S0-BLM-300

Jorge E. Garza
TX DL. # 08440044

COMMUNICATIONS ELECTRONICS
SYSTEMS, CIRCUITS, and DEVICES

Forrest Barker

Los Angeles City College

PRENTICE-HALL, INC., Englewood Cliffs, N.J. 07632

Library of Congress Cataloging-in-Publication Data

Barker, Forrest L.
Communications electronics.

Bibliography: p.
Includes index.
1. Telecommunication. I. Title.
TK5101.B3 1987 621.38 86-12347
ISBN 0-13-153883-7

Editorial/production supervision and
interior design: Erica Orloff
Cover design: Photo Plus Art
Manufacturing buyer: Carol Bystrom

Printed in the United States of America

10 9 8 7 6 5 4 3 2 1

ISBN 0-13-153883-7 025

PRENTICE-HALL INTERNATIONAL (UK) LIMITED, *London*
PRENTICE-HALL OF AUSTRALIA PTY. LIMITED, *Sydney*
PRENTICE-HALL CANADA INC., *Toronto*
PRENTICE-HALL HISPANOAMERICANA, S.A., *Mexico*
PRENTICE-HALL OF INDIA PRIVATE LIMITED, *New Delhi*
PRENTICE-HALL OF JAPAN, INC., *Tokyo*
PRENTICE-HALL OF SOUTHEAST ASIA PTE. LTD., *Singapore*
EDITORA PRENTICE-HALL DO BRASIL, LTDA., *Rio de Janeiro*

CONTENTS

PREFACE

This book is for the person who wishes to prepare for entry into an occupation in the field of electronic communications, a "technical" occupation, that is, one requiring the knowledge and skills associated with installing and maintaining communications equipment. It is intended, as well, for the generalist electronic technician student who recognizes the need to have a basic knowledge of the systems, circuits, and devices used in communications systems. Most of all, this book is for those who want to learn how to do hands-on work with electronic equipment, not pass an examination about it only.

Given the intended audiences of the textbook, its principal goal is to help its users learn "how to learn" about electronic communications systems, circuits, and devices. It is inevitable that the field of electronic communications will continue to change and change rapidly in the future, as it has in the past. If a person learns "how to learn" in this very technical field, he/she will possess the most important skill available for an interesting and rewarding career in an exciting field. Of course, we expect the book's users to obtain an extensive and in-depth knowledge of electronic communications systems and the building-block functions (circuits and devices) of which they are made. The book is dedicated to enabling them to do this as effectively, efficiently, and quickly as possible.

These goals are the basis of the distinctive characteristics of the book. It uses a from-the-top-down systems approach to provide integrated, meaningful explanations of communications systems and their circuits and devices. The emphasis is on "systems approach" and "explanation."

The systems approach first presents the reader with a general view of the system that he/she is going to learn about. Before looking at the bits and pieces of a system, a reader learns "what" the system is suppposed to do—why it exists at all. Then the reader learns, in a general way, "how" the system accomplishes what it is designed to do. It is the author's observation that with this general information as a background, a learner more easily recognizes his/her "need to know" the details of operation of the various functions which, in combination, accomplish the desired result: enable the system to "do" what it is supposed to do.

The "system approach" contrasts sharply with one in which the learner is confronted with information on an array of seemingly unrelated circuits and devices without having any clear idea why such functions need to be studied. One is more likely to be motivated to learn when there is strong evidence that what is being studied is related to a larger picture in which one is interested.

Presentations in this text are designed to provide "how" and "why" information. They are styled as user-friendly explanations—not simply statements of fact. Explanations are direct and based on physical concepts. They do not rely on abstract mathematical concepts. Wherever possible, explanations relate the new to the familiar and strive to evoke the reader's intuitive sense of how circuits and electronic devices work.

This emphasis on a tutorial approach is a consequence of two theories strongly held by the author: (1) a learner is more likely to accept and make use of a new concept if he/she has received a satisfying explanation of it, even if the explanation—the "why"—is soon forgotten; and (2) a concept is more likely to be remembered the better it is understood.

The book is nonmathematical in the sense that explanations are physical rather than mathematical. However, mathematical presentations do appear in conjunction with the calculations that a technician might be required to perform in the real, everyday world of electronic communications.

Although the book contains presentations relating to specific units of equipment (primarily in the chapter on two-way radio), it is not a goal of the text to prepare a user to be qualified to work on a particular piece or brand of equipment. Such a goal would be doomed to failure from the outset, the author believes. On the other hand, the author also believes that it is helpful in learning general, basic concepts to look at examples of actual, specific systems, circuits, and/or devices.

Further, it is not a goal of the book to present all the newest, latest ideas, equipment features, etc., in the world of electronic communications. That world is too broad and changing too rapidly for us to presume to strive for such a goal. As stated above, we believe that our most important goal is to help the user learn how to learn about communications electronics. If the user learns how to learn, he/she will be in a position to keep up to date with the numerous developments that will inevitably occur in this exciting, rapidly changing field during a lifetime career.

The book starts by presenting the reader, in Chapter 1, with the ideas of the systems approach. It proceeds immediately with a presentation of a generalized electronic communications system. Inasmuch as modulation is a unique and essential function in electronic communications, a presentation on the topic—the concepts, not the hardware—is given in Chapter 2. The material is mostly about amplitude modulation since AM systems are presented first in the subsequent chapters. However, a brief introduction to FM is included at the end of this chapter. Since most students using the book will, at the very least, have heard of FM, it was deemed desirable to mention it briefly, and somewhat out of context, at this early stage. A more thorough coverage of FM is included in Chapter 8.

Examination of AM systems begins with a comprehensive look at a relatively simple, basic AM broadcast-band receiver in Chapter 3. The reasons for beginning with the AM receiver are practical, pedagogical ones: (1) An AM receiver is a complete, operating subsystem: It "does something" readily observable by virtually anyone. (2) It can be used as a learning vehicle in the laboratory without concern for FCC regulations. (3) It is relatively inexpensive; in most situations it is possible to provide each student in the laboratory with a receiver with which to work individually. (4) The circuitry of a receiver is comparatively simple (i.e., is appropriate for beginners) and yet an AM receiver incorporates a high percentage of the basic functions of many electronic communications systems. Chapter 4 includes presentations on troubleshooting and alignment—laboratory and/or shop skills that can be practiced immediately on AM receivers.

AM transmitters are examined in Chapter 5. Included in this chapter are presentations on oscillation and oscillators, modulation, and transmitter power stages, among others.

Chapter 6 is the first of two chapters on transmission lines and antennas. In one sense the presentations in Chapter 6 are designed to complete the coverage of the AM system. However, they also have a much more general purpose than this. The second chapter of this pair, Chapter 10, presents more advanced coverage on the topics of transmission and propagation and includes a section on the Smith chart and one on waveguides.

Chapter 7 is an examination of the topic of noise. The terminology of noise and its consequences is introduced. Techniques of controlling noise are discussed.

Chapter 8 is a presentation on FM communications systems: angle modulation, FM transmitters and receivers, and an introduction to multiplexing. The very important topic, the phase-locked loop, is presented here.

Two-way radio systems and their circuits and devices are examined in Chapter 9. Presentations include a detailed look at specific examples of an FM transceiver and a single-sideband amateur transceiver. A brief look at the concept of a microprocessor-controlled transceiver is included. Troubleshooting and alignment procedures for radiotelephone equipment are included.

As noted above, Chapter 10 includes advanced topics on transmission and propagation.

Television stations and receivers are the topics examined in Chapter 11. Obviously, because of the limited space, the coverage can only be an introduction to basic concepts.

Extensive introductions to digital communications systems and light-wave-carrier systems are provided in Chapters 12 and 13, respectively.

In line with the goal of making this book of maximum usefulness to the person learning about communications electronics for the first time, the following features are included: (1) Worked-out examples are provided in situations where calculations are the topic to be learned. (2) A glossary of terms is provided at the end of each chapter. (3) From 6 to over 50 multiple-choice review questions are included at the end of each chapter. (4) Numerous essay questions and exercises are provided at the end of each chapter. (5) Each chapter contains numerous line drawings, circuit diagrams, and other figures to illustrate the concepts being examined.

Mine has been a lifetime of fascination with the "magic" of radio waves. In that lifetime I have learned about communication electronics, and "learned to learn" about it from numerous sources: students in classes I have taught, professors of classes I have taken, books and magazine articles, and personal experience. All of these deserve acknowledgment of a contribution to this book. However, the number is simply too large to recall, specifically, the names of all such sources, nor is there room to list them even if I could accomplish the recollection.

The following people assisted directly in the preparation of this book by way of their suggestions for the improvement of the manuscript and I am grateful for their contribution: Patrick J. O'Connor, DeVry Institute of Technology; Gary O. Coffman, ITT Educational Services; and Phillip J. Chiarelli, Electronics Institute. I would also like to acknowledge the assistance, encouragement, and support that I received from Gregory Burnell, Senior Editor, Electronic Technology, at Prentice-Hall. Finally, I wish to express my very great appreciation to my wife, Carolyn Berbower Barker, not only for her assistance with the manuscript but also for her supportive understanding, which made the whole project possible and worthwhile.

1 COMMUNICATION SYSTEMS

In a world that is constantly changing, there is no one subject or set of subjects that will serve you for the foreseeable future, let alone for the rest of your life. The most important skill to acquire now is learning how to learn. *If you know how to learn, you can adapt and change no matter what technological, social, or economic permutations occur.*

—John Naisbitt and Patricia Aburdene, Reinventing the Corporation, *Warner Books, September 1985*

1.1 INTRODUCTION

The field of knowledge and endeavor which we now call "electronics" was just getting started around the beginning of this, the twentieth, century. In the beginning, and for many years, the field was called "radio" because the only practical application of our very limited knowledge of electronics was in the field of radio communication. Of course, radio communication was also in its infancy.

One of the first applications of electronic devices to the field of radio was in the form of what would now be called a point-contact semiconductor diode. When first used it was called a crystal detector because its purpose was to "detect" or demodulate the radio signal and thus recover the information being transmitted by the radio equipment. The development and application of this device provided a very significant improvement in the performance of the radio equipment of the day. However, it would still be a number of years before any form of electronic amplification would be developed and utilized.

In the 80 or more years since the application of the crystal detector, our knowledge of electronics has grown astronomically. The application of this knowledge manifests itself in the form of the numerous modern miracles with which we are surrounded: radio, television, calculators, computers, space communication, microwave cooking, audio and video recording, and so on. The field of "radio" has expanded and is now more correctly called "electronic communication'' or ''telecommunication.'' But electronic communication, in turn, is only one of many areas of specialized knowledge in the larger field of ''electronics.''

The number of electron devices that have been developed over the years—the vacuum tube, the transistor, integrated circuits, etc.—is large. The circuits that have been and can be developed for utilizing these devices for worthwhile human purposes are virtually infinite in number. For example, a radio receiver, which in its earliest form contained relatively few components, is now a maze of circuits and devices.

The implication of this great progress in electronics for those of us who choose to be involved in this most fascinating of fields is that the task of gaining the knowledge necessary for successful work has grown larger if not more difficult. This calls for more efficient learning approaches. One such approach, and the one that we utilize in this book, is referred to as the *systems approach*. For example, a method for learning how a system such as an electronic communication system works might consist of laboriously determining the currents and voltage drops for all components in the system. But experience shows that this would not only be far too time consuming—it would not help us greatly to understand how the system functions. No, in the systems approach that we will use, we start our study at the system level rather than at the component or circuit level. First, we simply try to understand what the system, by design, is supposed to do for us as human beings. Next, we try to learn ''how'' the system accomplishes its purpose(s) in relatively simple, not-too-technical terms. Finally, we delve into the details of the many devices and circuits that must work together to accomplish the overall purpose.

1.2 THE SYSTEM CONCEPT

An electronic system is a combination of electronic circuits and devices working together to accomplish an overall purpose. Virtually all of the electronic appliances that we utilize in our daily lives, as well as the electronic equipment units used in all other applications, can appropriately be considered as systems, or at least as parts of systems, that is, *subsystems*. A list of examples of common systems, then, would include pocket calculators, tape recorders, computers, microwave ovens, and electronic games. Radio and television receivers are missing from this short list because, as receivers, they are really only part of a larger combination of elements—they are *subsystems*. If a radio or television receiver were completely isolated from the other essential portion of the communication system, the transmitter, it would not accomplish any useful purpose.

It is possible to represent any and all electronic systems (or subsystems) in a basic way by means of a simple block diagram (see Fig. 1.1). Next, it is valid and useful to say that every system has four important aspects to be considered in any attempt to understand how it accomplishes its purpose. These four items, shown in Fig. 1.1, are as follows:

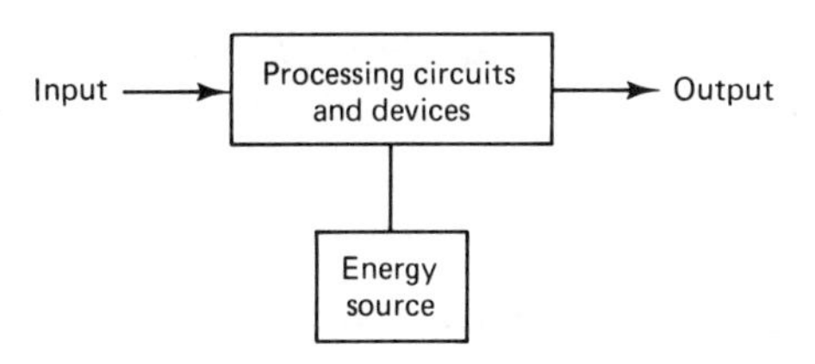

Figure 1.1 Basic block diagram of an electronic system.

1. An input
2. An output
3. Processing circuits and devices
4. A source of energy

It is now possible to give a somewhat more detailed definition of a system than that above: An *electronic system* is a combination of electronic circuits and devices which, working together under the control of the input, convert "raw" energy from the energy source into an output that is in some way useful to human beings. In this context, an ordinary flashlight can be analyzed as an electronic system. The input consists of the on-off switch. The energy source is, of course, the dry cells. The output is the light. The processing circuits and devices are the light bulb and the circuits connecting it to the dry cells.

The hand-held electronic calculator is an excellent example of a complete electronic system. It lends itself very readily to an analysis of the four essential elements of consideration of any system, as outlined in the preceding paragraph. In this case, it is obvious, again, that the battery is the energy source, an essential element in every system. The output is obviously the digital readout of the calculator. Although you may never have seen the working "insides" of a calculator, if you opened one up you would find that it contains one or more integrated-circuit chips (ICs) mounted on a printed-circuit board and interconnected with the key switches, the battery, and the output display unit. You would be viewing the "processing circuits and devices." Finally, the key switches are incorporated as part of the system to enable human beings to "input" information to the system.

1.3 HOW TO LEARN A SYSTEM

During the course of a career in electronics, you, as a technician, will probably have a need to learn a number of systems. You will find that different systems with very different purposes will contain, nevertheless, combinations of a relatively limited number of basic, building-block circuit functions. Examples of such basic functions include amplification, detection, signal generation (oscillators), modulation, and frequency control. Although the actual circuits performing a particular function may differ in the details of implementation, an understanding of the basic concept of a function can readily be used to gain an understanding of a particular circuit. Thus you need to have a reasonably good knowledge of the typical design details of the basic building-block circuits and a working understanding of how they function. With this you will be able to learn new systems relatively easily and quickly. Let us now become familiar with a logical, systematic approacn to facilitate the learning of any system. Let us explore in some detail the systems approach described briefly in Sec. 1.1.

Step 1. Find out what the system is supposed to do.

Typically, you will need to learn a new system because you are going to be involved with it by way of some process such as manufacturing, installation, calibration, operation, modification, or perhaps most commonly, troubleshooting and repair. Whatever the reason, it is extremely helpful, if not absolutely essential, to know what purpose the system serves. In many cases, as for example with a calculator, a radio receiver, or an audio tape recorder, it will be perfectly obvious what the unit is used for. Or you will know simply by observing the descriptive title of the unit what its function is. However, there are many specialized applications of electronic equipment which may require considerable effort to determine the intended function. Here, at the very least, it will probably be necessary to study an instruction manual.

In many instances, with very sophisticated equipment, your employer may send you to a training session on the system. A significant portion of the training may be devoted to helping you become familiar with the application and purpose of the equipment. Very often a part of the training will consist of providing you with the knowledge of how to "exercise" the equipment, that is, how to push the right buttons to make it do what it is supposed to do.

Consider the task of exercising a computer system for purposes of troubleshooting. This, generally, will require some skill in computer programming. A skill of this type will often be required as an entry-level skill at the time of hiring. In any event, an important requirement for success in performing any kind of technical tasks with an electronic system is a knowledge of what it is supposed to do and how to exercise it in that function.

Step 2. Using a block diagram and/or any other similar aids provided by the manufacturer, find out "how" the system accomplishes its purpose. Do not attempt to gain a detailed knowledge of circuits and devices at this level.

In the interest of saving time and effort in the learning of the essential features of a system (essential for purposes of one's involvement with a system), the best next step is usually to get a general, not-too-technical idea of how a system does what it is supposed to do. In most professional tasks it may be necessary to have a detailed knowledge of only a small portion of the circuits and devices of the overall system. Furthermore, experience indicates that it is easier to gain an understanding of a particular portion of the circuitry of a system if we already know how that portion contributes to the functioning of the larger system.

A well-constructed block diagram shows you at a glance what circuit functions are utilized and how these functions are interconnected to accomplish the overall result. In many cases, and especially with greater experience, you will already understand most, if not all, of the circuit functions.

The important task to be accomplished in this step is that of getting well in mind the overall plan of the system. What is the nature of the "input" to the system? Where does the input signal enter the system? What is the signal flow path? That is, through what function blocks is the input signal processed on its way through the system to the output? What is the nature of the processing performed on the signal? What auxiliary functions (as opposed to direct processing of the signal) does the system perform, and in what function blocks? What is the nature of the "output" of the system? Where is the output obtained? What is the nature of the energy source (or sources) for this particular system?

Step 3. Gain a knowledge of the details of the circuits and devices of the system and an understanding of how they function.

As suggested in step 2, in learning a new system you will no doubt discover that you already have an understanding of many, if not all, of the circuit functions you encounter. You may need only to familiarize yourself with the details of the particular circuits being used. In some instances, of course, it will be necessary for you to learn new functions and the details of the design of the circuits for performing them. And there will inevitably be new devices to learn about.

The challenge of learning about new circuits and devices is one of the major attractions of the field of electronics to those who have chosen it as their career. In learning about new circuits and devices it is absolutely essential that you effectively bring to bear on the task your knowledge and understanding of all the basic principles of electricity/electronics. For example, you must be prepared to make frequent use of Ohm's law and Kirchhoff's circuit laws, as well as your knowledge of basic electronic circuits such as amplifiers and rectifiers. Of course, never hesitate to make use of theory-of-operation explanations offered by manufacturers or other sources.

This three-step approach will be utilized throughout this text in presenting the information essential for an understanding of some basic electronic communication systems. The process begins in the next section with a presentation on a basic radio system. The presentation corresponds to steps 1 and 2 of the method outlined above.

1.4 THE BASIC RADIO SYSTEM

It would seem safe to say that the radio is the one item of electronic equipment familiar to the largest number of people in the world. The technical name for what is commonly referred to as a radio is *receiver* or *radio receiver*. A radio receiver is one of the two major subsystems of a radio system or an electronic communication system. The second subsystem is the *transmitter*. If there were suddenly no transmitters in the world operating at the appropriate frequencies, all of the radios of the world would be useless pieces of junk. They can perform their task of providing communication at a distance only if the other major portion of the system—a transmitter—is functioning correctly.

A very simple block diagram for a radio system is shown in Fig. 1.2. Notice that only one block for the processing circuitry is shown in this diagram. The block diagram of the system will be more helpful if the receiver and transmitter functions are shown separately (see Fig. 1.3).

Before proceeding further with the block diagram of the radio system, let's be sure that we have accomplished step 1 of the systems approach for learning the radio system. What does the radio system do for human beings? Although this question might be answered in some sociological or political way, we are, of course, interested here in a relatively technical answer. Our answer might be something like this: A radio system enables two or more persons to communicate with each other at relatively great distances, distances too great for communication by ordinary voice communication. Or a radio system enables one person to speak into a device called a microphone at one location and be heard by others at another location a great distance away.

Now, let us proceed to step 2 of the systems approach. In a general way, how does the system accomplish the stated result? The block diagram of Fig. 1.3 does not give a complete answer to this question, but it does facilitate an explanation of the operation of the system. Let us start at the left side of the diagram, the input side of the system, and discuss what steps are required in order to obtain an output from the system. Notice that the diagram is drawn so as to depict the output of the system on the right side of the diagram. This orientation, one in which the "signal" of the system passes from left to right on a diagram (either a block diagram or a schematic circuit diagram) is very popular. Although not a formally adopted convention or standard, it is used almost universally by persons and organizations involved in preparing such aids to information exchange.

Let us assume that this particular system is a radio broadcast system. Therefore, the diagram might represent your own transistor radio receiving a news broadcast from a local station.

The input to the system, then, is the sound energy from the announcer's voice. This sound energy is converted into electrical energy in the form of a voltage or a current by means of a *microphone*. The microphone is a *transducer*, a device for converting energy from one form to another. The electrical energy derived from sound energy is called the *information signal* of the system. It is often referred to simply as "the signal" or even "the information."

In this case, the information signal is composed of *audio-frequency* (AF) voltages (or currents) and, as such, is not capable of being transmitted, except

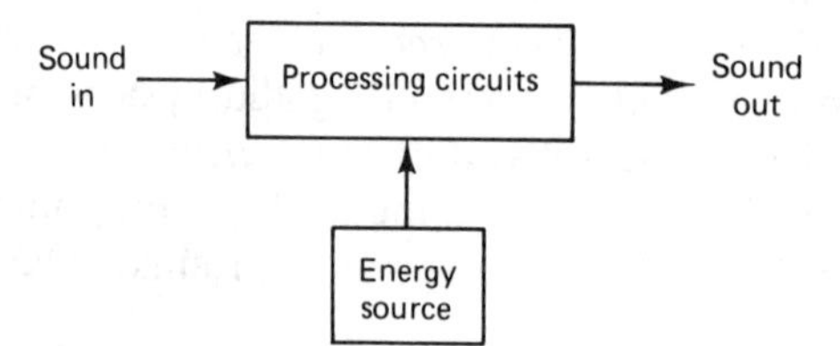

Figure 1.2 Block diagram of a radio system.

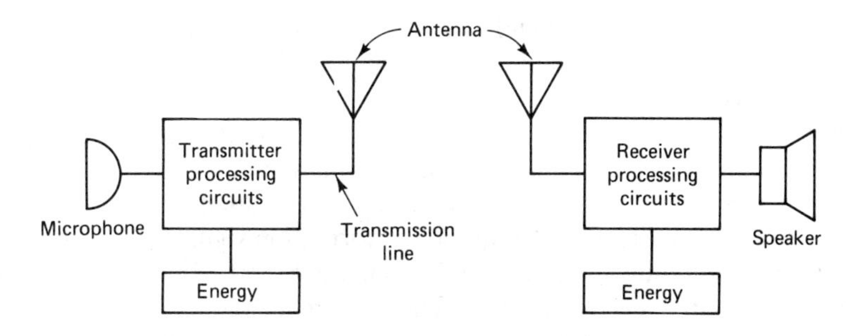

Figure 1.3 Revised block diagram of a radio system.

on conductors, over any except very short distances. Therefore, one of the principal purposes of the transmitter is to generate another signal, called the *carrier signal*, which can be transmitted over tremendous distances. Recall, for example, our communication with space probes hundreds of millions of miles distant from the earth. The carrier signal will be in the *radio-frequency* (RF) range of signal frequencies—20 kHz and higher. This signal is called the *carrier signal* because, as we shall see, it "carries" the information signal from the transmitter to the receiver.

For a signal to become the carrier, the transmitter must provide some means by which the information signal is "placed on" this higher frequency. The process is called *modulation*. "To modulate" means "to change." When a radio frequency is modulated by an information signal to produce a modulated carrier signal, the RF signal must be changed in such a way that the "result" of the process contains the "information" of the information signal.

There are a number of methods employed for modulating carrier signals. The two most common methods are termed *amplitude modulation* (AM) and *frequency modulation* (FM). These titles are aptly descriptive. Amplitude modulation means that a carrier's amplitude is changed so as to contain the information of the information signal. In frequency modulation, the frequency of the carrier is changed to contain the information of the information signal. The concepts and details of modulation will be presented in considerable detail in Chap. 2 and other subsequent chapters.

In addition to providing for modulating the carrier with the information signal, the transmitter will contain one or more stages of amplification. The purpose of the amplification is to increase the energy in the carrier so that it will be powerful enough to be transmitted over the distance between the transmitter and the receiver. When the energy in the carrier is at the desired level, this signal is transmitted by means of the *transmission line* from the output of the transmitter to the *transmitting antenna*. The transmission line, although it may just be a pair of conductors, possesses very special characteristics at radio frequencies. The concepts and characteristics of transmission lines are presented in detail in Chap. 6.

The purpose of the antenna is to facilitate the process called *radiation*. During the process of radiation the energy of the carrier detaches itself from the antenna and travels through space away from the antenna. (The word "radio" is derived from the same root as "radiation.") Antennas, like transmission lines, are relatively simple in their mechanical details. However, also like transmission lines, they are complex entities electrically. A discussion of the important concepts and characteristics of antennas is also presented in Chap. 6.

In summary, then, at the transmitter, sound energy is converted by a microphone to the audio-frequency information signal. The transmitter generates a radio-frequency signal which is modulated by the information signal to produce the modulated carrier signal. The transmitter amplifies the carrier signal and feeds it to the transmission line connected to the antenna. The transmission line conducts the carrier signal to the antenna. The antenna radiates the carrier signal into the surrounding space.

After traveling through some intervening space, the carrier signal arrives at the receiver antenna and induces an electrical potential in that device. This electrical potential, which is also called the carrier signal, is fed to the processing circuits of the receiver. These circuits must *select* this particular carrier and separate it from the signals of other transmitters which may be and usually are present. The receiver will increase the energy level of the carrier by amplification. Successful amplification maintains an acceptable ratio of carrier energy level to noise energy level—the "signal-to-noise ratio." Noise signals are always present in the atmosphere. The processing circuits themselves also introduce noise potentials and add them to the desired signal. The presence of a significant amount of noise in the output of the system would, of course, be unacceptable. The functions of the receiver just described are formally specified for a receiver by means of the technical terms of *selectivity* and *sensitivity*, respectively.

Next, an absolutely essential function which the receiver must perform is that of *demodulation*. Demodulation, also frequently called *detection*, is the process of recovering the information signal from the carrier signal. The process and the circuitry required are different for different types of modulation.

The signal recovered by demodulation of the carrier signal will be a replica of the information signal produced by the microphone at the input of transmitter (see above), if the system is working correctly. That is, the information signal present in the receiver, in a radio system, will be an audio-frequency signal. All but the most primitive of receivers will provide amplification for the information signal.

Finally, the audio-frequency signal will be fed to some type of device whose function is to *reproduce* the sound energy that was first introduced to the system by the announcer's voice. You know sound reproduction devices by the terms *loudspeaker*, or simply *speaker*, and *headphone*.

In summary, the receiver's antenna intercepts carrier signals from transmitters and converts these "air signals" to electrical potentials which are fed to the receiver's processing circuitry. This circuitry selects the potential of the desired carrier and rejects potentials of the undesired carriers. The receiver builds up the energy level of the desired carrier signal at the expense of undesired signals. The receiver demodulates the carrier signal—recovers the information signal. It amplifies (increases the energy level of) the audio-frequency signal (the information signal). Finally, it reproduces the original sound energy by way of a loudspeaker or headphone.

This completes a relatively nontechnical description of how a radio system accomplishes the task of providing for communication at a distance. The next step in learning the system is to study block diagrams of the system that are more detailed, which show how the two subsystems—transmitter and receiver—function in terms of discrete function blocks. Following that, the process includes gaining a thorough knowledge of the details of design and operation of each of the function blocks. The primary purpose of the remainder of this text is to provide such details.

1.5 OTHER ELECTRONIC COMMUNICATION SYSTEMS

The first steps of an effective approach for learning complex electronic systems, as they might apply to a relatively simple system, a radio broadcast system, have been previewed in the preceding sections. Much of the material presented could be applied with little change to more complex electronic communication systems. Of course, more complex systems utilize more electronic functions; the block diagrams for these would have to be expanded beyond that of the basic radio system.

An important, rapidly expanding application of the basic radio system is that of two-way radio, which enables two persons to communicate with each other, again, at a distance. A two-way radio system differs basically from a simple broadcast radio system only in that it requires both a transmitter and a receiver at each of the two locations involved in the communication. In modern practice in two-way radio, the transmitter and receiver of a station are, typically, incorporated into a single container package called a transceiver (transmitter/receiver). A simplified block diagram of a two-way radio system is shown in Fig. 1.4.

There are many applications of two-way radio systems. Noncommercial applications include the amateur radio service (''ham radio'') and citizens' band radio service (CB radio). Commercial applications include two-way radios in taxicabs, police and firefighting vehicles, gas, electric, and telephone utility vehicles, and many, many others.

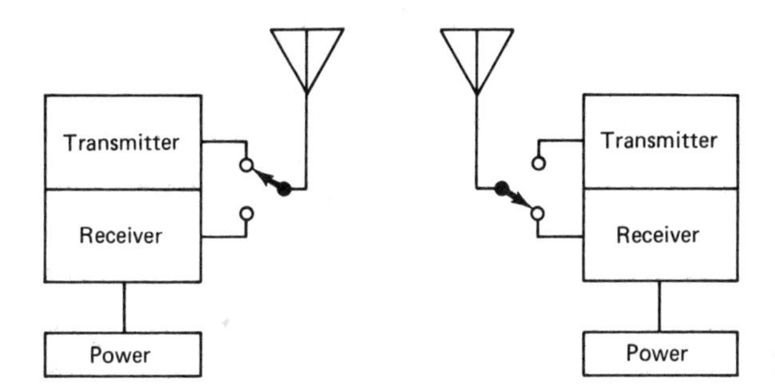

Figure 1.4 Two-way radio system.

Another extremely important example of an electronic communication system of the modern world is that of television. Even this system has many features and functions in common with the simple radio system presented above. In the case of television, there are two information signals that must be placed on carrier signals by way of modulation—the audio signal and the picture signal. The transmitters for television broadcasting are similar in many respects to those of a radio system. A significant part of a television receiver is similar to that of a radio receiver. Both television transmitters and receivers, however, must contain circuitry which makes it possible to convert a visual image into electrical signals, transmit those signals to a remote location, and then reproduce the image utilizing those signals. Electrical signals representing images are called *video signals*.

A relatively recent addition to the list of applications of electronic communication systems is in the transmission of data information between electronic digital computers. Because of the very rapid expansion in the utilization of computers, data transmission is one of the fastest-growing areas of electronic communication. Data communication equipment differs from voice communication equipment primarily in the modulation techniques used. Inasmuch as computer data are inherently in the form of electrical pulses, various forms of pulse modulation schemes are utilized for data communication, as opposed to the amplitude or frequency modulation of the analog signals of audio and video systems.

The principles of radio communication are utilized in the transmission of analog data in systems called *radio telemetry*. For example, a cardiac patient can be equipped with a miniature radio transmitter for sending electrocardiographic (EKG) data on his heart to a monitor in the intensive-care unit of a hospital. This, a radio-telemetry system, makes it possible for such a patient to walk about during recovery from a heart attack while still having his heart carefully monitored. Radio telemetry is used extensively in returning to earth stations the many types of physical data collected by space exploration vehicles. Radio telemetry is also used in many commercial and industrial applications.

SUMMARY

1. An efficient approach to the learning of electronic systems is called the systems approach. It proceeds from a study of what a system does and how it does this in a very simple, general way, to a study of the details of how each circuit and device in the system contributes to the overall function.
2. Virtually all electronic systems can be considered from the standpoint of an input, an output, processing circuits and devices, and a source of energy.
3. Studying the block diagram of a system is a very rapid way of becoming familiar with how a system functions without becoming prematurely involved in the intricate details of each function.

GLOSSARY OF TERMS

Amplitude modulation The changing of the amplitude of the radio-frequency signal of an electronic communication system in a way that incorporates the information of the signal to be transmitted.

Antenna An arrangement of wires, metal rods, etc., used to radiate and/or intercept the radio-frequency energy of an electronic communications system.

Audio frequency A frequency in the band of frequencies that are audible—about 20 to 20,000 Hz.

Block diagram A diagram composed of rectangles representing the functional divisions of an electronic system, showing how the functions are interconnected.

Carrier In an electronic communication system, the radio-frequency signal whose amplitude, frequency, or phase is modulated by the information signal.

Demodulation The process of recovering the information signal from the carrier signal.

Detection *See* Demodulation.

Frequency modulation The variation of the instantaneous frequency of the carrier signal in accordance with the information signal.

Information signal The electrical energy representing the information to be transmitted.

Microphone A device (a transducer) that converts the mechanical energy of sound waves into electrical energy.

Modulate In an electronic communications system, to vary the amplitude, frequency, or phase of the carrier signal in accordance with the information signal.

Pulse modulation The incorporation of information into a carrier signal by means of a scheme involving pulses—short, periodic bursts of energy.

Radiation The process in which energy is sent out through space, as if from a center.

Radio frequency A frequency in the band of frequencies between the audible frequencies and the infrared-light portion of the spectrum—approximately 20 kHz to 1,000,000 MHz.

Receiver An apparatus that intercepts carrier signals, selects a desired carrier, recovers the information signal from the carrier, and reproduces the information being transmitted by reconversion of the information signal.

Reproduce To make a copy, close imitation, duplication, etc., of a picture, sound, writing, etc.

Selectivity (of a receiver) The degree to which the receiver will reproduce the signals of a given transmitter while rejecting those of all others.

Sensitivity (of a receiver) The ability of a receiver to respond to incoming signals, usually expressed as the minimum input signal level required to provide a given output power level at a given signal-to-noise ratio.

Signal-to-noise ratio Ratio of energy level of desired voltage (or current) to energy level of noise (undesired voltage or current) in a device such as a radio receiver.

Transceiver An apparatus composed of both a receiver and a transmitter.

Transducer A device for converting energy from one form to another, for example, a microphone, which converts mechanical energy (sound waves) into electrical potential.

Transmission line An arrangement of wires, insulators, etc., used to conduct electrical energy from one location to another.

Transmit To send or cause to go from one person or place to another, especially across intervening space or distance.

Transmitter The apparatus that generates radio-frequency signals, modulates them in accordance with an information signal, and transmits them by means of an antenna.

Video signal The signal representing the picture portion of a television broadcast, as distinguished from the audio portion.

REVIEW QUESTIONS: BEST ANSWER

1. An electronic system is a combination of electronic circuits and devices working together to accomplish an overall purpose. **a.** True. **b.** False.
2. Which of the following is not an example of an electronic system? **a.** a transistor. **b.** a calculator. **c.** a tape recorder. **d.** a microwave oven. **e.** a digital electronic watch.
3. An important first step in becoming familiar with an electronic system is to: **a.** calculate currents in all resistors. **b.** find out what the system is supposed to do. **c.** write Kirchhoff voltage-law equations for all circuit loops. **d.** determine the function of each resistor and capacitor in the system. **e.** all of these.
4. A knowledge of the nature of the input and output of a system is of no value in learning how the system functions. **a.** True. **b.** False.
5. A knowledge of how to operate an electronic system is required of those who are paid to operate it but is of no importance to those who must repair it. **a.** True. **b.** False.
6. A block diagram of a system is useful in learning the: **a.** function of each device in the system. **b.** value of all resistors in the circuit. **c.** power rating of all transistors in the system. **d.** voltage rating of all capacitors used. **e.** in general, how the system functions.

REVIEW QUESTIONS: ESSAY

1. Choose an electronic system with which you are familiar, a system other than a radio system or an electronic calculator, and describe the four basic aspects of the system: the input, the output, the processing circuits, and the energy source.
2. Answer the question ''What is the system supposed to do?'' for each of the following systems: **(a)** television system; **(b)** calculator; **(c)** audio tape recorder; **(d)** public address system; **(e)** digital electronic watch.
3. Define *modulation.*
4. Define *information signal.*
5. Define *carrier signal.*
6. Define *reproduce* as it relates to a radio system.
7. Draw and label a simple block diagram that could be used to represent any electronic system.
8. Draw and label a simple block diagram for an electronic calculator.

2

MODULATION

2.1 INTRODUCTION

In Chap. 1 we were introduced to the idea that a radio-frequency signal can travel great distances through the atmosphere and/or outer space. We also learned that we can use an RF signal to "carry" information if we can somehow "place" a signal containing the desired information—an *information signal*—"on" the RF signal. An RF signal used in this way is called a *carrier signal* or simply the *carrier*.

The process of "placing" an information signal on a carrier is called *modulation*. The process involves using the information signal to *change* or *vary* (i.e., *modulate*) some property of the carrier. Properties, or more appropriately, *parameters* which are candidates for modulation are the carrier's *amplitude*, *frequency*, or *phase*. Schemes that involve the modulation of each of these parameters are in common use in the world of electronic communication. In this chapter we examine the basic ideas of *amplitude modulation* (AM) in considerable detail. We also take just a quick look at *frequency modulation* (FM) for comparison. We examine FM in greater detail in Chap. 8.

2.2 THE NEED FOR MODULATION

The mechanical energy contained in the sound waves of someone talking, or even shouting, travels only short distances, several hundred meters at most. If the mechanical energy of the sound is converted to an audio-frequency electrical signal, that signal can be radiated from an antenna and then will travel over a greater distance than as mechanical energy. However, there are two important problems involved in attempting to broadcast audio-frequency electrical energy.

Practical Antenna Length

First, detailed studies of antenna characteristics and operation have shown that in order to achieve significant amounts of radiation of electrical energy from an antenna, the antenna must be of a minimum length. The length required is inversely proportional to the frequency of the signal to be radiated. For example, if we wanted to broadcast an audio signal of 3.0 kHz, we would need an antenna approximately 16 miles long! Using an antenna 16 miles long is obviously ridiculous.

Utilization of the Frequency Spectrum

Even if it were feasible to radiate audio-frequency signals, a second difficulty would arise from the practice. Consider the situation in which several stations in a given area desire to broadcast audio-frequency programs. Since they would all be attempting to use the same range of frequencies, say 30 to 15,000 Hz, there would be nothing but interference, chaos. It would be impossible for listeners to select one station and reject the other undesired stations.

The solution is for stations to use carrier transmission: to modulate carriers of different frequencies which can then easily be sorted out by the selection circuitry of radio receivers. In fact, carrier transmission is also utilized in long-distance telephone communication, where signals are transmitted over wires or by microwave links (as distinguished from the wireless transmission of radio). Many telephone conversations can be transmitted simultaneously over a single line or microwave link when carriers are used.

2.3 AMPLITUDE MODULATION (AM)

Consider the waveform of an information signal composed of a single frequency, say 1000 Hz, as in Fig. 2.1(a). It is evident that the "information" of this signal is contained in two aspects of the signal: its frequency and its amplitude. Therefore, if the information of the signal is to be transferred by modulation to a higher-frequency carrier signal, such as that in Fig. 2.1(b), these characteristics must, in some way, be incorporated in the carrier after modulation. The waveform of Fig. 2.1(c) represents the carrier signal after it has been amplitude modulated by the information signal. Observe that if you focus on just the envelope of the modulated carrier, either upper or lower, you can see the information signal represented in the envelope. (The envelope of a waveform is an imaginary line joining just the tips of the waves.) The *amplitude* of the carrier cycles has been changed in a pattern of variation which is the same as the pattern of variation of the information signal.

Amplitude modulation (AM) is one of the most common forms of modulation

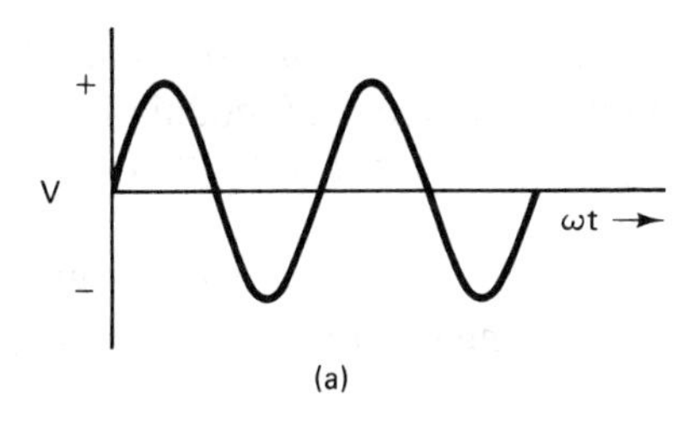

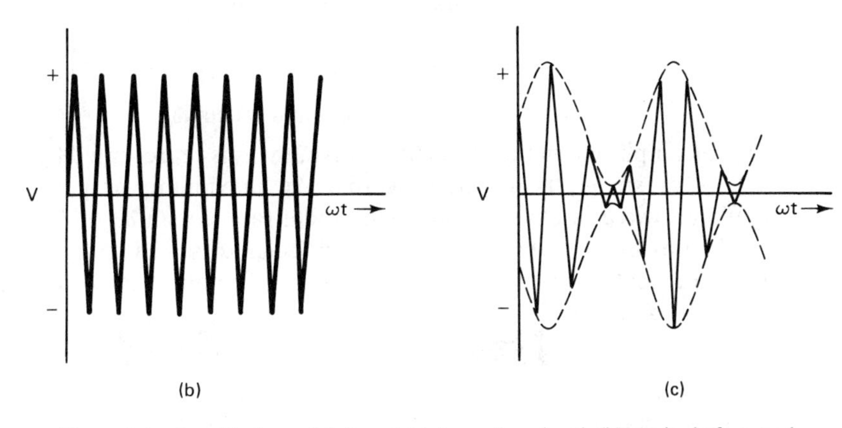

Figure 2.1 Amplitude modulation: (a) information signal; (b) carrier before modulation; (c) carrier after modulation.

employed in modern broadcasting. It is used in transmitting the picture information in broadcast television and is the form of modulation used in the so-called "AM band" of commercial radio broadcasting. The AM band includes carrier frequencies between 540 and 1600 kHz.

2.4 PERCENT MODULATION

Consider the effect of modulating a carrier whose instantaneous voltage amplitude is represented by v_c. The carrier is to be modulated by a sinusoidal voltage whose instantaneous value is v_m. The equation for the unmodulated carrier is

$$v_c = V_{cp} \sin \omega_c t \tag{2.1}$$

In this equation, V_{cp} represents the peak value of the carrier voltage wave and $\omega_c = 2\pi f_c t$, where f_c is the carrier frequency. The equation for the instantaneous value of the modulating signal is

$$v_m = V_m \sin \omega_m t \tag{2.2}$$

The peak value of the modulating signal is V_m and its frequency is f_m. There will be some relationship between V_{cp} and V_m. The relationship between the two values is conveniently expressed as the ratio

$$m = \frac{V_m}{V_{cp}} \tag{2.3}$$

The value m is called the *modulation coefficient* or *index* and, as will be seen, is an indication of the amount of modulation achieved. Solving Eq. (2.3) for V_m and substituting the value in Eq. (2.2) converts the equation for the modulating signal to

$$v_m = mV_{cp} \sin \omega_m t \tag{2.4}$$

It can be shown that the equation for the modulated carrier, instantaneous value, v_{mc}, is

$$v_{mc} = (V_{cp} + v_m) \sin \omega_c t$$

or

$$\begin{aligned} v_{mc} &= (V_{cp} + mV_{cp} \sin \omega_m t) \sin \omega_c t \\ &= V_{cp}(1 + m \sin \omega_m t) \sin \omega_c t \end{aligned} \tag{2.5}$$

The peak amplitude of the modulated wave, then, is represented by the expression $V_{cp}(1 + m \sin \omega_m t)$. In other words, the maximum, or peak, value of the modulated wave is changing at a sinusoidal rate ($m \sin \omega_m t$).

The extent to which a carrier is modulated is an important consideration in the practical operation of a radio station. The amount of modulation, called *percent modulation*, is obtained by multiplying the factor m of Eq. (2.3) by 100:

$$\text{percent modulation} = m \times 100\% \tag{2.6}$$

The value of m, and therefore percent modulation, can be determined from a plot of the modulated waveform, as, for example, in the display of the waveform on an oscilloscope (see Fig. 2.2). From the waveform in Fig. 2.2 it is evident that

$$V_m = \tfrac{1}{2}(V_{max} - V_{min})$$

and

$$V_{cp} = \tfrac{1}{2}(V_{max} + V_{min})$$

Therefore,

$$m = \frac{V_{max} - V_{min}}{V_{max} + V_{min}} \tag{2.7}$$

and

$$\text{percent modulation} = \frac{V_{max} - V_{min}}{V_{max} + V_{min}} \times 100\% \tag{2.8}$$

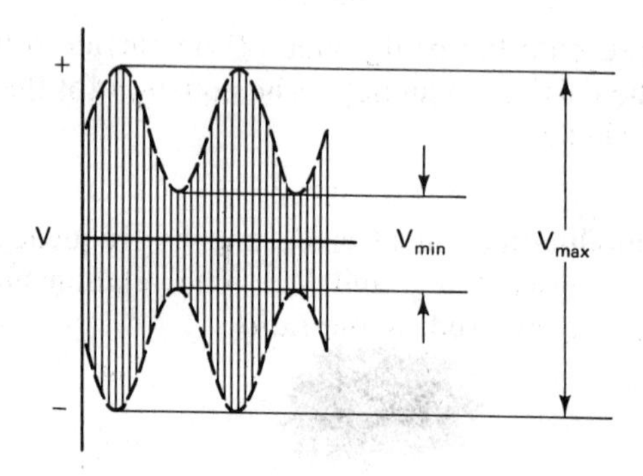

Figure 2.2 Determining percent modulation.

Example 1

The carrier $v_c = 150 \sin 2\pi 6.4(10^5)t$ is being modulated by the audio tone $v_m = 75 \sin 2\pi 800t$. (a) Determine the modulation coefficient. (b) Determine percent modulation. (c) What is the frequency of the carrier? Of the modulating signal?

Solution. (a) $m = \dfrac{V_m}{V_{cp}} = \dfrac{75}{150} = 0.50$

(b) Percent modulation $= 100m\% = 100 \times 0.50 = 50\%$

(c) $f_c = 6.4(10^5)$ Hz; $f_m = 800$ Hz

Example 2

Figure 2.3 represents the oscilloscopic display of a modulated RF carrier wave. Vertical scale factor: 0.2 V/div. Horizontal scale factor: 1.0 ms/div. (a) Determine the modulation coefficient. (b) Determine the percent modulation. (c) Determine the frequency of the modulating signal.

Solution. (a) $m = \dfrac{V_{\text{max}} - V_{\text{min}}}{V_{\text{max}} + V_{\text{min}}} = \dfrac{5 \text{ div} - 3 \text{ div}}{5 \text{ div} + 3 \text{ div}} = 0.25$

(b) Percent modulation $= 100m\% = 100 \times 0.25\% = 25\%$

(c) Period of $f_m = T_m = 3 \text{ div} \times 1 \text{ ms/div} = 3$ ms

$$f_m = \frac{1}{T_m} = \frac{1}{3(10^{-3})} = 333.3 \text{ Hz}$$

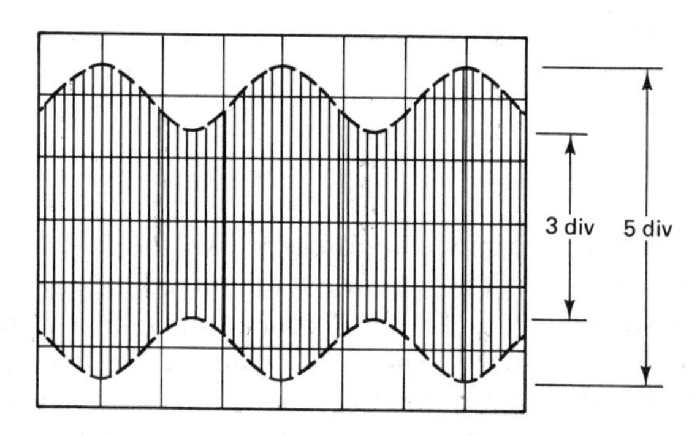

Figure 2.3 Modulated carrier wave.

2.5 FREQUENCY COMPONENTS IN AN AM WAVE

During the process of modulation of its amplitude, a carrier must undergo changes that distort its waveform. That is, the instantaneous values no longer conform to the equation of a pure sine wave,

$$v_c = V_{cp} \sin \omega_c t$$

In fact, as we will see later, *the process can occur only in a circuit that is operating in a nonlinear fashion*, thereby ensuring that the modulated signal at the output of the circuit will be distorted. The tools of higher mathematics reveal something extremely important about distorted sine waves: *A waveform with a pattern that repeats itself periodically, but which is nonsinusoidal, always contains frequency components other than the most obvious basic component.* An amplitude-modulated carrier wave is such a waveform. Consider a nonlinear circuit to which has been fed a carrier signal and a single-frequency information signal for modulation of the carrier (see Fig. 2.4). It can be verified experimentally that the output current in such a circuit will contain the following:

1. A dc component
2. A component of the original carrier
3. A component of the original information signal
4. Harmonics of the carrier
5. Harmonics of the information signal
6. A component whose frequency is equal to the sum of the carrier and information signal frequencies
7. A component whose frequency is the difference of the carrier and information signal frequencies

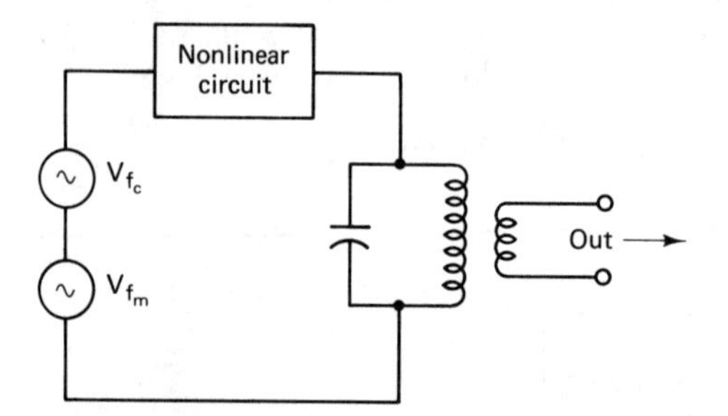

Figure 2.4 Circuit for amplitude modulation.

Side Frequencies

As shown in Fig. 2.4, the output of a modulation circuit will be fed to a tuned circuit. The function of the tuned circuit is to filter out (eliminate) all components except the carrier, the sum frequency, and the difference frequency (in the case of a single-frequency modulating signal). These components are the essential components for successful transmission of an amplitude-modulated signal. The current whose frequency is equal to the sum of the carrier frequency and information signal frequency is called the *upper side-frequency* (USF) component. That is,

$$\text{USF} = f_{\text{carrier}} + f_{\text{information signal}}$$

Similarly, the difference-frequency component is called the *lower side-frequency* (LSF) component,

$$\text{LSF} = f_{\text{carrier}} - f_{\text{information signal}}$$

A spectrogram of the products of amplitude modulation with a single frequency is shown in Fig. 2.5(a). A spectrogram of the modulated carrier, ready for transmission, is shown in Fig. 2.5(b). It is important to remember that the

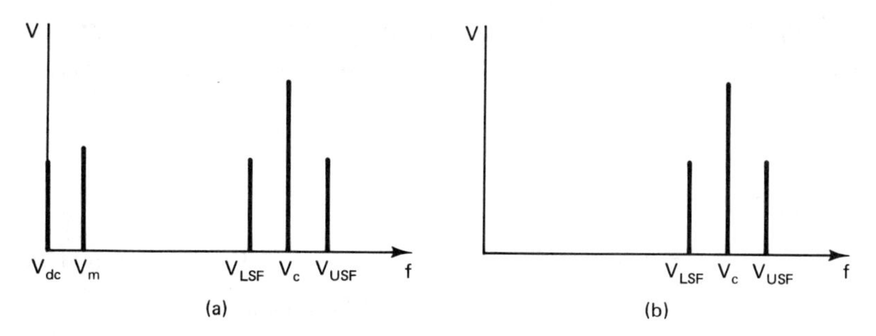

Figure 2.5 (a) Products of amplitude modulation process; (b) components of AM signal ready for transmission.

information signal is not present as a separate component in the composite signal that is to be transmitted. It manifests itself only as a "distortion" of the carrier signal, a distortion that produces side frequencies.

Sidebands

When an AM broadcast station's carrier is being modulated by typical program material, such as music, singing, or even someone speaking, the information signal will contain numerous frequency components. These will be constantly changing in both amplitude and frequency. Thus, as a result of modulation, there will be not merely an upper side frequency and a lower side frequency, but complete bands of sum and difference frequencies. During normal operation the signal broadcast from an AM station, then, consists of the carrier, the *upper sideband*, and the *lower sideband*. A spectrogram depicting this phenomenon is shown in Fig. 2.6.

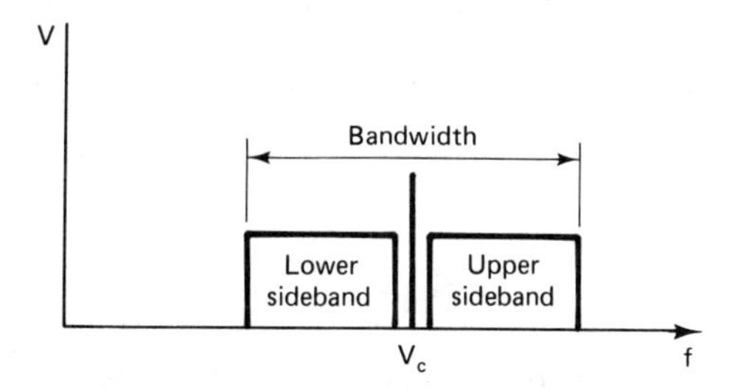

Figure 2.6 AM sidebands.

AM Bandwidth

The portion of the frequency spectrum utilized by a particular station is called its *bandwidth*. Mathematically, bandwidth is equal to the station's highest frequency minus its lowest:

$$\text{bandwidth} = f_{\text{highest}} - f_{\text{lowest}}$$

A station's highest frequency will be equal to its carrier frequency plus the frequency of the highest-frequency component in the information signal,

$$f_{\text{highest}} = f_{\text{carrier}} + f_{\text{highest information frequency}}$$

Similarly, a station's lowest frequency will be equal to the difference of the carrier frequency and the highest information signal frequency,

$$f_{\text{lowest}} = f_{\text{carrier}} - f_{\text{highest information frequency}}$$

Therefore,

$$\begin{aligned} \text{bandwidth} &= f_{\text{highest}} - f_{\text{lowest}} \\ &= f_{\text{carrier}} + f_{\text{highest information frequency}} - (f_{\text{carrier}} - f_{\text{highest information frequency}}) \\ &= f_{\text{highest information frequency}} + f_{\text{highest information frequency}} \end{aligned}$$

or

$$\text{bandwidth} = 2f_{\text{highest information frequency}} \tag{2.9}$$

That is, the bandwidth of an AM transmission is twice the highest frequency in the composite information signal to be transmitted. For example, stations in the commercial AM broadcast band are assigned channels that have a bandwidth of 10 kHz. On this basis, the highest audio frequency would be presumed to be 5,000 Hz. However, it is not correct to say that higher frequencies (up to 15,000 Hz) are never present in the modulating signal of AM stations.

Example 3

A 1.25-MHz carrier is being modulated by an audio signal whose frequency is 3000 Hz. (a) List the frequencies of the components of current that would be present in the nonlinear modulation circuit. (b) If the modulated signal were processed for broadcasting, what frequencies would be fed to the antenna?

Solution. (a) dc, 3 kHz, 1.25 MHz, LSF = 1.247 MHz, USF = 1.253 MHz; harmonics of audio: 6 kHz, 9 kHz, etc.; harmonics of carrier: 2.5 MHz, 3.75 MHz, etc.

(b) 1.247-MHz, 1.25-MHz, and 1.253-MHz signals

Example 4

At a particular moment, the program of an AM broadcast station is producing audio frequencies between 30 and 4000 Hz. The assigned carrier frequency is 890 kHz. (a) Assuming that there are no harmonic frequency components produced, what will be the highest frequency broadcast for the moment described? The lowest? (b) Again, for the moment described, what portion of the frequency spectrum will the station be using as it broadcasts its signal?

Solution. (a) $f_{\text{highest}} = 890\text{ kHz} + 4\text{ kHz} = 894\text{ kHz}$

$f_{\text{lowest}} = 890\text{ kHz} - 4\text{ kHz} = 886\text{ kHz}$

(b) Bandwidth $= 2f_{\text{highest information frequency}} = 2 \times 4\text{ kHz} = 8\text{ kHz}$

2.6 POWER DISTRIBUTION IN AN AM WAVE

For purposes of power distribution analysis, an operating transmitter feeding radio-frequency energy to an antenna can be compared to a simple source supplying energy to a dissipative load, a simple resistor, for example (see Fig. 2.7). When a simple source, whose voltage is V, is connected to a purely dissipative load, such as a resistor R, we say that the source supplies a power P, or the resistor dissipates that same amount of power, P. The formula for P is

$$P = \frac{V^2}{R}$$

When a transmitter is supplying energy to an antenna, most of that energy will be dissipated as a radiated carrier signal, although a very minor amount will be lost as heat. Again, it is possible to represent the situation with a formula:

$$P_c = \frac{V_c^2}{R_R}$$

where P_c represents the amount of carrier power supplied to the antenna by the transmitter (or the amount of carrier power "dissipated" by the antenna), V_c represents the effective (rms) value of the carrier voltage, and R_R represents a concept called the *radiation resistance* of the antenna. It is important to know that radiation resistance is not a simple physical property of the antenna which can be measured directly with an ohmmeter. Rather, at this point, to avoid an extensive delving into antenna theory, it is helpful to think of radiation resistance as a kind of imaginary property, a property that permits using a simple, familiar formula for power.

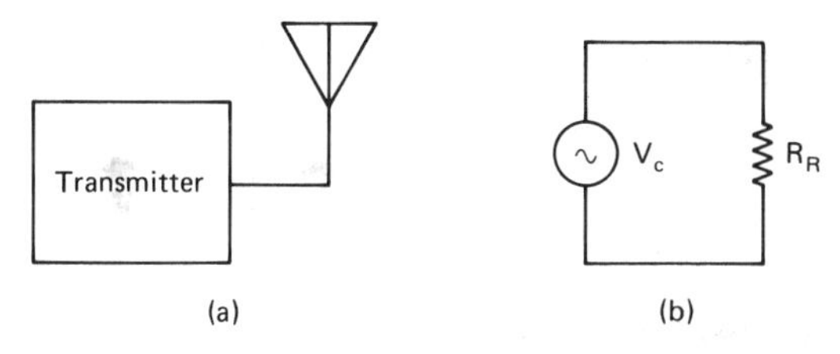

Figure 2.7 (a) Transmitter and antenna; (b) equivalent diagram.

Let's analyze power distribution for the case when the carrier is modulated by a single-frequency modulating signal. The transmitter now supplies, and the antenna dissipates, carrier power and also the power of the two side frequencies. To determine the amount of side-frequency power dissipated in the radiation resistance of the antenna, it is necessary to know the effective values of the side-frequency voltages. These values can be obtained if Eq. (2.5) (see Sec. 2.3) is multiplied out and expanded using the product of sines:*

$$v_{mc} = V_{cp}(1 + m \sin \omega_m t) \sin \omega_c t$$

$$= V_{cp} \sin \omega_c t - \frac{mV_{cp}}{2} \cos (\omega_c + \omega_m)t + \frac{mV_{cp}}{2} \cos (\omega_c - \omega_m)t \qquad (2.10)$$

* $\sin x \sin y = -\frac{1}{2} \cos (x + y) + \frac{1}{2} \cos (x - y)$.

In Eq. (2.10), the first term, $V_{cp} \sin \omega_c t$, represents the unmodulated carrier; the second term, $(mV_{cp}/2) \cos (\omega_c + \omega_m)t$, represents the upper side-frequency component in the modulated wave; and the third term, $mV_{cp}/2 \cos (\omega_c - \omega_m)t$, represents the lower side-frequency component. It is evident that the peak value of the voltage of each side-frequency component is equal to $m/2$ times the peak value of the carrier voltage. Hence the effective values of voltage for these components are also equal to $m/2$ times the effective value of the carrier voltage. Now, the equations for power in the side frequencies can be written

$$P_{\text{USF}} = \frac{V_{\text{USF}}^2}{R_R} = \frac{(mV_c/2)^2}{R_R} = \frac{(m^2/4)V_c^2}{R_R} \tag{2.11}$$

$$P_{\text{LSF}} = \frac{V_{\text{LSF}}^2}{R_R} = \frac{(mV_c/2)^2}{R_R} = \frac{(m^2/4)V_c^2}{R_R} \tag{2.12}$$

The total power in a modulated carrier signal, then, is

$$\begin{aligned} P_{\text{total}} &= P_{\text{carrier}} + P_{\text{USF}} + P_{\text{LSF}} = P_{\text{carrier}} + \frac{m^2}{4} P_{\text{carrier}} + \frac{m^2}{4} P_{\text{carrier}} \\ &= P_{\text{carrier}} + \frac{m^2}{2} P_{\text{carrier}} = P_{\text{carrier}} \left(1 + \frac{m^2}{2}\right) \end{aligned} \tag{2.13}$$

Equation (2.13) is the basis of several important observations regarding the distribution of power in amplitude-modulated carrier waves. The concepts contained in the following statements are utilized over and over again as further information on modulation and modulation circuits is developed. It is essential, therefore, that they be learned thoroughly:

1. The power in the carrier component of the modulated carrier wave is unchanged by modulation.
2. The total power content of a modulated carrier wave, compared to an unmodulated carrier wave, represents an increase of $m^2/2$ times the unmodulated carrier power.
3. The increase in power content of a modulated wave is obtained from the power in the side frequencies (sidebands, in the case of complex modulating signals).
4. The increase in the power content of a modulated wave is 50% when the modulation level is 100% ($m = 1$).

Example 5

At a particular moment, audio programming material is modulating a 250-W carrier at the 75% level. Determine: (a) the power in the carrier; (b) the power in the sidebands.

Solution. (a) $P_c = 250$ W (carrier power is unchanged)

(b) $P_{\text{sb}} = \dfrac{m^2}{2} P_c = \dfrac{(0.75)^2}{2} \times 250 = 70.31$ W

Example 6

(a) Determine the ratio of the effective value of the voltage of a transmitter signal modulated at the 80% level to that of an unmodulated signal (0% modulation). (b) By what percent does the voltage increase?

Solution. (a) In general,

$$P_{0\%} = \frac{V_{0\%}^2}{R_R} \qquad V_{0\%} = \sqrt{P_{0\%} R_R}$$

$$P_{m\%} = \frac{V_{m\%}^2}{R_R} = \left(1 + \frac{m^2}{2}\right)P_{0\%} \qquad V_{m\%} = \sqrt{\left(1 + \frac{m^2}{2}\right)P_{0\%}R_R}$$

Hence

$$\frac{V_{m\%}}{V_{0\%}} = \frac{\sqrt{\left(1 + \frac{m^2}{2}\right)P_{0\%}R_R}}{\sqrt{P_{0\%}R_R}} = \sqrt{1 + \frac{m^2}{2}}$$

and

$$\frac{V_{80\%}}{V_{0\%}} = \sqrt{1 + \frac{(0.8)^2}{2}} = \sqrt{1 + 0.32} = 1.1489$$

(b) Percent increase $= \dfrac{V_{80\%} - V_{0\%}}{V_{0\%}} \times 100\%$

$$= \frac{1.1489V_{0\%} - V_{0\%}}{V_{0\%}} \times 100\%$$

$$= 14.89\%$$

Example 7

Determine the percent increase in the effective value of the antenna current when the modulation of a transmitter is increased to 65% from 0%.

Solution. Since the transmitted power is proportional to the square of the antenna current (i.e., $P_{TRAN} = I_{ANT}^2 R_R$), then, as with the voltages (see Example 6),

$$I_{0\%} = \sqrt{\frac{P_{0\%}}{R_R}}$$

$$I_{m\%} = \sqrt{\frac{(1 + m^2/2)P_{0\%}}{R_R}}$$

and therefore,

$$\frac{I_{m\%}}{I_{0\%}} = \sqrt{1 + \frac{m^2}{2}} = \sqrt{1 + \frac{(0.65)^2}{2}} = 1.1006$$

percent current increase $= \dfrac{I_{m\%} - I_{0\%}}{I_{0\%}} \times 100\%$

$$= \frac{1.1006I_{0\%} - I_{0\%}}{I_{0\%}} \times 100\%$$

$$= 10.06\%$$

Example 8

Determine the percent of increase of transmitter power from 0% modulation to the following levels: (a) 25%; (b) 50%; (c) 75%; (d) 100%.

Solution. In general,

$$\text{percent increase in } P_{\text{tran}} = \frac{P_{m\%} - P_{0\%}}{P_{0\%}} \times 100\% = \frac{(1 + m^2/2)P_{0\%} - P_{0\%}}{P_{0\%}} \times 100\%$$

$$= \frac{m^2}{2} \times 100\%$$

(a) For 25% modulation, percent increase in $P = \frac{(0.25)^2}{2} \times 100\% = 3.1\%$

(b) For 50% modulation, percent increase in $P = \frac{(0.50)^2}{2} \times 100\% = 12.5\%$

(c) For 75% modulation, percent increase in $P = \frac{(0.75)^2}{2} \times 100\% = 28.1\%$

(d) For 100% modulation, percent increase in $P = \frac{(1.0)^2}{2} \times 100\% = 50\%$

2.7 FREQUENCY MODULATION (FM)

When the frequency, instead of the amplitude, of a carrier is changed, so as to incorporate the information of an information signal, the process is called frequency modulation (FM). A few of the important basic concepts of FM will be presented here. A comprehensive, detailed explanation of FM is presented in Chap. 8.

In order to grasp more easily the significance of changing the frequency of an RF signal, let us consider a situation in which we are displaying the output of an RF generator on the screen of an oscilloscope. The display would appear similar to that of Fig. 2.8(a), that is, a sine wave of constant amplitude and frequency.

Now, imagine that we can vary the frequency of the signal at a sinusoidal rate. That is, suppose that we changed the position of the frequency control knob on the generator, first clockwise to increase the frequency slightly, then counterclockwise to decrease the radio frequency by the same amount. We, in effect, "rock" the knob forward and back. The motion is sinusoidal, the "rate" of the rocking motion is, let's say, 400 times per second ($f = 400$ Hz).

Imagine that we measure and plot the position of the frequency control knob with respect to its beginning position. The plot would be similar to that of Fig. 2.8(b). The effect on the display of the signal would be similar to that of Fig. 2.8(c). Notice the "squeezing up" and "spreading out" of the display. Notice also that these phenomena track the position of the control knob. The display of Fig. 2.8(c) is that of a frequency-modulated signal.

Remember, modulation must transfer the information of the information signal to the carrier wave. The information of the information signal is contained in (1) its frequency, and (2) its amplitude. The modulated carrier must carry

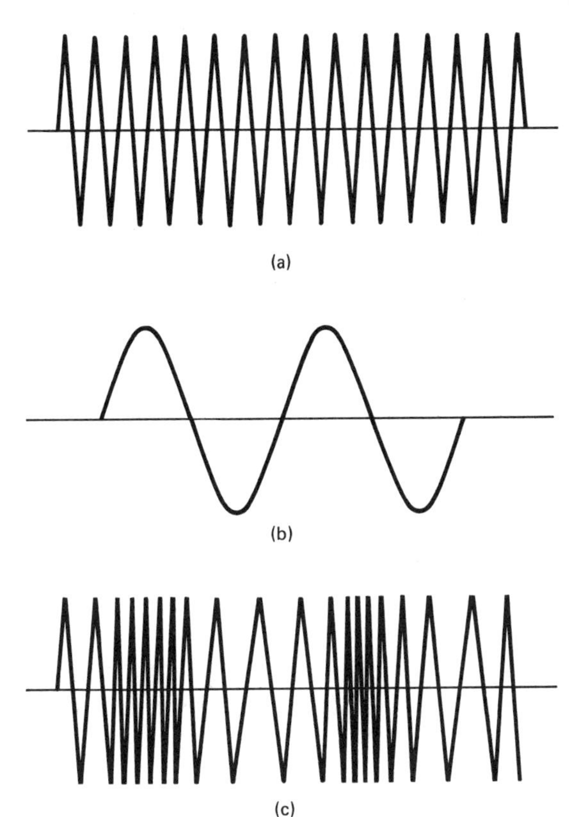

Figure 2.8 Concept of frequency modulation: (a) unmodulated RF signal; (b) information signal; (c) frequency-modulated RF signal.

these items of information. In an amplitude-modulated carrier, the presence of these items, in the envelope of the carrier, is obvious. In frequency modulation, information is carried as follows:

1. The amplitude of the information signal is represented by the amount of change in the frequency of the carrier. This change is called *frequency deviation*, or sometimes simply *deviation*. The symbol Δf ("delta f") is used to represent frequency deviation.
2. The frequency of the information signal is represented by the rate of the deviation, that is, referring to Fig. 2.8(c), by the number of times per second the carrier's frequency is increased and decreased (by the number of times per second the display is squeezed up and spread out). The rate of deviation, represented by f_m, is equal to the frequency of the information signal.

SUMMARY

1. Modulation is the process of varying some characteristic of a higher, steady frequency, called the carrier, by a lower-frequency information signal so that the carrier contains the information of the lower frequency. Amplitude, frequency, and phase may be changed.

2. Modulation is required because information, such as sound waves, cannot be transmitted over great distances. Further, modulated carriers permit much more efficient use of the frequency spectrum.
3. In amplitude modulation, the amplitude of the carrier is changed in accordance with the information signal. A replica of the information signal can be observed in the envelope of the modulated carrier.
4. In frequency modulation (FM) the frequency of the carrier is changed to incorporate the information of the modulating signal. The amount of frequency change, frequency deviation Δf, is directly proportional to the amplitude of the modulating signal. The rate of frequency deviation, f_m, is equal to the frequency of the modulating signal.

GLOSSARY OF TERMS

Bandwidth Of a station's signal, the range of frequencies within a group of related frequencies required to transmit a particular signal.

Envelope Of a waveform, the imaginary line drawn through the tips of the individual cycles of the wave.

Frequency deviation In frequency modulation, the amount by which the carrier frequency is changed (increased and then decreased) during modulation by the information signal.

Modulation coefficient (or modulation index) In amplitude modulation, the ratio of the amplitude of the modulating signal to the amplitude of the carrier.

Nonlinear circuit A circuit whose output is distorted and is not a good replica of the input; a circuit in which the ratio of output to input changes with the amplitude of the input.

Percent modulation In amplitude modulation, 100 times the modulation coefficient (see above). In frequency modulation, 100 times the ratio of the actual frequency deviation to the maximum deviation permitted by regulation.

Side frequency, lower In amplitude modulation, a signal that is produced as a result of the modulation of the carrier by a single-frequency information signal, and whose frequency is equal to the carrier frequency minus the information signal frequency.

Side frequency, upper In amplitude modulation, a signal that is produced as a result of the modulation of the carrier by a single-frequency information signal, and whose frequency is equal to the carrier frequency plus the information signal frequency.

Sideband The group of frequencies produced during the process of modulation.

REVIEW QUESTIONS: BEST ANSWER

1. The envelope of an amplitude-modulated carrier exhibits a replica of the: **a.** carrier. **b.** information signal. **c.** upper side frequency. **d.** lower side frequency. **e.** a and d.
2. A 350-V carrier is being amplitude modulated by a 230-V audio-frequency signal. The modulation coefficient or index is: **a.** 0. **b.** 0.25. **c.** 0.54. **d.** 0.66. **e.** 1.52.
3. An AM broadcast station is being modulated by a program whose frequency content is in the range 50 to 4000 Hz. The station's assigned carrier frequency is 1040 kHz. Assuming that there are no harmonics, the lowest frequency that the station will

broadcast during this particular program is: **a.** 990 kHz. **b.** 1036 kHz. **c.** 1039.05 kHz. **d.** 1040 kHz. **e.** none of these.

4. The bandwidth of the transmission of the station in Question 3 is: **a.** 100 Hz. **b.** 3950 Hz. **c.** 4050 Hz. **d.** 7900 Hz. **e.** 8000 Hz.
5. A 50-kW AM broadcast station is being modulated at the 90% level. The carrier power is: **a.** 20.25 kW. **b.** 40.5 kW. **c.** 45 kW. **d.** 50 kW. **e.** none of these.
6. The power in the sidebands of the station of Question 5 is: **a.** 20.25 kW. **b.** 40.5 kW. **c.** 45 kW. **d.** 50 kW. **e.** none of these.
7. The percent increase in the effective value of the antenna current when the modulation of an AM transmitter is increased from 0% to 75% is approximately: **a.** 10%. **b.** 13%. **c.** 20%. **d.** 37.5%. **e.** 50%.

REVIEW QUESTIONS: ESSAY

1. An information signal consists of a sine wave. What two aspects of the sine wave contain the information?
2. Define *envelope of a waveform.*
3. Define *modulation coefficent* or *index*. Give an example illustrating its use.
4. Define *percent modulation* as applied to amplitude modulation. Give an example of its application.
5. Describe briefly the type of circuit that is required to achieve amplitude modulation.
6. Describe briefly the nature of any nonsinusoidal, periodic waveform, according to mathematical analysis.
7. Define *upper side frequency* and *lower side frequency*. Give formulas. Define *sidebands*.
8. Define *bandwidth* of an amplitude-modulated transmitter.
9. Explain how the two basic aspects of an information signal are manifested in a carrier that has been frequency modulated. (*Hint*: Explain the appearance of the waveform as it might be observed on an oscilloscope.)

EXERCISES

1. Sketch the waveform of a carrier that has been amplitude modulated to each of the following levels: **(a)** 0%; **(b)** 25%; **(c)** 50%; **(d)** 100%.
2. A 280-V carrier signal is being amplitude modulated by an information signal whose amplitude is varying. Calculate the modulation coefficient for each of the following information signal amplitudes: **(a)** 20 V; **(b)** 70 V; **(c)** 150 V; **(d)** 200 V; **(e)** 280 V.
3. A 1400-kHz carrier is being amplitude modulated by an 1800-Hz tone. List the frequencies of all of the components produced by the modulation process. Ignore harmonics.
4. Calculate the side frequencies if a 760-kHz carrier is amplitude modulated by the following information signals: **(a)** single frequency of 1200 Hz; **(b)** composite signal of 800 Hz and 2.4 kHz; **(c)** composite of 400 Hz, 1.8 kHz, and 4.0 kHz; **(d)** composite of 30 Hz, 720 Hz, 8400 Hz, and 15 kHz.
5. Determine the bandwidth of each of the transmissions of Exercise 4.

6. Refer to Exercise 2. Assume that the specified carrier signal is being fed to an antenna whose radiation resistance, R_R, is equal to 50 Ω. Calculate the component of carrier power for each of the information signal amplitudes.
7. Calculate the total power in the sidebands for each of the information signals of Exercise 2. Assume that the antenna radiation resistance is 50 Ω.
8. Determine the percent increase in the effective value of the antenna current, compared to 0% modulation, for each of the modulation levels indicated in Exercise 2.

3

THE AM RECEIVER

In learning an electronic system, an effective approach is to proceed from the general to the specific. Chapters 1 and 2 have given an overview of electronic communication systems and some general theory about the process of modulation. It is now time to change the focus on the lens of our camera and begin to zoom in on the specifics of the system itself.

Focusing on the details of the receiver, before studying the circuits of the transmitter, may not appear to be the most logical next step in terms of how a communication system works. However, from a practical learning standpoint, approaching the receiver first has a number of advantages. Perhaps the most important advantage, particularly for persons who are studying the electronics of communications for the first time, is that it is much more practicable in the laboratory to work with receivers than with transmitters. Receivers are less expensive. They may be operated by anyone without any real concern for governmental regulations. On the other hand, a transmitter must be operated only in compliance with all regulations regarding such operation, to avoid interfering with others utilizing the frequency spectrum. This implies a knowledge that the learner may not have. Nearly all of the basic circuit functions required by transmitters are also found in an AM receiver. Thus practical work with typical communication circuits can proceed on a simple receiver in the laboratory concurrently with the study of the theory of electronic communication.

3.1 THE RECEIVER BLOCK DIAGRAM

In Chap. 1 the receiver was represented on the block diagram of an electronic communication system as a single block. Let us take a somewhat closer look at the receiver function by inspecting a block diagram of just the receiver subsystem, a diagram constructed using blocks representing the essential functions of a receiver (see Fig. 3.1).

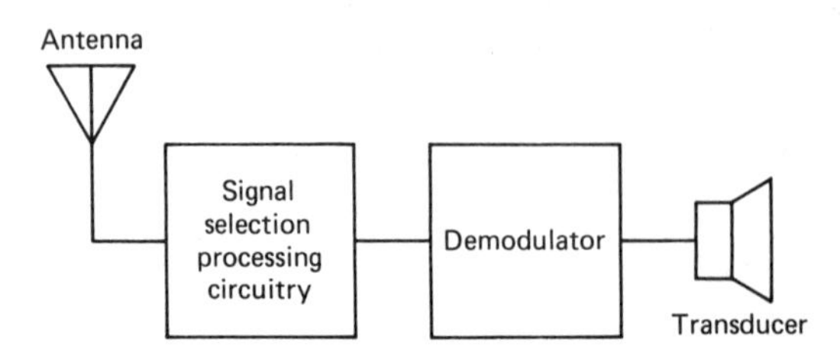

Figure 3.1 Block diagram of a receiver.

A receiver must include the following essential functions: (1) an antenna to intercept the energy from the transmitter and convert it to an electrical signal usable by the receiver circuitry; (2) processing circuits to select the desired carrier signal and reject all of the undesired carriers present at the antenna—to achieve the formal characteristic called *selectivity;* (3) a demodulator (or detector) circuit to recover the information content of the communication from the received carrier signal and convert it into a local information signal; and (4) a transducer—speaker or headphone—to convert the electrical energy of the information signal back into the mechanical energy of sound. Missing from this list of essential functions is that of amplification circuitry to achieve sensitivity. As will be seen, although all truly practical receivers incorporate amplification, it is possible for a simple receiver to function without amplification.

3.2 A PRIMITIVE RECEIVER

The dictionary definition of "primitive" includes such adjectives as "crude," "simple," "primary," and "basic." It is in the sense of these words that the diode-detector receiver is presented as a primitive receiver. This circuit is commonly packaged and marketed as a working toy. It is often referred to as a "crystal set," a "fountain pen radio," etc. It is the simplest circuit that can actually perform successfully all the essential functions of a receiver. Let us examine the circuit and the theory of its operation in detail (see Fig. 3.2).

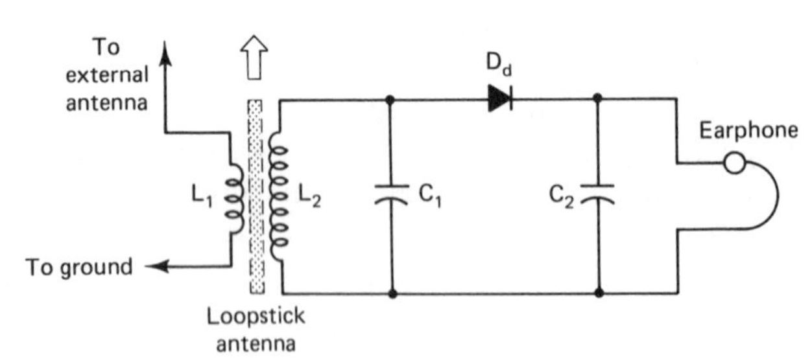

Figure 3.2 Schematic circuit diagram of diode-detector receiver.

The antenna and selection circuits of this receiver are often one unit, physically. This device is called a *ferrite-core loopstick antenna* (see Fig. 3.3). It consists of a small, movable core of material called ferrite inside a cardboard or plastic tube upon which is wrapped one or more multiturn coils of fine wire.

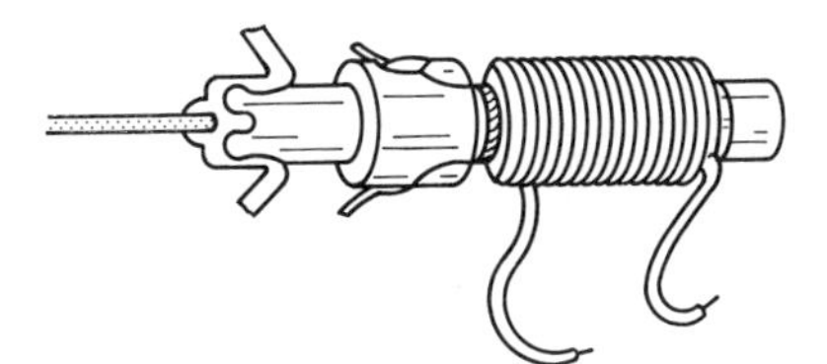

Figure 3.3 Ferrite loopstick antenna.

Ferrite is composed of a synthetic ceramic material which is also ferromagnetic. Initially, it is a fine, powder-like material of individually insulated granules. The bar is made by placing powder in a form and pressing under high pressure. The material has a very high permeability and, therefore, magnetic fields are induced in it relatively effectively by the passing electromagnetic fields of radio carriers. The induced fields, in turn, induce electrical currents in the coils wound on the bar. Because the individual granules of the core are insulated, there is very little loss of energy to eddy currents in the core. The ferrite loopstick, then, intercepts the energy of radio carriers from "air waves" and converts it into carrier signal currents at the input to the receiver.

As there are always many carriers competing to be received by any receiver, a next step in the process of successful reception is for the receiver to select the desired carrier and eliminate, or at least to drastically reduce the energy level, of the undesired carriers. In the case of the diode-detector receiver, this selection function is performed by a tuned (or resonant) circuit consisting of one of the coils (L_2) of the loopstick and the capacitor C_1 (see Fig. 3.2). Because the core of the loopstick is movable, the inductance of L_2 is variable, and therefore the frequency at which the circuit is resonant is variable.

Although, at first glance, the circuit of L_2C_1 appears to be a parallel resonant circuit, careful analysis will reveal that it is a series resonant circuit. (Recall that the voltages induced in the individual turns of a transformer winding are in series with each other as well as with the inductance of the winding itself. Therefore, the voltage induced in L_2 is in series with L_2 and C_1, and the circuit is a series resonant circuit.) Recall, also, that in a series resonant circuit at resonance the voltage across L or C is equal to Qv_a, where v_a is the voltage applied to the circuit and $Q = X/R$.

The action of the resonant circuit is such that when voltages (or currents) of various frequencies are present, the circuit discriminates in favor of the voltage at the resonant frequency and to the disadvantage of voltages of other frequencies. This is the selection process needed by a receiver.

The concepts involved are illustrated in Fig. 3.4. In Fig. 3.4(a) the antenna/input circuit of the receiver is depicted as consisting of a number of ac sources of various frequencies and amplitudes to represent the carriers present at the antenna. A spectrogram, showing the relative frequencies and amplitudes of

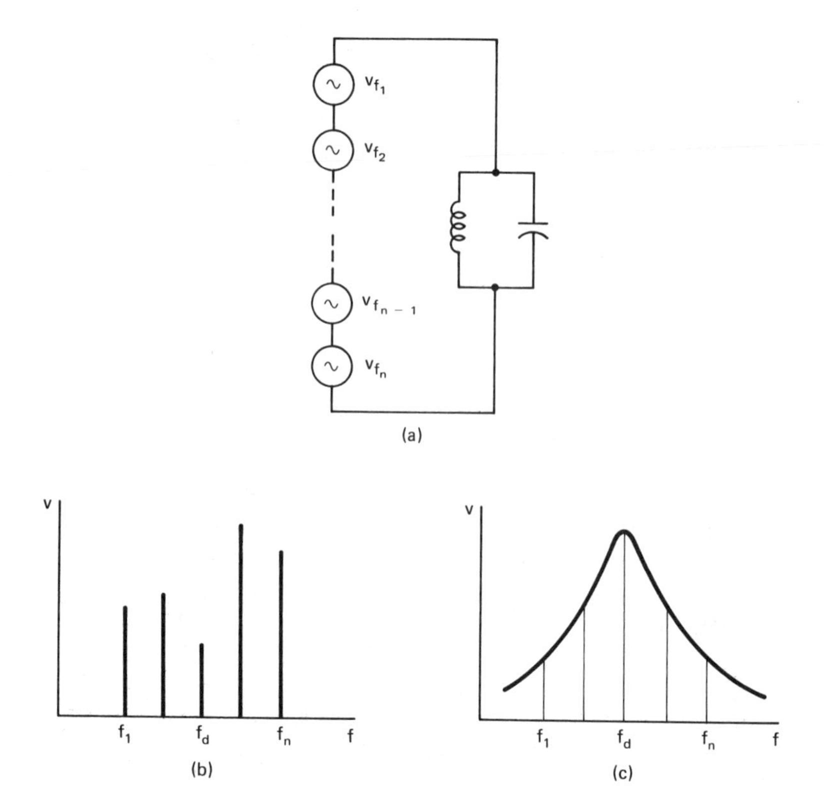

Figure 3.4 (a) Equivalent of antenna input circuit; (b) spectrogram of carriers present at antenna; (c) spectrogram of output signals.

these sources, is shown in Fig. 3.4(b). (The frequencies and amplitudes shown are arbitrary and do not purport to represent any real-life situation.)

If we assume that f_d is the desired carrier frequency, the effect of the tuned circuit could be that shown in Fig. 3.4(c), where f_d now has the greatest amplitude. Figure 3.4(c) depicts the result of tuning the L_2C_1 circuit so that the resonant frequency response curve is positioned to favor the desired frequency. This is exactly the function you perform when you turn the dial of a receiver to tune in a particular station. It is important to recognize that the signals of undesired stations are attenuated (reduced in amplitude) relative to the desired signal, but that they are not and cannot be eliminated completely.

The AM Detector Circuit

If the receiver works correctly, a signal voltage of significant magnitude representing the desired carrier will appear across the capacitor C_1 in the circuit of the primitive receiver. The next step in the process of "receiving" the desired carrier, in this receiver, will be to recover the information—to demodulate or detect the carrier.

The heart of a very common detector circuit for demodulating amplitude-modulated waves is a diode. Recall that a diode is a device that permits electrons to pass in only one direction. When used in a circuit energized by an alternating-current (ac) source, such as an ac power line source or an RF carrier signal, the diode will cause a pulsating unidirectional current—a "dc current"—to flow in the circuit. To complete the detector circuit, to provide a path for this dc current, there will be present a resistor or other circuit device called the *detector load*. Thus in the case of the primitive receiver (Fig. 3.4), the source for the detector is capacitor C_1, D_d is the detector diode, and capacitor C_2 is a filter in parallel with the headphone, the detector load.

Let us analyze in some detail what happens to a modulated carrier signal as it is processed by the detector circuit. Consider the simple circuit of Fig. 3.5(a), in which the capacitor portion of the detector load is missing. When an AM wave, such as that in Fig. 3.6(a), is applied to this circuit, the diode rectifies the RF ac cycles and the voltage across R_L might appear as in Fig. 3.6(b), if displayed on an oscilloscope. The voltage is that of a half-wave rectifier; the positive alternations are preserved and the negative alternations are eliminated.

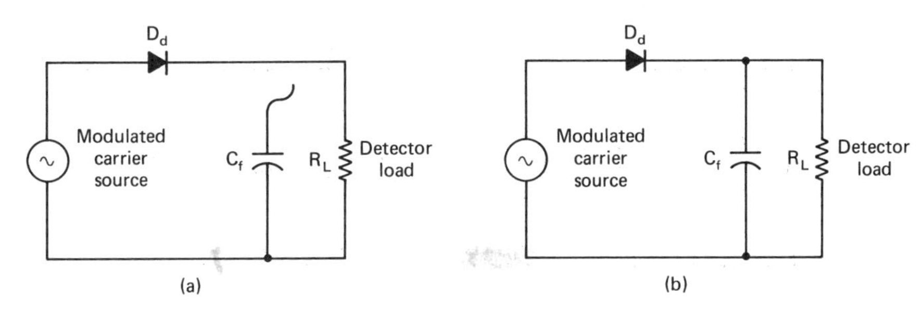

Figure 3.5 (a) Diode detector with unfiltered load; (b) diode detector with filtered load.

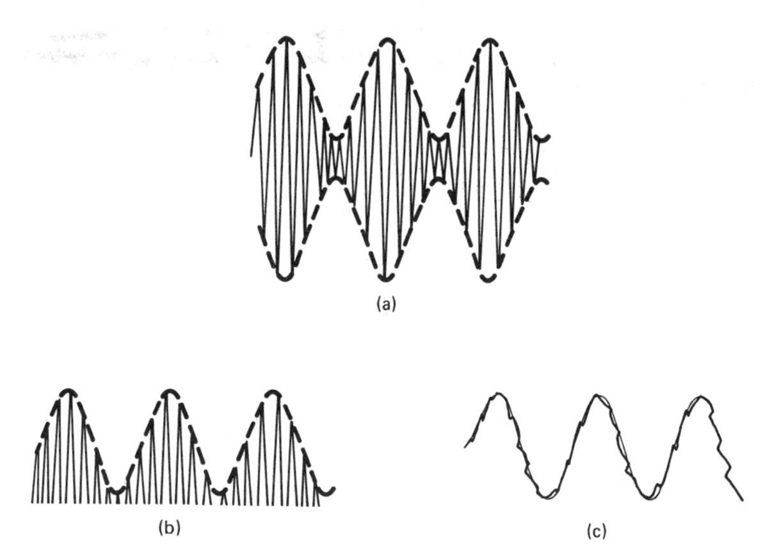

Figure 3.6 (a) Modulated carrier detector input; (b) unfiltered detector output; (c) filtered detector output.

Notice, however, that the envelope (the peaks of the individual cycles) of the waveform still provides a replica of the information signal. This waveform is called the *unfiltered output* of the detector; "unfiltered" because the filtering device, C_f, is disconnected, "output" because it is the waveform after the diode and across the diode load, R_L.

If the filter capacitor, C_f, is reconnected, the oscilloscope waveform display would appear as in Fig. 3.6(c). Here, because of the time constants of the *RC* charge and discharge circuits, the output of the detector is no longer individual cycles of rectified RF ac voltage. Because capacitor C_f charges rapidly and discharges much more slowly, its voltage dips only slightly between RF pulsations but follows relatively well the slower changes of the information signal. The output is a continuous dc voltage which varies in a pattern that is a replica of the original information signal (an audio signal in a radio system). The waveform in Fig. 3.6(c) is the filtered output of the detector.

An analysis of an AM detector is not complete unless its operation is also considered in the frequency domain, that is, unless the carrier is considered as a composite signal made up of an unmodulated carrier plus side frequencies (see Sec. 2.5). In this case, the detector circuit is considered from the standpoint of its being nonlinear and in which currents of different frequencies are present. A phenomenon similar to that which occurs during modulation (see Sec. 2.3) will occur. That is, sum and difference frequencies will be generated as a result of the interaction of the several currents present in the nonlinear circuit. Among other frequency currents generated will be the current at the difference frequency:

$$f_{\text{USF}} - f_{\text{carrier}} = f_{\text{information}}$$

when the carrier has been previously modulated by a single-frequency signal. There will also be generated a current at the frequency

$$f_{\text{carrier}} - f_{\text{LSF}} = f_{\text{information}}$$

Notice that in each case, the difference frequency is equal to the original information signal frequency. Thus a current at the original information frequency is present in the detector circuit and will produce a voltage across the detector load or will actuate the headphone in the case of the primitive receiver. The information signal is a relatively low frequency signal. All other signals present: f_c, f_{USF}, f_{LSF}, and all sum frequencies are much higher. The capacitor C_f in parallel with the detector load can now be considered as a low-pass filter circuit. This circuit produces a significant output at the low, information signal frequency and attenuates or bypasses to ground all the other higher-frequency currents.

Whether an AM detector is thought of as a circuit that converts the input RF signal to a pulsating dc current, or as a nonlinear circuit in which is generated the original information signal from the composite input signal, its purpose, again, is to recover the information signal from the modulated carrier signal. In each case, part of the task consists of attenuating the RF signals present. In each case, the output is an information frequency voltage across the detector load, or an information signal current in a transducer. The waveform of the output will be an exact replica of the envelope of the modulated carrier if all processing circuits are working perfectly—are not distorting either the carrier or the information signal. The envelope will be a perfect replica of the waveform of the original information signal if all relevant circuits are working perfectly in both the transmitter and the receiver.

Reproduction

The final function which the simple diode-detector receiver is capable of performing is that of converting the electrical energy of the information signal, as it appears at the detector output, to the mechanical energy of sound. This is called *reproduction.* Without amplification, however, the energy level is extremely low, of the order of microwatts. The only power available is that which the modulated carrier itself contained as it was intercepted by the antenna of the receiver. The only transducer that is capable of producing audible sound at this power level is an earphone. The current from the detector passing through a coil in the earphone produces a changing magnetic field, which, in turn, causes a small diaphragm to move in and out, creating sound waves. These sound waves are a reproduction of the original sound waves used to modulate the carrier at the transmitter. The process of "receiving" is completed with the reproduction of the sounds transmitted by the system.

3.3 IMPROVING ON THE PRIMITIVE RECEIVER: SOME HISTORICAL DEVELOPMENTS

Because of its simplicity, the primitive receiver described in Sec. 3.2 suffers from a number of shortcomings in its performance. Its sensitivity is very poor—it can receive, and reproduce audibly, the signals of only a few, strong local stations. Its selectivity also leaves a great deal to be desired. Often, even the few stations that a crystal set is able to receive cannot be separated cleanly—they interfere with each other at the listener's ear. And, of course, a receiver that can only operate an earphone has very limited acceptance in today's world. A modern radio receiver, then, in order to achieve its very much more satisfactory levels of selectivity and sensitivity, must consist of a more complex set of functions than the simple diode-detector receiver.

The story of the gradual evolution of today's electronic communication circuits is essentially identical to the early history of the entire field now known as electronics. A look at just a few of the early developments in the improvement of the receiver will provide at least a taste of the flavor of the history of electronics as well as an understanding of a few important, basic circuits.

As mentioned in Chap. 1, the first significant improvement in receivers was the development of a detector that operated on the basis of rectification, the same as the present-day detector. Prior to 1906, receivers used a device called a *coherer* in the detector circuit. The coherer consisted of a small glass tube containing a loose mixture of carbon and steel filings which normally presented a high resistance to the circuit. The coherer was connected as a resistive device in series with a tuned circuit, a battery, and a transducer or output device. The filings were tarnished with a thin film of oxide. When an RF signal was present, the films of adjacent particles were bridged, making them stick together or "cohere," thus lowering the resistance of the device and allowing current to flow through the transducer and produce an output. The coherer detector was very susceptible to static and electrical interference. It frequently had to be vibrated by hand to loosen the particles and restore its high resistance. All in all, the coherer was not a very satisfactory device, but was the best available at the time.

Attempts to improve the coherer led to the discovery of the concept of the rectifying effect of a sharply pointed conductor in contact with a semiconductor, such as the oxides of certain metals. This is the principle of the device known as a *point-contact semiconductor diode,* the forerunner of all semiconductor devices—the junction diode, the transistor, etc. The contact detector, as it was called, was a great improvement over the coherer. Attempts to improve the contact detector led to the discovery of the excellent properties of silicon as a semiconductor. Of course, work with silicon and other semiconductor materials ultimately led to the development of the bipolar junction transistor, the heart of many modern developments in electronics.

The Regenerative Receiver

The use of the contact detector improved the operation of early receivers primarily by reducing the detector's susceptibility to noise and increasing somewhat the efficiency of the recovery of energy from the RF wave. We could say that there was some improvement in the sensitivity of the receiver. The improved detector did virtually nothing to improve the selectivity—the ability to separate signals. Since this ability depends on the narrowness of the response of the basic tuning circuit of the receiver, a further narrowing of the response is required in order to improve selectivity. Before examining how such improvement might be achieved, let us review briefly some of the concepts and terminology related to the specification of the response of a tuned circuit.

Refer to Fig. 3.7, the familiar curve representing the response of a series resonant circuit. In this instance the response represents output voltage (voltage

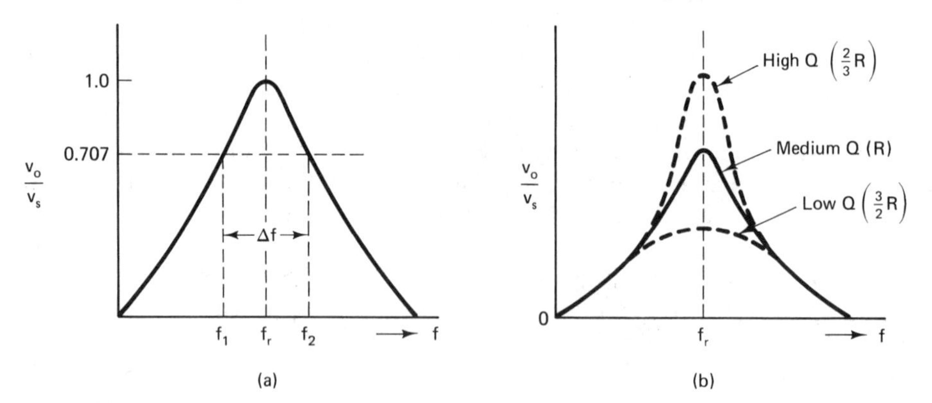

Figure 3.7 (a) Response curve of series resonant circuit; (b) effect of Q on bandwidth.

across either X_C or X_L), v_o, divided by the source voltage, v_s, plotted against the frequency of the source voltage. You will recall that the term *bandwidth* is used to specify the narrowness of the curve. The bandwidth is defined by the frequencies f_1 and f_2, which in turn are called *half-power points,* or points at which the voltage is down to 70.70% of the maximum value, that is,

$$\Delta f = f_2 - f_1$$

Furthermore, bandwidth is related to the Q of the circuit:

$$\Delta f = \frac{f_r}{Q}$$

where f_r is the resonant frequency of the circuit. The bandwidth of the circuit, the inverse of the narrowness of its response, is inversely proportional to the Q of the circuit. A high-Q circuit provides a narrow bandwidth, a low-Q circuit, a wide bandwidth. The effect of Q on the response of a tuned circuit is shown in Fig. 3.7(b).

To increase the selectivity of the tuning circuit of a receiver, then, it is necessary to increase its Q, or the equivalent. Since $Q = X_L/R$, the Q of a circuit may be increased by increasing X_L by making L larger and C smaller, or by decreasing R. (Recall that, at resonance, $X_L = X_C$ and $f_r = 1/2\pi\sqrt{LC}$.) To obtain high-Q circuits it is necessary to maximize the L/C ratio. However, unfortunately, in radio-frequency circuits, because of the phenomenon known as skin effect, increases in L beyond some point are accompanied by corresponding increases in R. There is thus a practical limit to the maximum value of Q obtainable simply by the choice of component values.

To obtain highly selective circuits, then, requires some approach other than that of choice of components. The formula for Q, $Q = X/R$ is derived from the ratio, reactive power/dissipation, or I^2X/I^2R. Thus it can be seen that Q could be increased if the dissipation (I^2R) in a circuit was decreased.

The apparent dissipation can be decreased in a resonating circuit if energy is injected into the circuit from an external source at the resonant frequency. That is exactly the principle utilized in one of the first improvements to receivers developed after the invention of the amplifying vacuum tube. The basic idea of the regenerative receiver is to reinject (regenerate) energy into the resonating tuning circuit to reduce the apparent dissipation, increase the Q, and thus increase the selectivity.

The schematic circuit diagram of a vacuum-tube regenerative receiver is shown in Fig. 3.8(a) and of the transistorized equivalent in Fig. 3.8(b). Let's examine the operation of the transistor version in detail.

The circuit composed of L_2C_1 is an input tuned circuit whose operation and function are essentially the same as those of the tuned circuit of the crystal receiver (see Fig. 3.2). Voltages of the various carriers whose air signals are intercepted by the antenna are induced in the coil L_1. The circuit discriminates in favor of a desired carrier voltage v_c, having been tuned to be resonant at the frequency of that carrier, f_c. That is, the circuit develops a "Q-multiplied" voltage, Qv_c, across C_1, for the carrier f_c. The signal Qv_c is fed to the base of the transistor through the parallel

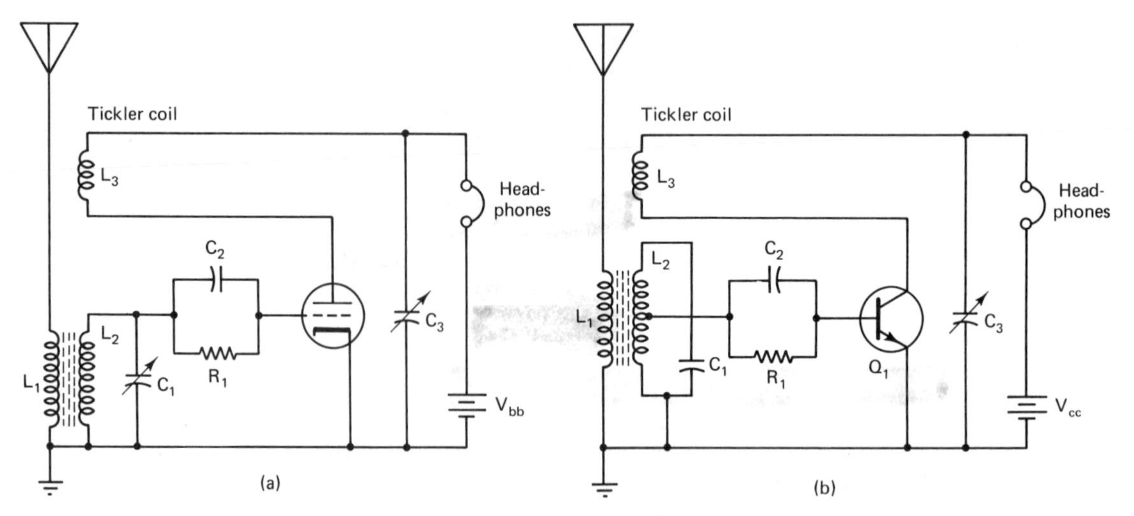

Figure 3.8 (a) Vacuum-tube regenerative receiver; (b) transistorized regenerative receiver.

bank, R_1C_2 (the R_1C_2 bank is in series with any current that flows from the input tuned circuit to the base of the transistor).

During the first moments of operation of the circuit several events happen which we analyze in a stepwise fashion:

1. On the first positive half-cycle of the voltage v_c, the base of the transistor is driven positive and the base–emitter junction is forward biased. Electrons flow from ground through the base–emitter junction, the R_1C_2 bank, and return to ground through the L_2C_1 circuit. This current, obtained by rectification of the input signal, begins the charging of C_2, with positive polarity on the left side of C_2. Let's call the voltage across C_2,V_2.
2. The RF ac voltage, v_c, reverses its polarity. The voltage on the base of the transistor is driven negative. At the negative peak of the cycle of the ac signal, this voltage is negative by an amount equal to the peak value of v_c plus V_2 (the dc voltage across C_2). The base–emitter junction is now reverse biased. Capacitor C_2 starts to leak (discharge) some of its charge through R_1. The amount of discharge is determined by the relative values of the time constant of the circuit, $t_{12} = R_1C_2$, and the period, $t_c = 1/f_c$, of the carrier signal.
3. When the signal voltage alternates again, making the top of the tuned circuit positive, the voltage at the base of Q_1 is equal to the instantaneous value of v_c minus V_2. Near the positive peak of the carrier wave, then, the base of Q_1 is again positive with respect to the emitter and additional current flows in the base–emitter circuit. A further charging of C_2 and an increase of V_2 (remember, positive polarity on the left side of C_2 in the circuit diagram) is the result.
4. This process of the charging of C_2 at the positive peaks and its discharging during negative alternations of v_c continues for several cycles of the carrier signal. Finally, the process stabilizes; the amount C_2 charges during positive peaks is just equal to the amount it discharges during negative alternations. The condition is characterized as a stable average charge on C_2 and a stable average value for V_2. The drawing in Fig. 3.9 depicts the development of the voltage V_2.

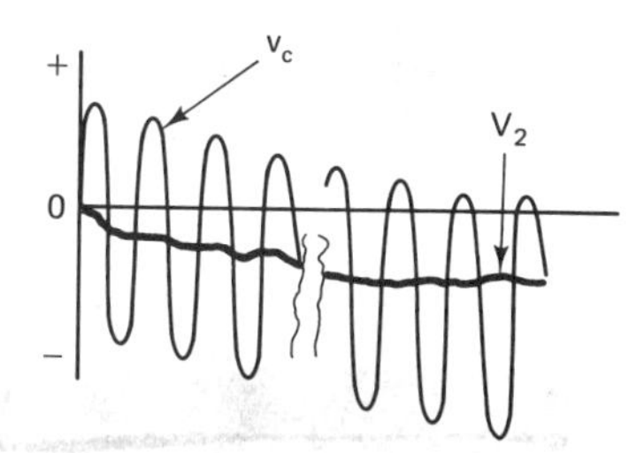

Figure 3.9 Development of grid-leak bias voltage.

The stable voltage V_2 (see step 4) represents a reverse bias for Q_1. The circuit, with the operation described, is often called a *clamping circuit*. The base is said to be "clamped" to the negative voltage V_2. Of course, the signal voltage is superimposed on (adds to and subtracts from) the dc voltage.

The dc voltage in circuits that operate in this fashion can be thought of as being *dynamically* generated: It is derived from the signal voltage and is proportional to it. Furthermore, if there is no signal, there is no clamping voltage.

In the tube version of this circuit (and many other similar circuits) the clamping voltage has long been called the *grid-leak bias voltage*. This voltage provides a negative dc bias voltage and biases the tube into cutoff.

As indicated, the relative values of the time constant of the clamping circuit, t_{12}, and the period, t_c, of the signal determine the average value of the clamping voltage: A relatively long time constant produces a greater magnitude of clamping voltage. The importance of this fact is that the time constant can be adjusted so as to determine how long collector current will flow during each positive alternation of the input signal. Time during which current flows is called the *conduction angle* of the circuit.

Remember, collector current flows only during the portion of the positive alternation when the instantaneous value of the signal exceeds the voltage across C_2. Typically, a regenerative circuit is designed and adjusted for class C operation, that is, for a conduction angle of considerably less than 180° (less than one-half of each cycle of the signal voltage). The collector current is simply short pulses of RF current. Of course, these pulses are *amplified* versions of the base current pulses.

The collector current pulses pass through coil L_3, called the *tickler coil*. The tickler coil is always in close proximity to L_2, and therefore the magnetic flux produced by the current through L_3 cuts the turns of L_2, inducing a voltage in L_2. If the turns of L_3 are oriented appropriately, the increasing current in L_3 will induce a voltage in L_2 which is in phase with the voltage induced by the passing carrier. In short, energy will be fed back to the resonant circuit at just the right moment to *aid* the action that is underway there.

The positive feedback of energy from the transistor circuit replaces energy dissipated (used up) in the resistance of the resonant circuit. The effect is that of a *negative resistance*. This negative resistance reduces the apparent R of the tuned circuit. The Q ($Q = X_L/R$) of the circuit is increased significantly. Since the bandwidth of a resonant circuit is inversely proportional to its Q (BW $= f_r/Q$), the bandwidth is narrowed. The selectivity of the circuit is increased! The primary purpose of the circuit—increased selectivity—is realized.

Of course, if the incoming carrier is modulated, the base and collector current pulses vary at an audio rate. The diaphragm of the headphone responds to the audio component of the current and reproduces the original sound transmitted by the system. In effect, the signal is demodulated in the base circuit. The circuit produces amplification of the demodulated audio signal as well as regeneration of the RF signal.

If the amount of energy fed back exceeds that which is dissipated, this circuit will *oscillate;* that is, it will still produce RF output current pulses even if the input signal voltage is removed. It will generate a signal by itself.

If the circuit were to oscillate, the input signal would be swamped (overwhelmed), and the circuit would cease to perform its intended function. The amount of energy fed back must be controlled. Capacitor C_3 is a variable capacitor that provides the return path for the ac components of the current through the tickler coil. Reducing the value of C_3 will increase the impedance of this ac return path and thus decrease the ac current. The reverse of this is also true.

When using this receiver the operator must not only adjust C_1 and/or the slug for L_2, to tune in the desired station, but for each station, must also adjust C_3 carefully to provide sufficient feedback for good selectivity but not so much as to put the circuit into oscillation. Although the receiver achieves significant improvement over the simple crystal detector, it is still inadequate for most applications of radio. It, too, is often packaged as a kit for assembly and use, especially by children or young people who are at the threshold of a hobby interest in electronic communication.

In summary, the regenerative receiver provides improvement over the simple diode-detector receiver because it (1) incorporates some audio amplification, and (2) improves selectivity by positive feedback of the RF signal which effectively reduces dissipation in the tuning circuit, thereby increasing its Q and therefore its selectivity. In addition to its amplification, the circuit incorporates two other important circuit principles: (1) The principle of positive feedback is the fundamental requirement of a class of circuits called *oscillators,* which will be presented in greater detail later. (2) The principle of generating a dc bias voltage, sometimes called *dynamic dc bias,* from a signal voltage is utilized extensively in electronic systems.

The Tuned-Radio-Frequency Receiver (TRF Receiver)

Another step in the evolution of the modern radio receiver was that of the *tuned-radio-frequency receiver,* or *TRF receiver.* The invention of the vacuum tube, of course, led to the development of the *radio-frequency amplifier,* or, as it is often called, the *tuned amplifier.* The schematic circuit diagram of a typical transistor-type tuned amplifier is shown in Fig. 3.10. Notice the presence of LC circuits on both the input and output sides of the amplifier. Normally, these circuits are designed to be resonant at the same frequency, the operating frequency of the amplifier. The signal must be processed by each of these circuits, in sequence, with the result that each circuit performs its selectivity function—

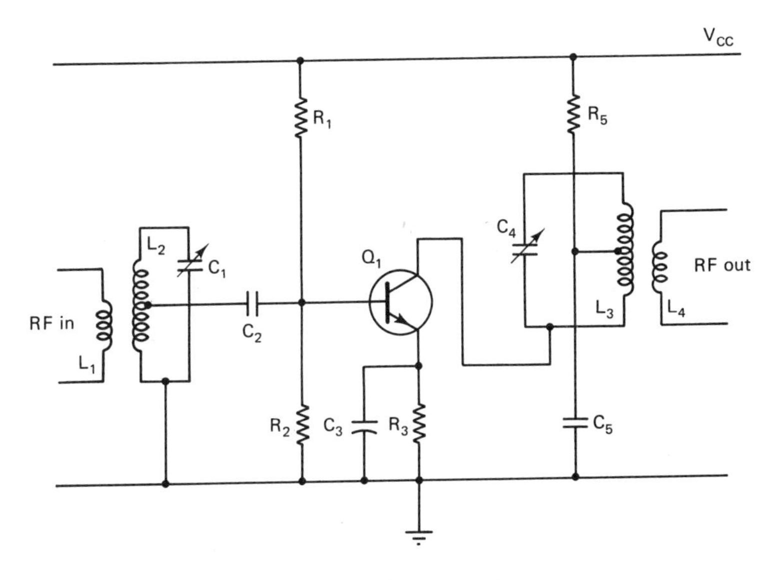

Figure 3.10 Tuned amplifier.

favoring the desired frequency and discriminating against undesired frequency components. The final result is a very great increase in selectivity—the desired frequency is amplified manyfold and the undesired frequencies are attenuated severely. Amplification increases sensitivity as well.

Further increases in both selectivity and sensitivity can be achieved by connecting two or more stages of tuned amplification in cascade (the signal is processed by the stages in series, the output of one stage is the input to the next stage, etc.). The block diagram of a complete receiver utilizing this principle is shown in Fig. 3.11.

Let us assume that the receiver is to be used in a service, such as the AM broadcast band, where a receiver must be designed to permit the operator to choose one station from a number of stations and to alter that choice at will. This feature would require that all of the tuned circuits include variable components, either L or C, which could be changed in unison and by exactly the same amount. For example, the tuning capacitors of all circuits could be mounted on a common shaft, and turning the shaft would change all capacitors, thus accomplishing the "in unison" requirement.

However, it has proven to be virtually impossible to construct such capacitors so identical that their capacitances will change by exactly the same amount when the position of the moving plates is changed. Of course, if all the capacitors in all the circuits are not changed in the same amount, the circuits will no longer be tuned to the same frequency and selectivity and sensitivity will suffer. We say that the circuits "do not track." The TRF receiver has achieved only a very limited acceptance, and that almost exclusively in applications where it is not necessary to tune to more than one station.

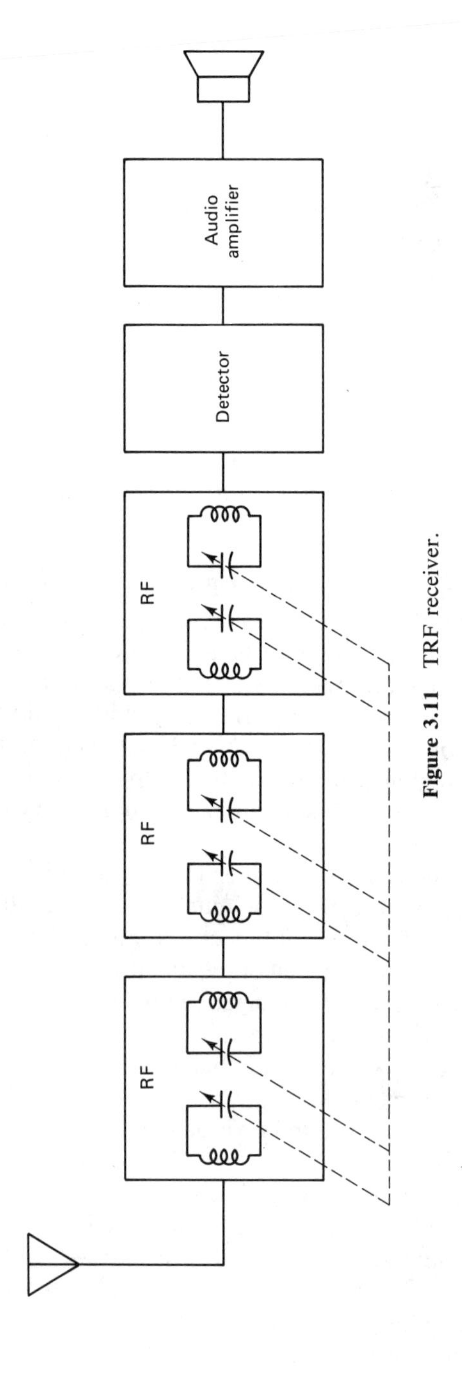

Figure 3.11 TRF receiver.

3.4 THE SUPERHETERODYNE AM RECEIVER: BLOCK DIAGRAM

The urge to build a better radio receiver led ultimately to a design whose performance is so superior that the basic concept of the design has not been changed since it was first conceived in the 1920s. All receivers utilizing the basic concept of this design are called *superheterodyne receivers.* The superheterodyne scheme is used in virtually all types of receivers: AM, FM, TV, radar, telemetry, etc. Let us begin the study of the superheterodyne idea by focusing on the block diagram of a superheterodyne receiver used to receive stations in the AM broadcast band, the most common type of household or car radio (see Fig. 3.12).

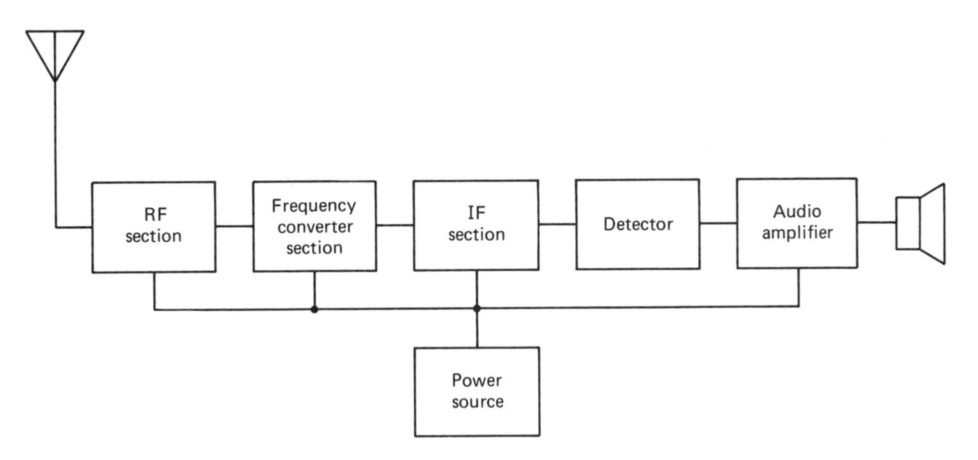

Figure 3.12 Block diagram of superheterodyne receiver for AM broadcast band.

RF Section

The purpose of the RF section is to achieve some preliminary, overall selectivity—to begin the process of pulling the signal of one station up out of the confusion of many signals. In many receivers the RF section may be nothing more complicated than a winding on an antenna loopstick connected in a series resonant circuit with a variable capacitor, a circuit virtually identical to the input circuit of the diode-detector receiver (Sec. 3.2). In receivers with more critical requirements because of their high price, or application (e.g., two-way radio communications systems), the RF section will include at least one stage of tuned RF (TRF) amplification; that is, an active device such as a transistor or tube will be included, along with both an input and an output tuned circuit.

Heterodyne Frequency Conversion Section

The heterodyne frequency conversion section provides a unique, essential function in superheterodyne receivers; its name is the source of the name for the receiver design. *Heterodyne* is the word used to describe the phenomenon in which the mixing of two frequencies, in an appropriate circuit, produces other frequencies, called *beat frequencies*. There are two beat frequencies. One is the sum of the two original frequencies. The other is the difference of the two original frequencies. The process of heterodyning is often called "beating." You may recall that the results of amplitude modulation (Chap. 2) are similar to those described above for heterodyning.

The frequency conversion function has two requirements: (1) there must be a locally generated RF frequency present to beat with the incoming carrier, and (2) there must be available a circuit which has the characteristic (nonlinearity) to cause the heterodyning to occur. The first requirement is met by a circuit function called the *local oscillator* (LO), the second by a function called the *mixer* (MIX).

Very briefly, the heart of the theory of operation of the unique feature of the superheterodyne receiver is this: a signal, frequency f_{LO}, from the local oscillator (LO) is beat with the desired carrier, frequency f_c, in the mixer circuit, producing sum and difference frequencies. The difference frequency, $f_{LO} - f_c$, is called the *intermediate frequency* (IF). The intermediate frequency is the same for all selected carriers. The intermediate frequency is "selected" in the intermediate-frequency amplifier; that is, the IF is amplified and other frequencies are discriminated against, or attenuated. Subsequent to the IF section, the selected IF signal is demodulated and the audio signal is processed and converted into sound energy.

The Local Oscillator (LO)

As indicated previously (see Sec. 3.3), when a circuit produces a signal within itself, it is said to be oscillating. When a circuit is designed to perform that function it is called an *oscillator circuit,* or simply an *oscillator*. The local oscillator (LO) in a superheterodyne receiver must be designed to meet certain requirements: (1) The frequency of operation must be variable. (2) The variable device that will be used to change the frequency must be available to the user. (3) The frequency of operation, at a given setting of the frequency control device, must be stable, not subject to significant change due to variations in temperature, humidity, etc. (4) The amplitude of the LO signal must always be larger than the amplitude of the carrier to ensure successful heterodyning.

The operating frequency of the local oscillator in AM broadcast superheterodyne receivers is designed to be higher than that of the desired carrier. The intermediate frequency of all AM radios, other than car radios, is practically, but not formally or legally, standardized at 455 kHz. Therefore, for most radios, for a given dial (station) setting, the local oscillator frequency is equal to the station frequency plus 455 kHz. Since the AM broadcast band includes the frequencies between 540 and 1600 kHz, the LO must be tunable between the frequencies of 995 and 2055 kHz.

Example 1

An AM broadcast receiver whose IF section operates at 455 kHz is tuning in a station with an operating frequency of 890 kHz. Determine the local oscillator frequency.

Solution

$$f_{LO} = f_c + f_{IF} = 890 + 455 = 1345 \text{ kHz}$$

Example 2

The local oscillator of an AM broadcast receiver is operating at 1850 kHz; its IF is aligned for 455 kHz. Determine the frequency of the station being selected.

Solution

$$f_c = f_{LO} - f_{IF} = 1850 - 455 = 1395 \text{ kHz}$$

Example 3

An AM radio is receiving a station whose carrier frequency is 1150 kHz. The local oscillator frequency is 1412.5 kHz. Determine the frequency to which the IF section is aligned.

Solution*

$$f_{IF} = f_{LO} - f_c = 1412.5 - 1150 = 262.5 \text{ kHz}$$

The Mixer (MIX)

To produce the sum and difference frequencies required for the superheterodyne principle, the signal from the local oscillator must be combined with the signal of the desired carrier in a certain way. The two waves must add phasorally and produce a resultant wave whose amplitude increases or decreases depending on the relative phases of the two component waves (see Fig. 3.13).

The production of the waveform of Fig. 3.13(c) requires that the two waves be mixed in a nonlinear circuit. The waveform in Fig. 3.13(c) is that of a distorted sine wave, a resultant wave very much like that produced by amplitude modulation. Just as in amplitude modulation, this waveform is made up of several components: the two original signals, a signal whose frequency is the sum of the two original

* The IF section of most car radios is designed to operate at 262.5 kHz.

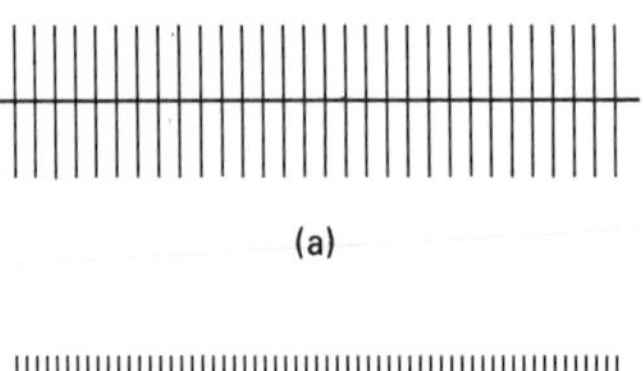

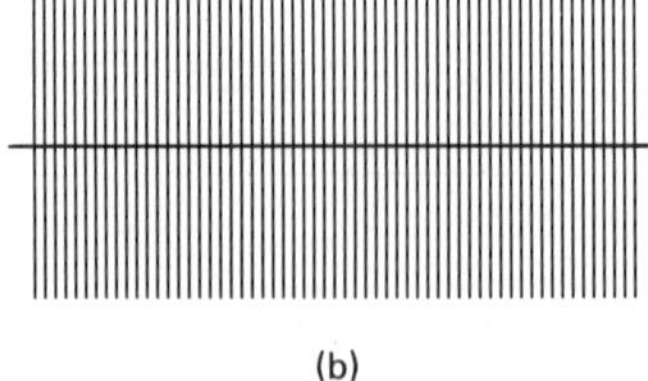

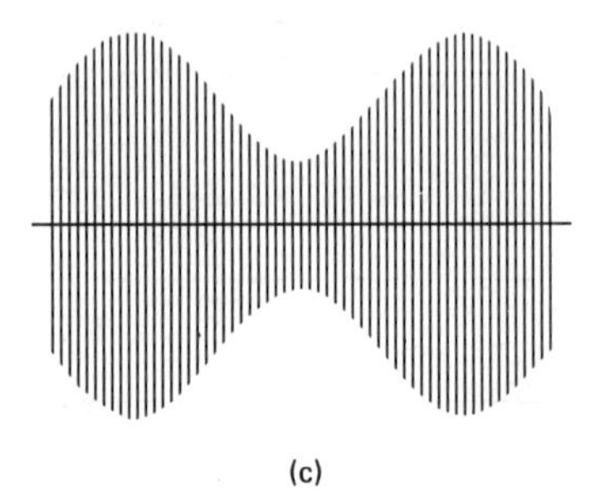

Figure 3.13 Signals of heterodyne mixer: (a) RF signal of lower frequency; (b) RF signal of higher frequency, (c) resultant RF signal of heterodyne mixer—contains sum and difference frequencies.

frequencies, and a signal whose frequency is the difference of the two original frequencies.

The mixer circuit is the nonlinear circuit required to produce the distorted sine wave. The most important output products of the mixer circuit, then, are signals whose frequencies are the local oscillator frequency, f_{LO}, the desired carrier frequency, f_c, the sum frequency, $f_c + f_{LO}$, and the difference frequency,* $f_{IF} = f_{LO} - f_c$ (see Fig. 3.14). Of course, there may be signals of undesired stations present in the mixer; the products of heterodyning for these will also be present at the output of the mixer.

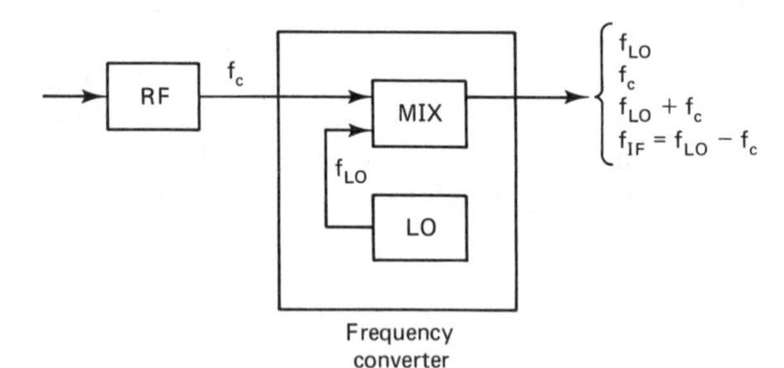

Figure 3.14 Output products of the mixer.

* The difference frequency is an intermediate frequency, intermediate between audio and radio frequencies; it is a super-audio heterodyne frequency, or simply, a superheterodyne frequency. It is this description of the output of the mixer that gave the receiver design its name.

The Intermediate-Frequency Amplifier

When an AM radio is tuned to receive a particular station, the only signal desired at the demodulator is that of the intermediate frequency produced by heterodyning the signal of the desired station with the appropriate local oscillator signal. However, as has been described, the output of the mixer (the frequency converter section) contains many signals in addition to this desired signal. It is the primary function of the intermediate-frequency amplifier section (IF section) to select and amplify the desired IF signal and attenuate the undesired signals.

The IF section is typically one or more stages of tuned amplification. It is considered to be a fixed-tuned amplifier since it can be designed to amplify a narrow band of frequencies centered around the IF frequency: 455 kHz for typical AM radios. Even though the IF circuit is designed as a fixed-tuned amplifier, the inductances in the resonant circuits are virtually always adjustable, by means of movable cores, to permit adjusting the resonant frequencies for optimum performance of the overall section. Because the section is fixed tuned, its frequency of operation can be chosen, as a part of the design process, for optimum selectivity and gain characteristics. It is this last feature that makes the superheterodyne receiver design so highly successful overall.

Remember, the IF section receives signals of various amplitudes and many different frequencies from the frequency converter section. Among these signals is the signal produced by heterodyning the local oscillator signal, of appropriate frequency, with the signal of the desired station. This signal, if the receiver is functioning correctly, is equal to the frequency to which the IF section is tuned, the intermediate frequency for the particular radio. The IF section will process this input from the converter section, amplifying the IF signal of the desired station and attenuating all other signals, and present the result to the detector.

The AM Detector Section

The AM detector in a modern AM radio, one employing solid-state devices such as transistors and integrated circuits, usually consists of a simple solid-state diode, a resistive load, and a capacitor filter. Thus the detector of almost any AM radio is very similar to that of the primitive receiver discussed in Sec. 3.2. The theory of operation of the diode detector of a superheterodyne AM receiver is the same as that presented in Sec. 3.2.

Audio Section

The amount of energy contained in the audio signal at the output of the detector is insufficient to operate a loudspeaker. The audio section of a modern receiver is basically an audio-frequency amplifier designed to amplify the audio signal so as to produce a loudspeaker output at some specified power level. This section almost always contains a volume control to regulate the level of sound output. In all but the least expensive radios, the section also includes a tone control to permit the user to adjust the relative balance in level between the low-frequency content of the program material and its high-frequency content.

Loudspeaker

The purpose of a radio is ultimately to reproduce the sound of the program material being broadcast by the radio station. The loudspeaker is a transducer that converts the electrical energy contained in the audio signal, which the receiver has finally produced from the received carrier, into the mechanical energy of sound. It accomplishes this by means of the interaction of the magnetic field of a permanent magnet and the field produced by the audio-frequency current flowing through a coil placed in the field of the permanent magnet.

3.5 THE SUPERHETERODYNE AM RECEIVER: CIRCUITS

The block diagram of the AM radio receiver has been presented in the preceding section, along with general information about the purpose of each section and how that purpose is accomplished. The next step in the systems approach (Sec. 1.3) is to look closely at the details of the circuits and devices of each block of the block diagram and gain an understanding of how these elements perform their functions. The schematic, or circuit, diagram of a typical transistor radio is shown in Fig. 3.15. This diagram will be examined in detail as a means of presenting information about the operation of typical receiver circuits.

Relating the Circuit Diagram to the Block Diagram

When learning an unfamiliar system (or subsystem) it is helpful, before digging into the circuits, to fit the block diagram to the schematic circuit diagram. This facilitates the study of the circuits by section or function block, a most effective approach. The generalized block diagram of an AM radio, presented in Sec. 3.4, can be fitted to the schematic of Fig. 3.15 very readily.

RF Section

Draw an imaginary, or preferably, an actual line with a pencil, down through the center of the core of transformer T_1 on the diagram of Fig. 3.15. Coil 1 of T_1 and capacitors C_1 comprise the RF section of this receiver. In this instance, the RF section is simply a tunable resonant circuit. As the tuning dial of the receiver is changed to tune in different stations, one section of C_1 is changed and this changes the resonant frequency of $T_1(1)C_1$ to match the carrier frequency of the desired station. Thus a small degree of selectivity is achieved in this simple, nonamplified RF section.

However, nearly all AM radios, other than the very inexpensive, utilize a stage of amplification in the RF section, ahead of the frequency converter. The addition of a stage of tuned RF amplification to the front end of a receiver improves both selectivity and sensitivity significantly. The details of this type of enhancement are presented in a later section of this chapter.

One of the capacitors labeled C_1 is one section of a multiple-plate variable

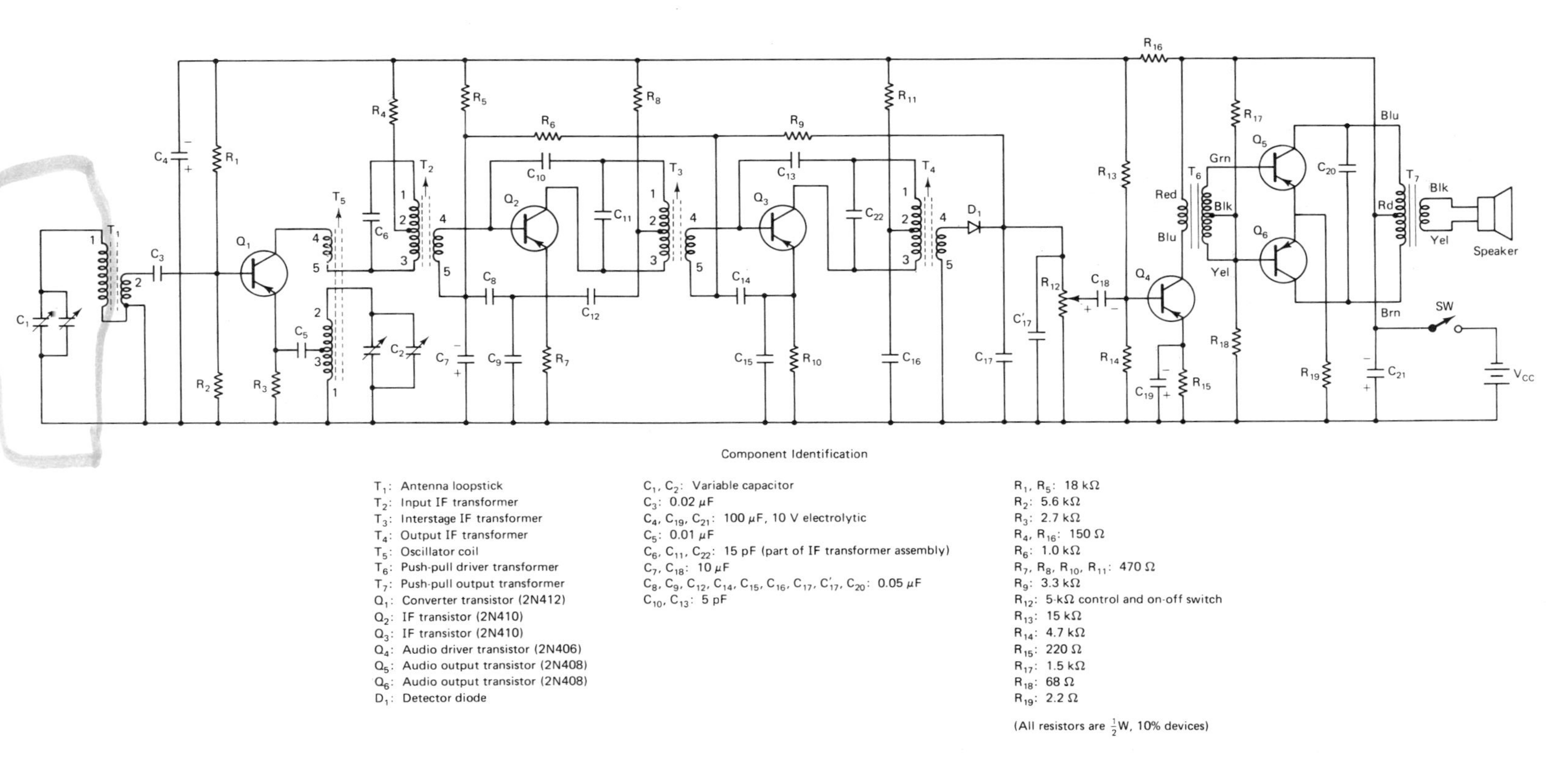

Figure 3.15 Six-transistor superheterodyne AM receiver. (Courtesy of Kelvin Electronics, Inc., 1900 New Highway, Farmingdale, NY 11735.)

capacitor with air dielectric (see Fig. 3.16). The oscillator variable capacitor, C_2, is the second section or "gang" of this two-gang capacitor. Each gang consists of a set of fixed plates, called the *stator,* and a set of movable plates, called the *rotor*. The movable plates of both gangs are mounted on a common shaft which turns when the tuning dial of the radio is changed. As the shaft is turned, the movable plates are moved into or out of the spaces between the fixed plates. The process is called *interleaving* or, more commonly, *meshing* the plates. When the rotor is turned so that the movable plates are filling the spaces between the fixed plates, the capacitor is said to be fully meshed. For the fully meshed condition, the area of exposure of the two types of plates is maximum, and therefore the capacitance is maximum. Thus, for example, when the tuning dial is changed to mesh more fully the two capacitors C_1 and C_2, the circuits of which these are a part are being changed to achieve lower resonant frequencies and the radio is being "tuned" to receive stations of the lower carrier frequencies of the AM broadcast band.

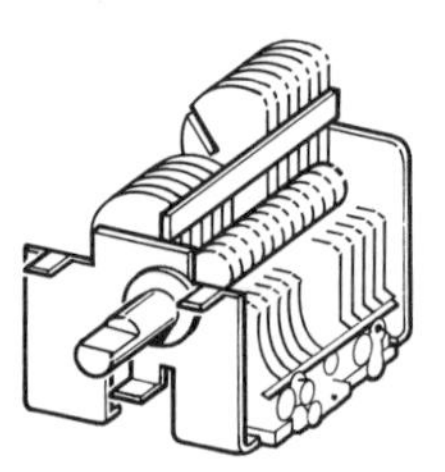

Figure 3.16 Two-gang, air dielectric, variable capacitor.

The capacitors C_1 and C_2 are mounted on a common shaft to permit the changes in the two circuits of which they are a part to "track" each other. ("Tracking" means that the changes occur simultaneously and by appropriate amounts for the respective circuits.) However, these capacitors do not track perfectly and the second parts of C_1 and C_2 shown in Fig. 3.15 represent extremely small, adjustable (as distinguished from variable) capacitors which are used to improve the tracking of the tuning circuits. The adjustment of these small capacitors, which are called *trimmers,* is performed by a technician as part of a process called *alignment*. An alignment is performed as one of the final steps in the manufacturing process, or when a receiver has been misaligned inadvertently, or when components that affect the alignment have to be replaced during the life of the radio.

The transformer T_1 is actually the loopstick antenna. The electromagnetic fields of the various carriers reaching the radio induce magnetic lines of force in the highly permeable ferrite core of the loopstick, and these lines of force, in turn, induce voltages in the primary and secondary windings (terminals 1 and 2 of T_1, respectively). The primary winding (terminal 1) is part of the preselection circuit discussed above. Its action favors the current of the one carrier to which it is tuned. This favorable discrimination provides for a stronger field for that signal and, therefore, an induced voltage of greater amplitude at that frequency in the secondary winding (terminal 2). The voltage across this secondary winding is the input voltage to the frequency converter section.

Frequency Converter Section

The transistor Q_1 is the active component of a combined local oscillator–mixer frequency converter section. That is, one transistor and its associated circuitry performs two functions. A circuit combining these two functions is called an *autodyne frequency converter.*

The frequency converter extends to the center of the core of transformer T_2. Draw a vertical line through T_2 and label the portion of the diagram between the lines you have drawn through T_1 and T_2 with the title "frequency converter section." Because this is a combined-function circuit, it is not possible to separate the circuit components clearly into two groupings and call one a "local oscillator" and the other a "mixer." It is possible, however, to analyze the circuit of Q_1 in terms of how it functions as an oscillator and as a mixer. A "how it works as an oscillator" explanation will be presented first.

For purposes of gaining an understanding of how the circuitry of Q_1 functions as an oscillator, let us imagine that the circuit is altered in two ways: (1) transformer T_2 is removed and the lower end of R_4 is connected to terminal 5 of transformer T_5 instead of to terminal 2 of transformer T_2; and (2) the left end of C_3 is connected directly to ground instead of to terminal 2 of T_1. The revised circuit would be as shown in Fig. 3.17. It can now be seen that the revised circuit will operate as a common-base circuit—the base is at signal ground because of the low-impedance path of C_3. The voltage-divider bias circuit for Q_1, R_1, and R_2 provides that Q_1 will be forward biased when power is turned on.

When power is turned on, a collector current will start to flow through winding 4–5 of transformer T_5, producing magnetic lines of force that will link with winding 1–2 of transformer T_5, injecting energy into the resonant circuit composed of winding 1–2 of T_5 and C_2. This circuit will start to oscillate with damped oscillations at its natural frequency, the resonant frequency.

This phenomenon is sometimes referred to as the *flywheel effect* of a tuned or resonant circuit. The flywheel effect is important in many applications of the resonant circuit.

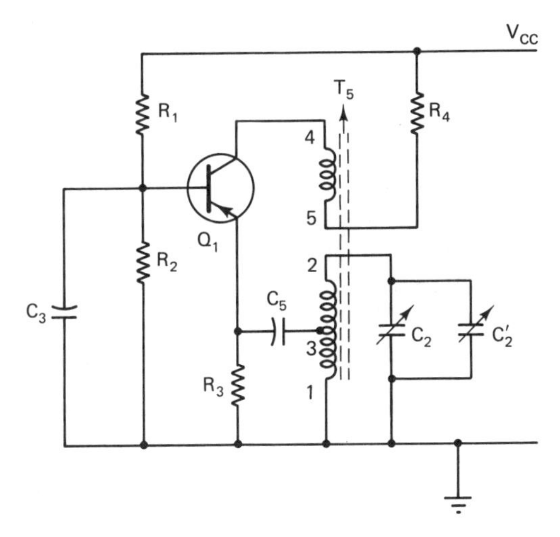

Figure 3.17 Stage Q_1 viewed as an oscillator.

If nothing further happened, the energy that was injected by the initial surge of collector current would soon all be dissipated in the resistance of the circuit and oscillations would cease. However, a portion of the voltage induced in the 1–2 winding of T_5 is applied to the emitter of Q_1 through the capacitor C_5. This voltage now becomes the input signal voltage for the stage acting as a common-base RF amplifier. The flow of emitter current will produce a voltage drop across R_3, reducing the forward bias of the transistor. The collector current will now flow only during a portion of the positive half-cycles of the oscillations.

The collector current, then, will be in the form of pulses and the pulses will occur at just the right moment to reinsert energy into the resonant circuit and keep the oscillations going. Looking at it another way, the collector current pulses will replace the energy lost in the resistance of the circuit and the damping of the oscillations will have been eliminated.

This circuit is called a *tickler-coil oscillator,* or an *Armstrong oscillator,* in honor of E. H. Armstrong, who devised it. The collector current in the *tickler coil*, winding 4–5 of T_5, "tickles" the tuned circuit with just the right amount of energy and at just the right moment to sustain oscillations.

The frequency of oscillation of this local oscillator is determined by the values of the inductance of the 1–2 winding of T_5 and total value of the two capacitors labeled C_2 (recall that $f = 1/2\pi\sqrt{LC}$). One section of C_2 is a gang of the two-gang variable capacitor described above in connection with C_1. Thus C_2 is the primary tuning device for this radio. The second part of C_2 is a trimmer used for altering the total value of C_2 to improve the tracking of the oscillator. The operating frequency of the LO must conform to the formula

$$f_{\text{LO}} = f_{\text{carrier}} + f_{\text{IF}}$$

The arrow shown near the top of the core symbol for T_5 indicates that the core slug for T_5 is movable and that the inductance values for T_5 are adjustable. This is a provision to facilitate the alignment of the receiver to achieve optimum tracking of its tuning circuits.

A mixer is simply a nonlinear circuit which has provision for the inputting of the local oscillator signal and the desired carrier signal from the RF section of a radio. The circuit of Q_1 operates in a nonlinear mode because its base bias is such that its collector current flows during less than 180° of a cycle of signal (class C operation). The carrier signal is fed into the stage through the secondary winding (terminal 2) of T_1 and C_3. As indicated above, the LO signal is present in the stage because the stage functions as the LO. Therefore, this Q_1 stage can, and does, function as a heterodyne mixer.

Let us return to the complete schematic of Fig. 3.15 and observe that the primary winding of T_2, between terminals 2 and 3, is connected in series in the collector circuit of Q_1. This means that the Q_1 collector current pulses also pulse through this winding. This collector current is not simply pulses of oscillator current, nor of RF signal current. It is a complex composite of currents—the output current of a heterodyne mixer. Remember, among other components it contains a current whose frequency is equal to the sum of the LO frequency and the selected carrier frequency. Most important, it contains a component whose frequency is equal to the LO frequency minus the selected carrier frequency, that is, the intermediate frequency—the IF signal current.

The entire primary winding of T_2 and capacitor C_6 across it are a resonant circuit which has been tuned, via the alignment process, to 455 kHz, the standard IF frequency. Therefore, this T_2 circuit will begin the process, which is to be carried out by the IF section, of discriminating in favor of the current of this 455-kHz signal. As a result of the flow of this favored component of current in the primary winding of T_2, a voltage whose frequency is also 455 kHz is induced in the secondary winding of T_2 (terminals 4-5) and is thus fed to the base of Q_2. This is the input signal to the IF section of the receiver. Resistor R_4 is added in series in the collector circuit of Q_1 to alter the operating point of the circuit and help ensure that the stage will operate in a nonlinear mode so that its function as a mixer will be achieved.

Notice that on IF transformers T_2, T_3, and T_4, the connections from the V_{CC} power supply is not made to the top of the primary windings, but, rather, to a tap slightly below the center of the primary windings. The reason for such tapped connections is to improve the impedance match between the relatively high impedances of the parallel resonant circuits of the transformer primaries and the relatively lower impedances of the collector circuits of the transistors involved. Tapping the primary winding of the transformer converts it to a mode of operation like that of a step-up autotransformer: there is a step up in voltage from the partial winding to the total winding; a step down in impedance from total winding to partial winding.

The IF (Intermediate Frequency) Section

The IF section of this receiver is, basically, a two-stage TRF (tuned radio frequency) amplifier. The IF sections of tube-type radios are typically single-stage amplifiers, whereas FM radios and TV receivers usually have at least three stages of amplification in the IF section. As indicated previously, the purpose of the IF section is to provide the major portion of the selectivity of the receiver, and to provide amplification of the desired signal. The typical IF section accomplishes this purpose very effectively and efficiently, primarily because it is a tuned amplifier that operates at a fixed frequency—the intermediate frequency.

When the operating frequency is fixed, the components of tuned circuits may be chosen to achieve maximum values of Q and therefore maximum selectivity. Furthermore, the selectivity is constant for all carrier frequencies since all are converted to the common, constant, intermediate frequency. Of course, the actual intermediate frequency, 455 kHz, which has become virtually a world standard, voluntarily, was chosen to permit optimization of the circuit as an electronic amplifier.

The two stages, Q_2 and Q_3, of the IF section of this receiver are nearly identical in configuration, circuit values, etc., and as indicated above, are stages of TRF amplification, or are RF amplifiers. Since RF amplifiers play an important role in both receivers and transmitters, let's look at a detailed explanation of their design features, theory of operation, and operating characteristics.

The designation "RF amplifier" is used by people in the world of electronics to indicate that an amplifier has a relatively narrow bandwidth rather than that it operates at a high frequency. For example, a "video amplifier," also called a wide-band amplifier, may operate at frequencies of 0 Hz to 6 MHz, or higher, whereas an RF amplifier might be used at frequencies as low as 40 kHz. Therefore, an RF amplifier, whether it is used in the "RF section" of some type of receiver, in the IF section of a receiver, or anywhere else, is, almost without exception, an amplifier that contains at least one resonant or tuned circuit. Observe that the collector load circuits of stages Q_2 and Q_3 are parallel resonant circuits (the primary windings of transformers T_3 and T_4 and the capacitors in parallel with them). During an alignment of the receiver, these circuits are adjusted to resonate at 455 kHz by adjusting the position of the ferrite slugs in the hollow cores of the windings.

When RF amplifiers are used to amplify unmodulated RF signals they are typically biased to operate in a class C mode. The class C designation indicates that the output current of an amplifier (the collector current, in the case of a transistor amplifier) flows for less than 180° of each cycle of the signal wave (see Fig. 3.18). Class C operation is highly desired in certain applications, for example, in the high-power stages of a radio transmitter, because it is much more efficient than any other mode of operation. This mode of operation is acceptable in RF amplifiers for unmodulated signals only because even though the output current is in the form of pulses—small portions of sine waves—the flywheel effect of the resonant circuit in the output will complete the cycles of the output voltage, with the result that it will be sinusoidal.

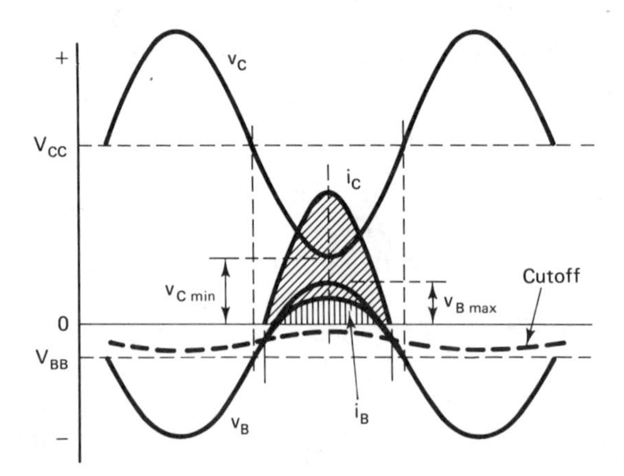

Figure 3.18 Output current in class C operation.

When the signal to be processed by an RF amplifier is an amplitude-modulated signal, however, the class C mode is not suitable. Figure 3.19 illustrates the effect on such a signal of processing it in a class C amplifier. Note that the portions of the signal wave which are below the cutoff level of the amplifier are missing in the output of the amplifier. This will cause the information signal content of the carrier wave to be distorted, an unsatisfactory condition. This kind of distortion can be avoided only if the output current flows over at least

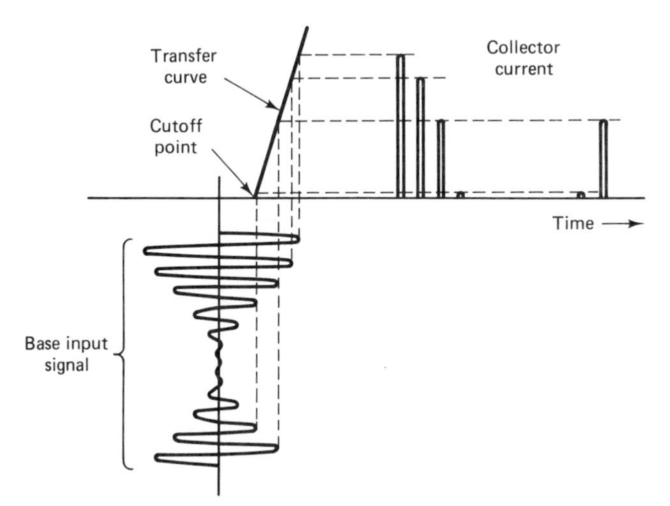

Figure 3.19 Effect on AM waveform of amplification in a class C amplifier.

180° of a cycle of the input signal waveform, that is, if the amplifier is operated in class B mode (180°), class AB mode (180° < AB < 360°), or class A mode (360°).

The base-bias circuits of stages Q_2 and Q_3 are modified forms of conventional voltage-divider bias circuits. Generally, a voltage-divider bias circuit has the appearance of that in Fig. 3.20(a). The relative values of the two resistors in the divider determine the value of the voltage at the base of the transistor. This conventional configuration has been modified in stages Q_2 and Q_3 to permit the bias voltages to be adjusted automatically by a voltage that is proportional to the amplitude of the signal being amplified [Fig. 3.20(b)]. The result of this adjustment will be an automatic change in the gain of the amplifiers—the gain is reduced when signal is strong, increased when signal is weak. This feature is called *automatic volume control* (AVC) in AM receivers and *automatic gain control* (AGC) in all other types of receivers (FM, TV, radar, etc.); it is the subject of another part of this section. Notice that the modification consists basically of disconnecting the lower resistor of the divider from ground and connecting that end to a variable voltage. Increasing or decreasing the variable voltage will make the base voltage more positive or less positive than its normal value. In any event, the net bias voltage, the voltage between base and emitter, must result in something other than class C operation.

To continue with the details of the IF stages, observe that each stage includes an emitter resistor (R_7 and R_{10}). The presence of an emitter resistor in common-emitter amplifier circuits is virtually universal since it serves to almost eliminate the effect of the beta (β) of the transistor on the gain of the stage. This is a vitally important characteristic for the circuit; it means that its operation will be relatively stable regardless of changes in operating temperature or transistors.

In each stage, the emitter resistor is bypassed by a capacitor. Without the capacitor, signal current flow in the emitter resistor produces a voltage across the resistor that will oppose, or reduce, the net signal voltage acting in the base circuit to change the base current. Therefore, this signal voltage across the

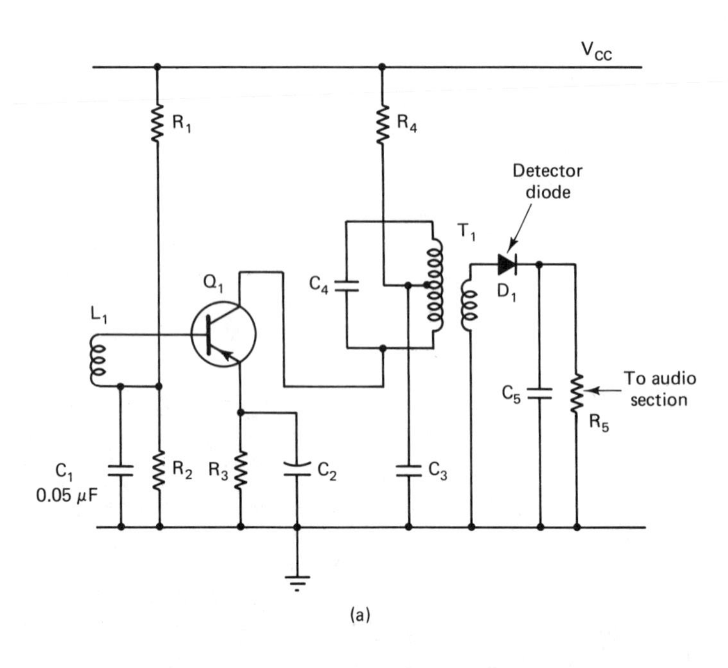

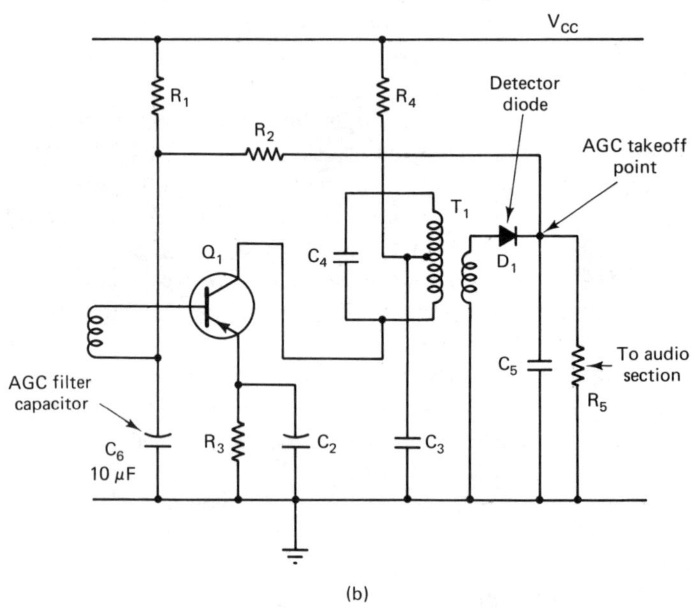

Figure 3.20 (a) IF amplifier stage with conventional voltage-divider bias circuit; (b) base circuit modified for AGC.

emitter resistor acts to reduce the gain of the stage. This phenomenon is a form of negative feedback; it is called *degeneration*. The purpose of the bypass capacitors C_9 and C_{15}, then, is to reduce the degeneration caused by the emitter resistors.

The base circuits of the stages must include provision not only for dc bias, but also for feeding in the IF signal voltage. The signal voltages appear across the secondary windings of the IF transformers (terminals 4-5) and thus are in series with the dc bias voltages. The signal voltages will add to or subtract from the dc voltages, causing the base currents, and, in turn, the collector currents to change.

It is important to minimize the possibility of any noise signals appearing across the dc portion of base circuits. The capacitors (C_8 and C_{14}) connected from terminal 5 of the IF transformer secondary winding to the top of the emitter resistor, in each stage, bypass the dc portion of the circuit to ground, ensuring that any undesired signals, such as noise voltages, appearing across the dc portion of the base circuit will be minimized. The capacitor C_7 connected from terminal 5 of T_2 (stage Q_2) to ground is part of the automatic gain control circuit. Details of its function are presented later in this section.

Tuned amplifiers have relatively high gains because of the effects of the amplification of the active device, as well as the multiplication effect of the Q factor of the resonant circuit. Because of this high gain, tuned amplifiers have a tendency to go into oscillation quite easily. This feature is an advantage when the desired outcome is oscillation, of course. When an amplifier, which is not intended to be an oscillator, goes into oscillation, the desired signal is lost, being overwhelmed by the stronger oscillation signal; the amplifier is useless as long as this condition exists. Therefore, it is essential that RF (or IF) amplifiers be designed so that oscillation cannot occur.

The mechanism that leads to oscillation is the feedback of signal voltage from the output side of the amplifier to the input side in such a way that at least a portion of the voltage fed back is in phase with incoming signal voltage. In this case, the phenomenon is called *positive feedback*.

In a transistor amplifier circuit, a portion of the output signal will be fed back from the collector circuit to the base circuit through what is called the interelectrode capacitance between the base and the collector. When amplifying a signal at its normal operating frequency, a tuned amplifier will usually feed back sufficient in-phase signal voltage for oscillation to occur. Therefore, it is essential that RF (or IF) amplifiers be designed so that this feedback will be minimized, if not totally eliminated.

One approach to the solution is to reduce the amount of the feedback voltage by simply reducing the gain of the stage. This can be accomplished by adding resistance in parallel with the tuned circuit, thereby reducing its Q. Reducing the gain of the stage, however, often is an unacceptable solution to oscillation.

Another, more common approach is to feed back an output voltage whose amplitude and phase will be such as to completely cancel out, or neutralize, the positive feedback signal. There are many possible ways of accomplishing this effect. In stages Q_2 and Q_3 it is accomplished by connecting capacitors (C_{10} and C_{13}) between the IF transformer winding and the base circuit. Notice that in each stage the transformer connection of the capacitor is at the end of the winding that is opposite to its connection to the collector. Because the ends of a transformer are out of phase, the voltages fed back to the base circuit will be out of phase. The capacitances of C_{10} and C_{13} are chosen to provide the proper amplitude of the negative feedback voltage, an amplitude that will neutralize the voltage fed back through the interelectrode capacitance. Capacitors C_{10} and C_{13} are called *neutralizing capacitors*.

A careful review of stages Q_2 and Q_3 in Fig. 3.15 will reveal that only resistors R_8 and R_{11} in the collector circuits, and the associated capacitors C_{12} and C_{16}, have not been discussed. These special circuits—a resistor in series with the V_{CC} supply connection and the stage, and a capacitor bypassing the connection point of the resistor to ground—are called *decoupling circuits*. They reduce the variations in the V_{CC} power supply that will reach the stages, as will be explained.

Decoupling circuits are frequently used for this purpose in receiver and transmitter stages of high sensitivity to minimize the introduction of noise to the desired signal. In a radio receiver, for example, a power supply that supplies dc voltage to all the stages of the radio may develop voltage variations at its terminals as a result of the varying current demand, especially of the audio power output stage. The current demand of this stage is very large relative to all other stages in the receiver. The variation of this high current demand through the internal resistance of the power supply produces the changing voltage observed at the power supply terminals. When this varying voltage is applied to a high-gain voltage amplifier, in the IF section, for example, the variations will cause all the dc voltages in the stage to vary. These variations will be amplified by the stage and added to the signal. Since they are not the signal, however, they are undesired noise and are to be avoided or minimized.

The decoupling circuit, in effect, adds additional internal resistance (in the form of R_8 and R_{11} in these stages) to the power supply, for the particular stage only. Then C_{12} and C_{16} present a low-impedance path to ground for any ac components of current produced by the varying dc voltage. Most of the variation in the power supply voltage, then, will be dropped across the decoupling resistors. We say that these variations are "decoupled," or, effectively, disconnected from the amplifier stage.

The Detector Section

The detector section of this receiver can be thought of as extending from the center of the core of transformer T_4 (the IF output transformer) to the right to the point where capacitor C_{18} connects to potentiometer R_{12}. The circuit is a conventional diode AM detector. The output of the IF section is coupled into the detector circuit as a voltage across the secondary winding (terminals 4-5) of T_4. Diode D_1 performs the rectification required for detection. Potentiometer

R_{12} is the resistive load for the rectifier. The two capacitors, C_{17} and C'_{17}, are the filter for the detector.

Recall (Sec. 3.2) that the filter capacitor(s) in an AM detector can be thought of as providing a low-impedance path for bypassing (shorting out) the higher-frequency components in the rectified wave, leaving only the audio signal component. A dc voltage, pulsating at the audio signal rate, will be produced across R_{12}, the detector load. This voltage is the result of the flow of a rectified current, a current caused by the application of the IF voltage to the detector circuit. The pulsating dc voltage can be thought of as consisting of two components: a pure dc component and the audio component. It is very important to be familiar with this manner of viewing the pulsating dc voltage, and each of its components will be considered in greater detail subsequently.

Like virtually everything else in the world, the AM diode detector is not perfect. There are several ways in which a diode detector may be imperfect. However, regardless of the imperfection, the result is the distortion of the information signal, either frequency or amplitude distortion. Distortion of the information signal by the detector means, for example, that its waveform after the detector is not an exact replica of the waveform as it was fed into the transmitter. The principal cause of amplitude distortion in a diode detector is the nonlinearity of the dynamic rectification characteristic of the diode itself. Frequency distortion is the result of a phenomenon called *diagonal clipping,* a result produced by the action of the filter circuit on the signal.

Let us compare the operating characteristic of an ideal, or perfect, diode with that of a practical diode, and also the effects of these characteristics on an AM signal during rectification. Refer to Fig. 3.21. A graph that is used to show the relationship between the input and output waveforms of a device, circuit, or system is called a *transfer characteristic.* Notice, in Fig. 3.21(a), that the line, or more correctly, the curve for the ideal diode is straight or linear once it starts to rise, beginning at the origin of the graph. For the practical diode, Fig. 3.21(b), the curve has some bend in it, especially between the origin and point 1. The active portion of the characteristic, the curve to the right of origin, for the ideal diode (a) is linear; for the practical diode (b) it is nonlinear. Both characteristics are nonlinear overall since they each have a sharp bend in them at the origin. Remember that, as was indicated in Sec. 3.2, to achieve demodulation, the carrier and its sidebands must be processed together in a nonlinear circuit (i.e., the diode rectifier circuit).

The expected output waveform, for a given input waveform, can be obtained from a transfer characteristic curve by sketching the input waveform on the input side of the graph and projecting a number of points from this waveform to the curve, and then to the output side of the graph. The output waveforms for the characteristics of Fig. 3.21 were obtained in this fashion.

Observe that the envelope of the rectified waveform in Fig. 3.21(a) is an exact replica of the envelope of the modulated input waveform. On the other hand, the envelope of the rectified waveform in Fig. 3.21(b) is a distorted replica of the envelope of the input waveform. The distortion is introduced because

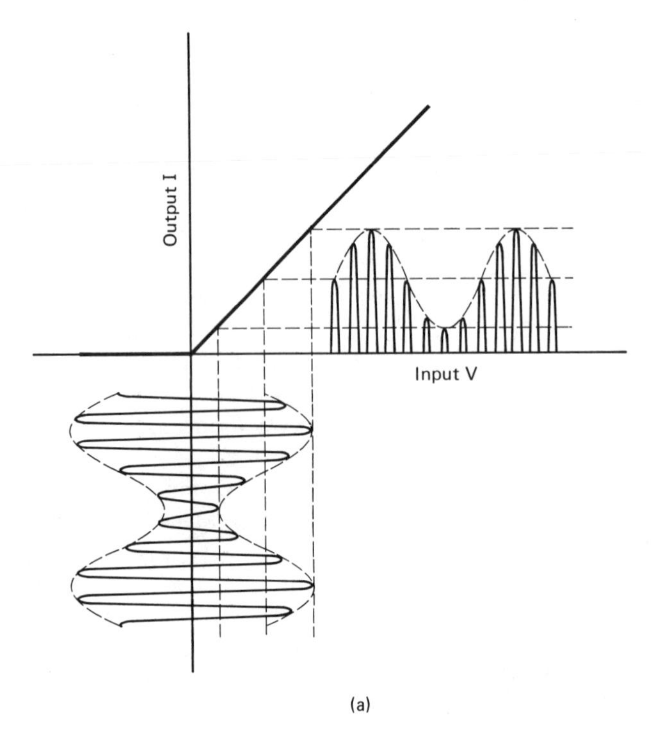

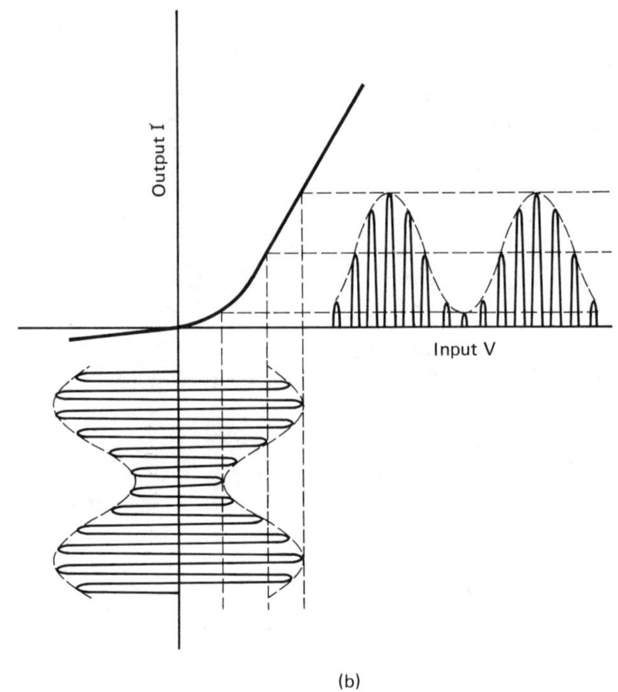

Figure 3.21 Transfer characteristics: (a) ideal diode; (b) practical diode.

the conduction characteristic of the practical diode is not linear over the range of the excursion of the applied voltage.

This distortion is obviously a distortion in the amplitude of the envelope, which, of course, represents the audio signal. It is often called harmonic distortion because it results in the introduction of harmonics. Harmonic distortion can be minimized, if not eliminated, if the amplitude of the signal input to a practical diode is sufficiently great to cause the more linear portion of the characteristic (to the right of 1 in the diagram) to be utilized. Also, distortion will be less if percent modulation does not approach 100%, if the "valleys" of the modulation envelope do not drop below the level of 1 V.

Another characteristic of the detector circuit produces an effect called *diagonal clipping*. The magnitude of this effect depends on the frequency of the modulating signal; the distortion it produces is therefore called *frequency distortion*. This phenomenon can best be described in relation to a diagram [see Fig. 3.22(a)].

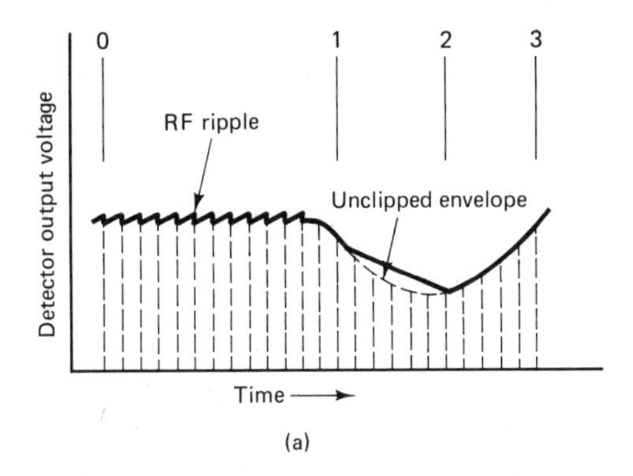

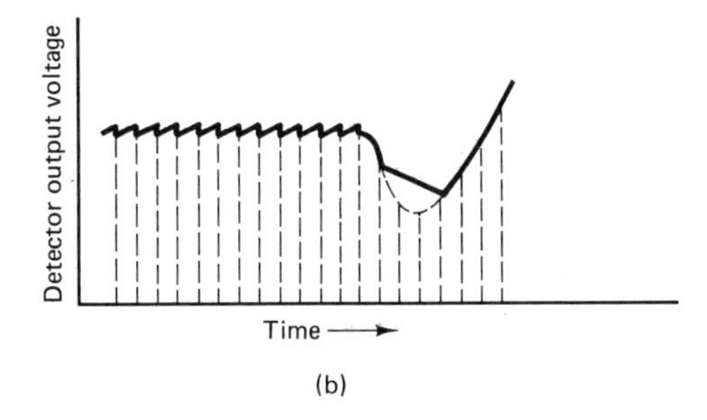

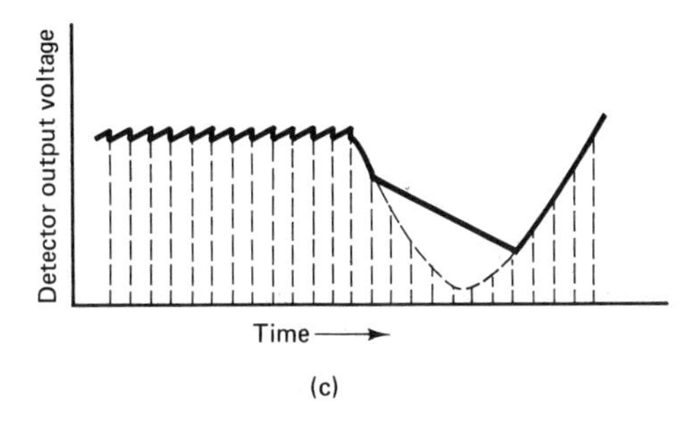

Figure 3.22 Diagonal clipping: (a) diagonal peak clipping; (b) effect of audio frequency; (c) effect of modulation level (audio voltage amplitude).

It is necessary to recall that when viewing the action of the detector in the time domain, the audio signal is produced by the action of the filter capacitor on the rectified RF (actually, the IF) wave. That is, the filter capacitor charges up quickly when the diode is conducting on positive half-cycles and discharges slowly when the diode is cut off. When the RF wave is unmodulated, region 0 to 1 in Fig. 3.22(a), the voltage across the capacitor reproduces the envelope of the wave accurately. Similarly, when the envelope is increasing in amplitude, region 2 to 3 of the diagram, the capacitor voltage is a relatively accurate replica of the envelope. However, in the region of the valley of the envelope, time interval 1 to 2 in the diagram, the capacitor does not discharge quickly enough (the RC time constant of the discharge path is longer than that of the charge path) and the voltage across it does not follow the envelope. The capacitor voltage does not equal what should be the rectifed envelope voltage until much later in the audio cycle. The audio waveform, the voltage across the filter capacitor, is clipped diagonally, and, of course, obviously distorted. Notice that the effect of the clipping is even greater when the audio frequency is higher, Fig. 3.22(b). Also, for a given frequency, the clipping is greater for a greater amplitude of audio (i.e., when percent modulation is higher) [see Fig. 3.22(c)].

Diagonal clipping, in summary, occurs when the RC time constant of the detector filter-load circuit is excessive. The problem can be minimized by choosing values for this circuit so that its time constant will be short compared to the period of the highest audio frequency and yet not so short as to eliminate the filtering out of the RF component of the rectified wave.

Let us return now to a concept stated earlier in this presentation on the detector. The filtered output (RF component attenuated) of the detector is a pulsating dc voltage whose pulsations are at an audio rate. This pulsating voltage can be viewed as being made up of two components: a pure (nonpulsating) dc voltage and an audio-frequency voltage (a pure ac voltage). The audio-frequency voltage is, of course, the information signal. It is separated from the pure dc voltage by coupling capacitor C_{18} and passed on to the base of transistor Q_4 for amplification in the audio amplifier section of the receiver. We say that C_{18} "blocks" the dc voltage and "passes" the ac voltage. A discussion of the audio section is presented below. Before proceeding to the audio section, however, it is essential that we examine one more detail of the IF section—automatic gain control (AGC).

Automatic Gain Control (AGC)

Automatic gain control is a means of changing the gain of an amplifier without human intervention. The gain of the IF and/or RF sections of a receiver must be changed to match the highly variable conditions of reception that most receivers are subjected to. Only if this is possible will reception be satisfactory.

In the first place, the energy levels of the various carrier signals that a particular receiver might be expected to receive may differ over a wide range—100:1 or more. This is true for several reasons, one of which is that different stations may broadcast their carrier signals at different assigned energy levels.

Carrier energy levels present at the antenna of a given receiver may differ also because the distances separating that receiver from the various stations that it is expected to receive differ widely in many situations. Moment-to-moment variations in a given signal level may occur when the efficiency with which radio waves travel through the atmosphere changes with atmospheric changes—changes in temperature, humidity, etc. When a receiver is mounted in a moving vehicle, the distance from receiver to transmitter changes with the location of the vehicle. Also, as the vehicle moves, barriers such as tall buildings or hills may come between the receiver and transmitting antenna, thus temporarily diminishing the energy level of the received carrier. Even an airplane flying overhead may momentarily intercept and diminish the signal level of a desired carrier.

To successfully receive and process weak signals—carrier signals having low energy content—and to reproduce the information contained in such signals, the RF and IF sections of a receiver must operate at or near maximum possible gain. On the other hand, if these sections operate at very high gain when stronger signals are being processed, the signal will be badly distorted. The signal will be distorted because one or more amplifier stages will be overdriven with too much signal. One outcome of an overdriven stage is the obvious one of clipping of the peaks of the waveform. A less obvious effect of driving an amplifier stage with too much signal is that it forces the stage to operate in a nonlinear portion of its transfer characteristic. This, by itself, causes distortion. In addition, however, another undesirable phenomenon may occur—*intermodulation distortion*.

If one or more undesired carrier or noise signals, as well as the desired carrier signal, are present in an amplifier stage that is operating in a nonlinear mode, the signals will modulate each other. They will produce the typical products of modulation (see Chap. 2), that is, sum and difference frequencies. One or more of the difference frequencies may be in the audio range. If this occurs, the detector will demodulate such a signal, producing an audio-frequency signal. This audio-frequency signal, the product of the intermodulation, will be heard from the speaker as a squeal or an annoying sound usually referred to as "static."

Manual gain control (control by the operator of a receiver) of the RF and/or IF sections of a receiver can be and has been provided in receiver designs. Manual RF/IF gain controls are almost always provided in the design of communications-type receivers. However, such manual controls would be unacceptable on receivers used by the general listener. Hence some form of AGC (automatic gain control) is provided in the design of virtually all radio receivers.

Automatic gain control (AGC) has two basic requirements: (1) The gain of the active elements (transistors, for example) in the amplifier of interest must be capable of being controlled by varying a dc voltage in the circuits of those elements. (2) It must be possible to provide a dc voltage whose amplitude is a measure of the strength (amplitude) of the signal being processed.

These requirements are easily met in the typical solid-state superheterodyne receiver. The gain of typical transistors used in RF/IF amplifier sections can be changed by shifting the quiescent operating point of the transistor stage (see Fig. 3.23). The operating point of a given transistor stage is determined by the dc emitter current, which in turn is determined by the dc base-to-emitter voltage.

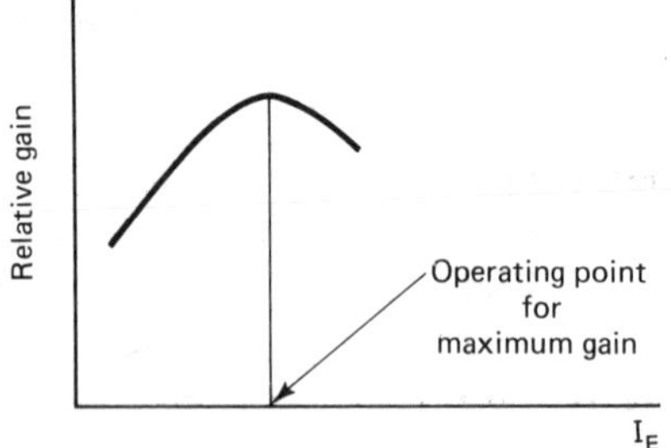

Figure 3.23 Variation of gain with emitter current for transistor amplifier.

Thus we need only be able to change the bias (base-to-emitter) voltage of a given transistor amplifier stage to change its gain. Furthermore, the output of the AM detector provides a dc voltage whose amplitude is in direct proportion to the amplitude of the carrier being processed by the IF section. (Of course, the carrier frequency has been converted to the intermediate frequency of the receiver.)

The AGC circuit, then, should be a circuit that samples the dc voltage at the output of the detector and uses that voltage to modify the dc emitter current of those amplifier stages in which gain is to be controlled automatically. Let us look now at the diagram (Fig. 3.15) of the receiver we have been studying to determine the exact nature of the AGC circuit used in it.

In this receiver, the two IF stages, Q_2 and Q_3, are the AGC-controlled stages. The AGC/bias circuit for these two stages consists of the branch, R_5–R_6–R_9, connected at one end of V_{CC} and at the other end to the detector output. The point where an AGC circuit connects to the detector circuit is commonly referred to as the *AGC takeoff point*. The AGC takeoff point here is where R_9 connects to the cathode side of detector diode D_1. Notice that transistor Q_2 receives its dc base voltage from the junction of resistors R_5 and R_6. Similarly, Q_3 receives its dc base voltage from the junction of R_6 and R_9. The end of the R_5–R_6–R_9 branch connected to V_{CC} will not change in voltage under normal conditions. However, the voltage at the other end of this branch, where R_9 connects to D_1, will change in accordance with the change in the strength (amplitude) of the received station carrier. Because of the direction of D_1, this voltage will become more positive as the amplitude of the carrier increases. With no signal being received, the voltage at the junction of R_9–D_1 is between -0.16 and -0.3 V, the junction voltage of D_1. Under this condition—no signal—D_1 is forward biased and the dc drop across the secondary winding of IF output transformer T_4 is virtually O V.

If an IF signal is being processed, that signal will appear across winding 4–5 of transformer T_4 as an ac voltage. The rectification of this voltage produces an electron current flow in a counterclockwise direction in the detector circuit, producing a voltage across R_{12} with its upper end (junction of R_{12} and diode D_1) positive. The greater the amplitude of the IF signal (i.e., of the received carrier), the more positive will the AGC takeoff point become. Thus one end of the AGC/bias circuit is anchored to V_{CC} but the other end is free to rise and fall with the amplitude of the received carrier. As the takeoff point becomes more positive it makes the bases of Q_2 and Q_3 somewhat more positive, thereby reducing the amplitudes of their base-to-emitter voltages. A reduction in these amplitudes reduces the dc (or quiescent) emitter currents of these two stages and thus shifts their operating points in the direction of reduced gain (see Fig. 3.23). If the amplitude of the IF signal decreases, the reverse of the above occurs. The overall net effect of the AGC action is to adjust the gain of the IF section so as to maintain a relatively constant level of IF output voltage over a relatively wide range of IF input voltage amplitudes.

The AGC voltage should be a pure dc voltage whose amplitude is directly proportional to the amplitude of the received carrier. The average amplitude of the dc voltage across the detector load is directly proportional to the strength of the received carrier. However, the detector load voltage is a pulsating dc voltage that contains an audio frequency ac component, when the carrier is modulated. If allowed to remain as a component in this dc voltage, the audio variation would vary the gain of the affected stages at an audio rate. This would be an unacceptable situation.

The AGC circuit must filter out the audio component. This is accomplished simply by connecting a relatively large capacitor to ground at some point in the AGC circuit, thus bypassing the audio current to ground. In the case of the receiver of Fig. 3.15, the AGC filter capacitor is C_7. Notice that it is a 10-μF capacitor, whereas RF bypass capacitors such as C_9, C_{15}, and C_{17} are only 0.05 μF. The AGC circuit is shown by itself in Fig. 3.24.

Although "AGC" is the label applied to this function in FM, television, and other types of receivers, the function has been traditionally referred to as "automatic volume control" (AVC) in AM broadcast band receivers. You may have the experience of seeing it so labeled still. This label is misleading, however, particularly to those who do not thoroughly understand the function. The word "volume" seems to imply that the volume of the sound coming from a receiver is controlled automatically. Indirectly, this is true when a change in carrier level causes a change in the level of the volume of the sound. However, for a given level of carrier the volume of sound must change with changes in the modulation level. The AGC and AVC circuits must not and do not interfere with this relationship.

The Audio Amplifier Section

The purpose of the audio amplifier section is to increase the energy level of the audio-frequency (AF) signal at the output of the detector so that a loudspeaker can be driven and thus reproduce the sound that was the original input to the system. The increase in energy level should be accomplished, optimally, without the introduction of noise or distortion. This performance characteristic requires that amplifier stages in the audio section operate in a linear mode so that the output signal of any stage is a perfect (or, at least, nearly perfect) replica of the input signal. Producing an output with a relatively high energy level is generally not compatible with a simple, so-called single-ended amplifier that must be operated with no or minimum distortion (i.e., class A and linear). Thus we find that the AF sections of most radio receivers include some form of the push-pull concept

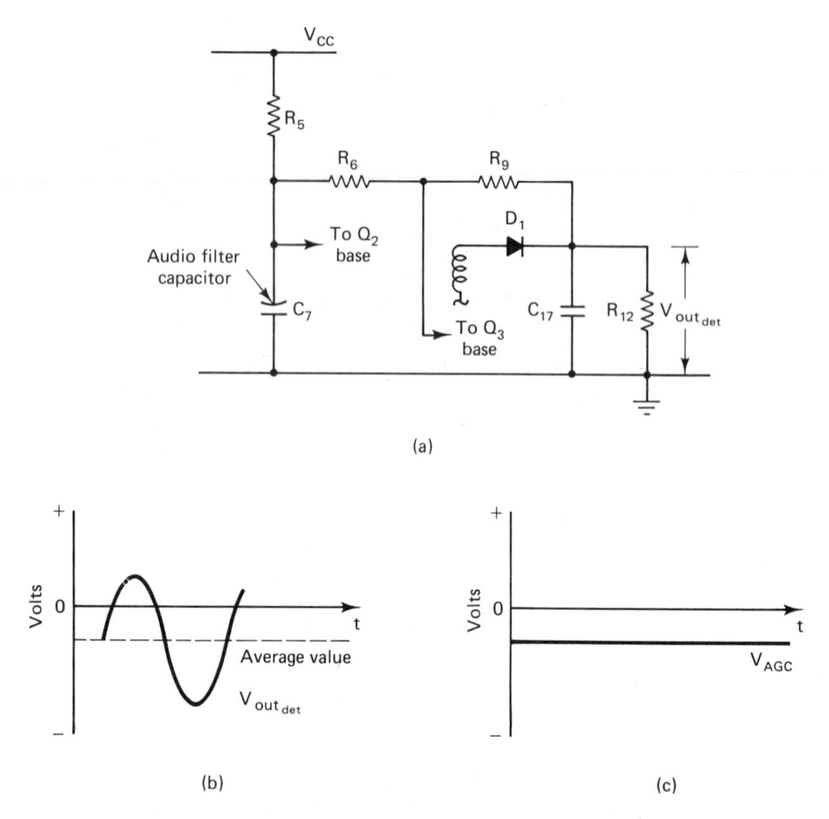

Figure 3.24 (a) AGC circuit; (b) detector output with RF filtering only; (c) detector output after audio filtering.

in the output stage. The AF section of the receiver of Fig. 3.15 employs a transformer-type push-pull output stage. Let us study the circuit of the AF section in some detail. The AF section, separated from the rest of the receiver circuitry, is shown in Fig. 3.25.

The circuit of Fig. 3.25 is that of a two-stage audio amplifier. It consists of a driver stage, transistor Q_4, and a push-pull output stage, transistors Q_5 and Q_6. The driver stage is biased for class A operation; that is, its collector current will flow during the full 360° of the input signal cycle. It employs the typical voltage-divider type of bias circuit—resistors R_{16}, R_{13}, and R_{14}.

Resistor R_{15} in the emitter circuit of this stage is used to swamp the emitter diode of Q_4 and thus minimize changes in gain of the circuit due to changes in temperature or the transistor itself. The emitter swamping resistor is bypassed with capacitor C_{19} to minimize attenuation of gain in the stage caused by negative feedback in R_{15}.

The driver stage is coupled to the output stage through transformer T_6. The secondary winding of T_6 is center tapped to provide two signals with a 180° phase difference between them. Such signals are required to drive transistors Q_5 and Q_6 in push-pull.

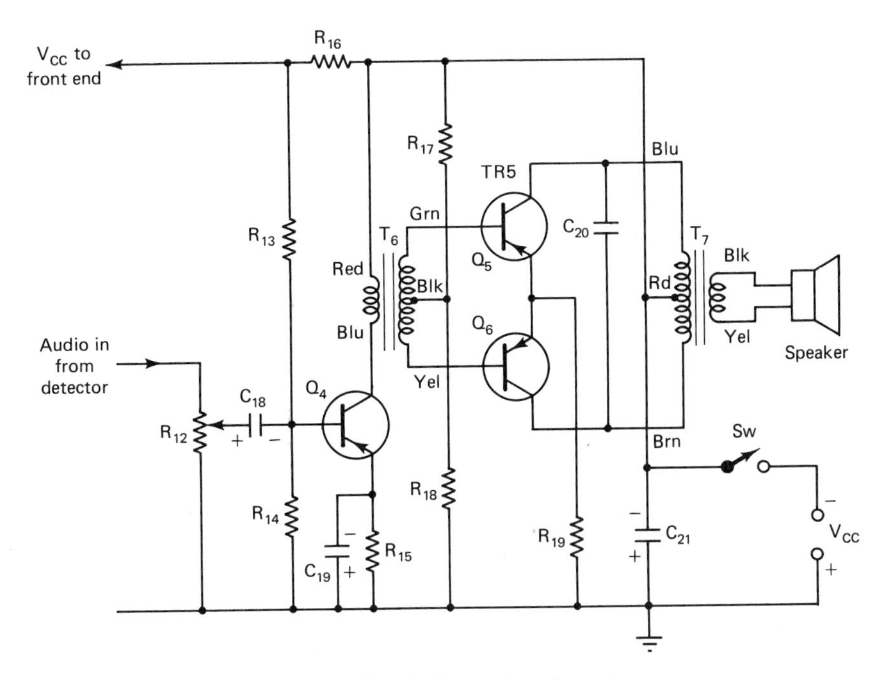

Figure 3.25 Audio section of receiver.

The output stage also uses a voltage-divider bias—resistors R_{17} and R_{18}. Emitter swamping resistor R_{19} is unbypassed. This provides some negative feedback. Negative feedback is useful in reducing distortion in an amplifier. This stage operates as a class AB amplifier. The collector currents flow for more than 180° but less than 360° of each cycle of the input signal.

One of the advantages of push-pull amplification is that it need not be class A in order to approach linear operation.

Two other advantages of the transformer-type push-pull amplifier are often mentioned. First, the push-pull connection eliminates the problem of transformer core saturation. This is a problem when a single transistor, operating class A, is used to drive a transformer-coupled speaker. This is the single-ended amplifier. Another advantage is that second and other even-numbered harmonic currents are attenuated in the output of a push-pull circuit. This minimizes the distortion normally present in a single-ended power amplifier.

Let us study the operation of a push-pull circuit in some detail in order to understand better the "why" and "how" of the foregoing advantages. Let us first look at dc operation with the aid of Fig. 3.26.

Transistors Q_5 and Q_6 receive dc base current in parallel from the base-bias network. The collector currents through the primary winding of the output transformer T_7 will be as indicated by the arrows I_{C5} and I_{C6} on the diagram. These currents are in opposite directions through the transformer winding. In a properly operating push-pull circuit the transistors are selected for matching characteristics. Therefore, the collector currents would be equal, or virtually so. The magnetic lines of force produced by the two dc collector currents would

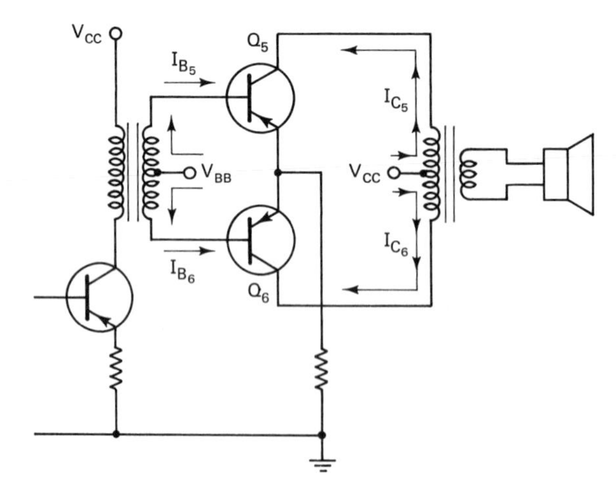

Figure 3.26 Dc currents in push-pull amplifier.

cancel each other. There would be no net magnetization of the transformer core by the quiescent current of the stage. It is important that you study the logic of this important feature of the circuit until you understand it clearly and can remember it easily.

Next, let us look at how the push-pull circuit processes the desired signal. This will provide a basis for more easily understanding how the circuit operates to minimize distortion by suppressing even-numbered harmonic currents.

The diagram of the circuit shown in Fig. 3.27 is labeled with waveforms to show the phase relationships of signal currents at various parts of the circuit. It is essential to be aware that the base signal currents of Q_5 and Q_6 are always 180° out of phase. Recall that an ac signal base current adds to and subtracts from the quiescent base current of a given transistor. The result is that there is a corresponding increase and decrease in the collector current of that transistor.

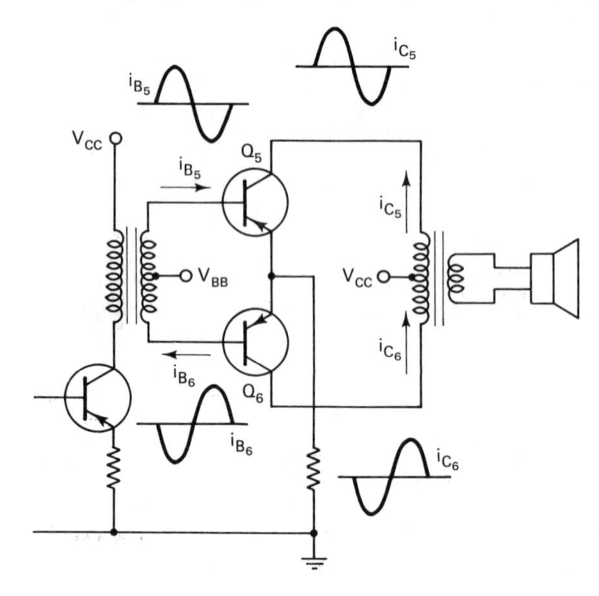

Figure 3.27 Signal currents in push-pull amplifier.

It follows that the collector currents of Q_5 and Q_6 will have ac signal components which are always 180° out of phase with each other. That is, the collector current of Q_5 will be increasing while the collector current of Q_6 is decreasing, and vice versa.

These collector currents flow in the two half-windings of the center-tapped primary of transformer T_7. An increase in I_{C_5} is equivalent to an electron current flow from the center tap of the winding toward the top of the winding. A decrease in I_{C_6} is equivalent to an electron flow up from the bottom of the winding toward the center tap. These currents produce magnetic lines of force which are in the same direction in the transformer core. They are, in effect, in phase as far as the transformer secondary winding is concerned. It is an easy step to the concept that one transistor is "pulling" while the other is "pushing" current through the transformer winding. This is the basis for the descriptive title of this circuit: *push-pull*.

Now, we want to look at how a push-pull amplifier minimizes distortion by suppressing second (and higher-order even) harmonics in its output. Let us review some of the facts about distortion in a single-ended amplifier. The typical transfer characteristic of such an amplifier is shown in Fig. 3.28(a). Observe that between points B and C on the curve the characteristic is straight. When biased and driven to operate in this region an amplifier's operation would be linear. This mode of operation is typical of a small-signal, class A voltage amplifier.

To produce a significant amount of output power, however, an amplifier must be driven over a greater portion of the transfer characteristic. This requires utilization of at least a portion of the curve between points A and B. The curve is nonlinear in this region. The effect on the output waveform of an amplifier operating in this fashion is shown in Fig. 3.28(b). Because the gain is less during the negative alternation of the input signal, the negative peak of the output waveform will be flattened, as shown. The positive alternation is also distorted by greater amplification of the positive peak.

A sine wave can be distorted in a manner identical to that shown if a second harmonic component of appropriate amplitude and phase is added to the sine wave [see Fig. 3.28(c)]. The conclusion, then, is that when the amplitude distortion of Fig. 3.28(b) is present in an amplifier, the circuit is introducing second, and higher, even-numbered harmonic currents in its output.

Observation: These currents are not found in the input side of the circuit; they are components of the output current only. They are generated by the circuit itself as a result of the distortion caused by the operation of the circuit. If the output circuit can be arranged so as to suppress these harmonic components of current, distortion will be reduced if not eliminated.

We are ready now to see how the push-pull circuit, by its very nature, suppresses the second harmonic component of current. The circuit of our discussion is shown again in Fig. 3.29. This time, waveforms to represent a component of second harmonic current have been added to the waveform diagrams on the collector side of the circuit. The waveforms are drawn as would be required to produce the distortion described in the preceding paragraph. Careful analysis

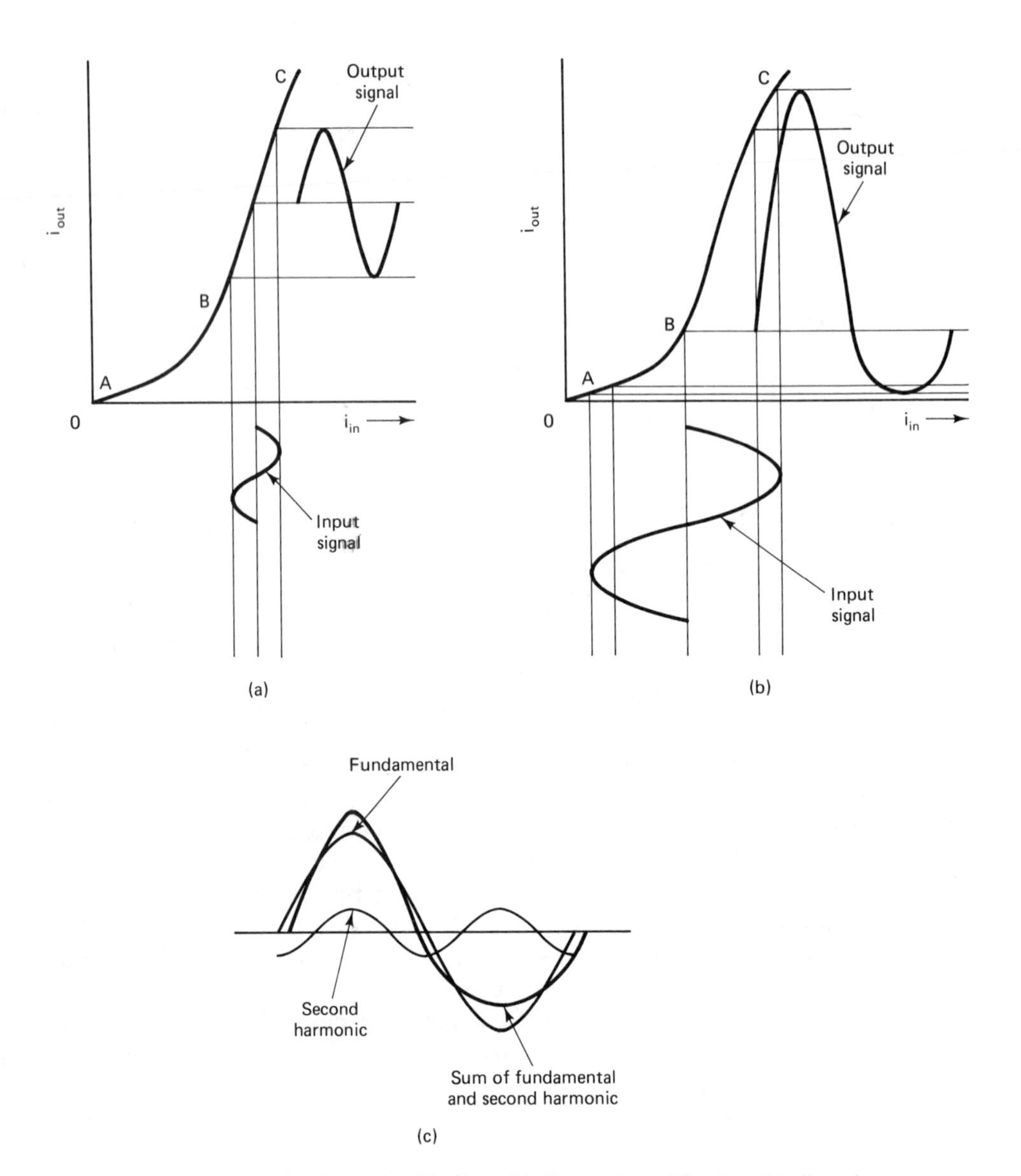

Figure 3.28 (a) Linear amplification; (b) distorted amplification; (c) distortion by addition of second harmonic to fundamental.

reveals that the second harmonic components of the two collector currents are in phase. That is, during their positive alternations, for example, these currents are flowing from the center tap of the primary winding, through their respective halves of the winding and toward the transistor collectors, and so on. The action is similar to that of the quiescent collector currents. The contributions to the magnetic field in the transformer core of these two components of harmonic current are equal but opposite. As a result, the effect of the second harmonic

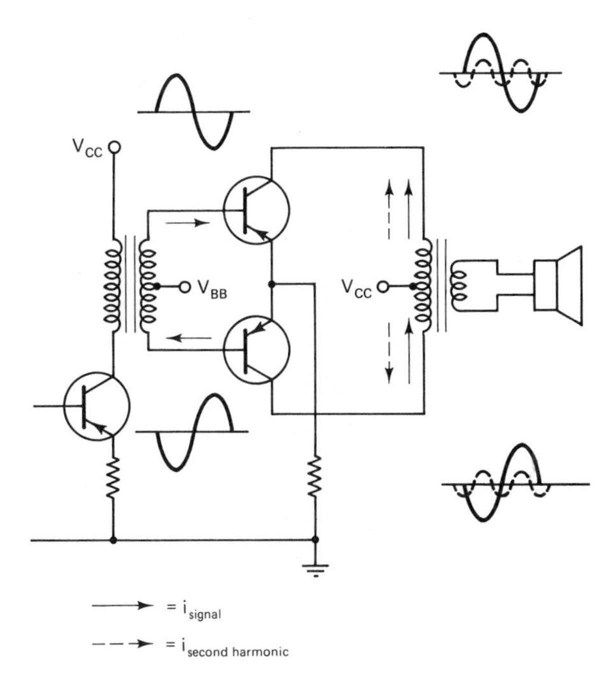

Figure 3.29 Signal and second harmonic currents in push-pull amplifier.

current generated by the amplifier on the current in the secondary winding of the transformer is suppressed. The distortion introduced by the nonlinear action of the circuit is diminished proportionately. In a similar manner it could be shown that the push-pull circuit suppresses all even-numbered harmonic currents, contributors to the distortion of the type described.

In a single-ended (see above) audio amplifier, there must be collector current flow at all times. This is necessary to provide both alternations of the ac waveform. Such operation is provided for by adjusting the bias of the stage so that the no-signal, or quiescent, base current is sufficient to provide some base current even at the negative peak of the signal base current. Operation in this fashion is called class A mode. However, in a push-pull audio amplifier it is not necessary that the circuit be operated class A. This is true because it is possible for one of the two transistors of the circuit to provide the current for the positive alternation of the output current. The other transistor would then provide current for the negative alternation of output current. If the bias for the circuit is arranged so that this sharing of currents is exactly true, the circuit is operating class B. Each transistor conducts during exactly 180° of the input signal cycle. Quiescent base currents are zero.

However, class B operation of push-pull amplifiers introduces another form of distortion, known as *crossover distortion* [see Fig. 3.30(a)]. Crossover distortion may be minimized, without operating the stage in full class A mode, if the bias is adjusted so that the quiescent currents are slightly more than zero [see Fig. 3.30(b)]. Operation under this condition is called class AB mode; the transistor conducts for more than 180° but less than 360° of the signal cycle. The efficiency

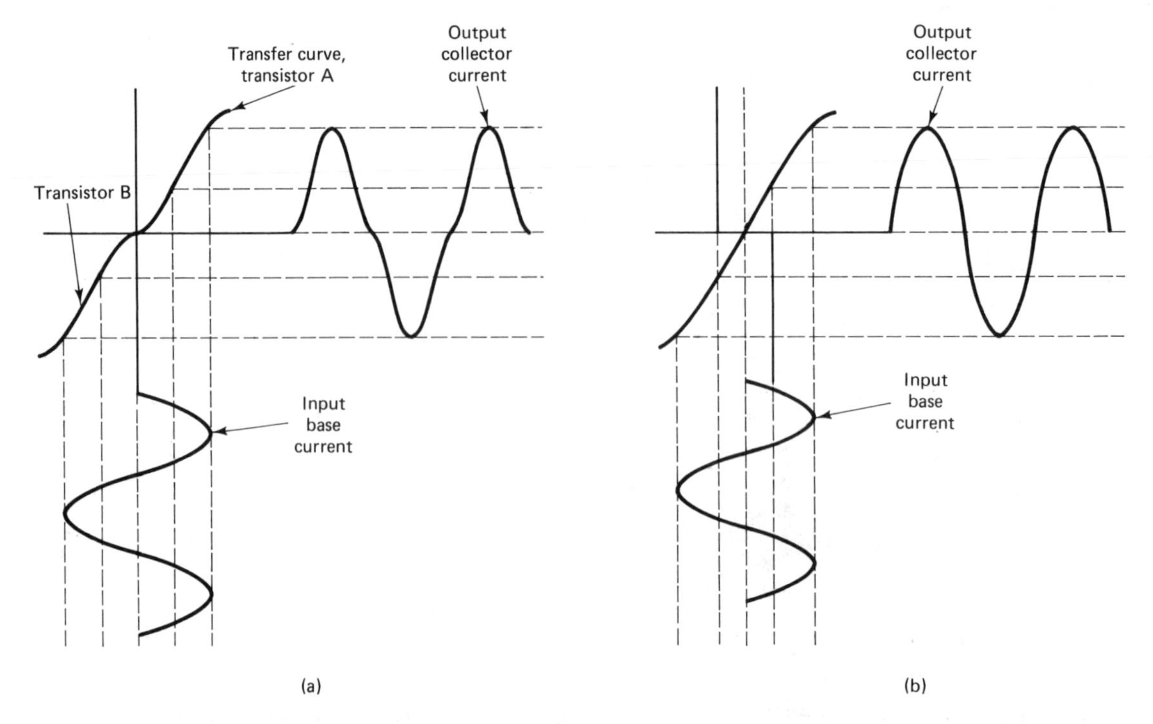

Figure 3.30 (a) Crossover distortion in class B push-pull amplifier; (b) crossover distortion eliminated by shifting operating point with trickle bias.

of an amplifier—that is, conversion of dc energy to ac energy—operating in the class AB mode is somewhat better than that of one in class A mode. Using transistors with a given power rating, more input drive can be used and more output can be obtained from an amplifier operating in class AB mode than from one operating in the class A mode.

In summary, push-pull audio output amplifiers, operating in class AB mode, are popular with radio receiver manufacturers. This design, compared to single-ended amplifier design, provides for greater output power for a given transistor power rating. The design also provides for less distortion with the same level of power output, or, alternatively, for greater power with the same level of distortion. These features are the result of the inherent action of the circuit, which minimizes transformer core saturation due to quiescent dc current and reduces distortion through the suppression of even-numbered harmonic currents in the output.

The Loudspeaker

Electrically and mechanically, a loudspeaker consists of a coil of fine wire attached to a paper cone in such a way that the coil is suspended in the field of a permanent magnet (see Fig. 3.31). The coil is free to move. It is caused to move when a

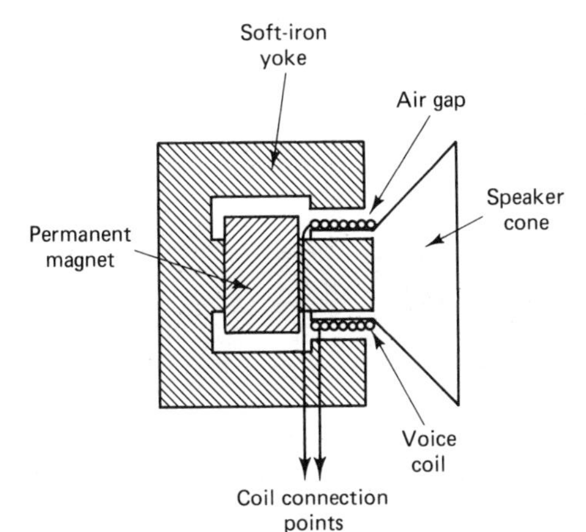

Figure 3.31 Permanent-magnet speaker construction.

current flows through the coil. The field of the permanent magnet interacts with the current in the coil and the coil moves, either in or out, depending on the direction of the current. The amplitude of the coil motion is directly proportional to the amplitude of the coil current. When the current in the coil is the current from the secondary winding of an output transformer of an audio amplifier, the movement of the coil and paper cone will be at an audio rate. This movement converts the electrical energy of the audio-frequency current into the mechanical energy of sound. The sound thus produced will be similar to that which was originally the input to the communications system involved. The more nearly all elements of the system operate without distortion, the more nearly will the output sound be just like that of the input sound.

The conversion of energy in the speaker represents the electrical equivalent of a resistive load. The ohmic value of the equivalent resistance is of the order of 4 to 16 Ω, with the most common value being 8 Ω. This range of values is much lower than the output impedance of many transistor power amplifiers. To achieve good power transfer from amplifier to speaker, therefore, the output transformer turns ratio is chosen so as to provide impedance matching between speaker and amplifier. The turns ratio can be calculated using the formula

$$\text{turns ratio} = \frac{N_2}{N_1} = \sqrt{\frac{Z_L}{Z_o}}$$

where N_1 = number of primary turns
N_2 = number of secondary turns
Z_L = impedance of the load (the impedance of the loudspeaker, in this context)
Z_o = output impedance of the output stage of the amplifier, as seen looking back into the amplifier from the primary side of the transformer

Example 4

A transistor amplifier whose output impedance is 1200 Ω is to be used to drive an 8-Ω speaker. Calculate the turns ratio of a transformer required to match the speaker to the amplifier for maximum power transfer.

Solution

$$\text{turns ratio} = \frac{N_2}{N_1} = \sqrt{\frac{Z_L}{Z_o}} = \sqrt{\frac{8}{1200}} = 0.08165 \simeq 1{:}12$$

SUMMARY

1. In its simplest form a radio receiver includes (a) an antenna, (b) a selection circuit, (c) a demodulation circuit, and (d) a transducer.
2. The superheterodyne receiver design is very effective. Its principles are used not only in AM broadcast band receivers, but also in FM, TV, and other types of receivers.
3. A basic AM superheterodyne radio receiver includes the following functions or sections: RF amplification, frequency conversion (local oscillator and mixer), intermediate-frequency amplification, detection, audio-frequency amplification, and reproduction.

GLOSSARY OF TERMS

Active device An electron device such as a diode, transistor, or tube.

AM broadcast band (commercial) The frequencies between approximately 540 and 1600 kHz used by commercial radio stations to broadcast programs to the general public using amplitude-modulated carrier signals.

Amplifier An electronic circuit, containing at least one active device, which is designed to increase the energy level of an input voltage or current called the signal.

Amplifier, class A An amplifier in which the amplifying device (transistor or tube) is active during the entire cycle (360°) of the signal waveform.

Amplifier, class AB An amplifier in which the amplifying device is active for more than 180° but less than 360° of the signal waveform. It is biased slightly above the point at which there would be no current flow when there is no signal. The point of no current flow is called cutoff.

Amplifier, class B An amplifier in which the amplifying device is active for exactly 180° of the signal waveform. It is biased at cutoff.

Amplifier, class C An amplifier in which the amplifying device is active for less than 180° of the signal waveform.

Amplifier, current An amplifier that is designed specifically to amplify an input current signal.

Amplifier, power An amplifier that is designed specifically to amplify the power level of an input signal and transfer the amplified power to the output circuit or device connected to the amplifier.

Amplifer, push-pull An amplifier, containing at least two active devices, which is designed so that the current in one device is increasing when the current in the other device is decreasing, and vice versa.

Amplifier, voltage An amplifier that is designed specifically to amplify an input voltage signal.

Amplifier stage An amplifier, or portion of an amplifier, which contains only one active device, except a push-pull amplifier stage, which contains at least two active devices.

Attenuate (an electrical signal) To reduce the amplitude of.

Automatic gain control (AGC) The control of the gain or amplification of an amplifier by electronic means, as differentiated from manual control by a person. This function has also been referred to as automatic volume contol (AVC) when related to an AM broadcast band radio receiver.

Beat frequency The frequency produced when currents of two different frequencies are present in a nonlinear amplifier circuit called a mixer. The beat frequency is equal to the difference of the other two frequencies.

Bias (of a transistor) The voltage across the base–emitter junction of a transistor, that is, the voltage measured from base to emitter of a transistor in an operating circuit.

Cutoff The condition of an active device in which there is no current flow through the device even though there is voltage applied to the device. For a transistor, this condition is controlled by the voltage across the base–emitter junction—the bias voltage.

Decoupling circuit A circuit used to reduce the effect on a given circuit of variations in the voltage supplied by a common power supply.

Degeneration The reduction of the gain of an amplifier caused by the output signal being combined with the input signal in a way that reduces the effect of that signal. *See* Negative feedback.

Demodulator A circuit for recovering the information signal from the carrier signal. (See the Glossary at the end of Chap. 1.)

Detector *See* Demodulator.

Diagonal clipping The clipping, at a diagonal, of the waveform at the output of an AM detector circuit. It is caused by excessive time constant of the detector load/bypass filter capacitor.

Distortion The altering (usually undesired) of the waveform of an electrical current or voltage by the circuit in which it is being processed. Alterations may be to the amplitude, frequency, or phase angle of the original waveform, or to any combination of two or more of these characteristics. Distortion is always accompanied by the introduction of one or more waveforms of frequencies different from that of the original waveform.

Distortion, crossover The distortion of the output waveform of a push-pull amplifier caused by the amplifier operation utilizing a nonlinear portion of the transfer characteristic when operation passes (or crosses over) from one transistor to the other, near the 0°, 180°, and 360° points in the signal cycle.

Distortion, nonlinear The distortion introduced by a device utilizing the nonlinear portion of its transfer characteristic.

Dynamic dc bias A voltage that functions to bias an active device and which is produced by the rectifying action of that device on an input signal. This voltage is called grid leak bias when produced in conjunction with an electron tube.

Feedback In a circuit processing a signal, the returning of a portion of the already processed signal to the input of the circuit, or to a point nearer the input of the circuit, and combining the returned signal with the signal in process.

Feedback, negative Feedback in which the returned signal is combined in opposition to the signal in process. *See* Feedback.

Feedback, positive Feedback in which the returned signal is combined in a way that adds to or reinforces the signal in process. *See* Feedback.

Ferrite loopstick An RF coil assembly containing at least two separate windings wound on a cylindrical form and containing a ferrite core.

Frequency converter Of a superheterodyne receiver, is a circuit function used to change the frequency of a selected carrier into the intermediate frequency (IF) of the receiver.

Frequency converter, autodyne A frequency converter in which the functions of the local oscillator and mixer are achieved in a single circuit with one active device.

Frequency domain As differentiated from time domain, a way of analyzing complex (nonsinusoidal) waveforms by considering them to be the sum of two or more sinusoidal waveforms.

Gain A measure of the performance of a circuit obtained by dividing the output (voltage, current, or power) of a circuit by its input. If the circuit is an amplifier, the gain is greater than 1. The gain for a circuit that attenuates the signal it is processing is less than 1.

Heterodyne To combine two or more frequencies to produce the sum and difference frequencies called beats.

Mixer, heterodyne An amplifier stage especially designed to operate in a nonlinear mode and with provision for inputting two separate signals. The output current includes components whose frequencies are: the two input frequencies, the sum of the input frequencies, and the difference of the input frequencies.

Oscillator An amplifier in which a portion of the output is fed back to the input in phase with the input signal causing the circuit to be self-generating. That is, it will produce an output signal without any external input signal.

Oscillator, Armstrong An oscillator in which the feedback takes place through a circuit arrangement called a tickler coil.

Oscillator, local (LO) In a superheterodyne receiver, the oscillator that supplies the local signal for converting the selected carrier to the intermediate frequency in the mixer.

Oscillator, tickler coil *See* Oscillator, Armstrong.

Spectrogram, frequency A bar graph that depicts the amplitudes of signals of various frequencies with bars.

Superheterodyne A frequency, produced by the heterodyne process, which is higher than what is normally considered an audio frequency. *See* Heterodyne.

Time domain A way of analyzing waveforms in which consideration is given to the values (amplitudes) of the waveform with respect to time. The normal display of a voltage waveform on an oscilloscope is in the time domain. *See* Frequency domain.

Transfer characteristic The manner in which a circuit produces an output as compared to the input. This characteristic of circuits such as amplifiers is commonly presented in the form of a graph in which the horizontal axis represents values of input and the vertical axis represents values of output. The graph is also called the transfer characteristic of the circuit.

REVIEW QUESTIONS: BEST ANSWER

1. The function that is not an essential of a radio receiver is: **a.** antenna reception. **b.** selection. **c.** amplification. **d.** demodulation. **e.** reproduction.
2. In a receiver, selection involves allowing the carrier of one station to be processed while attenuating the carriers of all other stations. **a.** True. **b.** False.
3. Virtually synonymous with selection is the: **a.** resonant circuit. **b.** antenna. **c.** detector circuit. **d.** oscillator circuit. **e.** none of these.
4. The circuit function that recovers the information signal from the carrier in a receiver is the: **a.** antenna. **b.** mixer. **c.** IF section. **d.** detector circuit. **e.** none of these.
5. The unfiltered output of an AM detector is a/an: **a.** pure audio waveform. **b.** pure dc voltage. **c.** half-wave rectified RF or IF waveform. **d.** unmodulated RF waveform. **e.** none of these.
6. The filtered output of an AM detector, when the input is an unmodulated carrier signal, is a/an: **a.** pure audio waveform. **b.** pure dc voltage. **c.** half-wave rectified RF or IF waveform. **d.** modulated RF waveform. **e.** none of these.
7. When an AM carrier signal is displayed on an oscilloscope, the part of the waveform that is a representation of the audio modulating signal is called the: **a.** phase angle. **b.** amplitude. **c.** frequency. **d.** envelope. **e.** none of these.
8. Changing the electrical energy of the information signal into sound energy is called: **a.** detection. **b.** reproduction. **c.** amplification. **d.** selection. **e.** none of these.
9. A resonant circuit with a high Q has: **a.** low selectivity. **b.** medium selectivity. **c.** high selectivity. **d.** average selectivity. **e.** none of these.
10. The positive feedback of energy to a resonant circuit will: **a.** lower its Q. **b.** raise its Q. **c.** have no effect on its Q. **d.** destroy the resonance. **e.** none of these.
11. The function that must be part of a superheterodyne receiver but not of other receiver designs is the: **a.** audio amplifier. **b.** frequency converter. **c.** detector. **d.** RF section. **e.** none of these.
12. The process of producing beat frequencies by combining two or more frequencies in a nonlinear circuit is called: **a.** heterodyning. **b.** detection. **c.** rectification. **d.** amplification. **e.** selection.
13. An oscillator is basically an amplifier that also incorporates: **a.** a rectifier. **b.** positive feedback. **c.** negative feedback. **d.** degeneration. **e.** a positive transducer.
14. A superhet receiver that has a 455-kHz IF section is tuning in a station whose carrier frequency is 890 kHz. The LO is operating at: **a.** 435 kHz. **b.** 672.5 kHz. **c.** 945 kHz. **d.** 1250 kHz. **e.** 1345 kHz.
15. The LO of a receiver is inoperative. The tuning dial is set for 950 kHz. **a.** The volume of the station at 950 kHz will be very weak. **b.** The receiver will tune in a station at 1405 kHz (950 kHz + 455 kHz). **c.** The frequency of the station received will be 495 kHz. **d.** The receiver will be inoperative. **e.** None of these will be true.
16. The ability of a resonant circuit to support oscillation at its resonant frequency for a brief moment (damped oscillation) is called: **a.** the tuning effect. **b.** the flywheel effect. **c.** the Edison effect. **d.** heterodyning. **e.** regeneration.
17. Degeneration, or loss of gain, in an amplifier utilizing an emitter resistor for stabilization, is reduced by means of a/an: **a.** decoupling circuit. **b.** tickler coil. **c.** local oscillator. **d.** bypass capacitor. **e.** neutralizing capacitor.

18. The tendency for high-gain RF (or IF) amplifiers to oscillate because of positive feedback through stray capacitance is minimized by means of a/an: **a.** neutralizing circuit. **b.** bypass capacitor. **c.** decoupling circuit. **d.** tickler coil. **e.** none of these.
19. Reducing the effect on an amplifier of variations in the dc power supply voltage is the purpose of a/an: **a.** neutralizing circuit. **b.** bypass capacitor. **c.** decoupling circuit. **d.** tickler coil. **e.** demodulator.
20. If the *RC* time constant of the diode detector load/bypass filter circuit is excessive, the result may be: **a.** diagonal clipping of the audio signal. **b.** oscillation of the detector. **c.** elimination of high-frequency audio. **d.** degeneration. **e.** regeneration.
21. The average value of the voltage at the output of the detector is a measure of the: **a.** amplitude of the received carrier. **b.** percent modulation of the received carrier. **c.** volume level of the audio signal. **d.** selectivity of the IF section. **e.** none of these.
22. The automatic gain control circuit (AGC) of a receiver: **a.** automatically adjusts the percent modulation. **b.** automatically adjusts the gain of the RF or IF sections, or both. **c.** depends on the percent modulation for its control voltage. **d.** depends on the carrier level for its control voltage. **e.** a and c are correct. **f.** b and d are correct.
23. A coupling capacitor is very commonly found between the output of the AM detector and the input of the audio amplifier to: **a.** block RF signal. **b.** attenuate low frequencies. **c.** block dc voltage. **d.** decouple high frequencies. **e.** a and c.
24. Nonlinear distortion in an amplifier is accompanied by the introduction of: **a.** degeneration. **b.** regeneration. **c.** detection. **d.** odd-numbered harmonic frequencies. **e.** even-numbered harmonic frequencies.
25. Even-numbered harmonic currents are in phase in the two halves of a transformer-type push-pull amplifier but are virtually eliminated in the output because the: **a.** transformer cannot respond to higher frequencies. **b.** harmonic currents flow in opposite directions in the transformer primary winding. **c.** speaker does not respond to the higher frequencies. **d.** harmonic currents flow in the same direction in the transformer secondary winding. **e.** none of these.
26. Matching the speaker impedance to the output impedance of the amplifier is important if: **a.** minimizing distortion is important. **b.** protecting the speaker is important. **c.** getting maximum power from amplifier to speaker is important. **d.** protecting the amplifier is important. **e.** none of these.

REVIEW QUESTIONS: ESSAY

1. Make a sketch of a block diagram of a radio receiver. Include in your sketch only the essential functions. Label each function with an appropriate word or phrase.
2. Describe briefly the function of the signal selection circuitry of a receiver. Why is it essential? Make a sketch of a simple circuit that could perform this function.
3. Discuss how a resonant circuit can perform the function of selection in a receiver. What is the significance of the bandwidth of the circuit in performing this function?
4. What is meant by the *envelope* of a modulated RF waveform? Make a sketch to illustrate your answer.

5. Describe the electrical nature of the unfiltered output of an AM detector. Is the voltage ac, dc, audio, RF, or . . .? Be specific.
6. List and describe the frequencies present in the output of a detector. Which frequency is selected for further processing in a receiver? Which frequency(ies) is/are not selected? What happens to those frequencies that are not selected?
7. Sketch and label the block diagram of a superheterodyne AM radio receiver.
8. Describe briefly the function of each of the blocks of a superheterodyne AM radio (see Question 7).
9. What is the basic requirement of the mixer in the frequency converter section of a superhet receiver? What frequencies are present in its output?
10. **(a)** Define *intermediate frequency* as used in relation to a superhet radio receiver. **(b)** Give the formula for IF in terms of received carrier frequency and LO frequency. **(c)** What is the value of IF for the typical AM broadcast band receiver?
11. **(a)** What is meant by *fixed tuned* when used in relation to IF amplifiers? **(b)** What is the advantage of a fixed-tune amplifier as compared to a variable-frequency amplifier?
12. **(a)** Describe a ganged capacitor. **(b)** What is meant by the *mesh* of a ganged capacitor? **(c)** Under what condition of mesh will a variable capacitor have maximum capacitance? minimum capacitance?
13. **(a)** Describe briefly what an oscillator is and how it functions. **(b)** What purpose does the tickler coil serve in a tickler coil oscillator?
14. Discuss the concept *tuned amplifier*. What is it, and why is it called a tuned amplifier?
15. **(a)** Describe a class C RF amplifier. **(b)** Why is a class C amplifier not suitable for amplifying an amplitude-modulated waveform?
16. **(a)** Discuss undesired oscillation in a high-gain amplifier. Why does it occur? What is its effect? **(b)** How can this problem be eliminated?
17. **(a)** Discuss the need for gain control in the RF/IF sections of a receiver. **(b)** Describe briefly a method for providing automatic gain control for a receiver. **(c)** What basic property of the average value of the detector output voltage makes it useful in an AGC scheme?

EXERCISES

1. Construct a spectrogram of the following signals: f_1 = 580 kHz at 8 mV, f_2 = 720 kHz at 4 mV, f_3 = 940 kHz at 12 mV, f_4 = 1140 kHz at 10 mV, and f_5 = 1560 kHz at 5 mV.
2. Frequency response measurements on several tuned circuits indicate that their output voltages are down to 70.7% of maximum (at 600 kHz) at the frequencies given. Calculate the bandwidth and Q for each circuit: **(a)** 570 kHz and 630 kHz; **(b)** 594 kHz and 606 kHz; **(c)** 596.7 kHz and 603.3 kHz.
3. Given: a superheterodyne radio receiver whose IF section operates at 455 kHz. Calculate the LO operating frequency for each of the following received-carrier frequencies: **(a)** 580 kHz; **(b)** 870 kHz; **(c)** 1270 kHz; **(d)** 1590 kHz.
4. Calculate the received-carrier frequency for each of the following LO operating frequencies (assume that IF = 455 kHz): **(a)** 1015 kHz; **(b)** 1345 kHz; **(c)** 1475 kHz; **(d)** 1.995 MHz.

5. A car radio is receiving satisfactorily a station whose carrier frequency is 1090 kHz. The LO frequency is 1352.5 kHz. What is the operating frequency of the IF section of this radio?
6. Construct a sketch showing the addition of a fundamental waveform and its second harmonic producing the waveform typical of nonlinear distortion in an amplifier.

4

TROUBLESHOOTING and CIRCUIT ALIGNMENT

Anyone who is successfully involved with electronic communication systems in a really hands-on technical way possesses certain laboratory and/or shop skills. The most important of these skills is that of *circuit diagnosis* or *troubleshooting*. Even a design engineer must be able to troubleshoot his/her prototypes. Consider the embarrassment involved in this situation:

> An engineer has designed a new radio receiver. The engineering laboratory technician has completed constructing the prototype. The unit does not work. Neither the engineer nor the technician knows how to troubleshoot. They call the local radio/TV repair shop and request the services of a technician to come and find the problem.

This story is not told to suggest that the situation has ever arisen—only to emphasize the importance of the skill of troubleshooting.

Another important skill in the field of communications technology is that of adjusting tuned amplifiers for optimum operation, a process generally referred to as *alignment*. Let's learn about these and certain other commonly used skills in communications work in the following sections.

4.1 TROUBLESHOOTING

The circuits of a complete electronic communication system contain literally hundreds of separate components. The failure of any one of these components may cause the system to fail to function at all or, at best, to function poorly.

The skill of troubleshooting is one of finding a defective component or other defect in a circuit when a system has failed. In the world of the late twentieth century the speed with which this task is completed is extremely important for at least two reasons: (1) In most cases it is important that the downtime of the equipment be held to a minimum; the equipment serves some critical need of its user. (2) The technician's time is expensive. For a technician to achieve a satisfactory level of productivity in troubleshooting, it is essential that he/she have a good grasp of the basic skills of a troubleshooting system. These can be studied, practiced, and mastered like other skills. Their mastery need not be left to the trial-and-error approach by which many have learned them.

A key phrase in the introductory paragraph above is "troubleshooting system." One of the most difficult problems in troubleshooting, for someone who is just beginning to learn the skill, is deciding where to begin. Then, after finally taking the first step, the novice is confronted with the problem of what to do next if the first step did not uncover the defective component, and so on. A good troubleshooting system, then, will be one that includes a plan, based on logical and scientific reasoning, which outlines how and where to start, and in general, how to proceed.

The troubleshooting system that we will study here is quite similar in concept to the systems approach we are using in learning a system. That is, we begin the task of trying to locate a single defective component by looking at the total system and observing what it is doing, or not doing. Then, step by step, we concentrate our attention on a smaller and smaller portion of the system. Ultimately, we end up at the defective component.

The process can be compared to viewing a scene on a television screen which is being picked up through a zoom lens: First, we see a long shot of the scene. Perhaps it is one containing people. Gradually, the camera zooms in closer and closer to the people. We are able to see more and more details of their features. Finally, the camera is focused very tightly on the face of a single person. We are able to discern all but the smallest of details of that face.

In troubleshooting a system, we first look at the system as a whole and attempt to determine in what next smaller portion is the malfunction occurring. Perhaps the system is composed of several relatively large subsystems. We will attempt to decide which subsystem the problem is in. Next, which section in that subsystem contains the problem? Having determined the section, the next step is to determine which stage in that section has the problem. Next, in what portion of the stage is the problem? Is it in the collector circuit, the base circuit, or the emitter circuit? Finally, if we have been successful in narrowing the area in which the defective component is likely to be located to this extent, the task of finding the actual component is relatively simple.

It is important to understand thoroughly that our approach takes us from the big picture of the system down through smaller areas until we finally focus on a small circuit containing several components at most. We do not start by making a guess as to which component is at fault, testing that guess, making another guess, testing it, etc. Neither do we start at one end of the system, calculate predicted currents or voltages, check against measured values, go on to the next circuit, calculate and measure, etc. The approach involves making a set of sequential decisions. Each decision reduces the size of the area in which the problem may be located and brings us closer to the defective component or circuit fault. Each decision is made on the basis of our analysis of data on the functioning of the system. The data are obtained through observations, tests, and/or measurements of the system.

Each step of the process relates to a subdivision of the system. The first step will be to obtain data and decide in which subsystem the problem is located. (This presumes that the system is sufficiently complex to contain subsystems.) The next step will be to gather the data and make the decision as to which section of the chosen subsystem contains the problem. The area of interest has been reduced to a single section. The third step involves obtaining the data and making the decision to point our finger at a particular stage in the designated section. Our area of focus is now a single stage: perhaps a transistor and several resistors and capacitors. The next step is to determine at which part of the stage to point our finger: the collector circuit, the base circuit, or the emitter circuit. A conclusive decision at this point leads to the fifth and last step: the determination of the defective component or other type of circuit failure. The process is summarized in the form of a flowchart in Fig. 4.1.

Utilizing the troubleshooting system described above requires that we view the total complex circuit which makes up a particular electronic system in a very different way than that which we have used on circuits in basic electronics courses. It is important now to divide the total circuit into sections and stages.

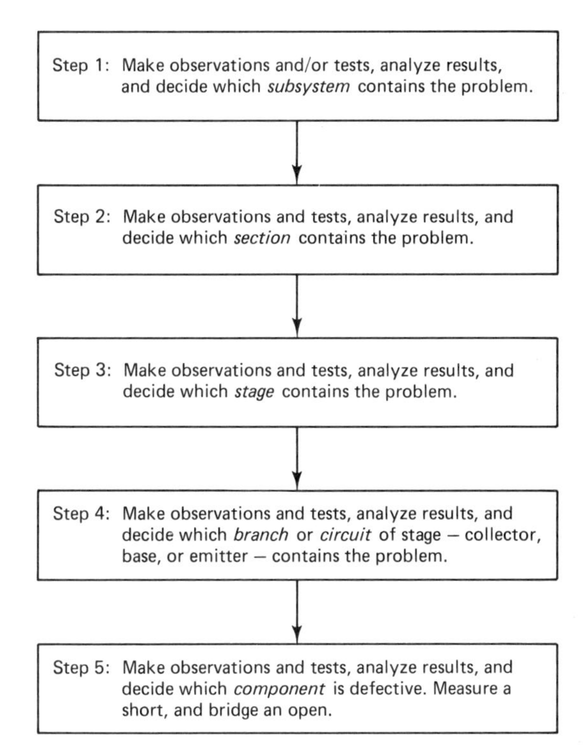

Figure 4.1 Troubleshooting system.

A section, in this view, is a portion of the total circuit which performs some unique function in terms of accomplishing the purpose of the overall system. Of course, this is precisely the view that we have taken in the systems approach to learning a new system.

In electronic communication systems as well as many other types of electronic systems, it is appropriate to consider that there is a signal to be passed through the system—a system signal. The system's basic function is to process that signal. Most sections of the system, then, process the system signal, each in its own unique way.

From the concept of system signal we jump to the idea that the signal follows some path through the system—a *signal flow path*. The signal flow path is similar to a chain. In this case the links of the chain are the sections. Within the sections there are often links of stages. As in any linked chain, if any one link is broken, the chain is broken, becoming ineffective.

We may think of the process of troubleshooting as one of finding the broken link in the signal chain. Implicit to this view is the idea that, generally, all but one of the sections of a faulted system are operable. This is the case, in fact. With appropriate testing we can identify the operable sections and/or stages and eliminate them from our area of concern. Performing this task systematically, as in the troubleshooting approach described above, leads us to the point where the signal path is broken.

To utilize this system we must perform a number of different tasks of measurement or testing, analysis, and decision making. These tasks involve techniques or skills of the electronic shop and laboratory. Let us now look at how we might go about applying the system, first, to a specific radio receiver and then to an electronic system in general. Let us imagine that we are troubleshooting the six-transistor receiver that we studied in Chap. 3.

We begin by assuming that the failure of the unit is due to a single defective component or circuit defect, such as an open or short. This assumption is one that is commonly made at the beginning of any troubleshooting procedure, even of a large system. Since the receiver is a subsystem of a communications system, step one of our troubleshooting system is already completed: the subsystem at fault is the receiver. The process of troubleshooting an actual receiver of this particular design might proceed as follows:

1. We power up the receiver and attempt to tune in a station in a normal fashion, that is, we turn the tuning knob across the dial and adjust the volume control for higher volume. If the receiver is completely inoperative, we will, of course, not be able to tune in any station. Nevertheless, we listen very carefully to the output from the speaker. Is there absolutely no sound whatsoever? Is the receiver completely "dead?" Or is there some noise from the speaker? If some noise can be heard from the speaker, can its volume be adjusted by the volume control? If the answer to the last question is yes, we can virtually eliminate the audio section as a possible location of the problem.

2. If the receiver is completely dead, a good next step is to check to see that dc power is being supplied correctly. Take a voltmeter and measure the V_{CC} voltage at two or three points in the receiver. Satisfy yourself that the

circuit is getting the proper dc voltage before going on to the next step. If the dc voltage is not correct, the next step is to investigate the power source.

3. If the receiver is dead and we are confident that it is obtaining power correctly, the next step is to make tests which will enable us to make a decision as to which section contains the problem. Here we will use a technique which is not explicit in the approach that has been described. We will mentally divide the receiver approximately in half and then make a test to reveal which half contains the problem. With one test we will have eliminated from suspicion approximately one-half of the receiver. This procedure has been referred to as the *half-split rule*.*

Let us assume that we have an oscilloscope available. We will split the receiver approximately in half by observing on the scope the IF output signal—the signal at the input to the detector. Adjust the sweep rate of the scope so that you display approximately three cycles of the IF signal (try 0.5 μs or 1.0 μs/div). Adjust the sensitivity of the vertical input of the scope for approximately 20 mV/div. Connect the ground side of the oscilloscope input to the chassis of the receiver and the active or ''hot'' side of the scope input to terminal 4, transformer T_4. Observe the scope display as you again attempt to tune in a station. Imagine that an IF signal display comes up at several points on the dial. You are correct to conclude that that portion of the receiver corresponding to the left half of the schematic diagram (see Fig. 3.15) is working. That portion of the receiver is commonly referred to as the *front end* of the receiver.

Thus the conclusion, as a result of this test, is that the front end—antenna/RF section, local oscillator/mixer, and IF section—is working. This conclusion also implies that the problem is located to the right of the test point on the schematic diagram.

It is also helpful for future reference to be thinking in terms of the block diagram of the receiver. Look again at Fig. 3.12. The signal path starts on the left side of the block diagram and proceeds to the right. On the block diagram, we have just made a test at a point between the IF section and the detector. As indicated, if we find a satisfactory signal here, one that can be tuned in at various points on the receiver dial, all the blocks to the left of the test point are working; all those to the right of the test point are still suspect. Of course, if we do not find a signal at this test point, we would conclude that the problem lies in one of the sections to the left of the test point.

We are going to imagine, then, that we have found a satisfactory signal at the input to the detector. What is our next step?

4. Using the results of step 3, we must conclude that the problem lies in the right half of the receiver circuitry, that is, in either the detector or the audio section of the receiver. It will again be productive to make a test at or near the center of this remaining circuitry. Displaying the signal on the base of transistor Q_4 would be a reasonable test. We must remember that a signal here would be an audio-frequency signal. We adjust the sweep rate of the oscilloscope accordingly.

* Donald H. Schuster, Ph.D., *Logical Electronic Troubleshooting*, New York, McGraw-Hill, Inc., 1963, p. 61.

Not knowing the expected amplitude of a signal here, if present, we increase the sensitivity of the vertical input to the scope to maximum, say 10 mV/div.

Let us imagine that we do find a signal here. It is not a sine wave but a waveform that changes rapidly since it is the representation of program material being broadcast by the station to which the front end of the receiver has been tuned. To ensure that what we are viewing is a real signal and not just noise, we turn the tuning knob across the dial. We see the displayed signal alternately disappear and then reappear in some slightly different form as different stations are received. As a result of this test our conclusion is that the problem is in the audio section. We are ready to find the defective stage.

5. For our next test we decide to "scope" the signal at the collector of Q_4 (i.e., connect the vertical input of the oscilloscope to the collector of Q_4 and the common to the chassis, and adjust the scope controls for a proper display). Imagine that we find no signal here. We expected an amplified version of the signal at the base. A signal at the base of a stage and no signal at the output (the collector) indicates that the stage is not amplifying. We have found the broken link! According to our troubleshooting system, the next step is to make tests to identify which part of the stage—collector side, base side, or emitter side—the problem is in, and then, finally, to determine which is the defective component.

Troubleshooting Using DC Voltages

Through step 5 of our imaginary troubleshooting exercise we have used a technique known as *signal tracing* as a means of obtaining data about the unit for the purpose of making decisions. Signal tracing simply means what the title suggests: We trace a signal along a signal path with the aid of an oscilloscope so that we can actually "see" the signal. In the example above the signal was that of a station or stations received and processed by the receiver. Of course, as such signals pass through the receiver their nature is changed—the RF carrier is changed to the IF signal, the IF signal is detected and "the signal" becomes an audio signal. In any case, in this example it is appropriate to think of the process described as that of tracing an "air signal" (a signal that has come through the air from a station) along the signal path through the receiver.

In some instances it is more effective to inject a signal from a laboratory generator than to rely on an air signal. If we inject an appropriate signal at various points in a system and observe the results of such injection, we are using the technique called *signal injection*. Both signal tracing and signal injection are effective techniques for determining which section and/or which stage contains the defect. However, once we have located the defective stage it is necessary to switch to another testing technique to determine where in the stage the defective component is located. This technique is one of measuring the dc voltage on the transistor terminals and analyzing the values so as to determine the location of the fault. Let us digress from the troubleshooting example long enough to study this technique.

This technique is based on the fact that in a normal, "healthy" transistor (or tube) stage, with no signal applied, there will attain a set of normal dc voltages on the electrodes or terminals of the transistor. That is, simply because of the ohmic values of components in the stage and the operating characteristic of the transistor, the dc voltages of collector, base, and emitter, as measured to ground will remain constant and predictable. This assumes that the power supply voltage, V_{CC}, remains constant. Further, the technique is based on the fact that when something happens to change the ohmic value of any of the components in the stage or the operating characteristic of the transistor, there will be a change in the dc voltages of the stage. In fact, the amount and direction of change of the voltages are a reliable clue as to what change has occurred to which component.

This is a powerful technique. It is popular with professionals whose responsibility it is to troubleshoot and repair electronic equipment. The technique is so effective, in fact, that equipment manufacturers go to great expense to measure and record normal transistor dc voltages on the schematic diagrams of their equipment as an aid to those who must service the equipment.

To use this technique successfully we must have a set of normal dc voltage values for the unit that we are going to troubleshoot. If we do not have a schematic with these values on it, we might measure the values on a unit of the same design which does not contain a defect. Of equal or greater importance than having the set of values is understanding why and how the values are what they are. Having this understanding prepares us for deciding what has caused a change from the normal voltages.

Let us examine the circuit of Q_4 in its normal state. It is a relatively typical transistor amplifier stage. For convenience, this stage is taken out of the overall receiver circuit and shown by itself in Fig. 4.2. Notice that the primary winding of transformer T_6 has been replaced with its dc resistance. Notice also that the diagram is labeled with normal dc voltage values for a V_{CC} of -5.8 V. Let us analyze the base circuit first.

Thévenize the base circuit. We Thévenize the base circuit to make it easier to understand the effect of changes in various parts of the stage. *Note*: Since the voltage at the junction of resistors R_{13} and R_{16} is determined primarily by the current drawn through R_{16} by stages Q_1, Q_2, and Q_3, we use the normal voltage for that point, -5.0 V, as the source voltage for the voltage-divider circuit, R_{13}–R_{14},

$$V_{BB_{\text{th}}} = \frac{R_{14}}{R_{14} + R_{13}}(-5.0) = \frac{4.7}{4.7 + 15}(-5.0) = -1.193 \text{ V}$$

$$R_{B_{\text{th}}} = \frac{R_{14} \cdot R_{13}}{R_{14} + R_{13}} = \frac{4.7 \cdot 15}{4.7 + 15} = 3.579 \text{ k}\Omega$$

The Thévenin equivalent circuit is shown in Fig. 4.2(b). The normal voltage on the base of Q_4 (with respect to ground) is -1.02 V. The Thévenin source voltage for the base, and, therefore, the voltage at the base if there is an open anywhere in the base–emitter junction circuit, is -1.193. The difference between these

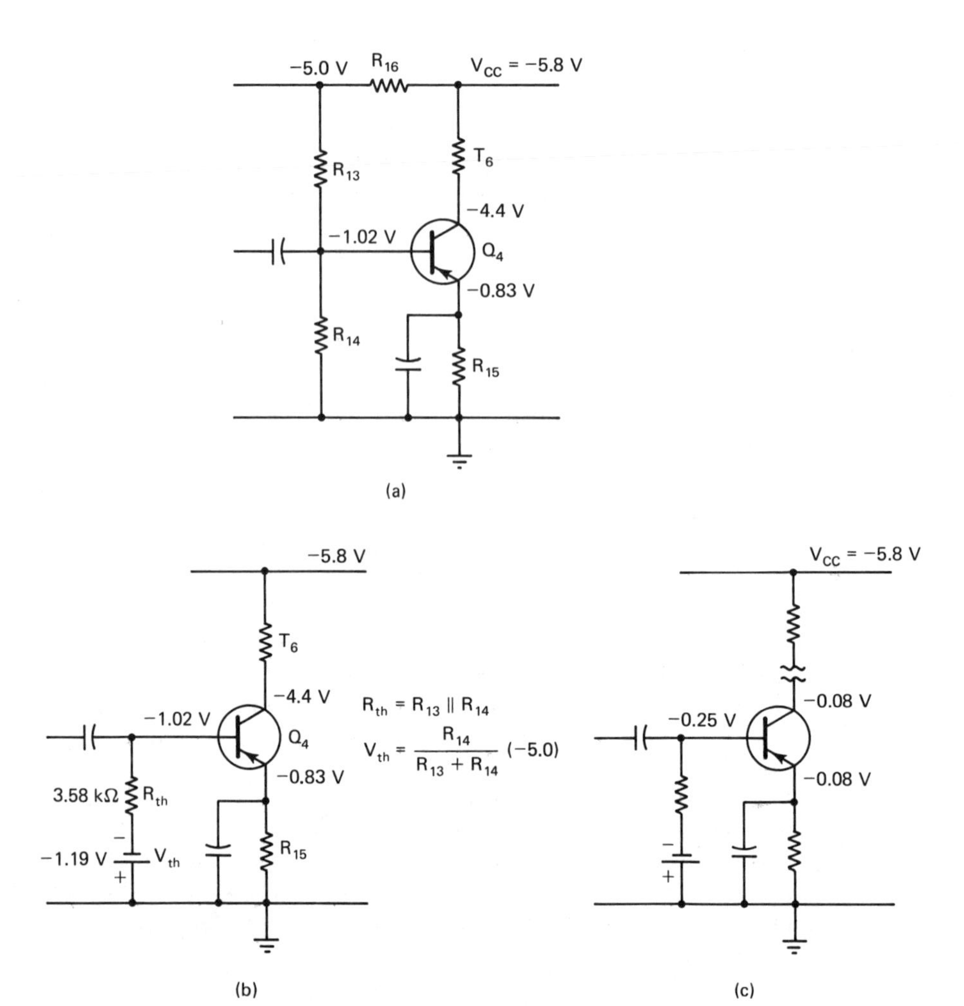

Figure 4.2 (a) Normal stage voltages; (b) Thévenin equivalent of base dc supply circuit; (c) dc voltages with collector circuit open.

voltages—the normal base voltage and the open-circuit base voltage—is small but very significant. We must be prepared, when troubleshooting, to check the base voltage carefully and take note of a small change.

Let us now make another calculation involving the base voltage. Imagine that something happens to cause an open in the collector side of the stage—the transformer winding opens or the collector connection inside the transistor case opens, for example. The collector current drops to zero. What will happen to the base voltage under this condition? Will it become more negative or more positive? How can we predict the probable result? The circuit would be similar to that of Fig. 4.2(c). Since there is now no collector current, the effect of the emitter resistor, R_{15}, on the base circuit will be determined by its actual value, not β times its actual value as is the case when there is collector current. Let us assume that the junction voltage will remain the same, -0.19 V, and calculate V_B for this condition.

$$V_B = \frac{R_{15}}{R_{15} + R_{B\text{th}}}(V_{BB\text{th}} - V_{BE}) + V_{BE}$$

$$= \frac{0.22}{0.22 + 3.579}[-1.193 - (-0.19)] + (-0.19) = -0.2481 \text{ V}$$

If there is an open in the collector side of the stage, we can expect the base voltage to become more positive or its magnitude to become less. Saying it another way, if we have a defective stage and we measure the base voltage and find its magnitude to be very low, one of the things we will want to check for is an open collector or collector circuit.

Let us look now at the collector voltage. As indicated in Fig. 4.2(a), the normal value for V_C is -4.4 V. It is important to understand why this value is what it is and how and why changes in it can be used to predict rather precisely certain types of defects in the stage. The equation for V_C is

$$V_C = V_{CC} - I_C R_L$$

This is a relatively simple equation, but it will be a valuable aid for you in troubleshooting if you learn to use it effectively. For example, we can use it to calculate normal collector current

$$I_C = \frac{V_{CC} - V_C}{R_L} = \frac{5.8 - 4.4}{370} = 3.784 \text{ mA}$$

Imagine that we are troubleshooting a defective stage and find that $V_C = -5.8$ V. What does the equation for V_C tell us to look for? Since $V_C = V_{CC}$ in this case, $I_C R_L$ must be zero. Therefore, either I_C is zero or R_L is zero. That is, either the transistor is in cutoff for some reason, or R_L is shorted. We have something to check on further.

Imagine now that in another instance, we find that V_C is approximately -0.08 V. What kind of problem does use of the equation suggest in this instance? The very low magnitude of V_C indicates that either V_{CC} is low or the $I_C R_L$ product is very large. In fact, this is the voltage, -0.08 V, that will typically be found at the collector if the collector circuit is open ($R_L = \infty$).

Looking now at the emitter voltage, we see from Fig. 4.2 that normally $V_E = -0.83$ V, approximately. The equation for this voltage is also a valuable troubleshooting aid: $V_E = I_E R_E$. Recall that $I_E = I_C + I_B$. Imagine that in troubleshooting a defective stage we find the magnitude of V_E greater than normal. What suggestions come to mind if we check the value against the equation and the normal value? Since the equation represents a simple product, one or the other of the two factors, I_E or R_E, must be larger than normal. Either the transistor is conducting more heavily than normal or the emitter resistor has increased in ohmic value, or it has opened, perhaps. Of if we find $V_E = 0$ V—what, then? Perhaps the emitter itself is open because there is apparently no I_E.

Looking at only one of the three voltages V_C, V_B, or V_E provides us with some clues as to what can be wrong in a defective stage. However, if we look at all three voltages and use a logical reasoning process in relating the amount and direction of the changes in each, we will be able to pinpoint precisely the circuit—collector, base, or emitter—which contains the fault. Let us look at some actual case studies of defective stages.

Example 1
A six-transistor radio like that of Fig. 3.15 was inoperative. Signal tracing led to the conclusion that stage Q_4 was not amplifying. The dc voltages on the stage were measured with the following result: $V_C = -5.8$ V, $V_B = -0.23$ V, and $V_E = -0.06$ V. What is the probable defect in the stage?

Solution. The increased magnitude of V_C suggests that either $I_C = 0$ or R_L is shorted. The bias voltage, $V_B - V_E = -0.17$ V, however, is about normal. This suggests that I_C should be about normal. However, since the magnitude of V_E is much less than normal, we have a confirmation that I_C is very low or zero. The defect must be an open in the collector circuit between the point where V_C is measured and the base–collector junction of the transistor. The open could be a cold-soldered joint at the collector lead, a poor contact between collector lead and socket receptacle, or an open somewhere inside the "can" (transistor enclosure). In fact, this defect was simulated by removing the collector lead from its socket.

Example 2
In another instance, when the six-transistor receiver of the preceding example was inoperative, the Q_4 stage was again found not to be amplifying. In this case measurement of the dc voltages on the stage yield the following values: $V_C = -0.14$ V, $V_B = -0.27$ V, and $V_E = 0$ V. What is the probable defect in the stage?

Solution. The magnitude of V_C is very small, indicating either that the transistor is saturated or that R_L is open. The bias voltage is greater in magnitude than normal (-0.27 V compared to -0.19 V). This suggests that the collector current should be found to be greater than normal. The emitter voltage is zero, indicating that either I_E or R_E is zero. All other indications are that I_E is not zero, but in fact, greater than normal. We must conclude that either R_{15} or C_{19} is shorted or that there is a short from the emitter of Q_4 to ground. In fact, this defect was simulated by connecting a jumper from the emitter to ground. This is a clear-cut example of a transistor in saturation: the base–collector junction is forward biased and the collector-to-emitter voltage drop is very small—0.14 V.

There is no magic key to sucessful diagnosis using dc voltage measurements. Success depends primarily on careful measurement of the dc voltages of a stage—V_C, V_B, and V_E—and a thoughtful analysis of these voltages. The analysis must include comparison with normal voltages for the stage, and consideration of the direction of change of the voltages using the basic equations for the voltages as a guide. These equations are simple Ohm's law equations or those of a series circuit. There is no substitute for using these very basic concepts. A 1–2–3 approach to diagnosis using dc voltages is as follows:

1. Use signal tracing, signal injection, or any other suitable technique to determine the defective stage. After deciding which is the defective stage, carefully measure the dc voltages of the stage—V_C, V_B, and V_E (these are the voltages of the respective leads of the transistor measured with respect to ground or chassis). Write these voltages down, one under the other, so that you can see all three at a glance. Do not attempt simply to keep them in mind. Calculate the bias voltage, $V_B - V_E$, and write that down near the other three voltages.

2. Begin the process of analyzing the meaning of the voltages by looking first at the collector voltage. Compare it with normal collector voltage for the stage. If it has changed from the normal value, does the direction of change suggest that the operation of the transistor has moved toward saturation or cutoff? For common-emitter *PNP* stages: An increase in magnitude (i.e., a more negative voltage) indicates movement toward cutoff; a decrease in magnitude (a less negative voltage) indicates movement toward saturation. For common-emitter *NPN* stages: An increase in magnitude, a more positive voltage, indicates movement toward cutoff; a decrease in voltage means a movement toward saturation. Draw a tentative conclusion, then, based solely on the direction of movement of the collector voltage: the transistor is conducting more than normal (collector voltage change suggests movement toward saturation), or the transistor is conducting less than normal (voltage change suggests movement toward cutoff). Be prepared to modify this tentative conclusion as you analyze other voltages in the stage.

3. Compare transistor bias with the normal bias for the stage. If the bias voltage has changed, what does the direction of change suggest: movement toward saturation, or movement toward cutoff? Draw a conclusion. Does your conclusion here agree with your conclusion in step 2?

4. Now compare the magnitudes of the base and emitter voltages with their normal values. Apply your knowledge of Ohm's law and/or series circuits to any variations from normal that you observe here. For example, if the magnitudes of both the base and emitter voltages have increased, the indication is that there is an open between the emitter transistor lead (the point where you measured the emitter voltage) and ground. The increased magnitude of the base voltage suggests that the base supply circuit has become unloaded. Recall the Thévenized version of the base supply circuit of Fig. 4.2(b). The increased magnitude of emitter voltage is the result of R_E in the equation $V_E = I_E R_E$ being very large. Of course, an increased magnitude of both base voltage and emitter voltage can also be caused by a reduction in the series resistance between the base and the base supply voltage, for example, R_{13} in Fig. 4.2(a). If this happens, base current will increase and the operation of the transistor will move toward saturation. There are several possibilities here; let's systematize the process of drawing a reasonable conclusion:

When magnitude of both base and emitter voltages have increased:	When magnitudes of both base and emitter voltages have decreased:
Look for open in emitter circuit if previous conclusion was that transistor has moved toward cutoff,	Look for open in collector portion of stage if previous conclusion was that transistor has moved toward saturation,
or	or
Look for reduced resistance in base supply circuit if previous conclusion was that transistor has moved toward saturation.	Look for reduction of resistance between base and ground in base voltage divider circuit if previous conclusion was that transistor has moved toward cutoff.

You will not be able to understand all of the ramifications of this dc voltage analysis troubleshooting technique simply by reading the foregoing material once. Study it carefully several times while referring to the circuit of an amplifier stage, for example, the circuit in Fig. 4.2. Add to your understanding of this technique and extend your knowledge of the symptoms of specific types of component failures by assuming a specific failure and predicting the consequences by means of pencil-and-paper analysis.

Of course, the most effective method of increasing your knowledge of the technique is to set up an actual amplifier stage and observe the effects of a variety of defects. Systematically observe the effects of opening and shorting each component in the stage.

This technique will not yield results on absolutely all component failures. For example, open capacitors, whether coupling or bypass capacitors, will not upset the dc voltages in a stage. Nevertheless, the technique will yield results in a very high percentage of problems. It is an indispensable tool.

Let's return now to the general process of troubleshooting the receiver, from which we have digressed. First, a summary of the steps taken so far:

1. Powered up the receiver and found it was completely "dead." There was no sound whatsoever from the speaker.
2. Checked to make sure that the unit was being supplied the correct voltage for dc power.
3. Dividing the circuitry approximately in half led to the decision to scope the output of the IF section. Found signal. Concluded that the front end of unit was functioning correctly, that the problem was in either the detector or audio sections.
4. Scoped at the input to the audio section. Found signal. Concluded that the problem was in the audio section.
5. Scoped at the output of the first audio stage. Found no signal. Concluded that the first audio stage contained the problem.

We continue and now use the dc voltage technique to find, first, the circuit of the stage at fault, and, finally, the defective component.

6. We measure the dc voltages at the leads of the transistor and record the values: $V_C = -5.8$ V, $V_B = -1.12$ V, V_E = a trace of voltage only. Bias voltage = -1.12 V approximately. The collector voltage indicates that the transistor is cut off. The bias voltage magnitude is excessively great. The magnitude of the base voltage is greater than normal. This suggests that the base supply circuit is unloaded, which, in turn, indicates an open in the circuit between the base test point, through the base–emitter junction, through the emitter resistor to ground. The trace of voltage at the emitter test point indicates that there is a small current flowing from the transistor through the emitter resistor. Hence the emitter circuit is not open. Conclusion: The open is between the base test point and the base–emitter junction. A test of the transistor itself revealed an open in the base. Replacing the transistor with a known good transistor restored the receiver to normal operation.

Let us work through another case. We begin with step 1. Imagine that during our first tests on the unit we observe noise from the speaker. We can adjust the volume of that noise with the volume control but are unable to tune in any station. Our conclusion based on this information is that the audio section is functioning—that the problem is not in that section. We go on to step 2 again.

2. We measure V_{CC} to make sure that the voltage is correct before proceeding.

3. Because of the noise test we are reasonably sure that there is no signal at the output of the IF section. We scope that output just to be sure and find no signal. Our conclusion is that the problem is located in the sections to the left of the detector on the block diagram: the RF circuit, the frequency converter section, or the IF amplifier section.

4. There are no powerful arguments for determining what the next step should be. Of course, we want to find the problem as quickly as possible, to make a minimum number of steps. As we move farther to the left on the block diagram, the signal amplitude will become smaller. It will require greater skill to observe the signal even when it is present. Checking to determine that the local oscillator is operating is a reasonable choice for a next test. We check for the presence of an oscillator signal by scoping at the emitter of transistor Q_1. It is essential that the scope be adjusted for an appropriate sweep speed and sensitivity. The frequency of the LO signal will be of the order of 1 MHz and the amplitude should be 100 mV or more. If we find a signal here it is important to assure ourselves that what we observe is actually the LO signal. We can do this by turning the knob while observing the scope display. We should observe a reasonably good sine wave whose frequency changes as we change the tuning knob setting. We will assume that we do find a healthy LO signal. This is not absolute proof that the frequency converter is doing its job, although, for the moment, a reasonable assumption is that it is. Hence we conclude that the problem is not in the frequency converter section but in the IF section.

5. Having decided that the problem is in the IF section, our next step is

to determine which stage, since the IF section has two stages. We scope the output of the first stage (connect scope test lead to the collector of Q_2). There will be no IF signal here unless the RF section and frequency converter section are adjusted to tune in a station. We adjust the tuning dial while observing the scope display. The IF signal comes up at several places on the dial. This stage appears to be working. Leaving the tuning dial set at a position that gives a strong IF signal, we move the scope input to the collector of Q_3. There is no signal. We check for signal at the base of Q_3 and find it. Stage Q_3 is not amplifying; it is the broken link in the chain.

6. Signal tracing has enabled us to conclude that stage Q_3 is the one with the problem. We turn now to dc voltage analysis in an effort to determine which part of the stage contains the defective component. The normal, no-signal dc voltages for the stage are: $V_C = -3.8$ V, $V_B = -1.11$ V, $V_E = -0.95$, $V_{BE} = -0.16$ V, and voltage to ground at the junction of R_{11}–$R_{16} = -4.8$ V. (This voltage is equivalent to V_{CC} for this part of the receiver.) We measure and record the voltages on the pins of the transistor socket: $V_C = -4.8$ V, $V_B = -1.23$ V, and $V_E = -1.30$ V. Bias voltage $= +0.17$ V. The stage is actually reverse biased, apparently! The collector voltage indicates that the transistor is cut off. The bias voltage confirms that the stage should be cut off. The magnitudes of both the base and emitter voltages are greater than normal, suggesting that the emitter resistor is open (R_{10}). We bridge R_{10} with a 470-Ω resistance from an R-box. The receiver immediately starts working.

It is important to understand why the normal voltages were given above as "no-signal" voltages: Stages Q_2 and Q_3 are AGC-controlled stages. Therefore, their dc voltages will vary with the level of signal being processed. A no-signal condition is one that can easily be duplicated by shorting out the local oscillator.

These examples of troubleshooting imaginary "bugs" in a receiver have been designed to demonstrate the specific steps that might be taken in following a troubleshooting system. They have demonstrated the application of the techniques of signal tracing and dc voltage analysis in very concrete examples. These descriptions will help you get started troubleshooting if you read them carefully several times and think about the logic of the conclusions as you read. You will become proficient at troubleshooting, however, only if you are able to practice it extensively with actual, or at least, simulated bugs in real equipment.

The examples presented here have been concerned with one particular electronic subsystem. Many of the principles demonstrated can be used on many other types of electronic systems. Before we leave the topic of troubleshooting, let's look at it in a more general way.

Troubleshooting in General

When we are confronted with the task of diagnosing trouble in an electronic system, one of the first things that we must do is to power up the unit and exercise it while we observe symptoms of malfunction. This requires that we know something about the system: what it is designed to do and how to operate it. Knowing how to operate a system may require extensive preparation over and above learning the technical details of how it works. For example, operating a computer system usually requires some knowledge of computer programming. Operating a system may sometimes involve observing very rigid procedures for the protection of the equipment and/or personnel, or for conforming to legal requirements. Testing an x-ray machine incorrectly can very easily destroy an expensive x-ray tube or radiate unnecessarily oneself or other persons. Operating a radio transmitter must be in conformance with federal regulations.

When we exercise a system as part of the process of troubleshooting it, we do so to observe symptoms of malfunction. It is important, then, to be extraordinarily alert while we are exercising the unit. We not only should be looking for the obvious signs of malfunction, but should be utilizing all of our senses to try to perceive any sign of abnormality about the operation of the system. Does the unit "sound right"? Is there any unusual vibration, or lack of normal vibration, detected as the unit operates? Is there any unusual odor emanating from the unit? Many components, in the process of failing, overheat, producing a peculiar odor and, sometimes, smoke. Such telltale signs have many times led an alert troubleshooter directly to the defective component. A cursory visual inspection of the circuitry of a unit may lead to immediate detection of an overheated or burned component, a broken conductor, or other obvious physical defect. If a system is completely dead, the first question should be: Is the unit getting ac power? Is the line cord plugged in? Or, when a system is wired directly to the ac circuit, is the line switch turned on?

The first, and even, second steps of our troubleshooting system—identifying the subsystem and section that contain the problem—can often be accomplished by using the features, controls, and indicator devices of the system itself. There may be no need for test instruments for these two steps. A televison receiver is an excellent example of a subsystem that contains many sections. Determining the section with the fault, in a malfunctioning TV receiver, can frequently be accomplished simply by manipulating the controls and carefully observing the results. "There is a picture but no sound," or "there is sound but no picture," are typical descriptions of TV receiver symptoms which, to an experienced TV service person, immediately reduces the possible location of a defect to one or two sections.

Knowing the block diagram of the system, or having one available, and being able to interpret it are essential for accomplishing these first two steps. Several different styles of block diagrams are illustrated in Fig. 4.3. Let's see how the configuration of the block diagram itself can help us in reducing the size of the area in which to look for the defect.

Figure 4.3(a) shows the block diagram of a system with a simple linear signal flow path. An AM radio receiver is an example of a subsystem with this type of diagram. The linear diagram, although straightforward, does not lend itself to the expedient technique being discussed here. The failure of any one section causes the failure of the whole unit.

Contrast this with the convergent type of flow path of Fig. 4.3(b). This

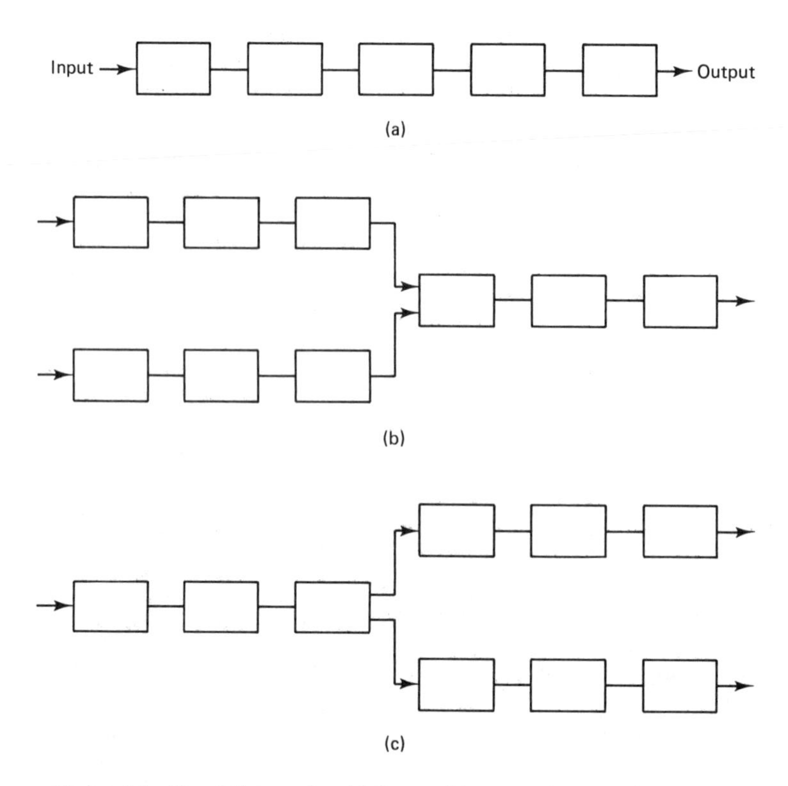

Figure 4.3 Signal flow paths: (a) linear; (b) converging; (c) diverging.

style is typical of the popular AM/FM high-fidelity receiver/record player units. In this case, switching between the various possible inputs will enable us to focus quickly on the area most likely to contain the problem. For example, if the receiver works on FM but not AM, it is apparent that the audio section functions satisfactorily and that the problem is located in some section unique to the AM function. This kind of information and conclusion can be gained through a simple process of informed switching, observation, and logical reasoning.

A TV receiver contains elements of the diverging flow path [Fig. 4.2(c)]. The IF section must process both picture and sound carrier signals. These are then separated and processed further in their own separate sections. The implications of the symptom of "picture but no sound" are obvious.

Once the point is reached in a troubleshooting process where it is necessary to use test equipment, the matter of personal knowledge of the equipment and skill in its use becomes extremely important. Signal generators, oscilloscopes, multimeters, etc., are simply tools. They are incapable of doing any thinking for us.

When injecting a signal, for example, it is crucial that we set up the generator to produce the correct frequency at an appropriate amplitude. Injecting an audio section with 455 kHz will not provide any useful information, nor will using a 10-V peak-to-peak audio signal when the typical signal is a nominal 1 mV.

Similarly, setting up an oscilloscope for meaningful observation of a signal requires considerable knowledge and skill and careful consideration of the frequency and amplitude of the signal we expect to observe. A sweep rate must be chosen and set which will provide a display of a few cycles of the desired signal. The vertical sensitivity must be chosen and set to provide an appropriate match to the amplitude of the expected signal. Failure to accomplish this can result in the conclusion that there is no signal present when in fact there is, or, conversely, mistaking a noise signal for the desired signal.

It is a valuable habit, when displaying a signal that we are not thoroughly familiar with, to verify that the signal we are viewing is in fact the signal we think it is. For example, if we are viewing a signal injected into a system from a signal generator, it is usually possible to adjust both its frequency and amplitude. Doing so and observing corresponding changes on the displayed signal are generally proof enough that what we are looking at is what we expected to be looking at.

We must know the limitations of the equipment we are using. For instance, most multimeters, even the electronic digital type, can measure accurately ac values of only several hundred hertz in frequency. Precise measurements of higher frequencies may require a special ac voltmeter or a special rectifying probe for RF frequencies.

Finally, when we arrive at the point where we think we have identified the defective component, there are two simple tests that can be of great help. These are useful in avoiding unsoldering a component until we are certain that we have correctly identified the defective component. Virtually all electronic equipment is now built utilizing printed-circuit techniques. According to the statistics of industrial quality control reports, one of the most common causes of failure in electronic equipment is the result of damage inflicted on components during a soldering or desoldering operation, especially on printed-circuit boards. Hence any unnecessary desoldering is to be avoided if at all possible.

A very useful troubleshooting policy is contained in the line:

Measure a short and bridge an open.

It is generally not feasible to measure the resistance of a component in-circuit (i.e., while the component is still connected in the circuit) if the circuit contains semiconductor devices. The reason for this is that there is nearly always a circuit path through a semiconductor junction which is in parallel with the component being measured. The voltage sources of many ohmmeters will forward bias such junctions. The measurement, then, would be of a parallel circuit, not of the resistance of the single component. The exception to this rule occurs if one has an ohmmeter that has a very low voltage source, low enough not to exceed the barrier voltage of the junction. However, if the component is suspected of having a dead short (R value $= 0\ \Omega$, approximately), the condition can be verified by an in-circuit ohmmeter measurement. Hence the rule: *Measure a short.*

On the other hand, if a resistor or capacitor is suspected of being open (open circuited), it is possible to test the conclusion by connecting a good component of the correct value in parallel with the suspected defective component. If the suspected component is indeed open, the circuit will function correctly. Hence the rule: *Bridge an open.* "To bridge" means "to connect in parallel with." Please be aware that this rule applies basically only to resistors, capacitors, and single-winding inductors. It definitely cannot be used with transformers unless both primary and secondary windings are bridged. It could be used with a transistor when one element, base, collector, or emitter, is suspected of being open if all terminals of the substitute are bridged to the old device. However, the result may not be conclusive if a remaining good part of the suspected transistor is still connected in the circuit.

4.2 ALIGNMENT

Many electronic systems include signal paths that incorporate tuned (resonant) circuits. Generally, the signal must pass through two or more such tuned circuits. The purpose of such an arrangement is to provide gain for a relatively narrow band of frequencies and attenuation of all other frequencies. In some instances the frequency response graph of the circuit resembles that of a simple resonant circuit with very high Q, as in Fig. 4.4(a). This response is typical of that of the IF section of an AM receiver. The response of the IF section of an FM broadcast band receiver is similar to that of Fig. 4.4(b). In this instance the bandwidth is greater; the top of the graph is relatively flat, or even dipped. The sides or skirts of the graph are still very steep so as to provide high attenuation of frequencies just outside the passband. The ideal passband of the IF section of a TV receiver is shown in Fig. 4.4(c). In this case, the response curve is said to be *shaped.* The gain levels at certain key frequencies must vary with only very limited tolerance fom specified values if the receiver is to perform at its optimum.

To achieve desired frequency characteristics the resonant circuits involved must be tuned very precisely to specified frequencies. In the case of the IF section of an AM receiver, typically three circuits are all tuned to 455 kHz. The TV IF section, on the other hand, requires many tuned circuits, some of which are used as traps or attenuation circuits at specified frequencies. In all cases it is virtually impossible to mass produce coils and capacitors so that the circuits will have the desired resonant frequencies after they are assembled. Therefore, it is necessary to be able to adjust the values of the L's and/or C's over a small range.

The process of adjusting the elements of circuits to achieve a desired passband characteristic is called *alignment.* An alignment is performed on new equipment just after it is constructed and before it is shipped. Alignment is generally necessary when a new L or C has been installed to replace a defective one in a tuned circuit. Alignment may be necessary because of the change of values of components due to aging. Someone may change the adjustment of one or more components in a unit inadvertently, thus necessitating an alignment.

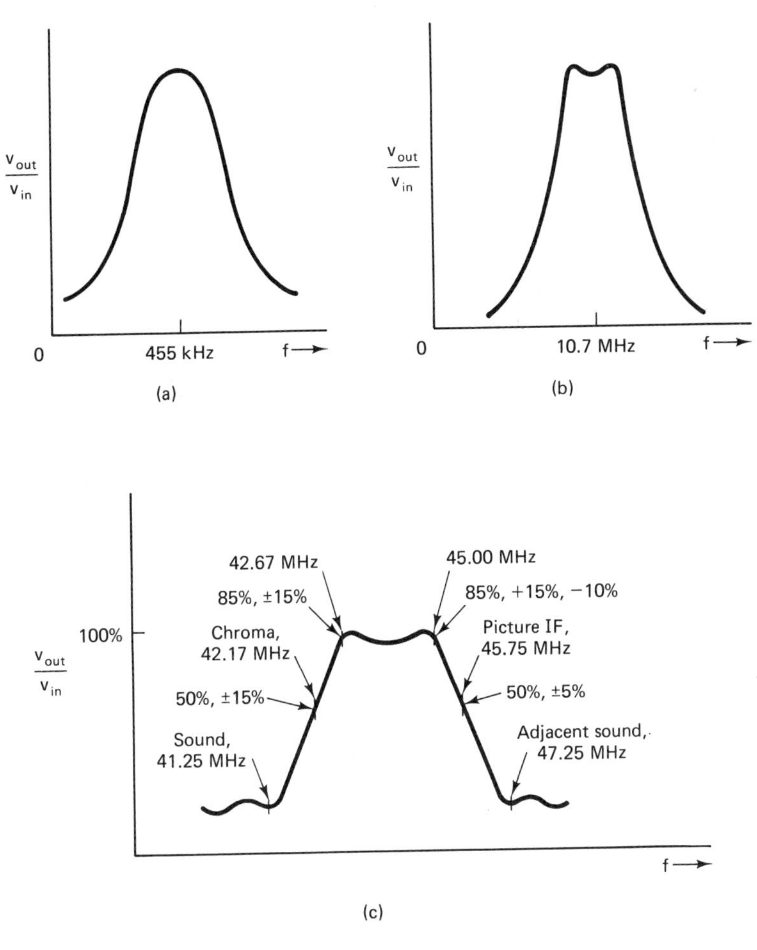

Figure 4.4 Frequency response graphs of tuned signal paths.

Aligning an AM Receiver

A superheterodyne AM broadcast band receiver incorporates tuned circuits in three of its sections: RF, frequency converter, and IF. Therefore, a complete alignment of this type of receiver requires the alignment of the circuits in each of these sections. The alignment begins with the IF section, proceeds to the frequency converter section, and concludes with the RF section. Although there are shortcut methods of alignment practiced by experienced technicians, let's look first at a formal procedure that makes use of typical laboratory or shop test equipment.

In simplest terms, the alignment of the IF section consists of adjusting the adjustable components in the section, the L's and/or C's, for maximum output of a signal whose frequency is precisely known to be the correct intermediate frequency for the particular receiver. To accomplish this result requires the following:

1. An RF generator with output amplitude adjustable
2. A means of measuring the amplitude of the signal after it has been processed by the IF section

Although it is possible to use a modulated signal for alignment and use the speaker output as an indication of the gain of the IF section, this technique should be used only in an emergency, when appropriate instruments are unavailable. The human ear is an unreliable means of assessing the amplitude level of a signal since its response is nonlinear.

The favored technique for alignment is to use an unmodulated signal and an oscilloscope to monitor the IF signal directly, at the output of the IF section. A dc voltmeter connected across the detector load can also be used reliably and conveniently with an unmodulated signal. An audio ac voltmeter connected across the detector load will serve as a monitor of alignment adjustments if a modulated IF signal is used. Generally, it is preferable to use an unmodulated signal or, at least, to monitor the audio signal before it is amplified in order to minimize disturbing others with an annoying audio test signal from a speaker.

The RF generator may be one with a fixed 455-kHz IF signal. However, since we will need other frequencies to complete the alignment, it is preferable to use a generator with the frequency, as well as the amplitude, variable. A typical hookup for performing an IF alignment on an AM receiver is shown in Fig. 4.5.

Choosing the method of injecting the IF signal will also require the exercise of some judgment on the part of the person performing the alignment. A requirement of the injection method is that it not disturb the electrical characteristics of the IF section, or, at worst, only minimally. One effective method is by means of an injection loop connected to the output of the RF generator. The loop may be made quickly from a short piece of hookup wire formed into a loop by wrapping around a pencil. The loop is then slipped off the pencil and placed in close

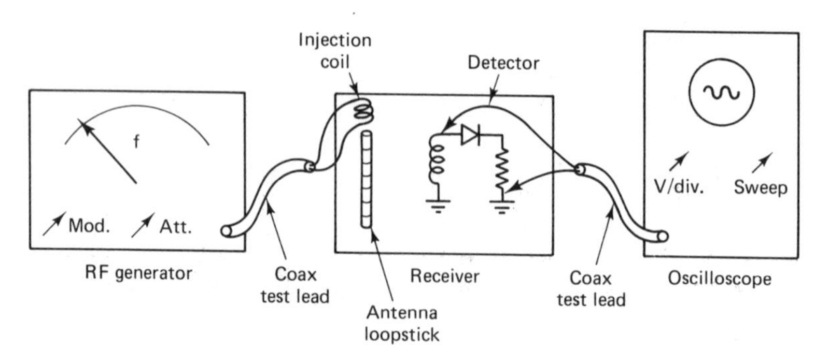

Figure 4.5 Setup for receiver alignment.

proximity to the antenna loopstick or coil. Or, alternatively, a few turns of hookup wire may simply be wrapped around the antenna loopstick and then connected to the generator. When this method is used, the IF signal, usually 455 kHz, is being coupled through the RF and frequency converter sections without being processed by them in the normal fashion. That is, the injected IF signal will not be converted by the frequency converter, but will be coupled through stray capacitance in the circuit to the input of the IF section, where it will then be processed in the normal way.

This technique may not work with some receivers, particularly those with "tight" front ends, for example those with an RF amplifier. In such cases it may be necessary to inject the signal at the input to the IF section through an isolating circuit such as a 0.01-μF capacitor and 1.0-MΩ resistor in series with the test cable from the RF generator. The capacitor will serve to isolate the generator output circuit as a potential dc load for any dc voltage that may be present at the connection point in the receiver. The resistor will minimize the ac loading effect of the generator on the IF circuit. Loading of the IF circuit will usually cause a change in the frequency response characteristic of that circuit with the obvious undesirable effect on the alignment process.

Let's assume that we are ready to begin the alignment of the IF section of a particular receiver. We have chosen to use an oscilloscope as a monitor and have connected it across the output of the detector in the receiver. We have wrapped a few turns of hookup wire around the receiver's loopstick antenna and connected a good RF generator to that loop. We have all the equipment powered up, including the receiver. We set the generator to provide an output of 455 kHz. We set up the oscilloscope to display the output signal by turning the sweep time and vertical sensitivity knobs to appropriate settings. For example, to determine what sweep-time setting would be required to display approximately three cycles of a 455-kHz waveform, we make these calculations:

$$T(\text{period of 455 kHz}) = \frac{1}{455 \times 10^3} = 2.198\ \mu\text{s}$$

$$\text{time for three cycles} = \text{time for 10 divisions} = 3 \times 2.198\ \mu\text{s} = 6.594\ \mu\text{s}$$

$$\text{sweep-time setting} = \text{time for 1 division} = \frac{6.594}{10} = 0.6594\ \mu\text{s/div}$$

We choose a setting of 0.5 μs/div for the sweep-time selector. This will give a display of slightly more than two cycles of the IF waveform. We set the vertical sensitivity selector to 10 mV/div as a start.

If the IF section is badly out of alignment, its output may not appear on the scope at first; we may have to "find" it by "rocking" the RF generator frequency dial. That is, we slowly adjust the frequency of the RF generator around 455 kHz while observing the display on the oscilloscope. Once we find the IF signal and are assured that the IF section is indeed functioning, we reduce the amplitude of the generator output by switching in attenuation at the generator. We alternately adjust attenuation and the frequency until the display has a peak-to-peak amplitude of two to three divisions when the generator frequency is adjusted for maximum output from the IF section. Repeating that last idea, we adjust the generator frequency for maximum output from the IF section. In doing this we determine to what frequency the IF section is tuned as we find it. For example, we could read that frequency off the generator dial. If the as-found frequency is 455 kHz, then, of course, the alignment will be very simple. We may need only to touch it up a bit to obtain optimum gain. On the other hand, if the as-found frequency is significantly different from 455 kHz, alignment is what the section needs.

In any case, what we are doing initially is to be sure that we have the oscilloscope adjusted so as to display the IF output and that what we are observing is actually that output. If the display changes as we change the amplitude and/or frequency of the signal at the generator, we can be reasonably sure that the display is the IF signal.

Let's assume that we have found the IF output and have it displayed in an appropriate fashion. We are ready to begin adjusting the adjustable components, say the transformers in the IF section, so as to maximize the IF output at the desired IF frequency. (For most AM broadcast band receivers the design IF frequency is 455 kHz, except for automobile radios, some of which use 262.5 kHz.) We set the generator for the correct frequency and begin to adjust each transformer, in turn, beginning with the final one and working forward to the first one in the section. As we are adjusting each transformer, we observe the output on the oscilloscope and stop adjusting when the output is a maximum.

Generally, as we proceed, the output will increase significantly because aligning an IF section that is out of alignment improves the gain of the section. This will require that we reduce the signal level from the generator. This is an important key to successful alignment—keeping the input near the minimum possible for a measurable output. If this is not done, there will be intermodulation distortion and consequently many spurious signals to confuse the process. Further, the AGC function will be activated. The action of AGC will be similar to that of reducing the overall Q of the section, making an accurate alignment impossible. We repeat this precaution: As the alignment proceeds and produces an increase in the gain of the section, reduce the amplitude of the signal from the generator by increasing the amount of attenuation used. Do this repeatedly. Keep the input level at the minimum required for a readable output. It is unnecessary to have as an output a clean sine wave of several hundred millivolts amplitude. In fact, it is very unlikely that a satisfactory alignment will be achieved if the input signal level is sufficient to produce that kind of output.

After we have adjusted each transformer once, we repeat the process. In fact, we continue the adjustments until no further improvement in the gain of the section can be achieved. Before leaving the IF section we again adjust the signal level from the generator to ensure that we have it at the lowest possible level while still producing a readable output on the oscilloscope. Next, we rock the generator frequency dial around the desired frequency and observe whether maximum output occurs at that frequency. If it does, we have completed the

IF alignment. If there is a maximum at some other frequency, we have probably not kept the input sufficiently low. We have aligned, not to the real signal, but to some spurious signal produced by intermodulation distortion as a result of excessive signal amplitude. The solution: Repeat the entire process, making sure to keep the input signal amplitude low.

The next step in the total alignment process for the receiver is the alignment of the frequency converter. The purpose of this alignment is to ensure that the frequency converter produces the reception of frequencies indicated on the tuning dial. This is called *tracking*. Visualize that turning the tuning knob changes not only the frequency indicated on the tuning dial, but also the position of the rotor of the ganged variable tuning capacitor and thus the frequency of operation of the local oscillator. The frequency of the local oscillator determines the frequency of the carrier that will be tuned in. It is necessary to make minor adjustments to the LO circuit components so that the circuit's operating frequency will be such as to produce the correct difference frequency (IF) for the carrier whose frequency is indicated on the tuning dial.

Although some receiver designs provide for three tracking adjustments in the local oscillator circuit, two is typical. One of these is a small adjustable capacitor with a single pair of plates separated by a mica insulator. Adjustment is achieved by means of a screw which, when tightened, reduces the separation between the plates and thereby increases the capacitance. This capacitor is in parallel with the main oscillator tuning capacitor. It is called the *trimmer capacitor*. The second adjustable component in the local oscillator circuit is the oscillator inductor. It is adjustable by means of a ferrite core made with threads. Turning the core moves it in or out of the coil, thus changing the inductance of the coil. With two adjustments, tracking can be adjusted at two settings of the dial—a low-frequency setting and a high-frequency setting. Unfortunately, because of the practical tolerances required during the manufacture of the oscillator capacitor and coil, the tracking of the tuning assembly is seldom perfect.

Let's now go through an imaginary alignment of the tuning assembly tracking. We start by setting the tuning dial to a high-end frequency, say 1400 kHz. Next we inject a 1400-kHz signal from the signal generator. The injection loop used in the procedure for the IF section is entirely satsfactory. We again monitor the output of the IF section by viewing the signal at the input of the detector on the oscilloscope. No change is necessary from the IF procedure. We adjust the trimmer capacitor until an IF signal appears on the oscilloscope. We attenuate the output of the generator until the waveform on the oscilloscope is just readable with the vertical input set for relatively high sensitivity, say 20 mV/div.

> *Caution*: Keeping the amplitude of the injected signal at a minimal value is even more important during this part of the process than it was during the alignment of the IF section.

After making sure that the signal we are viewing is, in fact, the real signal, and that the injected signal is at or near its minimal value for viewing, we now adjust the screw on the trimmer capacitor for maximum output signal amplitude—

we peak the output. This adjustment is very critical and, often, seemingly erratic. A useful hint here: Make the adjustment in such a way that you stop turning the screw as you are tightening it; it is less likely to be erratic. Before going to the next step, we rock the generator frequency dial through an arc of 300 or 400 kHz on either side of 1400 kHz while observing the scope display. There should be no signal at any other point on the dial. If there is, it means that the level of the input signal is too high, creating spurious signals. Finding another signal will require that we reduce the amplitude of the input signal and then repeat the adjustment just described.

Next, we set the tuning dial to a low-end frequency, say 560 kHz. Now we bring up the IF signal on the oscilloscope by adjusting the core of the inductor. We check to ensure that the amplitude of the injected signal is minimal and peak the output by adjusting the inductor core. We again rock the generator dial to check for other possible signals and to ensure that the peak output signal does occur at the frequency setting of the tuning dial (i.e., 560 kHz). This adjustment will, of course, upset the high-end operating frequency. We reset tuning dial and generator operating frequency to 1400 kHz and repeat the trimmer adjustment; return to the 560-kHz setting and repeat the core adjustment, and so on. We continue to go back and forth between these adjustments until no further adjustment is needed at either frequency.

The third and final phase of the receiver alignment is to make a trim adjustment to the resonant circuit(s) in the RF section of the receiver. In a simple receiver such as that of Fig. 3.15, the RF section may consist simply of a single, tuned circuit in which the inductor is a winding on the loopstick antenna core. In more sophisticated receivers the RF section is an amplifier with, typically, two tuned circuits and an active device. The resonant frequency of the circuit(s) in the RF section must change with changes in the tuning of the receiver.

For optimum performance, the RF section tuning should track the tuning of the frequency converter. That is, for each setting of the tuning dial, the frequency converter operates to convert a particular carrier frequency to the receiver's intermediate frequency—the RF section should be tuned to that same carrier frequency. As a practical matter, the typical RF section does not track the frequency converter perfectly.

Almost universally the RF tuning capacitor is fitted with a small trimmer capacitor very similar to the one on the local oscillator tuning capacitor. The alignment procedure for this section consists of adjusting the RF trimmer so that the RF section provides maximum gain for at least one setting of the tuning dial. Therefore, we set the receiver dial at, say, 1200 kHz and adjust the RF generator to inject a signal of the same frequency. We adjust the RF trimmer for a maximum on the monitor device. This completes the alignment of the receiver. Of course, the receiver should now be checked to see that it performs well in its function of "receiving" stations.

The procedure described is a general one and a composite of a number of specific procedures prescribed by receiver manufacturers for specific receivers. When you are required to perform the task of aligning a receiver, you should, if possible, obtain the manufacturer's recommended procedure for the particular receiver.

Aligning to Noise

Sometimes a receiver performs reasonably well but its sensitivity does not appear to be quite up to normal. This leads one to suspect that its alignment could stand a little "touching up." Both the IF alignment and the tracking of a receiver can be improved slightly without the aid of any instruments. The process involves tuning to the noise generated by fluorescent lamps. We describe the procedure.

Power up the receiver and bring it near an operating fluorescent lamp. Next, tune in a station near the high end of the dial and then off-tune the receiver from the station slightly. You should be able to hear a constant buzzing or grinding noise from the speaker, especially with the volume control turned all the way up. The receiver should not be too close to the lamp, only close enough to pick up the radiated noise signal. Now, touch up the IF transformer adjustments for maximum noise. Keep the noise level low by moving the receiver farther away from the lamp and/or by reducing the volume control setting. At a low level your ear will be more sensitive to changes in the level.

To check the tracking of the local oscillator, tune in a station near the high end of the dial and check for the match between its frequency and the dial setting. To improve the frequency match, touch up the oscillator trimmer slightly. (*Caution*: Do not make any major change in this adjustment.) Similarly, tune in a station near the low-frequency end of the dial and touch up the oscillator coil adjustment just slightly if an improvement in the accuracy of the dial setting is desired. Now, again tune in noise at a dial setting of approximately 1350 kHz. This time, touch up the trimmer capacitor setting on the RF coil for maximum noise output. If a receiver appears to be suffering from a major misalignment, a complete alignment with the use of instruments should be performed.

4.3 MISCELLANEOUS SHOP TECHNIQUES

Testing Semiconductor Devices with an Ohmmeter

It is often desirable in the everyday world of the shop to obtain a quick answer to the simple question: Is this transistor or semiconductor diode good or bad? There is no desire to determine the beta of a transistor or any other more comprehensive data about a particular device. The question can usually be answered with some simple tests with a 20,000-Ω/V volt-ohm-milliameter. Recall that an ohmmeter function, viewed in terms of its equivalent circuit, is a current-measuring meter in series with a voltage source and some resistance (see Fig. 4.6). When an ohmmeter is connected across a semiconductor junction, the meter reading can be interpreted as either a resistance value or a current value. It is okay to use the resistance-value interpretation if we remember that, in this test, it is an indication of a general opposition to current flow rather than a

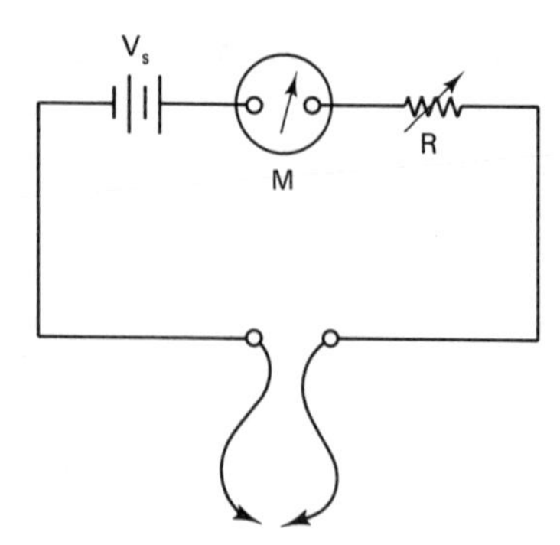

Figure 4.6 Equivalent circuit of an ohmmeter.

specific measure of ohmic resistance in the sense of the resistance of a resistor. The voltage source for the ohmmeter function has polarity since it is a dc source (a battery). Hence the ohmmeter circuit will either forward bias the junction or reverse bias it, depending on the way the meter is connected to the junction. Connected with one polarity, an ohmmeter test of any good junction should yield a relatively high resistance (low current) reading; a connection of the opposite polarity will yield a reading of relatively low resistance (high current). It is necessary to use the term "relatively" here since different types of devices will give different relative readings—it is not possible to give specific resistance values to expect. With experience in using the test you will develop a "sense" of the meaning of the readings obtained. Following are some suggestions for applying and interpreting the test.

1. Identify the polarity of the voltage on the leads of your ohmmeter. The "common" lead is not necessarily negative, the "hot" lead is not necessarily positive. You can do this identification by measuring the "voltage" on the ohmmeter leads with another dc meter set for "volts" function. Place a piece of masking tape with a notation concerning lead polarity on your meter so that you do not have to perform this step more than once.
2. To test a diode, set the ohmmeter to the $R \times 10$ range, connect the ohmmeter lead with the positive voltage polarity to the anode and the other lead to the cathode, and note the reading; reverse the leads and again note the reading. A good diode will give a low-resistance reading on the first test and a high reading on the second test. If both readings are low, the diode is shorted. If both readings are high, the diode is open. If these tests are performed on a diode connected in a circuit, the results will generally be the same. However, a low reading in both directions may now be obtained if there is a low-resistance circuit in parallel with the diode. Look at the circuit and determine if this is the case. This test can also be used to determine the anode and cathode of a diode whose polarity you are not sure of.
3. To test a transistor, set the ohmmeter to the $R \times 10$ range, and connect one test lead to the base as follows: lead with positive polarity to base if transistor is NPN, negative lead to base if PNP. Touch other lead to emitter and, then, to collector. Reverse the leads and repeat. The first part of this

test checks the base–emitter and base–collector diodes of the transistor in the forward direction; the second part checks these diodes in the reverse direction (see Fig. 4.7). Therefore, the first checks should give low-resistance readings; the checks with the leads reversed should give high-resistance readings. A high-resistance reading in the forward direction indicates an open in that circuit. A low-resistance reading in the reverse direction indicates a shorted junction.

For a final check of the transistor, connect one lead of the ohmmeter to the collector and the other lead to the emitter as follows: positive lead to collector, negative lead to emitter for an NPN transistor; the opposite of this for a PNP transistor. The resistance reading for this test should be infinity for a silicon transistor. The reading should be very high, but probably not infinity, for a germanium transistor. There is enough base-open leakage current in a germanium transistor to prevent the ohmmeter from reading infinity. If the resistance reading of this check is low, a defective transistor is indicated. If the resistance reading is not low, connect a 47-kΩ resistance between the collector and base. The resistance reading should now diminish to a low value corresponding to increased current flow in the collector–emitter circuit. This is characteristic of true transistor action (when the base–emitter junction becomes forward biased, current increases in the collector circuit). It is a reasonably reliable indication that the transistor is okay.

For one more check, reverse the ohmmeter connections to the collector and emitter of the transistor. The resistance reading should be infinity, or near infinity.

We have said that these tests are to be made with a conventional volt-ohm-milliameter (one of the nonelectronic types). The tests may or may not be valid when made with a solid-state electronic multimeter. To know whether a particular meter can be used, it is necessary to know something about the nature of its ohmmeter function. Some meters utilize an electronic constant-current source for the ohmmeter function, instead of a battery or dc voltage source. Generally, these do not perform the foregoing tests satisfactorily. However, most recent designs of electronic meters, such as digital multimeters, incorporate an ohmmeter function that can be (indeed, are designed to be) used to test semiconductor junctions as described above. Refer to the instruction manual for your particular meter to determine if and how it can be used to test diodes and transistors.

Special Techniques Appropriate for Working with Radio-Frequency Signals

Laboratory and shop tasks related to communications equipment are almost invariably involved with radio-frequency signals. Because of this, special care is required in performing such tasks. We mean that special care is required compared to tasks involving strictly audio-frequency equipment, industrial electronic equipment, and so on. In the first place, the amplitudes of typical RF signals are generally much smaller than the amplitudes of many other types of signals.

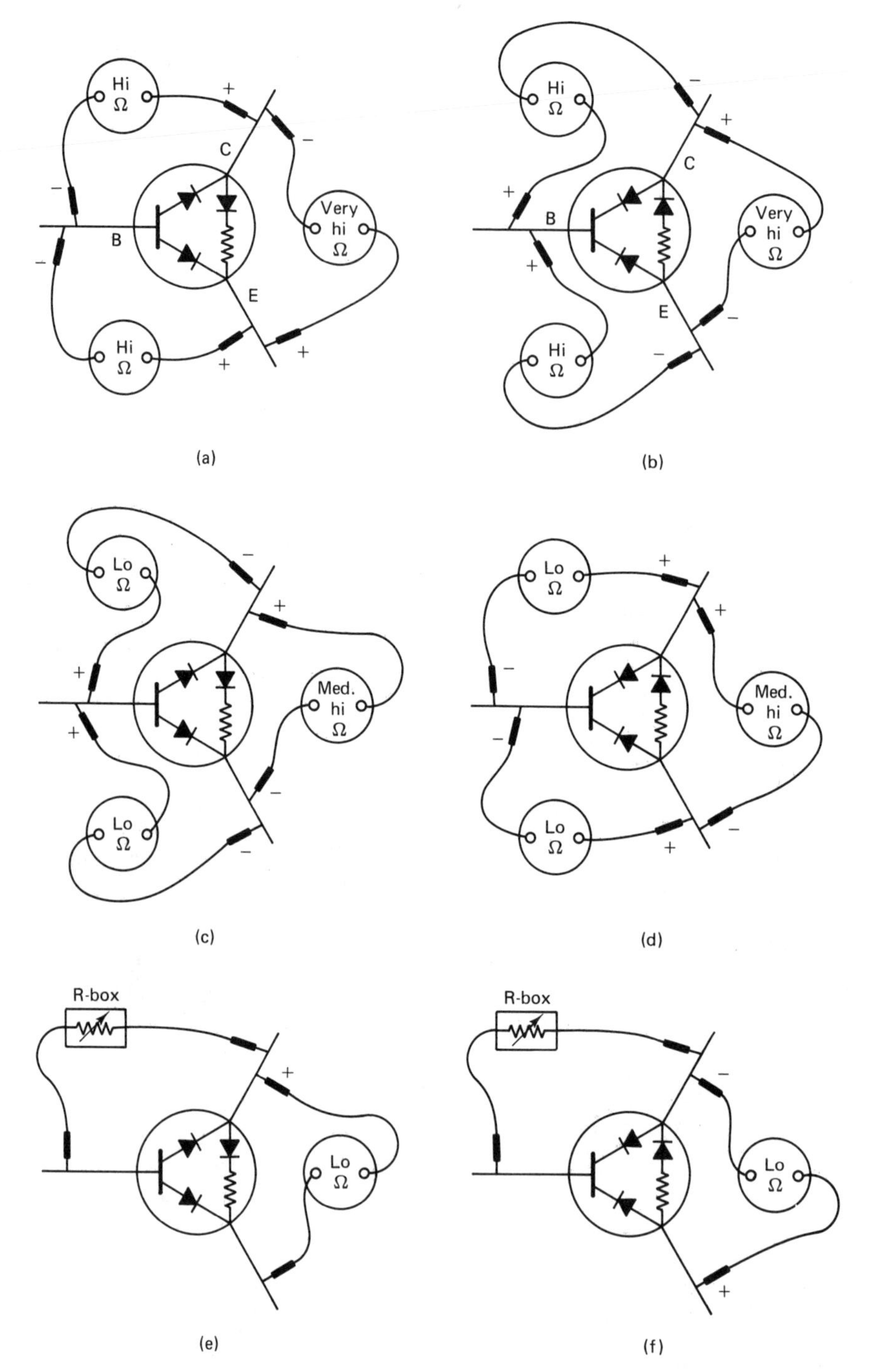

Figure 4.7 (a) NPN inverse diode tests; (b) PNP inverse diode tests; (c) NPN forward diode tests; (d) PNP forward diode tests; (e) NPN transistor action test; (f) PNP transistor action test.

Furthermore, by their very nature, RF signals are much more likely to "escape" ordinary conductors through radiation.

Many ordinary man-made devices produce RF noise signals: fluorescent lamps, automobile ignitions, thermostatically controlled heating devices such as heating pads, and many others. The result is that when we wish to obtain an RF signal from an item of communications equipment and display it on an oscilloscope, we must use reasonable precaution to prevent its being "polluted" by noise signals from a host of possible sources. We should always use a shielded test lead for conducting the signal from the device to the oscilloscope input. The shield portion of the test lead must be grounded (connected to the chassis of the device). The shield serves to intercept stray RF noise signals and prevent them from being added to the desired signal. The purpose of the shielded test lead should not be defeated by connecting a clip jumper lead to the center, "hot" conductor to extend its length or facilitate connecting it to the test point.

When attempting to display a desired RF signal, it is important to take steps, whenever possible, to assure ourselves that the signal we are viewing is indeed the desired signal. Because there is generally RF noise present "in the air," it is very easy to be fooled by a noise signal. In many instances we operate the vertical input of the oscilloscope at, or near, maximum sensitivity. This increases the chances that a stray noise signal will be displayed and interpreted as the desired signal. If there is any possible way to alter momentarily the frequency and/or amplitude of the desired signal so that the result of such alteration can be observed on the display, this should be done. Once we have assured ourselves that the displayed signal is the desired signal, we can proceed to perform the planned tests and measurements with confidence.

In summary, remember that RF signals are very "flighty" signals. They are everywhere at all times. We must always be at pains to ensure that we are not allowing an outsider to intrude on our attempts to display a particular signal.

The Significance of Oscilloscope Bandwidth in RF Tests and Measurements

When a signal is displayed in a normal way on an oscilloscope (the signal deflects the beam vertically), it is processed by an amplifier (called the vertical amplifier) in the scope. Like all other amplifiers, the vertical amplifier of an oscilloscope has a bandwidth limitation. Some typical vertical amplifier bandwidth specifications are: dc to 10 MHz, or dc to 50 MHz. What is the meaning of such a specification, and what significance does it have in the use of the oscilloscope?

In the notation "dc to 10 MHz," the items "dc" and "10 MHz" represent the lower and upper limits, respectively, to the band of frequencies over which the amplifier will amplify with constant-gain factor. The upper limit is defined as that frequency at which the gain has been diminished by the characteristics of the amplifier circuit to a value that is 3 dB (decibels) below the gain at other frequencies in the band. The gain at this limit frequency will be equal to 0.707 times the gain at other lower frequencies.

Stating it another way, let's visualize that we alternately display two signals, one whose frequency is equal to the upper limit frequency and one having a lower frequency. Let's say that each has an actual amplitude of 1.0 V peak-to-peak. If displayed at a sensitivity of 0.2 V (200 mV) per division, the lower-frequency signal display would be five divisions high (1.0 V/0.2 V/div = 5 div). The display of the signal at the limit frequency would be only 3.54 div (5 × 0.707). If we were using the oscilloscope display to determine the amplitude of the signal, we would be mistaken about the second signal. And, of course, if we attempted to display signals of higher frequencies, the error in the amplitude measurement would be even greater.

The significance of the bandwidth specification is that, for one thing, we cannot rely on the oscilloscope display as a means of measuring amplitude when the frequency of the signal being displayed is beyond the band limit. The amount of attenuation of the amplifier increases with the frequency once the frequency is beyond the band limit. Ultimately, the attenuation is so great that there is no usable deflection. Outside the band limit, the display is useful, if at all, only for purposes of waveform shape, phase angle, and frequency tests. We prefer to have an oscilloscope whose bandwidth includes the frequencies we wish to display.

SUMMARY

1. Effective, productive troubleshooting is facilitated by having an organized plan of approach to the task—a troubleshooting system.
2. Practical troubleshooting systems are based on the concept of limiting the probable area of the location of a defect to smaller and smaller subdivisions of the system. They proceed from a consideration of the entire system to a subsystem, a section, stage, circuit of the stage, and finally to specific components.
3. Troubleshooting is a process of gathering data about the operation, or lack of operation, of a system and its subdivisions. The process is guided by decisions made through the analysis of the data gathered.
4. Data are gathered by the use of techniques such as signal tracing, signal injection, and dc voltage measurements. Analysis of the data requires the use of Ohm's law, the principles of series and parallel circuits, and other basic circuit analysis concepts.
5. The dc voltages at the collector, base, and emitter of a transistor stage provide direct clues to the cause of most malfunctions in a stage. Understanding the reasons for the direction and amount of change of these voltages is the key to successful use of this technique.
6. "Measure a short and bridge an open" is a useful rule to remember and utilize when we think we have found the defective component. It assists us in confirming our diagnosis before unsoldering the component.
7. The alignment of a receiver or transmitter is the process of adjusting the adjustable inductances and/or capacitances in the resonant, or tuned, circuits in the equipment so that such circuits do, in fact, resonate at the intended frequencies.
8. The ohmmeter function of volt-ohm-milliameters and many electronic, solid-state digital multimeters can be profitably employed to test semiconductor devices. A knowledge and understanding of the circuit of the ohmmeter and the device is essential to success.

9. Great care must be taken in working wih RF signals to minimize the effect of noise signals on any ongoing test procedure. This is especially true of RF signals because they are normally very weak and radiate easily. Unshielded test leads are good receiving antennas or radio-frequency noise signals: Use shielded test leads!
10. The bandwidth specification of an oscilloscope is an important measure of its usefulness in RF test procedures. The bandwidth must include the frequencies that are intended to be displayed.

GLOSSARY OF TERMS

Bridge, to To connect in parallel with.

Defective component A component, such as a resistor, capacitor, inductor, transistor, etc., whose electric operating characteristics have changed in a way that causes it to fail to perform its function in a circuit. For example, a resistor may open up, become shorted, increase in ohmic value, or decrease in ohmic value.

Hot lead (the) The test lead of a meter, oscilloscope, or other instrument that is connected to the nongrounded side of the circuit.

Malfunction (of an electronic unit) The act of failing to function as it should.

Monitor, to To check the operation of, as with an oscilloscope.

Signal injection The technique of placing a signal from an external source into an operating circuit to test the performance of that circuit. The technique normally implies using some part of the circuit under test to monitor the response to the test signal.

Signal tracing The technique of using a test instrument such as an oscilloscope to follow a signal along the signal path of an electronic system.

Thévenize, to To convert a circuit into its Thévenin equivalent circuit, that is, to find its Thévenin equivalent voltage (V_{th}) and Thévenin equivalent impedance (R_{th}).

Tracking Of electronic circuits, the ability to perform functions in unison as some operating parameter, such as frequency, is changed.

REVIEW QUESTIONS: BEST ANSWER

1. A knowledge of a system in terms of its block diagram is of no value in troubleshooting that system. **a.** True. **b.** False.
2. An electronic system consists of several subsystems, many sections, stages, and thousands of components. In troubleshooting such a system, the first step should be to isolate the problem to: **a.** a particular component. **b.** one subsystem. **c.** a section. **d.** a stage. **e.** none of these.
3. An AM/FM hi-fi receiver with phonograph and cassette player inputs receives FM stations and works on phono and cassette inputs. It does not receive any AM stations. The section that definitely does not contain the problem is the: **a.** audio section. **b.** AM local oscillator/mixer section. **c.** AM IF section. **d.** AM RF section. **e.** none of these.
4. Noise can be heard from the speaker of an AM receiver when the volume control is turned, but it does not receive any stations, even with an external antenna. The

problem is likely in the: **a.** IF or audio sections. **b.** audio or mixer sections. **c.** mixer or IF sections. **d.** audio or detector sections. **e.** none of these.

5. When a modulated 455-kHz signal is injected at the input to the IF section of a malfunctioning AM receiver, a tone can be heard from the speaker and its volume level can be adjusted by the volume control. The problem is most likely in the: **a.** IF or audio sections. **b.** audio or mixer sections. **c.** mixer or IF sections. **d.** RF or mixer sections. **e.** none of these.

6. With an oscilloscope, a signal can be found at the output of the first IF stage of a malfunctioning AM receiver but not at the output of the second IF stage. The problem is most likely in the: **a.** first audio stage. **b.** second IF stage. **c.** LO stage. **d.** first IF stage. **e.** none of these.

7. There is no sound whatsoever from the speaker of a malfunctioning AM receiver when it is powered up. A signal is found at the output of the first stage of a two-stage audio section. The problem is in the: **a.** power supply. **b.** LO stage. **c.** audio output stage. **d.** first audio stage. **e.** a or d.

8. A signal is traced to the input of the second IF stage of an AM receiver. There is no signal on the output of that stage. A next logical step in troubleshooting the receiver is to: **a.** measure the dc voltages on the second IF stage. **b.** check for signal at the output of the detector. **c.** check for signal at the output of the first audio stage. **d.** check for proper operation of the LO. **e.** none of these.

9. In a defective NPN stage: (1) $V_C = V_{CC}$, (2) the base voltage is equal to its open-circuit Thévenin voltage, and (3) the emitter voltage is slightly greater than the base voltage. The defect is in the: **a.** emitter circuit. **b.** base circuit. **c.** collector circuit. **d.** a or b. **e.** b or c.

10. In a defective transistor stage the base-bias voltage is about normal, but the magnitudes of the voltages on base and emitter are much smaller than normal and the collector voltage is about equal to the emitter voltage. The defect is in the: **a.** emitter circuit. **b.** base circuit. **c.** collector circuit. **d.** a or c. **e.** none of these.

11. The bypass capacitor across the emitter resistor in a transistor audio stage is open. The result is: **a.** higher-than-normal emitter voltage. **b.** lower-than-normal emitter voltage. **c.** transistor cutoff. **d.** lower-than-normal gain. **e.** none of these.

12. One of the expected outcomes of aligning a receiver that needs an alignment is an improvement in its sensitivity. **a.** True. **b.** False.

13. A technician services CB (citizens' band) radio equipment in which the operating frequency is in the range 27 to 30 MHz. The most suitable bandwidth for an oscilloscope for viewing a carrier in this equipment is: **a.** dc to 5 MHz. **b.** dc to 10 MHz. **c.** dc to 15 MHz. **d.** dc to 50 MHz. **e.** none of these.

REVIEW QUESTIONS: ESSAY

1. You have been assigned the task of repairing an electronic system that is completely new to your department. No one else in your department has any knowledge of the unit. An excellent instruction manual, complete with block diagram, schematic diagram, operating instructions, and circuit description, is available. Describe how you would approach your task.

2. A malfunctioning TV receiver has a picture output but no sound output. You are aware that a TV receiver has an RF section, a frequency converter, an IF section for both picture and sound signals, a video detector, a sound detector, a sound amplifier section, and a video (picture) section. Name the section(s) that you think is/are most likely to contain a defective component and give the reasoning for your conclusions.

3. What is a signal flow path? What is the advantage of making a test for signal near the center of a signal flow path? Where should the next test be made? Why?
4. Describe the technique of signal tracing. Give a specific example of its application.
5. Describe the technique of signal injection. Give a specific example of its application.
6. What is the significance of noise heard from a speaker of an otherwise inoperable radio receiver?
7. Why is it important to check power supply voltage in a dead receiver?
8. Describe how you could use a dc voltmeter connected to measure a receiver's AGC voltage to check for operation of the front end of a receiver that is inoperable.
9. You have been given an AM/FM hi-fi receiver with built-in cassette player to troubleshoot. Describe what tests you would make on the unit, before ever removing it from its case, to begin the process of localizing the problem to a particular part of the subsystem.
10. Refer to Question 9. You discover that the unit will play a cassette but is inoperative on AM and FM. Describe the next step in your troubleshooting. Explain the reasoning for choosing that step.
11. Measuring and analyzing the dc voltages of an electronic stage is an important troubleshooting technique. Discuss the general theory underlying this technique. Why does it work? Give an example of a defect that will not be revealed by this technique.
12. Describe the effect an open collector circuit will have on the dc voltages of a stage. Will the effect of an open at the collector inside the transistor case be any different? Describe.
13. Describe the effect an open between the base and the base supply voltage will have on the dc voltages of a stage. Will an open base (an open inside the transistor case) have the same or a different effect? Describe.
14. Describe the effect an open emitter circuit (open between emitter and ground in a common-emitter connection) will have on the dc voltages of a stage. Will an open emitter (inside the transistor case) have the same or a different effect?
15. What is meant by *tracking*?
16. Discuss why a low ohmmeter reading for the inverse direction of a semiconductor junction is not necessarily an indication of a defective junction when the device is connected in an actual circuit.
17. Discuss why a "dc to 5 MHz" oscilloscope would be adequate for servicing AM broadcast band receivers but would have limited value in servicing FM receivers that have an IF of 10.7 MHz. The operating band for FM is from 88 to 108 MHz.

EXERCISES

1. Refer to the circuits shown in Fig. 4.8. The circuit in (g) is the normal circuit. The resistors are equal in value. The indicated voltages are with respect to ground. The other circuits are identical to (g) except that in each case there is one defect only:

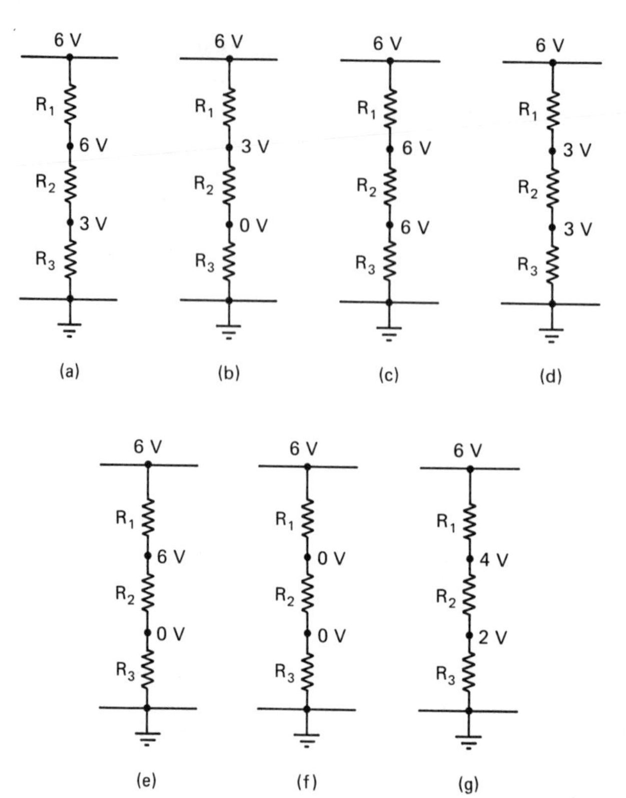

Figure 4.8

either one resistor is open or one is shorted. Analyze the voltage changes in each circuit and determine which resistor is defective and the nature of its defect.

2. Refer to the schematic diagram of the six-transistor receiver of Fig. 3.15. Thévenize the base supply circuit for the audio output stage of the receiver. Assume that $V_{CC} = -5.8$ *V* and that the dc resistance of the base windings of transformer T_6 is negligible.
3. Refer to Exercise 2. Assume that the transistors have a beta of 50, that the drop across the base–emitter junction is -0.15 V under no-signal condition, and that the dc resistance of each half of the primary winding of transformer T_7 is 5 Ω. Estimate the normal no-signal dc voltages for the stage.
4. Refer to Exercises 2 and 3. Calculate the base and emitter voltages (with respect to ground) for the open-collector-circuit condition. Assume that the base-to-emitter voltage is -0.15 V.
5. Refer to the schematic diagram of the six-transistor receiver of Fig. 3.15. Thévenize the base supply circuit of the second IF stage (transistor Q_3). Assume that the voltage at the junction of R_{13}–R_{16} (in effect, V_{CC} for stages Q_1, Q_2, and Q_{39}) is -5.0 V and is that of a power supply with negligible internal resistance. Further, assume that the base current of Q_2 is so small it can be ignored as a load on the R_5–R_6–R_9 voltage divider circuit. The no-signal AGC voltage, that is, the voltage at the junction of R_9 and D_1, is -0.2 V.
6. Refer to Exercise 5. Assume that transistor Q_3 has a beta of 100, that the drop across the base–emitter junction is -0.18 V under no-signal condition, and that the dc resistance

of the IF transformer winding is negligible. Estimate the normal no-signal dc voltages for the stage.

7. Refer to Exercise 5. Assume that the receiver is processing a signal such that the AGC voltage has become +0.20 V. **(a)** Thévenize the base supply circuit of Q_3 for this condition. Use the assumptions given in Exercise 5. **(b)** Calculate the normal dc voltages for the stage for the given signal level.

8. You are setting up an oscilloscope to display signals from an RF generator. For the given frequencies, calculate a sweep-time setting for the horizontal sweep of the scope which will give displays of not less than two cycles, nor more than five cycles; assume that the sweep-time dial settings are in the ratio: 1, 2, and 5 (e.g., 1 ms, 2 ms, and 5 ms): **(a)** 455 kHz; **(b)** 600 kHz; **(c)** 900 kHz; **(d)** 1200 kHz; **(e)** 1600 kHz.

5

AM TRANSMITTERS

The second major subsystem in an electronic communication system is the transmitter. Briefly, the functions of the transmitter are (1) to generate a high-frequency carrier signal, and (2) to provide the means for modulating that carrier with the information signal. If we count an antenna as part of the transmitter, a third function is to radiate the modulated carrier. When a transmitter utilizes amplitude modulation to place the information signal on the carrier it is referred to as an AM transmitter. Let us use the systems approach (see Chap. 1) to learn the details of construction and function of the AM transmitter subsystem.

5.1 TRANSMITTER BLOCK DIAGRAM

In Chap. 3 we saw that the receiver function could be performed with an extremely simple or primitive circuit. The transmitter function may also be performed very simply. As shown in Fig. 5.1(a), a transmitter may consist of only an oscillator, an antenna, a power supply, and a switch to turn the oscillator on and off in accordance with some code, for example, the Morse code. The diagram is that of a simple, continuous-wave (CW) transmitter, as might be used by a novice amateur radio operator (a ''ham'' operator). A slightly more sophisticated transmitter is shown in Fig. 5.1(b). In this case, provision has been made for modulating the oscillator with an audio signal from a microphone. This diagram is that of the simplest form of AM transmitter.

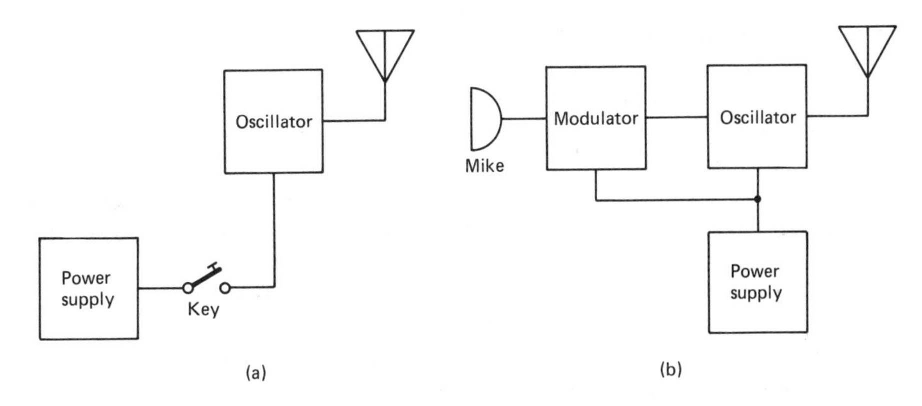

Figure 5.1 Transmitter block diagrams: (a) CW transmitter; (b) AM transmitter.

Although the AM transmitter of Fig. 5.1(b) can perform all of the essential functions required of a transmitter, it has a number of serious limitations, especially for use as a commercial broadcast transmitter. Transmitters as simple as that shown have been built for special purposes. For example, so-called "wireless" phonograph inputs for AM receivers are simple transmitters of the type shown. Very simple transmitters in the form of kit projects for young people are common. In these examples the amount of RF power produced is extremely small. In fact, this is one reason why such simple transmitters are permitted to be used by persons without an operator's license—there is virtually no danger of the radiation causing an interference problem with other users of the radio spectrum. The two most serious limitations of these simple transmitters for communications purposes are (1) low power output, and (2) poor frequency stability. It is not possible to get a great deal of power from a single oscillator stage. Modulating an oscillator causes a tendency for its frequency of operation to vary—to be unstable. These limitations can be overcome by using additional circuit functions.

If we add a stage of power amplification to the transmitter and move the point of modulation from the oscillator to the power amplifier, the block diagram would be that shown in Fig. 5.2. A transmitter of this type is referred to as a "MOPA" transmitter, for it consists of a *m*aster *o*scillator and a *p*ower *a*mplifier. The power amplifier, of course, increases the power level of the radiation. The

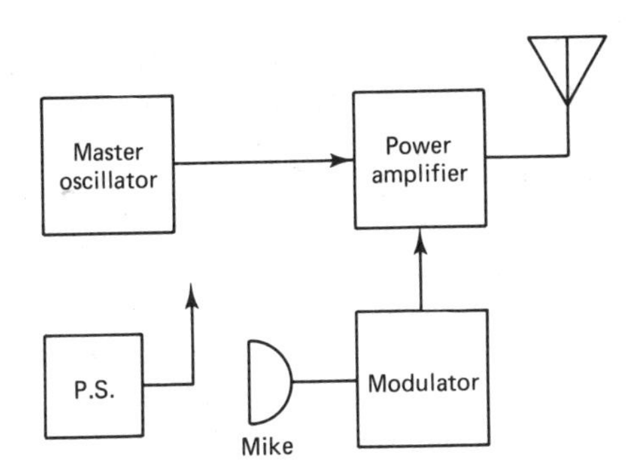

Figure 5.2 Block diagram of MOPA transmitter.

oscillator frequency stability will be improved because the oscillator is no longer being modulated. Although these are real improvements, the performance of the MOPA transmitter is still not adequate for many communications applications. Other improvements are needed or desirable.

To increase still further the power level of the carrier, a *driver amplifier* is needed to drive the power amplifier. To further protect the oscillator from the effects of modulation and lessen the tendency for its operating frequency to vary, a *buffer amplifier* may be included just after the master oscillator. In some instances it is not feasible to operate the oscillator at the final carrier frequency. It must be designed to operate at a lower frequency. In this case a *frequency multiplier* function is called for. The transmitter block diagram of Fig. 5.3 incorporates these additional functions.

We study the details of each of the circuit functions of a transmitter in the sections that follow.

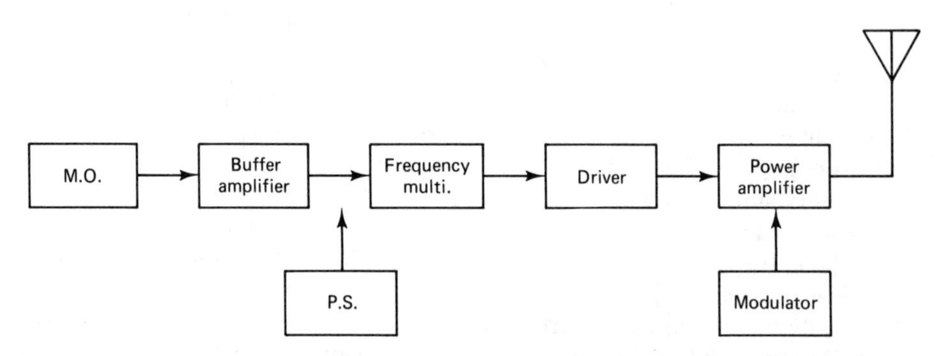

Figure 5.3 Transmitter block diagram.

5.2 THE RF AMPLIFIER

Most of the transmitter circuit functions are RF amplifiers of one form or another. This includes the oscillator. An RF amplifier contains at least one tuned circuit, or its equivalent, such as a quartz crystal. The bandwidth of an RF amplifier, when considered as the ratio of the highest-to-lowest frequencies normally processed, is narrow. This is in comparison to an audio or video amplifier. Let us look at the details of a basic RF amplifier as preparation for studying the special forms found in transmitters.

The mode or class of operation, as determined by the bias level, is of great interest in the utilization of RF amplifiers. Recall that in class A operation, the conduction angle (time during which output current flows) is a full 360° of the input signal waveform cycle. The conduction angle for class B operation is 180°, and for class C operation, less than 180°. The class C operating mode is of great significance in RF applications, for several reasons. First, unlike virtually all audio applications, it is feasible to use class C operation in many RF applications. We will see why shortly. Second, the conversion of dc power to ac signal power is much more efficient in class C operation that in either class A or B. This is important since a transmitter power amplifier may be called on to handle up to

100 kW of signal power. Third, amplitude modulation can occur only in a nonlinear circuit, and a class C amplifier is a nonlinear circuit. Let us look first, then, at a class C radio-frequency amplifier.

The schematic circuit diagram of a basic class C, tuned amplifier is shown in Fig. 5.4. Notice the absence of the usual voltage-divider bias network. Since we want the conduction angle of the collector current to be less than 180°, we can use the positive tips of the input signal to forward bias the base–emitter junction. In fact, the input circuit is an example of dynamic dc bias, as discussed in Sec. 3.3. Capacitor C_1 receives some charge during the positive alternation of the signal, while the emitter diode is forward biased. The capacitor discharges partially through R_1 during the negative alternation. The result is an average

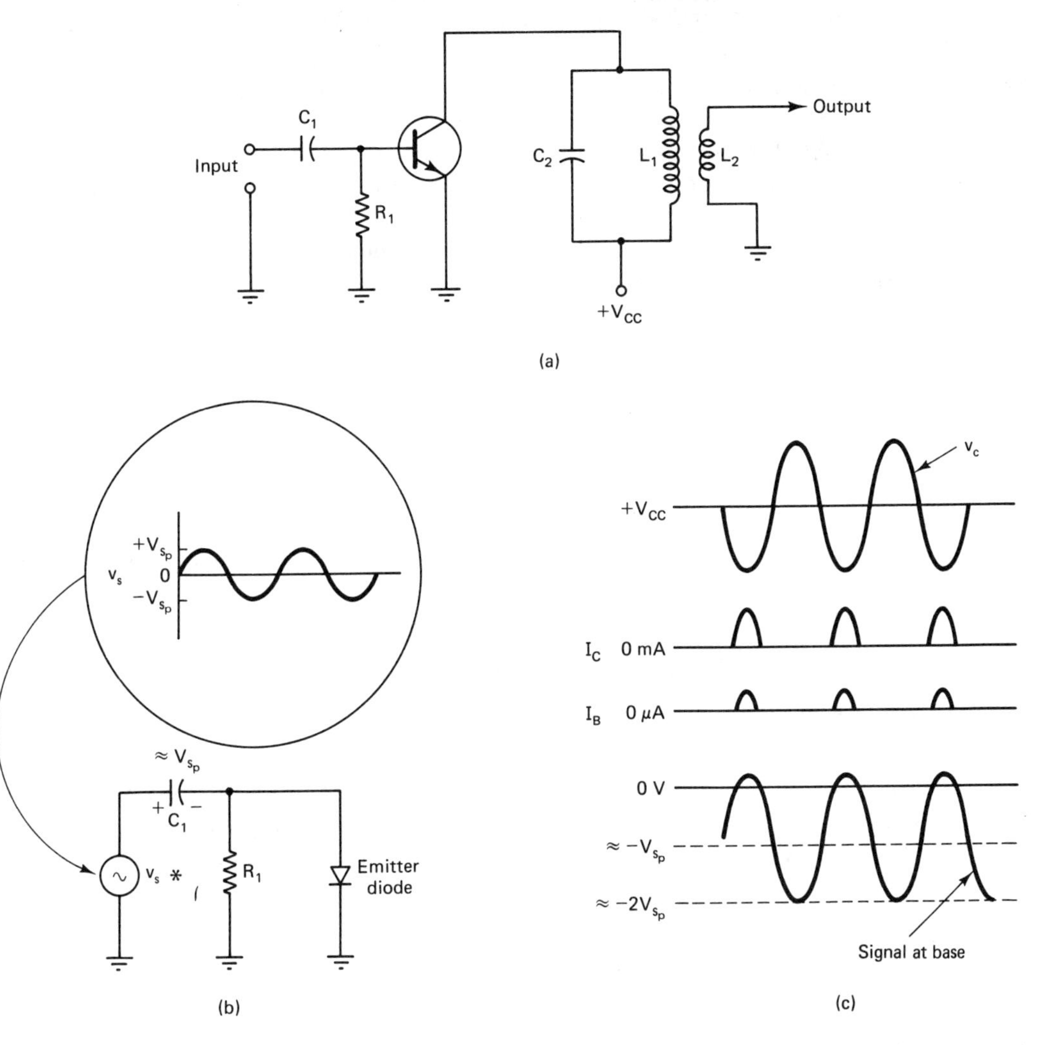

Figure 5.4 Class C amplifier: (a) typical class C amplifier schematic; (b) clamping in base circuit; (c) waveforms in stage.

voltage across C_1, with the right side—the side connected to the base of the transistor—negative. The voltage on C_1, then, actually provides a reverse bias for the transistor. Conduction of the transistor occurs only when the positive peaks of the signal exceed this reverse-bias voltage. The result is that base signal current is in the form of pulses of less than 180° duration. Collector current, similarly, is in the form of pulses of less than 180° duration. The waveforms for this stage are shown with the circuit in Fig. 5.4.

Let us digress for a moment and observe that the input circuit of the class C amplifier of Fig. 5.4 is also referred to as a *clamper*—a *negative clamper* in this particular case. A clamper is a circuit that adds a dc voltage to an ac voltage and thereby shifts the level of the ac voltage so that it is no longer alternating around a zero reference level. The ac voltage is "clamped" to some voltage other than zero. If the shift is to a positive voltage, the clamper is a *positive clamper,* if to a negative voltage, it is a *negative clamper.*

In the case of the circuit of Fig. 5.4, the incoming ac signal is clamped to a voltage almost equal to the negative peak voltage of that signal by the voltage across C_1. The actual clamping voltage depends on the peak-to-peak voltage of the signal and also on the relationship between the period of that signal and the time constant of the discharge path of C_1. If the RC time constant is long compared to the time when the base–emitter diode is cut off (approximately the period of the signal, or $1/f$), the capacitor will remain charged to a voltage very nearly equal to the negative peak voltage of the signal. On the other hand, if the RC time constant is relatively short compared to the period of the signal, the capacitor will discharge significantly between positive peaks and the average voltage across the capacitor (the clamping voltage) will be less.

TROUBLESHOOTING NOTE

Since a dynamic bias voltage (or clamper voltage) can be present only when a signal is present, testing for the presence of such a dc voltage can be used to test for the presence of a signal and/or the correct operation of the input circuit of a stage with this feature. That is, use a dc voltmeter (with high input impedance to avoid disturbing the time constant of the circuit) to test for the presence of signal in a circuit that generates dynamic bias, a clamper circuit.

Now, let us see how the class C tuned amplifier is able to provide an almost pure sine-wave output voltage when the output current is in the form of short pulses. First, we must recall from a previous study of resonant circuits, perhaps in an ac theory course, that a resonant circuit is capable of a phenomenon referred to as a *damped oscillation* (see Fig. 5.5). Visualize an LC circuit energized momentarily with a dc current, as in Fig. 5.5(a). Next, the connection to the dc source is broken and the coil-capacitor circuit is shorted on itself [Fig. 5.5(b)]. The collapsing magnetic field of the coil will attempt to keep the current flowing. This current is converted to charge on the capacitor. The energy given to the circuit from the dc source is being transferred from the magntic field of the coil to the electric field of the charging capacitor. Eventually, the magnetic field

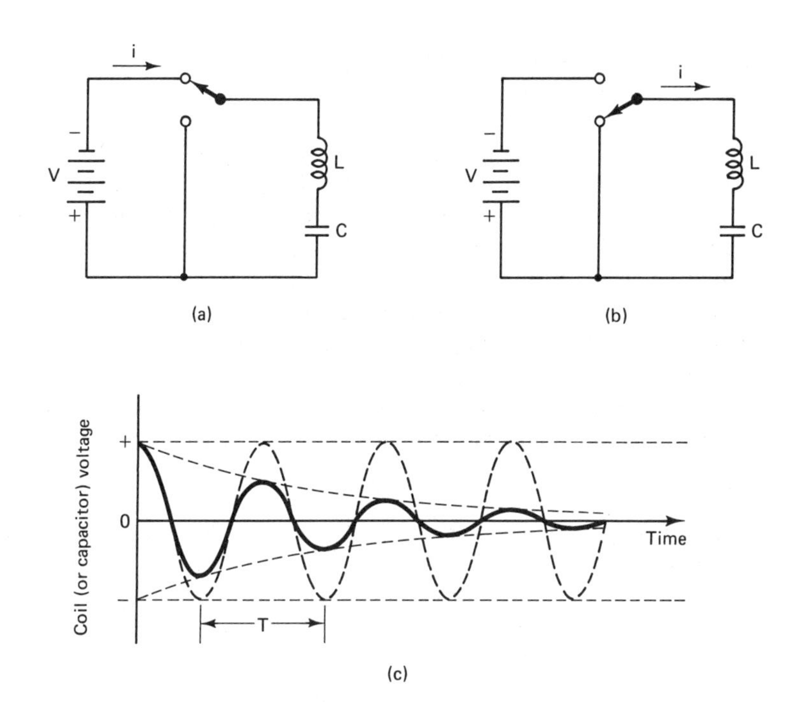

Figure 5.5 Oscillations in *LC* circuit: (a) charging circuit; (b) discharging circuit; (c) waveform of damped oscillation.

collapses completely. Current becomes zero. Most of the energy that was contained in the magnetic field of the current-carrying coil is now in the electric field of the charged capacitor. Some of the energy, however, has been lost to dissipation in the resistance of the circuit. Now, the capacitor starts to discharge, transferring the energy once again to the magnetic field of the coil. Eventually, the capacitor will be discharged and the process will again reverse itself. The energy "oscillates" between the coil and the capacitor until it has all been dissipated in the resistance.

The waveform of the voltage across the coil (or the capacitor) is that of Fig. 5.5(c), a typical damped oscillation. The waveform is not that of a sine wave, of course. However, the time between oscillations, the period of the waveform, is constant, as in a sine wave. The frequency of the damped oscillation, that is $1/T$, is the same as the resonant frequency of the circuit: $f_r = 1/2\pi\sqrt{LC}$.

Next, let us visualize what would happen in this circuit, and to its waveform, if we could inject, at just the right moment in each swing of the waveform, just the right amount of energy to make up the energy lost to dissipation. We would expect the peaks of the swings to be undiminished, or undamped. Observe the dashed-line extensions of the waveform peaks in Fig. 5.5(c). The voltage across the circuit would now be a sine wave. This is, in effect, what happens in a tuned, class C amplifier when the resonant frequency of the tuned circuit is exactly equal to the frequency of the input signal. The short collector current pulses are automatically timed to inject energy into the oscillating *LC* circuit at the correct moment.

The class C amplifier is especially desired as a power amplifier because its efficiency is significantly higher than that of either class A or class B operation. *Efficiency in an electronic amplifier is defined as the ratio of the ac output power to the dc input power*—it is a measure of the degree to which dc power is converted to signal power. Uncoverted dc power in an amplifier stage serves only to heat up the active device—the transistor, or tube. The more unconverted power a device must dissipate or get rid of, the higher must be its power rating. Higher power ratings imply larger and more expensive devices—both are undesirable characteristics.

The total dc power supplied to an amplifier stage is equal to the product of the dc supply voltage and the average current. For a transistor amplifier,

$$P_{dc} = V_{CC} \times I_C$$

where I_C is the average collector current. In a class A common-emitter amplifier, the average collector current is equal to the no-signal collector current. In a class B CE amplifier, the no-signal current is zero; the average current with a sine-wave signal is equal to 0.3183 times the peak value of the signal current since collector current flows for one-half of each cycle. It is obvious that for a given peak value of collector signal current, the average value of collector current in a class C amplifier will be less than that for a class B amplifier. By definition, collector current in a class C amplifier flows for less than one-half of the signal cycle. The actual average value of collector current for a given stage depends not only on the peak value of that current but also the conduction angle, or duty cycle in pulse circuit terminology. The conduction angle can be varied by changing the relationship between input signal amplitude and bias voltage. Thus the maximum theoretical efficiency of a class C amplifier approaches 100%. On the other hand, the maximum efficiency of a transformerless class A amplifier is only 25%, and for a class B amplifier it is 78.5%.

Another important characteristic of RF amplifiers that influences their design and application is that of maximum operating frequency. Every circuit, including amplifier circuits, incorporates stray capacitance. This is the capacitance that arises simply because two conductors are relatively close to each other. The two conductors could be a component lead lying close to the chassis. Much of the stray capacitance in an amplifier circuit is "across" (in parallel with) the signal path. Of course, in an absolute sense, the value of stray capacitance is small, of the order of a few tens of picofarads at the most. At low radio frequencies, the corresponding capacitive reactance of such capacitance is relatively large, and thus negligible. If the frequency of operation of a circuit is a high radio frequency (100 MHz or more), the reactance of stray capacitance is smaller and no longer negligible. The stray reactance that is in parallel with a signal path bleeds off or bypasses signal current. The net result can only be a reduction in gain of the amplifier. As the frequency goes higher, the undesirable effect is greater.

With careful design the amount of stray capacitance in an amplifier can be minimized. It can never be eliminated completely. Even the active device itself has inherent capacitance between its electrodes. Transistors are available which are designated "RF transistors." These have been manufactured so as to minimize the inherent capacitance between electrodes.

Every transistor has a specification called the *short-circuit limit frequency* or simply, *cutoff frequency*. This is the frequency at which the current gain of the transistor begins to diminish at a rate of -20 dB/decade because of the inherent shunting capacitance. It turns out that the inherent cutoff frequency for a given transistor is significantly higher for a common-base configuration than for a common-emitter configuration. Thus it is not uncommon to see the CB (common-base) connection used in RF amplifiers, especially in FM and television receivers.

A final consideration that we will mention in connection with RF amplifiers is that of noise. We are all familiar with noise in a radio receiver. There we actually hear something undesirable. We easily recognize that it is noise. The noise we hear from a receiver is the result of variations in the current through the speaker coil that are not intended to be there. Such variations are introduced into combination with the desired variations (signal) in two ways: from sources external to the circuitry, such as automobile ignition systems, etc., and internally as the result of phenomena caused by the particle nature of electrical current. (A more complete discussion of noise phenomena appears in Chap. 7.) Noise from external sources can be minimized, if not completely eliminated, by careful design, circuit shielding, and other techniques. Controlling internally generated noise is mostly a matter of selecting components that are known to have low-noise characteristics. For example, MOSFETs (metal-oxide semiconductor field-effect transistors) generate less noise internally than do conventional transistors. They are often used in very low signal RF amplifiers (FM and TV receivers).

5.3 THE OSCILLATOR

A radio transmitter is, first and foremost, a source of RF signal. An RF oscillator, as a generator of such a signal, then, is an absolutely essential function in a transmitter. An RF oscillator circuit, in turn, is an RF amplifier that has been modified to provide positive feedback. That is, first, it must be an electronic circuit with gain. Second, it must contain a circuit, called a feedback loop, which supplies a portion of the output signal to its own input terminals. The signal fed back must be in phase with what would be the normal input signal. In a true oscillator, however, no external input signal is required. The oscillator is said to be self-starting. As an RF amplifier, the RF oscillator will have a tuned circuit, or its equivalent, such as a quartz crystal. The amplifier favors one particular frequency—the resonant frequency of the tuned circuit. When power is applied, a current transient will occur in the circuit. The transient contains components of virtually all frequencies. The favored frequency will be amplified strongly, fed back to the input, amplified again, and so forth. As long as power is supplied, the circuit continuously generates a signal of the favored frequency.

We have looked previously at the oscillator function. In Chap. 3 we analyzed the operation of a tickler-coil oscillator used as the local oscillator in the heterodyne frequency converter of a receiver. The tickler-coil circuit lends itself to a certain type of analysis. In this approach, the damped oscillations of the tuned circuit are viewed as the most important and fundamental aspect of the circuit. The active portion of the oscillator stage is simply a means of reinjecting the energy lost through dissipation in the resistance of the circuit. Our understanding is that returning this energy to the tuned circuit enables it to operate continuously with undiminished oscillations. There are many different forms of oscillator circuits, however. Most other oscillator circuits are more easily understood when analyzed using a more general approach. Let us use that approach now to learn more about some of the oscillator configurations that are used in transmitters.

All oscillators can be represented in a general way by the simple block diagram of Fig. 5.6. As shown, an oscillator, in general, consists of an amplifier portion with a gain of G,

$$G = \frac{v_{\text{out}}}{v_{\text{in}}}$$

The feedback loop, which returns a portion of the amplified output to the input of the amplifier, has a transfer ratio of B,

$$B = \frac{v_{\text{fb}}}{v_{\text{out}}}$$

The term "transfer ratio" is used here instead of gain because the value of B is always less than 1. Analyzing the diagram, we write

$$v_{\text{out}} = Gv_{\text{in}} \qquad v_{\text{fb}} = Bv_{\text{out}} \qquad v_{\text{in}} = v_s + v_{\text{fb}}$$

Therefore, by substitution,

$$v_{\text{out}} = G(v_s + Bv_{\text{out}})$$

or

$$v_{\text{out}} = Gv_s + GBv_{\text{out}} \tag{5.1}$$

It is evident that if $GB = 1$, Eq. (5.1) will be true when $v_s = 0$. This defines the condition under which an amplifier with a positive feedback loop will become

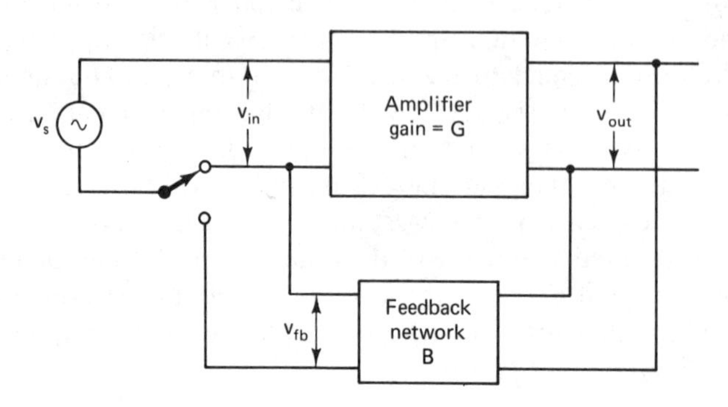

Figure 5.6 Block diagram of oscillator function.

a self-excited oscillator, a condition requiring no input but still providing an output. The condition is for the product, GB, to be equal to 1. If the oscillator is a single-stage, common-emitter amplifier, there is a 180° phase shift from input to output. In this case, $G = -v_{out}/v_{in}$. Therefore, in order for Eq. (5.1) to be true, B must also be negative. Or, more to the point, the feedback loop must also provide a 180° phase shift for there to be positive feedback.

The Hartley Oscillator

Let us apply this general approach now to a particular circuit configuration and observe how the circuit is arranged to provide for positive feedback. A very common oscillator circuit is shown in Fig. 5.7(a). As you will note, the coil in the collector circuit is tapped—to ground in this sample circuit. For this reason the circuit is sometimes called a *tapped-coil oscillator*. Generally, any oscillator circuit that has the tapped-coil feature is also called a *Hartley oscillator*. Since this circuit is a single-stage, common-emitter circuit, the feedback network must provide for a 180° phase shift. It may not be apparent to you how the 180° phase shift is achieved. To help in explaining this phenomenon, let us simply redraw the circuit slightly, as in Fig. 5.7(b). The oscillator will operate at the resonant frequency of the collector tank circuit

$$f_r = \frac{1}{2\pi\sqrt{(L_1 + L_2)C_3}}$$

At the operating frequency the feedback transfer ratio will be

$$B = \frac{X_2}{X_2 + X_3}$$

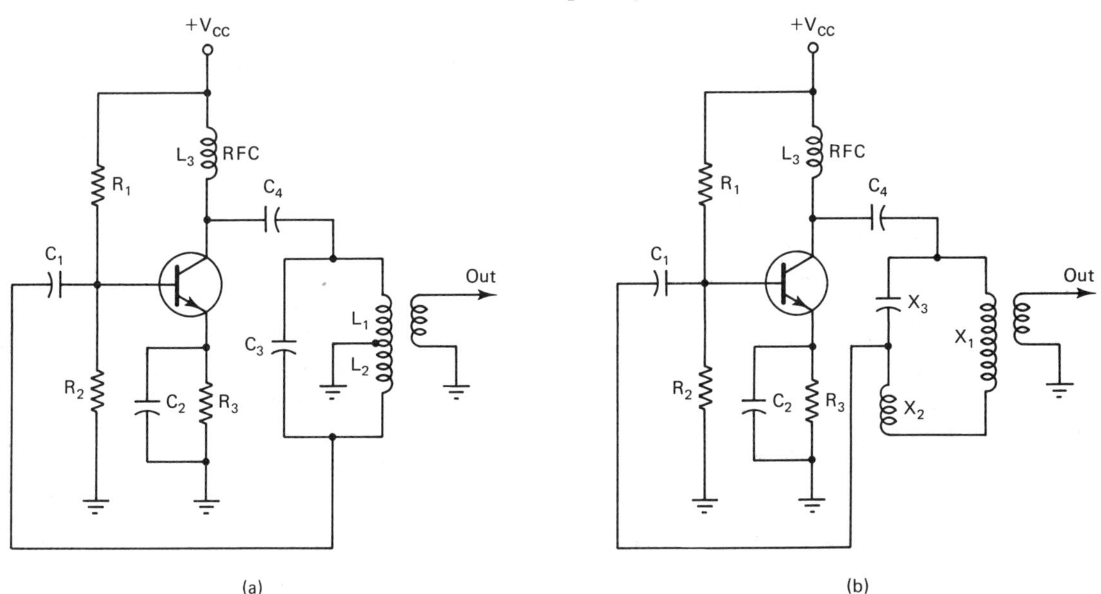

Figure 5.7 (a) Hartley oscillator; (b) Hartley circuit redrawn.

It is apparent that the relative values of L_1 and L_2 can be changed by changing the point of the tap on the coil. Therefore, the value of X_2 can be adjusted to achieve the required value for B. That is, since GB must equal 1, then

$$B = \frac{1}{G}$$

We want to understand how a 180° phase shift, between the signal voltage across the tuned circuit, and the feedback voltage, can be obtained. We call on our knowledge of some of the characteristics of parallel resonant circuits in which the Q is equal to or greater than 10. To the circuit external to the parallel circuit, that circuit appears as a very large resistance. The current supplied the tuned circuit is very small and is in phase with the voltage across the circuit. Each branch of the parallel circuit, however, can be analyzed separately. The inductive branch has a large current that lags the voltage across the circuit by almost 90°. (The series resistance of the coil prevents the phase angle from being exactly 90°.) The capacitive branch of a tuned circuit has a large current that leads the voltage across the circuit by 90°. The difference between these two branch currents is the very small line current that is in phase with the voltage across the circuit. Applying this information to the circuit of Fig. 5.7(b), we recognize that the branch with X_3 and X_2 appears as a net capacitive branch to the outside world. (It will be designed that way.) Therefore, the current in that branch leads the voltage across the tuned circuit by 90°. The voltage across the circuit is the output voltage of the amplifier. The current in X_2 leads the output voltage by 90°. The voltage across X_2 leads the current in the branch by approximately 90°. (At least there will be a very large component of the voltage across X_2 that leads the current by 90°.) Since the voltage across X_2 is the feedback voltage, the voltage to be supplied to the base, the feedback voltage leads the output voltage by 180°! The feedback loop accomplishes its purpose of providing a feedback voltage with a 180° phase shift from the output voltage.

The Hartley oscillator may be found with many different actual circuit configurations. The basic principles will be the same, however, no matter what the configuration. The circuit must provide a 180° phase shift for the feedback voltage.

The Colpitts Oscillator

Another common oscillator configuration is called the Colpitts oscillator. A typical circuit of this type is shown in Fig. 5.8(a). Note that the tuned circuit in the output of the stage again contains three reactances. In this circuit, however, there are two capacitive reactances and one inductive reactance. This is in contrast with the Hartley circuit, which has two inductive reactances and one capacitive reactance. This circuit is referred to by some as a "tapped capacitance" circuit. To facilitate visualizing the phase relationships in the circuit, let us again redraw the circuit slightly [see Fig. 5.8(b)]. The feedback ratio, B, is again given by

$$B = \frac{X_2}{X_2 + X_3}$$

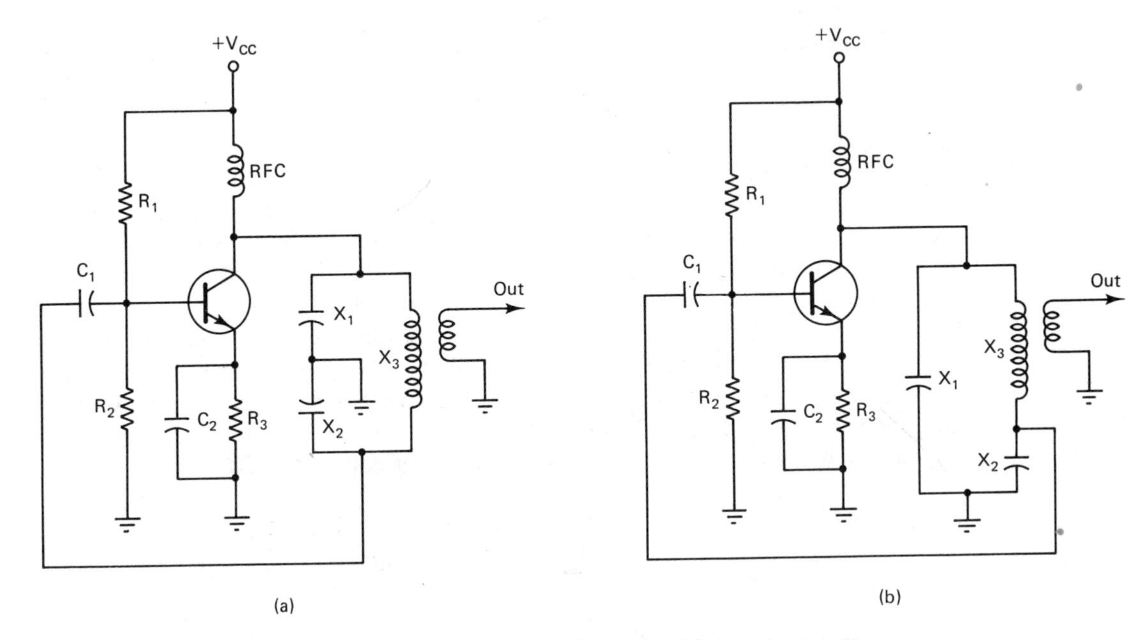

Figure 5.8 (a) Colpitts oscillator; (b) Colpitts circuit redrawn.

In this case X_2 is capacitive and X_3 is inductive. The nature of these reactances is the opposite of those in the Hartley circuit. Here the current in the X_2–X_3 branch lags the output voltage by approximately 90°.

The voltage across X_2 will lag that current by 90°. The feedback voltage will lag the output voltage by 180°. The feedback circuit again accomplishes the required 180° phase shift from output voltage to feedback voltage.

Let us look at some of the other details of the circuits of these two oscillators (Figs. 5.7 and 5.8). We note that in each case there is an RF choke between the V_{CC} supply and the collector. ("Choke" is another word for inductor. An "RF choke" is a coil of relatively few turns and low inductance. Still, it will have relatively high reactance at radio frequencies.) The RF choke has virtually zero dc resistance. It provides a path for the dc current required by the stage while blocking the RF current from the power supply. This particular circuit connection is referred to as "shunt fed." The direct current is supplied in shunt (in parallel) with the ac load (the tuned circuit). Another way to supply the dc current is in series with the ac load. In a series connection the power supply is connected directly to the coil of the tuned circuit and the direct current flows through part, or all, of the coil.

Another item of interest in both of these circuits is the circuit path for the feedback signal from output circuit to base. Note that in each case the path includes a series capacitor. The capacitor is required to block dc voltage and allow for different dc levels at the collector and base. The coil of the output tuned circuit is virtually a short for direct current.

Each of the circuits depicts voltage-divider bias and bypassed emitter swamping resistor. The swamping resistor provides for independence from the effects of changes in temperature and/or transistor characteristics (as, for example, when a transistor must be replaced). Oscillators will operate with either class A, B, or C bias.

As stated above, in order to operate continuously an oscillator must have a GB product equal to 1 ($GB = 1$). This product is called the loop gain, that is, the gain from input to output and back to input. In order to start oscillating the loop gain must be greater than 1, $GB > 1$. Generally, with $GB > 1$, a circuit starts to oscillate with a signal of very low amplitude. The amplifier is operating in the small-signal mode, and thus with maximum gain. As the signal amplitude increases, because of the feedback, the amplifier operates over a greater portion of its transfer or operating characteristic (which is always nonlinear). This results in a reduction of the gain, G. Then, either the GB product reduces to 1 or clipping of the signal occurs, which reduces G also. In any event, the operation adjusts itself, by one or a combination of these effects, until $GB = 1$. Oscillator operation is basically nonlinear, the waveform is not a perfect sine wave. However, the tuned circuit of an RF oscillator serves as a filter to eliminate most of the frequency components that distort the sine wave of the fundamental. The result is that the final output of an LC oscillator approaches that of a perfect sine wave.

Tuned-Input, Tuned-Output Oscillators

Oscillator circuits that incorporate tuned circuits on both the input and output sides are of special interest to us: first, because such circuits are found in actual practical applications, but more important, an understanding of their operation paves the way to an understanding of a serious problem found in most RF amplifiers—unwanted oscillation.

The circuit diagram of a tuned-base, tuned-collector transistor oscillator is shown in Fig. 5.9(a). The item of special interest in this circuit is the interelectrode

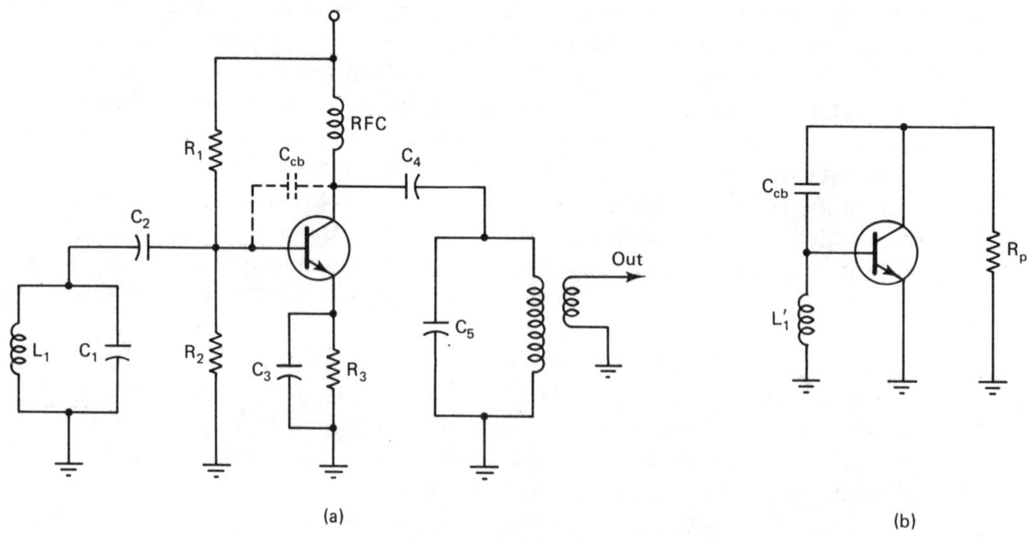

Figure 5.9 (a) Tuned-base, tuned-collector oscillator; (b) ac equivalent circuit.

capacitance between collector and base, C_{cb}. This circuit element is shown dashed on the diagram to indicate that it is not an actual, discrete component but rather an inherent property of the transistor and its connecting leads. The value of C_{cb} is very small, generally less than 10 pF. For this circuit to oscillate, a signal of the proper magnitude and phase must be coupled through C_{cb} from the collector to the base. The magnitude of GB for the circuit must be 1, at least. Since the circuit is a common-emitter circuit, the phase angle between the signals at the collector and base is 180°. Hence there must be a 180° shift in the phase of the signal coupled through C_{cb} from collector to base. How can this be accomplished?

Let us redraw the circuit as the ac equivalent circuit of Fig. 5.9(b). Note that the input tuned circuit is now represented as the inductance L_1. This means that the input LC circuit is tuned so that its resonant frequency is slightly higher than the operating frequency of the oscillator. (The net reactance of the input circuit is inductive.) Now, X_{cb} and X_1 form a series branch whose net reactance is capacitive ($X_{cb} > X_1$, because C_{cb} is very small). The current in this branch will lead the collector voltage by 90°. The voltage across X_1 (the feedback voltage to the base) will lead the branch current by approximately 90°. Thus there will be a large component of the feedback voltage that leads the collector voltage by 180°. The condition for oscillation will be obtained.

A straight RF amplifier is prevented from oscillating by feeding back a voltage from the output which is 180° out of phase with the signal voltage at the collector. Such a voltage is in opposition to the voltage fed back through C_{cb} and neutralizes it. This additional circuit to prevent oscillation is called a *neutralizing circuit*. In a circuit designed to oscillate, however, especially at a low radio frequency, it is sometimes necessary to connect an actual, discrete capacitor of very low value between the collector and base. This is required to provide sufficient feedback.

Crystal-Controlled Oscillators

We have seen that the condition for oscillation is very critical. In a tuned circuit oscillator the condition depends very much on the relative values of the components, including the operating parameters of the transistor (or tube, or other active component). The condition for oscillation also depends on the dc voltages in the circuit, to a degree. If any of the circuit parameters, including the load, change, the operating frequency will change. If the circuit changes are drastic, the circuit may not oscillate.

Changes in operating frequency are undesirable in the master oscillator of a transmitter. The radio-frequency spectrum is very crowded with users. If the operating frequency of a transmitter changes, it is very likely to cause interference with other users of the spectrum. When the operating power of a transmitter exceeds a few milliwatts it is essential that the operating frequency be maintained stable to within a few cycles per million. In fact, the degree of frequency stability required for the various communications services is defined by treaties negotiated in worldwide communications meetings, as well as by laws passed by the U.S.

Congress. A very high level of frequency stability can be achieved by the use of a quartz crystal as a replacement for the frequency-determining tuned circuit in the master oscillator of a transmitter.

Quartz is a naturally occurring mineral made of silicon dioxide, SiO_2. Silicon dioxide is the major constituent of ordinary sand. In quartz, the SiO_2 arranges itself in the form of a hexagonal (six-sided) crystal. In a crystal, the atoms or molecules of a substance arrange themselves in a definite pattern that is repeated regularly in three dimensions. The effect of the regular pattern is observable to the naked eye since a natural piece of quartz takes a form which is bounded by definitely oriented smooth planes that are in alignment with the internal structures [see Fig. 5.10(a)]. Quartz is clear and glass-like in appearance. A piece of natural quartz is made into an electronic component called a "quartz crystal" by, first, cutting the natural crystal into very thin slices with a high-precision saw. These slices are then mounted between two conducting plates and the assembly mounted ino a container with connecting leads so that the "crystal" can be connected into a circuit.

The operation of a crystal in an electronic circuit is called the *piezoelectric effect*. This term relates to the property of a substance of generating voltage when subjected to pressure, and conversely, undergoing mechanical stress when subjected to an electric field.

When a crystal is subjected to an ac voltage, it alternately expands and contracts, or vibrates. It has a natural frequency of vibration which is determined

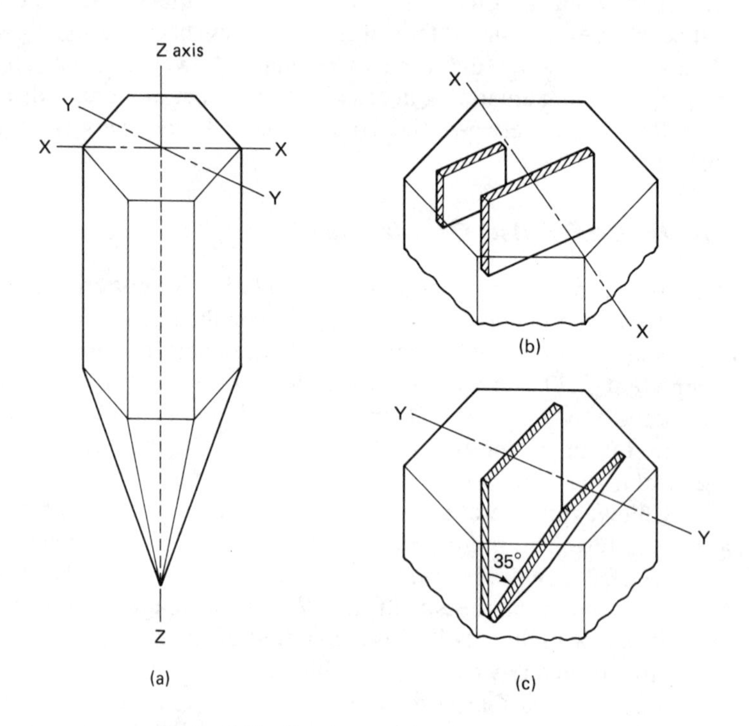

Figure 5.10 Quartz crystal and crystal cuts.

by its dimensions, and, in turn, its temperature. If its temperature is held constant, the operating frequency of a crystal is highly stable, hence its application in transmitters, watches, and other electronic devices.

In transmitters, provision is usually made to maintain the frequency-controlling crystal at a nearly constant temperature. In high-powered commerical broadcast transmitters the master crystal may be mounted in a small, insulated container called a "crystal oven." The oven is supplied with a heating element and the temperature is monitored and maintained by an elaborate electronic closed-loop control circuit. In low-powered transmitters, the temperature control is typically less elaborate. It may consist of a resistor mounted in contact with the crystal case. The current in the resistor is controlled by a simple thermostatic switch mounted somewhere in the transmitter case.

Electronic crystals are designated *X*- or *Y*-cuts according to the orientation of the cuts with respect to the axes of the natural crystal (see Fig. 5.10). There are many other types of cuts in actual practice, with designations such as *AT*, *BT*, *CT*, etc. Different cuts are obtained by rotating the plane around one or more of the axes, *X*, *Y*, or *Z*, of the crystal. The plane of the cut has a bearing on the natural frequency of vibration of the slice and on its susceptibility to temperature changes. Crystals are available with frequencies as low as about 6 kHz and as high as 30 MHz. The slice becomes extremely thin and fragile for the higher frequencies. The upper limit can be extended by selecting and mounting slices so that they are operated at harmonics of the fundamental, natural frequencies. Such special crystals are called *overtone crystals* and are available for operation up to 100 MHz.

As a piezoelectric device the action of a crystal is electromechanical. That is, it has mechanical as well as electrical characteristics. However, it can be represented with a purely electrical analog in the form of an equivalent circuit. For most practical purposes, the equivalent circuit is that shown in Fig. 5.11(b). In the diagram, the inductance, L represents the inertia of the mass of the crystal, its opposition to bending or change of shape during vibration. The capacitance C_C represents a mechanical property called elasticity or resiliance, the property that causes the crystal to restore itself to its original shape after being deformed

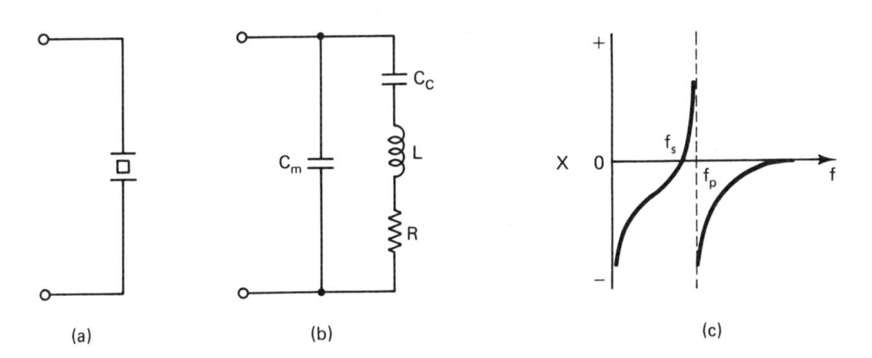

Figure 5.11 (a) Crystal circuit symbol; (b) equivalent electrical circuit; (c) crystal reactance.

by an electrical field or mechanical pressure. The resistance R is a representation of the energy losses in the crystal when it is being deformed. The capacitance C_m is the actual capacitance of the plates between which the crystal is mounted.

The crystal and its mounting can operate as a resonant circuit in both a series and a parallel mode. Figure 5.11(c) illustrates the net reactance of the unit as it relates to the frequency of vibration. The valuable property of the crystal, when viewed as a resonant circuit, is its very high Q, which is a result of a very high equivalent L/C ratio. Of course, these electrical properties are a direct result of the mechanical properties of the crystalline structure. The Q may range from a value of 10,000 up to 500,000 in some special designs. (Compare this with values of 10 to 100 for coil-capacitor tuned circuits.)

Crystals may be employed in many different ways in oscillator circuits. Some circuits use a crystal operating in its parallel resonant mode to replace a conventional LC parallel resonant circuit. Other circuits use a crystal operating in its series resonant mode strictly as the feedback component. In this case the crystal is effective in controlling the frequency of operation since the amount of energy fed back, and its phase angle, will be correct for sustaining oscillations only at the frequency of resonance of the crystal. Let us examine a number of different circuits employing crystals.

In Fig. 5.12(a) a crystal is used as the tuned-base parallel resonant circuit of a tuned-base, tuned-collector oscillator. Compare this circuit with that of Fig. 5.9(a). The tickler-coil oscillator of Fig. 5.12(b) is an example of the use of a crystal as the feedback connection component. See a similar application in the Hartley oscillator of Fig. 5.12(c) and the Colpitts circuit of Fig. 5.12(d). Remember, because of the very high Q of the crystal, changes in its equivalent series impedance will be extremely great for very small changes in the frequency of the signal it is passing. Therefore, in the circuits of Fig. 5.12(b)–(d), changes in the amplitude and phase angle of the feedback signal will be very large for even minute changes in the frequency of oscillation. In fact, the circuits will meet the condition for sustaining oscillation only at, or very, very near the series resonant frequency of the crystal.

In some circuits it is difficult to perceive how the crystal can be effective in controlling the frequency. For example, let us consider the circuit of Fig. 5.13(a). This is apparently a common-base Colpitts oscillator. How does the crystal function to control frequency? To help us understand the answer to this question, let us consider the noncrystal form of the circuit in Fig. 5.13(b). Since the crystal seems to replace C_5, what is the purpose of C_5? The circuit is a common-base circuit, it must have its base at signal ground, and C_5 provides the path to ground for the base (the reactance of C_5 is very low at the operating frequency and provides, effectively, a short circuit to ground for the signal). In Fig. 5.13(b) the reactance of C_5 is relatively low for a wide band of frequencies near the operating frequency of the circuit. Therefore, the operating frequency is determined by the components of the tuned circuit, C_1, C_2, and L_1. The interelectrode capacitances of the transistor also have some effect on the tuned circuit. It is these components that determine at what frequency the requirements for oscillation will be satisfied. The thing that is different about the circuit of

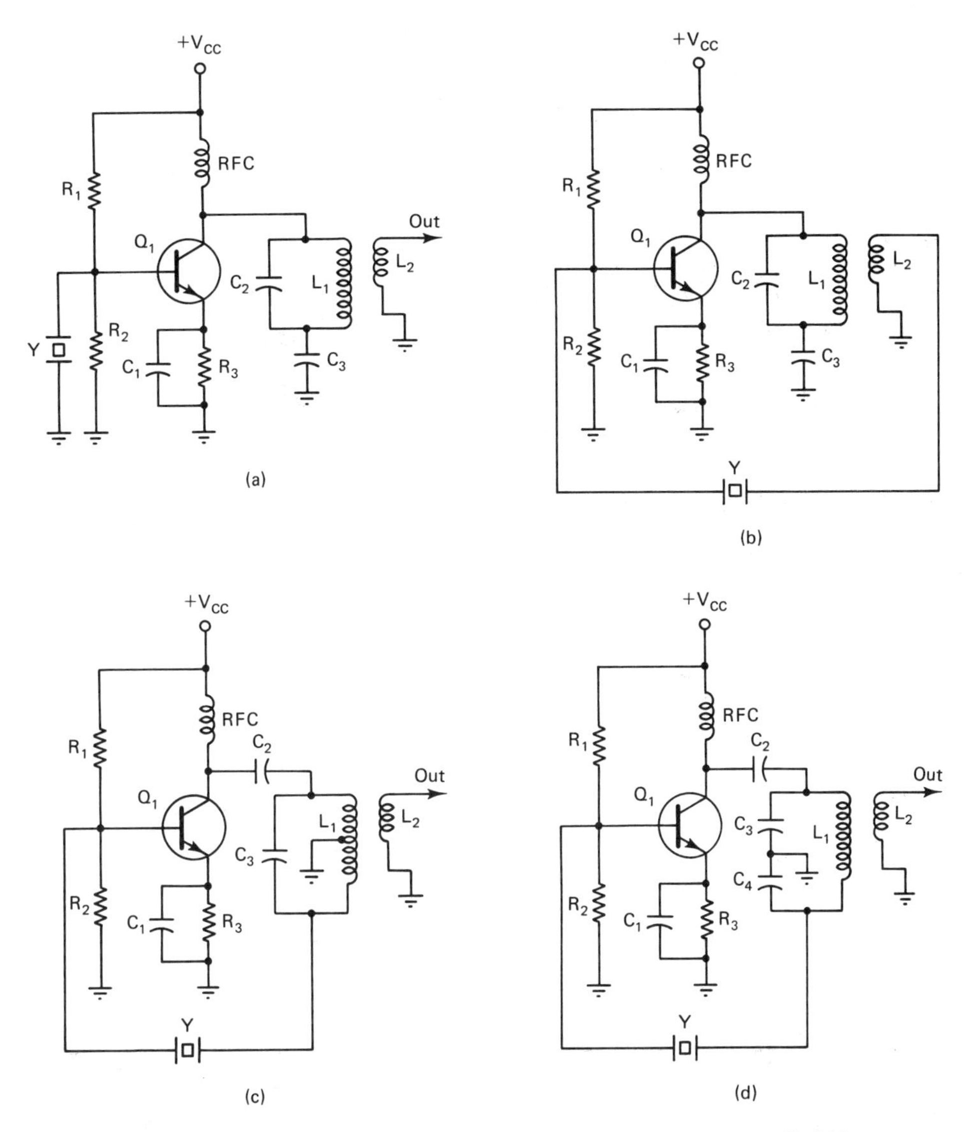

Figure 5.12 Crystal-controlled oscillators: (a) tuned-input, tuned-output; (b) tickler coil; (c) Hartley; (d) Colpitts.

Fig. 5.13(a) is that the impedance of the crystal will change radically with the operating frequency. Its impedance will be such as to satisfy the requirements for oscillation in the circuit at or near one particular frequency only. Obviously, the circuit will oscillate only at that frequency.

Another very important and interesting circuit is that of the Pierce oscillator (see Fig. 5.14). You will note that the circuits there do not include any *LC* resonant circuits. This permits the changing of the operating frequency of the circuit by changing only the crystal. This is an important feature in applications where it may be desirable to change the operating frequency of a transmitter

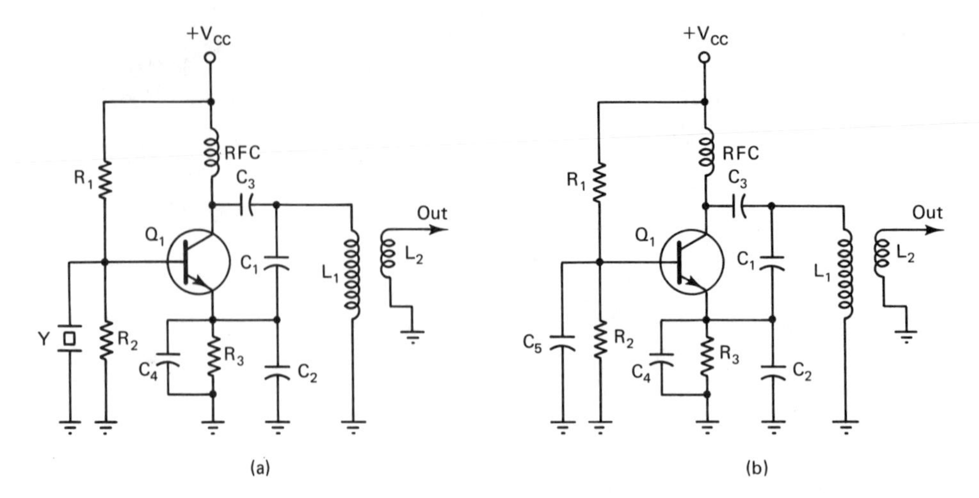

Figure 5.13 Colpitts oscillators: (a) Colpitts crystal-controlled oscillator (common base); (b) common-base Colpitts oscillator.

quickly or frequently. This is true, for example, of CB (citizens' band) two-way radio equipment. The circuit of Fig. 5.14(a) is another form of a Colpitts circuit and very similar to that of Fig. 5.13(b). The crystal replaces the collector tuned circuit. In the circuit of Fig. 5.14(b), the crystal must function on the inductive side of resonance. Viewed in this way, the circuit becomes very much like the common-emitter Colpitts circuit. There is a 180° phase shift in the feedback of the signal from collector to base (across the crystal).

A crystal-controlled oscillator that features current feedback instead of voltage feedback is shown in Fig. 5.15. The basic circuit is an emitter follower

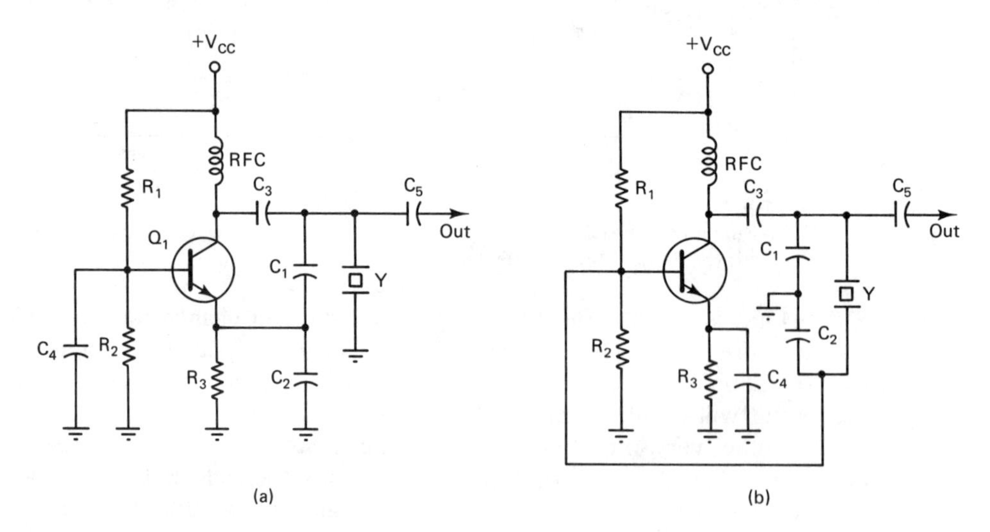

Figure 5.14 Pierce oscillators: (a) common-base Pierce oscillator; (b) common-emitter Pierce oscillator.

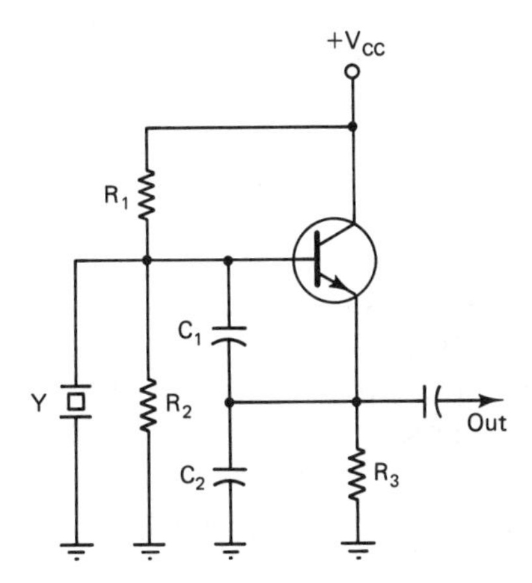

Figure 5.15 Emitter-follower Pierce oscillator.

which has current gain but no voltage gain. The crystal must function on the inductive side of resonance and forms a parallel resonant circuit with capacitors C_1 and C_2. Feedback is by means of charging current in C_2. The action is somewhat like a tickler-coil oscillator. Think of the parallel resonant circuit, of which the crystal is the inductive branch, as providing damped oscillations and a signal at the base of the transistor. The transistor amplifies the signal current. The amplified current is present in the emitter circuit. Some of this increased current is used to return energy to the oscillating tuned circuit. A small pulse of emitter current increases the charge on C_2 at just the right moment in each cycle.

Troubleshooting Oscillators

Troubleshooting an oscillator can be a difficult, frustrating experience. A successful diagnosis of an inoperative oscillator may require more ingenuity than that typical of troubleshooting a straight amplifier, for example. When we troubleshoot an amplifier section containing several stages, the IF section of a receiver, perhaps, it is a relatively straightforward procedure to trace a signal through the section and find the point at which the signal is lost. An oscillator, on the other hand, is normally the source of the signal. If it is not producing a signal, where do we start looking for the problem? The answer to this question depends to some degree on the situation. If we have just finished constructing the oscillator circuit and it does not work, the first thing to do is to check the circuit carefully to see that it is connected correctly. For example, if an oscillator incorporates a tapped coil as part of the feedback circuit, a reversal of two of the connections on the coil could give the circuit negative feedback instead of the positive feedback required for oscillation. This same error could be made when replacing a defective coil in an oscillator that had previously worked satisfactorily.

If we are troubleshooting an oscillator that has previously worked, however, we will not be looking for a wrong connection. We must look for a defective component. The use of dc voltage analysis (see "Troubleshooting Using DC Voltages" in Sec. 4.1) can often be effective in this type of diagnosis of an oscillator circuit. If dc voltages are normal, or within an expected range of nominal values, it will be necessary to check for defects in components that would not change the dc voltages of the stage. Examples of such defective components are: open bypass capacitors, open coupling capacitors, and, in certain instances, open coils.

An oscillator is basically an amplifier, of course. In cases especially difficult to diagnose, it may be expedient to inject a signal from an external source and probe the circuit with an oscilloscope. We would want to use this technique to determine whether the circuit is amplifying, for one thing. We can also use it to do some signal tracing within the stage. Finally, since an oscillator is usually a single stage, replacing every component would not be an impossible task if all other attempts to find the problem fail. The components should be replaced one at a time with circuit testing after each replacement. Replacement might best begin with the active device (transistor or whatever), proceed next with the crystal, if any, then capacitors, coils, and finally, resistors.

5.4 MODULATING THE AM TRANSMITTER

Let us review the meaning of amplitude modulation (see Sec. 2.3). In amplitude modulation the modulating circuit performs an operation on the radio-frequency signal in such a way that the amplitude of the RF voltage and current in the circuit are changed (modulated) in accordance with an information signal (audio or video, for example). The result is that the envelope of the RF signal is a replica of the information signal. The modulated circuit is typically an amplifier. To accomplish the modulation, then, the modulating (information) signal must, in effect, alter the gain of the amplifier. The gain must be increased momentarily to obtain the increase in amplitude of the modulated signal. The gain must be decreased momentarily to achieve the decrease in amplitude of the modulated signal. An amplifier with a nonlinear transfer characteristic provides the desired operating feature. The modulated circuit is generally a nonlinear RF amplifier. The modulating signal shifts the operating point between points of higher gain and lower gain.

An amplifier has an input side and an output side and modulation can be performed in either the input or output side. Let us begin our study of modulating circuits by examining an output-side modulation scheme. Refer to the circuit of Fig. 5.16. Observe that the modulation transformer is in series with the carrier output tuned circuit and the power supply. The amplifier is designated class C, which simply means that it supplies pulses of current of less than 180° duration to the carrier output circuit. The carrier output voltage is sinusoidal as a result of the flywheel effect of the tuned circuit. The amplitude of the carrier voltage is directly proportional to the energy in the circuit and, therefore, directly proportional to the current pulses supplied by the amplifier. For example, if the input drive signal is sufficient to drive the amplifier just to saturation at the peak of the signal, the peak-to-peak carrier voltage will be approximately equal to two times the dc supply voltage (i.e., $2V_{CC}$, in the case of a transistor circuit).

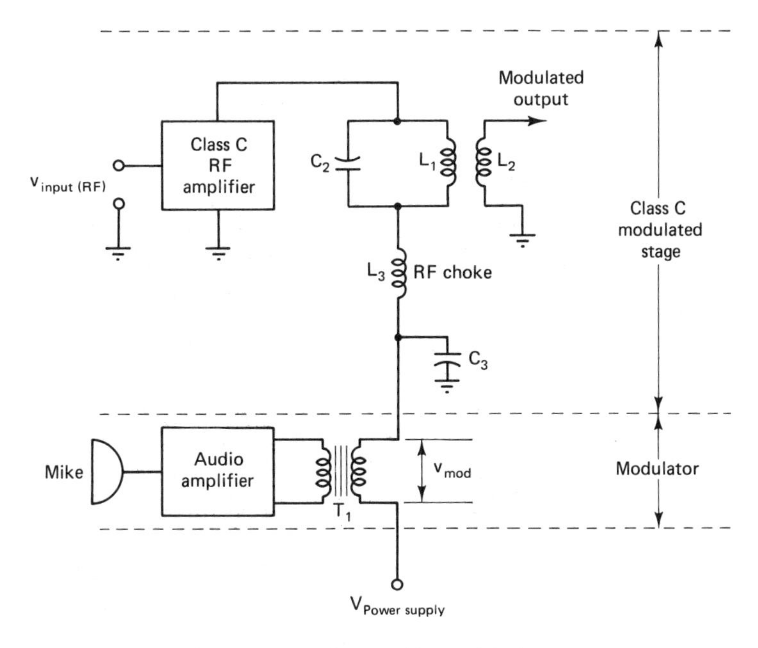

Figure 5.16 Output-side modulation.

At 100% modulation, the peak value of V_{mod} will be equal to the dc supply voltage ($V_{power\ supply}$). When the polarity of a V_{mod} peak is the same as that of $V_{power\ supply}$, the total dc voltage in the circuit will be doubled, saturation current will be doubled, and the output voltage will be doubled. When the polarity of a V_{mod} peak is opposite of that of $V_{power\ supply}$, the net dc voltage will be zero, current pulses will drop to zero, and the output voltage will be reduced to zero.

The modulating voltage (the information signal) is generally a slowly changing voltage (slowly changing compared to the carrier frequency). The net supply voltage will, consequently, be a slowly changing dc voltage. It will change in a pattern that is identical to the pattern of change of the modulating voltage. The pattern of change of the output current pulses, and therefore of the output voltage, will also be identical to the pattern of the modulating voltage. The carrier will have been amplitude modulated! These changes are illustrated in Fig. 5.17.

The modulation scheme explained above is applicable to both tube and transistor versions of RF amplifiers. The method is quite effective and popular in tube circuits and, there, is called plate-circuit modulation. The transistor version is called collector-circuit modulation, or, simply, collector modulation. The schematic diagram of a circuit utilizing collector modulation is shown in Fig. 5.18. This circuit is representative of those used in citizen's band transmitters which operate with an output of 5 W.

Unfortunately, transistor circuits have a tendency to compress the output waveform of a modulated signal as the level of modulation approaches 100%. One strategy for overcoming this deficiency is that of modulating the driver stage

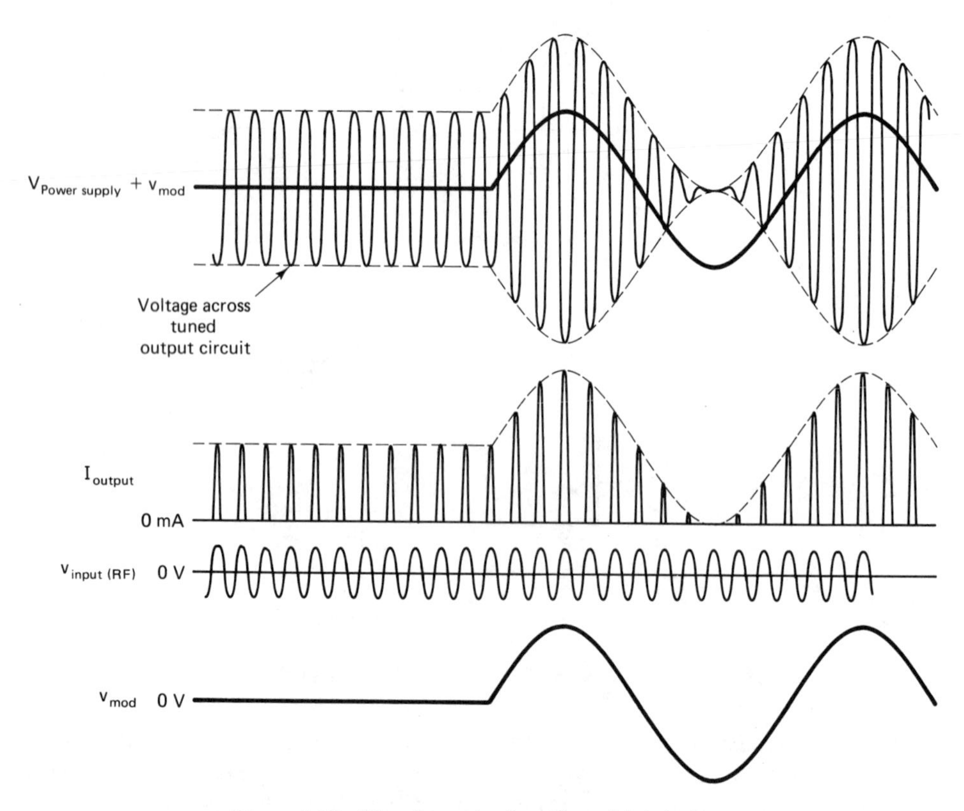

Figure 5.17 Waveforms in class C modulated stage.

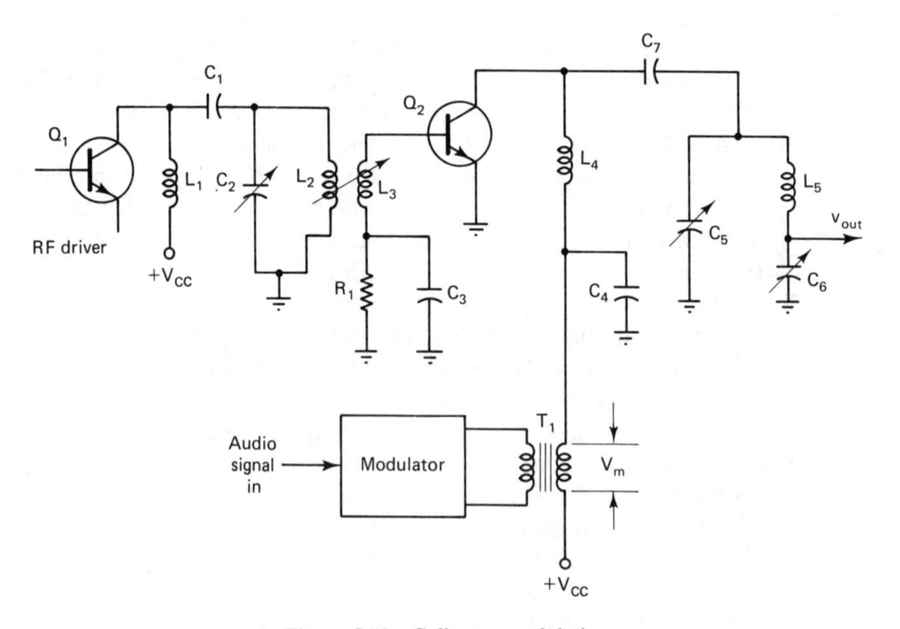

Figure 5.18 Collector modulation.

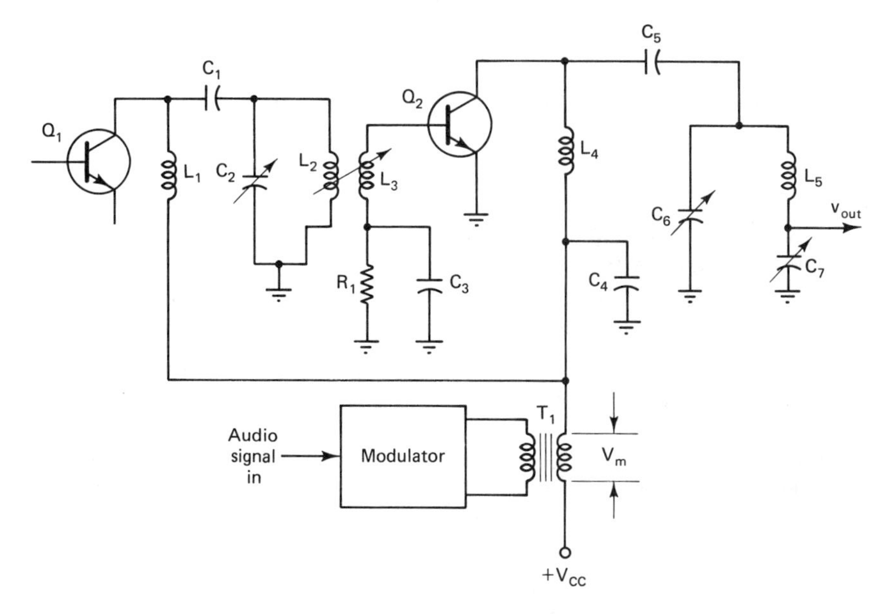

Figure 5.19 Circuit for improved collector modulation.

as well as the output stage. A circuit illustrating this technique is shown in Fig. 5.19.

High-Level versus Low-Level Modulation

A quick review of the facts concerning power distribution in an AM wave (see Chap. 2) reminds us that the power level of such a wave increases by the factor $m^2/2$ when it is being modulated. (The quantity m is the modulation index: V_m/V_c.) Thus, at 100% modulation, for example, the power is increased by 50%. In plate- or collector-circuit modulation this increased power must come from the modulator circuit, not from the dc power supply directly. The modulator circuit and its transformer must be capable of operating at and producing this level of power. For this reason, modulating the output side of the output stage of a transmitter is referred to as "high-level modulation." As an example, the modulator for a 50-kW radio station must supply 25 kW of audio-frequency power. In some applications, a transmitter design requiring such a high level of audio power is deemed inadvisable. In these cases the carrier is modulated on the input side of the power output stage, or in a low-power-level stage. Although modulating the grid circuit of a tube stage, or the base circuit of a transistor circuit is theoretically possible, this approach to modulation is not too common. Such circuits require very careful adjustment of circuit parameters for satisfactory operation.

A low-power-level stage may be modulated in the plate or collector circuit, as described above. The audio-frequency power level required of the modulator is correspondingly less. A transmitter utilizing this method is said to be using *low-level modulation.* The method is not without its drawbacks, however.

One of the major disadvantages of low-level modulation arises because of the fact that an AM signal must be amplified in a linear amplifier, not a class C amplifier (see Chap. 3). Hence, in a transmitter using low-level modulation, all amplifier stages following the modulated stage must be operated in a linear mode (class B generally). Class B mode is significantly less efficient than class C. If the output power level of the transmitter is more than a few watts, this requirement of linear operation may be a significant concern. The transmitter designer is always faced with the trade-off between the requirement for a high power level in the modulator (an audio-frequency circuit) versus low-efficiency operation of the high-power stages in the RF section.

Modulation Problems

There are several ways in which the modulation of an AM transmitter can be defective or unsatisfactory. The level (percentage) of modulation is very important since the "reaching" power of a station is proportional to the square of the modulation index (percent modulation/100). If the audio section of a transmitter is or becomes incapable of providing sufficient power for 100% modulation, the effectiveness of the transmitter is degraded significantly.

It is essential that the modulation process achieve modulation without distortion. Undistorted modulation means that the information content of the modulated carrier wave is a perfect replica of the information that was placed on the carrier by the modulation process. In practical terms, if a carrier wave is amplitude modulated by a single-frequency, sinusoidal information signal, then the envelopes of the carrier must be perfect sinusoids. This implies, for example, that the increase in the carrier amplitude at the peaks of the modulation cycle must be equal to the decrease in the carrier amplitude at the valleys of the modulation cycle. See the waveform in Fig. 5.20(a). This also means that the average peak (or peak-to-peak, or effective) value of the carrier must remain constant. If these conditions do not hold, the modulation envelope is not a perfect sine wave. It is distorted. The information content of the modulated carrier is distorted.

There are two ways in which the distortion just described can be introduced. First, the amplitude of the carrier can increase more at the peaks of modulation than it decreases at the valleys, as in Fig. 5.20(b). This results in an increase in the average of the peak or effective amplitude of the carrier. The phenomenon is called *positive carrier shift*. On the other hand, the carrier can increase less on the peaks than it decreases in the valleys, as in Fig. 5.20(c). In this case there is a decrease in the average peak or effective amplitude. This is called *negative carrier shift*, or, sometimes, *downward modulation*. The general problem referred to here is called *carrier regulation*. (You may recall from other studies or experience that the "voltage regulation of a source" is a measure of the change of the amplitude of the source voltage as the load on the source changes from no load to rated load.)

Any distortion of a radio signal ultimately results in the distortion of the sound which is heard from a receiver of that signal. A distortion of up to about

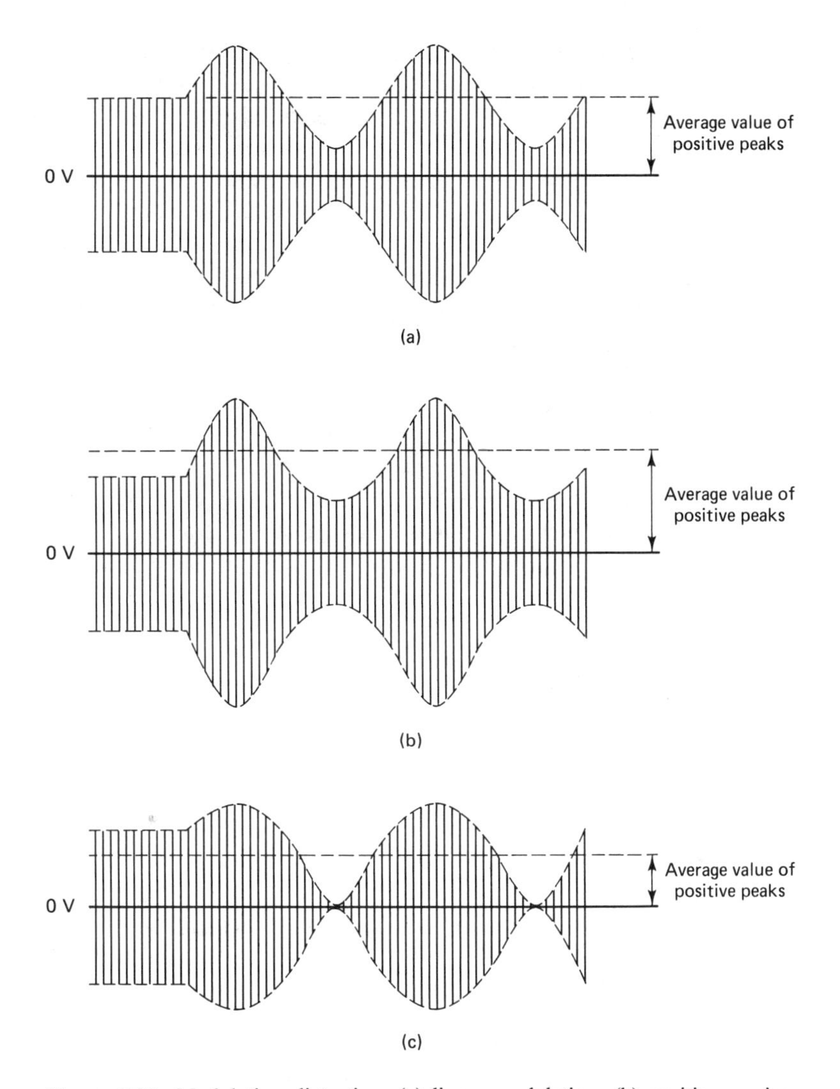

Figure 5.20 Modulation distortion: (a) linear modulation; (b) positive carrier shift; (c) negative carrier shift.

5% is hardly detectable by most persons. Positive carrier shift produces greater radiated power and may be considered to be advantageous by a transmitter operator if the distortion is not objectionable to the listener. Negative carrier shift, however, reduces radiated (reaching) power and produces sound that is compressed and weak. Negative carrier shift can be the result of a dc power supply having poor regulation. All of the energy produced by a transmitter must come, ultimately, from the dc power supply. At the modulation peak of a 100% modulated waveform, instantaneous power is four times that of the unmodulated carrier. The power supply must be capable of sustaining such a drastic change in power-level requirement.

A carrier-shift, or carrier amplitude regulation meter is a device designed to indicate a value that is related to the average value of the positive peaks of the carrier. (Please note that the statement is ". . . the average value of the positive peaks . . .," not ". . . the average amplitude . . ."). The meter is set up to indicate a convenient value, say half-scale, when monitoring an unmodulated carrier. Then, when the transmitter is modulated, if the meter does not change the modulation is symmetrical (the average peak value is unchanged). If the meter reads higher, the average peak value has increased. This is an indication of unsymmetrical modulation, specifically, positive carrier shift. Downward modulation, or negative carrier shift is indicated by a lower-than-normal reading on the carrier-shift meter.

As we have seen, the output of the modulator (the modulating stage) is used to achieve changes in amplitude of the modulated stage(s). The modulator is the final link in the audio chain of the transmitter. The freedom from noise and distortion, of the modulation, can only be as good as that of the signal arriving at the output of the modulator. The entire audio chain of the transmitter must be properly designed and operating correctly to achieve high-quality modulation.

5.5 BUFFER AMPLIFIERS

In order to be accepted by the Federal Communications Commission (FCC), for sale to the public, a radio transmitter must be able to meet certain requirements concerning its operating frequency. A transmitter must be able to operate while maintaining its operating frequency within prescribed limits. The limits have been set by law or regulation. Limits are different for different types of radio services—commercial broadcast, mobile communications, marine communications, amateur operations, etc. In all cases, however, limits are quite stringent. The purpose of such regulations is to help ensure that the utilization of the radio-frequency spectrum, a very limited resource in our modern world, is orderly and efficient. A transmitter whose frequency wanders will interfere unnecessarily with the transmissions of other services. In short, then, a high premium is placed on transmitter designs that produce stable operating frequencies. We have already studied the crystal-controlled oscillator (Sec. 5.3) and observed that its use enhances the stability of the operating frequency of a transmitter. The buffer amplifier is another function whose primary purpose is to help prevent the operating frequency of an oscillator from changing during operation.

A general definition of the term "buffer" implies the concept of lessening or absorbing the effect of a shock of collision or impact. A buffer amplifier placed after the master oscillator of a transmitter is there to lessen the effect on the oscillator of changes in the circuits downstream (in the signal flow path) from the oscillator. Amplitude modulation of a driver or power output stage, for example, can cause changes in the load seen by preceding stages, including the oscillator. Severe changes of this nature can be observed to "pull" (change) the frequency of operation of the oscillator. The buffer amplifier absorbs and lessens the effects of such load changes.

To achieve its purpose a buffer amplifier is typically a class A, RF amplifier. It is operated as a small-signal amplifier—it does not require a signal of great amplitude from the oscillator. It is lightly loaded—it is not expected to drive the amplifier placed after it with a large signal. Because of the way in which it is operated—class A, lightly loaded, lightly driven—the operating characteristics of the buffer amplifier change very little, if any, with changes in the operation of the transmitter stages downstream. It thus protects the master oscillator from those changes.

5.6 FREQUENCY MULTIPLIER STAGES

A frequency multiplier stage in a transmitter is a stage that increases (multiplies) the frequency of the already-generated RF signal by some integral multiple such as 2 or 3 (usually never more than 3 in a single stage). Frequency multiplication is required when it is impossible or technically unfeasible to operate a crystal oscillator at the operating frequency of the transmitter. For example, because crystals must be cut very thin to oscillate at high frequencies, practical crystal operating frequencies are limited to approximately 60 MHz.

A frequency multiplier stage is an amplifier operated in the class C mode. The output side must be a resonant circuit tuned to a frequency that is an integral multiple of the input signal frequency. For example, if a stage is to be a "×2" multiplier, and the frequency of the input signal is 45 MHz, the output tank circuit must be resonant at 90 MHz.

The circuit in Fig. 5.21 is that of a two-stage multiplier section. The first stage, Q_1, is a ×3 multiplier, the second stage, Q_2, is a ×2 multiplier. The overall multiplication is 6.

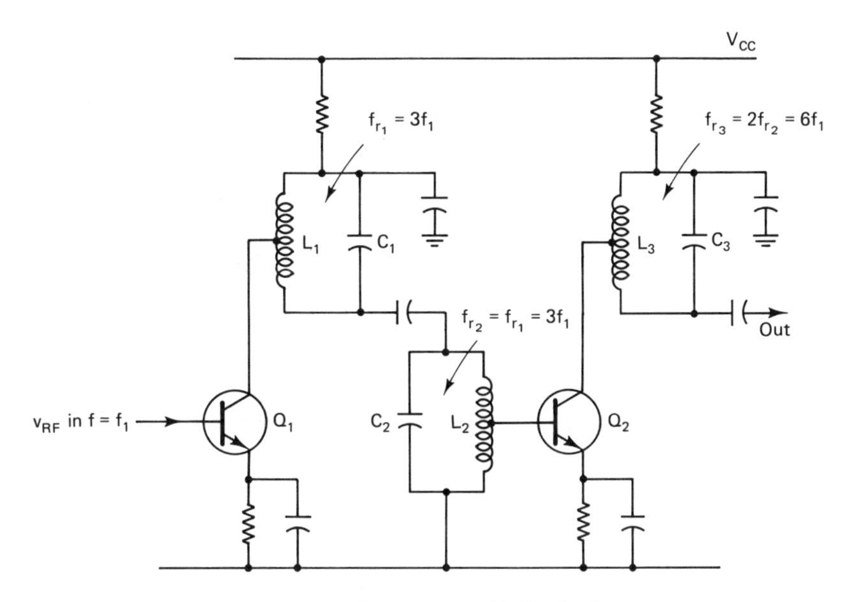

Figure 5.21 Two-stage multiplier (×6).

A frequency multiplier is able to accomplish its purpose by virtue of the flywheel effect of the tank circuit in the output side of the stage. Recall that in a typical class C RF amplifier, the output voltage waveform is virtually sinusoidal while the output current consists only of short pulses whose duration is less than 180° of the input signal cycle. In a nonmultiplying stage the current pulses occur for every cycle. However, the voltage is sinusoidal because of the flywheel effect of the tuned circuit [see Fig. 5.22(a)]. In a $\times 2$ multiplier, because the output tank circuit is tuned to two times the input frequency, the current pulses will occur in alternate cycles only. In the $\times 3$ multiplier, the current pulses occur in every third output cycle [see Fig. 5.22(b) and (c)]. As long as the multiplication attempted in a single stage is not greater than 3, the output voltage waveform is still substantially sinusoidal.

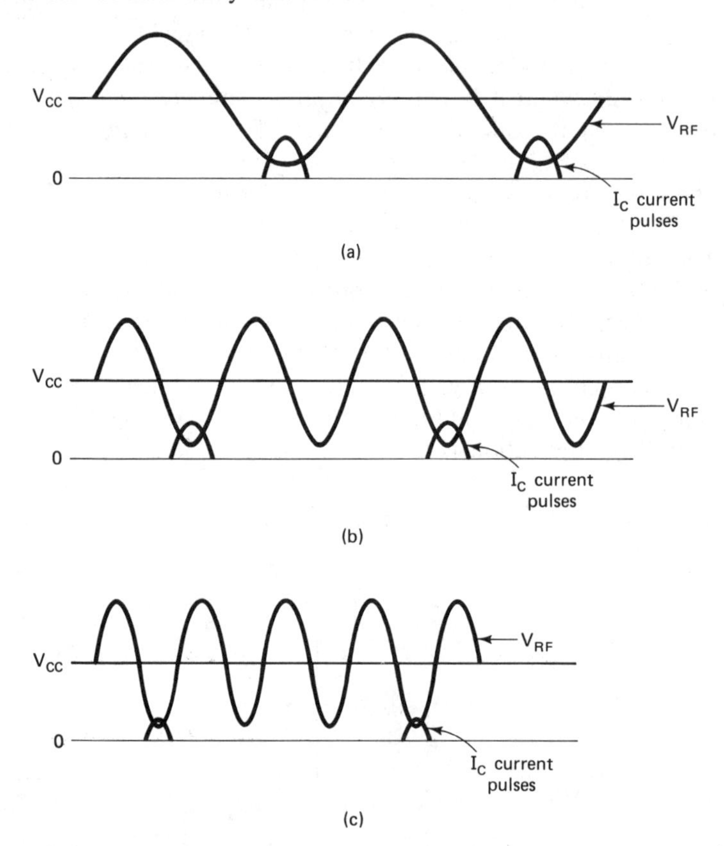

Figure 5.22 Frequency multiplier waveforms: (a) collector current and output voltage waveforms in class C amplifier; (b) waveforms for $\times 2$ frequency multiplier; (c) waveforms for $\times 3$ frequency multiplier.

5.7 TRANSMITTER POWER STAGES

Except for special-purpose units such as "wireless microphones" and micropower toy transmitters, AM transmitters, generally, have power ratings in the range 5

W to 50 kW. For example, the maximum rating for a CB transmitter is 5 W (dc input power to output stage); some so-called clear-channel AM broadcast stations are authorized to use transmitters rated at 50 kW (RF power into the antenna). All but the lowest-powered transmitters must utilize special amplifier stages, called power amplifiers and/or drivers, to increase the energy in the carrier signal to the desired power level. The last stage of a transmitter, the one that feeds RF power to the antenna, is called the power output stage. It is also called the transmitter final (stage) or simply the "final." The stage that drives the final is called a driver stage. It, too, is effectively a power amplifier since its typical power level is about 10% of that of the final.

RF power amplifiers incorporate special features to accomplish their purpose of producing high power levels. The finals of transmitters in the range 1000 W to 50 kW generally, if not exclusively, use electron tubes as the active devices. The output tubes of the higher-powered units require water cooling systems to carry away the heat produced by the amplification process.

Lower-powered AM transmitters, especially in the range 5 to 250 W, utilize power transistors in the power stages. Power transistors are larger and bulkier than the typical nonpower type of semiconductor devices. The terminals (connection leads), junctions, and the semiconductor die itself, of the power transistor, must be larger to be able to carry the high currents required to produce high power levels. A power transistor will be fitted with a heat sink to facilitate the dissipation of heat produced during operation. The heat sink must be mounted in intimate contact with the collector, or emitter, of a transistor for effective heat transfer.

Because of the amount of power it handles, it is important that a power stage be as efficient (see the definition for efficiency of an amplifier stage in Sec. 5.2) as possible. In an AM transmitter, the output stage can be a class C amplifier, the most efficient type of amplifier, only if it is also the modulated stage.

The schematic diagram of typical final and driver stages of CB transmitters is shown in Fig. 5.23. These are power stages that handle relatively low levels of power. A full-power CB transmitter is limited by regulation (law) to 5 W dc input power to the final stage. The RF power level is of the order of 2 to $2\frac{1}{2}$ W. You will recognize that both of these stages are designed to operate in the class C mode. There is no provision for providing any base current (or, therefore, collector current) when there is no RF signal present. That is, there is no provision for standby base or collector current. You will also recognize that collector modulation is used here, and that it is applied to the driver stage as well as the output stage.

Coupling circuits are extremely important parts of the power section of a transmitter. An interstage coupling circuit will be used to conduct, couple, or transmit RF power from the driver stage to the output stage. The coupling or matching circuit will provide for the transfer of power at a specific level. Similarly, a matching/coupling circuit will be used to transfer RF energy from the output stage to a transmission line and/or antenna.

The coupling circuit between final and antenna may also be designed to accomplish a high degree of attenuation of the energy of any undesired frequencies

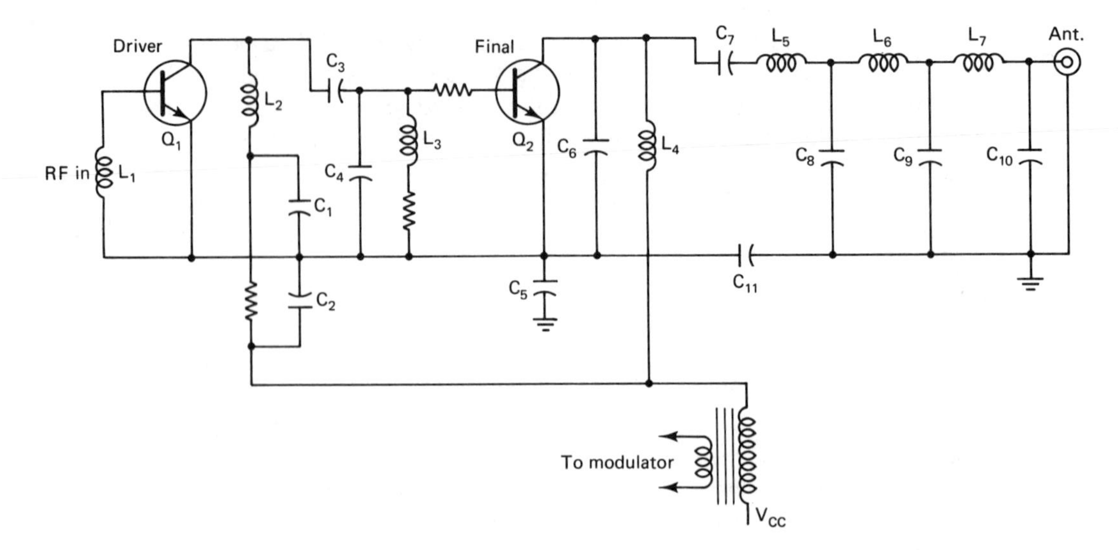

Figure 5.23 Typical power stages for CB transmitter.

present in the output stage. The output of a class C amplifier is high in energy components that are harmonics of the fundamental signal frequency. This is true simply because the operation is nonlinear. There may be other so-called *spurious* (unwanted) *signals* present in the output stage which must be attenuated so as to prevent interference with the signals from other radio stations.

In the design of the coupling circuit between amplifier stages—the interstage coupling circuit—consideration must be given to two aspects of amplifiers. From the point of view of the *driven* stage, the *driving* stage can be represented as a simple ac source with an internal impedance (as with the aid of Thévenin's theorem). This equivalent internal impedance is referred to as the "output impedance" of the stage. On the other hand, from the point of view of the *driving* stage, the *driven* stage can be considered as a load made up of a complex impedance. This equivalent impedance is called the "input impedance" of the stage.

Both input and output impedances of a given stage are conceptual representations of the operating characteristics of the stage. They are not actual, discrete components. The values of these impedances depend on both the physical characteristics of the transistor used and the operating conditions of the actual circuit. The design of the interstage coupling circuit must take these input and output impedances into account to achieve the desired transfer of power between stages. The relationship between the driving stage, the interstage coupling, and the input of the driven stage is depicted in Fig. 5.24.

In its function as a power transmitting unit, a coupling circuit is generally expected to "transform" a "load impedance" in order to accomplish a particular purpose. For example, in interstage coupling, a typical requirement is the transfer of a specified amount of RF power from the driver to the input of the power output stage. Given the power specification, say P_{out}, the relationship between

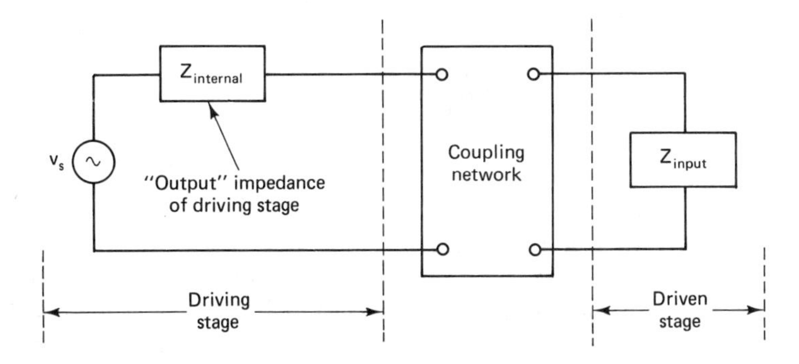

Figure 5.24 Relationship of output impedance, input impedance, and coupling network.

the RF voltage, V_{RF}, at the output of the driver stage and its (the driver's) load resistance, R_{load}, is the usual one, that is,

$$P_{out} = \frac{V_{RF}^2}{R_{load}}$$

However, if the driver stage is a class C amplifier, then, usually, the peak-to-peak value of the RF voltage is equal to $2V_{CC}$, or, $V_{RFp} = V_{CC}$. Therefore,

$$V_{RF}(\text{effective}) = \frac{V_{CC}}{\sqrt{2}}$$

and

$$V_{RF}^2 = \frac{V_{CC}^2}{2}$$

More precisely, if the collector saturation voltage, $V_{CE(sat)}$, is known, then

$$V_{RF}^2 = \frac{(V_{CC} - V_{CE(sat)})^2}{2}$$

Hence the amount of load resistance that must be presented to the driver stage in order to achieve the desired power level is

$$R_{load} = \frac{(V_{CC} - V_{CE(sat)})^2}{2P_{out}}$$

If the input impedance of the output stage (the driven stage) is not equal to this value of R_{load}, the coupling circuit is designed to transform that impedance to "match" this required value. Be aware that the match is not necessarily to achieve maximum power transfer, but to achieve the transfer of a specified amount of power.

The coupling circuit between the final stage of a transmitter and the transmitter load—usually an antenna at the end of a transmission line—must serve a similar purpose. The circuit is designed to transform the transmission line/antenna load so that it matches the value of an R_{load}, as seen by the output stage, required for the stage to produce a specified amount of RF power.

Coupling circuits of many different forms can be devised to accomplish the results described above. In fact, there is no limit to the number of circuits that could be constructed to do the job. It is outside the purpose of this book to present a comprehensive, detailed model of coupling circuit design. However, a general description of the approach used in designing coupling circuits can assist us in being aware of, if not fully understanding, the reasons that any particular coupling circuit has the configuration and circuit parameters with which we find it. Such a general description follows.

Among other facts, it is essential for us to recall that parallel ac circuits can be transformed to equivalent series circuits. The opposite transformation—series to parallel—is also possible. These circuits and their equations of transformation are shown in Fig. 5.25.

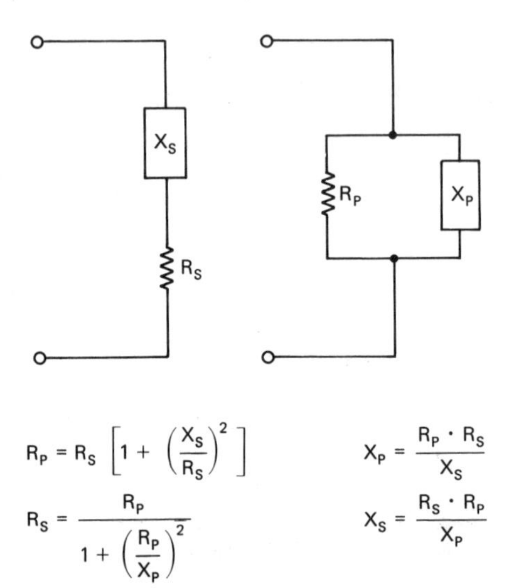

Figure 5.25 Series–parallel, parallel–series transformations.

When series circuits contain both inductive and capacitive reactances, there is one frequency that will make the magnitudes of the reactances equal. That frequency is called the *resonant frequency*. When the circuit is excited by a source operating at that frequency, the circuit will exhibit special characteristics. It is said to be "at resonance." At resonance the equivalent impedance of the circuit is simply R_s, the total series resistance of the circuit.

In parallel circuits, conditions are not so neatly distinguished at or near the frequency that makes the magnitudes of the reactances equal. The condition of importance in this discussion is called *antiresonance*. A circuit is in antiresonance when the net current to the parallel circuit is in phase with the voltage across it, that is, when the power factor is unity. This can also be described as the condition when the reactive components of the two branch currents are of equal magnitude. At antiresonance, the input impedance to the circuit is equal to R_p, the value of the total series resistance of the inductive branch transformed to its parallel equivalent, as in Fig. 5.25.

An "L" circuit (so-called because of its shape) composed of two reactances of opposite types is a simple and common coupling circuit used to transform the value of a component, or load, connected to its output terminals [see Fig. 5.26(a)]. When we redraw the circuit as in Fig. 5.26(b) it becomes obvious that the overall circuit, including R_L, is a parallel LC circuit. Let's assume that the resistance of the inductance L is negligible compared to R_L. This is a reasonable assumption for most practical cases. Then, at f_0, the frequency that makes the circuit antiresonant, the network will appear to be a pure resistance, R_{in}. The value of R_{in} is that of R_L transformed to its parallel equivalent [see Fig. 5.26(c)]. The most common form of the network is the one shown: X_L as the series element and X_C as the shunt element. This configuration also makes the circuit a low-pass filter that can attenuate the second and higher-order harmonic of the carrier and other spurious signals. Formulas for calculating values for X_L and X_C to obtain desired R_{in}/R_L transformation ratios are given below. These formulas are derived from the series-to-parallel transformation equations commonly found in textbooks on basic ac circuit theory.

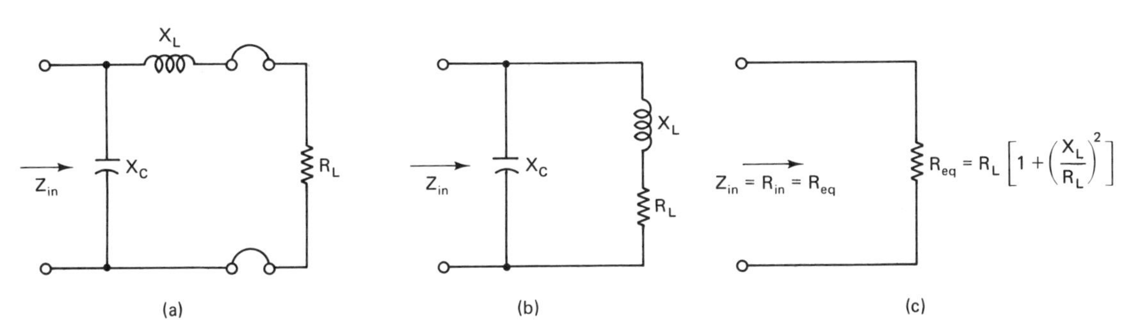

Figure 5.26 (a) L network; (b) L network redrawn; (c) equivalent impedance at resonant frequency.

Let $n = R_{in}/R_L$, $n > 1$; then,

$$X_L = \sqrt{R_{in}R_L - R_L^2} = \frac{R_{in}}{n}\sqrt{n-1}$$

$$X_C = \frac{R_{in}\,R_L}{X_L} = \frac{R_{in}}{\sqrt{n-1}} \qquad Q = \sqrt{\frac{R_{in}}{R_L} - 1} = \sqrt{n-1}$$

The L network, as described above, transforms the load resistance "up." That is, to a potential source (such as a transmitter output stage), a load resistance appears larger in value than it actually is, when seen "through" the network. A "down" transformation can be obtained with only a slight change in the circuit: Connect X_C in parallel with R_L, instead of at the input to the network [see Fig. 5.27(a)]. The X_C–R_L parallel circuit can now be transformed to an equivalent series circuit [see Fig. 5.27(b)]. This circuit has the potential to be series resonant. Thus, at f_0, the frequency at which the circuit is resonant, R_{in} for the network is equal to the transformed value of R_L, $R_{L\text{ser}}$ in the diagram, if we neglect the resistance of L. (Recall that Z_{in} for a series resonant circuit is equal to R_{ser}, the total series resistance of the circuit, because the reactances cancel each other.)

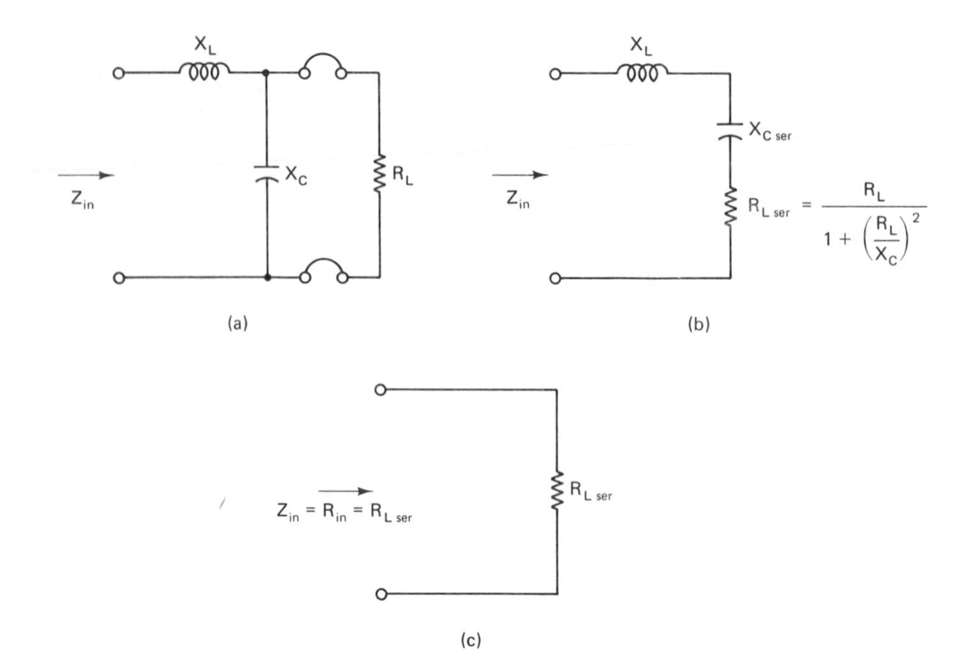

Figure 5.27 L network for "down" transformation: (a) L network for transforming an R_L to lower value; (b) L network transformed to equivalent series circuit; (c) equivalent L network at resonance.

The formulas for designing a step-down impedance transforming circuit of this type are as follows.

Let $n = R_L/R_{\text{in}}$, $n > 1$; then

$$X_L = \sqrt{R_{\text{in}}R_L - R_{\text{in}}^2} = R_{\text{in}}\sqrt{n-1}$$

$$X_C = \frac{R_{\text{in}}R_L}{X_L} = \frac{n}{\sqrt{n-1}}R_{\text{in}} \qquad Q = \sqrt{\frac{R_L}{R_{\text{in}}} - 1} = \sqrt{n-1}$$

As can be seen from the equation for Q of the L network, the Q cannot be controlled by the design of the network ($Q = \sqrt{n-1}$). It (Q) is a function of the transformation ratio only. This can be a distinct disadvantage. A simple modification of the circuit allows Q to be chosen by the designer. The modification is to connect a capactitive reactance in series with the series inductance of the circuit [see Fig. 5.28(a)]. Design equations for this circuit are as follows:

Select a desired value for Q. then

$$X_L = QR_{\text{in}}$$

$$X_{C1} = X_L - R_{\text{in}}\sqrt{n-1}$$

$$X_{C_2} = \frac{n}{\sqrt{n-1}}R_{\text{in}}$$

Various other coupling circuit configurations are in common usage. Several of these are shown in Fig. 5.28. The choice of one configuration over another

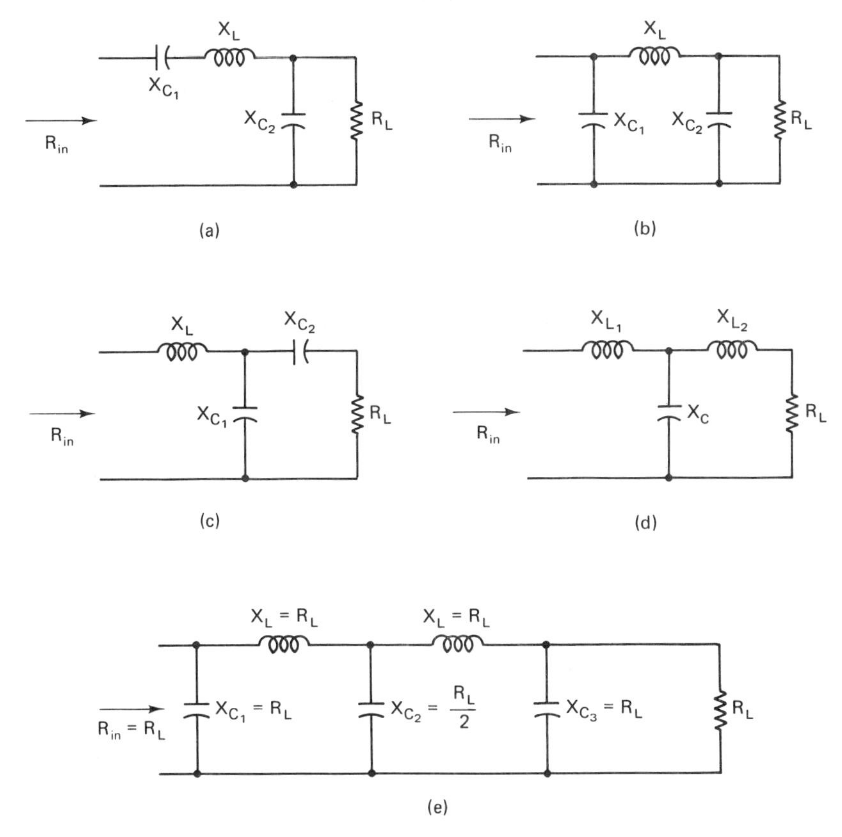

Figure 5.28 Networks for impedance matching and harmonic attenuation: (a) L network modified to permit choice of circuit Q; (b) pi network; (c) T matching network; (d) T network providing high harmonic attenuation; (e) network to provide additional harmonic attenuation.

is influenced by several factors: the transformation ratio required, the absolute values of the impedances R_L and R_{in}, whether the values of L and C components, as determined by the design equations, are practical, etc. For example, the pi network of Fig. 5.28(b) is more suited to the higher impedances of tube circuits than to the relatively lower impedances of transistor power stages. The T networks of Fig. 5.28(c) and (d) are more popular in the design of solid-state transmitters.

As has been stated previously, the attenuation of harmonic and/or other spurious signals is an essential requirement of the output circuts of a transmitter. Coupling circuits used to transform impedances invariably provide significant attenuation of this type. It is generally true that each separate reactive component in a low-pass filter (the typical coupling circuits presented above are all low-pass circuits) provides attenuation of 6 dB/octave. (A signal power level is attenuated by approximately 6 dB when the power is cut to one-fourth of the input value. Two frequencies are an octave apart when the higher frequency is twice the lower.) Hence, if a coupling network contains two reactances, as with the L network above, the attenuation is 12 dB/octave. When additional attenuation is required, additional reactances must be incorporated in the network between the output stage and the antenna. One simple solution is to place a pi network [see Fig. 5.28(b)] between the matching network and the antenna. The pi network would be designed now so that $R_{in} = R_L$. One interesting result of combining two pi networks is the network shown in Fig. 5.28(e). This network is easily obtained with components of practical values for the condition $R_{in} = R_L$. When designed for this condition it provides a phase shift of exactly 180° between input and output.

When one is engaged mainly in the task of troubleshooting and repairing transmitters, the details of design, especially of selecting one circuit configuration over another, are of no great concern. Even so, when confronted with a transmitter circuit containing several reactances in a ladder-type network between the output transistor and the antenna connection point, it is normal to be curious about the need for such a strange circuit. At worst, the situation can be intimidating if one is not familiar with the purpose, or design philosophy, of such circuits. The presentation in the preceding paragraphs is designed to assist in removing some of the mystery surrounding these circuits.

Example 1

A transmitter power output stage is to drive a 50-Ω antenna load. The transmitter will be supplied from an automobile electrical system with the result that V_{CC} = 12.5 V. It is desired to match the stage to the antenna load so that P_{out}(RF) = 2.5 W. The operating frequency is 27.5 MHz.

(a) Determine R_{in}, the equivalent load that the transistor must see in order to produce the desired amount of power for the given value of V_{CC} (assume class C operation). (b) Design an L network that will provide the required match.

Solution. (a) Assuming that $V_{CE(sat)}$ is negligible,

$$R_{in} = \frac{V_{CC}^2}{2P_{out}} = \frac{(12.5)^2}{2 \cdot 2.5} = 31.25\ \Omega$$

(b) Since R_{in} is less than R_L, we select a network of the type shown in Fig. 5.27. Then

$$n = \frac{R_L}{R_{in}} = \frac{50}{31.25} = 1.6$$

$$X_L = R_{in}\sqrt{n-1} = 31.25\sqrt{1.6-1} = 24.21\ \Omega$$

$$L = \frac{X_L}{2\pi f} = \frac{24.21}{2\pi \cdot 27.5(10^6)} = 0.14\ \mu\text{H}$$

$$X_C = \frac{n}{\sqrt{n-1}}R_{in} = \frac{1.6}{\sqrt{1.6-1}}(31.25) = 64.55\ \Omega$$

$$C = \frac{1}{2\pi f X_C} = \frac{1}{2\pi \cdot 27.5(10^6)(64.55)} = 89.66\ \text{pF}$$

The required network with components labeled with this respective values is shown in Fig. 5.29.

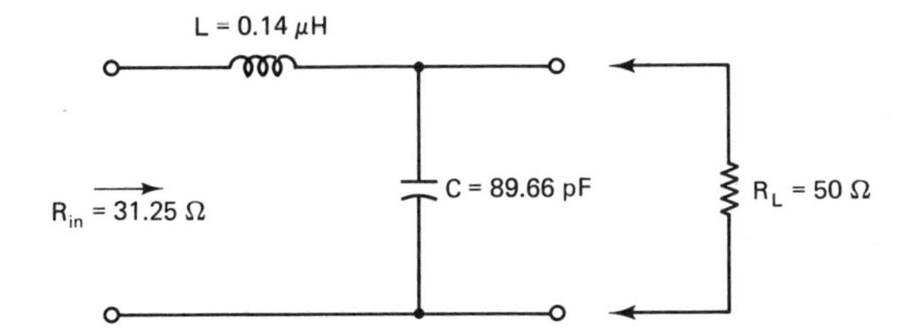

Figure 5.29 Network for transforming 50-Ω load to 31 Ω at 27.5 MHz.

GLOSSARY OF TERMS

Amplifier, buffer An amplifier placed between an oscillator, or any other critical circuit, and the remaining parts of a transmitter (or other system) to protect the critical circuit from the effects of changes in other parts of the system.

Amplifier, driver An amplifier that is specifically designed to provide the signal power required at the input of a power amplifier.

Carrier regulation The changes in the amplitude of a transmitter's carrier signal, other than the changes intentionally produced by amplitude modulation, as the amount of modulation changes.

Clamper A circuit that adds a dc component to an ac signal, usually as the result of rectification of the signal.

Clamper, negative A clamper circuit that adds a negative dc component to a signal.

Clamper, positive A clamper circuit that adds a positive dc component to a signal.

Cutoff frequency, of a transistor The frequency at which the current gain of a transistor begins to diminish at the rate of −20 dB/decade as the result of the shunting effect of interelectrode capacitance.

Efficiency, of an amplifier The ratio of signal output power to dc input power (multiplied by 100 if it is to be expressed in percent).

Final, the The output (and, therefore, "final") stage of a transmitter.

Frequency multiplier An RF amplifier whose output circuit is tuned to a multiple of the input signal frequency, enabling it to produce an output signal whose frequency is a multiple of the input signal frequency.

Negative carrier shift During amplitude modulation, the result when the increase in the amplitude of the carrier signal in the crests of the modulation pattern is less than the decrease of the carrier in the troughs of the pattern. The magnitudes of the increase, and decrease, in amplitude are equal in perfect or linear modulation.

Overtone crystal A quartz crystal that produces a strong harmonic of its fundamental or natural frequency.

Piezoelectric effect The property of a crystal of generating a voltage when subjected to pressure and, conversely, undergoing mechanical displacement when subjected to the field of an electric voltage.

Positive carrier shift During amplitude modulation, the result when the increase in the amplitude of the carrier signal in the crests of the modulation pattern is greater than the decrease in the troughs of the pattern.

RF choke An inductance coil, usually of relatively low inductance, used to impede (choke) the flow of RF current in a particular branch of a circuit.

Series feed A circuit arrangement in an RF amplifier or oscillator in which dc current passes through the RF component(s), such as inductance coils.

Shunt feed A circuit arrangement in an RF amplifier or oscillator in which dc current is conducted around (shunts) RF components, such as inductances.

Spurious signal Any unwanted signal.

REVIEW QUESTIONS: BEST ANSWER

1. A transmitter whose operating frequency is 1 MHz is specified as having a frequency stability of ±0.004%. The maximum permitted operating frequency error is: **a.** 4 kHz. **b.** 400 Hz. **c.** 40 Hz. **d.** 4 Hz. **e.** none of these.
2. One of the identifying features of an RF amplifier is: **a.** the lack of dc bias. **b.** direct coupling. **c.** low V_{CC} voltage. **d.** at least one tuned circuit. **e.** none of these.
3. The conduction angle for a class C amplifier is: **a.** 360°. **b.** less than 180°. **c.** between 180 and 360°. **d.** 180°. **e.** none of these.
4. A transistor amplifier stage without a provision for being supplied a dc bias voltage is probably designed to operate in class: **a.** A mode. **b.** B mode. **c.** C mode. **d.** D mode. **e.** none of these.
5. A circuit that creates a "dynamic dc bias" at the input of a transistor amplifier is called a: **a.** rectifier. **b.** clamper. **c.** clipper. **d.** peaker. **e.** none of these.
6. A pure sine-wave ac signal has a peak-to-peak value of 10 V. After passing through the input circuit of a class C RF amplifier, the peaks of the signal are at 9 V and −1 V. The input circuit of the amplifier is a: **a.** buffer amplifier. **b.** negative clamper. **c.** detector. **d.** positive clamper. **e.** none of these.
7. The signal at the input of the amplifier in Question 6 is clamped at (has a dc content of): **a.** 4 V. **b.** 5 V. **c.** 9 V. **d.** −1 V. **e.** none of these.
8. The transistor used in the amplifier of Question 6 is a type: **a.** NPN. **b.** PNP.
9. If the transistor in Question 6 conducts anytime the input voltage is more negative than −0.3 V, the conduction angle for the amplifier is: **a.** 59.32°. **b.** 61.37°. **c.** 30.68°. **d.** 90°. **e.** none of these.
10. The action that occurs in an *LC* circuit that is closed on itself after being energized briefly with a dc current is called: **a.** positive clamping. **b.** buffering. **c.** damped oscillation. **d.** frequency multiplication. **e.** none of these.
11. The period of the damped oscillation of an *LC* circuit is 25 ns. The resonant frequency of the circuit is: **a.** 400 kHz. **b.** 2.5 MHz. **c.** 4 MHz. **d.** 40 MHz. **e.** none of these.
12. In a transmitter power output stage, $V_{CC} = 13.6$ V and $I_C = 3.676$ A. The RF output power is 35 W. The stage efficiency is: **a.** 70%. **b.** 35%. **c.** 50%. **d.** 92.5%. **e.** none of these.
13. An essential requirement of an oscillator circuit is that it have: **a.** gain. **b.** low noise. **c.** high cutoff frequency. **d.** high efficiency. **e.** none of these.
14. Positive feedback in an amplifier means that the signal fed back: **a.** opposes the amplification action. **b.** adds a positive dc voltage to the input signal. **c.** aids the amplification action. **d.** reduces the gain of the stage. **e.** none of these.

15. If one-tenth of the output signal is fed back with appropriate phase to the input of an amplifier, in order to oscillate, the amplifier must have a gain of at least: **a.** 5. **b.** 10. **c.** 50. **d.** 100. **e.** none of these.

16. If an oscillator design is based on a common-emitter amplifier configuration, the phase angle between the feedback signal and the output signal must be: **a.** 0°. **b.** 45°. **c.** 90°. **d.** 180°. **e.** none of these.

17. A very small inductive reactance is part of a series circuit whose net impedance is virtually a pure capacitive reactance (the resistance of the circuit is negligible). The voltage across the inductive reactance will be out of phase with a voltage applied across the circuit by: **a.** 360°. **b.** 180°. **c.** 90°. **d.** 0°. **e.** none of these.

18. The arrangement for providing dc current in the circuit of Fig. 5.13(b) puts it in the class designated: **a.** shunt fed. **b.** series fed. **c.** negative feedback. **d.** neutralized. **e.** none of these.

19. One technique for troubleshooting an oscillator is to test its operation as simply a/an: **a.** amplifier. **b.** rectifier. **c.** voltage doubler. **d.** power source. **e.** none of these.

20. An essential requirement of an amplifier stage that is to be amplitude modulated is that its operation be: **a.** linear. **b.** without gain. **c.** nonlinear. **d.** neutralized. **e.** none of these.

21. One advantage of high-level modulation is that the stage handling the highest level of power in the transmitter can operate in class C mode and therefore with greatest: **a.** gain. **b.** efficiency. **c.** neutralization. **d.** linearity. **e.** none of these.

22. Negative carrier shift, or downward modulation, is undesirable, not only because it means that distortion is being introduced to the information signal, but also because: **a.** efficiency is reduced. **b.** detection is more difficult. **c.** radiated power is reduced. **d.** feedback is eliminated. **e.** none of these.

23. To achieve the desired goal of protecting, or isolating, critical circuits such as oscillators from the effects of load changes, etc., a buffer amplifier must: **a.** be lightly loaded and lightly driven. **b.** incorporate positive feedback. **c.** not be neutralized. **d.** be biased to cutoff. **e.** none of these.

24. Frequency multipliers are class C amplifiers in which: **a.** there is negative feedback. **b.** there is positive feedback. **c.** the input circuit is overtuned. **d.** the output circuit is resonant at the second or third harmonic of the input frequency. **e.** none of these.

25. The final (power output stage) of a particular transmitter produces 50 W of RF power and has a gain of 10 dB (dB $= 10 \log P_{out}/P_{in}$). The driver stage must produce RF power at the input of the final in the amount of: **a.** 1 W. **b.** 2 W. **c.** 5 W. **d.** 10 W. **e.** none of these.

26. The active devices of the finals of all but mini-powered transmitters require some form of: **a.** buffering. **b.** heat sinking or special cooling. **c.** negative feedback. **d.** positive clamping or peak clipping. **e.** none of these.

27. The collector supply voltage (V_{CC}) for a transmitter final is to be 28 V. The stage will operate in class C mode. It is desired to design the circuit so that the stage will produce 50 W of RF power into an antenna load of 50 Ω. The impedance-matching coupling circuit between the collector and antenna input must have what transformation ratio R_{in}/R_{load}? (Neglect collector-to-emitter saturation voltage.): **a.** 1 : 6.378. **b.** 1.852 : 1. **c.** 1 : 50. **d.** 28 : 50. **e.** none of these.

REVIEW QUESTIONS: ESSAY

1. Sketch and label the block diagram of a transmitter that has at least six function blocks in addition to the power supply.
2. Describe briefly the functions of a power amplifier and its driver.
3. Is a frequency multiplier required in all transmitters? Describe briefly the function of a frequency multiplier and its application. (When is it included in a transmitter?)
4. What is an RF amplifier? Why is it sometimes called a tuned amplifier?
5. Describe briefly the following terms applied to amplifiers: class A, class B, and class C.
6. Why is it possible for an RF amplifier to be operated satisfactorily as a class C amplifier but not an audio amplifier? When is it not appropriate to operate an RF amplifier in class C mode (for what kind of signal)?
7. Describe a clamper, indicating what components are essential to its operation. What is the basic electrical action of a clamper? Describe the difference between a negative clamper and a positive clamper.
8. Discuss the possible significance of a clamper voltage being less than normal (in magnitude) or completely missing. Assume that this information is discovered during a troubleshooting procedure on an inoperative transmitter.
9. Sketch the waveform of the damped oscillation of an *LC* circuit. Describe briefly the electrical action that produces such a waveform.
10. What is the relationship between the phenomenon of damped oscillation and the suitability of a class C amplifier for the amplification of an unmodulated RF signal?
11. In what way is the damped oscillation of an *LC* circuit similar to a rotating flywheel?
12. What is meant by the *conduction angle* of an amplifier?
13. Describe briefly the essential features and theory of operation of an oscillator circuit.
14. What characteristic of an oscillator circuit causes it to operate at one particular frequency?
15. Discuss the difference between positive and negative feedback.
16. Describe the two basic functions that the feedback network performs in an oscillator circuit.
17. Describe the difference between shunt feed and series feed, as applied to RF amplifiers and/or oscillator circuits.
18. What is a crystal-controlled oscillator? Why is it desirable (or, in most cases, required) for the master oscillator of AM transmitters to be crystal controlled?
19. An oscillator circuit has just been repaired by the replacement of a tapped coil that was known to be defective. The oscillator still does not work. What is a very likely cause for the oscillator not working, given this particular situation? (The tapped coil was part of the feedback network.)
20. Use information about power distribution in an AM signal to explain why modulation level is relevant when considering the "reaching power" of a radio station.
21. What is meant by *carrier regulation*?
22. What is the purpose of frequency multiplier stages? Describe at least two important characteristics of an amplifier circuit that enable it to be a frequency multiplier.
23. Define the efficiency of an amplifier stage. Why is efficiency of greater importance in a power amplifier than in, say, a voltage amplifier.

24. If a stage that produces 50 kW of RF power is 68% efficient, how much power is the cooling system for the stage required to dissipate?
25. Is the amplitude of the dc supply voltage for a transmitter final of any importance in determining the design of that stage? Discuss.
26. Discuss the purpose(s) of coupling circuits in the power section of a transmitter. Why are coupling circuits sometimes called impedance-matching circuits?
27. Why is harmonic suppression an important consideration in the design of the output circuits of a transmitter?

EXERCISES

1. The calibrated sweep-time control for an oscilloscope is set for 0.2 μs/div. The damped oscillation of an LC circuit is being displayed; three cycles occupy 8.5 divisions on the scope. The value of the capacitor in the circuit is 0.001 μF. Determine the value of L.
2. In a certain class A amplifier, the collector current is 80 mA, peak to peak. Signal load voltage is 12 V, peak to peak. Average collector current is 45 mA, and V_{CC} is 13.6 V. Find: **(a)** the dc power; **(b)** the rms signal power; **(c)** the efficiency of the stage. (Assume a sinusoidal signal.)
3. An amplifier is biased for class B operation (the conduction angle is exactly 180°). The collector supply voltage, V_{CC}, is 28 V. With a sinusoidal signal applied, the collector current has a peak value of 120 mA. An output of 680 mW is being produced in a load. Calculate the efficiency of the stage.
4. A class C RF amplifier is being supplied with a V_{CC} of 13.6 V. A dc current meter in the collector circuit reads 492.7 mA. A peak-to-peak signal voltage of 27.2 V is measured across a 15-Ω load (waveform is sinusoidal). Calculate the efficiency of the stage.
5. Refer to Fig. 5.7(a), the circuit diagram of a Hartley oscillator. It is desired to design an oscillator with an operating frequency of 4 MHz. A value of 300 pF will be used for C_3. **(a)** Calculate the value of the total inductance, $L_1 + L_2$. **(b)** Calculate the value of L_2 for the case when the open-loop gain of the stage is 10.
6. Refer to Fig. 5.8(a). A Colpitts oscillator is to operate at 4 MHz. A value of 5 μH has been chosen for L_3. **(a)** Calculate C_T for the resonant circuit. **(b)** If B is to have a value of 0.1, find the values of C_1 and C_2.
7. A high-level modulation scheme is used to modulate the power output stage of a 50-kW AM broadcast band transmitter. Calculate the amount of power the modulator section must supply for the following modulation levels: **(a)** 100%; **(b)** 80%; **(c)** 50%; **(d)** 25%. **(e)** What is the total power in the modulated signal for each of these modulation levels?
8. A transmitter class C final is to be supplied with 13.6 V dc. Calculate what equivalent R_L the transistor must see in order to produce an RF output of 7.5 W. If the transmitter is connected to a transmission line/antenna load of 50 Ω, what transformation ratio must the matching/coupling network provide? (Assume that the collector-emitter voltage at saturation is negligible.)
9. Design a simple L network to accomplish the matching required in Exercise 8. Sketch the circuit diagram of the network and label with the values of the L and C components. Assume that f = 27.5 MHz.

6

TRANSMISSION LINES and ANTENNAS

6.1 INTRODUCTION

In preceding chapters we have looked at the basic details of two important subsystems of an electronic communication system—the transmitter and the receiver. These two subsystems, however, are not the only essential elements of a radio or "wireless" communication system. In a true radio or wireless system, the carrier signal travels the major portion of the distance between the transmitter and receiver, not over wires, but through the earth's atmosphere, or in the free space above the atmosphere. Hence, the simplest system—one transmitter and one receiver—requires two antennas. An antenna can be thought of as any electrical element that is effective in either radiating electrical energy or intercepting electrical energy that has previously been radiated. The system, then, requires one antenna at the transmitter to "launch" the carrier into space, and another one at the receiver to intercept some of that carrier energy.

In many instances, it is not feasible to connect an antenna directly to the circuitry of the transmitter or receiver. In such instances a transmission line is needed to conduct the carrier energy to or from the antenna. In a basic sense a transmission line is simply two conductors used to transfer electrical energy between two points. However, as we shall see, at all but the lowest frequencies, a transmission line behaves like a relatively complex electrical network. It is the purpose of this chapter to present some of the fundamental concepts of transmission lines and antennas. More advanced presentations on these topics are given in Chap. 10.

Concepts relating to transmission lines will be presented before those on antennas. Several reasons could be cited for choosing that order. The decisive reason is that in some ways an antenna can be considered to be a special form of transmission line. Learning about antennas can be somewhat easier if we already know and understand some of the basic facts and principles of transmission lines.

6.2 CHARACTERISTIC IMPEDANCE OF A TRANSMISSION LINE

At first glance, it seems logical to assume that transmission lines should be a relatively simple topic. This assumption seems especially valid when a superficial comparison of transmission lines is made with many other aspects of electricity and electronics. However, learning about transmission lines requires learning a number of new, relatively abstract concepts. These concepts have become identified with a terminology that is used extensively in the everyday practical world of electronic communication, as well as in theoretical discussions. In fact, a good working knowledge of transmission lines could probably be gained simply by learning and understanding a well-chosen list of such terms.

Perhaps the most basic concept of transmission lines is that identified with the term *characteristic impedance,* or *surge impedance.* Characteristic impedance is almost universally represented by the symbol Z_0. One of many definitions of Z_0 is that it is the impedance seen looking into a line of infinite length. This idea is represented in Fig. 6.1(a). It is an idea usually difficult to accept at first. We reason that there can never be a line of infinite length and that, therefore, the definition is too imaginary to be of any practical significance. It helps the understanding only slightly if the definition is modified to a more practical ". . . very long line" from ". . . line of infinite length."

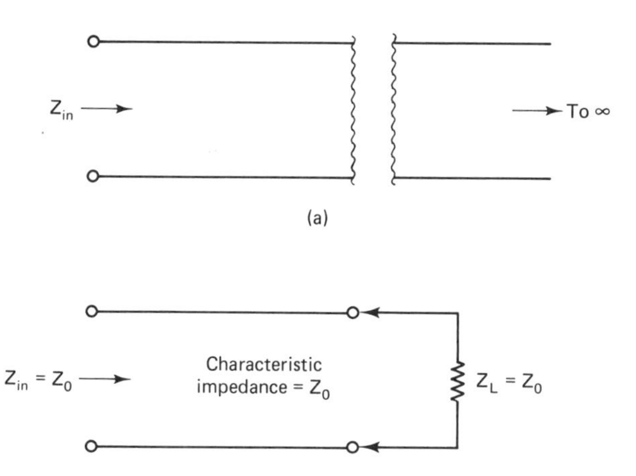

Figure 6.1 Characteristic impedance of transmission line: (a) characteristic or surge impedance, Z_0, equal input impedance, Z_{in}, of line of infinite length; (b) $Z_{in} = Z_0$ for line terminated with $Z_L = Z_0$.

Most of us are puzzled, in the beginning, by the specific values for the Z_0's of practical lines—50 Ω, 75 Ω, 300 Ω, etc. Another definition of Z_0 is that it is the input impedance of a line of any length that is terminated with a load impedance, Z_L, equal to its own characteristic impedance [see Fig. 6.1(b)]. That is a kind of circular definition that is little more comfortable than the first one. Let us look at some more elementary aspects of a transmission line in order to develop a basis for understanding Z_0.

Distributed Parameters versus Lumped Parameters

When we analyze a simple ac circuit, we use values for the R's, X_L's, and/or X_C's to determine Z_T, or I, or other circuit performance data. The concepts represented by these letters are said to be the *parameters,* or variables of the particular circuit. For a specific circuit they are fixed and can therefore be called constants. However, since specific components can be replaced with others of different values, they are, in a sense, also variable. It is in this sense that they are called parameters (of the circuit). When the circuit parameters are in their typical forms—resistors, wound coils, and/or capacitors—they are obviously *lumped parameters*. A network of lumped parameters is shown in Fig. 6.2(a).

Careful and detailed studies of sets of two conductors (transmission lines) reveal that such conductors, in reasonably close proximity to each other and of more than insignificant length, possess, in some degree, all the circuit parameters—resistance, inductance, and capacitance. These circuit values, however, are spread uniformly along the length of the conductors or transmission line. They are, appropriately, called *distributed parameters*. They are generally specified in values per unit length, for example, ohms per 1000 ft, or henries per foot, meter, or mile, or microfarads per meter, and so on. The unit length can be anything that is convenient; in a given situation, all values should be given for the same unit length.

It is virtually impossible to represent all of these distributed parameters in a single diagram. Distributed series resistance only could be represented as in Fig. 6.2(b). Distributed inductance would appear as in Fig. 6.2(c). It is more common to represent the parameters of a transmission line as lumped parameters per unit length, as in Fig. 6.2(d).

A transmission line must include the concept of shunt capacitance since a simple definition of capacitance is two conductors separated by an insulator. The representation in Fig. 6.2(d) also includes shunt resistance. This is required to explain the effect of a very small leakage current through any form of insulation separating the two conductors of a transmission line.

In effect, then, a transmission line can be thought of as a network with an infinite number of meshes [Fig. 6.2(d)]. Each mesh represents the lumped parameters of some unit length. Characteristic impedance is the input impedance to this network of infinite meshes. Let us analyze a network of similar type with explicit values of resistance only to see how the input impedance becomes, or at least approaches, a specific value.

First, to simplify analysis, we place all of the series parameters on one

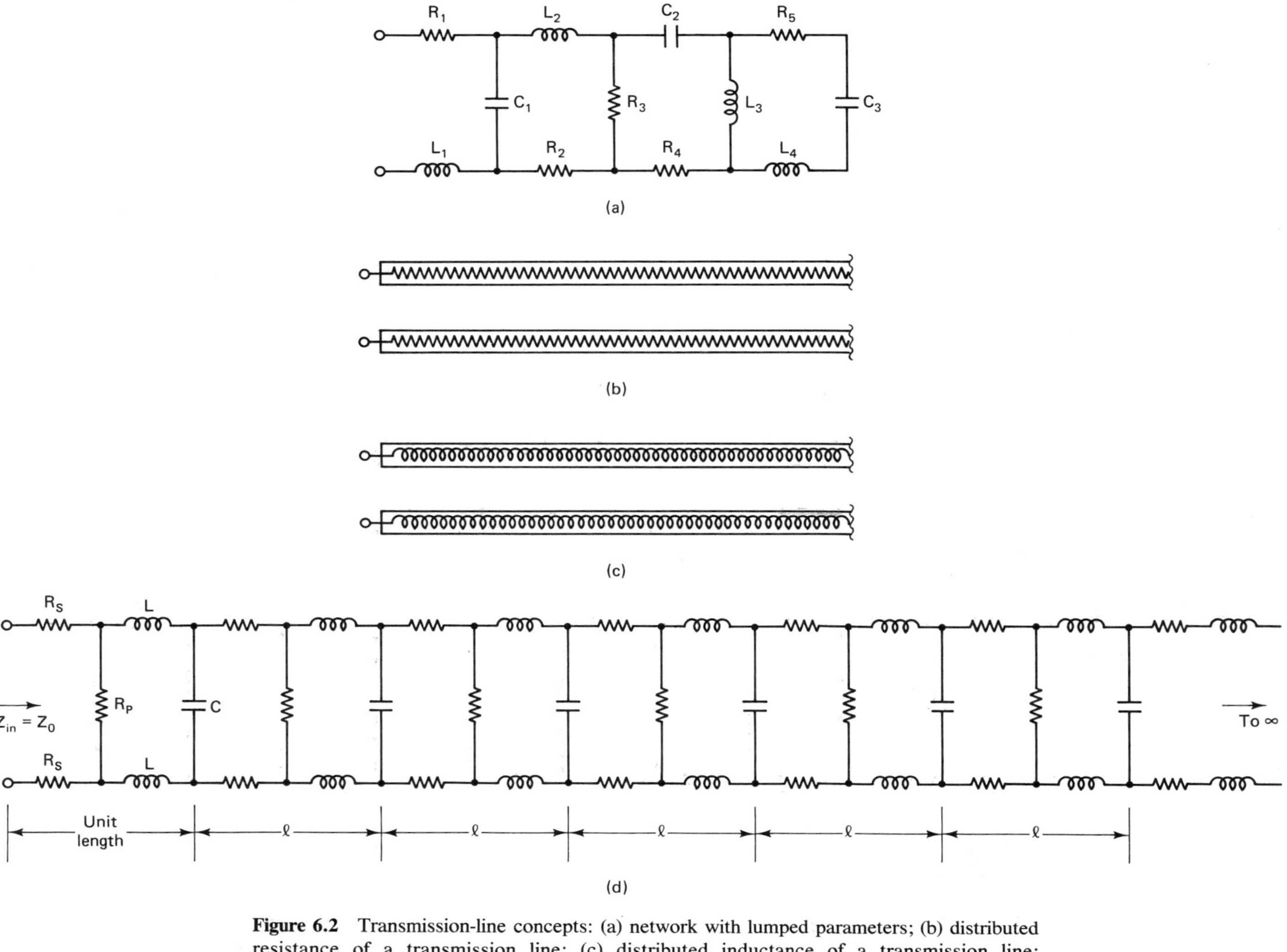

Figure 6.2 Transmission-line concepts: (a) network with lumped parameters; (b) distributed resistance of a transmission line; (c) distributed inductance of a transmission line; (d) transmission line as a network of lumped parameters: series R and L, and shunt R and C.

side of the circuit, as in Fig. 6.3(a). This does not destroy the equivalency of a network.

A step-by-step simplification of the network is demonstrated in Fig. 6.3(b). It is apparent, after only a few steps in the process, that the equivalent input impedance of even a finite number of meshes is approaching 300 Ω. Further, as can be seen in the last step shown in Fig. 6.3(b), if a load impedance of 300 Ω is connected to a single mesh the input impedance of that mesh is 300 Ω. Thus, if the input impedance of many sections is very near or exactly equal to 300 Ω, connecting those to another mesh simply makes the Z_{in} for that mesh equal 300 Ω. The more meshes included, the more nearly the input impedance becomes the specific value of 300 Ω. The 300 Ω is the characteristic impedance of the network.

Practical transmission lines for use in radio- and audio-frequency applications are manufactured in many different forms. A so-called balanced two-wire line is shown in Fig. 6.4. Transmission lines of this type are very often used in homes to connect television receivers to their antennas. The characteristic impedance of typical television "twin lead" is 300 Ω.

Another class of transmission lines is the coaxial line. Construction of lines of this type is diagrammed in Fig. 6.5. Coaxial lines are also used as television lead-ins. "Coax" is the most popular form of transmission line for connecting radio transmitters with their antennas. Typical values of Z_0 for coaxial lines are 50 Ω and 75 Ω.

In the preceding paragraphs we have looked at the characteristic impedance of a transmission line in terms of the input impedance of a line of infinite length. We have also spoken of characteristic impedance as that of the input impedance of a line of finite length terminated with a load impedance equal to Z_0. These are definitions based on the electrical performance of a line. However, characteristic impedance may be defined, measured, or calculated in other ways. For example, another very common definition is

$$Z_0 = \sqrt{\frac{L \text{ (of unit length)}}{C \text{ (of unit length)}}}$$

This definition reminds us that Z_0 is really determined by the physical characteristics of the line. It was derived by means of a very detailed electrical analysis of a line using the distributed parameters concept presented above. The formula represents a simplification based on the assumption that signal frequencies are high—in the RF range—and that the line losses are low. Line losses will be low if the conductors of the line are not so small that they have significant series resistance. The insulation must be of good quality to ensure that the shunt resistance per unit length is very high, thus minimizing losses in that medium. The values of L and C in the formula above are for a unit length of line. However, the value of Z_0 calculated is for a line of infinite length or a finite line terminated in its own Z_0.

It is also true that the Z_0 of lines for which the formula above applies is like a pure resistance. There is no net reactive component in Z_0.

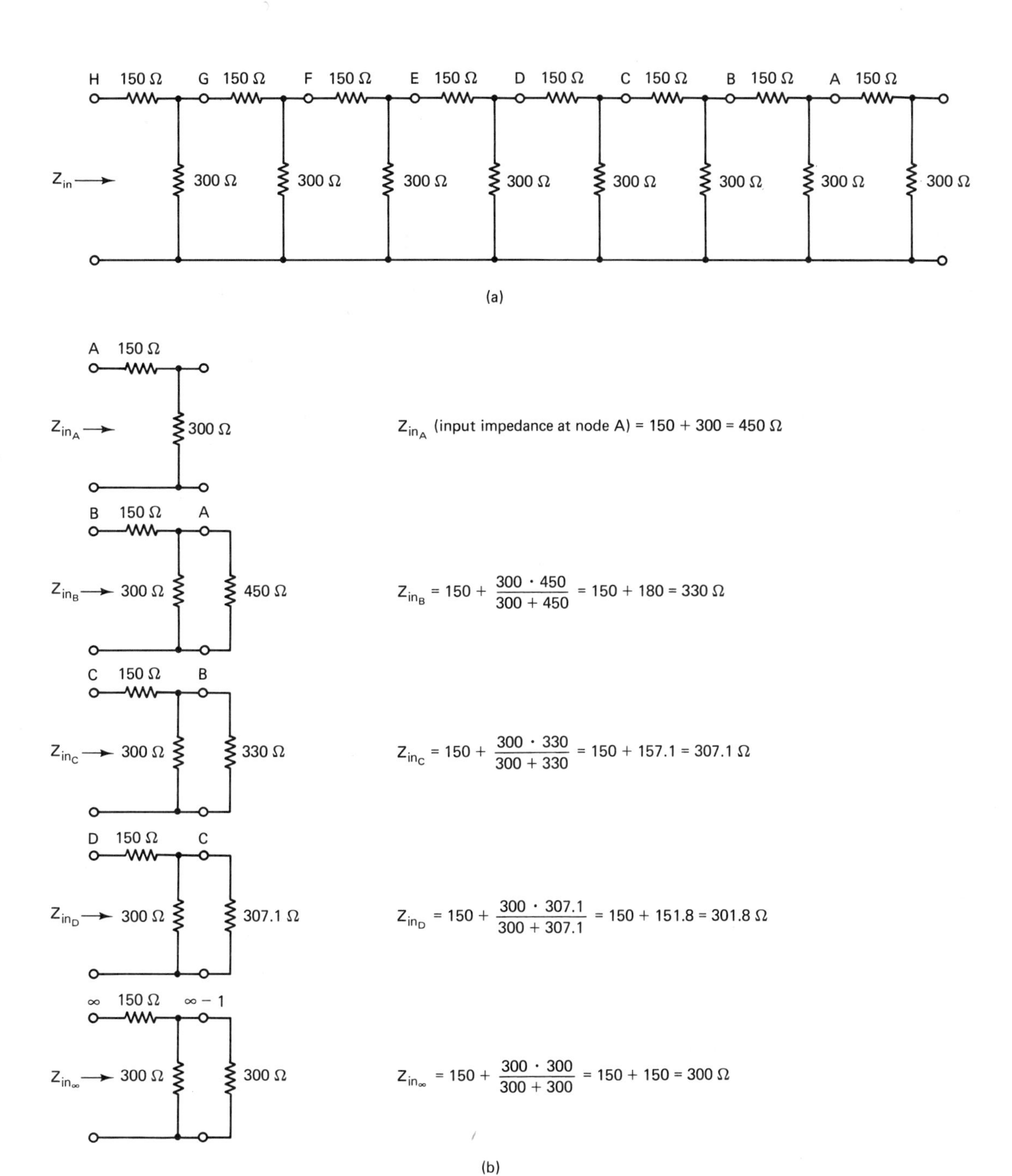

Figure 6.3 Characteristic impedance: (a) resistor network simulating transmission line; (b) reduction of network to constant, specific value.

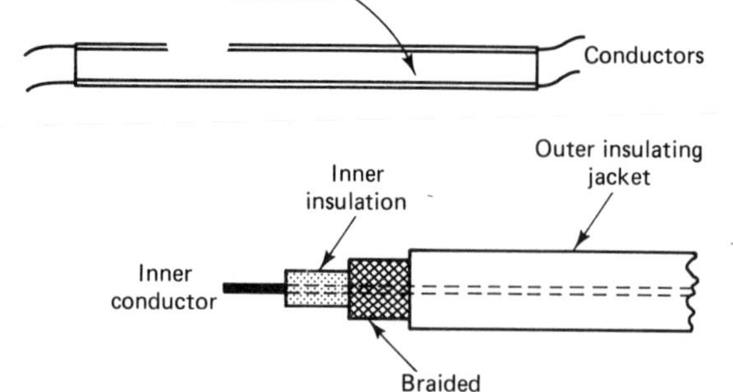

Figure 6.4 Parallel-wire ribbon line (TV twin lead).

Figure 6.5 Construction details of coaxial transmission line.

The self-inductance of a conductor is determined by its diameter. The capacitance of two conductors in close proximity is a function of (is determined by) the areas exposed to each other, the spacing between them, and the dielectric constant of the insulator between them. The net result is that for a line meeting the assumptions stated in the preceding paragraphs, L and C for a unit length are determined simply by the line's geometry. Thus it is not unexpected to find that formulas have been developed to calculate Z_0 for certain types of lines based on physical dimensions of the lines—conductor size and spacing. For example, the formula for a parallel, two-wire line with air dielectric is

$$Z_0 = 276 \log_{10} \frac{D}{r} \qquad \text{ohms}$$

where D represents the distance between the centers of the conductors and r is the radius of the conductors (see Fig. 6.6). The values for D and r may be in any units—inches, centimeters, etc.—as long as they are both in the same units. Although this formula is for a line with "air dielectric," it can be applied, with only slight error, to a line in which the conductors and the space between them are molded in a dielectric. (The familiar ribbon-type line used for television lead-in is an example of this kind of construction.)

Example 1

The diameter of the conductors in a ribbon-type transmission line is 0.0403 in. (No. 18 wire), the spacing is $\frac{11}{32}$ in. Calculate Z_0 for the line.

Solution

$$Z_0 = 276 \log \frac{D}{r} = 276 \log \frac{0.3438}{0.02015} = 276 \times 1.232 = 340\ \Omega$$

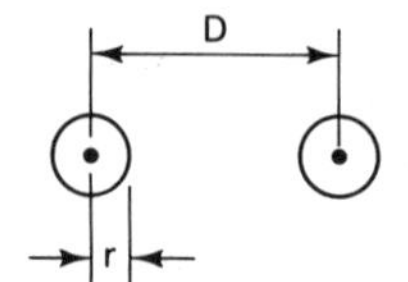

Figure 6.6 For a parallel, two-wire line, $Z_0 = 276 \log_{10} (D/r)$.

The characteristic impedance of a coaxial line is also related to its construction geometry. If a line is constructed with bead spacers instead of solid dielectric, an approximate formula for calculating Z_0 is,

$$Z_0 = 138 \log_{10} \frac{D}{d} \qquad \Omega$$

where D is the inner diameter of the outer conductor and d is the outer diameter of the inner conductor (see Fig. 6.7). The Z_0 of a line in which the space between conductors is filled with solid dielectric will be less than that given by the formula above. The formula must be divided by the square root of the dielectric constant, k, of the insulation (or, multiplied by $1/\sqrt{k}$).

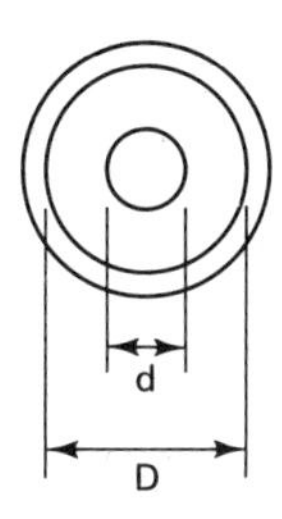

Figure 6.7 For a coaxial line, $Z_0 = 138 \log_{10} (D/d)$.

Example 2

A certain air-dielectric coaxial transmission line has the following dimensions: outside diameter of outer conductor, 0.375 in.; wall thickness of outer conductor, 0.032 in.; diameter of inner conductor, 0.081 in. Calculate Z_0 for the line.

Solution

$$D = 0.375 - 2 \times 0.032 = 0.311 \text{ in.}$$

$$Z_0 = 138 \log \frac{D}{d} = 138 \log \frac{0.311}{0.081} = 80.63 \ \Omega$$

Example 3

The measured Z_0 of the line in Example 2 is 74.15 Ω. Calculate the effective dielectric constant of the bead spacers used to hold the conductors apart.

Solution

$$k = \left(\frac{Z_{0 \text{ calculated}}}{Z_{0 \text{ measured}}}\right)^2 = \left(\frac{80.63}{74.15}\right)^2 = 1.182$$

Typical values for characteristic impedances of two-wire lines—250 to 650 Ω—are higher than those of coaxial lines—30 to 120 Ω.

It is important to know that the characteristic impedance of a line is similar to an inductance or a capacitance in its manifestation. The inductive effect of a coiled conductor is observed only when it is used in an ac circuit, or in a dc circuit while the current is changing. One does not expect to measure L directly with a dc ohmmeter. Similarly, a dc ohmmeter will not reveal the capacitance of a capacitor, although an ohmmeter can be used in a practical way to check for an open or short in a capacitor or coil. The characteristic impedance of a transmission line cannot be measured with a dc ohmmeter. It can be measured with an impedance bridge.

Summary: Characteristic Impedance

The characteristic impedance of a transmission line is an important, inherent property of the line. It must be taken into consideration in most applications of lines. Characteristic impedance is determined primarily by the geometry of construction of the line. The dielectric constant of the insulation between the conductors influences the value of Z_0 if that insulation is other than air and if it displaces most of the air surrounding the conductors. The input impedance of a transmission line is equal to its Z_0 if the line is extremely long or if it is terminated with an impedance equal to its Z_0.

6.3 WAVE PHENOMENA AND TRANSMISSION LINES

We have already seen how transmission lines are more than what they seem by virtue of that property called characteristic impedance. There are several other ways in which the approach to transmission lines must be different from that of circuits of direct current or low-frequency alternating current.

In many applications of electricity we think of the effects of an electric current as occurring simultaneously with the closing of the switch that starts the current flow. Of course, technically this is never true. The ''effect'' of electricity travels at a finite speed—the speed of light.

When circuits, such as transmission lines, are excited with radio-frequency energy the transit times of electrical phenomena become of great importance. At such high frequencies an ac voltage or current may pass through a significant portion of one or more cycles of variation between its point of origin and the point where its intended effect is produced. As a result, the performance of circuits at radio frequencies cannot be predicted adequately using only the simple theory of current as a flow of electrons. To explain and predict the performance of transmission lines, it is necessary to use many of the concepts of wave phenomena.

Traveling Voltage and Current Waves

Let us imagine that a parallel-wire transmission line is connected through a switch to a source of RF voltage, as in Fig. 6.8(a). Let's imagine, further, that the line is infinite in length. At the moment the switch is closed the effect of an electrical disturbance begins to be felt on the line. The effect is that of a radio-frequency sinusoidal voltage. This effect travels down (or along) the line at approximately the speed of light. (The speed may be somewhat less than the speed of light depending on the exact nature of the line.) At any given instant, because of the

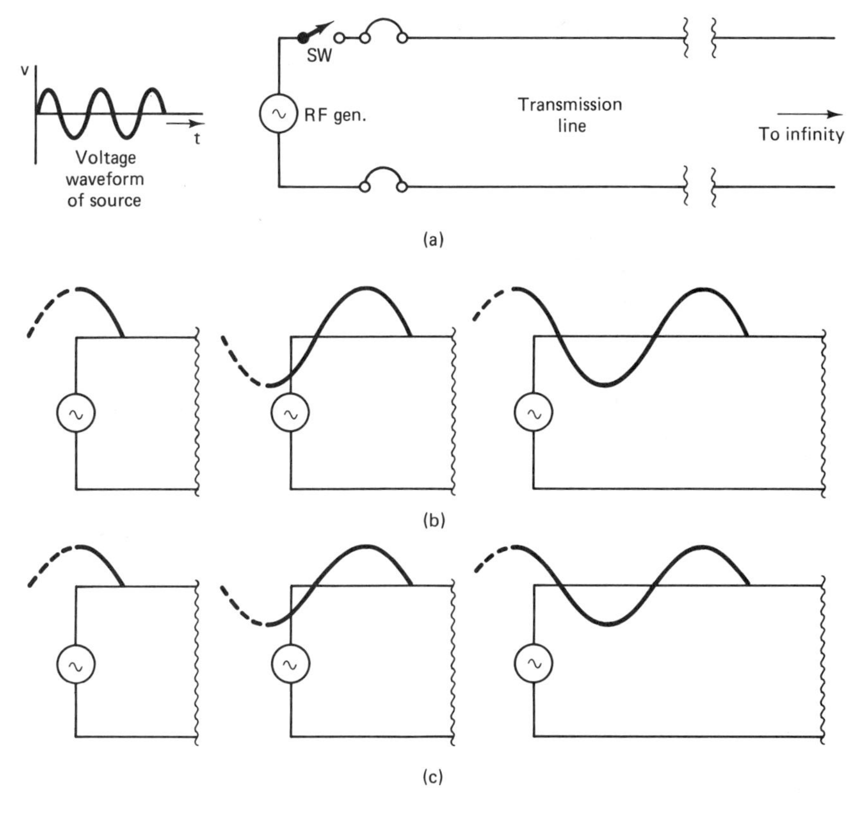

Figure 6.8 (a) Transmission line and RF source; (b) traveling voltage wave on transmission line; (c) traveling current wave on transmission line.

sinusoidal nature of the voltage, at some points along the line the voltage will be zero volts, at other points it will be maximum positive volts. At still other points the voltage will be equal to the peak negative amplitude, or it will be equal to anything in between these values. In other words, at any given instant there is a pattern of sinusoidal voltage variation along the line. And this pattern is traveling away from the source. We say that a voltage wave is traveling down the line from the source. The idea of a voltage wave is shown in Fig. 6.8(b).

Since the line is imagined to be one of infinite length, its input impedance will be equal to its Z_0. The source will supply a current to the line with a value given by $I = V_s/Z_0$. Because Z_0 is resistive in its nature, this current will be sinusoidal and in phase with the source voltage. The current effect will travel down the line just as the voltage effect did. We say that there is a current wave traveling down the line. The current wave is depicted in Fig. 6.8(c).

In summary, when a line of infinite length is connected to a source of ac voltage there is produced on the line a traveling voltage wave and a traveling current wave. These waves travel away from the source toward the opposite end of the line. The current wave is in phase with the voltage wave at every point along the line; its amplitude is determined by $I = V_s/Z_0$.

Concept of Wavelength

When we view a sinusoidal wave pattern on an oscilloscope, the pattern varies with time, or, we say it is a function of time. The horizontal length of one cycle of variation is the time for one cycle of variation to be completed. On the "scope", this distance can be converted to time by multiplying the distance by the scale factor of the time-base setting. The time so obtained is the period of the waveform.

When a sinusoidal voltage or current travels down a transmission line, the pattern of variation along the line can only be imagined or depicted as a drawing from the imagination. In this case the pattern of variation is a spatial one—its variation is a function of distance along the line. The distance over which one cycle of variation occurs is an actual length. This length is called the *wavelength* of the wave—voltage or current. Wavelength is generally represented by λ, the Greek lowercase letter lambda.

Determining the wavelength of a wave is somewhat like determining the length of the cars in a train. If we know the total length of the train and the number of cars, we divide that total length by the number of cars and get the length of one car. A voltage or current wave travels at the speed of light: approximately 186,000 mi/s, 3×10^8 m/s, or 9.843×10^8 ft/s. Hence the "train" that passes a given point in 1 s is the speed-of-light long (186,000 mi, etc.). The number of "cars" (cycles or waves) in that train is equal to the frequency, f, of the wave. The formula for wavelength, then, is

$$\lambda = \frac{c}{f}$$

where c is the speed of light in free space. The unit for λ will be the same as that used for c. (Later we will observe that the speed of light is different in a medium such as the insulation of a line. We will modify formulas involving c accordingly.)

Example 4

Determine the wavelength of a voltage wave whose frequency is 10 MHz. Find the length in both feet and meters.

Solution

$$\lambda = \frac{c}{f} = \frac{9.843(10^8)}{10(10^6)} = 98.43 \text{ ft}$$

$$= \frac{300(10^6)}{10(10^6)} = 30 \text{ m}$$

Electrical Length

When considering the various aspects of transmission lines and antennas it is often more convenient to work with what is called "electrical" length or distance rather than actual physical distance—meters, feet, or similar units. For example, a short length of line may be said to be "$\frac{1}{4}\lambda$" (one-quarter wavelength) long, at a given frequency. This means that the actual physical length of the line, in meters, feet, or whatever, is equal to a quarter of the length of a wave of the specified frequency. The electrical length of a line can be found by dividing its actual length by the wavelength of the given frequency. Both lengths must be in the same unit.

Example 5

A piece of transmission line is 7.5 m in length. What is its electrical length at 10 MHz?

Solution. The value of λ for 10 MHz was determined in Example 4 to be 30 m. Hence

$$\text{electrical length} = \frac{7.5 \text{ m}}{30 \text{ m per } \lambda}$$

$$= \tfrac{1}{4}\lambda \text{ or } \lambda/4$$

Example 6

What is the physical length of a $\frac{1}{2}\lambda$ antenna if the design frequency is 30 MHz?

Solution

$$\lambda \text{ for 30 MHz} = \frac{300}{30} = 10 \text{ m}$$

$$\tfrac{1}{2}\lambda = \tfrac{1}{2} \times 10 = 5 \text{ m}$$

Reflection of Voltage and Current Waves on an Open Line

A source has no way of knowing whether a line is infinite or finite when it begins to supply current and voltage waves to the line. If the line is terminated (connected to a load) in a resistance whose value is equal to Z_0, the voltage and current waves will "enter" that resistance and be dissipated. The energy that the waves represent will be taken off the line by the terminating device (the resistance); none of the energy will be returned to the line.

On the other hand, if the line is simply an open line of finite length, something must happen to the waves when they reach the end of the line. Since there is nothing connected to the line to absorb them, they will be reflected back from the end of the line and will travel along the line toward the source. On the line there will now be voltage and current waves coming from the source, and voltage and current waves traveling back from the end of the line. The waves from the source are called *incident waves,* those reflected from the end are called *reflected waves*. As with any ac voltage or current, the two sets of waves will combine phasorally at each point along the line—the incident voltage wave with the reflected voltage wave, incident current wave with reflected current wave. As a result of the waves combining, there will be established on the line, patterns of voltage and current variations. It happens that these patterns do not move or travel. These are, therefore, *standing waves* and are known by that name.

The standing waves of voltage and current of an open line are depicted in Fig. 6.9. The points in a standing-wave pattern where a voltage or current is a maximum are called *loops,* the points where values are minimums are called *nodes*.

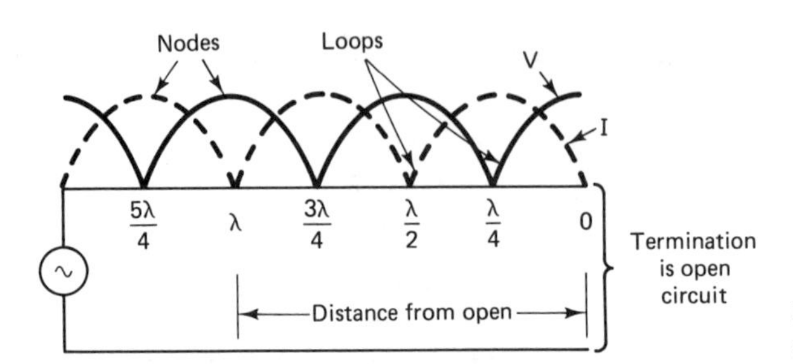

Figure 6.9 Standing waves on an open line.

You will observe that on an open line, the voltage standing wave has a loop at the open end, and the current standing wave has a node at that end. This is as we would expect from the theory of ordinary circuits—the voltage is maximum across an open, the current is zero at the open.

It is of importance to understand how this can be explained using the concept of reflected waves. For the voltage wave on an open line, the explanation is that the wave is reflected *as it would have continued.* The result is that at the end of the line, the incident and reflected voltage waves are *always in phase*. The voltage standing wave, at the end, varies between twice the positive peak of the incident wave and twice the negative peak. On the other hand, at exactly $\frac{1}{4}\lambda$ back from the end, the incident and reflected waves are *always exactly 180° out of phase*. Thus if the reflected wave is not attenuated due to line loses, the two waves will cancel at $\frac{1}{4}\lambda$ from the open end. The way in which incident and reflected voltage waves combine, for various phases of the incident wave, is shown in Fig. 6.10. Spend some time studying the way in which the wave is reflected at the end. Note the phase relation of the reflected wave with the incident wave along the line. Verify that the standing wave (the heaviest line) is the result of adding the instantaneous values of the incident and reflected waves.

The current wave at the end of an open line must be reflected in a completely different fashion than the voltage wave. The incident and reflected waves must always cancel at the open end. The current standing wave has a node at the open end. This effect will be achieved if the current wave is reflected 180° out of phase with the wave that would have continued (if the line had not ended). Thus, at the end of the line, the reflected current wave will always be equal and

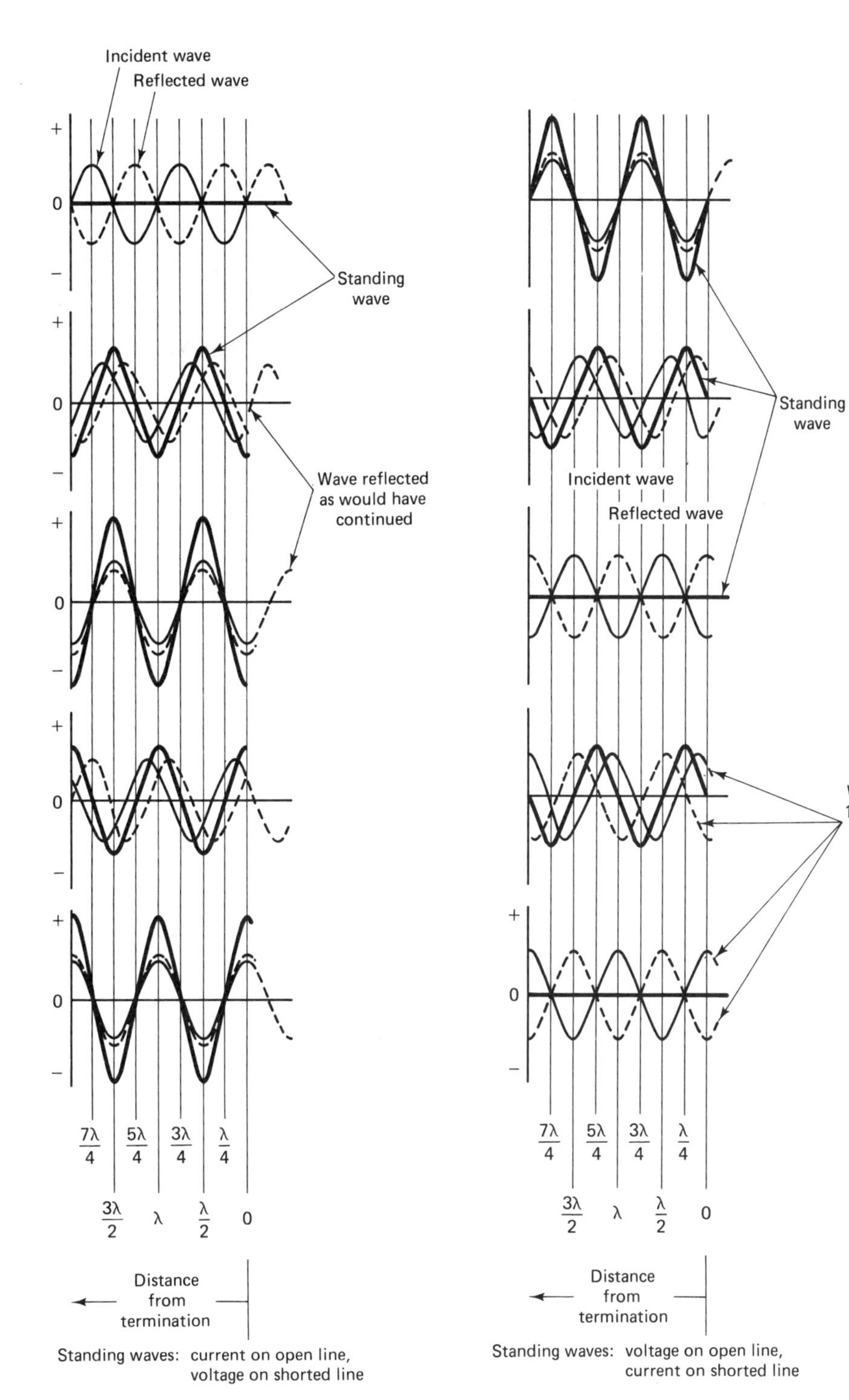

Figure 6.10 Standing waves: voltage on open line; current on shorted line.

Figure 6.11 Standing waves: current on open line, voltage on shorted line.

opposite to the incident current wave. The instantaneous sum of the two waves will always be zero at that end, no matter what the phase of the incident wave. How current waves are reflected and combined with incident waves, for various phases of incident current waves, is shown in Fig. 6.11. Again, study these diagrams in detail until you are familiar with the manner in which the current waves reflect and combine to form the current standing-wave pattern.

Now that we have a familiarity with the way in which voltage and current standing waves are formed, let's look again at Fig. 6.9 and study other aspects of the pattern of these waves. First, note that loops and nodes are a quarter-wavelength apart (two adjacent nodes, or loops, are a half-wavelength apart). For example, at a point $\lambda/4$ from the open end of the line there is a voltage node and a current loop.

The input impedance of a transmission line is determined by the familiar formula for impedance: $Z = V/I$. In the case of a transmission line with standing waves, V and I in this formula take on the values of the standing waves at the input of the line. Since on an open line there is a voltage node and a current loop $\frac{1}{4}\lambda$ from the open end, consider the effect on the input impedance of the line if in fact the line is only $\frac{1}{4}\lambda$ long: V will be 0 V approximately, and I will be at its maximum value. The input impedance of an open transmission line which is exactly $\frac{1}{4}\lambda$ long is virtually 0 Ω!

At a distance of $\frac{1}{2}\lambda$ from the end of an open line the condition of the standing waves is the reverse of that at a $\frac{1}{4}\lambda$ from the end: There is a voltage loop and a current node. We conclude that the input impedance of an open line $\frac{1}{2}\lambda$ long is infinity (for a lossless line).

Reflection of Voltage and Current Waves on a Shorted Line

When a transmission line is shorted at its load end, the situation involving the voltage and current at that point is just the reverse of that for the open line: the voltage must be zero, the current a maximum value. Again, it must be possible to show how these values attain using the concept of waves.

In the case of a shorted line, the current wave will be reflected as it would have continued (if line had continued). You will recall that this was true of the voltage wave on an open line. On the shorted line, the voltage wave will be reversed (or shifted in phase by 180°) before it is reflected. Study the waveforms of Fig. 6.12 in detail and verify that reflections of the types just described will produce the standing-wave patterns for voltage and current depicted in Fig. 6.12.

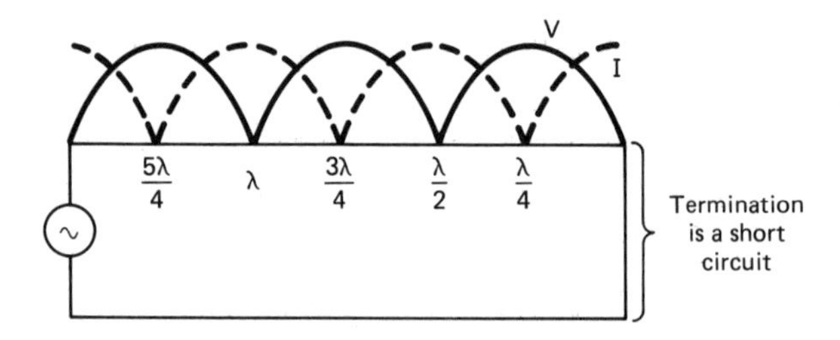

Figure 6.12 Standing waves on shorted line.

Note carefully that at $\frac{1}{4}\lambda$ from the shorted end there is a voltage loop and a current node. As a result, the input impedance of a shorted line whose length is $\frac{1}{4}\lambda$ will be infinity! In fact, the input impedance of a shorted transmission line of any odd multiple of a quarter-wavelength will be infinite if the line is lossless. If the line is not lossless, the input impedance will be very high but not infinite.

We have been looking at the effects of opens or shorts as the terminations of transmission lines. We have noted how these terminations produce standing waves and have studied briefly the nature of such waves. We have seen how standing waves can be used to predict the input impedances of open or shorted lines whose lengths are integral multiples of a quarter-wavelength. In some instances, the information on input impedance is hardly believable: for example, the input impedance of a $\frac{1}{4}\lambda$ shorted line is infinite!

Earlier, we noted that if a line is infinite in length, or if it is terminated in its own characteristic impedance, its input impedance is equal to its characteristic impedance. In the latter two cases there are no standing waves because the energy the current and voltage waves represent either continues traveling down the line or is completely absorbed by the load (or terminating) impedance.

There are two other general classes of conditions relating to transmission lines that are important. The first of these classes relates to terminations. If a line is terminated in its own Z_0, there will be no standing waves and the input impedance of that line will be Z_0 regardless of its length. Any other termination for a line produces what is called a *discontinuity* for the voltage and current waves. A discontinuity always causes standing waves. The input impedance of a line with standing waves is virtually always something other than Z_0. Furthermore, the input impedance of a line whose termination produces a discontinuity varies with the length of the line. For a given termination, input impedance may be resistive, inductively reactive, or capacitively reactive, depending on the length of the line. The variation of input impedance with length represents the second class of conditions mentioned above.

As might be guessed from the preceding paragraph, predicting the input impedance of a transmission line with a discontinuity is not a simple matter. Fortunately, a special tool for analyzing transmission lines has been devised. It is called the *Smith chart*. Although the Smith chart is an excellent aid for transmission-line calculations, becoming proficient in its use requires an extended effort. A detailed presentation of the use of this device is given in Chapter 10.

Standing-Wave Ratio

As has been stated above, when voltage and current waves are reflected on a line due to a discontinuity, standing waves are produced. Standing waves are the result of the summing of instantaneous values of incident and reflected waves at every point along a line. The summing process produces a pattern of variation (the standing wave) along the line. It is of the nature of this pattern that there are points of maximum and minimum values. A quantitative indication of the nature of a particular standing wave is given by the *standing-wave ratio* (SWR). The standing-wave ratio is defined as the ratio of the maximum value of a wave to its minimum value. For a voltage standing wave,

$$\text{VSWR} = \frac{V_{\max}}{V_{\min}}$$

and for a current wave,

$$\text{ISWR} = \frac{I_{\max}}{I_{\min}}$$

When there is no standing wave, the parameter (voltage or current) is varying to the same degree all along the line. Variations continue to move down the line as traveling waves. There are no points along the line where the wave is always zero or a minimum value. Or, to say it another way, the maximum and minimum values are the same. Hence the SWR for a line that has no reflections is 1.

It is helpful to recall that a maximum value is the value of a loop of a standing wave, a minimum value of a standing wave is the value of a node. On very short lines and/or lossless lines with standing waves, the value of a node is virtually zero. The SWR for such a line approaches infinity (the calculation for SWR involves division by zero, or a very small value).

Values of SWR greater than 1 indicate that there is a discontinuity on the line. In most cases, the discontinuity is a mismatch between the load impedance, Z_L, and Z_0. A discontinuity may also be caused by a physical distortion of the line geometry (e.g., the spacing of a parallel-wire line is changed because of physical damage to the line).

The SWR is an excellent guide to the degree of mismatch of a line termination—the larger the value of the SWR, the greater the discrepancy between Z_L and Z_0.

Effect of Length of Line on Input Impedance

When a transmission line has standing waves (i.e., is open, shorted, or terminated with a Z_L not equal to Z_0) its input impedance will vary with its length. In general, when the length of the line is an even multiple of a quarter-wavelength ($\frac{2}{4}\lambda$, $\frac{4}{4}\lambda$, $\frac{6}{4}\lambda$, etc.) the input impedance is equal to the terminating impedance: a short looks like a short, an open looks like an open. This is true because the phase relationship between voltage and current waves at even multiples of $\frac{1}{4}\lambda$ are the same as at the termination end of the line. On the other hand, there is a phase change of 90° in voltage and current standing waves between the end of a line and a point an odd multiple of a quarter wavelength ($\frac{1}{4}\lambda$, $\frac{3}{4}\lambda$, $\frac{5}{4}\lambda$, etc.) away from the end. On open and shorted lines, this means that maximums become minimums, and vice versa—an open looks like a short, a short looks like an open. When a mismatched load is reactive, it appears as the opposite form of reactance at odd multiples of $\frac{1}{4}\lambda$ away from the end—an inductance looks like a capacitance, a capacitance appears as an inductance.

Let us examine standing-wave patterns, again and in detail, to determine the changes that occur in the phase relationships of voltage and current standing waves at increasing distances away from the terminated end toward the input end of a transmission line. To make the procedure as straightforward as possible, we will examine the standing-wave patterns of open and shorted lines. The patterns are reproduced here for convenience.

A detailed inspection of the SW patterns on an open line, Fig. 6.13(a), reveals that between the $\frac{1}{4}\lambda$ point and the end of the line, the current wave decreases from its maximum to its minimum. At the same time, the voltage wave is increasing from its minimum to its maximum. Examine the drawing now, and confirm for yourself this characteristic of the two wave patterns.

This relationship is the same as that which exists between the current and voltage in a purely capacitive circuit where current values lead voltage values by 90°. Thus it is not surprising that actual tests of open transmission-line stubs of less than a $\frac{1}{4}\lambda$ show that the input impedance of these lines is purely capacitive (for lossless lines). Further analysis of the standing-wave patterns of Fig. 6.13(a) reveals that the same situation exists as any odd multiple of $\frac{1}{4}\lambda$ is approached from the terminated end of the line.

We now state a generalization: *The input impedance of an open, lossless transmission line stub whose length is greater than an even multiple of* $\frac{1}{4}\lambda$, *but is less than the next odd multiple of* $\frac{1}{4}\lambda$, *is capacitive in character.*

Let's now examine the voltage and current standing-wave patterns for an open line for distances from the terminated end which are greater than $\lambda/4$ but less than $\lambda/2$. That is, examine Fig. 6.13(a) again carefully, especially between $\lambda/4$ and $\lambda/2$. Note that at $\lambda/2$ the voltage wave is a maximum, the current wave a minimum. As we move from the $\lambda/2$ point toward the $\lambda/4$ point, the voltage values are decreasing, the current values are increasing. Between $\lambda/2$ and $\lambda/4$ the voltage values "lead" the current values. It is correct to conclude that the input impedance of transmission line stubs of this length—less than $\lambda/2$ but greater than $\lambda/4$—will be inductive in nature. Input impedance measurements of actual transmission lines confirm this to be true. Again, it is true that this generalization applies to longer stubs of appropriate length as well.

The input impedance of an open, lossless transmission line stub whose length is greater than an odd multiple of $\frac{1}{4}\lambda$ *but less than the next even multiple of* $\frac{1}{4}\lambda$ *is inductive in character.*

It is now appropriate to state another important generalization about open

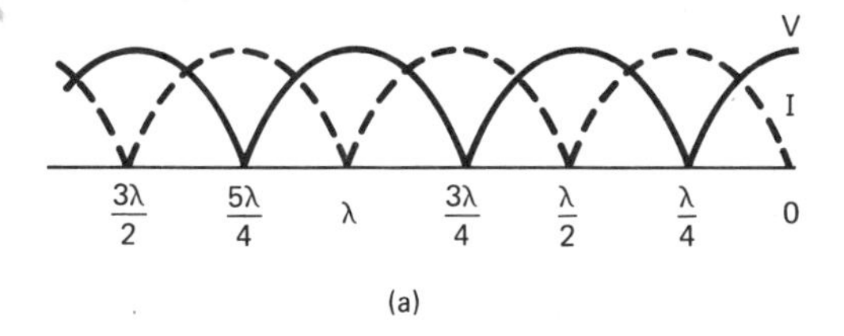

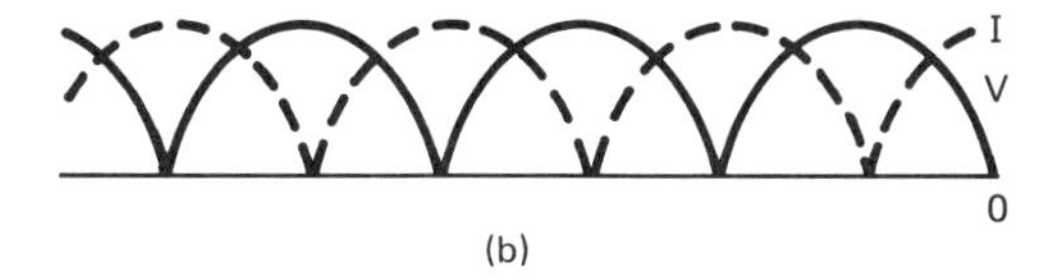

Figure 6.13 (a) Standing-wave patterns on open line; (b) standing-wave patterns on shorted line.

transmission line stubs: *The input impedance of an open transmission line resembles that of tuned (or resonant) circuits, parallel and series, at resonance and off resonance.*

At $\lambda/4$ on an open line the voltage and current standing waves are such that the input impedance of a $\lambda/4$ stub is very low, similar to that of a series resonant circuit. The input impedance of a $\lambda/2$ stub is similar to that of a parallel resonant circuit—very high. The input impedance of a stub less than $\frac{1}{4}\lambda$ in length is capacitive, and, therefore, similar to a series resonant circuit below resonance. The input impedance of stubs between $\frac{1}{4}\lambda$ and $\frac{1}{2}\lambda$ in length is inductive, similar to the input impedance of a series resonant circuit above resonance, or a parallel resonant circuit below resonance.

Now, examine again the standing-wave patterns for voltage and current on a shorted line. These are illustrated in Fig. 6.13(b). Note that compared to the open line, the positions of voltage and current standing-wave patterns are simply reversed. This means that at the end of the line and at even multiples of $\lambda/2$ the voltage value is very small, the current value relatively large—the equivalent of low resistance. Hence the input impedance of shorted transmission-line stubs, whose lengths are even multiples of $\lambda/4$, is a very low resistance (or a short). This is similar to the input impedance of a series resonant circuit at resonance.

By similar reasoning we conclude that the input impedance of shorted line stubs, whose lengths are odd multiples of $\lambda/4$, is a very high resistance, a condition similar to that of a parallel resonant circuit at resonance. Further, even more detailed study of the standing waves of Fig. 6.13(b) reveals that the pattern of changes of the voltage and current waves between the $\lambda/4$ point and the end of the line is like that of a circuit with lagging current—an inductive circuit. Stubs whose lengths are greater than an even multiple of $\lambda/4$ but less than the next longer odd $\lambda/4$ have input impedances of this same nature. They are inductive, like series resonant circuits above their resonant frequencies or parallel resonant circuits below resonance.

By similar reasoning we conclude that shorted stubs whose lengths are greater than an odd multiple of a $\lambda/4$ but less than the next even multiple are capacitive in nature, like series resonant circuits below resonance and/or parallel resonant circuits above resonance.

These are important basic facts about the input impedances of open and shorted transmission-line stubs. It is helpful to summarize these facts with a graphical representation so as to make them more easily perceived and remembered. The diagrams of Fig. 6.14 are a popular way of doing just that.

Because the input impedances of open and shorted transmission-line stubs resemble the input impedances of resonant circuits, stubs are very commonly used in place of more conventional circuit elements at very high radio frequencies. It is practicable to use stubs as circuit elements above, say, a few hundred megahertz, because quarter-wavelengths at such frequencies are beginning to be short enough to be manageable in small spaces ($\lambda/4$ at 500 MHz $= 0.25 \times 300/500$ m $= 0.15$ m, or approximately 6 in.). A presentation on several typical applications of transmission-line stubs as circuit elements appears as a subsequent section in this chapter.

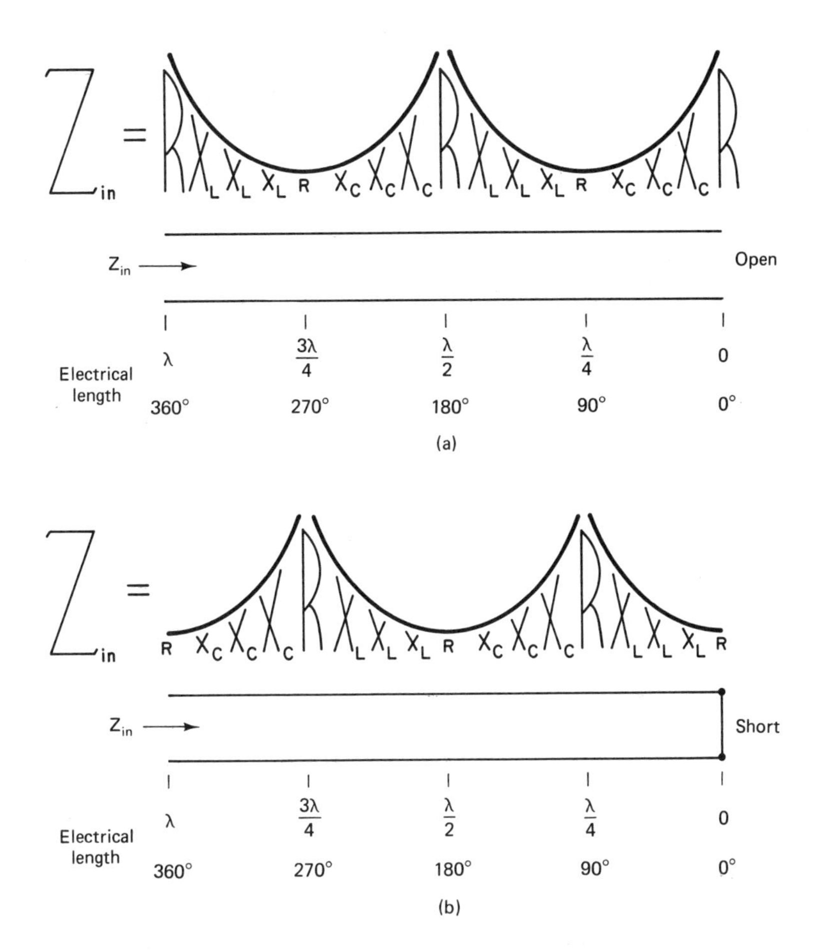

Figure 6.14 (a) Z_{in} for open line versus length; (b) Z_{in} for shorted line versus length.

Effect of Load Impedance, Z_L, on Input Impedance, Z_{in}

Let us recall that when $Z_L = Z_0$, the input impedance, Z_{in}, of a transmission line is also equal to Z_0. Further, when a transmission line is open or shorted, the Z_{in} of the line changes with the length of the line. What is Z_{in} if Z_L is neither a short, an open, nor equal to Z_0? The answer is that Z_{in} depends on both Z_L and the length of the line.

There will be reflections and standing waves on any line that is neither infinite in length (infinite lines are seldom realized) nor terminated in a load impedance equal to Z_0. The magnitude of the input impedance of a line with standing waves is determined by the amplitudes of the voltage and current waves at the input to the line. The phase angle between the voltage and current standing waves, at the input to the line, determines the nature of the input impedance: that is, whether it will be resistive, capacitive, or inductive.

As we have seen with open and shorted lines, the phase angle between voltage and current at the input to a line with standing waves depends very much on the precise electrical length of the line—the length specified in wavelengths (λ's). The amplitudes of the voltage and current standing waves at the input of a given line, for given values of voltage and current input, depend on the proportions of the incident or transmitted waves that are reflected. These amplitudes are affected also by the losses along the line—the greater the losses, the smaller the reflected waves. The proportion of energy reflected, and the way in which it is reflected is determined by the load impedance. Let's examine several specific cases.

Z_L resistive and greater than Z_0. When the termination of a transmission line is resistive and has an ohmic value significantly greater than Z_0, voltage and current waves will be reflected in a fashion very similar to that for an open line. The termination, when resistive, will absorb some of the incident energy (but not all of it as a matched load would). Hence the reflected waves will not have relative amplitudes as great as they would have been if the line were open. Nevertheless, as with an open line, the voltage wave will be reflected so as to be in phase with the incident voltage, producing a voltage loop at the end of the line. The current will be reflected out of phase and will produce a current wave node at the end.

The input impedances of lines terminated as described will be equal to Z_L when line lengths are multiples of $\lambda/2$ (i.e., even multiples of $\lambda/4$). For lines whose lengths are odd multiples of $\lambda/4$, input impedances will be the equivalent of small resistances, but will not be zero (as with open lines).

When line lengths are neither even nor odd multiples of $\lambda/4$, the input impedance will be reactive in nature. The exact nature of the input impedance can be determined only through a laborious analysis procedure or by the use of a Smith chart (see Chap. 10).

The standing wave ratios, voltage or current, for lines terminated as described will vary with the ratio of mismatch between Z_L and Z_0. Measuring the SWR on a line is a very popular technique for determining the degree of mismatch of a line and its load.

Z_L resistive and less than Z_0. When Z_L is resistive and less than Z_0 the conditions on the line will be similar to those for shorted lines. When lines so terminated are even multiples of $\lambda/4$ in length, their input impedances will be resistive and less than Z_0. When lengths are exactly an odd multiple of $\lambda/4$, input impedances will be resistive and much greater in magnitude than Z_0. Again, when line lengths are not precise multiples of $\lambda/4$, the input impedance will be reactive in nature and will be dependent on the exact length of the line. Precise values for Z_{in} can be obtained with the aid of a Smith chart.

Z_L reactive. When Z_L is reactive the task of determining Z_{in} becomes laborious indeed. However, as with the other examples cited above, Z_{in} for lines that are multiples of $\lambda/2$ is equal to Z_L. A presentation on determining the input impedance of lines with reactive loads and of other lengths is included with the discussion of the Smith chart.

Resonant and Nonresonant Lines

When a transmission line is mismatched with its load impedance, or is open or shorted and of finite length, standing waves will be created on the line if it is excited or energized by a source. When there are standing waves, a line appears very much like a resonant circuit to the source. The line may appear like a circuit at resonance, either series or parallel, depending on its exact electrical length. Or it may appear to be a circuit off resonance, and therefore reactive, depending on the length. Of course, the electrical length of the line changes with the frequency of the source. Hence its impedance changes with frequency, as does the impedance of a resonant circuit. For these reasons a line with standing waves is commonly referred to as a *resonant line*.

If a line is infinitely long or is terminated with a load impedance equal to its own characteristic impedance, no standing waves will be produced when it is excited. The input impedance will be equal to Z_0 and therefore equal to Z_L also. It is as if the line were not there, it produces no consequences as long as the matched conditions exist. The frequency of the source can be changed with no change in the input impedance. In this case the line is referred to as a *nonresonant line*. It is also called a *flat line*. When the purpose of the line is simply to transmit energy from a source to a load, the flat, or nonresonant, line is the preferred condition, by far. It transmits the maximum possible energy.

6.4 SOME APPLICATIONS OF TRANSMISSION LINES

The most important and obvious application for transmission lines is the one of conducting electrical energy from one place to another—from a source to a load. In electronic communication systems, transmission lines usually connect transmitters and antennas, or antennas and receivers. Transmission lines are also used extensively in long-distance telephone or data transmission systems to conduct signals over great geographic distances. In all these instances it is desirable to achieve maximum effectiveness in the transfer of energy. Consequently, considerable effort is expended to ensure that lines and their loads are matched, producing flat or nonresonant lines (see the preceding section) with virtually no standing waves.

However, there is another, large class of applications of lines in which the resonant properties of lines are utilized. In these applications, lines are used entirely for their resonant circuit-like properties and not at all to transmit energy. In this section we examine some of these special applications of transmission lines. When used for such special purposes, lines are typically short in electrical length, and hence are called *transmission-line stubs* or simply *stubs*.

The Metallic Insulator

One of the most startling and, seemingly unlikely, applications of a transmission-line stub is as an insulator (see Fig. 6.15). Recall that the input impedance of a shorted $\lambda/4$ stub (see Sec. 6.3 under the heading "Reflection of Voltage and

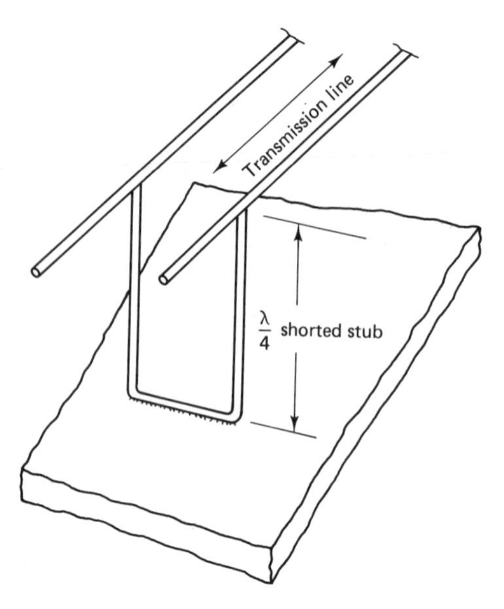

Figure 6.15 Metallic insulator.

Current Waves on a Shorted Line'') is infinite when the line is lossless. A short length of line is virtually lossless if its conductors are of average size, say No. 14 wire (diameter = 0.064 in.). In an actual application, a parallel-wire line used for transmitting energy could be supported with strong, U-shaped steel supports if the dimensions of those supports made them equivalent to a shorted $\lambda/4$ stub at the operating frequency of the main line.

Transmission-Line Stubs as Filters

A second application of transmission-line stubs is as filters: low pass, high pass, etc. You will recall from your course in basic ac theory that a filter is a circuit which is designed to be frequency selective. A filter allows certain frequencies, of a signal which contains many frequencies, to pass virtually with no attenuation. It severely attenuates other frequencies. A filter accomplishes these results by virtue of combinations of inductances and capacitances whose effects in a circuit are frequency dependent. At audio and lower radio frequencies the L's and C's of filters are actual coils and capacitors. At higher radio frequencies it is feasible to use resonant transmission-line stubs, which, as we have seen above, behave like resonant circuits. Let us examine the use of a stub as a low-pass filter at the output of a transmitter.

The output of any radio transmitter is distorted to some degree. Since a distorted signal inherently incorporates spurious (undesired) signals which are harmonics of the fundamental signal(s), the output of any radio transmitter will include at least some harmonic content. The second harmonic of a station's carrier frequency is the strongest of the undesired signals. If radiated, it is likely to interfere significantly with the carrier signal of a station operating at or near double the frequency of the station producing the interference. It is a universal practice to use some means to suppress harmonics and other spurious signals at the output of a transmitter. This frequently takes the form of a simple low-pass filter—a series inductance and a shunt capacitance, for example.

However, for second harmonic suppression especially, it is feasible to use transmission-line stubs at frequencies of a few hundred megahertz and above. For example, if an open $\lambda/4$ stub is inserted in series in one side of the transmission line connecting the output of a transmitter to its antenna [see Fig. 6.16(a)], it appears as a short to the carrier and produces no attenuation of it. However, for the second harmonic of the carrier this same length of line appears as an open $\lambda/2$ stub in series with the transmission path. The input impedance of a $\lambda/2$ stub is infinite—the second harmonic is greatly attenuated.

An alternative approach is to connect a shorted $\lambda/4$ stub in parallel with the antenna, as in Fig. 6.16(b). As a shorted $\lambda/4$ stub this transmission-line section has infinite impedance. At double the frequency (the second harmonic) the same section is a shorted $\lambda/2$ stub having zero input impedance. The stub's input impedance is in parallel with the antenna, is infinite at the carrier frequency, and is a short at the second harmonic frequency. It is in effect a low-pass filter that passes the carrier unattenuated but suppresses the second harmonic with significant attenuation.

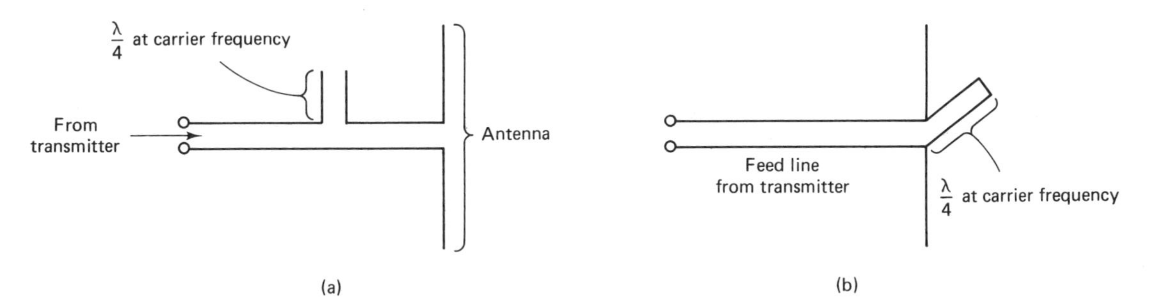

Figure 6.16 (a) $\lambda/4$ open stub as low-pass filter; (b) $\lambda/4$ shorted stub as low-pass filter.

Parallel-wire transmission lines (Fig. 6.16) have been used to illustrate this discussion. It is important to be aware that the principles involved are applicable to all other types of transmission lines—coaxial lines and waveguides.

Transmission-Line Stubs as Oscillator Tank Circuits

One of the most common features of noncrystal RF oscillator designs is the presence of a parallel-resonant *LC* circuit—the so-called tank circuit. The number of workable oscillator circuit configurations is almost without limit. However, almost invariably, the *LC* tank circuit is used in a way that its resonant frequency determines, and is approximately equal to, the operating frequency of the oscillator.

You will recall that the resonant frequency of a tuned circuit (and therefore the operating frequency of an associated oscillator) is inversely proportional to the square root of the *LC* product. Hence, as the design frequencies of oscillators increase, the values of L and C must decrease. In fact, for oscillators in the UHF range, L is so small that it becomes a single-turn coil, and C is so small that it is less than the normal stray capacitance in a circuit. (The designation "UHF" means *ultrahigh frequency*. The UHF range includes the frequencies between 300 and 3000 MHz, by agreement of the nations of the world.) Required values for C as a discrete component in the tank circuit of a UHF oscillator are impracticable! The use of a resonant transmission-line stub as the frequency-determining tank circuit in a UHF oscillator is a necessity, not just an interesting or novel application.

The use of transmission-line stubs as circuit elements is illustrated in the UHF oscillator circuit of Fig. 6.17. You will observe that stubs are being used both as frequency-determining resonant circuits and as RF chokes. (The chokes permit the passage of dc current but block the RF current.) A stub is also shown as an inductive trimmer element for fine tuning the operating frequency of the circuit. In this example, the transmission lines are of the so-called "stripline" construction: they are fabricated as foil patterns on the two sides of a printed circuit board.

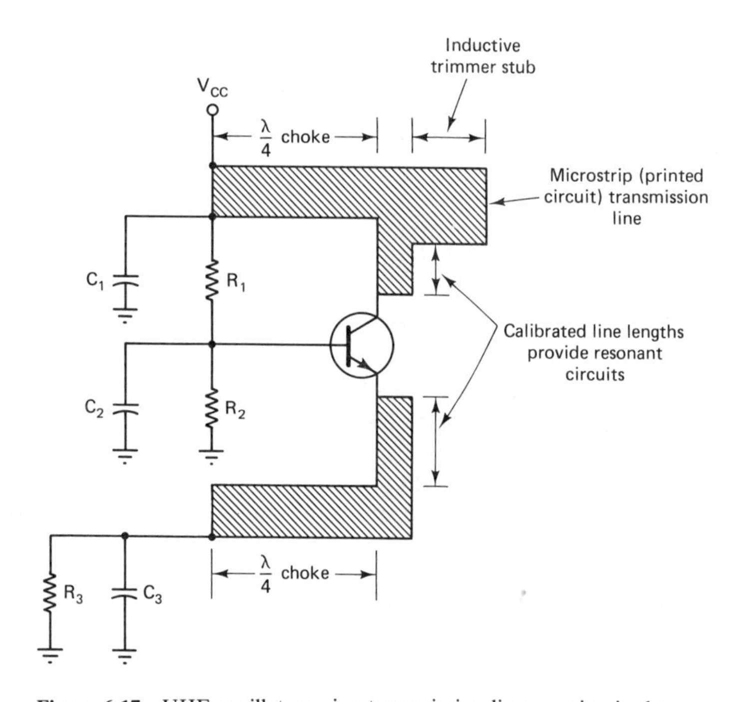

Figure 6.17 UHF oscillator using transmission lines as circuit elements.

The UHF tuners of home television receivers contain UHF amplifier, oscillator, and mixer sections. [TV receivers utilize a heterodyne design similar to that of simple AM radio receivers (see Chaps. 3 and 11).] Transmission-line stubs are universally used in these sections. They replace the discrete components of tuned circuits of lower-frequency systems.

6.5 ANTENNAS: BASIC PRINCIPLES

Let us imagine that we have all the facilities for exciting a transmission line with RF energy and for carefully measuring transmitted and reflected power on the line. The line, a parallel-wire line, is open and is slightly longer than one wavelength at the exciting frequency. Excited, it develops standing waves because it is neither infinite in length nor terminated in a load that will remove all of the RF energy transmitted. If not removed, energy reaching the open end of the line will be reflected. The reflected energy, measured at the sending end, will be equal to the energy fed to the line from the source minus the energy lost during its trip from the generator, along the line to the open end, and back.

We expect some energy to be lost—the usual I^2R loss due to current flowing in the resistance of the conductors, and a much smaller loss in the dielectric between the conductors. We can predict these losses quite accurately, however, using facts about conductor size, measured current, and so on.

We are puzzled, therefore, when we discover that the reflected energy, as measured at the sending end of the line, is significantly less than the transmitted energy minus the predicted losses. What can explain the greater-than-expected loss of energy? The answer is *radiation!* A small but significant amount of RF energy has simply left the transmission line and is traveling away from it. This phenomenon is the one that makes possible all wireless communication. An antenna is a device that enhances the process of radiation of RF energy from a system.

The Phenomenon of Radiation

The radiation of electrical energy is very common. The science of measuring and predicting radiation is highly developed. However, because of its nature—it is silent, invisible, and odorless—it does not lend itself to an explanation in simple physical terms. The explanation that follows is a common one. It does not tell the entire story, but is useful in providing a working understanding of radiation independent of the very elegant but highly sophisticated mathematical treatments.

To continue with the example of the transmission line of the preceding paragraph, let us imagine the conductors of the last $\lambda/4$ of the line being spread apart slightly, as in Fig. 6.18(a). The standing waves of voltage and current produce an electric field between the conductors and a magnetic field around each conductor. These are represented by the lines of force of Fig. 6.18(b). Notice the *fringing* (bowing out) of the electric field lines at the end of the line. (Fringing of electric field lines is a common phenomenon at the boundaries of an electric field between two conductors.) Fringing occurs because field lines running in the same direction exert a repulsive force on each other. At a field boundary, this force produces the spreading out and bowing of the lines.

When the electric field lines are the product of an ac voltage, the lines must be produced, collapsed, and reproduced in a reverse direction at a rate equal to the frequency of the voltage. We can imagine the lines being sent out from, and withdrawn to, the conductors. It is useful to theorize that, when the

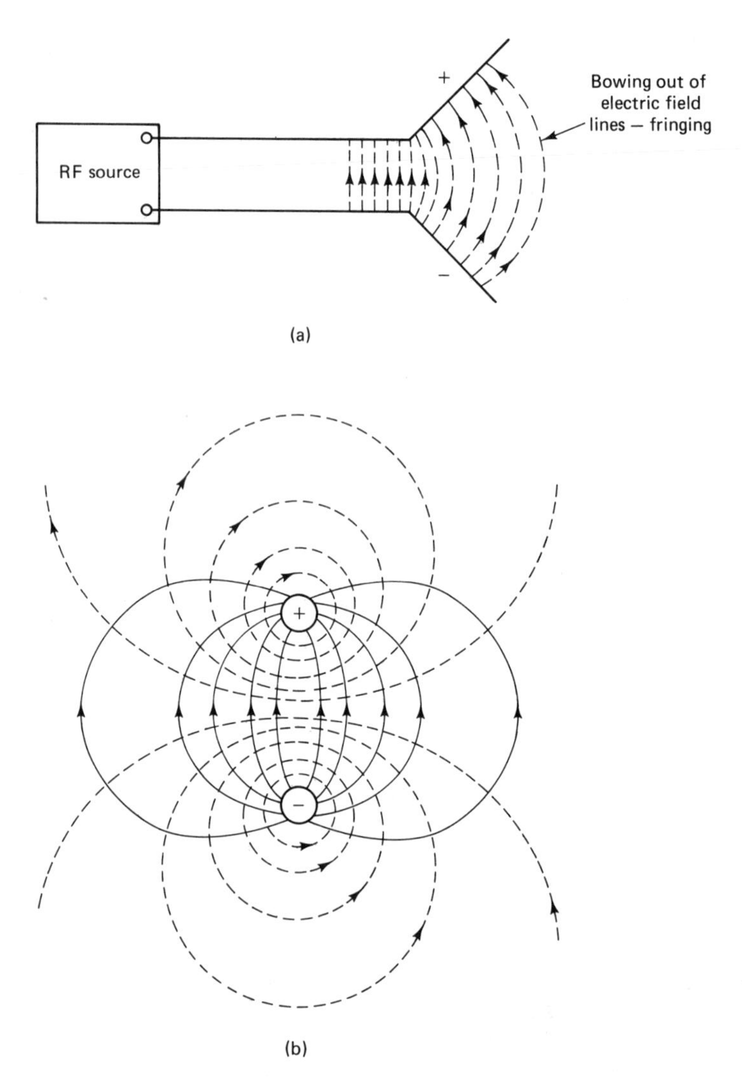

Figure 6.18 (a) Fringing of electric field lines at end of transmission line; (b) cross-sectional view of twin-lead transmission line with electric and magnetic field lines.

frequency exceeds approximately 20,000 cycles (40,000 reversals) per second, the outermost line of the field simply cannot keep up with the process of reversal. Being unable to return to its conductor, it closes upon itself, forming a closed loop. This loop is repulsed by the outermost line of force produced by the next alternation of the ac voltage. That line of force subsequently fails to make it back to the conductor and forms another closed loop, etc. The process repeats itself during each cycle. The closed loops are driven farther and farther away from the conductors. The result is a continuous wave train of energy being discharged (radiated) and repulsed from the conductors. The drawing of Fig. 6.19 is a conventional method of depicting this theory of the process of radiation.

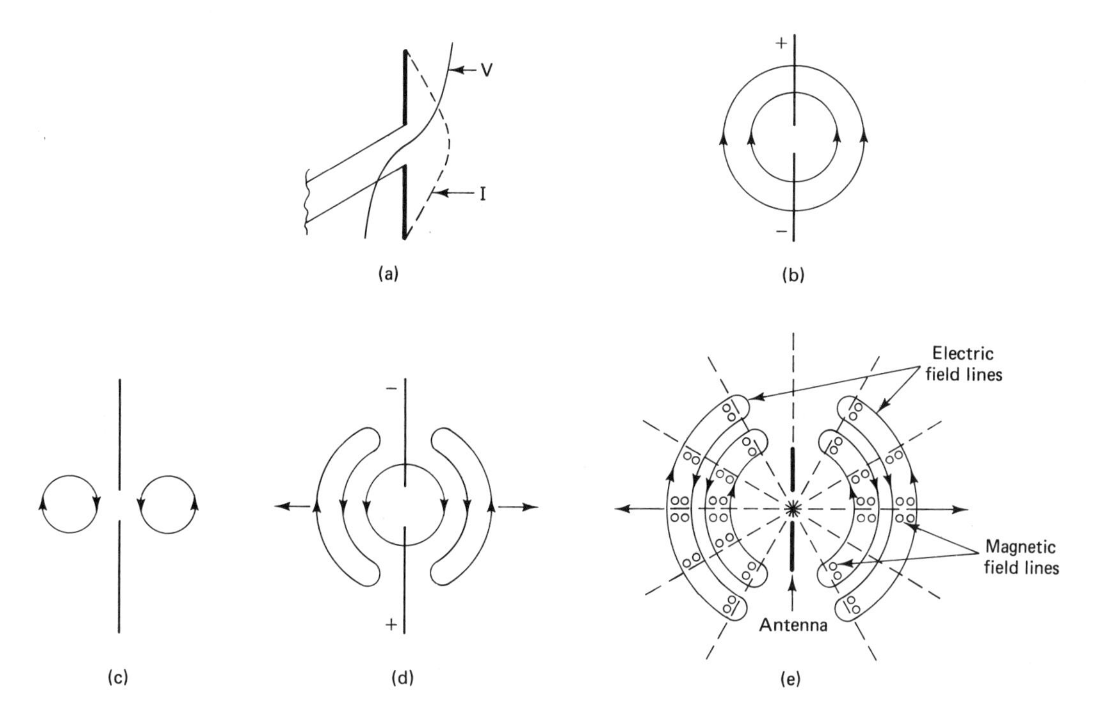

Figure 6.19 Process of radiation: (a) voltage and current standing waves on half-wave dipole; (b) electric field lines on antenna; (c) field lines closing on themselves after voltage collapse; (d) old field lines forced away by new lines after voltage reversal; (e) electromagnetic waves traveling away from antenna.

A number of observations about the nature and behavior of radiated energy in accordance with the theory just related are listed below. It is essential that you understand and become thoroughly familiar with these statements as a background for the study of antennas and RF energy propagation.

1. The RF electrical energy released from conductors in the process called radiation is a wave-like disturbance. The disturbance travels through the atmosphere, or space, at the speed of light (approximately 186,000 mi/s or 3×10^8 m/s). The wavelength of radiated energy is calculated in the same way as the wavelength of voltage or current waves on a transmission line, that is, λ = speed of light/frequency.
2. Radiated electrical energy contains energy in the form of both an electrical field and an associated magnetic field. These fields are the "disturbance" that travels at the speed of light, as described in statement 1. The fields always act at right angles to each other. That is, their directions as indicated by arrows on lines of force in field drawings, for example, are perpendicular to each other. The fields are inseparable in the same sense as are the electric and magnetic fields produced on a transmission line excited by an electrical source. The electric field component is oriented, or polarized, in the same direction as the electric field between the conductors from

which radiation occurred. The fields vary in amplitude and polarity at the same frequency as the source that produced them.

3. The "front" of the "disturbance," the *wavefront,* travels outward from the source as if carried on the surface of an expanding sphere (e.g., as if on the surface of a balloon in the process of being inflated) (see Fig. 6.20). A given amount of energy, therefore, is spread out over an ever larger area as the distance from the point of radiation is increased. The energy level in the propagating (or traveling) disturbance or wave is inversely proportional to the distance from the point of radiation. This is contrasted with the strength of the magnetic field surrounding a current-carrying conductor—the *induction field*—which is inversely proportional to the *square of the distance from the conductor*. The induction field thus diminishes in strength very rapidly with distance from its source. Its effect is felt only at very short distances from a conductor. The *radiation field,* on the other hand, is capable of being felt at astronomical distances, as evidenced by our ability to communicate via radio with space probes millions of miles from the earth.

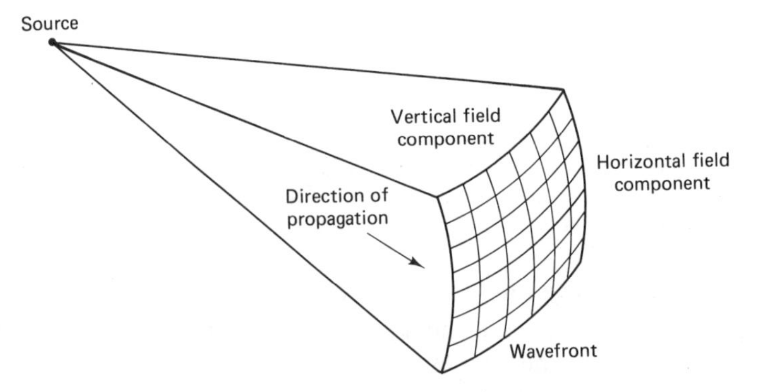

Figure 6.20 Spherical wavefront of propagating electromagnetic wave.

4. The radiation field, once detached from the radiation source, exists independently of the source. It continues to move away from the source indefinitely. This is in dramatic contrast to the induction field, which collapses and ceases to exist soon after the electrical source producing it is turned off.

5. The level of energy radiated from a conductor is directly proportional to the frequency of the electrical source energizing it. There is some radiation from every circuit no matter what the frequency, even 60-Hz power circuits. However, the amount of energy radiated from a circuit is negligible if the frequency of excitation is not 20,000 Hz or higher. At frequencies of several hundred megahertz and higher it is seemingly impossible to contain the energy in circuits made of simple conductors because radiation occurs so freely.

A Basic Antenna: The Half-Wave Dipole

Let us return to the picture of energy being radiated from the slightly spread end of a parallel-wire transmission line. Refer to Fig. 6.18 again. What might we do to maximize radiation, since that is what is desired in an antenna? The answer is to bend further the final quarter-wavelength of each conductor until each is at right angles to the line (see Fig. 6.21). Since the length of each conductor bent is $\lambda/4$, the overall length of the perpendicular portion of the line is now $1/2\lambda$. The device thus produced is called a *half-wave dipole*. The half-wave dipole is a simple, basic antenna. It is commonly used as a basis for comparison for more complex antennas.

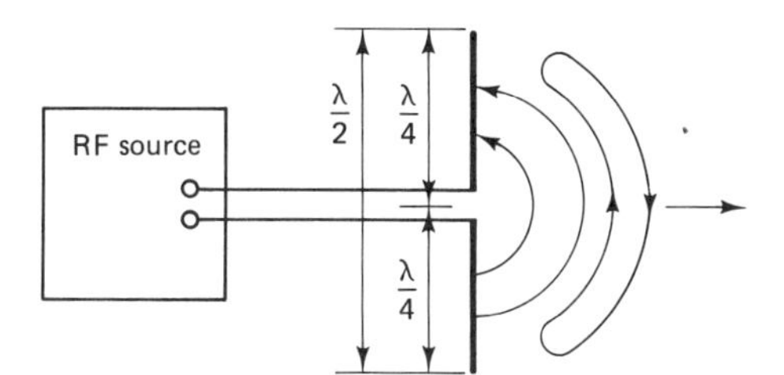

Figure 6.21 Half-wave dipole antenna made by spreading apart conductors of $\lambda/4$ of transmission line.

Information about standing waves on transmission lines can be transferred to the half-wave dipole antenna. It is like an open line. The distance from the tip of each pole to the center feed point is $\lambda/4$. Hence there will be voltage standing-wave loops at the tips of the dipole (like the loop at the end of an open line). There will be a voltage wave node $\lambda/4$ away, at the center feed point. Similarly, there will be current standing-wave nodes at the tips, and a current loop $\lambda/4$ away, at the center feed point (see Fig. 6.22). The standing-wave pattern just described is exactly the one needed to maximize radiation.

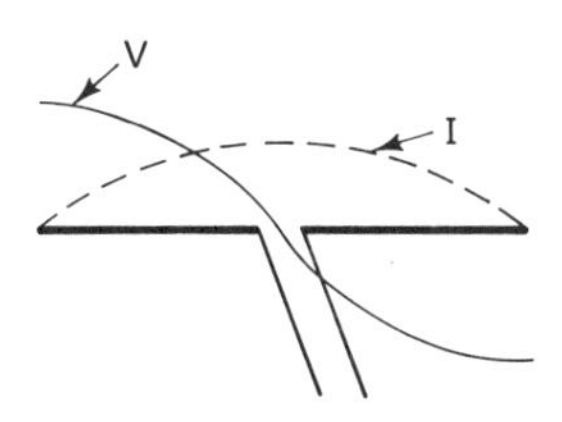

Figure 6.22 Voltage and current standing waves on half-wave dipole antenna.

Polarization

An electromagnetic wave radiated from an antenna is characterized in many ways, one of which is its *polarization*. The term "polarization" is used to describe the orientation of a wave with respect to the earth's surface. Specifically, the direction of polarization of a wave is defined by the direction of the electric field component of the wave. If the direction of the electric field component is parallel to the earth's surface, the wave is *horizontally polarized*. If the electric field is perpendicular to the earth's surface, the wave is *vertically polarized*.

As stated previously, the direction of the electric field, as the wave leaves the antenna, is parallel to the length of the antenna. Hence the term "polarization" is also used to describe the orientation of an antenna. For example, a half-wave dipole mounted so that its length is parallel to the earth's surface (like home TV antennas) is horizontally polarized. A "vertical antenna" or usually, simply "a vertical" is an antenna that radiates an electromagnetic wave whose electric field component is perpendicular to the earth's surface. The antenna's length is perpendicular to the earth's surface.

Radiation Patterns

Let us envision a half-wave dipole antenna positioned at the center of an imaginary sphere. It is mounted so that nothing interferes with the radiation emanating from it. It is being excited with a radio signal of the appropriate frequency. Let us further imagine that we can move to any point on the sphere and measure the level of radiation reaching that point from the antenna. (Remember, all points on a sphere are equidistant from its center.) Would the radiation measure the same at every point on the sphere? The answer is no.

If we very carefully measured the level of radiation at every point on the sphere and plotted this information on circular graphs (graphs with so-called polar coordinates), we would be producing a representation of the *radiation pattern* of the antenna. The graphical presentations, also called radiation patterns, indicate the levels of radiation at equal distances from an antenna. A particular graph presents radiation intensity data for a single plane with respect to the antenna. The data are plotted so as to represent the antenna at the center of the graph. Intensity is represented as distance from the center—the greater the distance of a point from the center, the greater the intensity. The second coordinate of each point on the graph is the angle a radius through that point makes with a beginning or reference radius or direction.

For a simple antenna, such as a half-wave dipole, a relatively complete representation of the radiation pattern can be made if it is plotted in just two planes. One is a plane through the center of the antenna and perpendicular to its axis (length). The second plane should be one parallel to the axis of the antenna. One of these two planes will generally be a horizontal plane, the other a vertical plane. Which is vertical or horizontal depends on the polarization of the antenna.

The typical radiation pattern of a horizontally polarized half-wave dipole antenna is depicted as a three-dimensional sketch in Fig. 6.23(a). Notice that the representation is doughnut-shaped. The distance away from the center of any point on the surface of the doughnut represents the relative intensity of radiation in the direction indicated by a line drawn from the center to the point.

If we cut through the center of the doughnut with a horizontal plane, a plane parallel to the earth's surface, we see the pattern of Fig. 6.23(b). This then is the radiation pattern, in the horizontal plane, of a horizontally polarized half-wave dipole antenna. It is sometimes described as a "figure 8" pattern, for an obvious reason. Analyzing the pattern, we draw the conclusion that maximum

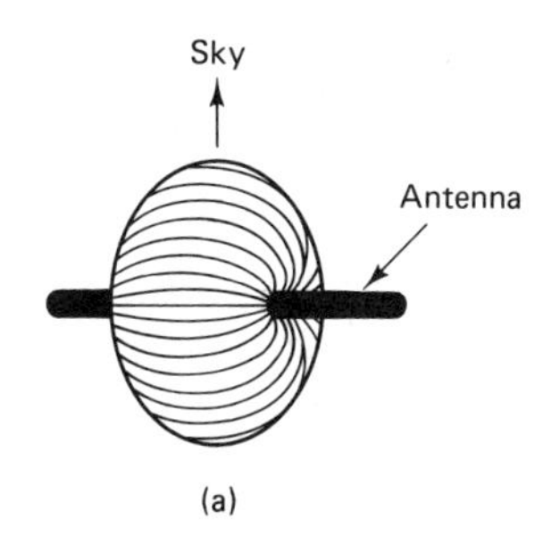

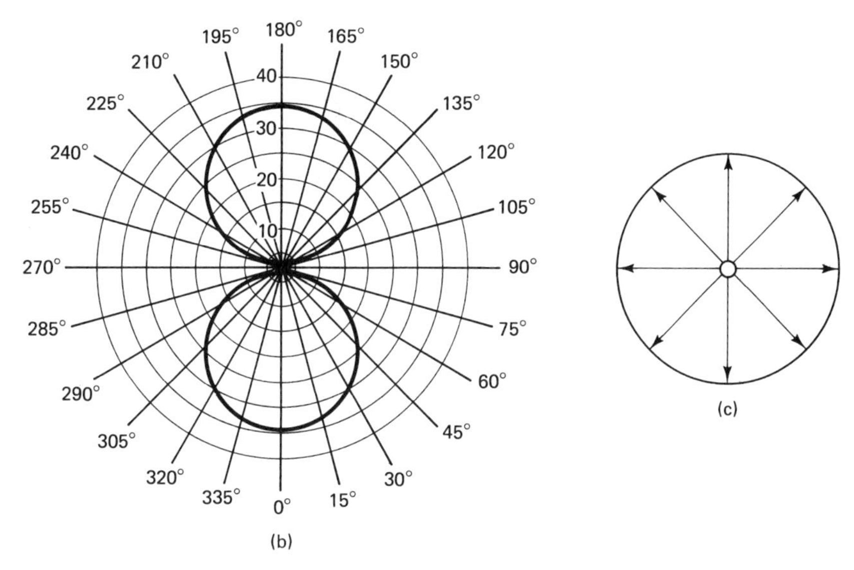

Figure 6.23 Radiation pattern of horizontally polarized half-wave dipole antenna: (a) three-dimensional pattern; (b) horizontal pattern plotted on polar-coordinate paper; (c) radiation pattern in vertical plane.

radiation occurs in the direction of a line drawn through the center of the antenna and at right angles to it. Further, radiation is virtually nonexistent off the ends of the antenna, that is, in directions in line with the axis of the antenna.

If the doughnut is sliced with a vertical plane through its center, we are presented with the pattern shown in Fig. 6.23(c). This indicates that radiation is uniform in all directions at right angles to the axis of the antenna. Incidentally, the representations given here are true only if the antenna is mounted at sufficient distance from the earth's surface so that reflections from the earth (or any other object of significant size) do not alter the amount of radiation at any given point. The radiation pattern of an isolated dipole, in a plane perpendicular to its axis, is a circle.

If a dipole is mounted with vertical polarization, its radiation pattern in the vertical and horizontal planes will be just the reverse of those for the horizontally polarized dipole (see above). The pattern for such an antenna is presented in

Fig. 6.24. The pattern is now a horizontal "doughnut." The pattern in the horizontal plane is a circle, the pattern in the vertical plane is now the figure 8.

Every antenna has its radiation pattern. The pattern is determined by the form or configuration of the antenna. The pattern is also influenced by reflections from nearby objects such as the earth's surface, buildings, other antennas or antenna elements, a vehicle's surface (in the case of mobile antennas), and so on. Radiation patterns are extremely important in the intelligent selection and utilization of antennas.

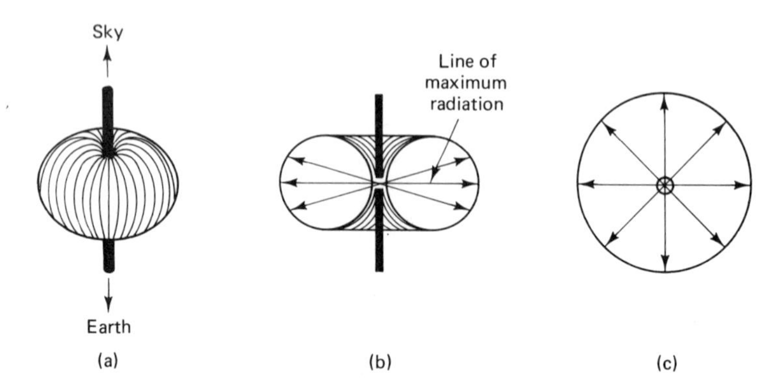

Figure 6.24 Radiation patterns of vertically polarized half-wave dipole: (a) three-dimensional pattern; (b) pattern in vertical plane; (c) pattern in horizontal plane.

Antenna Types

Antennas of numerous sizes, shapes, and configurations have been built and utilized. Amateur radio operators (hams), especially, have enjoyed experimenting with many different types of antennas. Details of design and construction of several special types appear in Sec. 6.8. At this point the names and a brief description of two basic types of antennas will be presented. Reference to these two is frequently made. The lack of knowledge of the titles and description of these types becomes a handicap to anyone required to study or work with antennas.

The first of the two types to be presented is the *Hertz antenna*. [This antenna is named after the German physicist Heinrich Rudolph Hertz (1857–1894), an early investigator of radio waves.] A Hertz antenna is simply an isolated half-wave dipole antenna. The discussion of the preceding paragraphs was about a Hertz antenna. *Isolated,* in the sense used here, means that the antenna is mounted away from all objects so that its performance, radiation pattern, etc., will not be altered by radiation or reflections from such objects.

A second basic type of antenna is the quarter-wave, grounded vertical antenna, also called a *Marconi antenna*. [Guglielmo Marconi (1874–1937), an Italian physicist, is sometimes referred to as the father of modern radio because of his pioneering work in developing a working radiotelegraph system.] A quarter-wave antenna whose lower end is connected to an earth ground of good conductivity behaves very much like a vertical half-wave antenna. Its radiation pattern is like a horizontal doughnut split in half at the earth's surface. That is, the pattern in the horizontal plane is omnidirectional (see Fig. 6.25). When the antenna is well grounded, the pattern of voltage and current standing waves is identical to that of one-half of a half-wave antenna (see Fig. 6.26). It is as if the conductivity of the ground supplies the other half of a half-wave vertical antenna. The effect produced is referred to as an *image antenna*.

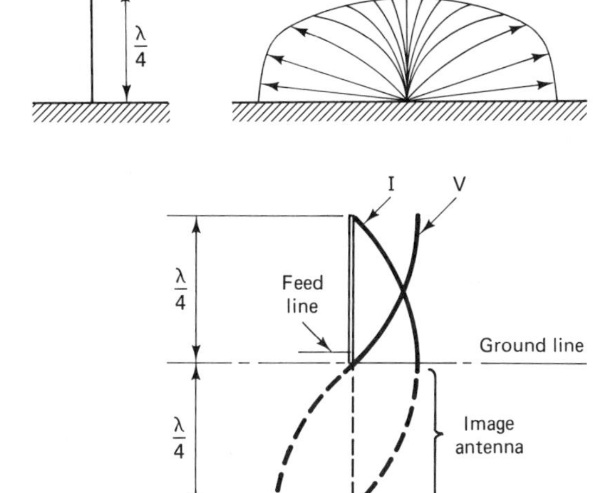

Figure 6.25 Quarter-wave grounded vertical antenna and its radiation pattern.

Figure 6.26 Grounded vertical with image antenna and standing wave pattern.

Because there is a current loop at the center of the actual antenna plus image antenna, there will be high current flow in the ground at the antenna base. For effective operation of this type of antenna, steps must be taken to ensure that the earth in the vicinity of the antenna has high conductivity. In some installations, such as commercial AM broadcast stations, a grid of heavy conductors, called a *counterpoise,* is placed in the ground surrounding the area where a grounded-vertical antenna is mounted.

The quarter-wave grounded vertical antenna is the antenna of choice for commercial broadcast stations in the AM band (540 to 1600 kHz). Its radiation pattern is ideal for such applications. The radiation is primarily of the ground-wave type, meaning that reception is very strong within a radius of 50 to 100 mi of the antenna and attenuates rapidly beyond that. A quarter-wave antenna mounted vertically, in this frequency range, is feasible—a full half-wave vertical antenna is hardly feasible because of its excessive height. (A half wavelength at 1000 kHz is over 490 ft!)

6.6 PROPAGATION OF RF ENERGY

The phenomenon of radio-frequency energy traveling through the earth's atmosphere, as well as through the ''empty'' space above the atmosphere, is called *propagation*. Recall that after radio-frequency energy has been emitted from an antenna (as a disturbance in the form of a wave) it travels away from the antenna at the speed of light. Once radiated, the wave exists and propagates independently of the system that produced it. That is, changes that occur at the transmitter or antenna have no effect on energy that has already been radiated. The front of the wave is spherical in nature, as if carried on the surface of an expanding balloon. The direction(s) of propagation of a wave is determined primarily by the nature of the antenna from which the wave was radiated.

For purposes of describing, analyzing, and predicting propagation, this phenomenon is generally considered to take place in two fundamentally different ways, or modes: *ground-wave propagation mode* and *sky-wave propagation mode*. These terms are aptly descriptive. Ground-wave propagation refers to the propagation effects which are most directly determined by the earth's presence. Sky-wave propagation is affected more by the nature of the sky high above the earth's surface. We will examine each of these modes of propagation in some detail. Before proceeding to propagation modes, let us first look at two basic effects that are fundamental in determining the behavior of propagating waves. It is essential that we have a working knowledge of the phenomena of *reflection* and *refraction*.

Reflection

It is without risk to say that everyone who reads this book is familiar with the phenomenon of reflected light. We have all looked at ourselves in a mirror.

Radio waves are, in a valid sense, simply long light waves. As such they are also capable of being reflected. As with visible light, the amount and quality of reflection of radio waves depends on the nature of the reflecting surface. First, to reflect a radio wave the dimensions of a surface must be large compared to the electrical dimensions of the wave (i.e., its wavelength). The significance of this fact is easily recognized if one visualizes the effects on an ocean wave of two objects of significantly different dimensions: a single post, 2 ft in diameter, or a partially submerged rock 30 to 50 ft in length. The post deflects (reflects) the wave hardly at all; the rock reflects the wave and creates an area of quiet water (a shadow) behind it. The microwaves of radio signals of very high frequencies are reflected by virtually any object large enough to be seen by the naked eye. Only the earth's surface, or the surfaces of other planets are large enough to reflect the very long waves of very low frequency radio signals.

Although light is reflected from many different surfaces, certain surfaces reflect better than others. Some surfaces reflect hardly at all. Smooth water and glass come to mind as reasonably good light reflectors. In fact, almost anything with a very smooth surface reflects at least some light.

Mirrors are made by coating one surface of a smooth, clear glass with an amalgam—a mixture of mercury with another metal, or metals. In a similar way, certain surfaces are much better radio-wave reflectors than others. Some surfaces reflect radio waves hardly at all. Surfaces that are also good electrical conductors make the best radio-wave reflectors—seawater with its salt content is a good conductor and a good radio-wave reflector. Almost any metal object, if large enough, will reflect radio waves since metals are conductors. Smooth metal surfaces reflect radio waves better than rough surfaces. Surfaces with poor conductivity, such as wood, reflect radio waves very poorly.

When light strikes a reflecting surface, the angle between the direction of the light and a line perpendicular to the surface is called the *angle of incidence*. The light which leaves that surface, by reflection, has a certain direction. The angle between a line perpendicular to the surface and the direction which the light takes is called the *angle of reflection*. As you are probably already aware, the angle of reflection is always equal to the angle of incidence. Exactly the same observation can be made about reflected radio waves—*the angle of reflection equals the angle of incidence* (see Fig. 6.27). This principle must be used in predicting the behavior of reflected waves.

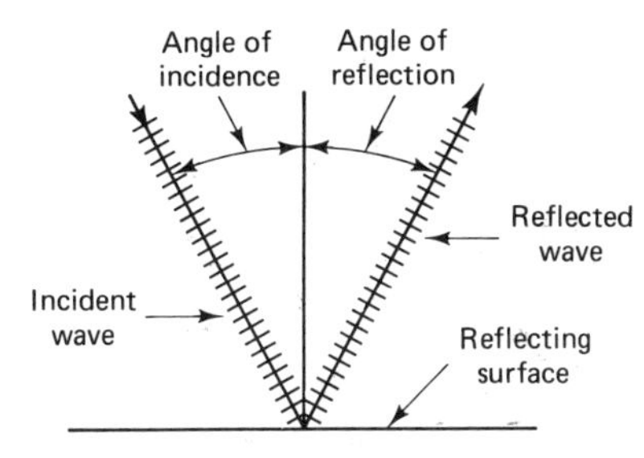

Figure 6.27 Reflection of radio waves.

Refraction

When a wave of energy, such as visible light, passes from one medium into another the direction of the wave is generally changed. We experience this effect when we try to grasp an object below the surface of water and discover that the object—like a bar of soap in a tub of bathwater—is not where we think we see it to be. In this case the light is said to be *refracted*. To refract means to bend. We are accustomed to seeing things with light rays that travel in straight lines. When the light rays from the soap reach the surface of the water they are bent and we are fooled about the location of the soap bar (see Fig. 6.28). A common explanation for refraction depends on the knowledge that light, and other electromagnetic waves (including radio waves), travel at different speeds in different media. As we shall see, radio waves are refracted (bent) as they pass through different layers of the earth's atmosphere. The layers are of differing densities and composition and thus cause radio waves to change direction when passing through the boundaries between them. This phenomenon is like that of light passing through the boundary between water and air.

Ground-Wave Propagation

Visualize an antenna that is located relatively near the earth's surface. Imagine that it is "aimed" or positioned so that its radiation will travel near the earth's surface rather than into the sky, toward a satellite, for example. The propagation of the radio-frequency energy from such an antenna would be observed to conform with one or several of a group of general observations describing the phenomenon called *ground-wave propagation*.

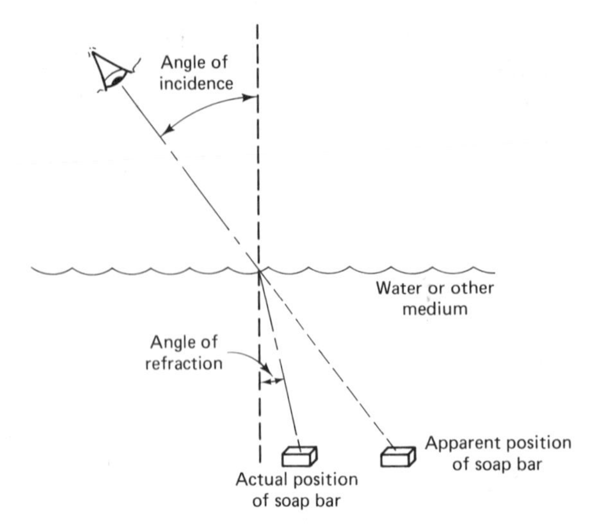

Figure 6.28 Refraction.

Ground-wave propagation includes several components (see Fig. 6.29). The *direct-wave* component includes the radiation that travels through the atmosphere on a direct line from the sending antenna to the receiving antenna. The direct-wave phenomenon is sometimes referred to as line-of-sight propagation since it is limited, as is sight, by the curvature of the earth's surface (see Fig. 6.30).

In actual fact, radio waves propagating near the earth's surface bend and follow the surface to a small degree. This bending is a refraction caused by the atmosphere. As a result of the refraction, the "radio horizon" is slightly beyond the visual horizon.

The maximum distance over which a radio wave, propagating as a direct wave, can be expected to be received is calculated by the formula $d = \sqrt{2h}$, where d is the distance to the radio horizon in miles and h is the height of the transmitting antenna in feet. The distance to the visual horizon, about 15% less, is given by the formula $d = 1.23\sqrt{h}$. If both receiving and transmitting antennas are elevated above the surface of the ground in the area adjacent to the mounting site, the radio horizon is $d = \sqrt{2h_T} + \sqrt{2h_R}$, where h_T represents the height of the transmitting antenna and h_R is the height of the receiving antenna.

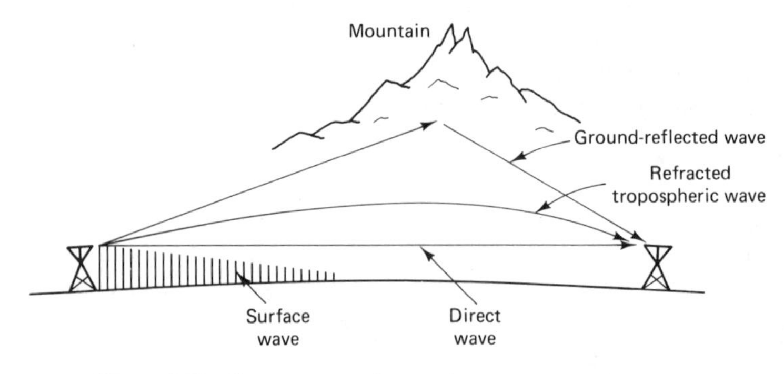

Figure 6.29 Components of ground-wave propagation mode.

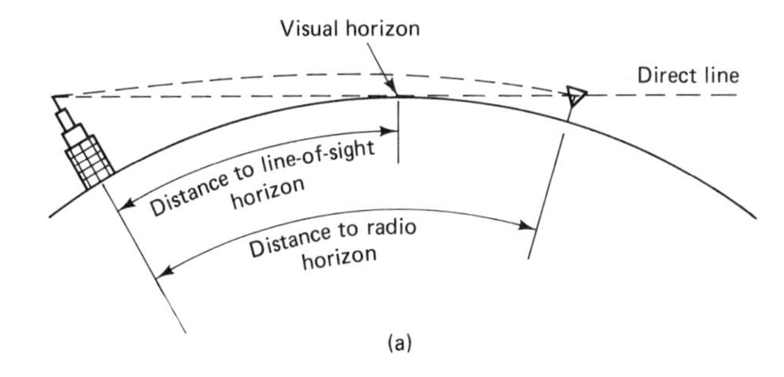

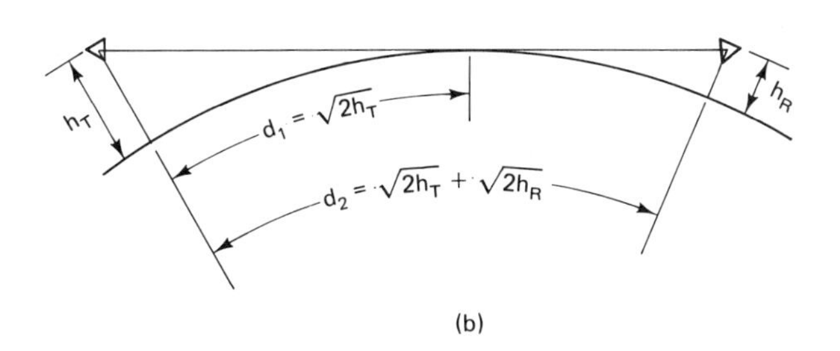

Figure 6.30 (a) Visual horizon versus radio horizon; (b) extending propagating range of direct waves by elevating antenna(s).

Direct-wave propagation is especially important when the frequency of radiation is above approximately 30 MHz. At such frequencies the direct wave is the major component of ground-wave propagation. For example, signals for home TV and FM broadcasts are receivable, without the aid of cable or satellite transmission systems, at maximum distances of 100 mi, approximately.

Example 7

A transmitting antenna is mounted at a height of 750 ft. What is the approximate range at which a signal can be received by direct-wave propagation?

Solution

$$d = \sqrt{2h} = \sqrt{2 \times 750} = 38.73 \text{ mi}$$

(The visual horizon is at $d = 1.23\sqrt{h} = 1.23\sqrt{750} = 33.68$ mi.)

Example 8

A television antenna is mounted on a peak with an elevation of 4250 ft. At what maximum distance is the signal likely to be received if the receiving antenna height is 100 ft?

Solution

$$d = \sqrt{2h_T} + \sqrt{2h_R} = \sqrt{2 \times 4250} + \sqrt{2 \times 100} = 106.34 \text{ mi}$$

It is also important to know that "line-of-sight" transmission means that signals which are propagated primarily as direct waves do not "go around" or through mountains or any other obstructions of large size (very large buildings). Many home TV antennas are located within a few miles of a TV transmitting antenna but provide very poor (if any) reception of the signal from that antenna because they are in the "radio shadow" of an intervening mountain peak, or large building.

Another important component of ground-wave propagation is the *surface wave*. Surface waves propagate along the surface of the earth almost as if the earth's surface were a single-conductor transmission line. The propagation is greatly affected by the surface. A surface with poor conductivity attenuates the wave more than a surface with good conductivity (e.g., the ocean's surface). The signal strength at a distance from a transmitting antenna is less than inversely proportional to the distance because of the absorption of signal by the surface. Absorption of energy by the surface increases with the frequency of the wave. Signal strength associated with a surface wave diminishes rapidly with height above the earth's surface. Best propagation is obtained with vertically polarized waves.

Surface-wave propagation is the means by which the vast majority of listeners of AM broadcast stations receive their signals. This is especially true during daylight hours. Within their range of effectiveness—100 to 150 mi—surface-wave signals are generally very reliable and not affected by fading. Grounded-vertical quarter-wave antennas are effective producers of surface waves, hence their popularity as the antenna of choice for AM broadcast stations.

Another important component of the ground wave is one called the *tropospheric-wave component*. The *troposphere* is the lower portion of the earth's atmosphere. (The earth's atmosphere can be thought of as being made up of two parts: the troposphere and the *stratosphere*. These are separated by a surface called the *tropopause*.) The troposphere is the part of the atmosphere in which "weather happens." It is also characterized by a relatively uniform decrease in temperature with increasing height above the earth's surface. Relatively sharp changes in atmospheric density with altitude occur in the troposphere. In short, the troposphere is not a homogeneous mass of gas by any measure. Rather, it is often many masses of differing temperature, humidity, and density characteristics. Refraction of radio waves occurs at the boundaries of these masses. The concept of tropospherically refracted waves is depicted in Fig. 6.31.

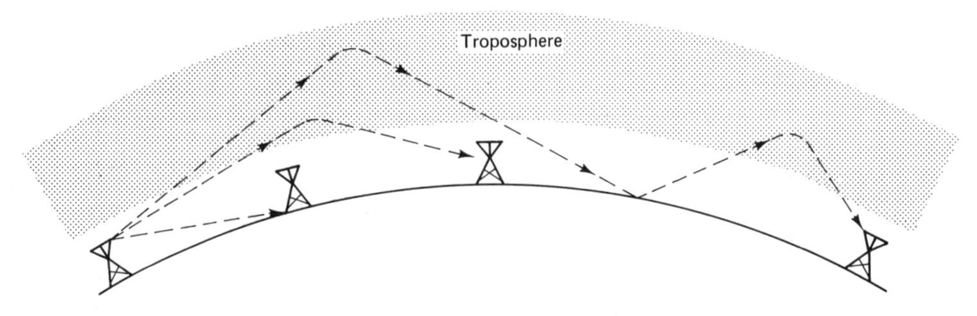

Figure 6.31 Tropospheric propagation.

When weather conditions are right, and tropospheric refraction is occurring, the range of ground-wave propagation is extended significantly. Since it happens that shorter waves are more susceptible to bending than long waves, radio transmissions at frequencies of 50 MHz and higher are sometimes received at amazingly great distances.

An atmospheric condition that is perhaps the most significant cause of tropospheric refraction is called *temperature inversion*. This is the phenomenon which is also an important cause for "smog," an unwelcome condition of atmospheric pollution which occurs in most of the highly populated and industrialized areas of the world of the late twentieth century.

As the name implies, temperature inversion is a condition in which the usual decrease of temperature with altitude does not obtain. Specifically, temperature inversion (with respect to the atmosphere) refers to a condition in which a warm layer of air exists above a cooler layer of air. This condition can arise as a result of one or a combination of several other phenomena: A mass of warm air may be blown or drift over a body of cool water (above which the air is cool). The air nearest ground level may cool rapidly after sunset leaving a mass of warmer air at a higher elevation. A mass of cool air over a nearby ocean surface may be blown onshore and under a warmer, higher air mass. The air above a cloud or ground fog layer may be heated by the sun's rays reflected from the top of the cloud/fog layer.

Whatever the cause, a temperature inversion produces layers of differing densities and thus boundaries between media with differing refraction indexes. These are the conditions for significant refraction of radio waves.

Weather conditions often change rapidly, sometimes within minutes. Thus the conditions for good tropospheric refraction are unstable and unpredictable. Radio-wave propagation by tropospheric refraction is open for use by anyone desiring to take advantage of its possibilities, primarily remarkable range. But it is only as reliable as the weather. It is subject to fading and sudden variations in received signal strength.

As indicated in the diagram of Fig. 6.29, ground-wave propagation includes a component called the *ground-reflected wave*. The ground-reflected wave is similar to the direct wave. It is a direct wave that is reflected from the ground or any other large object. If a receiving antenna is excited only by a ground-reflected wave, there is no complexity introduced. However, in many instances an antenna will intercept both a ground-reflected wave and a direct wave. This phenomenon usually introduces some undesirable effects.

When a wave is reflected, its polarity is changed by 180°. If a reflected wave arrives at the receiving antenna at exactly the same time as the direct wave, the two will cancel each other. Or if the reflected wave has been attenuated slightly, it may simply cancel part of the direct wave. Generally, the path taken by the reflected wave is longer than that taken by the direct wave. Its time of arrival at the receiving antenna is slightly later. The two waves, now with a phase difference between them, will combine, producing a resultant wave. The resultant wave may vary in amplitude from zero to twice the amplitude of one wave alone. That is, the waves combine just like phasor voltages. Exactly how they combine at any specific location depends on the path taken by the reflected wave in comparison with the path taken by the direct wave.

The effects of this phenomenon on the users of a propagation path depend on the nature of the resultant wave. In rare cases the waves combine to produce a stronger signal—usually a desirable effect. When the waves cancel, producing what is called a "null," the effect is usually felt as very undesirable. In the case of television signals, a reflected wave is generally the source of the very distracting effect known as a "ghost image."

In summary, although ground reflected waves may sometimes produce desirable effects (they are important in the propagation of radio waves between low-flying aircraft), in general this component of ground-wave propagation is more troublesome than beneficial.

In the discussion on the direct wave, above, there is an implication that there is no radio-wave reception when the line-of-sight view from receiving antenna to transmitting antenna is obstructed by a building or hill. This is not entirely true. Just as the line between light and shadow in a "light shadow" is never sharp, there is never a sharp cutoff of radio reception behind an obstruction. In either case the effect is caused by a phenomenon called *diffraction*. Whenever light, or any other form of electromagnetic energy (e.g., radio waves), passes through a narrow slit or by the edge of an opaque object, an interference with the normal transmission of the energy occurs. The practical effect, as far as radio-wave propagation is concerned, is a bending of the path of the wave (see Fig. 6.32). A radio wave does "go around" an obstruction to a limited degree.

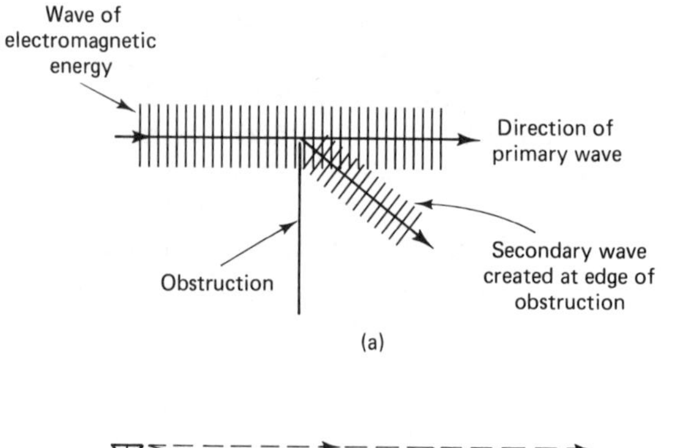

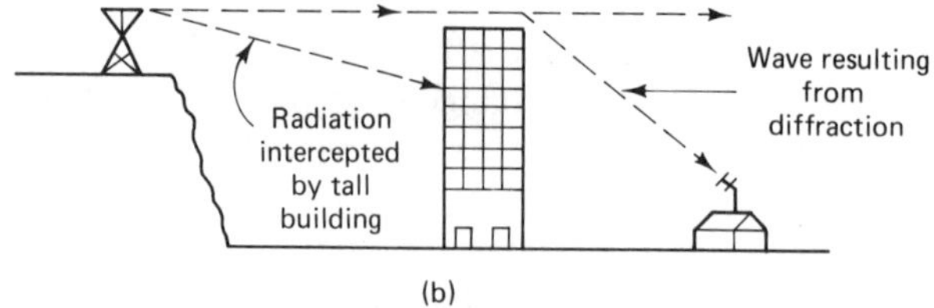

Figure 6.32 (a) Basic concept of diffraction at edge of obstruction; (b) partial elimination of "radio shadows" by diffraction.

Summary: Ground-Wave Propagation

1. A ground wave includes these components: surface wave, direct wave, tropospheric wave, and ground-reflected wave.

2. A surface wave propagates near the earth's surface and is significantly affected by that surface. This wave is most effective with vertical polarization and relatively low frequencies, say below about 2 MHz. Absorption by the earth's surface limits the range of the wave.
3. Direct wave propagation is utilized for transmissions of frequencies higher than those for which the surface wave is used. Polarization may be vertical or horizontal—both are used. Vertical polarization is used commonly for two-way radio communication—police vehicles, taxicabs, service trucks, CB radio on private vehicles, and so on. Horizontal polarization is used for commerical broadcast FM and television transmissions. The range of direct wave propagation is limited to line of sight, which, because of some small degree of atmospheric refraction, is approximately 15% greater than visual line of sight.
4. The tropospheric wave is a direct wave that has received significant refraction by the troposphere, the lower atmosphere. Refraction occurs when weather conditions, such as temperature inversion, produce layers of differing refractive indexes in the atmosphere. The boundaries of such layers produce refraction. Tropospheric wave propagations achieve remarkable ranges at times, but may also be unreliable.
5. Ground-reflected waves may enhance a direct wave but often cause undesirable effects such as television ghost images.

A Word about Terminology

The terms used to classify various forms of radio-wave propagation have not been standardized. There is no single set of terms that everyone in the world concerned about these matters has agreed to use. As might be expected, different people use different terms, although the variation is not great. Perhaps the most significant difference in terminology from that used in the preceding paragraphs is that the tropospheric wave is categorized as a distinctly separate mode of propagation rather than as a component of the ground wave. The direct wave and ground-reflected wave are often, together, referred to as a *space wave,* and as the second of two components of the ground wave. The other component is the surface wave, as used above.

Sky-Wave Propagation

A second major category of wave propagation includes what is called a *sky wave*. The sky wave is the result of radiation whose initial direction is above the horizontal, toward the sky. It is a relatively complex phenomenon, there being several factors that determine the ultimate nature of this mode of propagation. Figure 6.33 depicts the several paths of transmission commonly associated with sky waves. The one most important factor influencing propagation by sky wave is the *ionosphere*. Before proceeding further with an examination of the sky wave phenomenon, let us look at the ionosphere—what it is and what determines its characteristics.

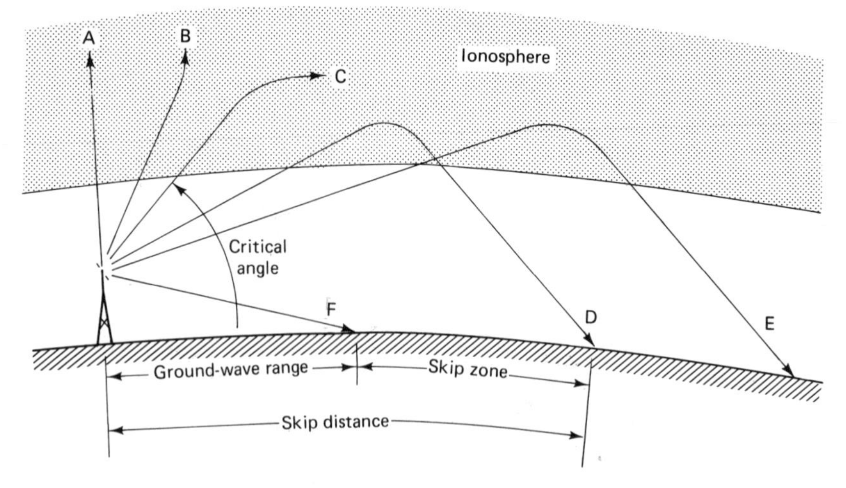

Figure 6.33 Sky-wave propagation.

The Ionosphere

The term *ionosphere* literally means a *sphere of ions*. It is a sphere of ionized gases that surrounds the sphere on which we live, the earth, in somewhat the same way that an orange peeling surrounds the more edible "meat" of that fruit. Of course, an ion is an atom or molecule (group of atoms) which has a net electrical charge—it has given up, or taken on, one or more electrons. The sphere of ions has four relatively distinct layers. These are designated the D, E, F_1, and F_2 layers. The layers are at different heights above the earth's surface. The order of the heights is the same as that listed, with the D layer nearest the earth's surface. The concept of the ionosphere with its four layers is shown in Fig. 6.34.

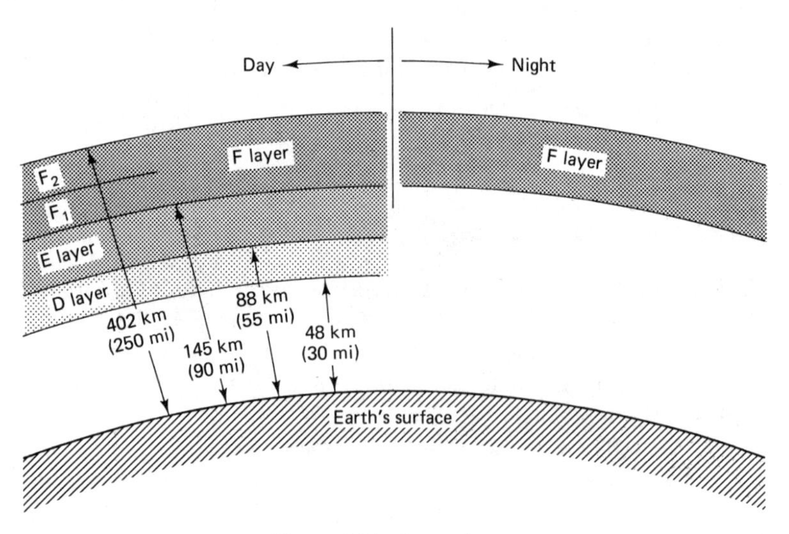

Figure 6.34 Ionosphere.

The molecules of the gases at the upper limits of the earth's atmosphere are ionized by radiation from the sun. (A particle of energy, called a *photon,* interacts with an electron in the outer orbit of a gas atom and increases its energy level. With increased energy the electron is able to escape from its orbit, producing a free electron and an ion.) This ionization is maintained for long periods because the low density of the gases in the upper atmosphere reduces significantly the probability of deionization, the recombination of electron and atom. Although, in all probability, most of the ionizing energy comes from the sun, it is believed that there are other forms of energy that contribute to the ionization: cosmic rays from unknown sources in the universe, as well as a constant bombardment of fast-moving electrons and positrons from unknown sources.

The amount of ionization, as could be expected, depends on the intensity of the ionizing energy. Since the vast majority of that energy is ultraviolet radiation from the sun, the character of the ionosphere, and its effect on radio-wave propagation, are determined virtually totally by the character of the sun's radiation striking the earth's atmosphere. For example, when a region of the earth and its atmosphere is experiencing nighttime darkness, the ionosphere consists of little more than a weakly ionized F layer—the D and E layers disappear.

Because of the variable nature of the sun's energy reaching the earth, the intensity of ionization and the number and altitude of the layers of the ionosphere undergo constant change. These changes follow repeating patterns, to a very large extent. The patterns can be called cycles—they repeat themselves over certain periods of time. Some patterns have periods of only an hour. That there would be a cycle with a period of a day is obvious. Another cycle has a period that corresponds to the period of the seasons of the year—the seasons are determined by the changing positions of sun and earth.

There is another cycle of changing radiation of a much longer period—the *sunspot cycle*. Sunspots, which manifest themselves as dark spots on the sun's surface, are perceived to be tremendous storms, or localized disturbances. The disturbances consist of eruptions of incandescent gases which flare out from the sun's surface to distances of half a million miles. These eruptions increase the level and proximity to the earth of the sun's ionizing energy. Sunspots therefore increase the intensity of ionization in the ionosphere and enhance the ability of the ionosphere to influence radio-wave propagation. Sunspots appear, on the average, about every 11 years.

In addition to these more or less regular, or cyclical, changes, the ionosphere is also subject to sudden, unpredictable changes that alter its effect on the propagation of radio waves. These irregular variations involve changes in the degree of ionization or the density of the layers of the ionosphere. Like storms in our lower atmosphere, these disturbances often appear suddenly, are localized geographically, and disappear in due time.

An ionized gas is a relatively good conductor. The layers of the ionosphere affect the propagation of radio waves, then, because they are large conducting surfaces. Depending on the exact circumstances, an ionospheric layer may *absorb, reflect,* or *scatter* electromagnetic radiation. The overall effect of the ionosphere on the transmission path of a particular radio wave depends on several factors, including: height of layer, frequency of wave, angle between direction of propagation and surface of layer, intensity of ionization of layer, and density of layer. Let us examine some of the more common effects of the ionosphere and see how these are used in achieving reliable, long-distance radio communication.

For long-distance communication, the most useful effect of the ionosphere is the reflection of a wave. If a wave is reflected, the well-known rule about reflections is observed: the angle of reflection is equal to the angle of incidence. A reflected sky wave is capable of traveling much further around the earth than a surface wave or direct wave (see Fig. 6.32).

Whether or not a wave is reflected from an ionospheric layer depends on its frequency and the *ionization density* of the layer. This density determines the distance between the ionized molecules in a gas. In a high-ionization-density gas, the distance between ionized molecules is relatively shorter than in a low-ionization-density gas. A high-density gas will be able to "turn back" (reflect) a shorter wave than a low-density gas. A low-density gas will allow a short wave to pass through (between the molecules, as it were) and turn back only long waves.

For a given ionization density of layer, there is a highest frequency that will be reflected; waves of higher frequency will simply pass on through. This frequency is called the *critical frequency*. Since the ionospheric layers nearest the sun—the E and F (F_1 and F_2) layers—have the greatest ionization, their critical frequencies are greater (they reflect higher frequencies than does the D layer). Of course, as described above, the ionization of the various layers is constantly changing, sometimes in predictable patterns, sometimes unpredictably. Therefore, not only does each layer have a different critical frequency, but those critical frequencies change with changes in ionization.

The concept of critical frequency, as described in the preceding paragraph, applies to radiation that is beamed directly at the ionosphere from an antenna on the earth (the direction of travel makes a 90° angle with respect to the earth's surface). Reflection of waves whose frequencies are higher than the critical frequency occurs when the angle of travel is smaller (see Fig. 6.32). Here again, however, as the angle between the direction of travel and the earth's surface increases (approaches 90°), an angle is reached at which the direction of travel of the radiation is not bent sufficiently to permit return to the earth. The radiation is effectively lost in, or absorbed, by the ionospheric layer (see line *C* in Fig. 6.33). This angle is called the *critical angle*.

The critical angle also depends on the ionization density of the layer and the frequency of the wave. The critical angle is smaller for higher frequencies; it is smaller for lower ionization densities. For a given frequency, the critical angle is larger for the E layer than for the D layer, for example. When the direction of travel that a wave makes with the earth's surface is greater than the critical angle, the wave simply passes on through a given layer. It may or may not be reflected by a higher layer—it depends on whether the critical angle is exceeded for such a layer.

Before looking at the overall effect of the ionosphere on radio-wave transmission let us examine specific details concerning the ionospheric layers. The *D layer* has the least ionization density. It is further from the sun. The higher layers are between it and the sun, intercepting and diminishing the energy available for ionization at the lower level. The D layer is present only during daylight hours and has its greatest significance at midday when the sun's effect is greatest. Its ionization is insufficient to affect any significant bending of wave directions except for the lowest frequencies. It reflects only waves whose frequencies are approximately 500 kHz or less. It has a detrimental effect on the propagation of sky waves of higher frequencies because it absorbs energy from them, and thus attenuates their field strength, as they pass through. The D layer extends from approximately 30 to 55 miles above the earth's surface. The exact thickness and height of the layer varies with the conditions described above.

The band of ionization extending from about 55 to 90 miles above the earth is designated the *E layer*. This region of ionization is also called the Kennelly–Heaviside layer, in honor of the two men who discovered the effects of the ionosphere on radio transmissions (Sir Oliver Heaviside of England and Professor Arthur Kennelly of the United States, 1902). Like the D layer, the characteristics of this layer are subject to variation for the various reasons outlined above. It, too, almost completely disappears at night; it has its greatest effect at noon. When effective, its ionization is greater than that of the D layer. It reflects waves of higher frequencies—up to 20 MHz. Because of its height and ionization density it is useful in providing radio transmission paths whose effective range is up to approximately 1500 miles.

The effective range of a wave reflected from an ionospheric layer is determined by the basic geometry involved: the layer is at some definite distance from the earth's surface; the direction of travel from the antenna must take into account the limitation imposed by the critical angle concept. The travel path of a wave forms two sides of a triangle (see Fig. 6.35). The effective communication distance—the distance from transmitter to receiver, measured along the earth's surface—is the third side of the triangle. The length of that third side is determined by the lengths of the other two sides and the angle between them. These dimensions and angle, of course, are determined by the height of the point of reflection and the angle the path makes with the surface of the earth. Since a wave is not turned around sharply at the surface of a layer but travels along an arc (see Fig. 6.35), the effective height of the point of reflection is slightly greater than the height of the lower surface of the layer.

To obtain greater transmission distances it would be necessary to decrease the angle the direction of travel makes with respect to the earth's surface (see path *B* in Fig. 6.35). However, such a move is not effective since now the wave travels a considerable distance through the ionized layer and is significantly attenuated by it. Greater range for higher frequencies is obtained by utilizing the F layer for reflection.

The most highly ionized layer of the ionosphere is the *F layer*. It is also, therefore, the most useful for long-distance transmission of the higher radio

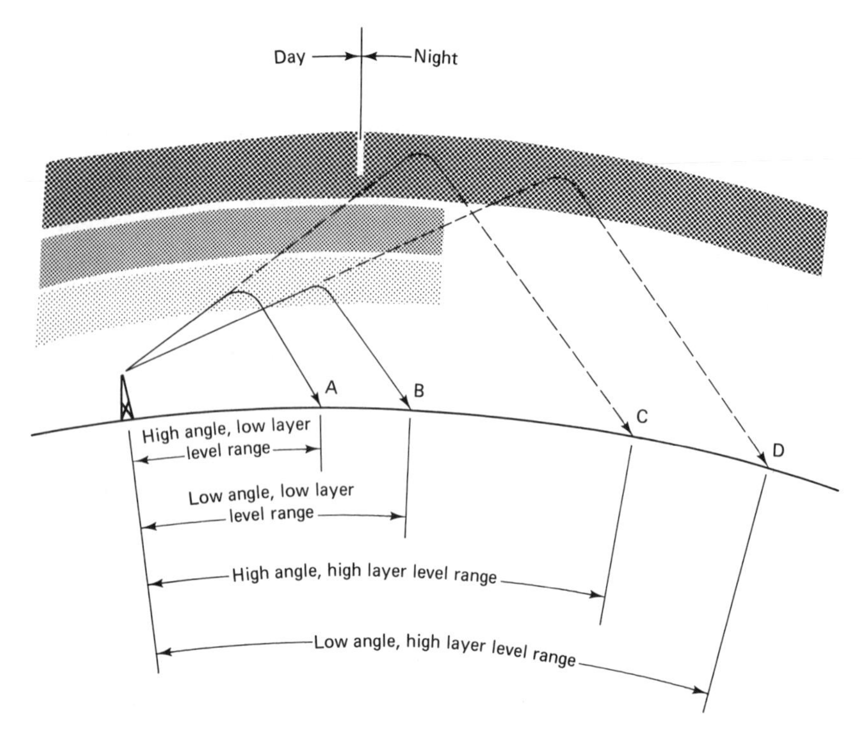

Figure 6.35 Effect of radiation angle and height of ionospheric layer on radio transmission range.

frequencies. The F layer exists from about 90 mi above the earth's surface to an altitude of approximately 250 mi. During times of maximum ionization in the cycles of variation (see above) the F layer is perceived as splitting into two layers: the F_1 layer extending from 90 to about 150 mi above the earth, and the F_2 layer, which exists between distances of approximately 150 and 250 mi above the earth. The F_1 layer is lowest during daylight hours; the two layers appear to merge into one at night and the lower surface is somewhat higher than during the day.

Let us visualize there being many receiving stations located at various distances from a transmitting antenna (refer to Fig. 6.33 again). It is apparent that if we left the transmitter and visited the various stations (imagine being able to monitor all stations at the same moment) we would find that the signals at the various locations would vary over a wide range of amplitudes of field strength. The field strength near the antenna would be relatively high because of the surface wave. Beyond the ground-wave range, a distance of 100 mi or so from the transmitting antenna, receivers would be in what is called the *skip zone,* a region in which the transmitted signal is highly attenuated, or virtually too weak to be received satisfactorily at all. At some considerable distance from the transmitter (a distance determined by many different conditions described above) we would find another area in which the reception is very good as a result of the reflection of the transmitted wave from an ionospheric layer. The distance from the transmitting antenna to such reception points is called the *skip distance*. Skip distance, and the transmission range due to skip propagation, varies with the conditions that alter the characteristics of the ionosphere (see discussion above). Variations in skip range between day and night are illustrated in Fig. 6.35.

Conditions that determine the skip distance for a particular transmission can change very rapidly—within minutes. The result is a sudden attenuation of the received signal, a phenomenon called *fading*. Signals may fade in and out of reception.

Unfortunately, fading does not simply affect the amplitude of a received signal. Because a modulated RF signal represents a band of frequencies, fading may cause distortion of the information signal as well. This is a phenomenon called *selective fading*. Different frequencies are reflected by different amounts. Thus different frequencies, even in a given transmission, may be attenuated differently, producing selective fading and the resulting distortion. Of course, communication under such conditions is undependable, at best.

Reception by way of skip (reflection) is limited to a relatively small region at just the right distance from the transmitter. The signal is again too weak to receive satisfactorily just beyond that location. Under optimum conditions—a signal transmitted parallel to the earth's surface and reflected at maximum height from an F_2 layer—a maximum range of about 2500 mi is possible. How, then, are transmissions regularly made over much greater distances achieved? The answer is *multihop transmission* (see Fig. 6.36).

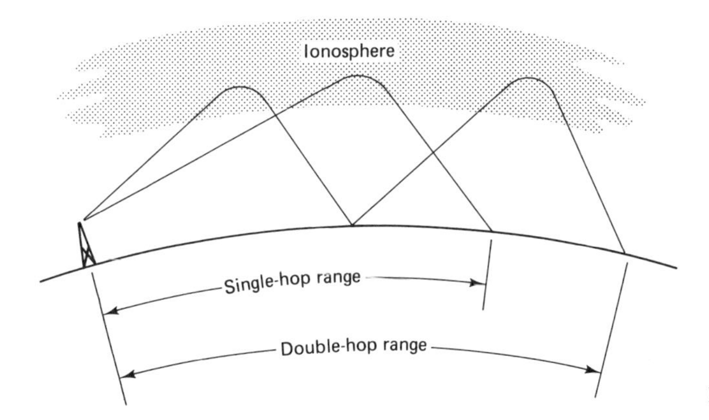

Figure 6.36 Multihop transmission.

In multihop transmission a radio wave is reflected from the ionosphere to the earth and then back to the ionosphere one or more times. There is significant attenuation of a wave when it is reflected; hence there is a very practical limit to the number of hops that can be utilized.

All of the propagation phenomena which are dependent on the reflection of radio waves from the ionosphere are, of course, limited to those frequencies that can be reflected. As a result, the term *maximum usable frequency* is frequently used in discussions involving ionospheric propagation. The maximum usable frequency (MUF) is the highest frequency at which signals will be reflected by the F_2 layer, the layer of greatest ionization (or highest capability to reflect radio waves). It is important to note that the MUF is not the same as the critical frequency (see above). The critical frequency is lower than the MUF because it relates to signals which are sent straight up from the earth and are not reflected. The term "MUF" defines the highest frequency that will be reflected when the angle of incidence is less than 90°. When the ionization of the F_2 layer is low, the critical frequency may be as low as 4 or 5 MHz and the maximum usable frequency as low as 7 or 8 MHz. On the other hand, at noon on a summer day the critical frequency may be up to 8 to 10 MHz and the MUF as high as 70 MHz.

As you have no doubt already realized, there are many factors to be considered if you are involved in "bouncing signals off the ionosphere" (i.e., utilizing transmission paths made possible by reflections from the ionosphere). In addition to the more-or-less predictable, cyclical variations of characteristics of the ionosphere, there are numerous sudden, unpredicted variations. All of this makes working with long-distance radio communication interesting, to say the least. Many amateur radio operators (hams), in fact, do specialize in "DX" operation, as it is called, with great pleasure. These, and others, find satisfaction in studying and predicting propagation conditions.

Auroral Propagation

An aurora consists of luminous (giving off light) bands or streamers of light in the night sky. It is sometimes described as being like draperies of light waving in the wind. A more scientific description is that it is a region of the atmosphere which temporarily has a very high ionization density. The phenomena seem to occur mostly, if not totally, at the poles of the earth. The northern aurora is called the *aurora borealis* and the southern, the *aurora australis*.

A theory of auroras is that the ionization is caused by charged electrical particles (electrons and/or positrons) from the sun that are diverted toward the magnetic poles of the earth. The aurora activity is greatest when the sunspot cycle is at its peak.

As an ionized region of the atmosphere, an aurora is capable of reflecting radio signals. If two stations aim their antennas at an aurora, communication is possible. However, because the nature of an aurora is one of extremely rapid change in quality and position, communication by means of auroral propagation is hardly acceptable. The quality of voice communication by way of a modulated radio signal bounced off of an aurora has been described as similar to that of someone talking through an electric fan.

Propagation via Meteor Trails

A promising new communications technique makes use of the ionized trails made when small meteors enter the upper atmosphere. It is called *meteor-burst communications*. Trails of ionized gas, lasting for only a fraction of a second, occur at a rate of 2 to 8 billion a day. Such trails occur in a region approximately 20 mi (30 km) deep beginning at a height of about 52 mi (85 km). (This is a region centered approximately at the boundary of the D and E layers.) The trails, when they are of high ion density, reflect radio signals, as from ionospheric layers. When the density is low, the signals pass through the trail but in doing so excite the free electrons, which then act as small dipole antennas and reradiate the energy back to the earth.

The technique of meteor-burst communication is successful and economically feasible only because microcomputer technology makes it possible for the communications portion of the system to locate and switch rapidly from one short-lived trail to another. The technique is being used successfully in applications where data (such as snow cover, temperature, precipitation, etc.) are gathered in remote locations and transmitted by way of unmanned stations and meteor trail propagation to a master collection station.

Moonbounce Propagation

Radio amateurs in different parts of the world have communicated with each other by bouncing their radio signals off the moon. [The moon's average distance from the earth is 238,855 mi (384,218 km).] The equipment used was not unusual—1000 W. A critical requirement for moonbounce propagation is that the signal be above the critical frequency to permit it to pass through the ionosphere. Using higher frequencies, however, also permits the use of smaller antennas which can be designed for high gain and aimed more easily than large antennas—a distinct advantage in this application.

6.7 RADIATION MEASUREMENTS AND THEIR UNITS

In order to operate communications systems effectively, or analyze and compare them, it is desirable, if not indeed necessary, to have a means for measuring the performance of such systems. One measure of the overall performance of a transmitter–antenna–propagation system is the signal level receivable at some distance from the radiating antenna. It is desirable that the measured signal level be an *absolute value*. That is, the measured value should be a value which is independent of the characteristics of the antenna used in the measuring process. Thus *the field intensity or field strength of an electromagnetic (radio) wave at a given point in space is equal to the amount of voltage induced in a wire antenna exactly 1 m (39.37 in.) in length located at that point.* In actual measurements it is not necessary that the antenna used for the measurement be a 1-m antenna. If the antenna has been calibrated against a standard 1-m antenna, a correction factor can be applied which will convert a voltage meter reading to an absolute field intensity value. The meter reading may incorporate the conversion factor.

The voltages that are induced in receiving antennas are generally very low—of the order of microvolts. Hence field strength readings are most commonly given in *microvolts per meter*. Commercially available field strength meters are calibrated in this unit. A reading of 38 μV/m taken from a field strength meter means that a voltage of 38 μV would be induced in a wire antenna 1 m in length.

For future reference it is important that you recognize and remember that a field strength specification is like a voltage specification in conventional circuits. For example, consider the situation in which signal power is one of the items of concern and field strength is a given, or measured, quantity. You must remember that signal power is proportional to the square of the field strength. If one field strength value is one-half of a second, the signal power for the first is only one-fourth of that of the second. Conversely, field strength is proportional to the square root of power. If the power being radiated by a transmitter–antenna system is doubled, the field strength of the signal from that system at a given point in space would be increased by a factor of 1.414, and so on.

Example 9

Calculate the percent increase in field strength for an AM radio station when modulation is increased from 0% to 80%.

Solution. The formula for output power for an AM transmitter with modulation index m (where m = percent modulation/100) is

$$P_{\text{modulated}} = \left(1 + \frac{m^2}{2}\right) P_{\text{carrier}}$$

Therefore,

$$\frac{P_{\text{modulated}}}{P_{\text{carrier}}} = 1 + \frac{m^2}{2}$$

$$= 1 + \frac{0.8 \times 0.8}{2}$$

$$= 1.32$$

and

$$\frac{\text{field strength (80\% modulation)}}{\text{field strength (0\% modulation)}} = \sqrt{1.32} = 1.1489$$

$$\text{percent increase in field strength} = (1.1489 - 1) \times 100$$

$$= 15\% \text{ approx.}$$

Field Strength Meters

A commercial *field strength meter* is a portable instrument which is designed to enable the user to quickly and easily measure the field strength of a signal of a specified frequency. The instrument includes a readout meter which may be of either the analog (pointer and scale) or digital type. Instruments are designed to provide either absolute field strength or relative field strength value. The relative field strength instrument (see Fig. 6.37) is the simplest type. It consists of a whip antenna and a receiver tunable to the desired frequency. The receiver is of the superheterodyne type. The readout meter is connected to monitor some point in the receiver, such as an AGC line, which provides a voltage that is proportional to the amplitude of the signal voltage induced in the antenna. The

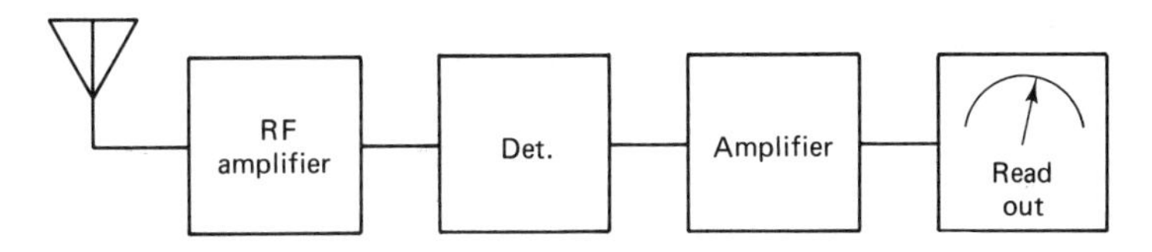

Figure 6.37 Block diagram of field strength meter.

meter is typically labeled to indicate that its readings are to be interpreted as microvolts per meter. If it is a relative field strength meter the readings are meaningless except as a means of comparison.

An instrument capable of providing absolute field strength readings must incorporate some means of calibrating its antenna and meter against standards. The antenna may be calibrated by comparing its output to that of a standard 1-m antenna. The indicating meter itself may be calibrated by a process in which an appropriate signal from a calibrated RF signal generator is injected into the instrument's receiver. A block diagram for such an instrument is shown in Fig. 6.38.

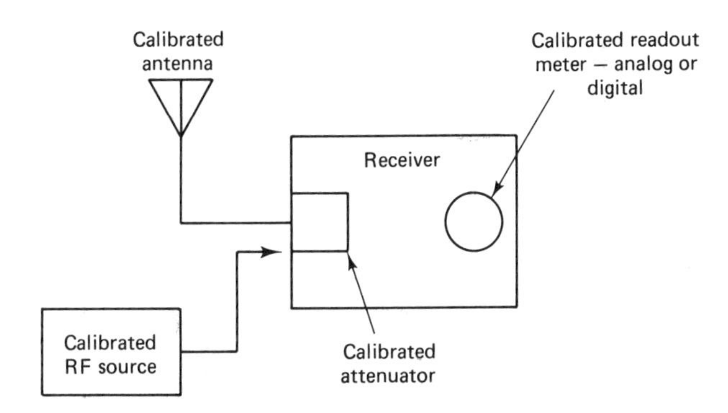

Figure 6.38 Absolute field strength meter.

Field strength meters are very popular test instruments for engineers and technicians involved in installing and servicing television antenna systems and/or cable TV systems. For example, when installing a directional TV antenna it is desirable to aim the antenna so that it provides maximum signal output to the receiver for the available broadcast signals. By connecting the antenna's output to a field strength meter, in place of the instrument's normal test antenna, the best aim of the antenna can be ascertained by rotating the antenna while visually monitoring the meter for maximum output. You can obtain a quick check of the overall performance of a TV cable system by connecting the cable's output, the fitting normally connected to a customer's receiver, as the input to a field strength meter.

Antenna Gain Measurements

An antenna is classified as a *passive device*. This means that it cannot add any energy to a signal that has been fed to it for processing. A circuit containing a transistor or electron tube is an *active circuit* because, if it is designed appropriately, it will convert dc energy to signal energy and thus add energy to the signal it is processing. Active circuits generally have the characteristic called *gain* because they are capable of providing an output signal which has more energy than the input signal. Consequently, passive devices or circuits such as antennas are generally not considered to have gain. They may cause a loss in the signal they are processing and hence be said to have negative gain. However, in spite of the fact that an antenna is a passive device, it is a useful and common practice to use the concept of gain with respect to antennas.

By virtue of its design and physical "architecture," one antenna may be more effective than another in converting and radiating the RF energy fed to it from a transmitter. Such an antenna is properly said to have gain as compared to the less effective one. Or, as we shall see in the next section, it is possible to design an antenna so that it effectively "directs" the energy it is processing in a narrow beam. The result is that compared to an antenna which radiates energy equally in all directions (an omnidirectional antenna), it provides a gain in a desired direction, the direction of its beam.

Principle of Reciprocity

Before proceeding further with the concept of gain as applied to antennas, let us become familiar with an important principle, the principle of *reciprocity*. Reciprocity applies to many electrical circuits, especially simple ones. This principle states that in a network containing one or more sources and one or more impedances, reciprocity exists if a voltage V applied in branch 1 causes a current I to flow in branch 2, and the same V applied in branch 2 also causes current I to flow in branch 1.

Reciprocity is true of all *linear, passive* networks. It is true of antennas. When applied to antennas it means that a given antenna will have the same characteristics—gain, directivity, input impedance, etc.—whether used as a receiving or a transmitting antenna. One of the practical implications of this principle is that the characteristics of a given antenna may be ascertained by testing it as a receiving antenna. This may be far simpler and more practical than testing it as a radiating antenna.

We return to gain measurements. It is possible to measure the performance of any two antennas with a field strength meter and compare the readings. If the antennas are mounted and oriented to receive exactly the same radiation, then if one provides a higher reading than the other, it "has gain" with respect to the second. This *relative gain* of the one over the other would be calculated as follows:

Let V_1 be field strength reading for antenna 1 and V_2 be field strength reading for antenna 2; then the relative gain of antenna 1 with respect to antenna 2 is

$$\text{gain (dB)} = 20 \log \frac{V_1}{V_2} \tag{6.1}$$

Remember, gain in decibels is by definition a comparison of two power levels, say P_1 and P_2:

$$\text{gain (dB)} = 10 \log \frac{P_1}{P_2} \tag{6.2}$$

Observe that in Eq. (6.2) the multiplier of the logarithm (to base 10) of the power ratio is 10. In Eq. (6.1) the multiplier is 20. The multiplier must be doubled when using field strength readings because signal power is proportional to the square of the field strength. (For a comprehensive presentation on decibels, see Appendix A.)

Example 10

While receiving exactly the same radiation, antenna 1 provides a relative field strength reading of 27 μV/m and antenna 2 a reading of 3.9 μV/m. Calculate the gain of antenna 1 with respect to antenna 2.

Solution

$$\begin{aligned}\text{gain (dB)} &= 20 \log \frac{V_1}{V_2} \\ &= 20 \log \frac{27}{3.9} \\ &= 16.81 \text{ dB} \quad (= 17 \text{ dB approx.})\end{aligned}$$

More common than comparing just any two antennas is the practice of comparing the performance of an antenna of interest with that of a basic *isotropic antenna*. An isotropic antenna is one that radiates equally in all directions. It is impossible to physically achieve an antenna which behaves in this way, so that it is said to be a *hypothetical* (assumed or supposed) antenna. Therefore, the field strength of an isotropic antenna cannot be measured. However, it can be calculated.

In technical literature—journal articles, catalogs, etc.—an antenna will often be described as having x dB gain. At first glance, we are apt to conclude that a decibel is an absolute unit, like volt or ampere. It is important to know that such is not the case. *A decibel value always represents a comparison.* It is true that there are decibel units that are used as if they were absolute. For example, *dBm* is one such unit. It means the number of decibels a given power level is above or below an accepted zero level of 1 mW. This is not a standardized unit but it is accepted and used extensively in certain industries, namely, telephone and broadcasting.

The gain unit for antennas is often written "dBi," the "i" being used to designate "isotropic." Therefore, an antenna with a gain of, say 10 dBi, is an antenna whose field strength is 3.16 times that of an isotropic antenna (N dB $= 20 \log 3.16 = 10$). (A half-wave dipole in free space has 2.1 dBi.)

Unfortunately the utilization of "dbi" is not universal in the world. Because the isotropic antenna is hypothetical, it is a popular practice to use antenna dB-gain figures which compare a given antenna with the half-wave dipole. When an antenna gain figure is given simply in "dB" it is difficult to decide what is meant—dBi, or comparison with a half-wave dipole? Sometimes the author's intent can be determined from the context. In any event, if one assumes that dBi is meant, when comparison with a half-wave dipole is actually the case, the gain figure will be low by 2.1 dB (see above).

Example 11

Under identical conditions the output of antenna X is 75 μV/m when the output of a half-wave dipole is 12.6 μV/m. Using the popular convention (gain over $\lambda/2$ dipole), what is the gain of antenna X? What is the gain of X in dBi?

Solution

$$N \text{ dB} = 20 \log \frac{\text{field strength of } X}{\text{field strength of dipole}}$$

$$= 20 \log \frac{75}{12.6}$$

$$= 15.49$$

The antenna gain would be specified as 15 dB.

$$N \text{ dBi} = N \text{ dB} + 2.1 = 15.49 + 2.1 = 17.59 \text{ dBi}$$

6.8 DIRECTIONAL ANTENNAS

Some radio station operators wish to communicate with the entire world. Hence they use *omnidirectional antennas* to radiate their signals. It is more common for a broadcast service—radio or TV—to be interested in reaching (or be required to reach) a limited audience. In this case, an antenna that "beams" a radiated signal in a limited number of directions is more appropriate. Thus we have the need for *directional antennas.* If you think about it you will realize that you are already familiar with directional antennas, at least in a limited sense. You probably see them in many different shapes mounted on the roofs of the homes in your city or town. Most TV receiving antennas are *directional.* That is, they are constructed so that they receive, very efficiently, signals coming from one direction and, very poorly, signals coming from all other directions. You may also have noticed the directional array of a ham operator mounted on the roof or in the backyard of a house in your neighborhood. Let us examine the principles of why and how radiation can be directed: the fundamentals of directional antennas.

At frequencies below those which produce microwaves (say 1 GHz), directional (or directive) antennas are generally made by adding one or more elements to a basic element, usually a half-wave dipole. The additional element(s) may be *passive* or *driven.* A passive element simply "sits" there and does its "thing." It intercepts radiated energy from the original active element and reradiates that energy in a way that reinforces, or adds to, the radiation from the main, driven element. When it is of the appropriate length and mounted in a particular location with respect to the driven element, the additional element enhances the radiation intensity in one direction at the expense of radiation in the opposite direction. When mounted on the side away from the direction of maximum intensity, the passive element is called a *reflector.* If the passive element is mounted on the side of the direction of maximum signal, it is called a *director.* Examples of a simple half-wave dipole with a reflector and a director are shown in Fig. 6.39. When an element is connected directly to the feed transmission line (from a transmitter or to a receiver) it is called a *driven element.*

Let us consider the arrangement in Fig. 6.39(a) to learn how directional performance is obtained in an antenna array. Recall that the radiation pattern of an isolated (from the earth or surrounding objects), horizontally polarized half-wave dipole is like a doughnut placed around the antenna with the axis of the doughnut coinciding with the axis of the dipole. The direction of maximum radiation is at right angles to the length of the antenna. Looked at in a horizontal plane, the radiation pattern includes two lobes of maximum intensity.

If we now place another 0.5λ element at a distance of 0.25λ from the main, driven element, and parallel to it, the following happens when the driven element is excited by an RF signal. First, some of the radiation from the excited element reaches the passive element (also called a *parasitic* element) and *induces* a voltage in that element as it cuts across it. The induced voltage is 180° out of phase with the wave (recall the phase relationships in a transformer). A current, in phase with the voltage, flows in the element producing a reradiated wave.

Meanwhile, at the driven element, the phase of the exciting signal has advanced by 90°, a change that occurs in the time required for the wave to travel the distance of 0.25λ from driven element to parasitic element. At the parasitic element, the reradiated wave, which is nearly equal in amplitude to the original wave, virtually cancels the original wave on the side away from the driven element. Remember, the reradiated wave is 180° reversed from the original wave.

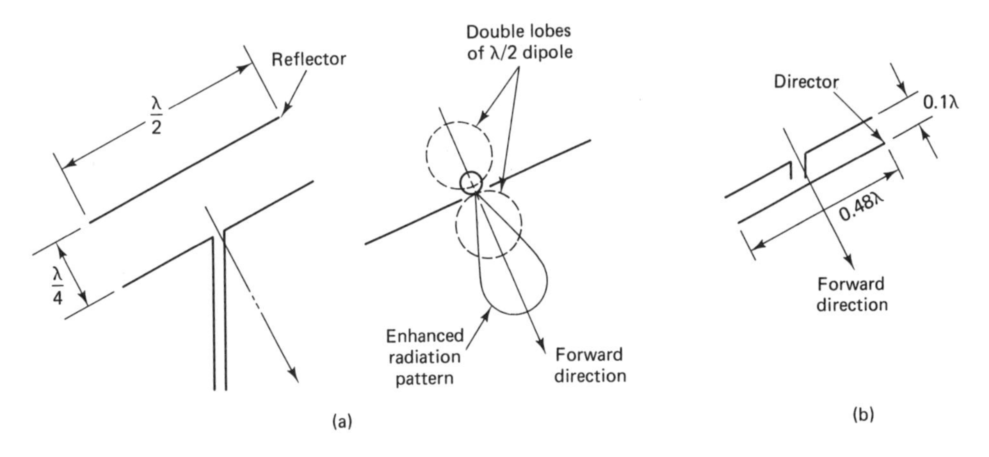

Figure 6.39 (a) Half-wave dipole with reflector; (b) half-wave dipole with single director.

There is very little radiation, now, in that direction. However, there is a reradiated wave traveling back across the 0.25λ space to the driven element. It arrives there just at the moment that the exciting signal has advanced another 90° in time phase. The wave being radiated fresh and the reradiated wave are now in phase, and, of course, combine to produce a wave with nearly twice the amplitude of a wave from an antenna without a reflector element.

The direction of the enhanced wave is called the *forward direction* of the antenna. Please take note that forward direction is toward the side *without* the reflector. As indicated previously, reciprocity applies to an antenna. Although the explanation here is for a radiating antenna, a receiving antenna made of a half-wave dipole and a reflector is also directive.

The explanation in the preceding paragraphs is a theoretical explanation of why a simple directive antenna performs as it does. In actual practice, however, we find that slightly different dimensions provide a greater gain in the desired direction. Best results are obtained when the length of the reflector is about 5% longer than 0.5λ. Similarly, an actual spacing of approximately 0.18λ to 0.20λ gives optimum results.

If a passive element is made about 5% shorter than a driven element and placed approximately 0.1λ away from it, the radiation toward the side of this parasitic element is enhanced. The element is called a *director* [see Fig. 6.39(b)]. Further directivity can be achieved if both a director and a reflector are used (see Fig. 6.40). In fact, the typical home TV antenna for relatively weak signal areas includes one reflector element and several director elements.

Directional antennas are popular as home TV antennas for two reasons: Directivity increases the effectiveness (gain) of an antenna in its forward direction, and signal amplitudes are increased at the expense of the random noise signals of the environment. Directivity is helpful in reducing signals that can cause ghosts, signals that arrive at the antenna via paths other than the direct path from the radiating antenna (see ground-reflected waves above).

When an antenna contains several passive elements it is called a *yagi array* (in honor of its inventor, a twentieth-century Japanese engineer). Yagi arrays are found of many shapes and forms, both in TV antennas and in antennas used by ham radio operators. As more elements are added, greater directivity and forward-direction gain are achieved, but at a cost. The cost is a reduction in an antenna property called *radiation resistance*.

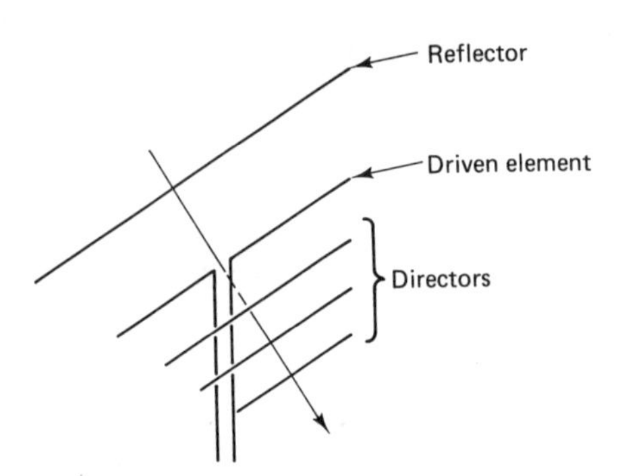

Figure 6.40 Directional antenna with reflector and three directors.

The term "radiation resistance" does not describe a real ohmic resistance at all but is a kind of imaginary property of an antenna used in explaining the loss of energy from an antenna system by way of radiation. For example, it is used in the formula for radiated power,

$$P_{rad} = I_{ant}^2 R_{rad}$$

Additional elements added to an antenna can be likened to additional branches added to a parallel circuit—they reduce the total resistance. To achieve the same amount of radiated power with reduced radiation resistance, R_{rad}, the antenna current, I_{ant}, must be increased. Increased antenna current means increased I^2R losses in the transmission line and antenna elements. The result is that, practically, the losses soon approximate the gain achieved by adding the additional elements. There is a practical limit, then, to the number of elements that can be used.

Another limit to the application of a yagi array is an antenna property called its *bandwidth*. Bandwidth describes the set of frequencies over which the other important properties of the antenna remain usable in the sense of its being effective in radiating a signal. For example, we have stated the dimensions of a yagi array in terms of the wavelength of the signal it is radiating. Obviously, if the antenna is excited by signals whose frequencies differ greatly from the frequency used in stating its dimensions, it is not likely to function as intended. Such frequencies are outside its bandwidth.

Driven Arrays

In another class of antennas in which there are a number of elements arranged in an array, the elements are all driven, or excited, by the signal from the transmitter. Such a system is called a *driven array*. Driven arrays are generally classified as *collinear arrays, broadside arrays,* or *end-fire arrays*. Antennas using combinations of two or more of these arrangements are not uncommon, particularly in the amateur radio service.

Collinear arrays. If we again start with a simple half-wave dipole and add other elements, but place the elements on a line with the original element instead of in parallel with it, we will have created a *collinear array*. ("Collinear" means *in the same straight line*.) In Fig. 6.41, an additional $1/4\lambda$ element has been connected to the end of each of the original elements of the dipole. In effect, there is now a $\lambda/2$ element on each side of the transmission-line feed point. There is one current in each element. If we think of an element as two $\lambda/4$ elements in series, the currents in the two $\lambda/4$ elements are in phase. Carrying this idea a little further, we have two dipole antennas mounted on the same line and excited with in-phase currents. Their radiation patterns will be superimposed. The resultant pattern will be of the same general shape as that for a single dipole—a circle in the plane perpendicular to the line of the antenna, a "figure 8" pattern in the horizontal plane and a doughnut in the three-dimensional view (see Fig. 6.23). The resultant pattern will be different in that the doughnut will be flatter but bigger. The radiation intensity in the direction of maximum

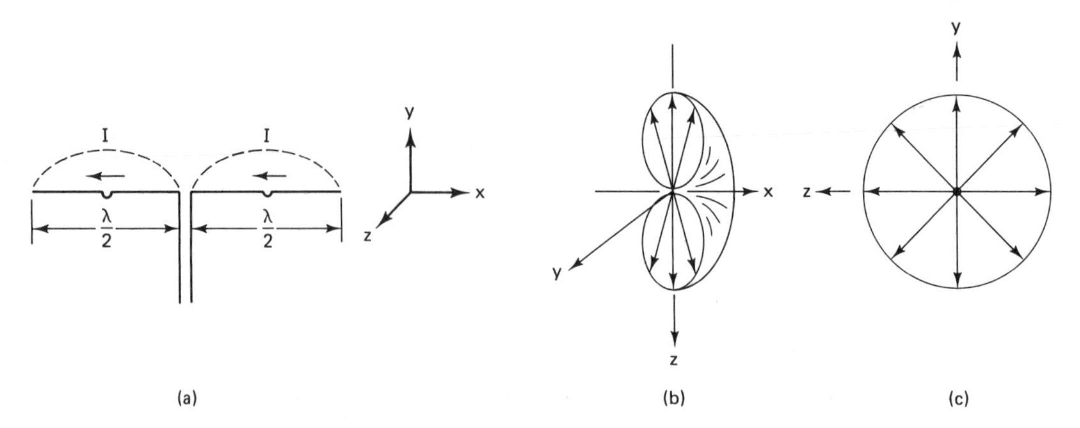

Figure 6.41 Collinear antenna: (a) collinear array; (b) radiation pattern in *x-z* plane (horizontal plane for horizontally polarized array); (c) radiation in *y-z* (vertical) plane.

radiation will be doubled, approximately (making the doughnut bigger). The radiation in other directions will be diminished relative to the maximum radiation (making the doughnut flatter).

Most applications of this array will use horizontal polarization because the additional length makes vertical mounting less practicable. With horizontal polarization, the radiation pattern in the horizontal plane is bidirectional (instead of unidirectional as with a yagi array). The array achieves directivity in the sense that it produces a narrower beam with a gain of about 2 (6 dB) as compared to the radiation of a simple dipole.

Broadside arrays. The title "broadside array" is so descriptive it hardly seems to require further elaboration. In a broadside array, elements are mounted in parallel, physically, just as in a yagi array. However, in this array it is arranged so that all elements receive direct excitation from the source, all are driven. With appropriate phasing of the excitation, the direction of maximum radiation will be at right angles to the "broad side" or plane of the array rather than perpendicular to the line of the elements in the plane of the array, as with a yagi.

A broadside array with four parallel, vertical half-wave elements is illustrated in Fig. 6.42. Notice that all four elements are connected to the feed transmission line. Since the spacing between elements is $\lambda/2$, the connections between elements is crossed, thereby providing in-phase signals at each element (while a signal is

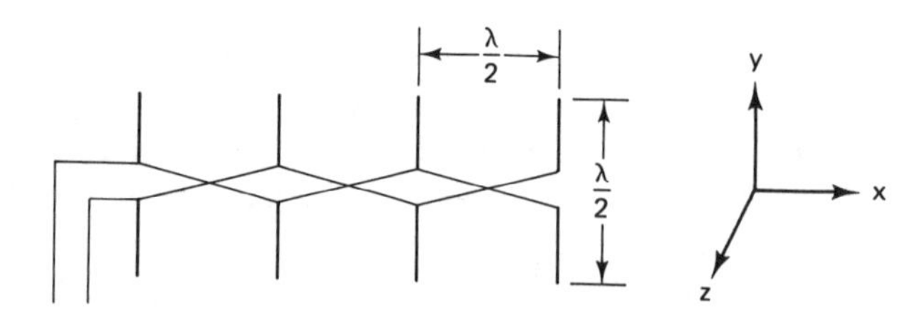

Figure 6.42 Vertically polarized broadside array.

traveling a half-wavelength along a line, the input signal is advancing 180° in time phase).

We can imagine each dipole producing its own circular pattern in the horizontal plane. Consider the radiation that would reach any point on the z-axis (see Fig. 6.42), in front of or behind the array, and at a significant distance from it. Since the waves leaving the elements are in phase, and since the distances traveled are equal, at least when the total distance is significantly greater than $\lambda/2$, the waves will be in phase on arrival. The radiation intensity at any such point will be a summation of the intensities contributed by the individual elements. Along the z-axis the gain will be approximately 12 dB.

At any point along the x-axis, a wave from one element will cancel the wave from the next element because the difference in distances traveled by waves from adjacent elements is $\lambda/2$, the distance between elements. The radiation from the edges of the array will be minimal.

At locations on the horizontal plane between the x- and z-axes, there will be partial cancellation (or partial summation) of waves. The difference in distances traveled by waves from adjacent elements to points at such locations will be equal to $\lambda/2$ times the cosine of the angle between the x-axis and the direction of travel.

Radiation from the broadside array in the vertical direction (up and down) will be minimal. The radiation off the ends of any half-wave dipole is very small. The overall pattern of radiation from an array of the type shown in Fig. 6.42, then, is one with a relatively narrow, bidirectional beam along the z-axis of Fig. 6.42. A three-dimensional view of the pattern is shown in Fig. 6.43.

The discussion on the broadside array in the preceding paragraphs centered on an example made up of vertical, half-wave elements. The technique is also very commonly applied to three or four grounded, quarter-wave verticals of AM

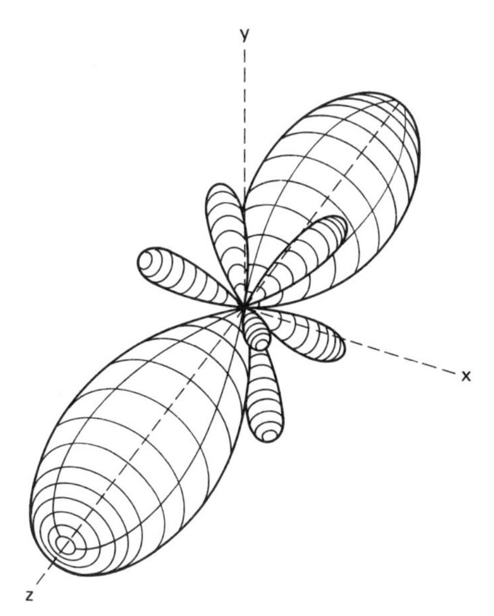

Figure 6.43 Three-dimensional radiation pattern of broadside array.

broadcast stations. By adjusting the phasing of the signals fed to the various elements of an array, it is possible to achieve any number of different radiation patterns. For example, some stations are required, by terms of their licensing permit, to have a different radiation pattern at night from that used during daylight hours. The antenna systems of such stations include provisions for quickly and easily changing the phasing of signals fed to the antenna towers in the array. Arrays of this type, although a form of the broadside category, are often called *phased arrays*.

The major factor influencing the making of agreements that require a station to change its daytime antenna patterns to nighttime patterns is one of changing propagation conditions. The daytime range of a station may be only 100 mi. Its nighttime range, because the ionosphere has shifted to a greater elevation, may be several hundred miles. The probability of one AM station interfering with another station, even when they are several hundred miles apart, is increased significantly at night because of the increased nighttime range.

End-fire arrays. In physical appearance, an end-fire array is quite similar to a broadside array. See Fig. 6.44(a), the diagram of an end-fire array with two half-wave dipoles mounted vertically and parallel to each other. They are separated by a space of $\lambda/2$ and are fed from the same transmission line. Since there is no twist in the line between them, the actual signals at the two elements are shifted 180° with respect to each other. The waves leave their respective elements 180° out of phase. Let's look first at what happens to the waves along the x-axis, both to the left of the array (negative portion of x-axis) and to its right (positive portion of x-axis). Along the positive x-axis, the waves from two adjacent antenna elements will be in phase. Of two adjacent elements, the wave from the one on the right travels $\lambda/2$ less in distance, but it leaves its element 180° later in time phase. Going to the left, a wave from the element on the right leaves 180° later than a wave from the element on the left and travels a half-wavelength farther—a total lag of 360°. As a result, the two waves are also in phase to the left of the array. With a two-element array, the gain in either direction along a line off the end of the array will be 2, or 6 dB, compared to a half-wave dipole.

There will be virtually no radiation off the top or bottom of the array because these areas are equivalent to the ends of the individual dipoles. As we have seen previously, there is no radiation off the end of a dipole. Radiation will cancel along the z-axis in front of or in back of the broad side of the array.

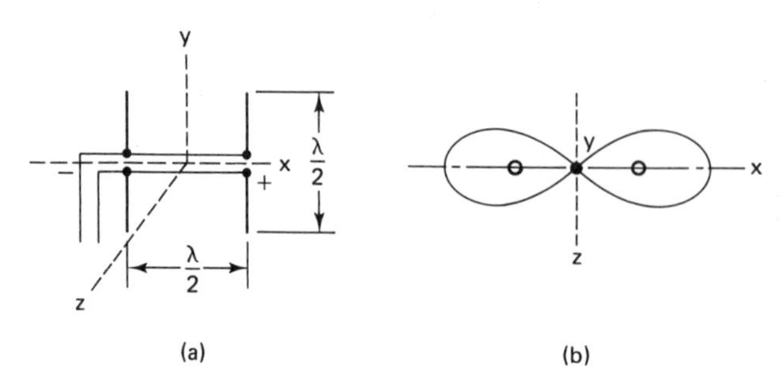

Figure 6.44 (a) Vertically polarized end-fire array; (b) radiation pattern, horizontal plane, of end-fire array.

This is true because waves arriving at points on this axis are 180° out of phase. The overall radiation pattern of this array is a relatively narrow, bidirectional beam along the x-axis of the diagram—off the ends of the array. This, of course, is why the array is called an end-fire array. [See the radiation pattern in Fig. 6.44(b).] It is not uncommon for an array to contain several elements instead of the two illustrated in this discussion.

The end-fire array can also be used to produce a heart-shaped, or *cardioid,* radiation pattern. Consider the array of Fig. 6.45. The spacing between the elements is now $\lambda/4$. The phase delay between elements is 90°, because of the time required for the signal to travel the quarter-wave distance between elements. To the right of the array, along the positive x-axis, the waves from the two elements will arrive at a given point at the same time, and thus reinforce each other. To the left of the array the effect is now quite different. The wave from the element on the right is displaced by 180° with respect to the wave from the element on the left—it leaves its element 90° later in time and has to travel a quarter-wavelength farther to arrive at a given point. The radiation in the direction to the left of the array is virtually eliminated. The overall pattern for a two-element array is that shown in Fig. 6.45(b). End-fire arrays with several (four to eight) elements are used to achieve unidirectional patterns with high directional gain.

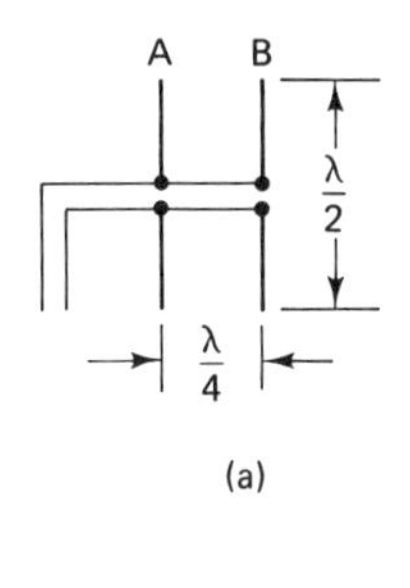

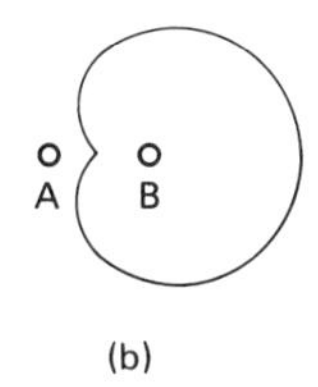

Figure 6.45 (a) End-fire array with 90° phasing; (b) cardioid radiation pattern for (a).

Nonresonant Antennas

All of the antenna systems that have been described in this section are called *resonant* antennas or systems. This term is properly applied to the units discussed because each has been based on an element or elements that are $\lambda/2$ in length. The properties that have been described generally apply only when the physical lengths of the elements correspond to the electrical $\lambda/2$. A given antenna or array of the types described is useful only for a limited range of frequencies (a narrow bandwidth). In many applications, that limitation is completely acceptable.

Some applications, however, are better served by an antenna or antenna system that has good radiation characteristics over a broad range of frequencies. Antennas possessing desirable characteristics over a wide range of frequencies must be *nonresonant;* they are also *broadband devices.*

Consider a dipole antenna in which each of the two elements is one wavelength long (see Fig. 6.46). The currents in the two elements are now 180° out of phase. There will be cancellation of radiation in any direction perpendicular to the axis of the antenna (the direction of maximum radiation for a half-wave dipole) because waves traveling in such a direction are also 180° out of phase. However, at angles of less than 90° with respect to the axis there will not be complete cancellation because, even though the waves leave their elements 180° out of phase, distances traveled are unequal. It has been found that this antenna provides a radiation pattern of the type shown in Fig. 6.46. Maximum radiation occurs at an angle of 54° with respect to the axis. The pattern shown is for the horizontal plane when the axis is also horizontal. The pattern, however, is three-dimensional—it is identical in the vertical plane of this antenna. Of great practical significance is the fact that the angles of the lobes of maximum radiation decrease as the length of the antenna is increased. The result is that a so-called *long-wire antenna* is a directive antenna with a relatively strong bidirectional pattern. The axis of maximum radiation approaches that of the antenna. The pattern of the long-wire antenna can be made to be unidirectional simply by terminating the free end with a resistance that has a value which minimizes standing waves on the wire. The radiation is directed toward the terminated end.

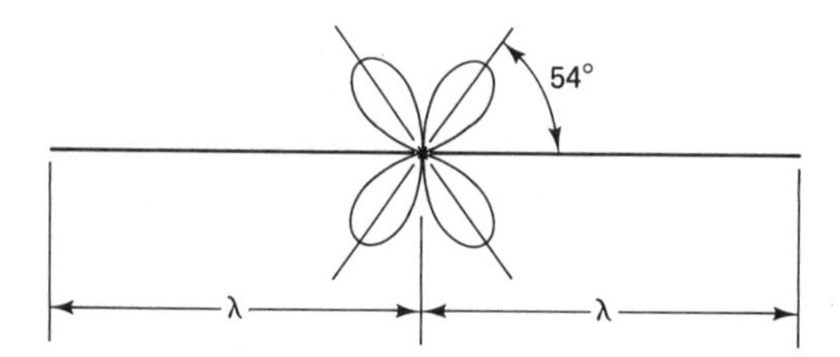

Figure 6.46 Long-wire antenna.

A long-wire antenna which is several wavelengths long for a given frequency is, of course, even longer for all higher frequencies. The greater the electrical length of such an antenna, the more pronounced is its distinctive characteristics. The long-wire antenna is, therefore, a broadband antenna.

Rhombic antennas. There are many forms of long-wire antennas, with many varied characteristics. Because of the limitation of space, we will examine only one of the special forms of long-wire antennas—the *rhombic antenna.* This antenna takes its name from a geometric figure, the *rhombus.* A rhombus is a four-sided figure in which the sides are all of equal length and opposite sides are parallel. The definition especially applies to such figures in which one pair of angles is greater than 90°. (A square is a special form of rhombus.) A drawing of a rhombic antenna is shown in Fig. 6.47. You will note that the diagram includes the radiation pattern for each arm of the antenna, considered as a separate long-wire antenna.

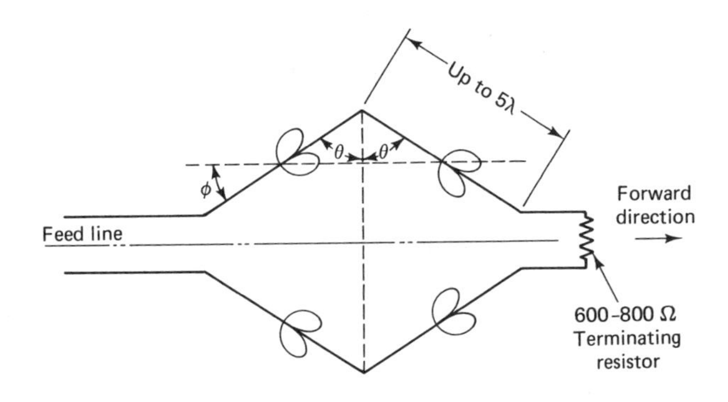

Figure 6.47 Rhombic antenna.

Rhombic antennas are used most commonly in the frequency range between 30 and 70 MHz, although some have been used for frequencies as low as 1 MHz. Each arm of the rhombus may be made up to five wavelengths long at the design frequency. Thus the antenna typically occupies a large space. For example, at 30 MHz the overall length may be of the order of 80 m (over 240 ft). The antenna is nearly always constructed in the horizontal plane. Because of the size, rhombic antennas are typically mounted between poles such as are used for power or telephone lines.

If constructed with appropriate proportions, a rhombic antenna has a very strong directional characteristic. The line of maximum radiation coincides with the longer diagonal of the rhombus (see Fig. 6.47). When terminated in a resistance of appropriate value (usually 600 to 800 Ω), the antenna has a unidirectional pattern with an excellent front-to-back ratio. (*Front-to-back ratio* is the ratio of radiated power density from the side of greatest radiation of an antenna, to the power density from the side of least radiation.) When operated in the unidirectional mode, a rhombic antenna with arm lengths of four to five wavelengths has a gain of up to 40 (16 dB) over a half-wave dipole.

The directivity characteristic of a rhombic antenna depends greatly on the angle, called *tilt angle,* between the short diagonal and the arms of the figure. The tilt angle is designated θ on the drawing of Fig. 6.47. When the tilt angle is adjusted so that it is equal to 90° minus ϕ (see Fig. 6.47), the angle between the major lobes and the desired line of radiation, maximum directional gain will be obtained. This is true because the major lobes in the forward direction are then in the same direction, the axis of the antenna. The sum of their power densities will be the arithmetic sum of the full value of each, not just a vector sum—the sum of in-phase components. At the same time, radiation in all other directions will be canceled almost completely.

Loop Antennas

Loop antennas are important to the study of communications systems if only because there are literally millions, perhaps billions, of them in the world. Some form of loop antenna has been an integral part of virtually every AM broadcast band (the 0.5- to 1.6-MHz band) receiver built since the 1930s. The loop antenna is popular as a receiving antenna for medium-wavelength signals because it has reasonably good signal gathering characteristics even when implemented in a very compact package. It is not a good transmitting antenna and is used for radiating a signal only for special applications.

A loop antenna can be constructed in many different forms. It consists of a coil of wire wound in the shape of a circle, square, diamond, etc. (see Fig. 6.48). A basic feature of many loop antennas is that the diameter (or diagonal) of the loop is small compared to the wavelength of the signal for which it is being used. When this is true, the currents in the coil are in phase. This condition creates a radiation pattern which is almost identical to that of a dipole. The equivalent dipole is positioned at the center of the loop, with its length perpendicular to the plane of the loop. The radiation pattern [see Fig. 6.49(b)] is doughnut shaped—the central plane of the doughnut coincides with the plane of the loop, and the centerline of the doughnut is perpendicular to the loop and through its center. The directional qualities of a loop antenna are utilized in one special application called a *radio direction finder*.

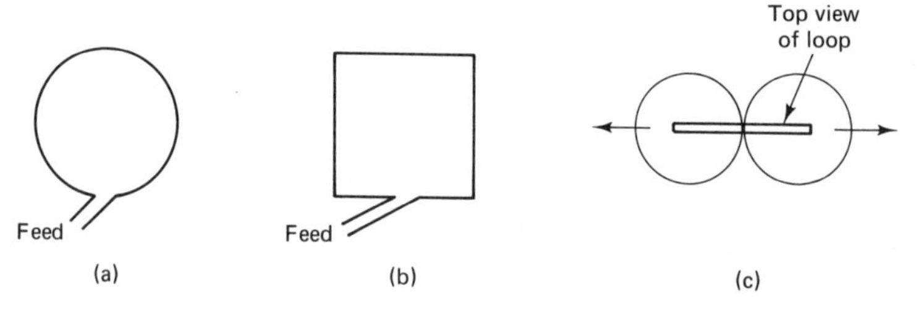

Figure 6.48 (a) Circular loop; (b) square loop; (c) radiation pattern of loop antenna.

The *loopstick antennas* used in AM broadcast band receivers consist of a length of conductor wound on a ferrite (powdered-iron) core (see Fig. 6.49). The core improves the efficiency of the loop in intercepting the energy of radio waves that pass over it. In most applications, the core also serves as the core of a transformer. The primary of the transformer is the antenna coil and the secondary connects to the input of an RF amplifier or selection circuit, coupling the antenna signal to the receiver. The antenna pattern of the loopstick can be visualized by considering the ferrite bar as a dipole. When the dipole (bar) is horizontal, the pattern is a figure 8 in the horizontal plane. It is directional; changing the orientation of the receiver/bar changes the received signal level. When the dipole

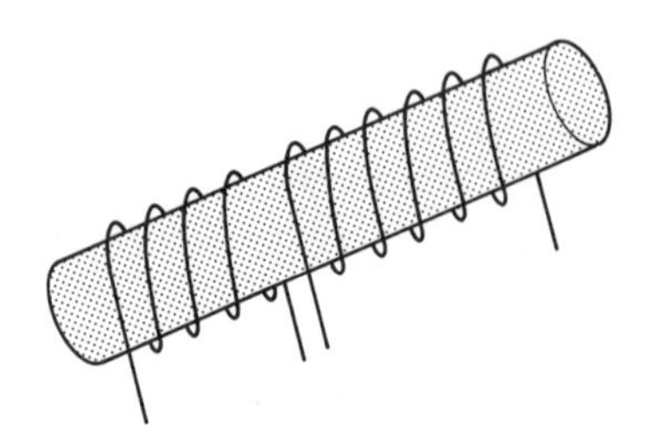

Figure 6.49 Loopstick antenna.

(bar) is vertical, the pattern in the horizontal plane is a circle, the circumference of the doughnut—the pattern is nondirectional (see Fig. 6.50).

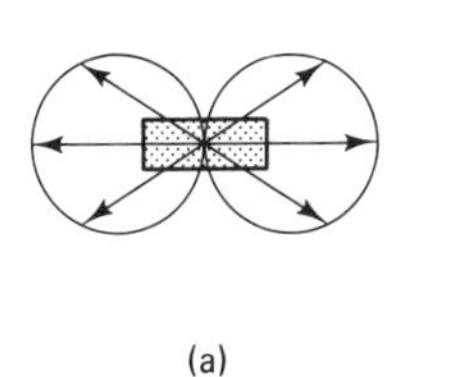

(a)

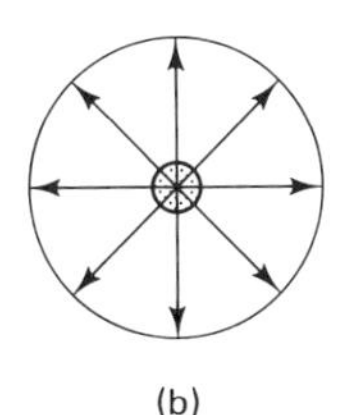

(b)

Figure 6.50 Horizontal radiation patterns of loopstick antenna: (a) core of loopstick horizontal; (b) core vertical.

6.9 ELECTRICAL PARAMETERS OF ANTENNAS

In the two preceding sections we have considered antennas in terms of their actual radiation effects. However, to produce radiation, an antenna has to be excited or driven by an RF signal. This means that the antenna is part of an electrical circuit; it is a circuit component with its own peculiar parameters. Like other components, a given antenna will have boundary characteristics such as the maximum voltage at which it can be operated without arcing or some other form of breakdown. An antenna will also have a maximum current rating. However, of greater importance than these are an antenna's input impedance and radiation resistance.

Input Impedance

When connected to an antenna, a transmission line "sees" the equivalent of a simple two-terminal impedance. This two-terminal impedance seen by the transmission line is called the *input,* or *driving-point impedance* of the antenna. It is also sometimes called the *characteristic impedance* of the antenna. In any case, it is equal to the voltage at the terminals of the antenna divided by the current at that point.

For a given driving-point signal voltage, the current taken by an antenna depends on many factors, for example, the electrical length of the antenna and its proximity, and hence electrical relationship to ground and/or other nearby conductors, to name just two. The input impedance, then, is influenced by these same factors. The influence of nearby conductors, such as other antennas or antenna elements, is similar to the effect of the mutual inductance of a coil that has both self- and mutual inductance. In fact, an antenna that is not completely isolated from all other conductors is said to have both a self-impedance and a mutual impedance. The influence the length of the elements of an antenna has on its impedance is analogous to the effect of the electrical length of a transmission line stub on its input impedance.

Let us first consider the driving-point impedance of a simple half-wave dipole which is remote from any objects that could influence its operation. The transmission line is connected across a gap of insignificant width at the center of the antenna—a typical mode of connection of practical lines and antennas (see Fig. 6.51). The distribution of voltage and current standing waves is depicted

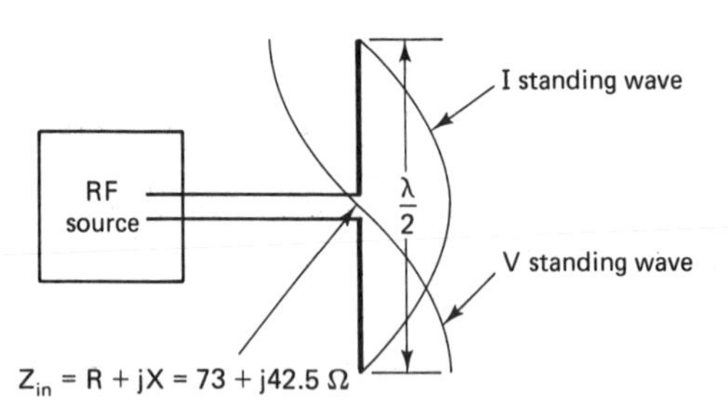

Figure 6.51 Driving-point impedance of half-wave dipole antenna.

superimposed over the antenna in Fig. 6.51. You will note that the current is zero at each end and is symmetrical with respect to the center of the antenna. The wave has a loop at the center. The voltage is a minimum at the center and a maximum at each end. The voltages at the ends are of opposite phase. The impedance of such an antenna, using the sophisticated methods of engineering and higher mathematics, has been calculated to be

$$Z_0 = R + jX = 73 + j42.5\ \Omega$$

Since the value of X is 42.5 Ω and not 0 Ω, it is apparent that a dipole which is exactly $\frac{1}{2}$ wavelength long is not a resonant antenna. Practical, so-called half-wave antennas are made a few percent shorter than $\lambda/2$ to make them resonant. The parameter R is called the *self-resistance* of the antenna. It is also the characteristic called *radiation resistance* (see above). The value of R is slightly less than 73 Ω when the antenna is shortened to make it resonant.

Radiation resistance is an important characteristic or parameter of antennas. It determines the amount of power radiated for a given signal voltage, or antenna current,

$$P_{\text{rad}} = \frac{V_{\text{sig}}^2}{R_{\text{rad}}} = I_{\text{ant}}^2 R_{\text{rad}}$$

The radiation resistance is also a key factor in determining the electrical efficiency of an antenna, that is, the ratio of power radiated to input power. Input power is equal to power radiated plus power losses. Power losses in an antenna/transmission line system are due to signal currents flowing in the ohmic resistances of the line and antenna elements. When radiation resistance is high, these resistances are comparatively small. However, if radiation resistance is reduced and the ohmic resistances are not reduced proportionately, the efficiency suffers.

The input or driving-point impedance of an antenna depends on numerous factors, including the following: length of the antenna, size and shape of the conductor of which the antenna is constructed, feed point of the antenna with respect to its length, and proximity of ground and/or other antennas or conductors. Knowing or being able to determine the impedance of an antenna is desirable because it permits an operator to exercise control in matching the components of a system—transmitter or receiver, transmission line, and antenna—for maximum performance of the system.

For radiating systems, knowing the radiation resistance is especially important because that value may be used in determining the power being radiated by a station. In fact, an accurate knowledge of the radiation resistance of a station's antenna is required by governmental regulation for certain types of stations. It is required of AM broadcast stations in the United States, for example. The radiation resistance is used by such stations in the required process of monitoring station power. The process includes continuous monitoring of antenna current. Station power is determined by the so-called *direct methods:* $P_{rad} = I^2_{ant}R_{rad}$.

There are two basic approaches to the determination of input impedance of an antenna. One method makes use of an *impedance bridge* to measure the impedance of the antenna, just as if it were another two-terminal impedance component. A second approach uses measured standing-wave ratios on the transmission line feeding the antenna.

Impedance Bridge Method

An impedance bridge is very much like a Wheatstone bridge. You will recall that a Wheatstone bridge is an instrument for measuring the dc resistance of circuit components. The "bridge" is essentially two circuit branches in parallel, with each of the branches containing two components in series. When the bridge is "balanced" there is no voltage difference between the nodes (points of connection of the two components) of the parallel branches. This condition can be "detected" by a meter connected between the two nodes. Also, when the bridge is balanced, the ratio of the two components in one of the branches is equal to the ratio of the corresponding components in the second branch. If the values of three components are known, the value of the fourth can be calculated.

An impedance bridge is a "Wheatstone bridge for ac impedances." The circuit configuration of an impedance bridge is identical to a Wheatstone bridge except that the arms of the bridge may contain reactance as well as resistance, instead of resistance only. Further, the source activating the bridge must be a sinusoidal ac source. An unknown impedance, such as an antenna, is connected to the appropriate terminals and becomes one of the arms of the bridge. The bridge is balanced with the aid of some form of sensitive ac meter. The unknown impedance is calculated using the known values of the other three arms of the bridge. Commercial impedance bridges are now available which contain microprocessors to facilitate the selection of known impedances and to test oscillator frequencies, and balancing. These instruments calculate the R and X values of the impedance under measurement and provide a direct readout on a digital scale.

If the radiation resistance of an antenna is to be used to determine radiated power, the impedance must be measured at the point that the antenna current will be measured. For greatest accuracy, the impedance should be measured with several test frequencies near the actual operating frequency of the antenna, the values plotted on graph paper, and the value for the operating frequency read from the graph.

Standing-Wave Ratio (SWR) Method

A very common way of measuring the input impedance of an antenna makes use of standing-wave measurements on the feedline and is therefore called the SWR method. It is sometimes called the transmission-line method. Standing waves, as we have seen previously, exist on a transmission line when the termination of the line does not take, or remove, all of the energy being sent down the line. Energy not removed is reflected and combines with the incident energy to form patterns of voltage and current standing waves. The ratio of the loops (maximum values) of such waves to their nodes (minimum values) is called the standing-wave ratio (SWR) on the line. The SWR is an indication, then, of the mismatch between the input impedance of the line termination, an antenna for example, and the characteristic impedance of the line. Thus, if the SWR can be determined and the characteristic impedance of the line is accurately known, the input impedance of an antenna as the termination of a line can be calculated. It is common for station operators to have SWR meters permanently connected to monitor the SWR on the line between transmitter and antenna. The calculation of antenna input impedance from SWR is best made with the aid of a Smith chart.

The SWR Meter

A meter for measuring the standing-wave ratio (a SWR meter) on a transmission line is useful not only for determining the actual input impedance of an antenna but is indispensible in monitoring the operating condition of the transmission line–antenna system in general. The readings of a SWR meter are used by operators to adjust the tuning of antenna matching networks for optimum operating conditions.

At the heart of a SWR meter is a circuit function called a *directional coupler.* A directional coupler is a simple circuit, in component count, which is able to sense the flow of power in both directions on a transmission line. Recall that when standing waves exist on a line, it is useful to consider that there are two power flows on the line: forward, or incident or transmitted power, and reflected power. *Forward power* is the energy transmitted from the transmitter in the direction of the load or antenna—the *forward direction* on the line. When a load or antenna is not matched to a transmission line, some portion of the energy is removed, but that which is not is turned around and sent back toward the transmitter end and becomes *reflected power. The voltage and current components of forward power are always in phase; the voltage and current components of reflected power are always 180° out of phase!* A directional coupler, then, is a circuit that combines a line voltage reading with a sample of line current in the forward direction to give a forward power reading. It combines the same line voltage reading with a sample of line current in the *reverse* direction to give a reflected power reading.

The voltage standing-wave ratio (VSWR) of a line is the ratio of the voltage of a loop of the standing wave (V_{max}) to a node (V_{min}). It is also

$$\text{VSWR} = \frac{V_f + V_r}{V_f - V_r}$$

$$= \frac{1 + V_r/V_f}{1 - V_r/V_f}$$

where V_f is the voltage amplitude of the forward wave and V_r is the voltage amplitude of the reflected wave. But since for a given load resistance, line voltage is proportional to the square root of power,

$$\text{VSWR} = \frac{1 + \sqrt{P_r/P_f}}{1 - \sqrt{P_r/P_f}}$$

Thus it is practicable to use the same meter to indicate either forward power, reflected power, or SWR, simply by providing a selector switch to change the connection to the current sensor and/or the calibration circuit of the meter. This is, in fact, the way SWR/RF POWER meters are commonly arranged.

A very much simplified diagram for the circuit of a SWR/POWER meter is shown in Fig. 6.52. Current sampling is provided by transformer T_1, a ferrite-core transformer with a single-turn primary (the transmission-line conductor). The circuit connected to the right end of the secondary winding of T_1 produces

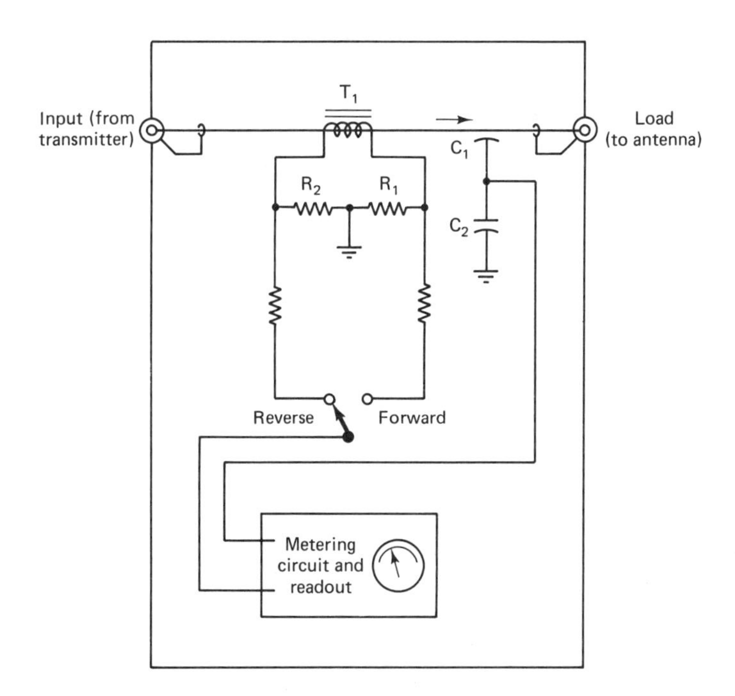

Figure 6.52 Bidirectional RF power/SWR meter.

a voltage across R_1 which is proportional to the forward line current. A sample of line voltage is provided by the C_1–C_2 voltage divider. These two voltages are fed to a metering circuit which combines, then rectifies them and provides an indication on a dc current meter. It is possible, then, to calibrate the circuit and meter scale so as to give a forward power reading *for a given load, say 50 Ω*. Calibration is usually performed with a so-called *dummy antenna,* a noninductive resistor load carefully constructed to contain the desired amount of resistance.

The circuit connected to the left end of the secondary winding of transformer T_1 produces a voltage across R_2 which is proportional to the inverse of the line current. This voltage is combined in the metering circuit with the line voltage sample from C_1–C_2. The metering circuit for this connection, in conjunction with the meter scale, is designed to indicate the reflected power. The indication is a comparison with conditions when a matched load is connected to the system and there are no standing waves (for this condition the metering/sensing circuit would be adjusted to give a reading of zero).

Please note that this metering system is not a true wattmeter in the sense that it measures actual in-phase voltage and current and indicates the product of the two. Rather, its indications are based on samples of voltage and current for a specified normal loading condition. Its ability to give useful readings of power and SWR is dependent on the metering circuit being designed for use with the specified load and calibrated with that load in place. If the normal loading condition is changed, the metering circuit must be changed and recalibrated.

Permanently connected SWR/power meters are widely used in radio stations. The SWR meter is used to make adjustments in antenna matching networks to achieve SWR values of unity or near unity. In routine station operation, the SWR/power meter is used to indicate the "health" of the antenna/feed line system.

6.10 DANGERS OF RADIATION

The effect of x-rays and other types of radiation associated with the energetic particles of atomic nuclei on the human body has been studied extensively in the years since World War II. Although much has been learned of the biological effects of these forms of electromagnetic radiation, there is still a great deal that is not understood in this regard. However, enough is known to indicate the very great danger of such radiation. The message is simple. The human body can stand very little x-ray or nuclear radiation without serious harm or death. Appropriate steps must be taken to prevent unnecessary exposure to these forms of radiation. When exposure is necessary or desirable, as with making x-ray film for medical diagnostic purposes, the amount of the exposure to radiation must be carefully controlled and limited.

The radiation from a radio antenna is different from x-rays primarily in the wavelength. Both are forms of electomagnetic radiation. X-rays have an extremely short wavelength and therefore penetrate the body readily. Extremely short radio waves, microwaves, can also penetrate the human body and may cause some harm. They must be of very high intensity—one must be in close proximity to a source of significant power (in the kilowatt range). Microwave cooking is possible because microwave radiation produces cellular changes in organic tissue. Appropriate precautions are taken in the construction of microwave ovens to contain the radiation within the oven housing.

The radiation from the antennas of the numerous radio and television stations in the world has not been found to be harmful to persons in any location normally occupied by human beings. This is not to say that you would not be harmed if you put yourself within inches of an antenna being energized with a 100-kW signal. A prudent policy is one of an awareness of the potential hazards of any form of electromagnetic radiation (one can get a lethal dose of radiation with too much exposure to the sun's rays at the beach). Appropriate steps to minimize the exposure of one's body to radiation is always in order.

GLOSSARY OF TERMS

Angle of incidence The angle that an electromagnetic wave (light or radio, for example) striking a surface makes with a line perpendicular to that surface.

Angle of reflection The angle that an electromagnetic wave leaving a surface makes with a line perpendicular to that surface.

Antenna, directional An antenna that is designed to increase its effectiveness in some directions at the expense of other directions.

Antenna, isotropic An imaginary antenna that sends or receives radio waves equally in all directions.

Antenna, omnidirectional A simple antenna which, at least in one plane, sends and receives radio waves equally in all directions.

Aurora A region of the atmosphere that temporarily has very high ionization density, usually manifesting itself as luminous bands or streamers of light in the night sky.

Coax A popular expression for coaxial transmission line.

Coaxial transmission line A pair of electrical conductors fabricated with one conductor at the center of and insulated from a second, outer conductor.

Counterpoise (ground, of an antenna) A grid-like structure of radial and concentric conductors placed in the ground with an antenna at the center to increase the conductivity of ground at the base of the antenna.

Critical angle of radiation The maximum angle, with respect to the earth's surface, at which radiation can take place and still be reflected from and not go through the ionosphere.

Critical frequency The maximum frequency of radiation that will be reflected from an ionospheric layer of given ionic density for any angle of radiation.

Diffraction The disturbance to a wave phenomenon when it is deflected at the edge of an opaque object or passes through a narrow slit.

Dipole, half-wave A simple antenna constructed of a single, straight conductor one-half wavelength long and broken at the center for connection to a transmission line.

Direct wave The radio wave, a component of the ground wave, which travels the line-of-sight path from sending antenna to receiving antenna.

Discontinuity On a transmission line, any change in the basic characteristic of the line, as compared to a line that continues unchanged to infinity.

Field intensity The amount of voltage induced in an antenna of standard dimensions, usually expressed in microvolts per meter.

Field strength *See* Field intensity.

Fringing The distortion of an electric or magnetic field, perceived as a set of lines, at its boundaries, manifested as an increase in space between the lines (bowing) and caused by the repulsive force between adjacent lines.

Ground wave The portion of the radio-frequency energy leaving an antenna which remains closest to the earth's surface.

Ground-reflected wave A radio wave that has been reflected from the earth's surface or an object on or near the earth's surface.

Ham An amateur radio operator.

Hertz antenna An isolated half-wave dipole antenna.

Incident wave A wave on a transmission line that travels from the sending end toward the receiving end.

Induction field The magnetic field immediately adjacent to a conductor carrying an electric current.

Ionosphere The region of the upper portion of the earth's atmosphere in which the relatively fewer atoms and molecules of the atmospheric gases are in an ionized state part of the time as a result of bombardment by photons (energy particles) from the sun and other cosmic phenomena.

Isolated antenna An antenna that is sufficiently far removed from the earth, other antennas, or any other form of conductor that its characteristics are not influenced in any way by external objects.

Load impedance Of a transmission line, the circuit or antenna connected to the end of a transmission line opposite the sending end.

Loop A point on a standing wave where incident and reflected waves add to produce the maximum amplitude of the standing wave.

Lossless (antenna or transmission line) Loss due to current flowing in the ohmic resistance of a unit is insignificant in comparison to the desired effect.

Marconi antenna A grounded-vertical antenna one quarter-wavelength long.

Maximum usable frequency (MUF) The maximum frequency that can be used to provide radio communication between two specified points on the earth's surface, as determined by ionospheric propagation conditions.

Node A point on a standing wave where incident and reflected waves cancel, producing a standing-wave minimum.

Nonresonant line A transmission line that has no standing waves because its load is matched to the line in impedance value.

Parameter Any property of a circuit, transmission line, etc., which may vary and which is used to determine other characteristics.

Parameter, distributed The properties, particularly of a transmission line, such as inductance, resistance, and capacitance, which are perceived as being uniformly present throughout a device, as opposed to being present in discrete quantities or lumps.

Parameter, lumped The properties of a circuit that is made up of discrete components separated by conductors whose resistance, inductance, or capacitance is insignificant.

Polarization Of an antenna or radio wave, the direction of the electric field with respect to the earth's surface, and therefore the direction of the length of an antenna.

Propagation Of a radio wave, the spreading or travel of the wave away from the sending antenna.

Radiation The process of energy leaving a source and traveling through space as opposed to being removed by conduction or the convection currents of a medium, as with heat convection by air.

Radiation field The region of influence of energy released from a source by the process of radiation.

Radiation pattern Of an antenna, the way in which the radiated energy manifests itself, in terms of intensity or strength, in the three-dimensional space around the antenna, commonly portrayed in the form of a curve on a graph using polar coordinates—amplitude versus angle.

Radiation resistance A hypothetical concept used to provide an analogy between the energy "lost" from a system due to radiation and that lost due to the true ohmic resistance of current-carrying components; the in-phase component of the input impedance of an antenna.

Reciprocity Of an antenna, the ability to present identical characteristics, whether being used to send or to receive.

Reflect To bend, or turn around and throw, or send back.

Reflected wave A wave that has been bent or turned around and sent back.

Refract As of a light or radio wave, to bend the direction of travel when passing from one medium to another or through layers of different density of the same medium.

Resonant line A line with standing waves whose input impedance therefore changes with the frequency of the input signal.

Skip zone An area of the earth's surface where the radio reception from a given station is weak or nonexistent because its distance from the station is greater than the range of surface or direct waves but short of the range of waves propagated by ionospheric reflection.

Sky wave That portion of the radiation from an antenna which is directed away from the earth and toward the sky.

Standing wave A stationary wave-like pattern of voltage or current variation on a transmission line produced by the phasor summing of incident and reflected traveling waves.

Standing-wave ratio The ratio of maximum amplitudes of voltage or current to minimum values of amplitude on a transmission line with standing waves.

Stub A common expression used in place of the complete expression: transmission-line stub, meaning a piece of transmission line whose electrical length is not more than a half wavelength.

Sunspot cycle A pattern of variation of approximately 11 years' duration of the spots (giant storms) visible on the surface of the sun.

Surface wave That component of radiation which propagates very close to the earth's surface virtually as if that surface were a transmission line.

Temperature inversion An abnormal condition of the earth's lower atmosphere in which a layer of cool air exists above a region of warm air, thus restricting the warm air from rising, as in a normal situation.

Transmission line stub A piece of transmission line whose electrical length is generally not greater than a half-wavelength.

Troposphere The lower portion of the earth's atmosphere.

Wave phenomenon Any "disturbance" in the physical world, such as sound, light, electricity, etc., whose variation over time and/or space is wave-like and whose behavior can be analyzed or predicted with the aid of a number of principles generally true of any wave-like pattern.

Wavefront The leading edge of a traveling wave.

Wavelength The distance between two adjacent, exactly similar points on a traveling wave; also, the distance found by dividing the speed of the wave by the number of waves per second (i.e., the frequency).

Yagi antenna (or just "yagi") A directional antenna array (named after H. Yagi, a twentieth-century Japanese engineer) in which a basic $\lambda/2$ dipole is reinforced by several reflector and director elements.

REVIEW QUESTIONS: BEST ANSWER

1. The characteristic impedance, Z_0, of a transmission line is equal to the input impedance of a/an: **a.** shorted line. **b.** few meters of open line. **c.** line terminated with resistance. **d.** line of infinite length. **d.** none of these.
2. The input impedance of a line of any length that is terminated with an impedance equal to the Z_0 of the line is equal to: **a.** Z_0. **b.** infinity. **c.** zero. **d.** $Z_L/2$. **e.** none of these.
3. The statement that No. 14 gauge copper conductor has a resistance of 2.575 Ω per 1000 ft is an example of a/an: **a.** lumped parameter. **b.** distributed parameter. **c.** input impedance. **d.** infinite line. **e.** none of these.
4. The input impedance of a two-terminal network with an infinite number of meshes is likely to be: **a.** a finite value. **b.** zero. **c.** infinite. **d.** very large. **e.** none of these.
5. The characteristic impedance of a transmission line is determined by its: **a.** length. **b.** construction geometry. **c.** dielectric. **d.** a and b. **e.** b and c.
6. Transmission lines considered as electrical circuits are not simple because at radio frequencies: **a.** electrons travel faster. **b.** electron transit times become important. **c.** electrons produce light. **d.** electrons are no longer involved. **e.** none of these.
7. The fact that, at radio frequencies, voltage and current may go through several cycles of variation between source and load makes it useful to use the concepts of: **a.** electron flow. **b.** conventional current flow. **c.** wave phenomena. **d.** pulse technology. **e.** a and d.
8. Except when the dielectric of a transmission line is not air, voltage and current waves travel along the line at a speed of approximately: **a.** 1086 ft/s. **b.** 3×10^8 m/s. **c.** 186,000 mi/s. **d.** infinity. **e.** b or c.
9. The current and voltage waves on an infinite line are in phase because: **a.** Z_0 is purely resistive in nature. **b.** the waves do not yet know what Z_L will be. **c.** they are in phase at the source. **d.** there are no reactances in a transmission line. **e.** none of these.
10. A 300-Ω transmission line is terminated with a load equal to its own Z_0. The voltage is 75 mV at a point 3 m away from the source. The current at that point is: **a.** cannot be determined. **b.** 75 mA. **c.** 0.3333 mA. **d.** 0.25 mA. **e.** none of these.

11. The wavelength of a current wave whose frequency is 27.58 MHz is: **a.** 100.8 m. **b.** 10.88 m. **c.** 356.9 ft. **d.** 35.69 ft. **e.** b or d.

12. A wavelength of 80 m corresponds to a frequency of: **a.** 2.325 MHz. **b.** 12.30 MHz. **c.** 47.35 MHz. **d.** 3.750 MHz. **e.** none of these.

13. The electrical length of 4.5-m transmission line at 15 MHz is: **a.** 0.225λ. **b.** 4.444λ. **c.** 3.333λ. **d.** 0.30λ. **e.** none of these.

14. When all of the energy on a transmission line is not removed by a properly matched resistive load, the effect is that the voltage and current waves are: **a.** diffracted. **b.** refracted. **c.** repelled. **d.** reflected. **e.** none of these.

15. Reflected waves combine phasorally with incident waves to produce: **a.** standing waves. **b.** square waves. **c.** radio waves. **d.** short waves. **e.** none of these.

16. The points where standing waves are at their maximum and minimum values, respectively, are called: **a.** nodes and loops. **b.** valleys and peaks. **c.** loops and nodes. **d.** peaks and crests. **e.** none of these.

17. In order for a standing wave, voltage or current, to have a loop at the end of a transmission line, the incident wave must be reflected: **a.** 180° reversed. **b.** with a 90° phase shift. **c.** with a 45° phase shift. **d.** as it would have continued. **e.** none of these.

18. When a transmission line is shorted, the voltage wave is reflected: **a.** 180° reversed. **b.** with a 90° phase shift. **c.** with a 45° phase shift. **d.** as it would have continued. **e.** none of these.

19. A termination or condition on a transmission line that causes standing waves is called a/an: **a.** matching impedance. **b.** discontinuity. **c.** short. **d.** open. **e.** none of these.

20. On a line with standing waves, the ratio of maximum amplitude to minimum amplitude of either voltage or current is called the: **a.** input impedance. **b.** characteristic impedance. **c.** standing-wave ratio. **d.** radiation resistance. **e.** none of these.

21. If a transmission line with standing waves is an even number of quarter-wavelengths long, its input impedance looks like: **a.** a short. **b.** an open. **c.** its termination. **d.** its characteristic impedance. **e.** c or d.

22. If a transmission line is an odd number of quarter-wavelengths long and is open, its input impedance looks like: **a.** a short. **b.** an open. **c.** a capacitive reactance. **d.** an inductive reactance. **e.** none of these.

23. The input impedance of a line with standing waves depends on its termination and electrical length, and therefore on: **a.** distance between conductors. **b.** the frequency of the exciting signal. **c.** its height above ground. **d.** the amplitude of input voltage. **e.** none of these.

24. A transmission line with standing waves is called a/an: **a.** flat line. **b.** nonresonant line. **c.** resonant line. **d.** floating line. **e.** none of these.

25. A transmission line with a properly matched terminating impedance has no standing waves and is called a/an: **a.** flat line. **b.** nonresonant line. **c.** resonant line. **d.** a and b. **e.** none of these.

26. An electrically short transmission line, especially when used for a special purpose other than transmitting power, is called a/an: **a.** flat line. **b.** resonant line. **c.** stub. **d.** a and c. **e.** none of these.

27. The process by which electrical energy detaches itself from a conductor and, then, moves away from the conductor is called: **a.** standing waves. **b.** attenuation. **c.** fringing. **d.** radiation. **e.** none of these.

28. A device that enhances the process of radiation of electrical energy is called a/an: **a.** antenna. **b.** transmission line. **c.** stub. **d.** radio. **e.** none of these.

29. Radiated electrical energy is a wave-like phenomenon and contains two field components, namely: **a.** right and left fields. **b.** positive and negative fields. **c.** vertical and cylindrical fields. **d.** electric and magnetic fields. **e.** none of these.

30. The fields of a radio wave are always perpendicular to each other and the polarization (or direction with respect to the earth's surface) of the wave is identical with that of the: **a.** negative field. **b.** magnetic field. **c.** left field. **d.** electric field. **e.** none of these.

31. The magnetic-field component of a radio wave is horizontal; therefore, the wave is polarized: **a.** horizontally. **b.** vertically. **c.** positively. **d.** negatively. **e.** a and c.

32. If a radiation field has diminished to one-tenth of its original strength in a given distance from a conductor, the strength of an induction field at the same distance, compared to its original strength, will have diminished to: **a.** $\frac{1}{10}$. **b.** $\frac{1}{2}$. **c.** $\frac{1}{1000}$. **d.** $\frac{1}{100}$. **e.** none of these.

33. A plot of the level of radiated energy from an antenna, at points of equal distance from the antenna, is called a/an: **a.** doughnut. **b.** circle. **c.** figure 8. **d.** radiation pattern. **e.** none of these.

34. An isolated half-wave dipole is horizontally polarized. Its radiation pattern in a horizontal plane is a/an: **a.** doughnut. **b.** circle. **c.** figure 8. **d.** cardioid. **e.** none of these.

35. An antenna whose radiation pattern is omnidirectional in the horizontal plane is the: **a.** Marconi antenna. **b.** Hertz antenna. **c.** grounded quarter-wave vertical antenna. **d.** grounded horizontal half-wave antenna. **e.** a and c.

36. The phenomenon of a radio wave traveling through space and/or the earth's atmosphere is called: **a.** radiation. **b.** sky waves. **c.** polarization. **d.** propagation. **e.** none of these.

37. Propagation modes are sometimes categorized as either: **a.** vertical or horizontal. **b.** ground wave or sky wave. **c.** ground wave or ionospheric. **d.** tropospheric or reflective. **e.** none of these.

38. Two important physical phenomena that are true of both radio waves and visible light waves are: **a.** speed and velocity. **b.** reflection and refraction. **c.** phase and reciprocity. **d.** linearity and portability. **e.** none of these.

39. Compared to the distance to the visible horizon, the distance to the radio horizon is slightly: **a.** less. **b.** higher. **c.** lower. **d.** greater. **e.** none of these.

40. The range and quality of tropospheric propagation can change frequently and rapidly because of: **a.** changes in height of ionosphere. **b.** weather changes. **c.** wind changes, primarily. **d.** ocean temperature changes. **e.** none of these.

41. "Ghost images" on TV screens are often caused by: **a.** ground-wave reflections. **b.** tropospheric refraction. **c.** ionospheric propagation. **d.** temperature inversion. **e.** none of these.

42. One of the reasons that a radio signal is never totally absent behind an obstruction is because of the phenomenon called: **a.** refraction. **b.** field intensity. **c.** diffraction. **d.** translucence. **e.** none of these.

43. If the angle of radiation (angle between earth's surface and direction of travel of a radiated wave) is greater than the critical angle, for a particular wave, that wave will: **a.** be diffracted. **b.** be reflected back to earth. **c.** continue through the ionosphere into space. **d.** be absorbed in the ionosphere. **e.** none of these.

44. If the frequency of a radio signal exceeds the critical frequency for ionospheric propagation, this propagation mode is: **a.** excellent for achieving great range at that frequency. **b.** unsuitable for communication at that frequency. **c.** the best mode for the frequency. **d.** probably used normally for propagation of signals in a band of frequencies near that frequency. **e.** a, c, and d.

45. The greater the skip distance of a particular signal the: **a.** greater is its range. **b.** less satisfactory is the propagation path. **c.** lower the quality of the transmission. **d.** more likely it is to be used by an amateur operator. **e.** a and d.

46. Because an antenna is a passive device, the gain of an antenna refers to its ability: **a.** to increase the radiated signal level. **b.** to increase radiation in one or more directions at the expense of other directions. **c.** to amplify a signal. **d.** to convert dc energy to RF energy. **e.** none of these.

47. An antenna can be tested as a receiving antenna and the evaluation obtained can be used to judge its suitability as a transmitting antenna. This property of antennas is called: **a.** propagation. **b.** indifference. **c.** reciprocity. **d.** radiation resistance. **e.** none of these.

48. An antenna that radiates equally in all directions, but which, for that reason, is impossible to obtain, is called a/an ______________ antenna. **a.** Hertz **b.** isotropic **c.** Marconi **d.** half-wave dipole **e.** none of these.

49. While receiving exactly the same radiation, antenna *A* provides an output of 87.53 μV/m and a half-wave dipole antenna has an output of 6.385 μV/m. The gain of antenna *A* over the half-wave dipole is: **a.** 22.74 dB. **b.** 13.71 dB. **c.** 11.37 dB. **d.** 26.18 dB. **e.** none of these.

50. An antenna that is designed to receive, or radiate, signals better in certain directions, at the expense of other directions is called a/an ______________ antenna. **a.** isotropic **b.** Hertz **c.** directional **d.** passive **e.** none of these.

51. Because several of their dimensions are critically important, directional antennas have very limited: **a.** usefulness. **b.** impedance. **c.** gain. **d.** bandwidth. **e.** none of these.

52. Collinear arrays, broadside arrays, and end-fire arrays are examples of: **a.** passive-array antennas. **b.** theoretical antennas. **c.** driven-array antennas. **d.** obsolete antennas. **e.** none of these.

53. Antennas that have electrically long elements are nonresonant and have improved: **a.** bandwidth. **b.** isolation. **c.** impedance. **d.** cardioids. **e.** none of these.

54. An important specification of antennas that are directional in one primary direction is: **a.** wavelength. **b.** ground isolation. **c.** front-to-back ratio. **d.** parasitic elimination. **e.** none of these.

55. The most popular antenna used with AM broadcast band receivers is a form of ______________ antenna. **a.** grounded-vertical **b.** isolated half-wave **c.** yagi **d.** loop **e.** none of these.

56. It is important to know the input or driving-point impedance of an antenna to: **a.** prevent overdriving the antenna. **b.** facilitate impedance matching with the feed line. **c.** maximize front-to-back ratio. **d.** minimize parasitic oscillation. **e.** none of these.

57. The RF voltage at the input of a half-wave dipole antenna ($R_R = 73\ \Omega$) is 270.2 V.

Determine the radiated power. **a.** 50 W. **b.** 100 W. **c.** 278.9 W. **d.** 1000 W. **e.** none of these.

58. Measurement of forward and reverse power on a transmission line are possible because, in the reverse direction, voltage and current are: **a.** in phase. **b.** 180° out of phase. **c.** dc values. **d.** lower in frequency. **e.** none of these.

59. All forms of electromagnetic radiation are potentially hazardous to the human body: **a.** is a false statement. **b.** if absorbed in excessive amounts. **c.** and RF radiation more than any other form. **d.** especially x-radiation or the radiation from nuclear activity. **e.** b and d.

REVIEW QUESTIONS: ESSAY

1. Give two definitions of *characteristic* or *surge impedance* of a transmission line.

2. Describe a line of infinite length.

3. Describe the difference between distributed and lumped parameters.

4. Describe the phase relationship between the voltage and current waves on an infinite transmission line. Explain why this relationship occurs.

5. Make a sketch of a voltage wave on a transmission line. Label your sketch so as to show the meaning of a wavelength. Describe wavelength in terms of your sketch. Give the definition of wavelength in mathematical terms.

6. Explain what is meant by *electrical length.* Describe how to express the length of a transmission line in the units of electrical length. Make up and give an original example of physical length converted to electrical length.

7. Incident voltage and current waves on a transmission line are always in phase (see Question 4). What is the phase relationship between reflected voltage and current waves? Why does this relationship occur?

8. What are standing waves? How are they produced?

9. Explain *loop and node* as these terms are used in connection with standing waves. What is the spacing, in electrical length, between adjacent loops? between adjacent nodes? between an adjacent loop and node?

10. Explain *discontinuity* on a transmission line.

11. Describe in words the meaning of *standing-wave ratio.* Give a mathematical formula for this concept. What is the abbreviation for standing-wave ratio? for voltage standing-wave ratio? for current standing-wave ratio?

12. Under what condition(s) is a transmission line called "resonant"? What does this classification mean? Why is it appropriate?

13. Describe a flat line. What is another term for the same line?

14. Discuss and compare an induction field and a radiation field.

15. Explain what is meant by the polarization of a radio wave. Describe the relationship between the polarization of a wave and the polarization of the antenna from which the wave was radiated.

16. Why is a radio wave called an electromagnetic wave?

17. Given: an isolated, horizontally polarized half-wave dipole antenna. Sketch the radiation pattern of this antenna: **(a)** in the horizontal plane of the antenna; **(b)** in the vertical plane that passes through the center of the antenna.

18. Give two major modes or categories of radio-wave propagation.
19. Describe the ground-wave propagation mode and list its components.
20. Describe the sky-wave propagation mode and list its components.
21. What is meant by line-of-sight propagation? visual horizon? radio horizon? Is the distance to the radio horizon equal to the distance to the visual horizon? If not, why not?
22. What is the troposphere? Why is the troposphere a consideration in the study of radio-wave propagation?
23. Explain in detail the connection between ground-reflected waves and TV ghosts.
24. What is the significance of diffraction in radio-wave reception?
25. Describe the phenomenon of ionospheric propagation. Include in your description mention of the basic physical phenomenon involved in this form of propagation and an explanation of why the ionosphere is capable of influencing propagation.
26. Explain the difference between critical frequency and maximum usable frequency (MUF).
27. What is the meaning of the term *critical angle?*
28. Explain *skip zone* and *skip distance*.
29. What are *fading* and *selective fading?*
30. List and describe briefly three types of propagation, other than ionospheric propagation, which involve the reflection of sky waves.
31. Give the definition of field intensity or field strength. Give the unit of this measurement and explain the significance of the unit.
32. Explain what is meant by the *gain* of an antenna. What is meant by *absolute gain* of an antenna? What is the common, practical reference for gain values for antennas?
33. Describe an isotropic antenna. Where can it be purchased?
34. Describe briefly the details of construction of a directional antenna. Why is a directional antenna used? Give an example of a directional antenna.
35. Explain the difference between passive and driven elements in directional antennas. What is another term used to describe a passive element?
36. Why does a directional antenna typically have a limited bandwidth?
37. What is a nonresonant antenna? Why is it nonresonant? What are the advantages, if any, of a nonresonant antenna?
38. Define the input or driving-point impedance of an antenna. What factors can change the input impedance of an antenna? Why is knowing the input impedance of an antenna important to someone designing or operating a radio station?
39. Discuss *radiation resistance,* its meaning and importance.
40. Explain how you would calculate SWR if you knew only forward and reflected power values for a transmission line.

EXERCISES

1. Draw the circuit diagram of a network made up of numerous two-resistor sections. Each two-resistor section has a series resistance of 15 Ω and a shunt resistance of 300 Ω (see Fig. 6.3). What is the characteristic impedance of this network, assuming

that there are an infinite number of sections? (That is, what is the input impedance of the network when there are a very large number of sections?)

2. A parallel-wire transmission line has the following parameters: 1.8 μH/m and 20 pF/m. Calculate the characteristic impedance of the line.

3. A coaxial line has a characteristic impedance of 75 Ω. If its inductance parameter is 1.2 μH/m, what is its capacitance per meter?

4. A TV twinlead line is made of No. 18 gauge wire spaced at 0.6351 cm. The diameter of No. 18 gauge wire is 0.1024 cm. Calculate the Z_0 of the line.

5. A coaxial transmission line is constructed with an inner conductor of No. 12 gauge wire and an outer shield whose inside diameter is 0.5461 cm. If the diameter of No. 12 gauge wire is 0.2053 cm, what is the characteristic impedance of the line? Neglect the effect of the dielectric supporting the inner conductor.

6. If the measured Z_0 of the line of Exercise 5 is 50 Ω, calculate the effective dielectric constant of the insulation between the inner and outer conductors.

7. Calculate the wavelength of signals having the following frequencies. Express answers both in meters and in feet. **(a)** 1.8 MHz; **(b)** 3.0 MHz; **(c)** 14 MHz; **(d)** 20 MHz; **(e)** 50 MHz; **(f)** 250 MHz; **(g)** 875 MHz.

8. Calculate the frequency of signals with the following wavelengths. **(a)** 750 ft; **(b)** 275 m; **(c)** 185 m; **(d)** 80 m; **(e)** 35 m; **(f)** 20 m; **(g)** 10 ft; **(h)** 25 cm; **(i)** 2 cm.

9. Make a sketch showing the pattern of voltage and current standing waves on an open line. Label the sketch to show multiples of a quarter-wavelength from the terminated end of the line.

10. Make a sketch showing the pattern of voltage and current standing waves on a shorted line. Label with quarter-wavelength points.

11. Calculate the SWR corresponding to the following values of maximum and minimum voltages, respectively, on a line: **(a)** 25 V, 25 V; **(b)** 25.6 V, 24.4 V; **(c)** 26.1 V, 23.9 V; **(d)** 240 V, 220 V; **(e)** 400 V, 100 V; **(f)** 500 V, 100 V.

12. Refer to Fig. 6.14. Using expressions such as "inductive reactance," "low resistance," etc., characterize the input impedance of transmission-line stubs with the indicated terminating conditions and lengths: **(a)** open, $\frac{3}{4}\lambda$; **(b)** shorted, $\lambda/2$; **(c)** open, between $\lambda/4$ and $\lambda/2$; **(d)** shorted, between $\lambda/2$ and $\frac{3}{4}\lambda$; **(e)** open, less than $\lambda/4$; **(f)** shorted, less than $\lambda/4$.

13. **(a)** Make a labeled sketch showing the electric and magnetic fields of an excited transmission line. **(b)** Make a sketch to show fringing.

14. Make a labeled sketch depicting **(a)** horizontally and **(b)** vertically polarized half-wave dipole antennas.

15. **(a)** Assuming a clear day and no obstructions, how far could you see from the top of a 1700-ft peak? **(b)** What is the range of a direct-wave radio signal from that peak? **(c)** What is the range if the receiving antenna is mounted atop a 150-ft tower?

16. At a certain location 50 km from an AM radio station, the measured field strength of the station's signal is 375 μV/m while the station is unmodulated. What will be the field strength when the modulation level is at 95%?

17. When a TV antenna was replaced with a more sophisticated one the signal level measurement at the receiver increased from 35 μV to 285 μV. Calculate the gain in decibels of the new antenna with respect to the old.

18. While receiving exactly the same signals, a half-wave dipole antenna produced an output of 47.5 μV and a directive antenna produced an output of 127.5 μV.

(a) Calculate the gain of the directive antenna. Express your result in dB. (b) What is the gain of the antenna in dBi? (*Hint:* A half-wave dipole antenna has a dBi gain of 2.1; and total gain in dB $= dB_1 + dB_2 + \cdots$.)

19. A given antenna is specified as having a dBi gain of 10. What is its gain with respect to a half-wave dipole?
20. A special directional antenna receives a given signal at a level of 375 μV. When the orientation of the antenna is changed by 180°, the received level of the same signal is 2.57 μV. What is the front-to-back ratio of the antenna? Express your answer as a pure ratio and in decibels.
21. A directional TV antenna, designed to minimize ghost images caused by the reception of ground-reflected waves, was tested under special conditions to determine the amount of rejection of off-end signals (signals whose direction is off the end of the antenna, i.e., 90° with respect to the desired signal direction). When directed at the transmitting antenna the antenna under test received at a level of 535 μV. When the test antenna was rotated 90°, the received signal measured 3.25 μV. Calculate the off-end rejection ratio.
22. When excited by an RF voltage of $75 \angle 0°$ V, the input current to an antenna was measured as $1.15 \angle 15°$ A. (a) Determine the input impedance of the antenna. (b) What is its radiation resistance? (c) What input voltage would be required to produce 100 W of radiated power?

7

COMMUNICATIONS SYSTEMS and NOISE

7.1 INTRODUCTION

One of the most significant factors influencing the design, operation, and utilization of all forms of communication by electronic means is that of *noise*. Of course, noise is a negative influence—it always degrades the quality of communication or limits the amount of communication that can be accomplished over a given facility.

Anyone who has ever listened to a radio or watched a television screen is familiar with the effects of noise on those two particular communications media. On AM radio, especially, we may hear the crack of static during a thunderstorm, the crackle caused by the ignition system of an automobile, the buzz of an electric motor running nearby, or the interference of another station. When watching television we may experience the effects of noise as "snow" in the picture, or as some other form of reduction in the technical quality of the picture. These are common manifestations of electrical noise in communications systems.

When communication involves the transmission of the data used by computers, noise manifests itself as a polluting of the complex waveforms that a system is normally capable of interpreting as discrete elements of data. The electrical pollution may result in extraneous data being added to the transmitted data, thereby rendering it erroneous. Or the quality of the waveforms may be sufficiently lowered that no interpretation at all is possible.

The study of noise and its effects on communication systems has become a highly developed and sophisticated scientific discipline. The study of noise is part of what is called *information theory*. Entire books have been written on the subject. It is the goal of this book, and this chapter in particular, to provide information on noise sufficient to the needs of persons for whom the overall book is intended. For example, we (the users of this book) need to know some general, basic terminology of noise. We need to know something about how attempts to minimize the limitations imposed by noise affect the design of circuits and construction of equipment. This is important, first of all, to help satisfy the curiosity that all of us have, especially in this field. In working with electronics equipment, we are frequently asking "Why was this built this way instead of that?" But more important, we need to know enough about design for noise reduction to avoid destroying the effects of that design when we repair or modify equipment—replace parts, make adjustments, etc.

7.2 NOISE TERMINOLOGY AND BASIC CONCEPTS

In developing a knowledge of the technical terms used in connection with electrical noise, it is well to start with the word *noise* itself. Although anyone who hears distracting sounds coming from a radio's speaker is able to identify those sounds as "noise," something more precise than that is meant in the context of a technical discussion of noise.

The "noise" that is heard from a loudspeaker is produced by an electrical current. Such a current, of course, is undesired and is one not intended to be present at the speaker. A more technical definition of electrical noise, then, is that it is a voltage or current present, but not intended to be present, in a circuit. Where do such voltages or currents come from? The answer is: from many sources. Attempts to categorize such sources leads to other terminology.

Noise signals (voltages or currents) can originate within the circuits of a system. In this case, the noise is called *internal noise*. If the noise originates from outside the system circuits, it is called *external noise*.

External Noise

Man-made noise. A number of different types of equipment used in great numbers every day are sources of radiated electrical energy. The ignition systems of automotive engines, fluorescent and neon lighting systems, and certain types of electric motors are all sources of relatively potent radiated energy. The frequencies of the energy radiated by these sources are spread randomly across most of the radio-frequency spectrum. The radiated impulses enter communications systems and produce undesired currents. Thus we have *man-made noise*.

The common factor in man-made noise sources, the reason that they produce energy that can be radiated, is that each periodically produces sharp interruptions in relatively high currents. When the electric and magnetic fields associated with the circuits carrying such currents attempt to collapse rapidly, some of the energy contained in them is radiated, just as from an antenna. When the energy waves from these sources are intercepted by the antenna of a radio or television receiver, noise currents are induced in the antenna and some undesired effect is usually observable—audible noise from the radio's speaker or distracting disturbance to the image on the television screen.

Some of the radiated noise energy may induce currents directly in the conductors and coils of a receiver. Careful shielding of portions of receiver circuitry where desired signal levels are extremely low is effective in minimizing noise due to this cause. The only sure cure for noise induced in a receiver's antenna is the geographic separation of the antenna and noise source. This is the rationale for locating the earth stations for interplanetary space communications systems in desert areas far from industrial centers.

Atmospheric and space noise. All other noise that originates external to an electronic system is generally placed in one or the other of two categories: *atmospheric noise* or *space noise*. Atmospheric noise is composed of undesired electrical impulses that originate from natural phenomena in the earth's atmosphere. The electrical discharges of lightning are the major source of atmospheric noise. Lightning discharges generate frequencies across the RF spectrum. However, the energy level of this disturbance is inversely proportional to its frequency. It is hardly a problem at frequencies above 20 MHz. Lightning storms that occur geographically near a receiver will produce the greatest amount of interference. The noise signals of lightning discharges may propagate over great distances if conditions are also favorable for the propagation of radio signals over those distances.

Space noise is a classification representing the electrical signals that originate in the universe beyond the earth's atmosphere. The sun and the other stars are huge masses of ionized gases. By definition, the electrons and ions of such gases are in motion. Such motion represents electrical currents, currents that can produce radio-frequency emissions. These emissions propagate in all directions and some of them reach the earth. However, only those whose frequencies are above about 8 MHz are able to penetrate the ionosphere and reach the earth's surface. Emissions that originate from the sun are called *solar noise,* all other noise from space is called *cosmic noise*.

A great deal of effort in the latter half of the twentieth century is being expended in the monitoring and analyzing of cosmic emissions. There are many who believe that there is a possibility that all such emissions are not simply "noise," but that some may be the result of the activities of intelligent beings in universes other than our own. The belief is that some of the cosmic radio-frequency emissions may be the result of intentional radio broadcasts by other human-like beings somewhere in the cosmos. The desire to explore the possibility of such broadcasts has provided the incentive for major advances in the development of low-noise receivers and antenna systems.

Internal Noise

Unintended and undesired voltages and currents arise in circuits for several different reasons. Perhaps the most common cause of internal noise is that of the constant, random motion of electrons in all electrical components. Electrons in any form of electrical conductor possess the energy to be in motion at any temperature above what is called absolute zero. Absolute zero, 0°K (K is for Kelvin), is equivalent to −273°C. (A working definition for absolute zero is that temperature at which electrons cease to move.)

Taken over a significant time period, the effect of electron motion is totally random—there is no net motion in any one direction. However, over extremely short periods of time, there may be net motion of electrons in one or another direction in a component. Net motion means an electrical current—a current produced entirely as the result of the thermal activity (activity produced by the heat in the material) of the electrons. Such currents are unintended and undesired and are called *thermal noise*. Thermal noise currents are extremely small and, therefore, are of significance only in high-gain amplifiers.

The frequencies of thermal noise signals are spread relatively uniformly over the entire radio-frequency spectrum. This is analogous to the phenomenon of white light, which includes light of all wavelengths (frequencies). Hence thermal noise is also called *white noise*.

As might be expected, thermal noise is related to the temperature of a conductor or component: the higher the temperature, the greater the noise. In fact, J. B. Johnson, a U.S. physicist, in 1928 was able to show a precise relationship between the power in a thermal noise signal and the temperature of the device in which it is produced

$$P_n = kT\,\Delta f$$

where k = Boltzmann's constant = 1.38×10^{-23} J/°K
T = temperature of the device (°K)
Δf = bandwidth over which the circuit will operate

The result in this formula, P_n, is in watts and is the maximum possible power that can be generated by thermal noise for a given condition.

This formula enables us to estimate the noise voltage that could be generated by a component, in terms of the resistance, R, of that component. Refer to Fig. 7.1, in which a thermal noise source is represented as a generator with voltage V_n and internal resistance R (the equivalent resistance of the circuit in which

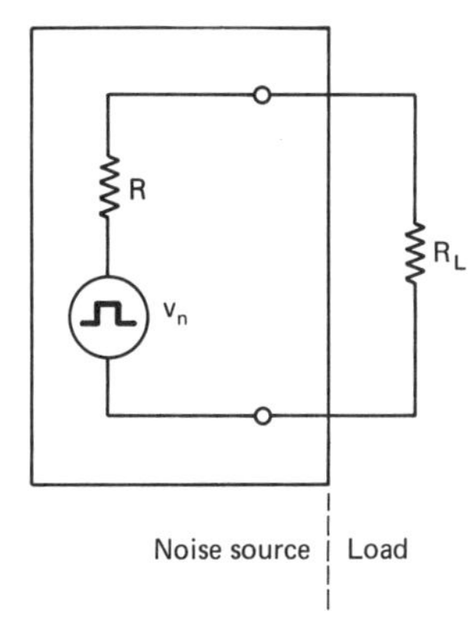

Figure 7.1 Equivalent diagram of noise source.

the noise is being generated). The generator is connected to a load R_L. Maximum power will be transferred to the load when $R = R_L$. Thus

$$P_L = \frac{V_L^2}{R} = \frac{(V_n/2)^2}{R} = \frac{V_n^2}{4R}$$

and hence

$$V_n = \sqrt{4kT\,\Delta f\,R} \qquad \text{volts (rms)} \tag{7.1}$$

The noise voltage, V_n, obtained in this way is an rms value because Eq. (7.1) is for average noise power generated, not for an instantaneous peak value.

Example 1

The input circuit of a high-gain RF amplifier has an equivalent resistance value of 1000 Ω. The amplifier has an overall voltage gain of 1000 and a bandwidth of 200 kHz. Estimate the maximum noise voltage amplitude at the input to the amplifier due to thermal noise only when the operating temperature is 29°C.

Solution

$$V_n = \sqrt{4kT\,\Delta f\,R} = \sqrt{4 \times 1.38 \times 10^{-23} \times (29 + 273) \times 200{,}000 \times 1000}$$

$$= \sqrt{3.334 \times 10^{-12}} = 1.826 \times 10^{-6}\text{ V} = 1.826\ \mu\text{V}$$

The implication of the estimate of 1.826 μV in this example is that this amplifier would not begin to be useful until the input information signal voltage is several times 1.826 μV.

All devices that have ohmic resistance produce thermal noise. (Thermal noise is also called *Johnson noise,* in honor of the person who first provided extensive information about it. The terms *thermal noise, white noise,* and *Johnson noise* are interchangeable.) Thus, not only resistors, but also capacitors, inductors, and even electronic devices such as tubes, diodes, and transistors introduce thermal noise voltages into the circuits in which they are connected. Most components have relatively low resistance, compared to most resistors, and therefore their contribution to the total thermal noise in a circuit is usually negligible. Noise voltages are additive, just as are the voltages of any other sources in series.

When it is highly desirable to produce an amplifier with low internal noise, it is obvious that decisions concerning the resistors in the circuit become exceedingly important. In the first place, since noise voltage is proportional to the square root of the resistance in the circuit, designs that minimize the resistance values required will minimize noise. Further, the type of resistors used is significant in low-noise design. Equation (7.1) is for a resistor made of copper wire. The noise power for resistors made of any other material is greater. The very common carbon-composition resistors tend to be the noisiest. So-called *low-noise resistors* are simply resistors that are known to be less noisy than common types. They are usually of the wire-wound type and are thus bulkier and more expensive than common types. Significant reductions in thermal noise can be achieved with careful design and attention to selection of components.

Shot noise. Another type of noise that is generated within an electronic circuit is called *shot noise*. Sources of shot noise are the electronic devices themselves: electron tubes, transistors, and semiconductor diodes. Shot noise gets its name from the fact that when a shot noise current flows in a speaker coil producing sound, the sound is similar to that of lead shot falling on a drumhead or sheet of metal. This, in turn, suggests that the noise signal is generated by a nonuniform flow of current.

Investigation has revealed that on a moment-to-moment basis, the current flow in electron devices is spasmodic, even under pure dc operating conditions. To understand why this can be, we first recall that current flow in electron devices is the result of the movement of a discrete, or limited, number of carriers. In arriving at a given destination, if some of the carriers take longer than others, there is a momentary reduction in current. When the tardy carriers finally arrive, there is a momentary increase in current. In the case of transistors, for example, different carriers have different transit times because they have different distances to travel. We contrast this with current flow in a good conductor. There we have a massive overabundance of carriers for current, somewhat like a pipe full of molasses. With a constant pressure (voltage) at one end of the pipe, there is a constant, given amount of carriers emerging from the other end.

As can be expected, the amount of shot noise produced by different devices depends in some measure on the way they are manufactured. Unfortunately, a simple means for estimating the noise contribution of a particular device has not been discovered. Generally, device manufacturers make extensive tests on device types and publish typical shot-noise information together with other technical information about each device type. Since shot noise is additive with thermal noise, shot-noise information is usually published as an *equivalent noise resistance*. In predicting expected noise potentials in a circuit, a designer simply includes the noise resistance of each device in the analysis of the "noise circuit" to obtain a total equivalent noise resistance. This value is then used in Eq. (7.1).

7.3 SIGNAL-TO-NOISE RATIO (S/N)

When the time comes to apply information about noise to actual electronics systems, such as receivers, one basic concept in common use is the *signal-to-noise ratio* (S/N). Let us examine what the term means and how it is used.

The signal-to-noise ratio (S/N ratio) at any given point in a system is obtained, as the title suggests, by dividing the signal power at that point by the noise power at the same point. That is,

$$\frac{S}{N} = \frac{\text{signal power}}{\text{noise power}}$$

The result is a pure number; it has no units. Signal-to-noise ratios are often expressed in decibels:

$$\frac{S}{N}\,(\text{dB}) = 10 \log \frac{\text{signal power}}{\text{noise power}}$$

or

$$\frac{S}{N}\text{(dB)} = 20 \log \frac{\text{signal voltage}}{\text{noise voltage}}$$

(assuming that both signal and noise voltages are measured or calculated across the same resistance, as is usually the case).

Example 2
The voltage across a receiver's speaker coil is 890 mV when the receiver is adjusted for normal operation while receiving a station. The voltage across the coil when the receiver is not receiving a station (hence the noise voltage) is 12 mV. Determine the S/N ratio in decibels.

Solution

$$\frac{S}{N} = 20 \log \frac{\text{signal voltage}}{\text{noise voltage}} = 20 \log \frac{890}{12}$$

$$= 37.4 \text{ dB}$$

A moment's thought about Example 2 will raise a question about the validity of the S/N ratio as calculated. Is the voltage across the speaker, when the receiver is receiving a station, signal voltage only? No, the voltage is actually signal-plus-noise voltage. The ratio is better designated the "signal-plus-noise-to-noise ratio" [$(S + N)/N$ ratio]. In practical work it is generally very difficult to measure a pure signal voltage. If a pure signal voltage and noise voltage can be measured or calculated, it is possible to determine a pure S/N ratio; otherwise, as is usually the case, an $(S + N)/N$ ratio is what is determined. In most cases, as in Example 2, the noise voltage is so small compared to the signal voltage as to be negligible in the $S + N$ sum. The result is that the S/N and $(S + N)/N$ ratios are virtually identical, especially for practical purposes.

The *sensitivity* of a radio receiver is an indication of the minimum input signal level required to produce a usable output. Producing a receiver with excellent sensitivity is not simply a matter of incorporating circuits with abundant gain or amplification. The amplifying circuits must produce gain while minimizing the amount of internal noise which they introduce. Obtaining adequate gain is a relatively easy matter, obtaining low-noise gain is more difficult. In fact, there are practical limitations to improvements in the gain versus noise battle.

Objective, technical receiver sensitivity specifications always include two parts: (1) a specified power output level at a specified signal-to-noise figure, and (2) an input signal level required to produce the prescribed output level. In the final analysis, receiver sensitivity is an indication of how well the introduction of internal noise has been minimized.

SINAD Ratio

In the industry that manufactures and services the equipment for two-way radio communications systems there is a standard performance specification for communications-type receivers. The performance standard is based on the SINAD ratio. The expression "SINAD" is an acronym for "*si*gnal plus *n*oise plus *d*istortion." The SINAD ratio is

$$\text{SINAD} = \frac{\text{signal} + \text{noise} + \text{distortion}}{\text{noise} + \text{distortion}}$$

In the SINAD ratio, it is recognized that a receiver may introduce some distortion to the signal. Measurements for the ratio do not attempt to separate either noise or distortion from the signal.

The so-called "12-dB SINAD ratio" is a sensitivity measurement/specification for communications receivers. It is a means of providing a standard for the comparison of the useful sensitivity of various receivers. The 12-dB SINAD sensitivity is obtained by testing a receiver with a calibrated, modulated RF signal source. With a 100% modulated input, the receiver's audio output voltage is measured both with a signal output and with the test audio signal filtered out by a precise, "notch" filter. (The audio signal is the modulating signal of the RF generator after it has been processed by the receiver.) The minimum RF voltage that will bring the output (signal + noise + distortion) to 12 dB above the "notched out" output level is the 12-dB SINAD sensitivity.

7.4 NOISE FIGURE (NF)

How much noise does a given component or circuit add to the signal being processed? The answer to this question can be very useful. *Noise figure* (NF) is a concept used in providing an answer to the question. Noise figure (or sometimes, *noise factor*) is a ratio of ratios: the ratio of the signal-to-noise ratio at the input of an electronic unit to the signal-to-noise ratio at the output of that unit. Thus

$$\text{noise figure} = \frac{\text{input } S/N}{\text{output } S/N}$$

The result for NF using this formula is a pure number. The ratio can also be expressed in decibels:

$$\begin{aligned}\text{NF (db)} &= 10 \log \frac{\text{input } S/N}{\text{output } S/N} \\ &= \text{Input } S/N \text{ (dB)} - \text{output } S/N \text{ (dB)}\end{aligned}$$

You will note that if an amplifier adds no noise, S/N at input and output will be the same, and NF = 1 or 0 dB. An NF of 1 or 0 dB characterizes an ideal amplifier and is not something that is found in actual devices.

Noise figure data for specific electronic devices, such as RF transistors, is often made available by device manufacturers. The data are usually presented in the form of families of curves since NF varies with frequency and bias level, among other factors.

Example 3

An amplifier using a transistor with an NF = 4 dB has an input S/N of 40 dB. Estimate the output S/N.

Solution

$$\begin{aligned}\text{Output } S/N\text{ (dB)} &= \text{Input } S/N\text{ (dB)} - \text{NF} \\ &= 40 - 4 \\ &= 36\end{aligned}$$

7.5 NOISE AND RECEIVER AUTOMATIC GAIN CONTROL CIRCUITS

Assume that a receiver is to have several stages of amplification in its RF/IF sections and it is desired that it have high sensitivity. Is it necessary that all of the amplifier stages be designed specifically as low-noise stages? If not, which one(s) should be so designed? How should automatic gain control be applied to the stages to achieve best performance, including high sensitivity? To help us answer these questions, let us refer to the block diagram in Fig. 7.2. Each stage injects a noise voltage peculiar to that stage. The input voltage of all but the first stage is the output voltage of the preceding stage. The output voltage of a stage is equal to the gain of the stage times the sum of the input voltage and the noise voltage of the stage. The output voltages of the three stages are as follows:

$$\begin{aligned}v_{01} &= G_1(v_{\text{in}} + v_{n1}) \\ v_{02} &= G_2(G_1v_{\text{in}} + G_1v_{n1} + v_{n2}) = G_2G_1v_{\text{in}} + G_2G_1v_{n1} + G_2v_{n2} \\ v_{\text{out}} &= v_{03} = G_3(G_2G_1v_{\text{in}} + G_2G_1v_{n1} + G_2v_{n2} + v_{n3}) \\ &= G_3G_2G_1v_{\text{in}} + G_3G_2G_1v_{n1} + G_3G_2v_{n2} + G_3v_{n3}\end{aligned}$$

As you will note, both the input signal, v_{in}, and the noise voltage of the first stage, v_{n1}, are amplified by all three stages. The noise voltage of the second stage is amplified by two stages and the noise voltage of the third stage is amplified by only one stage. The noise voltage introduced by the first stage, then, is the largest contributor to the noise component of the final output voltage, v_{out}. The first stage is the one that must be designed for low noise if a favorable, overall S/N ratio and high sensitivity are to be achieved. If the first stage can be designed

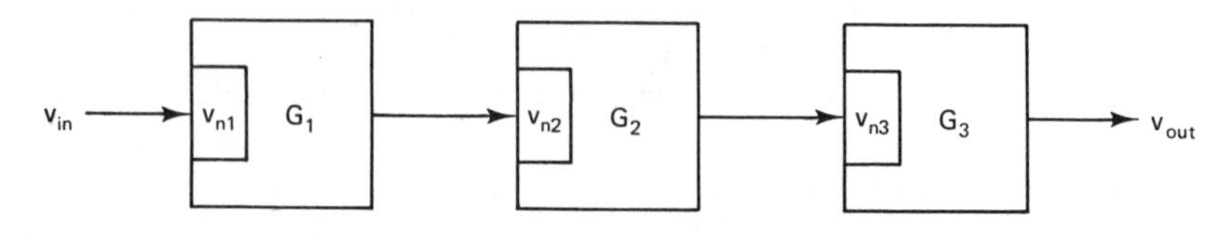

Figure 7.2 Effect of internal noise voltages in multistage amplifier.

to provide good amplification of the signal while adding minimal noise, the following amplifiers need not require excessive attention to minimization of noise.

Automatic gain control circuits, ideally, match the overall gain of an amplifier section to the incoming signal level. They automatically provide maximum gain for low-level signals and decrease the gain when signal levels are higher than the minimum.

If there are three stages in an amplifier section, as in Fig. 7.2, is it good strategy to reduce the gain of all three stages by the same amount for stronger-than-minimum signals? Assume that we can decrease the value of G_1, G_2, and G_3 individually, but that some overall gain product is needed.

For minimum output noise it is apparent that the gain of the third stage should be reduced first. If more reduction in gain is needed, the second stage gain should be reduced. The gain of the first stage should not be reduced unless further reduction in gain is required to process the incoming signal satisfactorily without overdriving any of the stages. The AGC circuitry of some receivers, particularly TV receivers, is arranged to operate in just this fashion. Such circuits are referred to as "delayed AGC" circuits. The word "delay" usually suggests a postponement as a time-related phenomenon. In this case it is being used to indicate a "postponement" related not to time but to signal level: "Delay reducing the gain of the first stage until the signal amplitude exceeds a certain level."

The expression "low-noise design" generally refers to techniques that are or can be taken to reduce the effect of internal noise in electronic systems. Low-noise techniques are particularly relevant in the design of receivers—radio or television. This is true because, often, even with the best of antenna systems, the signal that must be processed to provide a useful output is of an extremely low amplitude. We must do the best we can with that signal. The alternative is no signal and no output at all.

However, internal noise is not the only enemy of reception. External noise is also a serious obstacle to acceptable reception especially when the desired signal amplitude is small. By its nature, most external noise is amplitude-modulated electrical impulses, or it produces amplitude modulation of a desired signal within the receiver itself. Thus an AM detector demodulates noise energy and outputs it to a speaker together with the desired signal. One answer to external noise is a communications system using frequency modulation (FM). As we shall see in Chap. 8, FM is inherently a low-noise system, and, in this case, external noise is included in the kinds of noise whose effects are minimized.

7.6 NOISE AND INFORMATION TRANSMISSION

As we have seen, most types of noise are spread across the radio-frequency spectrum. It is logical, then, that the greater the bandwidth of a particular communications system, the more susceptible the system is to noise. Bandwidth is somewhat analogous to size of opening for admitting light: the larger the opening, the more light admitted, but also, the more easily dust can enter.

It also appears logical that the greater the bandwidth, the more information that can be transmitted in a given time. This relationship has been studied carefully to determine its validity. In 1928, R. V. L. Hartley of the Bell System reported the results of his investigations which verified that validity. His report resulted in the acceptance of the relationship as a fundamental principle of information theory. It is now known as the *Hartley law of information theory*. This principle, of course, presents a dilemma: greater bandwidth permits the transmission of a greater amount of information in a given time, but also opens the system to the introduction of more noise. Noise is indeed a serious limiter to the transmission of information.

Systems that can successfully limit noise while providing abundant bandwidth are definitely superior to narrow-bandwidth systems for the transmission of large amounts of information. These are characteristics which are true of the FM standard broadcast system. It has several inherent noise-limiting features, and its bandwidth, 200 kHz, is 20 times that of the standard AM broadcast service. It is for this reason that FM is known as the "high-fidelity system." It transmits more information in that, for example, it passes all audio frequencies up to 15 kHz while AM broadcasts generally include frequencies up to 5 kHz only.

The relationship between bandwidth, information transmission, and noise can be used in a different way to enhance communications systems. For example, if a limit is placed on the amount of information that will be transmitted, the bandwidth can be limited and the signal-to-noise ratio will be improved, or it can be maintained at a satisfactory level without excessive attention to low-noise design. Most two-way radio systems are so-called narrow-band systems. Since they are to be used for two-way voice communication, a limit of 3 kHz is placed on the audio frequencies transmitted. This reduces the bandwidth requirement of the system, inherently reduces noise levels, and provides a system with acceptable signal-to-noise values at reasonable cost. Narrow-band systems also economize on the utilization of the radio-frequency spectrum.

GLOSSARY OF TERMS

Atmospheric noise Electrical impulses produced by natural phenomena in the atmosphere (e.g., lightning discharges) and which are detrimental to the operation of electronics systems, particularly communications systems.

Delayed AGC A form of automatic gain control for an amplifier section designed so that the gain of the stage nearest the source of the incoming signal is not reduced unless reduction of gains of other stages is insufficient to provide overall optimum operation.

Emission The radio-frequency energy radiated by an antenna.

External noise Noise that originates outside an electronic system.

Hartley law A statement of the physically observable principle that the amount of information that can be transmitted in a given time (the maximum information rate) is proportional to the range of frequencies used.

Internal noise Noise (unwanted electrical impulses) that originates within electronic components themselves.

Johnson noise *See* Thermal noise.

Low-noise design The design of electronic equipment which includes careful attention to techniques known to minimize the potential for noise to appear in the signals being processed by the equipment (e.g., the selection and utilization of low-noise resistors).

Noise figure A measure of the noise added to a signal by a component or circuit; equal to the ratio of the input signal-to-noise ratio to the output signal-to-noise ratio.

Sensitivity, receiver The minimum input signal to a receiver required to produce an output of specified power level and signal-to-noise ratio.

Shot noise Electrical impulses that arise in electron devices such as tubes and transistors as a result of the discrete particle nature of the current carriers in such devices.

Signal-plus-noise-to-noise ratio The ratio of the sum of signal and noise powers to the noise power, at a given point in an electronic circuit; usually abbreviated as $(S + N)/N$.

Signal-to-noise ratio The ratio of signal power to noise power at a given point in an electronic circuit; usually abbreviated as S/N.

SINAD The ratio of signal-plus-noise-plus-distortion to noise-plus-distortion.

Thermal noise Noise generated in resistors and other circuit components as a result of the motion of the electrons and/or other current carriers not being purely random as theorized.

White noise *See* Thermal noise.

REVIEW QUESTIONS: BEST ANSWER

1. A current or voltage that appears in a circuit, but is unintended or undesired, is called: **a.** emission. **b.** signal. **c.** information. **d.** noise. **e.** none of these.
2. The buzz heard from a radio's speaker while a fluorescent light is operating is a form of: **a.** thermal noise. **b.** man-made noise. **c.** space noise. **d.** white noise. **e.** none of these.
3. A technician replaces a burned-out wire-wound resistor in a TV tuner with an inexpensive carbon resistor.
   a. If the correct value was used, the receiver should work normally.
   b. The replacement resistor will introduce more noise, which may be observable as "snow" in the picture.
   c. The picture will be better than before because of a reduction in noise.
   d. The receiver will not work because the proper resistor was not used.
   e. a and c.
4. A voltage of 0.926 V is measured across an 8-Ω resistor connected to temporarily replace a radio's speaker. The test is made while the receiver is being fed an RF signal from a laboratory generator which is being modulated with a 1000-Hz signal. When a 1-kHz "notch filter" is placed between the resistor and the audio amplifier, the voltage drops to 7.35 mV. The signal-plus-noise-to-noise ratio is: **a.** 126 dB. **b.** 2.10 dB. **c.** 42 dB. **d.** 21 dB. **e.** none of these.
5. A signal with an S/N of 38 dB will be fed to an amplifier with an NF of 18 dB. The estimated output S/N in decibels is: **a.** 56 dB. **b.** 38 dB. **c.** 20 dB. **d.** 18 dB. **e.** none of these.
6. The input signals required to produce rated output power and S/N for four receivers are as follows: A: 5 μV, B: 5 mV, C: 200 μV, D: 35 μV. The receiver with greatest sensitivity is: **a.** A. **b.** B. **c.** C. **d.** D. **e.** none of these.

7. While producing rated outputs with identical input signals, four receivers exhibited the following output signal-to-noise ratios: A: 12 dB. B: 25 dB. C: 40 dB. D: 55 dB. The receiver with the best (highest) sensitivity is most likely receiver: **a.** A. **b.** B. **c.** C. **d.** D. **e.** none of these.
8. In a multistage amplifier, for best overall signal-to-noise performance the stage that should be especially designed for low noise is the: **a.** output stage. **b.** middle stage. **c.** input stage. **d.** a and c. **e.** none of these.
9. The performance of a "noisy" communications channel might be improved by reducing its bandwidth, but that would also: **a.** lower the maximum frequency of operation. **b.** lower the rate of information transmission. **c.** increase the amount of gain required. **d.** minimize the SINAD. **e.** none of these.

REVIEW QUESTIONS: ESSAY

1. Discuss the reasons for the importance of noise in the design and operation of communications systems.
2. Describe shot noise.
3. What is meant by *low-noise design*?
4. Discuss the term *signal-to-noise ratio*. What does it mean, and how can it be useful to persons working with communications equipment?
5. If you were a communications electronics technician, why might you consider the $(S + N)/N$ a more practical ratio than S/N?
6. Give the meaning of *noise figure* (NF). Describe how data on noise figures, as given by device manufacturers, could be useful in a practical way.
7. What does SINAD mean? What is the SINAD ratio? Where is it used?
8. What is the relationship between noise, channel bandwidth, and rate of information transmission on a given channel?

EXERCISES

1. The output of a receiver is measured as 763 mV across an 8-Ω dummy speaker load. With the signal filtered out, the voltage measured 86.4 mV at the same place. Calculate the $(S + N)/N$.
2. A video amplifier with a bandwidth of 4.5 MHz has an input impedance of 3600 Ω. **(a)** Estimate the noise voltage at the input to the amplifier due to thermal noise if the operating temperature is 28.5°C. **(b)** If the output signal voltage is 600 mV across a typical load, what is the S/N at the output? Use a gain of 100 (voltage) for the amplifier and assume that the input thermal noise is the only noise present.
3. An RF amplifier has a bandwidth of 200 kHz and a gain of 50 dB. The total resistance present in the input circuit is 20,000 Ω. **(a)** Estimate the thermal noise for an operating temperature of 31°C. **(b)** What is the input signal in volts if it is 10 times the noise voltage introduced by the input circuit? **(c)** What is the S/N at the output for the input signal of part (b)?

4. A receiver has an output S/N of 43 dB when the input S/N is 55 dB. What is the noise figure for the receiver? Express the result in two forms.
5. What will be the output S/N of a receiver when the input S/N is 57 dB if the NF for the receiver is 6 dB?
6. What S/N is required at the input of a receiver if an S/N of 48 dB is desired at the output and the receiver has an NF of 9 dB?

8

FM COMMUNICATIONS SYSTEMS

Modulation in the first radio communications system consisted simply of turning the carrier on and off in accordance with a predetermined code. This was a *radiotelegraph* system and the emission it produced is now called "CW" for *continuous wave*. Subsequently, equipment was developed which provided for modulating the amplitude of a carrier. This was the beginning of *radiotelephone* systems—two-way radio communications using the human voice—and led ultimately to radio broadcasting.

These early developments took place in a world in which very little was known about electronics in general, or about the principles of radio in particular. As experience was gained with AM (amplitude modulation) systems, it was discovered that there are several drawbacks to such systems. One of the disadvantages of an AM system is that it is inherently noisy: Most electrical noise, from whatever source, is either amplitude modulated itself, or can easily cause amplitude modulation of the desired signal in a radio receiver. Another disadvantage of AM is that with its sidebands, it is not economic in its use of the spectrum.

Increased knowledge of the principles of radio led to the development of a method for transmitting information by modulating the *frequency* of a carrier rather than its amplitude. Furthermore, it was discovered that frequency modulation (FM) has an inherent characteristic of actually suppressing the effects of noise, as well as escaping the effects of AM-type noise. These discoveries led to the development of the standard FM broadcast service as well as other applications of FM. Unlike AM applications which mostly "grew" instead of being carefully engineered, FM systems were developed as systems. A great deal of attention was given to engineering into the systems as many advantages of the modulation method as possible. As a result, the FM broadcast service, especially, is generally accepted as a low-noise, high-fidelity broadcasting system.

8.1 PRINCIPLES OF ANGLE MODULATION

We have learned previously that the operating frequency of an oscillator can be controlled by an *LC* tank circuit—the frequency of oscillation being equal, or approximately equal, to the resonant frequency of the *LC* circuit. Obviously, if the value of either *L* or *C* changes, the oscillator's operating frequency will change accordingly. For example, if *C* increases, the operating frequency will decrease, and vice versa. This is true because of the relationship of resonant frequency to *L* and *C*:

$$f_r = \frac{1}{2\pi\sqrt{LC}}$$

Let us refer to the circuit of Fig. 8.1, then, and imagine what will happen to the operating frequency of the oscillator as someone speaks into the capacitor microphone: The movable diaphragm of the mike will alternately move closer to, and farther away from, the fixed plate of the mike, increasing and decreasing the capacitance that the mike represents. Since the capacitance of the mike is in parallel with the *C* of the oscillator tank circuit, the net *C* of the tank circuit will alternately *increase* and *decrease*. And the operating frequency of the oscillator will alternately *decrease* and *increase!*

The operating frequency of the oscillator will change at a rate equal to the frequency of the sound wave producing the change in capacitance. The amount of change of the operating frequency will be proportional to the amplitude of the sound wave. Observe the waveforms of Fig. 8.1(b) and note how the cycles of the oscillator waveform are alternately compressed and spread out, depicting changes in its frequency.

What we have just described is true frequency modulation: The *rate* of frequency change is equal to the *frequency of the modulating force; the amount* of frequency change (the deviation) is proportional to the *amplitude* of the modulating force. We do not wish to imply that the circuit of Fig. 8.1 is a typical frequency modulation scheme. The circuit illustrates simply and directly the idea of FM; we will be studying more practical modulating circuits in detail in this chapter.

Before going on to the details of true, or direct, FM, let us become aware of a slightly different, but related form of modulation: phase modulation (PM), or indirect FM. When we investigate, in detail, the effects of true FM, we find that the relative amount of change in a typical FM carrier during modulation is very small: at most, 100 kHz in 88 MHz, or about 0.1%. The change is virtually imperceptible to the human eye, for example, when a modulated FM carrier is displayed on an oscilloscope at a sweep rate that permits viewing individual

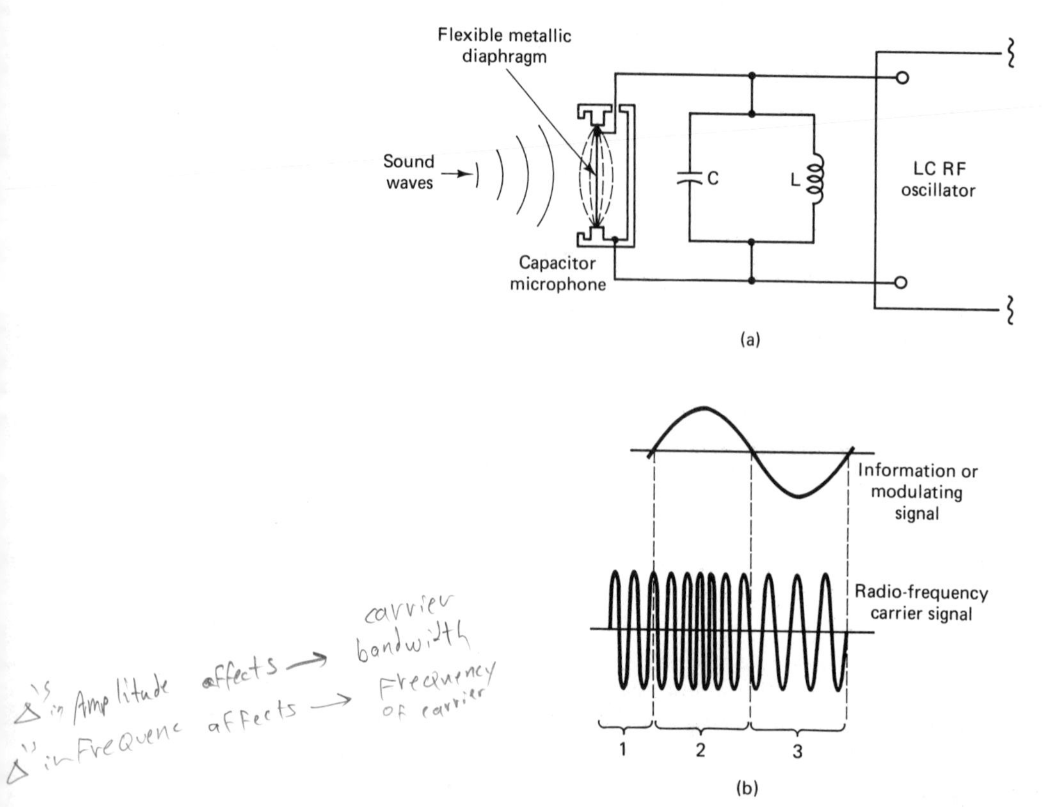

Figure 8.1 (a) Capacitor microphone modulating oscillator. (b) Waveforms of frequency modulation: 1, carrier at center, or unmodulated frequency; 2, carrier frequency increased as a result of modulation; 3, carrier frequency decreased by modulation.

cycles of the carrier. To electric circuits, for example a resonant circuit, the momentary change in frequency of FM is equivalent to a momentary change in the phase angle of the sine wave of the carrier: the phase angle goes leading, then lagging, and then back to normal as the carrier frequency increases, decreases, and returns to its normal or *center frequency*. This being the case, since it is possible to modify, or modulate, the phase angle of a signal *after* it has been generated, it is possible to achieve a form of FM without actually changing the frequency of the oscillator. This form of modulation is called *phase modulation* or *indirect FM*. As we will see, although PM is not true FM, it can be corrected, by appropriate circuit design, so that it becomes completely equivalent to true FM. Because both FM and PM are equivalent to changes in the *phase angle* of a carrier, together they are generally referred to as *angle modulation*.

The Terminology of FM

There are a number of terms relating to frequency modulation which have very explicit meanings. Knowing and understanding the meaning of these terms facilitates the study of FM.

The amount by which the frequency of a carrier is changed is called *deviation* and is generally represented by Δf (pronounced "delta f"). In a frequency-modulated carrier, the amplitude of the modulating signal is represented by the amount of deviation, or

$$\Delta f \propto v_m \qquad \text{or} \qquad \Delta f = kv_m$$

where v_m represents the amplitude of the modulating signal and k is a constant of proportionality.

How often, or how frequently, is the frequency of an FM carrier changed? The frequency at which the carrier is changed is called the *deviation rate*, or simply, the *rate*. The rate is the number of times per second the carrier goes through a cycle of frequency change: increase, decrease, back to center frequency. The rate is equal to the frequency of the modulating signal:

$$\text{deviation rate} = f_m$$

MEMORY JOGGER

1. Amount of frequency change, deviation or Δf, is proportional to the amplitude of the modulating signal.
2. The deviation rate, the number of times per second the carrier goes through a cycle of frequency change, is equal to the frequency of the modulating signal.

In amplitude modulation, the amount by which a carrier can be modulated is limited in a sort of natural way: the amplitude can only be reduced to zero. That is equivalent to 100% modulation in AM. Theoretically, an FM carrier could be deviated until its frequency is reduced to zero. (What would that do to the signals of all stations whose frequencies are less than that of the given carrier?) In the development of FM systems it became desirable to define a *maximum deviation*, $\Delta f_{\max}$. A maximum deviation has been prescribed by FCC regulations for each communications service that employs FM. For standard FM broadcast, $\Delta f_{\max} = \pm 75$ kHz, for the aural carrier of television broadcasts, $\Delta f_{\max} = \pm 25$ kHz. Other maximum deviations have been prescribed for other services. Frequency deviation is sometimes referred to as "frequency swing." Hence it is often said that standard FM broadcast is permitted a maximum swing of ± 75 kHz.

It is now possible to state the definition of percent modulation in FM: *Percent modulation is equal to 100 times the ratio of actual deviation to maximum permitted deviation*

$$\text{percent modulation} = \frac{\Delta f_{\text{actual}}}{\Delta f_{\max}} \times 100$$

For example, if an FM broadcast carrier is being deviated ± 50 kHz, the modulation level is 67%. Remember, just as in AM, percent modulation is an indication of the amplitude of the modulating signal; at the speaker of a receiver, sound level is proportional to percent modulation.

Another way in which FM is different from AM is in the number of side frequencies produced by the modulation process. You will recall that when a carrier is amplitude modulated by a single-frequency signal, only an upper side frequency and a lower side frequency are produced. In FM, the modulation process "distorts" the carrier to a much greater extent and the number and amplitude of side frequencies is in accordance with a complex mathematical relationship known as the *Bessel function*. One of the parameters of an FM wave used in conjunction with Bessel functions to determine the makeup of FM sidebands is called the *modulation index,* m_f. The FM modulation index is equal to the ratio of the actual deviation, Δf, to the modulating frequency, f_m:

$$m_f = \frac{\Delta f}{f_m}$$

Table 8.1 lists the number of significant side frequencies corresponding to various modulation indexes. Side frequencies always come in pairs; the numbers in the table indicate pairs.

TABLE 8.1

Modulation index	Number of side frequencies (pairs)
0.25	1
0.5	2
1.0	3
1.5	4
2.0	4
3.0	6
5.0	8
10.0	14
15.0	16

The side frequencies in FM are separated from the carrier frequency (also called the center frequency), f_c, and each other by an amount equal to the modulating frequency, f_m. Thus, if a carrier were being modulated by a frequency of 8 kHz with an index of 1 (three side-frequency pairs), there would be side frequencies at f_c plus 8, 16, and 24 kHz as well as at f_c minus 8, 16, and 24 kHz.

The amplitudes of side frequencies in FM also vary in a complex pattern that can be predicted, again, with the aid of Bessel functions. The total power in an FM wave does not vary with the level of modulation, although the power in the carrier itself does vary with modulation. What happens is that power i transferred from the carrier to the side frequencies. The number and amplitude of side frequencies for various modulation indexes is shown in the spectrograms of Fig. 8.2. As you study the diagrams, note how the amplitude of the carrier is different for various modulation indexes. Make sure that you are aware that the modulation index is a function, not simply of actual deviation, but of the deviating frequency as well.

It is interesting to take note of the conditions depicted in Fig. 8.2(e): f_m = 15 kHz, the maximum audio frequency for FM broadcast, and Δf = 75 kHz,

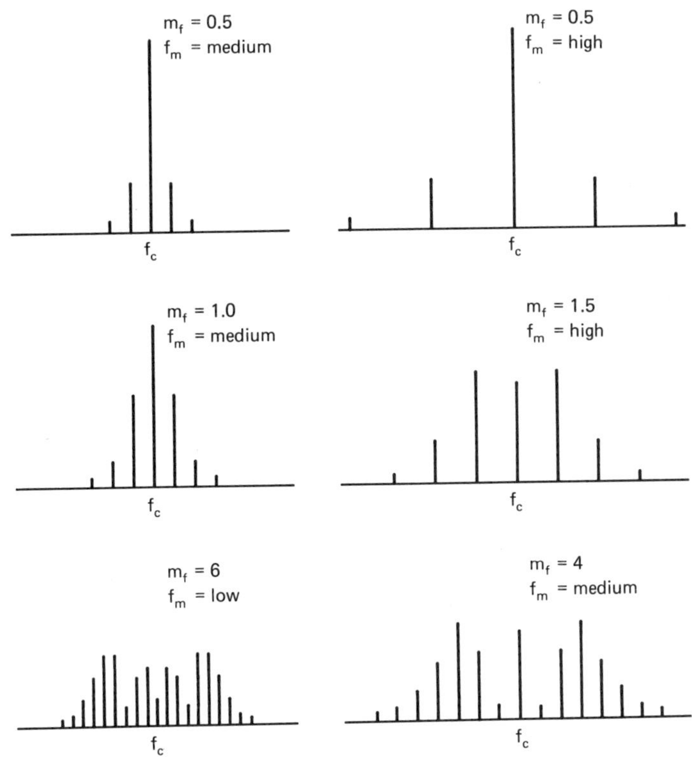

Figure 8.2 Effect of modulation index and modulating frequency on side frequencies and carrier amplitude in FM.

the maximum permissible deviation. These conditions produce a modulation index of 5, the extreme side frequencies are at ±120 kHz (8 × 15 kHz). Thus an FM carrier modulated in this fashion would occupy a bandwidth of 240 kHz (2 × 120 kHz); the authorized bandwidth is 200 kHz. This is not a serious problem, however, because the amplitude of the eighth side frequency is only about 2% of that of the unmodulated carrier. Furthermore, very little program material produces a modulating frequency of 15 kHz.

Example 1

An FM broadcast station with f_c = 96.1 MHz is being modulated with a 12-kHz test signal so that Δf = 60 kHz. Determine: (a) percent modulation; (b) modulation index; (c) number of side frequency pairs; (d) extreme upper and lower side frequencies; (e) bandwidth of modulated carrier.

Solution. (a) Percent modulation $= \dfrac{\Delta f_{\text{actual}}}{\Delta f_{\max}} \times 100 = \dfrac{60}{75} \times 100 = 80\%$

(b) Modulation index $= \dfrac{\Delta f}{f_m} = \dfrac{60}{12} = 5$

(c) From Table 8.1, for a modulation index of 5 there are eight significant side-frequency pairs.

(d) Extreme upper side frequency $= f_c + 8 \times 12 \times 10^3$
$= 96.1 \times 10^6 + 8 \times 12 \times 10^3$
$= 96.196$ MHz

Extreme lower side frequency $= f_c - 8 \times 12 \times 10^3$
$= 96.1 \times 10^6 - 8 \times 12 \times 10^3$
$= 96.004$ MHz

(e) Bandwidth = extreme USF − extreme LSF = 96.196 − 96.004
= 0.192 MHz
= 192 kHz

FM Bandwidth Requirements

The bandwidth of a transmitter refers to the portion of the frequency spectrum occupied by all of its components of emission of significant energy level. The FCC prescribes the bandwidth limitations for each of the numerous radio services authorized for operation. Prescribing bandwidths and making a reasonable effort to ensure that such regulations are being complied with by operating stations is part of the FCC's effort to provide for maximum utilization of the radio-frequency spectrum and minimum interference between stations. The authorized bandwidth for *standard FM broadcast stations* is given in the FCC definition of an *FM broadcast channel:* "A band of frequencies 200 kHz wide and designated by its center frequency." FCC regulations also state that a standard FM broadcast transmitter shall be able to process audio signals up to 15 kHz and shall be able to achieve a deviation of the carrier of ±75 kHz.

The bandwidth specifications for other radio services which utilize frequency modulation differ from those stated above for standard FM broadcast stations. Such specifications, along with other relevant information, for several other services will be presented and discussed in other sections of this book.

Power Distribution in FM Emissions

When a carrier is amplitude modulated, the power in the emission is increased. The additional power is entirely accounted for by the power in the side frequencies which are produced by the modulation—the power in the carrier remains unchanged. The situation is completely different in frequency modulation: The total power of the emission remains unchanged; power is shifted from the carrier to the side frequencies. As we have indicated previously, the amplitudes of the side frequencies cannot be determined in a simple direct way, only by use of the Bessel function. Since for a given load resistance, power is proportional to voltage amplitude squared, there is no simple direct way to state how power is distributed between the carrier and side frequencies of an FM emission.

Since total RF power in an FM emission does not change with modulation, antenna current is unaffected by modulation. This is in sharp contrast to the situation at an AM station where current must change with the modulation level. One of the advantages of the constant-power-level phenomenon is that the regulation of the transmitter power supply is less critical—the current it is required to supply does not fluctuate wildly.

Example 2

The deviation of an FM broadcast station is changed from 0 to 48 kHz. Determine: (a) the percentage modulation; (b) the percent change in antenna current.

Solution. (a) Percent modulation $= \dfrac{\Delta f_{\text{factual}}}{\Delta f_{\text{max}}} \times 100 = \dfrac{48}{75} \times 100 = 64\%$

(b) Since there is no change in total RF power with modulation, in FM, there is no change in antenna current.

Noise Suppression in FM Systems

As we have stated previously, an FM system, by its very nature, has less noise than an AM system. There is more than one reason for this fact. One reason has to do with the nature of noise energy itself. Most noise "signals" are amplitude modulated. Or, noise signals mix with desired signals in circuits operating in a nonlinear fashion in AM receivers to produce the equivalent of an AM signal, which is detectable. The latter process is called *intermodulation distortion,* generally abbreviated *IM.*

As we shall see, typical FM receivers are designed to eliminate, prior to detection, virtually all variations in the amplitude of an FM carrier signal. To the extent that this process (which, incidentally, is called *limiting*) is successful, noise caused directly by variations in the amplitude of an FM signal is eliminated.

However, eliminating all variations in amplitude just prior to detection does not eliminate all *effects* of changes in the amplitude of the FM carrier. Let's look at an important basic statement and then see why it is true: *Changes in the amplitude of a carrier signal produce changes in the phase of the signal and, therefore, the equivalent of frequency modulation. Or, amplitude modulation causes frequency modulation!*

A diagram will help to demonstrate what happens when the amplitude of a signal changes. Refer to Fig. 8.3. The left end of the waveform depicts a carrier that is neither amplitude nor frequency modulated. In this portion of the waveform, the fact that the frequency is constant could be determined by noting that the distances between the peaks (points 1) of the cycles are constant. Or constant frequency could be verified by checking distances between other similar points on each cycle, say points 2. For example, if we were special "beings" that could check these points "as they come by" for their time separation, we could determine whether the frequency of the waveform is constant. Either points 1 or 2 would be OK for this purpose. Or would they?

Consider the waveform at the center and to the right side of the diagram. The frequency of the signal has not been changed. However, as you can see, the amplitude has: it was first increased and then decreased from its normal value. What effect, if any, does an increase in amplitude have on the arrival

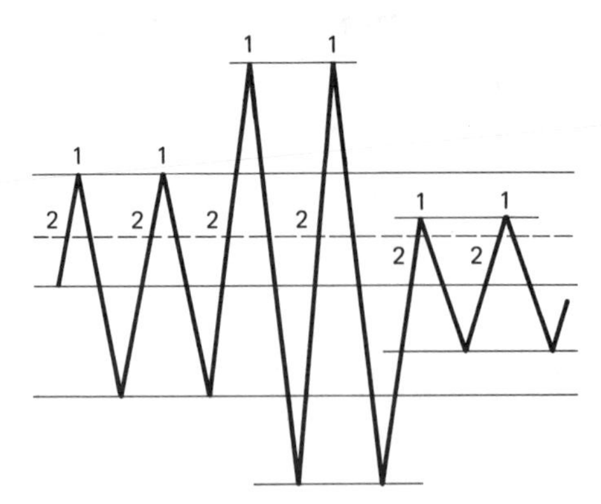

Figure 8.3 Effect of amplitude changes on phase of sine wave.

time of points 1 and 2 of the waveform? If the frequency has not been changed, the *peaks* arrive at the predicted time, their time separations are unchanged. What about points 2? Isn't it true that if the peak of a wave is higher and "arrives" at its "place" at the correct instant, other levels of the waveform must "arrive" at their places "sooner?" The time phase, not of the peak, but of other points on the waveform must advance slightly. The greater the increase in the amplitude, the greater must be the advance in phase. Of course, if the amplitude stabilizes at a new, higher value, the distance between points 2 will again stabilize and be a reliable indicator of frequency.

Looking at the right side of the diagram of Fig. 8.3, where amplitude has been decreased, we would reason that since *peaks* are arriving at their level, a lower level, at their usual times, other points will arrive later. The phase of other points will appear to *lag* while the change in amplitude is occurring.

The reactive circuits of an FM detector in an FM receiver might be thought of as "beings" that can detect the changes in time phase just described. These phase changes are indistinguishable from the phase changes that occur when the frequency of the waveform is changed. Such changes are detectable by an FM detector. Since they are the result of unintentional, and therefore, unwanted effects, when reproduced they are perceived as noise. This noise is an FM noise and is to be distinguished from "AM noise."

In true FM the *amount of deviation* is proportional only to the amplitude of modulating signal: $\Delta f \propto v_m$. When the frequency is changed, in the modulation process of FM, it is changed by a comparatively small amount. And the change does not persist—the frequency is increased, then decreased, increased, and so on, at a rate equal to the frequency of the modulating signal. Therefore, the actual change in frequency appears as a change in the phase angle of the carrier. The phase angle changes in the leading direction, then in the lagging direction, etc. The amount of the phase-angle change is, like the frequency change, proportional to the amplitude of the modulating signal.

In the case of phase-angle changes produced by changes in the amplitude of the carrier, *the amount of phase angle change is directly proportional to the frequency of the amplitude change!* If the amplitude is increasing slowly (the frequency of the change is low), the FM carrier values will advance only slightly.

If we call the change in phase "ϕ," then ϕ is small. If the increase in amplitude is rapid (the frequency of amplitude change is relatively high), ϕ will be larger.

Remember, there is a "ϕ" in FM also. There ϕ is strictly proportional to the amount of deviation, which is proportional to the amplitude of the modulating signal. The FM detector interprets the size of ϕ as indicative of the amplitude of the information signal. Thus the amplitude of noise, produced as described in the preceding paragraphs, is proportional to the frequency of the disturbance producing it! Let's state this another way: The noise heard from the speaker of an FM receiver will be louder for higher frequencies produced!

The phenomenon we have been studying in the preceding paragraphs—FM noise produced by amplitude modulation of the carrier signal—is *phase modulation*. In this case, it is noise produced by phase modulation. Phase modulation can, and is, used to actually achieve desired frequency modulation. Therefore, a knowledge of phase modulation—what it is, how it can be achieved, and its similarities and differences with pure FM—is an extremely important facet of the study of the general topic of angle modulation. You are urged to study the explanation above intensively. Generally, it is not possible to gain an adequate understanding of phase modulation simply by reading over an explanation such as the one above, or even by reading it over several times. You must study an explanation of phase modulation until you "see" what is happening yourself.

There is another, more technical explanation for phase modulation produced by amplitude modulation of the carrier with which you should be familiar. The explanation involves the use of phasors.

Refer to Fig. 8.4. The diagrams of Fig. 8.4 are phasor diagrams of a carrier, C, an interfering or noise signal, N, and the resultant, R, of these two. Think of the phasors as rotating counterclockwise. If we assume that the frequency of N is greater than the frequency of C, we can represent various time conditions by drawing C in a fixed position and N at various angles with respect to C Then R is constructed as the phasor sum of the other two, for each condition.

On the diagram, ϕ represents the deviation of R with respect to the carrier, and therefore means the same as ϕ in the paragraph above. With careful study it can be seen that R goes from one side to the other of C; ϕ, the phase deviation, can be positive or negative. There is a maximum ϕ, and it occurs when N and R are perpendicular to each other.

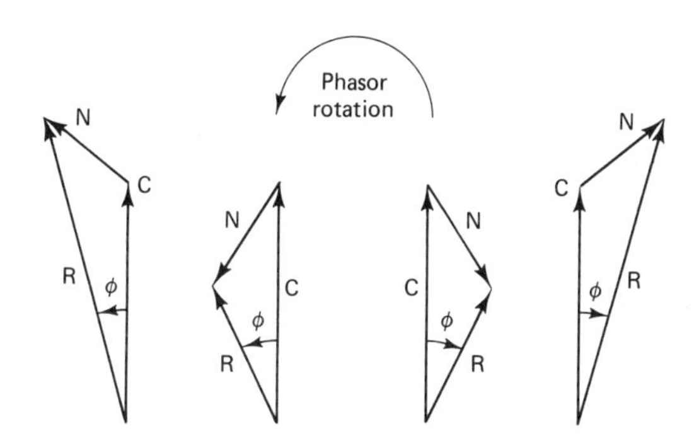

Figure 8.4 Phase modulation as a result of amplitude modulation.

However, the diagram does not indicate, in any direct way, the relationship between the actual equivalent frequency deviation produced and the frequency of the force causing the modulation. Unlike true, or direct FM, there is a relationship between Δf and the modulating frequency in phase modulation. It is known that in indirect FM, or phase modulation, the amount of equivalent frequency deviation, Δf, is proportional to the frequency of the force producing the deviation. The frequency of the force, in the case of Fig. 8.4, is the difference in frequency between C and N. That is, the modulating frequency is the beat frequency of C and N ($f_c - f_N$).

How can we explain, with the use of the diagram, why the Δf produced is proportional to the modulating frequency, the beat frequency in this case? Let's start by reminding ourselves that when we use a phasor diagram we are thinking of it as a "freeze photo" of rotating phasors (arrows). In the phasor diagram of phase modulation (Fig. 8.4) the rotational speed of C, representing the frequency of the carrier, is constant, so that its position remains fixed. The position of R, the phasor sum of C and N, changes. Sometimes R is leading, or is ahead of C; sometimes it is lagging C. The position of R changes because N is rotating faster than C and when we find their phasor sum, at various times, the sum is in different places. We can interpret this changing of the position of R as a variation of its speed (frequency)—a speeding up and slowing down. In terms of frequency, it is a plus and minus Δf. But R is the resultant carrier; it is the signal the receiver will process as the received carrier. The phase, and thus indirect frequency modulation, of R is the equivalent of frequency modulation of the received carrier.

The angular separation of R and C, in Fig. 8.4, is indicated as ϕ; thus ϕ is a measure of the change in position of R (with respect to C). What determines the value of ϕ? Of course, for one thing, the relative positions of C and N are determining factors. But ϕ is also dependent on the relative values of C and N. If N equals ϕ, then R equals C and $\phi = 0$. If N is small compared to C, ϕ is small no matter what the positions of C and N; if N is larger, ϕ is larger. The value of ϕ, then, is proportional to the magnitude of the phenomenon producing it.

What is the significance of the value of ϕ when interpreted as a frequency deviation for R? Take the example of R leading C. In order for one rotating element to get ahead of another, the first obviously has to rotate faster. What if ϕ is negative? R has to slow down to attain a position "behind" C. If we consider two different values of ϕ, two values that have been attained in *the same amount of time,* then, the greater the value of ϕ, the greater is the difference in speed between R and C. Or, relating it to frequency, the greater the value of ϕ, the greater is Δf. An important condition here is pointed to in the phrase "in the same amount of time."

What happens if we try to compare two values of ϕ, two equal values, for example, which have been attained in different amounts of time? This is what we are doing when we consider the significance of the value of ϕ for different frequencies of the modulating force. (In true FM the amount of frequency deviation, Δf, must be completely independent of the frequency of the modulating signal.) Visualize two situations represented by two phasor diagrams: the value of ϕ is the same for each diagram; however, the time required to achieve ϕ in the first situation is twice the time required in the second. It is obvious that the amount of change in rotational speed (or the Δf) for the second situation must be twice that of the first. The time (period) of a periodical phenomenon is inversely proportional to its frequency. Thus, in the two imaginary situations just described, the frequency of the modulating force in the first situation is one-half the frequency of the second. The equivalent Δf of the first is one-half the equivalent Δf of the second.

We now restate a very important principle of phase modulation: The equivalent frequency deviation caused by a change in the phase of a signal is directly proportional to the frequency of the force that is causing the change. With phase modulation, then,

$$\Delta f = \phi \times f_m \tag{8.1}$$

where ϕ = phase displacement of the carrier signal produced by the force causing the modulation (rad)

f_m = frequency of the force causing the modulation

When phase modulation is the result of amplitude modulation produced by a noise impulse signal, or the interfering signal of another station, the amount of phase modulation is directly proportional to the relative amplitudes of the noise or interfering signal, and the carrier signal. The equivalent Δf is also directly proportional to the difference in frequency of the two signals, their beat frequency.

Suppression of Noise from Impulse Signals

Let's compare the signal-to-noise ratios of AM and FM systems for noise caused by impulse signals. An example of an impulse signal is the radiation from the ignition system of a passing automobile. Let's say the radiation has an energy component whose frequency differs from that of the desired carrier (AM or FM) by an amount that is in the audio-frequency range. The radiation enters our receiver's IF section and mixes with the desired carrier, producing a change in the amplitude of that carrier of 50%. In the case of the AM system, assume that the information signal is modulating the carrier 100%. The signal-to-noise ratio is 2 to 1, or 6 dB. The AM receiver will demodulate the noise-modulated carrier and pass the noise on to the speaker. The S/N at the speaker will also be 6 dB.

Now consider the FM system. The direct change of the amplitude of the FM carrier will be removed by the amplitude limiter circuit of the receiver. Any noise reaching the speaker will be the result of phase modulation. Consider the phasor diagram of Fig. 8.5. Noise will be maximum when Δf is maximum and Δf is directly proportional to ϕ. When is ϕ maximum? ϕ is maximum when the angle between N and R is 90°. The maximum value of ϕ is obtained using the relationship

$$\phi_{\max} = \arcsin \frac{N}{C}$$

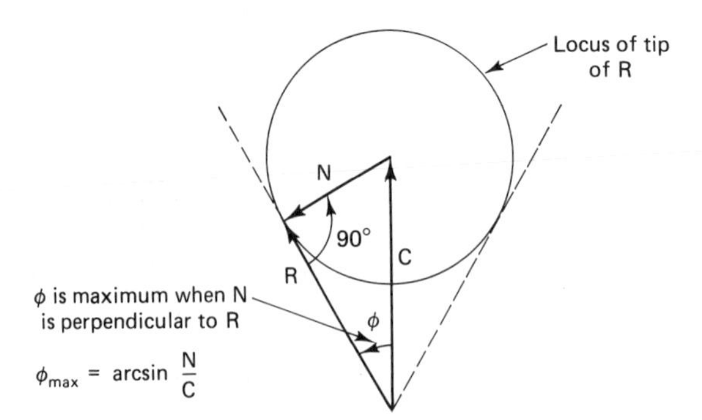

Figure 8.5 Condition for maximum ϕ.

In this example, then

$$\phi_{max} = \arcsin\frac{1}{2} = 30.00° = 0.5236 \text{ rad}$$

To calculate the frequency deviation caused by the noise impulse, we must decide on a frequency to use. Let's assume the worst case and use $f_{noise} = 15{,}000$ Hz. Thus

$$\Delta f_{noise} = \phi \times f = 0.5236 \times 15{,}000 = 7854 \text{ Hz}$$

If we assume that the FM carrier is being modulated 100% ($\Delta f = 75{,}000$ Hz), as we did with the AM carrier, then

$$\text{FM}\frac{S}{N} = 20\log\frac{\Delta f_{signal}}{\Delta f_{noise}} = 20\log\frac{75}{7.854} = 20 \text{ dB}$$

The S/N of the FM receiver is 14 dB better than that of the AM receiver, for the same condition of external noise. Or putting it another way, the FM system has suppressed the noise-to-signal ratio from $\frac{1}{2}$ to almost $\frac{1}{10}$.

We say that FM has a noise-suppression characteristic. That is, in comparison with an AM system under identical conditions, the noise-to-signal ratio of an FM receiver is less than that of an AM receiver. This happens, not because of any special effort to design in quality in the FM receiver, but simply because of the way an FM system functions.

Let's study this example carefully. It demonstrates two important principles relating to the noise-suppression characteristics of FM systems. First, noise suppression is inversely proportional to frequency. This conclusion is drawn from an analysis of Eq. (8.1). Since frequency deviation produced by phase modulation (almost all noise in FM enters by way of phase modulation) is directly proportional to frequency, the output noise level is lower for lower frequencies: lower frequencies are suppressed more than higher frequencies. Second, the FM system permits trading off bandwidth with signal-to-noise ratio.

In determining the FM signal-to-noise ratio in the example above, we compared the values of Δf_{max} for each. In the case of the FM signal, if we divide Δf_{max} by the maximum modulating frequency, f_{max}, prescribed for the system (15 kHz for standard FM broadcast service), we obtain the modulation index for the maximum factors. We also obtain a modulation index for the noise if we solve Eq. (8.1) for ϕ: $\phi = \Delta f_{noise}/f_{max}$. If we divide the index for signal by the index for noise, we arrive at the same ratio we used previously:

$$\frac{\Delta f_{max}/f_{max}}{\Delta f_{noise}/f_{max}} = \frac{\Delta f_{max}}{\Delta f_{noise}} = \frac{S}{N}$$

The significance of this result is that, in FM, signal-to-noise ratios can be obtained simply by dividing the modulation index of maximum factors by the phase deviation produced by a noise, that is, ϕ_{noise}. Let's use this idea in seeking understanding of the statement above about bandwidth–noise trade-off.

The phase deviation of noise is constant for a constant relative amplitude of noise (i.e., a constant AM signal-to-noise ratio); therefore, the signal-to-noise ratio in an FM system can be altered by changing the modulation index of the system. For example, the signal-to-noise ratio can be improved (increased) by increasing the maximum permissible deviation (and thereby, the bandwidth required). On the other hand, a good signal-to-noise ratio can be attained, with much lower bandwidth, if the maximum permissible modulating frequency is reduced. In the so-called "narrow-band FM" applications of several two-way radio services, values of ±5 kHz and 3 kHz for maximum deviation and maximum audio frequency are used. This yields a lower modulation index—5/3 = 1.33—and, therefore, a poorer signal-to-noise ratio than that of the FM broadcast service. However, the bandwidth occupied—specified as 16 kHz by regulations—is also very much less. Even with a modulation index of 1.33, however, the worst-case S/N for narrow-band FM is 1.33/0.5236 = 2.54 when the AM S/N is 2.

Example 3

The signal voltage at the input terminals of an AM radio, for a certain station, is 87.5 mV. The input for a distant station operating on the same frequency is 28.7 mV. (a) Determine the S/N ratio for the stated condition, at the receiver input and at the speaker. (b) Assume that an FM receiver is experiencing exactly the same signal and interfering signal conditions as stated for the AM receiver. Calculate S/N for the FM receiver, at input and at the speaker. In both cases, assume that there is no other noise present.

Solution. (a) For AM receiver

$$\text{Input } \frac{S}{N} = \frac{87.5}{28.7} = 3.049$$

$$\frac{S}{N}\text{(dB)} = 20 \log 3.049 = 9.68$$

$$\frac{S}{N} \text{ at speaker} = \text{same as input } \frac{S}{N} \text{ for AM receiver}$$

(b) For FM receiver

$$\text{Input } \frac{S}{N}\text{: same as for AM receiver, 3.049 or 9.68 dB}$$

$$\frac{S}{N} \text{ at speaker: noise modulation index} = \phi_{max} = \arcsin \frac{\text{noise}}{\text{signal}}$$

$$= \arcsin \frac{28.7}{87.5} = 0.3342 \text{ rad}$$

$$\frac{S}{N} = \frac{\text{modulation index of system}}{\text{noise modulation index}}$$

$$= \frac{5}{0.3342} = 14.96$$

$$\frac{S}{N}(\text{dB}) = 20 \log 14.96 = 23.5$$

Capture Ratio

A peculiar, and often puzzling to the unwary, manifestation of the noise-suppression characteristic of FM occurs when an FM receiver is processing the signal of one station and the signal of an interfering station is present with an amplitude approaching 50% of that of the desired station. This condition is not uncommon in regions where there are several large metropolitan areas in relatively close proximity. In such a setting there are, usually, many FM stations. Consequently, there is a fair probability of stations being assigned to the same or adjacent channels, creating a potential for interfering signals under unusual propagation conditions. By a listener of a particular station, what is experienced is a sudden "capturing" of the channel by another station, an interfering station. That is, without touching the tuning dial, a listener suddenly realizes that the receiver is receiving a completely different station from that of a moment before. The phenomenon is called the *capture effect* of an FM receiver.

The capture effect can be explained as follows: As long as the signal of the desired station is, let's say, twice that of the interfering station, the system effectively suppresses the interfering station and the listener is unaware of the interfering station. This is in contrast to the AM system where such suppression does not occur—an interfering station with a signal much less than 50% of that of the desired station can be heard along with the desired station. In FM, if the desired signal decreases momentarily, or the interfering signal increases, the receiver is "captured" by the interfering signal. That is, the receiver switches from processing the desired signal and suppressing the interfering signal to the reverse: processing the interfering signal and suppressing the desired signal.

The situation that results in capturing is most commonly experienced with receivers in automobiles. If a driver is listening to an FM station while in transit between two adjacent cities, the signal from the station in the home city is steadily growing weaker, that of a possible interfering station in the destination city is growing stronger. At some point, a capturing is likely to be experienced. In fact, within a certain region along the journey, a capturing/recapturing phenomenon is likely to continue until the station in the destination city clearly has the stronger signal.

The capture ratio, the ratio of the amplitude of desired signal to amplitude of interfering signal at which capturing takes place is determined, primarily, by the noise-suppressing function of the FM system. Please recall that noise suppression is an inherent feature of frequency modulation and is not a special enhancement achieved by a conscious design effort on a particular item of equipment. Nevertheless, the capture ratio may vary slightly between receivers of different manufacturers.

The capture ratio of a particular receiver design is usually measured by the manufacturer and included in the list of performance specifications of the receiver. A typical capture ratio is 2:1. A capture ratio of 2:1 means that when the amplitude of the wave of the interfering station is 50% or greater of the amplitude of the wave of the primary station, the receiver may be captured by the interfering station.

Preemphasis and Deemphasis

Although the superior, low-noise characteristics of frequency modulation are primarily inherent, as we have seen, the designers of the "FM system" did "design in" a feature to further enhance the noise-suppressing capability of FM systems. Please recall that we have learned that when an FM carrier is phase modulated, as it is by an interfering or noise impulse signal, the equivalent frequency deviation produced is proportional to the frequency of the phenomenon producing the phase modulation. Since audio output amplitude is proportional to frequency deviation, the higher frequencies of a desired modulating signal are more likely to be interfered with by noise than are lower frequencies.

Designers of the FM system largely overcame this weakness of the system by writing into FCC regulations a requirement that all FM transmitters incorporate a *preemphasis network* and all FM receivers include a *deemphasis network*. These circuits are simple filters which: (1) at a transmitter, provide increased gain for the higher audio frequencies of the modulating signal at the expense of lower frequencies; and (2), at a receiver, attenuate higher frequencies more than lower frequencies by an amount that reverses the preemphasis of highs and provides, overall, equal gain for all frequencies. By design, then, the system gives higher frequencies an extra boost at the transmitter. After the signal has been processed by the system, including being subjected to the possibility of pollution with noise, the deemphasis network, in the audio section of the receiver, cuts the higher frequencies, including the higher-amplitude noise which may have invaded the signal, back down to normal level.

To work properly, all transmitters and receivers must incorporate preemphasis and deemphasis networks which have identical electrical characteristics. Regulations state that these networks shall have a *time constant* of *75 microseconds*. Networks could be as simple as those shown in Fig. 8.6(a) and (b)—a single resistor and a single capacitor. In actual systems, pre- and deemphasis networks, although equivalent to those shown, are found in many different forms. The 75-μs specification refers to the RC time constant of the circuits shown—any combination of R and C whose product is 75 μs is, by regulation, permitted. The actual choice of component values is determined mostly by the parameters of the circuit with which the network is to be combined.

You will note that the preemphasis network is a high-pass filter with the frequency response characteristic shown in Fig. 8.6(c). The deemphasis network

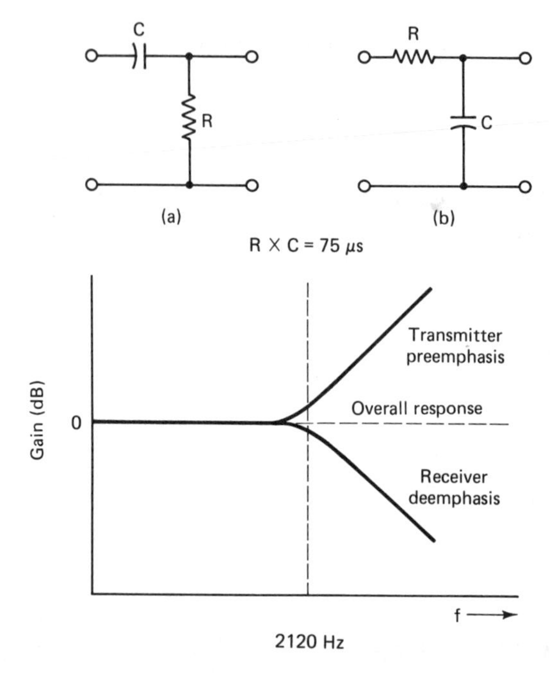

Figure 8.6 (a) Preemphasis network; (b) deemphasis network; (c) frequency response characteristics of pre- and deemphasis networks.

is a low-pass filter. Its frequency response graph is the lower curve in Fig. 8.6(c).

As you will observe in Fig. 8.6(c), the preemphasis network starts to increase the amplification (or decrease the attenuation) of an input signal at a frequency of 2120 Hz. That is, the amplification has risen by 3 dB from a flat, 0-dB level at that frequency. The 2120-Hz frequency is not determined by accident—it is the frequency whose period is 75 μs (i.e., $1/2\pi RC = 2120$ Hz).

The frequency response graph of the deemphasis network [see lower curve in Fig. 8.6(c)] is down by 3 dB at 2120 Hz. The slope of this curve is the negative of the preemphasis curve. When added, as decibels can be to determine total gain, the sum of the two curves is a straight, horizontal line indicating *equal overall amplification* for all modulating frequencies. All noise at frequencies above 2120 Hz is attenuated—the higher the frequency of the noise, the more the attenuation.

Remember, pre- and deemphasis is a characteristic designed into the FM system—it is not simply inherent. In recent years, other schemes for improving the signal-to-noise characteristics of FM have been devised and experimented with. One of these, called the *Dolby system* after its inventor, has been authorized for use by FM stations on an optional basis. The Dolby system is more complex in concept than simple emphasis, but does include pre- and deemphasis network requirements. Since it is an optional, not a required feature, stations that employ it generally do so only for some programs and not others. Transmitters are equipped with modifications which can quickly be switched in or out depending on the owner's choice. Similarly, some receivers may be purchased with switchable circuitry which provides the user with the option of using the Dolby system, as and when desired. The emphasis networks in Dolby have time constants that differ from the standard 75 μs.

8.2 FM SYSTEMS: TRANSMITTERS

The overall block diagram of an FM communications system is identical to that of the AM system presented in preceding chapters. An FM system consists, on one side, of a transmitter, an associated antenna, and a transmission line connecting the two. On the other side is at least one receiver with its associated antenna and connecting transmission line. If the FM system is a broadcast system, there would be numerous receivers. If the system is a two-way communications system, there would be at least two transmitters and two receivers.

In general conception, an FM transmitter is virtually identical to an AM transmitter (see Chap. 5). That is, an FM transmitter must include a source of RF signal (oscillator), a means of modulating the RF signal (modulator), and circuits to increase the energy level of the signal (RF voltage and power amplifiers). An FM station, like an AM station, will have an antenna and a transmission line to connect transmitter and antenna. FM and AM transmitters are different in the details of their design mainly in the modulator and in the sections, if any, which process the RF signal after modulation.

There are two, basically different, ways of modulating a transmitter to achieve a frequency-modulated signal. In the first method, the frequency of the oscillator is changed by the modulator producing *true* or *direct FM*. In the second method, the phase of the RF signal is changed—the operating frequency of the oscillator is left untouched—producing *phase modulation* or *indirect FM*. Each of these methods has advantages and disadvantages. Each has its peculiar requirements for transmitter functions and circuits. We will study the block diagrams and circuits of each in this section.

Direct FM Transmitters

In its simplest form, a transmitter for direct FM consists of an oscillator and a modulator connected to alter the frequency of the oscillator. (A complete station, of course, also includes an antenna.) The block diagram of such a simple, direct FM transmitter would appear as in Fig. 8.7. For any application other than as a toy, this transmitter would be inadequate and for several reasons. First, most applications require more RF power than can be obtained from an oscillator. Most FM transmitters have one or more stages of RF amplification. If we increase the power level of the output of the transmitter, we must ensure that the *at-rest,* or *center, frequency* of the oscillator does not drift. (For example,

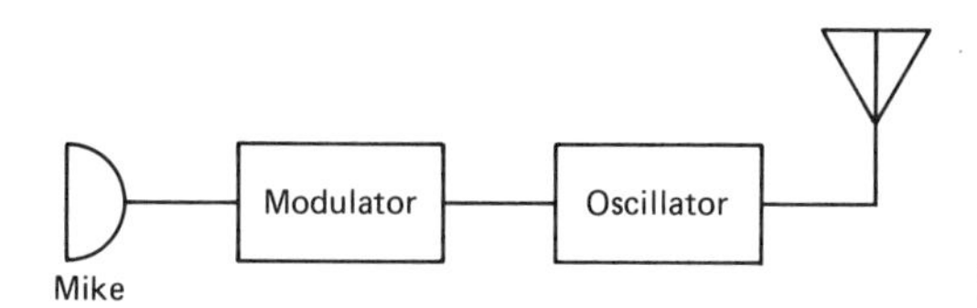

Figure 8.7 Simple FM transmitter.

FCC regulations require that the center frequency of an FM broadcast station of more than 10 W not depart from its assigned frequency by more than 2000 Hz.) Of course, by definition, the frequency of a direct FM carrier will deviate as a result of modulation. But the modulation deviation must be balanced—the frequency swing must be the same above and below the carrier frequency—so that the average or center frequency remains constant.

Since, in this scheme, the oscillator frequency itself is being modulated, it is not possible to use a crystal to maintain oscillator frequency stability. Some other means of maintaining center-frequency stability, while allowing that frequency to be modulated, must be utilized. The function being referred to here is called an *automatic frequency control* (AFC) circuit. (The operating frequency of an *LC* oscillator tends to change, or drift, with time, even if not modulated. There can be several reasons why this occurs. For example, changes in temperature and humidity change the spacing of the turns of a coil slightly, changing its inductance, and, in turn, the operating frequency of the oscillator of which it is a part.)

Let us look at the block diagram of an FM transmitter that incorporates the functions required for providing adequate output power and frequency stability of the carrier. Refer to Fig. 8.8. The block diagram depicted in Fig. 8.8 is typical of transmitters that produce direct FM and of what is called the *Crosby FM system*.

You should find several familiar labels in the diagram: power amplifier, intermediate power amplifier, master oscillator, and frequency multiplier. These are labels of circuits/circuit functions which, upon examination, you would also find familiar even if not identical to circuits previously studied. All FM transmitters, broadcast or otherwise, operate at higher frequencies than AM broadcast transmitters, for example (Chap. 5). Inductors are smaller for higher frequencies. Lead "dress" (or placement) is more critical at higher frequencies. The need to protect low-signal, high-gain sections from induced noise signals and to prevent stray radiation from such circuits becomes critical at higher frequencies. Such circuits are typically completely enclosed in metallic containers, for example. Beyond such details, however, many circuit functions used in FM transmitters and receivers are, indeed, very similar to their counterparts in AM systems. Let us examine the details of the functions that are new and/or different for FM transmitters: the frequency modulator and automatic frequency control circuit.

Modulator for FM

Modulator circuits for FM have changed as new electronic devices have been developed and made available in mass-produced quantities. Whatever the device used, the net effect of an FM modulator must be that of a *variable reactance,* either X_L or X_C. The variable reactance is connected in parallel with the tank circuit of the master oscillator. The equivalent reactance of the modulator circuit is changed in accordance with a changing dc voltage. The dc voltage varies at an audio rate since it is derived from the audio modulating signal. As the equivalent reactance changes with the modulating signal, the overall equivalent

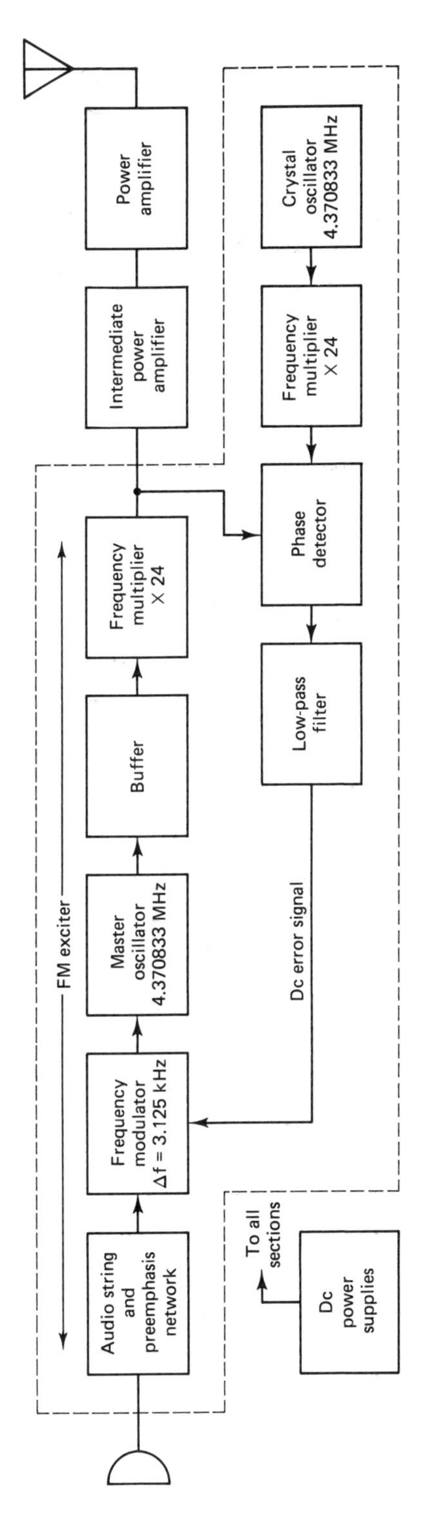

Figure 8.8 Block diagram of Crosby-type FM transmitter.

L or *C* of the oscillator tank is also changed a proportional amount and at the same rate. The consequence is the direct frequency modulation of the oscillator output. All direct FM modulators act in this fashion. Various circuits designed to accomplish this result differ only in the details of achieving the variable reactance.

One of the newest devices for obtaining electronically variable reactance is the *varactor.* (The word "varactor" is formed from two words: *var*iable re*actor*.) In simplest terms, a varactor is a junction diode and is usually made from silicon. Among other characteristics, all junction diodes have some inherent series resistance, junction capacitance, and reverse (or, back) resistance. In general-purpose diodes, junction capacitance is a nuisance to be reduced as much as possible by design and fabrication techniques. In varactor diodes, junction capacitance is increased, again, by design and fabrication techniques, to a useful value. Special effort is also extended to maximize back resistance and minimize series resistance in varactors in order to make them more like capacitors.

Junction capacitance is a phenomenon that is inherent in the nature of a semiconductor junction. You will recall from the theory of semiconductor devices that a junction consists, first, of two opposite types of semiconductor materials: *N*-type and *P*-type. When the junction is formed during fabrication, a so-called *space charge* region is created when electrons from the *N* material drift across the junction and recombine with holes in the *P* region of the junction. Because of this action, the *P* region is charged slightly negatively and the *N* region is charged slightly positively. These charged regions prevent further migration and recombination of current carriers and thus create the space charge region and the accompanying potential barrier of the junction. The potential barrier can be overcome, allowing motion of current carriers across the junction, only by an externally applied voltage.

Of interest here, in a consideration of varactor diodes, is the space charge region, which is also called a *depletion layer* since no free carriers exist in this region. The depletion layer is effectively an insulator, an insulator sandwiched between two conductors (actually, semiconductors). By definition, then, the junction is a capacitor: two conductors separated by an insulator. Furthermore, it is a voltage-variable capacitor because the thickness of the depletion layer (the separation between the plates of the capacitor) can be changed by changing the amount of voltage impressed across the junction.

A diode junction is forward biased when an external voltage is connected to oppose the potential barrier of the space charge region. This voltage effectively reduces the width of the depletion layer. When an external voltage that has the same polarity as the potential barrier of the junction is connected to a diode (− to *P*, + to *N*), it is reverse biased. The depletion region is widened, and current carriers find it even more difficult to move across the junction. Varactor diodes, when used as voltage-variable capacitors, are connected so that they are reverse biased. The depletion layer width varies directly with the reverse voltage. The equivalent capacitance of a junction varies inversely with the applied reverse voltage since capacitance is inversely proportional to the thickness of the dielectric between the conductors (plates).

Varactors are available with nominal ratings of from 0.1 to 2000 pF. Nominal capacitance ratings are usually stated for a specific voltage. An important characteristic of a varactor is its *capacitance ratio or range* (ΔC). This ratio is stated with respect to a specific bias voltage range: from a designated level, such as -1 V, for nominal capacitance, to a maximum safe working voltage, usually, -50 or -100 V. Capacitance ratios range from 0.2:1 to 12:1, depending on the type of varactor. Some schematic circuit symbols in use for varactors are shown in Fig. 8.9.

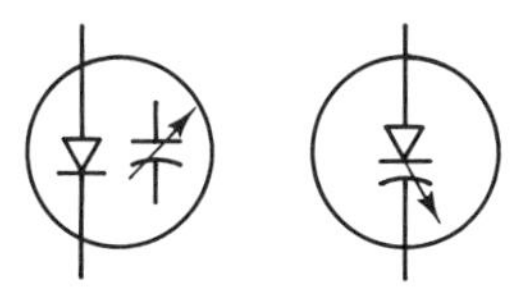

Figure 8.9 Varactor-diode schematic circuit symbols.

Let's now look at a possible circuit for using a varactor to vary the equivalent C of an oscillator tank circuit and thus produce frequency modulation of the oscillator's output (refer to Fig. 8.10). The oscillator circuit could be one of those presented in Chap. 5. The important idea to keep in mind about the oscillator in the discussion here is that the frequency of the oscillator's output is determined by the values of L_1 and C_{eq}, where C_{eq} is the equivalent of C_1 and C_V (the equivalent capacitance of the varactor, D_1) in parallel.

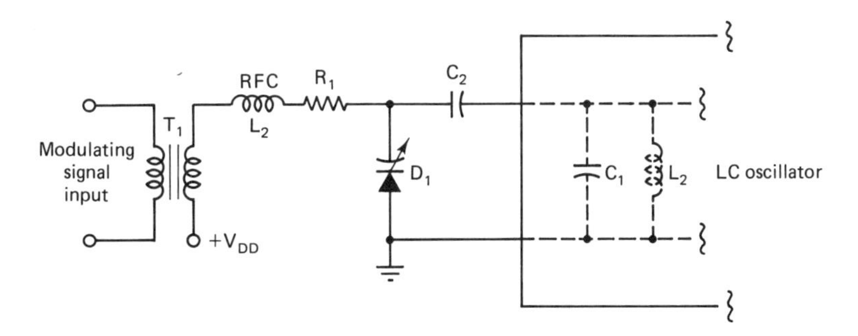

Figure 8.10 Varactor-diode modulator circuit.

The varactor, D_1, is a voltage-variable capacitance. Therefore, it requires a dc voltage source to set its unmodulated value and to place the operating point approximately at the center of a linear portion of the capacitance-voltage characteristic of the particular device. The dc voltage source for this purpose is designated V_{DD} in the diagram. Observe that the dc source is connected to reverse bias D_1. It is highly likely that the dc level of the modulator circuit will be different from that of the LC tank circuit of the oscillator. Hence a dc blocking capacitor is needed; C_2 is that capacitor in this circuit.

The net voltage across the varactor is the sum of the voltage of the dc power supply, V_{DD}, and the voltage across the secondary of modulation transformer, T_1. The primary of the modulation transformer is fed an audio signal from the audio section of the transmitter. The secondary voltage of T_1, then, is an ac voltage that varies in amplitude and frequency with the program material of the station. When the upper end of transformer T_1 is positive, the two voltages add; the reverse voltage across the varactor is increased; the depletion layer is widened; the capacitance of the diode is decreased; and the frequency of the oscillator is increased. When the upper end of the transformer winding is negative, the voltages are opposed: the net voltage is their difference; the capacitance of the varactor is increased; the output frequency of the oscillator is decreased. The output frequency of the oscillator varies by an amount that is directly proportional to the *amplitude* of the modulating signal; the output frequency of the oscillator varies at a *rate* which is the same as the *frequency* of the modulating signal. That is frequency modulation, true and direct! An oscillator whose frequency can be changed by changing a dc voltage is called a *voltage controlled oscillator* (VCO).

Other components in the circuit of Fig. 8.10 have the following functions: L_2 is an RF choke (RFC) preventing RF energy from the oscillator circuit feeding into the audio transformer. Resistor R_1 limits the current in the circuit in the event that peaks of audio voltage exceed the voltage of the dc source and momentarily forward bias the varactor diode.

A Transistor Reactance Modulator

Another circuit that can be used to produce a voltage-variable reactance is a single-stage amplifier called a reactance transistor circuit. This is a solid-state version of a classic circuit—the reactance tube circuit. The circuit is more than worthy of the effort required to understand it because it demonstrates so clearly how controlling the phase angle of a circuit can produce distinctive results.

A model of a reactance transistor circuit that could be used as a frequency modulator of the master oscillator of an FM transmitter is shown in Fig. 8.11. The circuit functions basically like an RF amplifier with a JFET as the active device. Overall, the effect of the circuit is that of a variable reactance in parallel with the tank circuit of the master oscillator. With the appropriate selection and

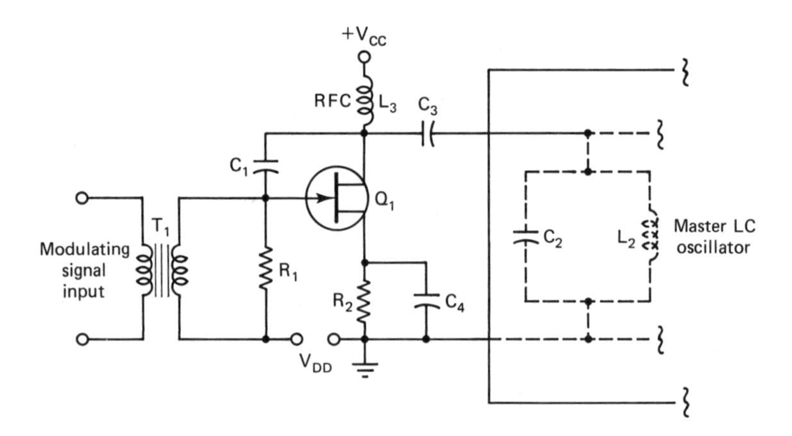

Figure 8.11 Transistor reactance modulator.

positioning of resistor R_1 and capacitor C_1 it can be made to appear either as an inductive reactance or a capacitive reactance. The stage is connected across (in parallel with) the oscillator tank circuit through the common ground connection and capacitor C_3. C_3 is necessary for keeping different dc levels separated. It is large enough that its reactance has no effect on the circuit as a reactance.

The RF "signal" for the stage is obtained from the R_1C_1 network which is also across the tank circuit. This network is the heart of the circuit—its design is crucial to the circuit's performing as a reactance. The values of the components in this network must be chosen so that X_{C1} is very much larger than R_1. This causes the current in this network to lead the voltage across it by 90°, and that, in turn, produces a voltage across R_1 which leads the voltage across the network, and across the tank circuit by 90°. The voltage across R_1 is the signal input voltage to the gate of the JFET. The RF component of the drain current is in phase with the signal input voltage; it, therefore, leads the voltage across the tank circuit by 90°. Note that carefully: the RF component of the drain current leads the voltage across the oscillator tank circuit by 90°, which is exactly what the current in a capacitive reactance connected across the tank circuit would do! The stage, then, affects the tank circuit the same as a capacitive reactance would.

Now, if the RF component of the drain current can be made to vary in accordance with an audio signal, the circuit will appear as a variable capacitive reactance and will modulate the frequency of the oscillator output signal. Of course, all we have to do to vary the amplitude of the drain RF current is change the operating point of the stage by changing its bias. This we do by connecting the secondary of a transformer (T_1) in series with the bias circuit of Q_1 and feeding the station's audio signal to the primary of that transformer. Changing the bias of the stage in effect changes its gain and thus changes the amplitude of the RF component of drain current.

In summary, the stage appears as a variable capacitive reactive load to the oscillator tank circuit. An audio modulating voltage varies the gain of the stage: on positive alternations of the audio the gain is greater; RF drain current is greater; the circuit appears as a smaller reactance—a larger C; the frequency of the oscillator is reduced. On negative alternations of the audio the gain is reduced; RF drain current is less; the circuit appears as a larger reactance—a smaller C; the frequency of the oscillator is increased.

If the positions of R_1 and C_1 and their relative values are reversed, the stage will appear as a variable inductive reactance to the oscillator tank circuit. This mode of operation is explained as follows. The current in the R_1C_1 branch is in phase with the tank circuit voltage; the voltage across C_1 which is the input signal voltage to the JFET lags the tank circuit voltage by 90°; the RF component of the drain current is in phase with the input voltage and therefore lags the tank circuit voltage by 90°, which is what the current in an inductive reactance would do. Changing the bias has the same effect as in the previous case.

Before leaving the topic of direct frequency modulation let's observe one additional point about the operation of the two circuits that have been presented here. First, let's review the basic idea of the circuits: the audio voltage changes the apparent reactance of a circuit that is connected as a load across the *LC* tank circuit of the master oscillator; this change in apparent reactance changes the operating frequency of the oscillator. Now, the audio voltage is an ac voltage: the changes in the oscillator frequency are balanced; that is, for each increase in frequency there is an associated decrease in frequency. Frequency deviations due to modulation are both positive and negative. If there are no long-term variations in circuit components, the oscillator frequency returns to a constant, center frequency.

But what if there are long-term changes in components? The oscillator's center frequency will change, the carrier frequency will drift off the assigned frequency. If we knew the amount of the drift, could we not correct for it, bringing the operating frequency back to the assigned value, simply by adjusting the steady, dc level of the modulation circuit? Yes, we can, of course, and if we assign the task of determining the amount of frequency drift to an electronic genie, we will have an *automatic frequency control* circuit, the topic of the next section.

Automatic Frequency Control (AFC) Circuits

As stated previously, one of the potential problems with direct FM is that the master oscillator cannot be crystal controlled since it must be directly frequency modulated. With a temperature-controlled crystal as its frequency setting element, an oscillator, with reasonable attention to design and construction, can be built that will operate well within the frequency departure tolerances set by FCC regulations. Without direct crystal control, however, an oscillator must be provided with some form of electronic (automatic) frequency control in order to operate within specified standards. Conceptually, an AFC scheme requires (1) a means of determining the amount and direction of the long-term departure (drift) of the oscillator frequency from its assigned value, (2) a way to convert frequency error information into a dc voltage whose amplitude is proportional to the amount of the frequency error and whose polarity indicates the direction of that error, and (3) a means of using the dc voltage to correct the operating frequency of the oscillator. As indicated in the preceding paragraph, the modulation schemes presented inherently provide a means for correcting oscillator frequency error under the control of a dc voltage input. Let us now investigate how elements 1 and 2 of an AFC scheme can be implemented.

The circuit function that can be used to determine the departure of a carrier from its desired frequency is called a *phase discriminator* or *phase detector*. (A form of this circuit function is also the heart of an FM detector in an FM receiver.) As we shall see, a phase discriminator is also capable of providing a dc voltage indicative of the amount and direction of a frequency change or error.

At the heart of a phase discriminator is a resonant circuit. You will recall that at resonance, a resonant circuit has zero power factor. Or, stating it another way, considered as a single equivalent impedance, a resonant circuit has a zero phase angle at resonance. Considered at a frequency slightly lower than the resonant frequency, a series resonant circuit has a positive, or leading phase angle—its net effect is as a capacitive reactance. At a frequency slightly higher than the resonant, a series circuit behaves like an inductive reactance; its phase angle swings in the other direction and becomes negative or lagging. For a parallel resonant circuit these changes in equivalent impedance and phase angle for off-resonance frequencies are the opposite of a series circuit—below resonance the circuit is inductive with a lagging phase angle, above resonance the parallel circuit is capacitive with a leading phase angle. A resonant circuit, then, can be used to "generate" a phase-angle change as a function of frequency, a change that is indicative of both the magnitude of frequency change and the direction of that change.

How can we use a phase change to generate a voltage? Consider the three ac sources shown in Fig. 8.12(a). Imagine that sources A and B are arranged so that their phase angles can be varied from 0° with respect to a reference, to any angle, plus or minus, with respect to that reference. However, the two sources are interconnected so that their phase angles change together. Source C is a source with fixed phase angle with which sources A and B will be combined. Refer to the phasor diagram of Fig. 8.12(b) for the relative phase angles of the three sources for the beginning condition. Let's agree that this initial condition corresponds to the condition in a phase detector when the signal being monitored is at its correct frequency (when the center frequency of an FM transmitter is equal to its assigned value, for example). Note that the phasor sum of the voltages of C and A, V_{AC} is equal in magnitude to the phasor sum of C and B, V_{BC}.

Now consider the effect of adjusting the phase of A and B to +45°. The phasor diagram for this condition is shown in Fig. 8.12(c). Now we observe

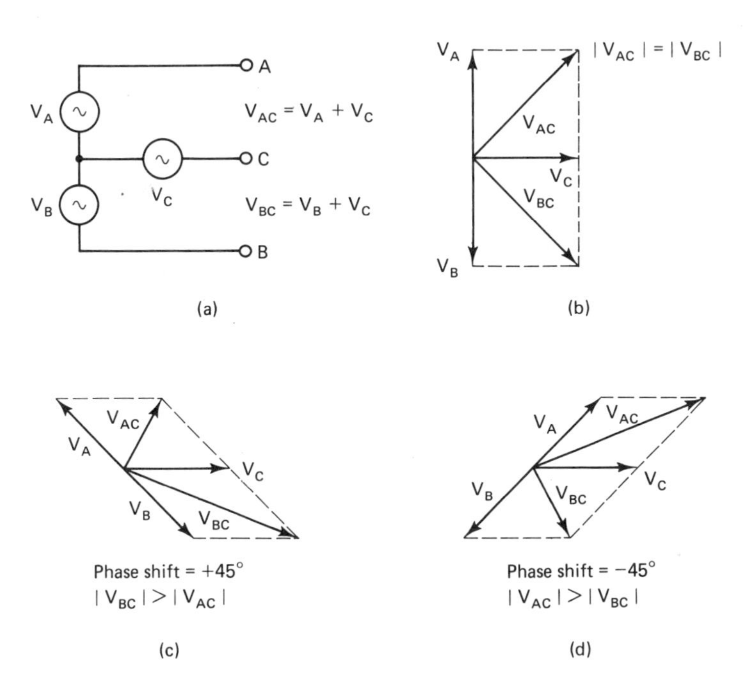

Figure 8.12 Effect of phase angle on phasor sums.

that the magnitude of phasor V_{AC} is less and that of phasor V_{BC} is greater than for the initial condition.

Next, we consider the condition when the outputs of A and B are made to lag by, say 45°. Phasor V_{AC} is now greater and V_{BC} is less than for the initial condition [see Fig. 8.12(d)]. In summary, simple changes in the phase angle of two sources can be converted into changes in the magnitudes of the phasor sums of those two sources with a third, reference source.

The next step is to devise a means of converting the change in the two resultant phasors into a single dc voltage whose polarity will indicate the direction of change of the phase angle of the two sources, A and B, and whose amplitude will be indicative of the magnitude of the phase-angle change. The desired result can be achieved with the aid of two diodes and two resistors (see Fig. 8.13). Diode D_1 will rectify the current which is proportional to phasor V_{AC} and produce voltage V_1 across resistor R_1 with the polarity as marked on the circuit diagram of Fig. 8.13. The amplitude of V_1 will be proportional to V_{AC}, and thus proportional to the amount of phase change of A.

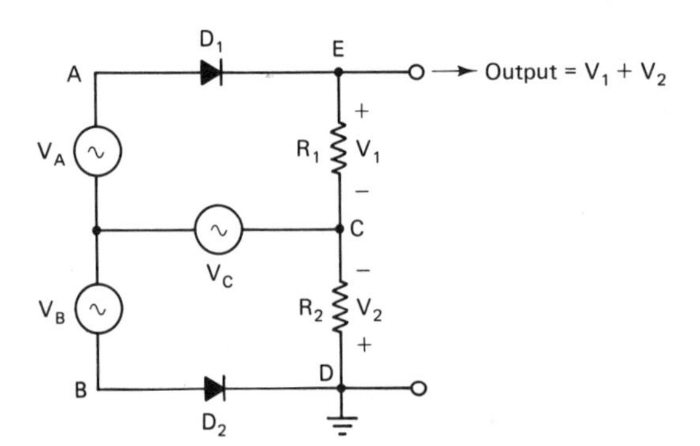

Figure 8.13 Changing dc voltage from phase-angle changes.

In a similar way, diode D_2 will produce a voltage V_2, with a polarity as marked on the diagram and with an amplitude indicative of the direction and magnitude of the phase change of sources A and B. Note that the circuit is grounded at D and an output from the circuit is taken at E. The final, most important outcome of this entire circuit concerns the *amplitude* and *polarity* of the dc voltage of E with respect to D, V_{ED}.

In the circuit of Fig. 8.13, V_{ED}, the "output voltage" of the phase discriminator, is equal to the algebraic sum of voltages V_1 and V_2. The output voltage will be positive when the phase angle of sources A and B is adjusted to lag its initial state; the output voltage will be negative when the phase angle is leading its initial state. This circuit, then, produces a dc voltage (1) whose polarity changes with the direction of phase change (+ for lag, and − for lead), and (2) whose amplitude is determined by the amount of phase change (if amplitudes of ac sources are constant).

Please note that the relationship of the polarity of the dc voltage and the direction of change of the phase angle can be changed by reversing the connections of the diodes. That is, if the diodes were reversed, a leading phase angle would produce a positive dc output voltage and a lagging phase angle would produce a negative dc voltage.

Now let's put together a circuit that uses these two important ideas: (1) a resonant circuit can be used to produce either a leading or lagging phase angle as a function of the degree of off-resonance of the signal being processed; and (2) in an appropriate circuit, a phase-angle change can be converted into a dc voltage whose polarity and amplitude are a function of the direction and magnitude of the change.

Refer to Fig. 8.14. Notice the similarity between this circuit and the circuit of Fig. 8.13. The sources A and B have been replaced by the center-tapped secondary winding of transformer T_1. This winding and C_1 form a series resonant circuit that is tuned to the frequency to be controlled (e.g., the center frequency of an FM transmitter). Source C has been replaced by the secondary winding of transformer T_2. This winding inputs a precisely controlled reference signal from a crystal-controlled oscillator.

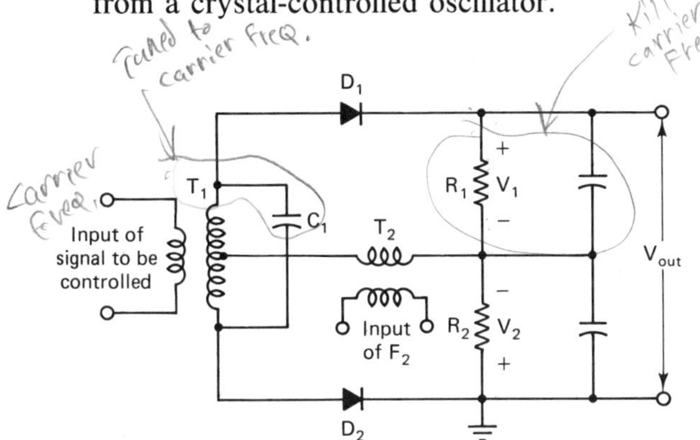

Figure 8.14 Phase discriminator circuit.

If the signal whose frequency is to be controlled is on frequency, the circuit of the secondary winding of T_1 and C_1 is resonant and its phase angle is zero. This sets the initial or on-frequency condition for the circuit. By appropriate means it would be possible to arrange the circuit so that, for this condition, the voltages at the top and bottom of the winding lead and lag, respectively, the input reference voltage by 90°. (This creates a condition like that with the three sources in Fig. 8.12.)

If the carrier frequency drifts higher, the resonant circuit will become inductive (see the discussion above). The current will lag its initial phase position. Since the voltages at the top and bottom of transformer T_1 are the phasor sums of voltages induced from the primary and the IX_L voltage drops across the inductive reactances of the windings, these voltages will also lag their initial phase positions. The condition is now like that of the phasor diagram of Fig. 8.12(d). The dc output voltage of the circuit, V_{out}, is positive and proportional to the magnitude of the error of the carrier frequency.

If the carrier drifts to a lower frequency, the resonant circuit becomes capacitive; the current in that circuit becomes leading in phase; the phase of the transformer terminal voltages will undergo a similar shift in the leading direction;

the phasor diagram of voltages in the circuit will be similar to Fig. 8.12(c); the dc output voltage will be negative and proportional to the amount of frequency error.

The phase discriminator circuit is found in numerous forms. Regardless of the details of such circuits, however, the basic theory of their operation, and the basic function performed is as described here. It is very easy to show that the polarity of the dc output voltage of a discriminator is determined by the phase angles of the sources involved, and hence by the frequency error of the signal to be controlled. It is not a simple matter to demonstrate that the amplitude of the dc output voltage is directly proportional to the amount of frequency error. It is not one of the goals of this book to prove that relationship. We will stop with stating that within relatively narrow limits of frequency departure, a phase discriminator circuit can provide a dc output voltage which is linearly proportional to the magnitude of frequency departure.

Thus far we have examined a means (1) of determining the magnitude and direction of the frequency error of an FM carrier, and (2) of converting that information into a dc voltage of appropriate polarity and amplitude. An important part of this scheme is the crystal-controlled reference signal used in the phase discriminator. The discriminator, in effect, compares the carrier frequency to the frequency of this reference and provides information, in the form of a dc voltage, about the difference between the two, or the error of the carrier.

There is one other function that is required to make the AFC work smoothly and effectively: a low-pass filter between the output of the phase discriminator and the control input point of the modulator circuit. It is important to recognize that the FM carrier is subject to two types of frequency changes: (1) desired changes which are the result of frequency modulation by an audio-frequency signal representing program material and whose average value is 0 Hz, and (2) undesired changes in the form of a slow drift off frequency caused by environmentally induced changes in component values. Let's refer to the first of these types as *short-term changes* and to the second as *long-term changes*.

We do not wish for the AFC to affect the short-term changes because they are what the whole system is about. The period of these changes, at the most, might be about 30 ms corresponding to a lowest audio frequency of 30 Hz. Long-term changes have periods of several seconds or more, and thus are equivalent to very low frequencies. We block the varying dc voltage caused by modulation by passing the dc "error signal," as it is called, through a low-pass filter before it is fed to the modulator circuit. The low-pass filter also enables the system to correct frequency error in a stable manner, that is, without overcorrection and "hunting." ("Hunting" is a term used to describe the action of an unstable, self-correcting system when it first overcorrects, must correct for overcorrecting, and so on, and therefore is constantly "hunting" for the correct operating point.)

Let us now review the automatic frequency control (AFC) "system" as a complete unit so as to gain an overall view of its composition and operation. A block diagram showing the basic functions (see Fig. 8.15) is of great help in this regard. You will note that in the block diagram of Fig. 8.15, there is identified

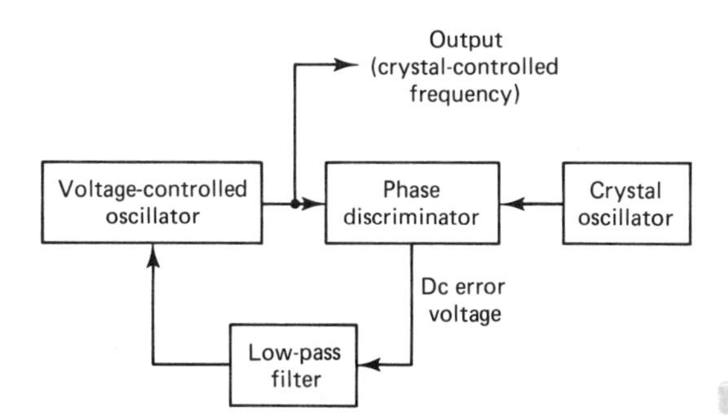

Figure 8.15 Automatic frequency control function (AFC or PLL).

a "voltage-controlled oscillator" (a VCO), but there is no master oscillator and frequency modulator. If you think about it, a master oscillator and a modulator circuit, taken as a unit, are, in fact, an oscillator whose frequency is controlled by a dc voltage. Next, there is a phase discriminator which has two signal inputs: an input from the VCO and another input from a reference source—a crystal-controlled oscillator. The output from the phase discriminator (or phase detector) is the "dc error voltage" which is fed to the low-pass filter and from there, to the VCO. The overall function is a complete loop—from VCO to phase detector, to low-pass filter, and back to the VCO.

This complete function has become extremely popular with electronic designers; it is utilized in numerous applications other than FM transmitters and is referred to as a *phase-locked loop,* whose abbreviation is PLL. Complete PLLs, minus the crystal, are available as integrated-circuit chips (ICs). The output (the real purpose) of a phase-locked loop is a signal whose frequency is "phase locked" to another, a reference, frequency. If the reference frequency is produced by a crystal-controlled oscillator, the output of the PLL is effectively crystal controlled. However, the frequency of the master oscillator can be modulated, and with the appropriate circuitry between the reference source and the loop, can even be varied, thus providing virtually a crystal-controlled, variable-frequency source.

FM Exciter

The functions (of an FM transmitter) that we have examined thus far are often considered part of a subsystem called the *FM exciter.* We have noted that the master oscillator could take the form of one of a number of the different oscillator types discussed in Chap. 5. We have examined two types of direct FM modulators and an automatic frequency control system. In addition to these functions which generate and/or process the RF carrier signal, an FM exciter generally includes one or more stages of frequency multiplication.

Frequency multiplication in FM transmitters may be utilized for one or the other or both of two reasons: (1) All services employing frequency modulation utilize relatively high carrier frequencies, making it expedient to operate the master oscillator at a frequency well below the desired carrier frequency and then multiply that signal to the desired frequency. (2) The amount of frequency deviation obtainable by any method is so limited that it is virtually impossible to obtain the maximum allowable deviation by direct modulation of the assigned carrier frequency.

Frequency multipliers for FM are not unlike those for AM, which were described in Chap. 5. Since FM frequencies are generally higher than those of AM systems, FM multipliers will be designed for higher operating frequencies. You will recall that a multiplier stage is basically an RF (tuned) amplifier whose input circuit is tuned to (is resonant at) the incoming signal frequency and whose output tuned circuit is resonant at a multiple (often two and seldom more than three) of the input frequency. A multiplier circuit multiplies any frequency deviation present in an incoming signal, as well as its frequency.

Consider the example of a 45-MHz signal with a deviation of 25 kHz. At maximum positive swing the frequency is 45 MHz + 25 kHz = 45.025 MHz. If passed through a frequency doubler, the center frequency is 90 MHz and, at the height of a positive swing, the frequency is 90.050 MHz—the deviation is doubled to 50 kHz.

In the block diagram of Fig. 8.8, multiplication of $\times 24$ is indicated for obtaining a carrier frequency of 104.9 MHz from a master oscillator operating at 4.370833 MHz. A maximum deviation of ± 3.125 kHz is required at the oscillator frequency to achieve a maximum deviation of 75 kHz of the final carrier frequency.

Example 4

An FM broadcast exciter is to have an output frequency of 97.5 MHz. The exciter provides frequency multiplication of $\times 18$. Determine (a) the operating frequency of the master oscillator, and (b) the amount of deviation required at the oscillator frequency. (c) Since, by FCC regulation, departure of the carrier from the assigned frequency due to drift is limited to 2000 Hz, what is the maximum departure permitted at the oscillator?

Solution. (a)
$$f_{osc} = \frac{f_C}{\text{multiplication factor}} = \frac{97.5\text{ MHz}}{18} = 5.416667\text{ MHz}$$

(b)
$$\Delta f_{osc} = \frac{\pm f_{max}}{\text{multiplication factor}} = \frac{\pm 75\text{ kHz}}{18} = \pm 4.167\text{ kHz}$$

(c)
$$\frac{2000\text{ Hz}}{18} = 111.1\text{ Hz}$$

The output of the exciter is a modulated FM carrier. It is of the assigned frequency and frequency deviation. It is ready for power amplification before it is finally fed to an antenna. Amplification of a frequency-modulated signal is simpler than the amplification of an amplitude-modulated signal. Amplification of an FM signal may take place in a class C amplifier; an AM signal requires linear amplification. The power involved in the amplification of an FM signal is constant; the power involved when an AM signal is processed varies with the amount of modulation. The design characteristics for intermediate and power amplifiers in FM transmitters are similar to those of AM transmitters. You will recall that the topic of RF power amplifiers was covered extensively in Chap. 5. It is recommended that you review that material again at this point. A fresh look at it will assist you to gain an understanding of a complete FM transmitter. You must remember, of course, that the power amplifier stages in an FM transmitter will not be complicated by the modulation circuits discussed in Chap. 5.

Let us look again at the block diagram of Fig. 8.8. If you will examine it carefully you will conclude that all functions have now been discussed with the following exceptions: the audio section and the preemphasis section. Actually, information on the preemphasis circuit was presented in Sec. 8.1. Recall that a preemphasis circuit is a high-pass filter that gives the higher frequencies in program material a boost at the transmitter. The extra gain for high frequencies is removed, or equalized, by the deemphasis circuit of FM receivers. The audio section of a transmitter contains one or more stages of audio amplification. If the transmitter is for the broadcast service, the audio section will include provision for selecting one of a number of audio sources: microphones, for live programming; pickups, for disk-type recordings; playback heads on tape decks; and telephone lines, for program material from remote locations. A brief introduction to audio amplification was included in Chap. 3 as part of the presentation on AM receivers. Audio amplification will be treated more extensively in the section on FM receivers in this chapter.

Transmitters Utilizing Phase Modulation: The Armstrong System

As we have seen, a serious drawback to the method of producing a frequency-modulated carrier by direct modulation is that the carrier center frequency is subject to drift. An alternative to using an elaborate AFC system to obtain the required stability is phase modulation of the RF signal. Phase modulation is performed on the carrier *after* it has been generated; the master oscillator can be crystal controlled. This method is generally referred to as the *Armstrong system*. (Edwin H. Armstrong, an American electrical engineer, is credited for the invention of wide-band frequency modulation. He was born in 1890.) A block diagram of a phase-modulated (also called indirect FM) transmitter is shown in Fig. 8.16.

If you will study the diagram of Fig. 8.16 carefully you will see that this transmitter is different from the Crosby-type transmitter of Fig. 8.8 in three ways: (1) It uses a crystal-controlled master oscillator (there is no AFC circuit since none is needed). (2) The modulator is located after the oscillator, in the signal flow path, and is a phase modulator. (3) The audio section includes a $1/f_M$ frequency-compensating filter. Let's examine each of these three features.

Since the carrier is phase modulated, and this occurs downstream from the

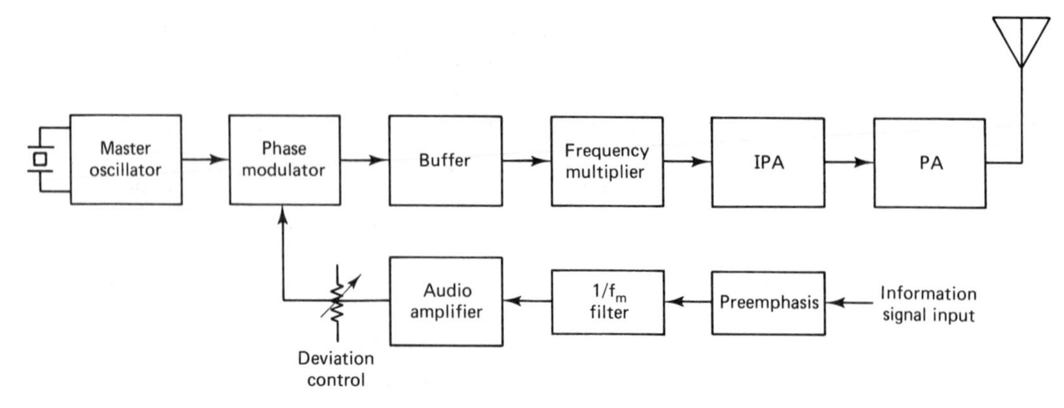

Figure 8.16 Block diagram of Armstrong-type phase-modulated FM transmitter.

oscillator, the oscillator can be crystal controlled. The oscillator circuit can be any one of numerous possible configurations that utilize the stabilizing feature of a crystal. A crystal-controlled oscillator is highly frequency stable and requires no additional frequency control circuitry.

A phase modulator is any circuit that can be used to change the phase of the RF signal coming from the master oscillator. This circuit must be able to change the phase angle of the carrier so that it becomes, alternately, leading and lagging. The circuit must be arranged so that phase changes are produced by the audio (the modulating) signal and so that the amount of the change is proportional to the amplitude of the modulating signal. As could be expected, there are numerous circuits that can be devised to accomplish phase modulation. In fact, any circuit that can be used to modulate the frequency of an oscillator directly, by changing a parameter of its tank circuit, can be used to achieve phase modulation: For example, if the total equivalent capacitive reactance of an RF amplifier's tank circuit load is modified by a varactor diode, the resonant frequency of that tank circuit will be changed and the phase angle of the current in the circuit will be altered.

A phase modulator using a varactor diode is shown in Fig. 8.17. Varactor D_1 is reverse biased by the voltage at the junction of R_1 and R_2. This produces a depletion layer in the junction region of the diode and a nominal value of capacitance that is in parallel with capacitor C_1 of the tank circuit. The tank circuit, including the capacitance of the varactor, is designed so that its resonant frequency is equal to the operating frequency of the oscillator. For this initial condition of the varactor, the tank circuit operates at unity power factor.

An audio signal processed through the audio section of the transmitter will cause the voltage across *deviation control* R_3 to vary at an audio rate. A portion of this varying voltage is coupled to the varactor through C_2. The capacitance of the varactor will be altered in accordance with the amount and rate of this varying voltage. The resonant frequency of the tank circuit will vary with the change in the capacitance of the varactor. This change in f_r will produce a variation in the phase angle of the tank circuit current with respect to the applied voltage, and this alters the phase of the output voltage of the tank.

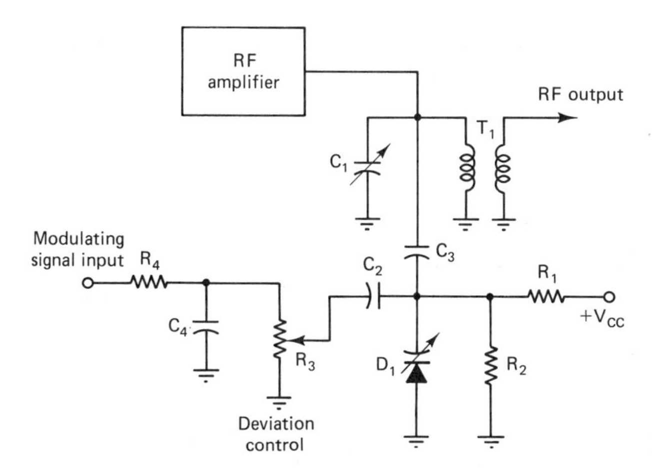

Figure 8.17 Phase modulator.

The overall net effect of this circuit is that the phase angle of the voltage out of the modulator varies at a rate that is equal to the frequency of the modulating signal (the audio input to the transmitter). The amount or magnitude of the phase angle variation is directly proportional to the amplitude of the modulating signal. This magnitude of variation can be adjusted by the *deviation control,* which simply adjusts the amplitude of voltage variation applied to the varactor.

The modulator circuit presented in the previous paragraphs produces *phase modulation* of the carrier. But that is not exactly what we want. If you will recall the discussion of phase modulation produced by noise or an interfering signal (see Sec. 8.1), you will also recall that although phase modulation produces frequency modulation, the frequency deviation produced is *directly proportional to the frequency of the modulating signal.* And that is not acceptable; something must be done to the circuit to eliminate that characteristic of phase modulation. The correction is very simple: we add a low-pass filter in series with the audio signal so as to cause the amplitude of the varying voltage fed to the varactor diode to be inversely proportional to the frequency of the audio. This is referred to as the "$1/f_M$" filter, and in Fig. 8.17, R_4 and C_4 provide this function.

The amount of frequency change per unit of input voltage change (of modulating signal) is called the *sensitivity* of a frequency or phase modulator. Unfortunately, it is impossible, with all known methods of modulation, to achieve good sensitivity while maintaining linearity of modulation (frequency deviation precisely proportional to amplitude of modulating signal for all levels of amplitude). It is difficult to achieve sufficient deviation.

In the so-called narrow-band FM employed generally in two-way mobile communications, audio input is typically limited to the range 200 to 3000 Hz, and maximum deviation is 5 kHz. With these limitations it is possible to achieve the desired deviation levels by modulating the carrier at a relatively low frequency and then using total frequency multiplication of the order of 12 to 48.

However, in standard broadcast FM where audio signals up to 15 kHz must be processed and maximum deviations of 75 kHz produced, a method other than direct frequency multiplication must be employed to achieve the desired frequency deviation. This method uses frequency conversion as well as multiplication. The block diagram of Fig. 8.18 shows the functions required.

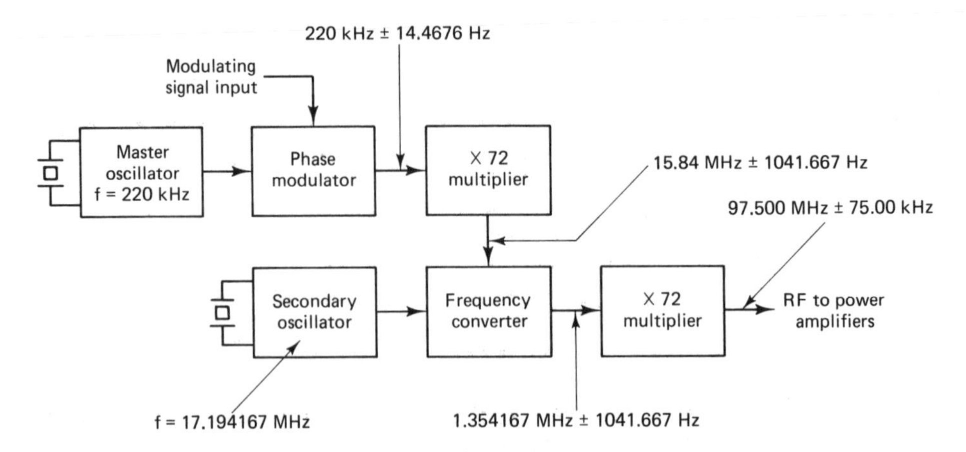

Figure 8.18 Scheme for increasing frequency deviation.

In the example of Fig. 8.18, a carrier of 220 kHz is generated by a crystal oscillator. A phase modulator produces a maximum deviation of that carrier of ±14.4676 Hz. A ×72 multiplier section increases the carrier frequency to 15.840 MHz and deviation to 1.041667 kHz. The modulated carrier is beat with the signal from a second crystal oscillator operating at 17.194167 MHz in the frequency converter section. The output of the frequency converter—the difference-frequency signal—is a modulated carrier at 1.354167 MHz with a deviation of ±1.041667 kHz. Note that the converter does not change the amount of the deviation, only the frequency. When this carrier is again multiplied by 72, the result is a 97.5-MHz carrier with a maximum deviation of ±75 kHz.

MEMORY JOGGER

1. Frequency multiplication increases both the carrier frequency and the amount of frequency deviation.
2. Frequency conversion changes the carrier frequency to a new frequency but leaves unchanged the amount of frequency deviation.

8.3 FM RECEIVERS

In general terms, FM receivers are very similar to AM receivers (see Chap. 3). For example, essentially all FM receivers utilize the superheterodyne principle to achieve good selectivity.

However, there are several significant differences between FM and AM receivers. Some differences in the two types of receivers are strictly the result of the two forms of modulation. For example, even though each type of receiver must have a detector or demodulator, there is no real similarity between an AM detector and an FM detector. Other differences are, in a sense, more accidental. Most radio services that employ frequency modulation utilize much higher carrier frequencies than is used in AM services: FM receivers typically process higher frequencies; their reactances are smaller; shielding, to prevent noise pickup, is more critical; lead "dress" is more critical because the reactance of an "accidental capacitor" (two wires in close proximity) is less at higher frequencies.

FM broadcast receivers tend to be somewhat more expensive than AM receivers because satisfactory operation requires slightly more attention to design and careful construction. Perhaps because they are slightly more expensive in basic form, FM receivers are generally significantly more expensive than common AM receivers. This is because manufacturers attempt to build in additional quality and eye-catching extra features.

The block diagram of an FM broadcast receiver is shown in Fig. 8.19. Study the names of the functions and you will discover that you are already familiar with the names and circuit operation of most of the functions. Let's start at the beginning of the signal path—the antenna—and examine the functions one by one, learning the details of operation and circuit construction of any that are new. The antenna itself will provide a new experience.

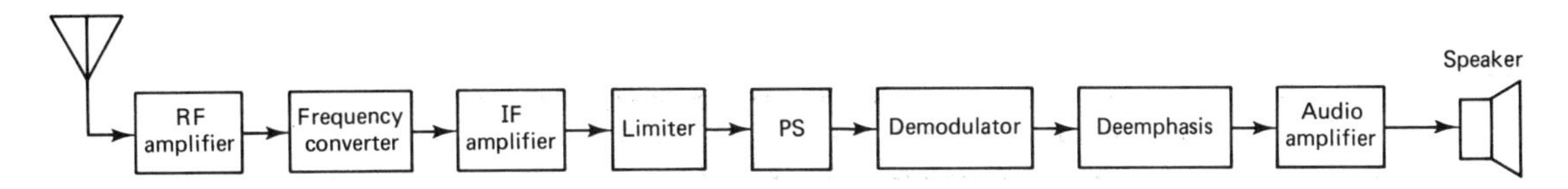

Figure 8.19 Block diagram of FM receiver.

The small loopstick antenna used for AM receivers does not function effectively at frequencies typical for FM. Installation instructions that come with expensive FM receivers generally specify that an external antenna be used with the receiver. Some receivers include a form of antenna made by wrapping a piece of hookup wire around the ac line cord. The line cord, and even house wiring, serves as the pickup element in such an arrangement. The signal picked up in this way is capacitively coupled to the hookup wire "gizmo" which is connected to the receiver's antenna input terminal. Portable FM receivers generally include a whip antenna—a telescoping, tubular aluminum rod approximately 1 m in length when fully extended.

The block diagram indicates an RF section. In FM receivers the RF section usually includes an active device; it is a true RF amplifier with gain. An RF amplifier preceding the frequency converter in a superheterodyne receiver serves at least two important functions. It improves the sensitivity of the receiver by building up the signal energy level in a circuit especially designed for low noise amplification. Second, since the RF amplifier is a tuned amplifier, it improves the selectivity of the receiver. Specifically, it improves the selectivity by reducing the probability of a problem that is inherent in superheterodyne design, namely, interference produced by an *image-frequency signal*.

Consider the frequency relationships involved when a receiver is tuned to receive an FM station with an assigned frequency of 88.1 MHz. The standard intermediate frequency for FM broadcast receivers is 10.7 MHz. If the local oscillator is designed to operate at a frequency higher than the received frequency, its frequency in this instance would be 98.8 MHz. But this means that if the signal of a station operating at a frequency of 109.5 MHz enters the frequency converter it could also produce an IF signal of 10.7 MHz. The 109.5-MHz signal is the image-frequency signal of the desired 88.1-MHz signal. The superheterodyne image frequency is equal to the desired frequency plus or minus two times the intermediate frequency:

$$f_{\text{image}} = f_{\text{desired}} \pm 2f_{\text{intermediate}}$$

Now, 109.5 MHz is outside the FM broadcast band, and the image frequencies for all FM channels are outside the FM broadcast band. However, any signal of the appropriate frequency can cause an image-frequency interference if that signal is not attenuated before it reaches the frequency converter. This statement is true for any superheterodyne receiver: AM, FM broadcast, narrow-band FM, or television. A tuned RF stage between a receiver's antenna and its frequency converter provides very high attenuation of image frequencies since such frequencies are well outside the pass band of such a stage. Of course, the stage's tuning must change as the receiver is tuned (dial setting changed). Image frequency interference is more likely to be observed on AM broadcast-band receivers than on FM broadcast band receivers because the front ends of such receivers are not as "tight" (narrow-band selective). Image-frequency interference is very common in economy-model shortwave receivers because the design of such receivers does not include a highly selective RF section. In summary, a major purpose of a stage of RF amplification in a superheterodyne receiver is to suppress image-frequency signals.

Example 5

The local oscillator (LO) of an FM communications receiver operates at 137.3 MHz to receive a carrier signal of 148.0 MHz. Determine: (a) the intermediate frequency of the receiver; (b) the image frequency of the desired frequency, 148.0 MHz.

Solution. (a) $f_{\text{intermediate}} = f_{\text{carrier}} - f_{\text{LO}}$

$$= 148.0 - 137.3 = 10.7 \text{ MHz}$$

(b) $f_{\text{image}} = f_{\text{desired}} - 2f_{\text{IF}}$

$$= 148.0 - 2(10.7) = 126.6 \text{ MHz}$$

Be sure you note that in this case with the LO frequency lower than the carrier frequency, the image frequency is equal to the desired frequency *minus* twice the intermediate frequency.

Return now to the block diagram of Fig. 8.19 and proceed to the next function along the signal flow path: the frequency converter. You will recall that the AM receiver of Chap. 3 also contained this function. The frequency converter, using the principle of the heterodyne mixer, converts an incoming carrier to the intermediate frequency of the receiver. The converter requires a local oscillator (LO) function and a mixer function. In AM receivers, which are frequently designed to be priced very competitively, these two functions are generally performed by a single device. FM receivers, on the other hand, usually utilize a separate active device for each function. Also, because of the much higher frequencies involved, local oscillators in FM receivers typically operate at frequencies which are below those of received carriers. The higher the operating frequency of any circuit, the more difficult it is to design and construct it for stable operation: the effects (reactance) of stray capacitance and inductance are more significant at higher frequencies.

Next in line in the signal flow path of an FM receiver is the IF amplifier section, or simply, the IF section. Although there is no legal standard requiring it, virtually all receiver manufacturers in the world design FM receivers to use 10.7 MHz as the intermediate frequency. (For this reason, 10.7 MHz could be called the *de facto standard* intermediate frequency for FM receivers.)

As we will see in a subsequent chapter, receivers used in narrow-band FM two-way communications systems often have two IF sections. One of these operates at 10.7 MHz and the second one operates at a lower frequency, 455 kHz, for example. Such receivers are called *double-conversion* receivers because the carrier is twice converted to a lower frequency before it is finally demodulated.

Although an FM IF section is like the IF section of an AM broadcast receiver in function, it will usually be different in several physical respects. But the differences are not readily seen.

In the first place, the IF section of a broadcast-band FM receiver must have a much greater bandwidth than that of an AM receiver: 200 kHz for the FM receiver compared to about 10 kHz for an AM receiver. In relative terms these two specifications are not too different; each is about 2% of the center frequency of the band. However, the broader bandwidth requirement will result in some small differences in the design of the FM section as compared to an AM section.

One technique used to achieve broader bandwidth is the use of *double-tuned transformers* in the interstage coupling circuits. A double-tuned circuit is one in which both the primary and secondary sides are designed to be resonant circuits. In an actual receiver, each circuit would have its alignment slug: an IF transformer would have two alignment slugs. If each circuit is tuned to a frequency slightly different from that of the other, and from the center frequency, the overall result is a wider bandpass with some loss in gain. Achieving a broader absolute bandpass (as compared to AM IF sections), using whatever method, is at the expense of gain. Consequently, FM IF sections typically have three stages of amplification.

Amplitude Limiters

A function usually performed by an FM IF section (and not found in an AM receiver) is that of *amplitude limiting.* If the section is designed with stages using discrete components, the third and final stage of the section is usually an amplitude limiter stage.

The purpose of amplitude limiting is to get rid of amplitude variations in the signal (the IF signal) which is going to be demodulated. Variations in the amplitude of the IF signal are the result of the injection of noise or interfering signals. This "pollution" of the desired signal occurs within the receiver itself. Of course, we are speaking of pollution (amplitude variations) which occurs at an audio rate and thus has the potential of being heard as noise from the speaker. (You are urged to review the discussion of FM noise in Sec. 8.1 under the heading "Noise Suppression in FM systems.") If the IF signal is first amplified generously, then limited in a limiter stage, the noise-induced amplitude modulation will not reach the detector or speaker. The amplitude limiter function is thus an essential part of the overall scheme that makes FM significantly more noise free than AM.

An amplitude limiter stage is shown in Fig. 8.20(a). The action of a limiter is depicted in the waveform diagram of Fig. 8.20(b). The major difference between the limiter stage, Q_2, and the preceding nonlimiter stage, Q_1, in Fig. 8.20, is the addition of the 470-Ω resistor in the collector circuit of Q_2. The effect of the 470-Ω resistor is to reduce the range of excursion of the collector voltage of the stage. The input signal required to drive the stage between cutoff and saturation is less. Note the flattening of the positive and negative peaks of the output waveform.

The cost of achieving an output of constant amplitude (within obvious limits) is a reduction in gain. This clipping-limiting action also causes a distortion of the waveform being processed, an introduction of harmonics. However, since the signal is passed through another resonant circuit before demodulation, the harmonics will be severely attenuated. We draw to your attention the fact that the circuit of Fig. 8.20 utilizes double-tuned IF circuits to achieve the bandwidth desired of broadcast band FM receivers.

It is important you know that removing noise-causing amplitude modulation by limiting does not eliminate all of the potential noise of a particular noise-inducing incident. As was pointed out in Sec. 8.1, amplitude modulation of an FM signal also produces a phase modulation of that signal. And phase modulation produces the equivalent of frequency modulation. The action of a limiter does not remove the phase-modulation effect of the disturbance of the amplitude of the desired signal. Nor is there any way yet devised to eliminate the noise so caused. However, since the amplitude of that noise is proportional to its frequency, the preemphasis/deemphasis functions serve to reduce significantly the effect of such noise.

The IF sections of FM receivers often incorporate automatic gain control (AGC) circuitry. You will recall that this topic was explored in conjunction with the IF section of an AM receiver (see Chap. 3). An AM detector provides a

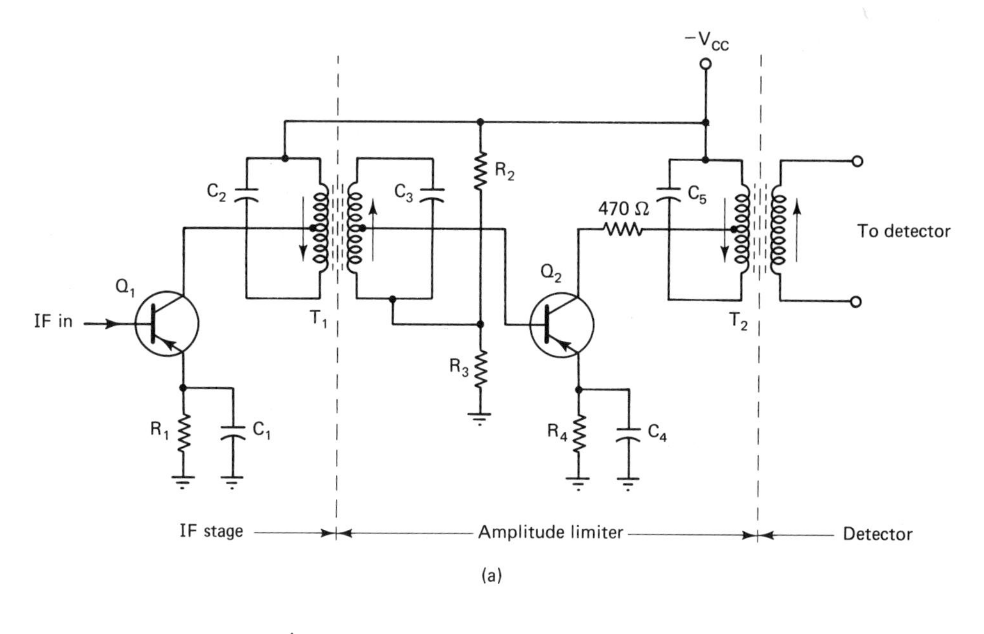

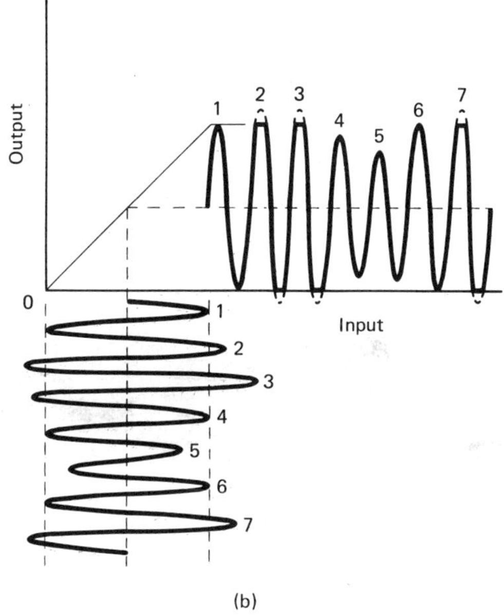

Figure 8.20 (a) Amplitude limiter stage; (b) effect of limiting on waveforms.

readily available source of a dc voltage whose amplitude is directly proportional to the amplitude of the IF signal, and thus of the received carrier signal. Such a voltage is required to perform the function of automatically controlling the gain of an IF or RF section. You will discover in the presentation on FM detectors that all FM detectors do not provide such a voltage. Hence a separate diode, to rectify a portion of the FM IF signal and provide a dc voltage for

AGC, may be incorporated in the IF section. When used, AFC is generally applied to at least one stage of IF and to the RF amplifier, if one is included in the design.

When an FM system is working properly, the output of a receiver is quite free of noise. If a receiver is functioning but its output is noisy, this is a clue to the potential source of trouble.

Excessive noise often indicates that the signal input to the limiter is not sufficient for limiting to occur. The problem then is between the input to the limiter and the station. There could be something wrong with the receiving antenna, the antenna-to-receiver lead-in, the RF section, the frequency converter section, or the first stages of the IF section. Of course, the problem may simply be that the field strength of the signal at the receiving antenna is not sufficient to provide the noise-free operation normally expected with an FM system. This is very often the case with FM receivers mounted in motor vehicles. It is very easy for a vehicle to move into locations where large buildings or hills are located in the line of sight between the receiving and transmitting antennas. Such obstructions intercept an emission and produce a "radio shadow" area of severely attenuated field strength.

FM Demodulators

Moving along the block diagram of Fig. 8.19, the next function is that of the demodulator. The demodulator, or detector, is the function in an FM receiver which is most unlike the corresponding function in an AM receiver. The FM demodulator function must detect a change in frequency (or phase) and convert that change into an audio signal: The amplitude of the audio signal is to be derived from the *amount* of frequency change (the deviation) and the frequency of the audio signal is to be derived from the *rate* of the frequency change (in the incoming signal).

When constructed with discrete components, FM detectors are usually one of two basic types: the *frequency discriminator* (also called the Foster–Seeley discriminator after its inventors) or the *ratio detector*. When integrated circuits are used in the design of a receiver, FM detection may be by means of a phase-locked loop (PLL) or some form of circuit referred to as a *synchronous detector*.

The Frequency Discriminator

Perhaps the first circuit built specifically for the demodulation of an FM signal was some form of the one now called the *Foster–Seeley discriminator*. (It is possible to achieve a degree of FM demodulation using an AM IF section and AM detector by operating the section slightly off frequency. Off frequency, the section processes a signal on the slope of the response curve; the output varies with the frequency of the IF signal. Hence this scheme is sometimes called a "slope detector." It has serious limitations and is never incorporated into the design of a receiver intended to be marketed.)

The Foster–Seeley (F-S) discriminator is very similar in concept and theory of operation to the phase discriminator described in connection with the automatic frequency control circuit of the Crosby-type FM transmitter. A typical F-S discriminator circuit is shown in Fig. 8.21.

Compare the circuit of Fig. 8.21 with that of the phase discriminator of Fig. 8.13 and you will discover basic similarities in the two circuits. The major difference of the two is that the reference voltage in the circuit of Fig. 8.13 is obtained from a separate source—the crystal-controlled oscillator; the reference voltage in the circuit of Fig. 8.21 is obtained from the top of the IF transformer and is coupled into the circuit through C_1 and appears across L_1. The output of the demodulator corresponds to the error signal of the phase discriminator of the AFC. The output is now the desired audio signal: Its amplitude is proportional to the amount of frequency deviation; its frequency is the same as the rate of frequency deviation.

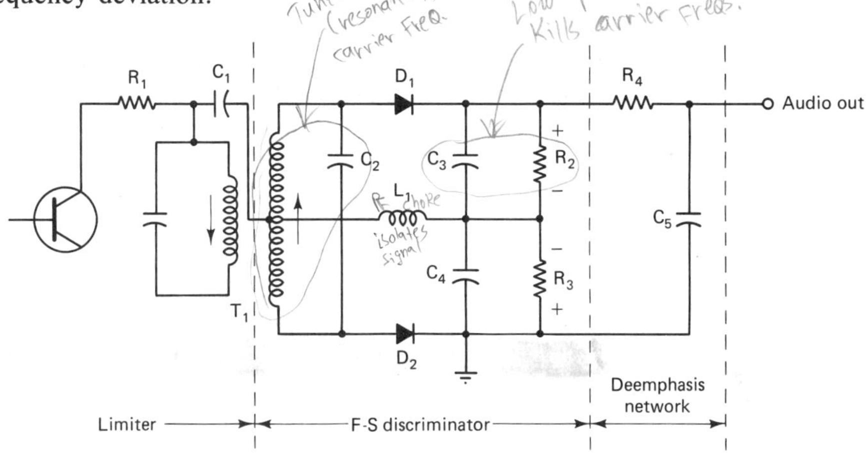

Figure 8.21 Foster–Seeley discriminator.

The theory of how the F-S discriminator derives the audio signal from the frequency-modulated wave is identical to that described for the development of the error-signal output of the phase discriminator of a PLL (see above). A discriminator can be tested by applying an appropriate signal to its input from a variable-frequency RF generator. The dc output of the circuit is measured and recorded in a table of values for a number of different input frequencies. If these values are plotted on a graph—dc output voltage versus input frequency—the result would be similar to that of Fig. 8.22. Such a graph is called the *response curve* of the discriminator.

Study the curve of Fig. 8.22 carefully and observe the following:

1. The curve is linear between points 2 and 6; this is the portion of the response characteristic that provides a high-fidelity output. The receiver should be aligned so that the IF signal is centered to this region. The frequency span (or bandwidth) of the linear response should be a minimum of 200 kHz to correspond to the active bandwidth of an FM broadcast carrier.
2. The response changes from a negative output to a positive output at point 4; this point is called the *crossover point* of the discriminator. The crossover

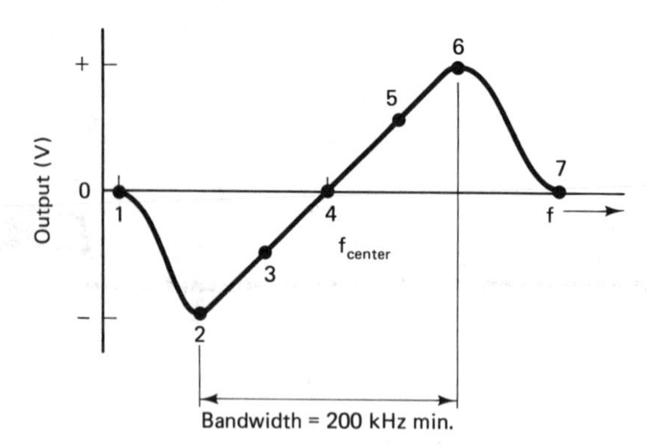

Figure 8.22 Frequency response curve of FM detector.

point corresponds to zero modulation and the zero level of the ac audio waveform. For best results, the discriminator should be designed and adjusted so that its crossover point coincides with the center frequency of the system (the IF of the receiver). Alignment of the IF section/discriminator includes adjusting the operating frequencies of these two sections so that the crossover of the discriminator coincides with the intermediate frequency.

3. Demodulation of a carrier will occur when a receiver is tuned to place the IF at either point 1 or 7. However, the quality of the audio will be extremely poor. Many FM receivers incorporate a tuning indicator of some type to assist the user to avoid tuning the receiver for reception at one of these two points. The response curve of an FM detector is often referred to as an "S curve."

The Ratio Detector

Another, very popular, FM detector circuit is one known as a *ratio detector*. The ratio detector was developed to overcome one of the shortcomings of the discriminator, and at a lower cost. The discriminator responds to small changes in the amplitude of the IF signal as well as to its frequency changes: In a word, it demodulates AM as well as FM. Amplitude modulation of an FM signal always represents noise and is therefore to be eliminated if possible. In short, a discriminator requires a limiter to provide the noise-free quality possible with FM.

Incorporating a limiter in the design of a receiver generally implies the need to add two additional stages—one as the limiter, and another stage to provide the additional amplification required. When receivers were built utilizing electron tubes, adding two stages added significantly to the cost of the overall receiver. As we shall see, a ratio detector provides a degree of limiting as an inherent function of its operation; a separate limiter stage is unnecessary, and can be eliminated for a savings in cost, if an output of average quality is acceptable. The ratio detector is the demodulator of choice in receivers manufactured to be priced competitively. Let us look at the circuit of a ratio detector (see Fig. 8.23).

If you will examine the circuits of Fig. 8.23 carefully, you will note that they are quite similar in arrangement to that of the discriminator. One very

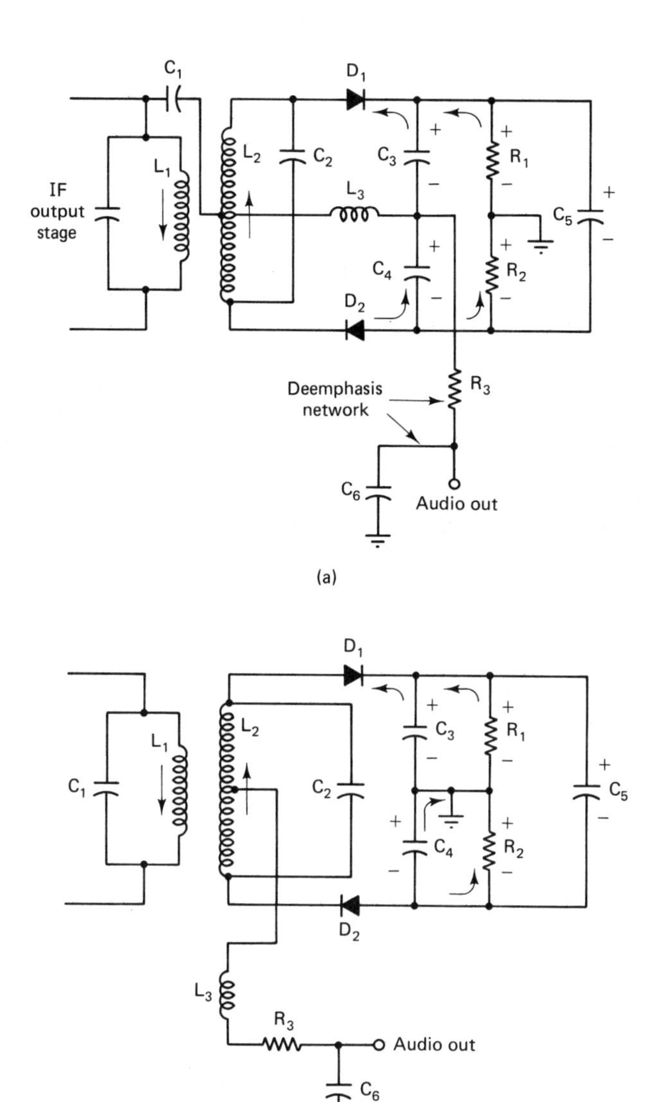

Figure 8.23 Ratio detector circuits.

obvious difference is the connections of the diodes: The diode connection of the discriminator (Fig. 8.21) is series opposing; the diodes are connected series aiding in the ratio detector (Fig. 8.23). (As you look at the diagram of the discriminator, you perceive the diodes as "pointing" in the same direction; they point in opposite directions in the ratio detector.) A second significant difference between the two types of circuits concerns the method of obtaining the reference voltage. The reference voltage, the voltage across L_3, may be obtained by capacitive coupling from the previous stage as in the circuit of Fig. 8.23(a). Or, as is more common, L_3 may be inductively coupled directly with the primary of the IF transformer, L_1, as in Fig. 8.23(b).

Let us examine the operation of the circuit of Fig. 8.23(a). The ac voltages across the upper and lower halves of L_2 in conjunction with the voltage across L_3 provide potentials, which when rectified by diodes D_1 and D_2, produce dc current flows in the directions indicated by the arrows on the diagram. Directly as a result of these currents, dc voltages are developed across C_3, C_4, R_1, and R_2 with polarities as indicated on the diagram.

It is an important characteristic of the ratio detector that the sum of the voltages across C_3 and C_4, or R_1 and R_2, is constant as long as the amplitude of the signal being demodulated is constant. In fact, short-term fluctuations in this total voltage will be virtually eliminated by the charge-storage effect of C_5. (C_5 typically has a value of the order of 5 μF, whereas C_3 and C_4 are of the order of 300 pF.) That is, the total voltage across the circuit will not change significantly with the short-term changes caused by noise-induced amplitude modulation of the IF signal! The effect is equivalent to amplitude limiting: The ratio detector, within limits, is *self-limiting*.

Of course, the voltage must change somewhere in the circuit with changes in the frequency of the IF if the circuit is to perform its function of detecting FM. In the circuit of Fig. 8.23(a), the phasor sums of the voltages across L_3 and, either the upper half of L_2, or its lower half, change with changes in the frequency of the IF; and the changes are for the same reasons as the changes noted in regard to the phase discriminator. These voltage changes cause corresponding changes in the amplitudes of the rectified currents in the upper and lower diodes of the circuit. There are corresponding changes in the voltages across C_3 and C_4. If the voltage across C_3 increases by, say 2 V, the voltage across C_4 will decrease by 2 V. The sum of the two voltages remains the same, except for long-term changes in the amplitude of the received signal. However, the *ratio* of the two voltages changes, as does the voltage at the junction of the two capacitors, measured with respect to ground. The voltage at this junction point, which is a function of the ratio of the two capacitor voltages, is the audio output of the detector. The circuit is named "ratio detector" because of its dependence on the changing ratio (with frequency) of two voltages rather than the changing sum of two voltages, as in the discriminator.

The circuit of Fig. 8.23(b), although obviously structured differently, produces the same result as that of Fig. 8.23(a)—an audio output that is a function of the ratio of the currents in diodes D_1 and D_2. As described previously, the reference voltage for the circuit (the voltage across L_3) is obtained through inductive coupling with the primary of the IF output transformer. Another difference between the circuits of Fig. 8.23(a) and (b) is in the circuit paths for diode currents. In the second circuit, the currents of both diodes must pass through R_3, the input circuit of the audio amplifier and the common ground. It is more relevant to consider the audio output as a function of the ratio of these two currents.

Whatever the configuration of the particular ratio detector circuit, the response curve is very much the same as that shown in Fig. 8.22 for the discriminator—an S curve.

The Quadrature Detector

A third type of FM detector is known as the *quadrature detector*. This method of detection was originally used mostly in television receivers to demodulate the FM sound signal of television programs. However, it is experiencing renewed popularity as the detector of choice in FM receivers built using integrated circuits (ICs). A circuit that can perform this function lends itself to integration. Only one resonant circuit, external to the IC chip, is required for operation and tuning of the detector, thus simplifying alignment.

"Quadrature" is a fancy word that implies a relationship of 90° between two quantities. In the case of a quadrature detector, the word is appropriate because the detector requires two RF input signals which are at 90° with respect to each other when the carrier is on its center frequency.

A circuit incorporating the essential functions of a quadrature detector is shown in Fig. 8.24. The triangle-shaped symbol on the diagram represents an *operational amplifier,* or "op amp." Its specific function is to provide an output signal whose amplitude is proportional to the *mathematical product* of its two input signals. Its input signals are two versions of the receiver's frequency-modulated IF signal.

At the center frequency, the network consisting of C_1, C_2, L, and R_1 provides a signal at input terminal B which is exactly 90° out of phase with the signal at input A. (The signals are said to be "in quadrature," hence the title of the detector.) The detector output for this condition is zero. (You will recall from your study of power in ac circuits, for example, that the product of two quadrature waveforms of the same frequency is zero. Or, more specifically, power in an ac circuit is proportional to the product of current and voltage and the cosine of the angle between them. When the angle is 90°, of course, the product is zero.)

When modulation swings the frequency of the input signals slightly off center frequency, the angle between the two input signals also changes. In fact,

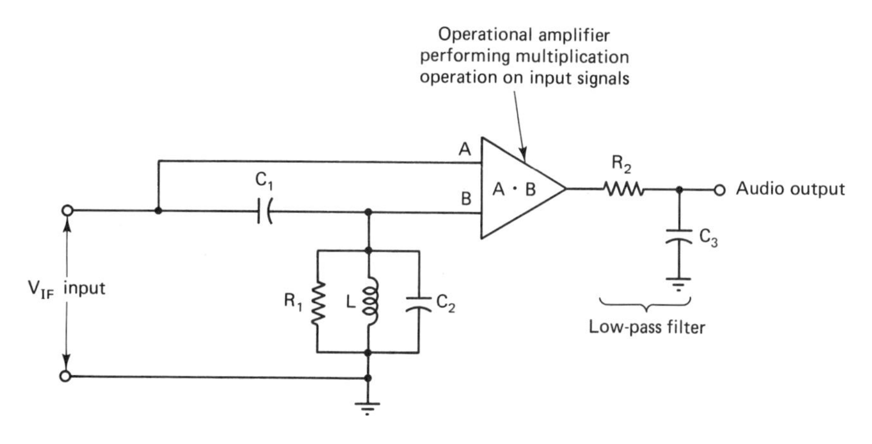

Figure 8.24 Quadrature detector circuit.

when frequency changes are relatively small, as they are in FM, the change in the angle between the two signals is proportional to the amount of frequency change. Furthermore, for small angular changes around 90°, *the cosine of that angle is equal to the amount of the angular change in radians*. Therefore, since, in effect, the amplitude of the output of the detector is proportional to the cosine of the angle between the two input signals, it is proportional to the change in frequency of the input signals.

The output of the multiplier also includes a component of the original signal frequency (the IF signal). This is easily filtered out by the low-pass filter, consisting of R_2 and C_3 in Fig. 8.24. The final output of the circuit represents a recovery of the original modulating signal: the amplitude of the output signal is proportional to the amount of frequency deviation (or its equivalent); the frequency of the output signal is equal to the rate of the frequency deviation.

An example of this function as manufactured on an integrated-circuit chip is that of the Motorola MC1357. The *LC* circuit that is required to provide the in-phase and quadrature versions of the signal at the input to the detector is external to the chip. An important feature of this particular circuit is that only a simple, low-cost, easily aligned single-winding coil is required for tuning the detector.

A quadrature detector is also included on the CA3089E chip, which is manufactured by RCA, among others. A block diagram depicting the application of this chip in an FM broadcast receiver is shown in Fig. 8.25. You will note that this subsystem-on-a-chip includes other functions as well: a three-stage FM IF amplifier/limiter with level detectors (for AGC), a low-level audio amplifier with provision for muting, and an AFC output. The muting feature permits the system to mute the audio output during tuning when the noise level is extremely annoying. It will also permit rejecting stations whose signals are too low to permit satisfactory reception.

Automatic Frequency Control (AFC)

As we have seen, FM demodulator circuits have limited bandwidths—approximately 200 kHz. For best results—maximum undistorted audio output—the signal fed to an FM detector must be centered to the operating band of the detector. The incoming signal is the receiver's IF signal, generally 10.7 MHz. Even when tuned properly for best results with the aid of a tuning meter, a receiver's local oscillator may soon drift off of the correct frequency required to convert the incoming carrier frequency to the 10.7-MHz center frequency of the detector. When this happens, one side or the other of each audio cycle is "clipped" by the detector. The sound emanating from the speaker is distorted and/or reduced in volume.

Drifting of the oscillator frequency occurs because of changes in the values of the oscillator's *LC* frequency-determining circuit. Such changes can be the result of heating, humidity changes, vibrations, etc. Correcting the unsatisfactory operating condition of the receiver requires either manual retuning by the user, or automatic frequency control of the LO. Of course, manual retuning is an unacceptable option.

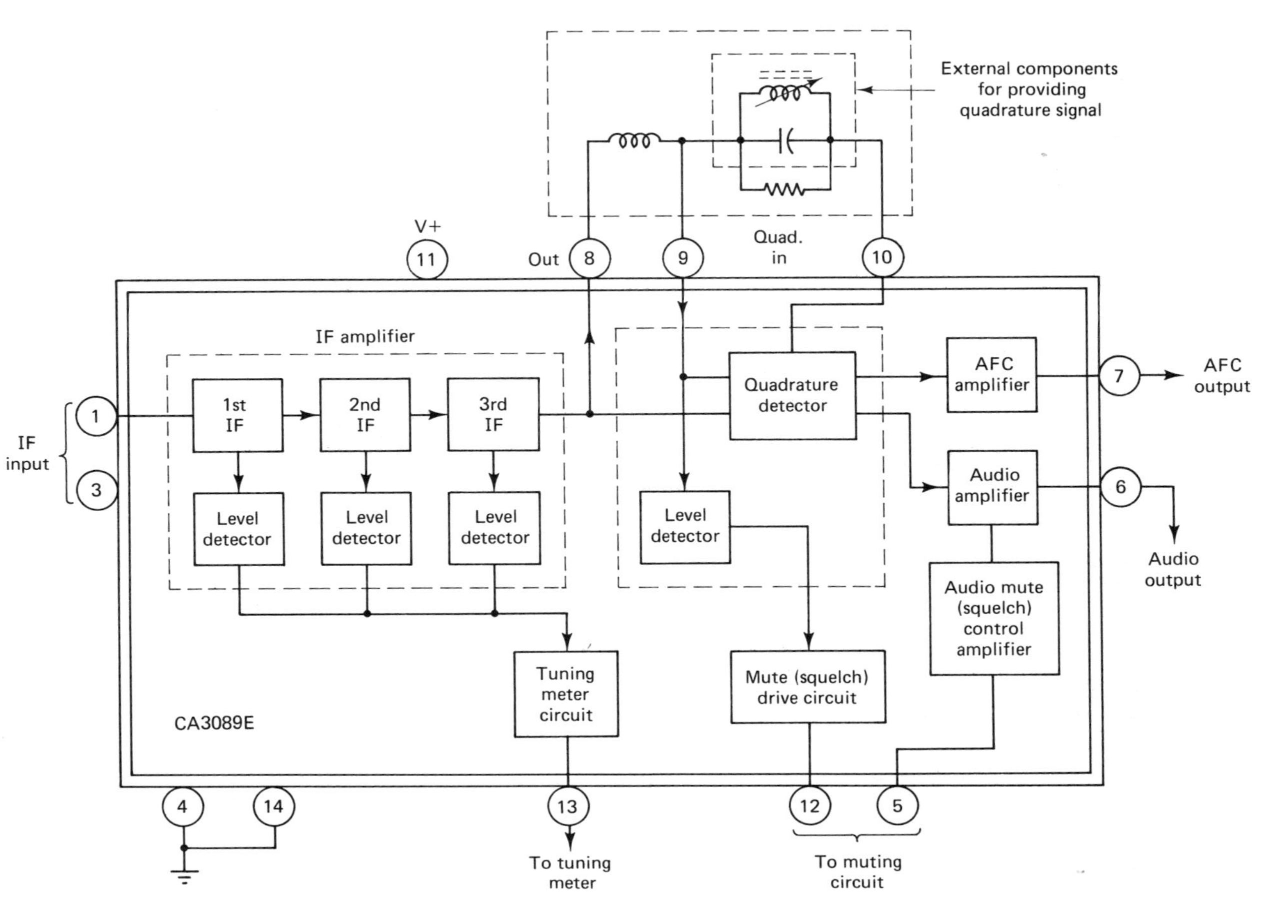

Figure 8.25 Integrated circuit for FM-IF, including quadrature detector. (Courtesy of RCA Corporation.)

Review the discussion of the automatic frequency control for FM transmitters in Sec. 8.2. You will recognize that with its LO and FM detector, an FM receiver has nearly all the functions it needs for automatic frequency control of the LO. In fact, only two additional functions are needed: (1) a varactor circuit in parallel with the oscillator's tank circuit to make the LO a voltage-controlled oscillator (a VCO); and (2) a feedback path, including a low-pass filter, from the output of the detector to the varactor circuit to provide a dc "error" voltage. The average value of the output of an FM detector is typically zero volts, even when the carrier is being modulated. That is, this is true when the signal being fed the detector is equal to the center frequency of the detector response. If the incoming signal is off frequency, the detector output will have a dc component called an "offset." The amplitude and polarity of the dc offset will reflect the amount and direction of the "error" of the frequency of the incoming signal. This offset voltage can be used to control the frequency of the LO and thus keep the IF centered to the detector's response curve.

The basic scheme for an automatic frequency control circuit for an FM receiver is shown in Fig. 8.26. You will note that it incorporates all of the basic functions of the phase-locked loop (see Sec. 8.2) except for a crystal-controlled reference frequency. In the case of the circuit of Fig. 8.26 there is no precision reference signal. The error output voltage is generated simply on the basis of the change in circuit parameters with changes in actual center frequency. If the actual center frequency (of input signal) is equal to the functional center frequency of the circuit, the average output is zero volts. For any other center frequency, within the functional bandwidth of the detector, the output voltage has a dc offset. In actual operation, the AFC is constantly "monitoring" the detector output and "correcting" the LO frequency so as to maintain the error voltage (dc offset) at a minimum level, if not, in fact, at zero volts.

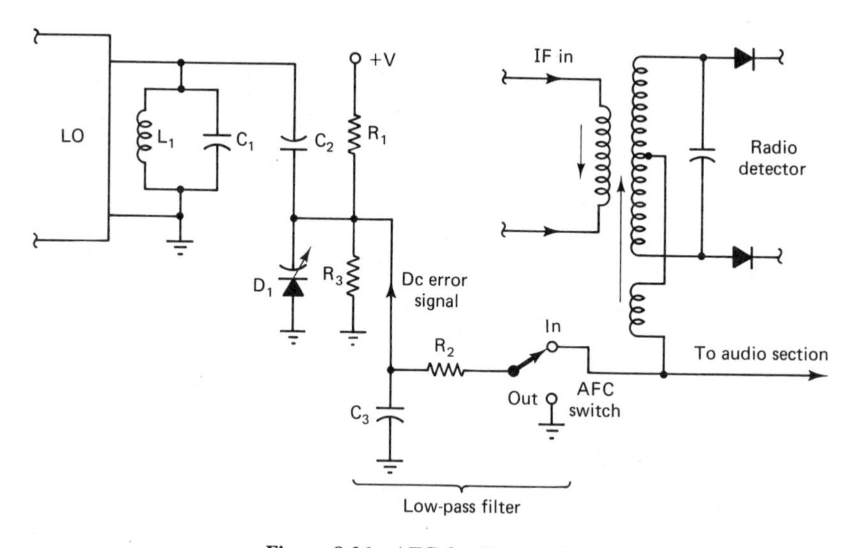

Figure 8.26 AFC for FM receiver.

An AFC function in an FM receiver affects the "feel" of the tuning of such a receiver compared to one without AFC, or one in which the AFC is temporarily defeated. When one tunes an FM receiver, one is trying to center the IF signal for a given station to the center of the detector response curve. If the AFC is not active, it is immediately apparent when the limits of the response curve are approached: the output sound is distorted and/or reduced in volume.

One technique for precise tuning is to "rock" the tuning knob (turn the knob from side to side in the vicinity of the desired station's position on the dial). One can sense the limits of good tuning on either side of the response curve, and then, finally, "center" to that imaginary curve. However, when an AFC function is enabled (is in operation), as soon as one's tuning places the LO frequency in close proximity to the correct frequency for a particular station, the AFC will "pull" the LO into the frequency that produces zero error signal at the output of the FM detector.

This setting of the tuning system may or may not place the RF tuning section at its best position for that particular station. Furthermore, as one tries to tune to another station the AFC will pull the LO frequency along, trying to maintain zero error voltage. For a short distance along the tuning dial, the "feel" is as if the receiver does not wish to let go of the station that it has been receiving. All in all, somewhat more precise tuning can be achieved with AFC disabled. Many receivers include a switch with which a user can place AFC "in" or "out" of operation. With this provision, the user is advised to switch AFC "out," tune for best reception to the desired station, and then switch AFC "in."

The phenomenon described in the previous paragraphs has a descriptive title in technical terminology: *lock*. That is, when the VCO frequency is sufficiently close to the desired frequency, the closed loop pulls the VCO into the desired frequency and *locks* it: The error in frequency, and consequently, the dc error voltage are maintained at an irreducible minimum. The range of frequencies over which the phase-locked loop can maintain lock is called the *lock range*. In the case of the AFC function of an FM receiver, lock range would refer to the range of frequencies over which the LO would operate if it were not in lock by the AFC loop. The lock range is greater than another range of frequencies called the *capture range*. The capture range is the range of frequencies over which the loop will "capture" the VCO and pull it into lock.

The Audio-Frequency Section

Moving further along the signal path of an FM receiver brings us next to the audio-frequency (AF) section. The AF section of all but the least expensive of FM receivers is almost always more sophisticated than that of an AM-only receiver (see Chap. 3). Almost all FM broadcast receivers incorporate the circuitry for decoding and reproducing stereophonic programs. (AM stereo is just beginning to become available as this is being written—1985.) A very large percentage of receivers with good-quality FM sections also provide for amplifying signals from phonograph and tape inputs. All of these additional functions—stereo, and phono and tape amplification—make the AF section more complex. However, aside from the stereo decoding circuitry, an FM audio section is simply an audio amplifier designed to reproduce the AF signals which it receives with maximum fidelity. More expensive versions of radio/audio systems separate the AM/FM tuning and audio amplification functions and place them into completely independent housings. The hardware units so produced are called "tuners" and "audio amplifiers." Let us look first at the basic portion of an audio section: the audio amplifier.

The block diagram of a basic audio amplifier typical of the quality and sophistication found in medium- to higher-priced AM/FM tuners or receivers is shown in Fig. 8.27. Let's examine the block diagram and, subsequently, some of the circuit details, considering only a single channel, a monaural amplifier. Meanwhile, keep in mind that all but the least expensive broadcast receivers with FM capability incorporate audio sections for stereo sound. That is, they include the equivalent of two complete audio amplifier strings.

You will note that the block diagram includes the following functions: preamplifier, tone amplifier, predriver, driver, and power amplifier. Not all of these functions are essential. An audio section might include only a driver and a power amplifier.

The preamplifier is a low-gain, low-noise, small-signal amplifier. Its purpose is to provide only a modest gain, but in a circuit that will minimize the injection of noise. It is often a two-stage amplifier in which the stages are direct coupled. A typical setup is one in which the collector of the input stage is connected directly to the base of the second stage. Direct coupling minimizes the attenuation of low frequencies. (When a signal is coupled through a coupling capacitor, the amount of signal passed from one stage to the next is diminished at lower frequencies because more signal is lost across the higher reactance, at lower frequencies, of the capacitor.)

Direct coupling introduces a problem of bias stability. Direct-coupled stages generally incorporate some form of negative dc feedback to overcome the stability problem.

A preamplifier section may incorporate what are called *equalization networks*. An equalization network is a circuit that changes the frequency response of the

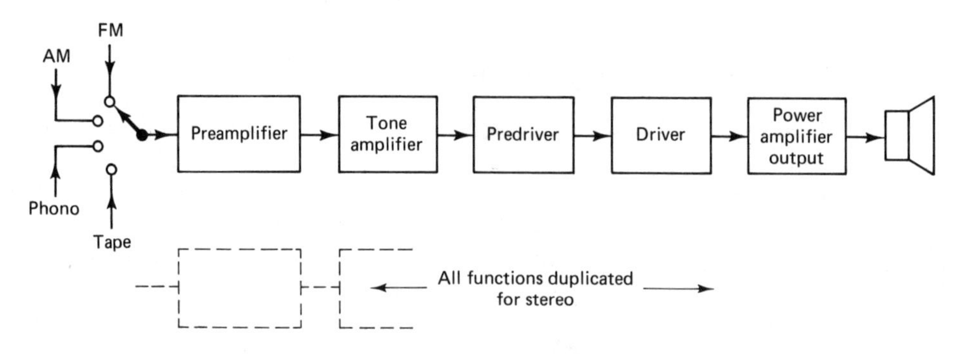

Figure 8.27 Block diagram of audio section of AM/FM receiver.

amplifier to reverse or undo the distortion intentionally added to a signal during the process of its being recorded on a phonograph record or magnetic tape. Such predistortion is a technique used to enhance the quality of the recorded signal by compensating for the known limitations of the recording medium. The concept is somewhat like that of preemphasis and deemphasis in FM broadcast systems.

A tone amplifier, if present, is a stage that incorporates special frequency-variable impedances. These impedances, generally combinations of resistors and capacitive reactances, shape the frequency response characteristic of the overall amplifier. The frequency response characteristic determines the relative amplification of low frequencies versus middle and/or high frequencies, etc. Front panel controls enable a listener to vary some of the resistive elements in this section. Labels such as "tone," "bass," "treble," and "mid" are familiar titles of such controls. Tone controls may be part of a preamplifier or a predriver; a special tone amplifier is not always included in audio sections.

A *volume control* for controlling the overall gain of the audio section (and the level of sound out of the speakers) is typically located between the preamp and the first driver amp.

At this point in the signal flow path the process of significantly increasing the energy level of the signal begins. The "voltage amplifiers" of the preamp or the tone amp require relatively little power to drive them. The power output stage, on the other hand, requires a significant level of "driving power." Therefore, depending on the power rating of the amplifier, one or more stages of "driver amplifiers" will be included. Predriver and driver stages are simply amplifiers that are designed to boost the power level of the signal sufficiently to drive the power output stage. These stages process larger signal amplitude swings than a preamp. They are more likely to introduce distortion into the signal; they typically incorporate negative feedback to counteract the distortion. The transistors have higher power ratings so as to be capable of dissipating more power without being destroyed in the process.

The power output stage provides the power to drive the speaker or speakers. Power ratings of 200 W per channel for living-room-type audio equipment is not uncommon. Assuming a speaker impedance of 8 Ω, output currents of the order of 5 to 10 A are to be expected. Transistors capable of handling such currents come in large packages. They require appropriate heat sinks to carry away the heat generated during the processing of currents of this order of magnitude. Output stages generally incorporate some form of push-pull operation to permit the production of high power levels with minimum distortion. A given output stage may include operation of transistors in parallel to achieve the desired power level. Power output stages are generally included in a negative feedback loop designed to minimize distortion of the signal during amplification.

Direct-Coupled Preamp

A circuit to demonstrate several basic concepts found in direct-coupled preamplifiers is shown in Fig. 8.28. This circuit is from a Heathkit Model AR-27 solid-state

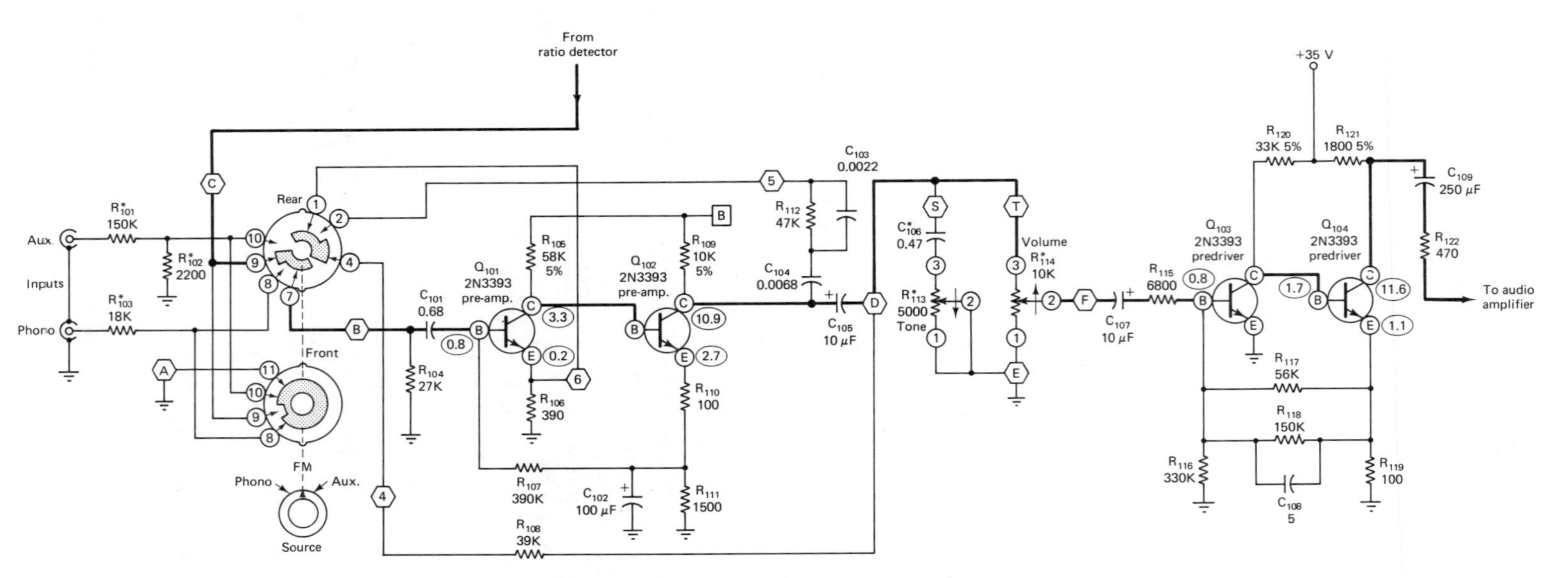

Figure 8.28 Direct-coupled audio preamplifier. (Reprinted by permission of Heath Company.)

monophonic FM receiver. Study the diagram carefully. Observe the following facts that affect the biasing of the two stages: (1) The base of Q_{102} is connected directly to the collector of Q_{101}. (2) The resistance in the emitter circuit of Q_{102} is split between two resistors, R_{110} and R_{111}. Only R_{111} is bypassed (C_{102}). The voltage across R_{111} is applied to the base of Q_{101}, through R_{107}, to supply bias for that stage. This arrangement—obtaining bias voltage for the first stage from across an emitter resistor of the second stage, when the second stage is direct coupled to the first stage—is a means of providing a degree of bias self-regulation for the combination. It stabilizes the bias against the effect of changing temperature, for example. The arrangement helps ensure that the circuit will function satisfactorily when one or both transistors have to be replaced after becoming defective. Without self-regulation, new transistors with somewhat different characteristics could require a change of the bias parameters to achieve satisfactory operation.

To understand why the bias arrangement of the circuit of Fig. 8.28 is said to be self-regulating, consider the following. Imagine that for some reason, a higher-temperature or whatever, transistor Q_{101} starts to conduct more (even with no input signal). The voltage across R_{105} in the collector circuit of Q_{101} will increase, the voltage applied to the base of Q_{102} will decrease. Q_{102} will conduct less, producing a smaller drop across R_{111}. The voltage providing forward bias for Q_{101} will be reduced, reducing the conduction of Q_{101}. The opposite effect will attain if Q_{101} drifts toward lower conduction (aging, perhaps): The voltage at the collector of Q_{101} will increase; the voltage at the base of Q_{102} will increase, causing greater conduction of Q_{102} and a rise in the voltage across R_{111}. An increase in the voltage across R_{111} increases the forward bias on Q_{101} and an accompanying increase in its conductivity, etc.

A defect in direct-coupled stages causes extreme changes in the dc voltages in such circuits. The normal dc voltages for the transistor terminals in the circuit of Fig. 8.28 are noted on the diagram. These voltages are the result of a dynamic balance, as described in the preceding paragraph. Consider the effect of an open Q_{101} or its emitter resistor, R_{106}: Without Q_{101} collector current the voltage drop across R_{105} will diminish significantly; the voltage at the collector of Q_{101}, and the base of Q_{102}, will rise significantly. Q_{102} will probably be driven into saturation. The voltage at the collector of Q_{101} will approach that of its collector supply—28 V. The voltage at the collector of Q_{102} will probably drop to around 4 V, the voltage predicted across the series combination of R_{110} and R_{111} with the equivalent resistance of the transistor assumed to be zero.

An open in Q_{102} or its emitter circuit will cause both stages to go to cutoff; both would have high collector voltages. Troubleshooting direct-coupled stages is often simpler than that of *RC* coupled stages, provided that you have an understanding of the interactions of such circuits.

The circuit of Fig. 8.28 also illustrates very basic applications of two other functions typically found in preamps: phono equalization and tone control. Study the circuit again and you will observe that R_{106}, the emitter resistor for Q_{101} is unbypassed. However, it has something connected to it through the rear deck of the *source* selector switch. When the source switch is set for "FM" (as shown) or "AUX," the junction of R_{106} and the emitter of Q_{101} is connected through switch contacts 4 and 1 and R_{108} to the output of the preamp. This is a feedback circuit; the voltage fed back is in phase with the signal voltage across R_{106}. That voltage is the result of emitter current flowing in the resistor. It is degenerative: it reduces the gain of the amplifier. Therefore, the feedback from the output of the preamp is also degenerative; it is a *negative feedback.*

Negative feedback is a way of attenuating, if not eliminating, undesirable "additions" to a signal, particularly additions made by the amplifier itself. Negative feedback, then, can greatly reduce distortion to a signal caused by any nonlinearity of operation of the amplifier. It can reduce noise introduced by the components of an amplifier.

When the "source" switch is selecting the "phono" input, the feedback to the emitter of Q_{101} is through the circuit consisting of C_{104}, C_{103}, and R_{112}. Since this circuit contains capacitance, the amount of feedback will be different for different frequencies. This is a circuit that provides equalization or compensation for the preemphasis (abnormally high amplitude level) given the high frequencies during the making of a phonograph record. For example, the Record Industry Association of America (RIAA) prescribes a standard preemphasis curve which is used by most record makers. A feedback or deemphasis circuit that corrects for the preemphasis, a form of predistortion, of the RIAA standard is called an *RIAA equalization* or *compensation circuit.*

A very simple tone control circuit is part of the preamp of Fig. 8.28. You will note the circuit consisting of C_{106} and R_{113} (connected as a rheostat) is in parallel with the volume control, R_{114}. It is a shunt path to ground. Its impedance is inversely proportional to frequency: it bypasses higher frequencies more than lower frequencies. When R_{113} is adjusted to reduce its value, the circuit bypasses higher-frequency currents even more readily. Therefore, when the listener turns the TONE control counterclockwise (opposite to the direction of the arrow on the diagram), he/she will perceive the speaker sound as having "weaker highs" and "stronger lows." If he/she turns the TONE control clockwise, the circuit will bypass or attenuate high frequencies less. The perceived effect will be that of "stronger highs."

Generally, a tone control section is significantly more complex than that shown in Fig. 8.28. In some cases tone control units contain numerous capacitors, resistors, and adjustable resistor controls. "Graphic equalizers" are a popular innovation on home entertainment units. A graphic equalizer is a tone control system which in essence consists of a number of adjustable, bandpass filters. These enable one to "shape" the overall audio passband by adjusting relative amplification levels at say, 50 Hz, 250 Hz, 1 kHz, 4.5 kHz, 10 kHz, and 15 kHz. The adjustments are by means of slide-type rheostats. The slides are mounted side by side. The pattern of the slide positions suggests the shape of the passband that will be produced by the particular settings; hence the title: graphic equalizer.

The circuit of Fig. 8.28 also shows the volume control, a variable resistance connected as a potentiometer. The volume control, depending on its position, "picks off" a portion of the total signal available across it. It is thus a manually adjusted gain control. If all of the available signal were fed to the amplifier following the control, the gain would be some specific amount. When the volume control is set to feed only a portion of the total available signal to the amplifier following, the output will be less and the gain will be less.

Audio Power Amplifiers

The presentation on the audio section of the AM receiver in Chap. 3 included a discussion of a transformer-type, solid-state push-pull amplifier. However, transformers are now virtually never seen in the audio amplifiers of radio receivers of the latest design. When electron tubes were the only practical amplifying device available, transformers were indispensable in the design of audio amplifiers. They were needed to achieve a match between the relatively very low impedance of speakers and the relatively high output impedance of tube amplifier stages. Power transistor amplifier stages typically have much lower output impedances than tube stages and can be designed to drive speakers directly. Most transistor audio power amplifier sections are now designed to be transformerless. At least, they have no output transformers; some may still have input transformers for phase shifting. Nevertheless, they are designed to take advantage of the benefits of push-pull operation—greater efficiency and power-handling ability with less distortion. Let us examine how a transformerless push-pull design can be evolved from one that utilizes a transformer.

Review the discussion of the transformer-type push-pull amplifier in Chap. 3 and be reminded that a push-pull amplifier is less subject to nonlinear distortion because all even harmonics cancel out. In that discussion, the cancellation was seen to occur in the primary of the output transformer; the second-harmonic currents entered the primary winding at opposite ends but *in phase*. Their magnetic effects, and hence their effect on the load, were canceled. If the transformer is eliminated, the circuit must be arranged in a different way to achieve similar results.

Consider the circuit of Fig. 8.29(a). You will note several differences from the transformer-type circuit: The transistors are complementary types—Q_1 an NPN, Q_2 a PNP. There are two power supplies. The speaker coil is connected directly into the circuit between ground and the common point between the power supplies. The input signal is applied in phase to both transistors. An analysis of the dc currents in the speaker coil reveals that their effects cancel because of the complementary nature of the transistors.

Now, consider the effect of a signal at the input to the stage. Since the same signal is applied to both bases, a positive-going signal will, for example, cause the current in Q_1 to increase and that in Q_2 to decrease. We would say that the collector currents are 180° out of phase. But in the speaker coil these effects are in the same direction: The signal currents are additive in the speaker coil. On the other hand, the even-harmonic components of the collector currents are in phase in their respective circuits; they are out of phase and cancel in the speaker coil. The disadvantage of this evolutionary circuit is that it requires two power supplies.

With a few more evolutionary changes the circuit can be made to operate with one power supply [see Fig. 8.29(b)]. In this circuit the speaker is connected

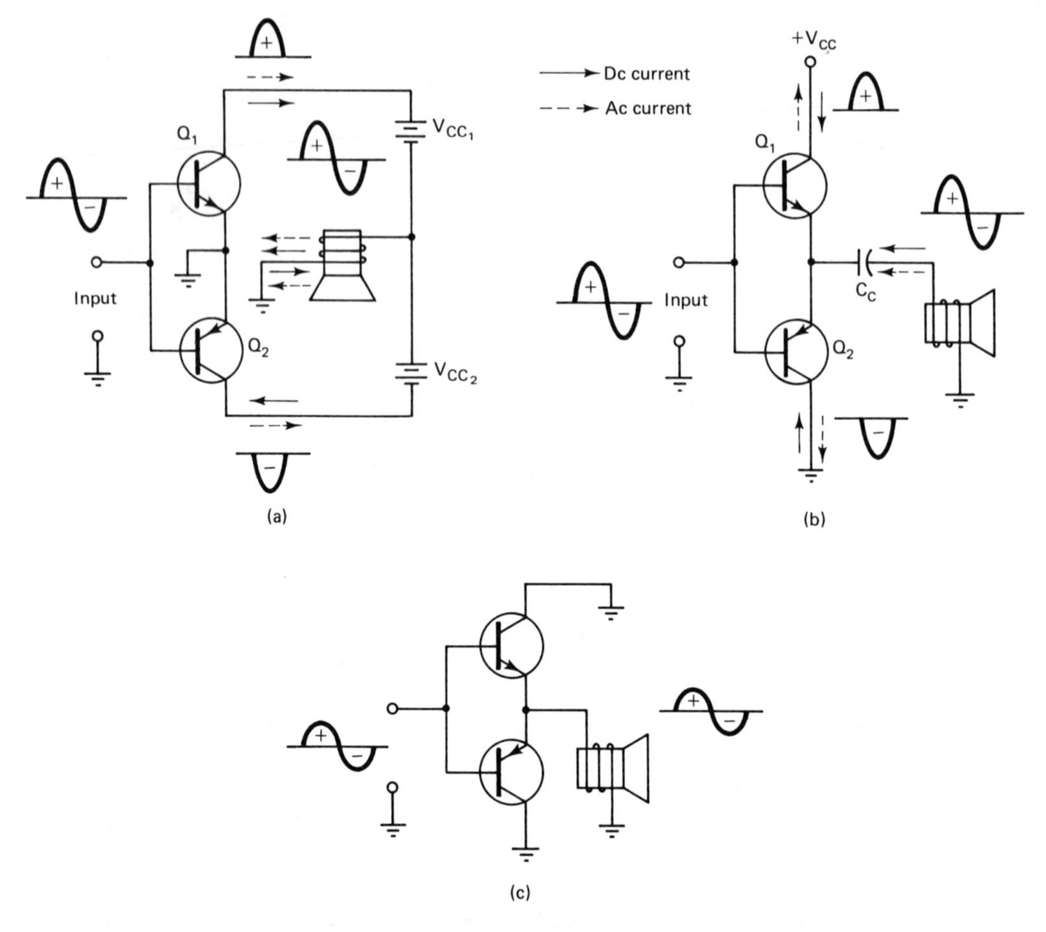

Figure 8.29 Evolution of transformerless push-pull output stage.

between the common point of the emitters and ground, through a large capacitor. Here the dc circuit consists of the two transistors in series, and in series with the single power supply. With no signal, the collector-to-emitter voltage across each transistor is equal to one-half of the supply voltage. The coupling capacitor, C_C, is charged to this voltage. When a signal is applied to the stage, during a positive alternation, Q_1 conducts more and Q_2 conducts less. The voltage across Q_1 decreases; the voltage across Q_2 increases. The emitter common point moves toward V_{CC}. Capacitor C_C will try to charge more—positive on the left side, negative on the right. This draws current through the speaker coil.

During a negative alternation of an input signal the situation is reversed. The emitter common point will be drawn down toward ground potential. The coupling capacitor will try to give up charge, driving discharge current through the speaker coil. The net effect is that the speaker coil is driven by an ac current which is a replica of the input signal current. The net dc current in the speaker coil is zero: it is blocked by C_C.

An ac equivalent of the circuit is shown in Fig. 8.29(c). The even-harmonic currents produced by the two transistors, as before, are in phase but drawn through the speaker coil in opposite directions, and therefore their effects cancel.

This circuit is a form of emitter follower and exhibits the characteristics of that type of circuit. The important characteristic in this application is low output impedance. This permits the circuit to drive the low impedance typical of speakers without resorting to the impedance-matching capability of a transformer.

We generally think of an audio amplifier as requiring operation in the class A mode. That is, collector current must flow the full 360° of each cycle to provide an undistorted output, thus producing a replica of the input waveform. This is contrasted with an RF amplifier, in which the conduction angle can be less than 180° per cycle. The flywheel effect of the resonant circuits in RF amplifiers "completes the cycle" and produces a good waveform.

However, a class A amplifier is an inefficient energy converter. A class B amplifier is significantly more efficient. The push-pull amplifier offers the promise that an audio amplifier can be operated successfully in class B mode: Set the bias so that both halves of the circuit are at cutoff with no signal; let each half circuit supply half of the output waveform. There is a hitch, however. Because the signal voltage must overcome the junction barrier potential before normal conduction starts, the two halves of the circuit do not produce undistorted half sine waves (when the input signal is a pure sine wave) (see Fig. 8.30). The result is called *crossover distortion*. The problem can be overcome by arranging the circuit so that the transistors will be supplied a small amount of no-signal base current, say 5% of that required for saturation. This decreases the efficiency of the circuit only slightly. This low-amplitude bias current is called "trickle bias." It places the operation of a stage in a mode technically described as class AB.

The circuit of Fig. 8.31 incorporates several features typical of the design of transformerless push-pull audio amplifiers. It is variously called a "stacked circuit" and a complementary-symmetry output circuit. Let us examine it in some detail.

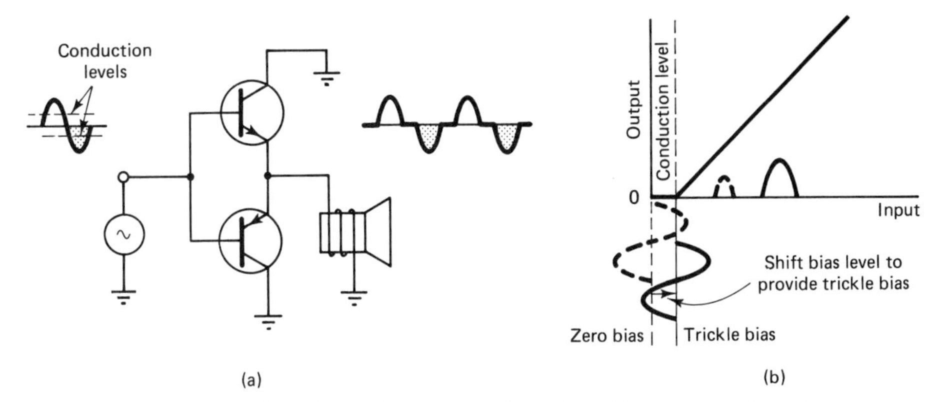

Figure 8.30 Trickle bias and crossover distortion: (a) crossover distortion; (b) transfer curve for amplifier.

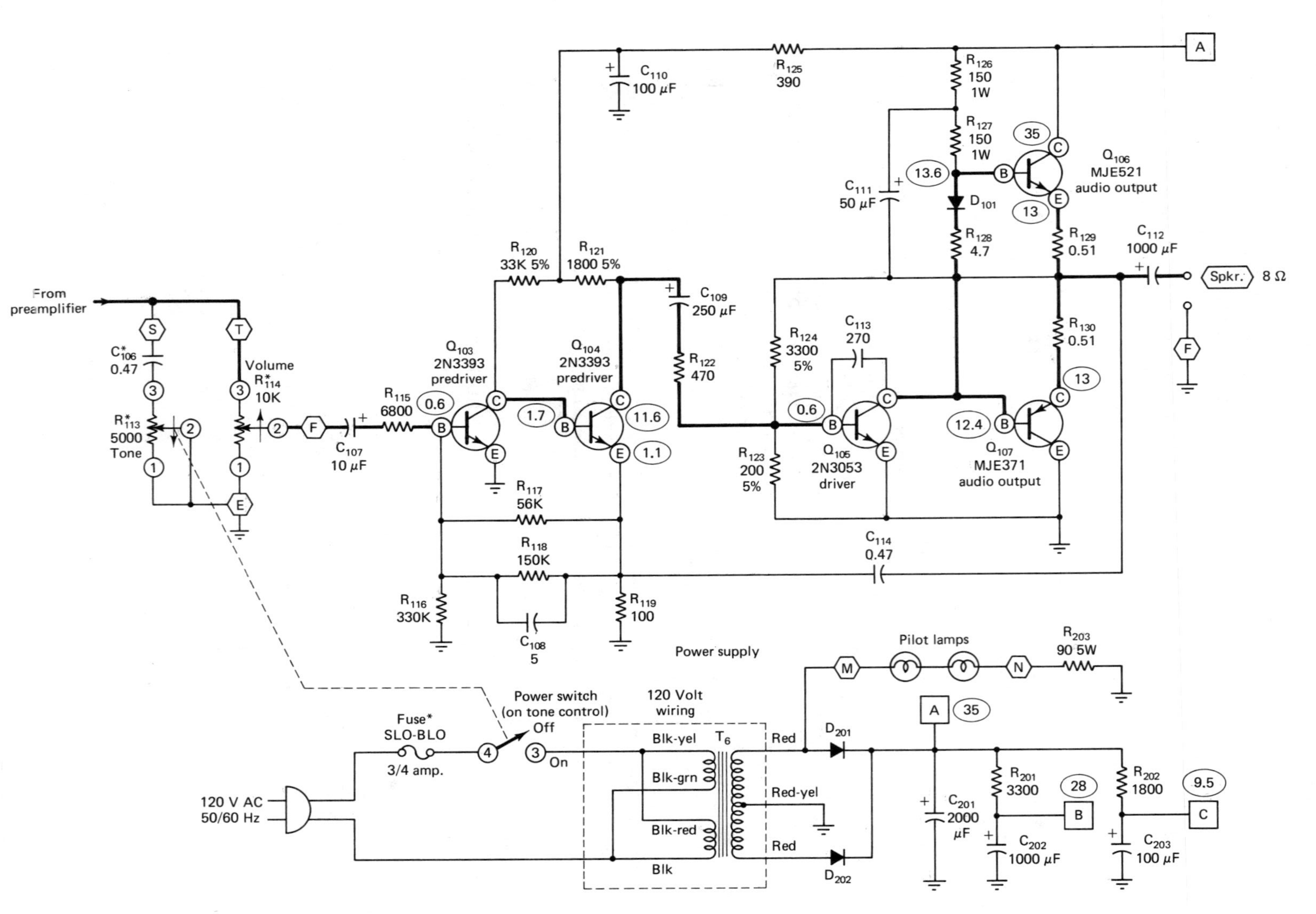

Figure 8.31 Transformerless audio amplifier output stages. (Reprinted by permission of Heath Company.)

Obtaining the desired performance from a class AB push-pull audio amplifier requires a relatively precise control of bias current. If there is too much of a trickle, the operation will be less efficient. Transistors will tend to overheat. As the temperature of transistors increases, they tend to conduct more readily. This produces more heating, then more conduction, and so on. If the bias current is not sufficient, some crossover distortion will occur. It is important to build into the circuit some means of compensating for this chain-reaction type of phenomenon.

A diode is very commonly placed in the circuit between the bases of the two halves of a push-pull circuit to improve stability. (See diode D_{101} in the circuit of Figure 8.31.) A voltage difference of about 1 V between the bases of the output transistors is normally required. (In the circuit of Fig. 8.31 this difference is indicated as $13.6 - 12.4 = 1.2$ V.) A diode and a resistor, as part of the voltage divider supplying the bias, provide this difference.

The diode is there to provide temperature compensation. It works like this. First, the diode will be specially selected so that its thermal characteristics will be similiar to those of the output transistors. For example, it will be chosen to be of the same material as the transistors, silicon, usually. For best results, the diode will be thermally coupled to the transistors so that its temperature changes will match closely those of the transistors. Thermally coupled, practically, means that the diode will be mounted on the same heat sink as the transistors. As the unit, transistors and diode, heats up, the conductivity of all increases. With increased conductivity, the equivalent of a lower junction resistance, the voltage drop across the diode decreases. In the circuit of Fig. 8.31, the voltage at the base of Q_{106} decreases slightly, lowering its collector current slightly. The voltage at the base of Q_{107} increases slightly, reducing its forward bias and collector current slightly.

Let us examine other features of the circuit of Fig. 8.31. First, tracing the signal flow path, we find that the output of Q_{103}, the first predriver, is direct coupled to Q_{104}. The output of Q_{104} is coupled through C_{109} and R_{122} to driver Q_{105}, which is direct coupled to the output stage—Q_{106}, Q_{107}. Note that Q_{103} receives its bias from Q_{104} as the voltage across R_{119} in the emitter circuit of Q_{104}. This represents dc feedback and is used to stabilize the dc operating point of these two direct-coupled stages. Q_{103} also receives negative signal feedback—from the junction point between output transistors Q_{106}–Q_{107}, through C_{114} and the parallel combination of R_{118} and C_{108}. This connection is provided to minimize the distortion introduced by the amplifying action of the various stages in the section.

There is also dc and ac feedback from the output stage to the driver stage. The connection from the output stage through R_{124} to the base of Q_{105} provides bias for Q_{105}. However, changes in the dc operating level of the output stage will be fed back to Q_{105}: For example, if Q_{107} starts to conduct more, the voltage across it will decrease, lowering the bias on Q_{105}. Q_{105} will then conduct less, raising the voltage on its collector and the base of Q_{107}. The bias of Q_{107} is lowered, thereby, and it will conduct less. The feedback provides a measure of dc operating point stability. Since R_{124} is unbypassed, signal voltage from across the speaker drive circuit is fed back to the base of Q_{105}. This is a negative feedback, and, again, is incorporated in the design to reduce distortion. Resistors R_{125} and R_{126} and capacitors C_{110} and C_{111} provide decoupling between the power supply and the circuits in which they are connected.

Troubleshooting Audio Amplifiers

You will undoubtedly encounter audio amplifiers that differ in specific details from the examples used above. Stereo amplifiers, for example, contain two complete identical sections. More expensive and/or higher-powered units are generally more sophisticated than has been illustrated. Nevertheless, when taken stage by stage, there are many similarities in all audio amplifiers, wherever they may be found.

Troubleshooting audio amplifiers is generally easier than troubleshooting RF or IF amplifiers. This is true, for one reason, because testing techniques are less critical when working with lower frequencies. The probability of capacitively loading a circuit, and changing its operating characteristics, by clipping on a test lead, is less. When a speaker is connected to the output of an audio amplifier, the combination provides its own test monitor. Generally, some noise can be heard from a speaker attached to an operable amplifier. If the noise amplitude can be varied by adjusting the volume control, there is an excellent chance that the amplifier is functioning correctly. If there is no noise at all, a problem in the amplifier is highly likely. Testing an audio amplifier for operation by injecting a signal from an audio generator is a simple and effective technique. In fact, any source of audio frequency is a candidate for providing a test signal if an audio generator is not available. Analyzing dc voltage changes (see Chap. 4) is an effective method in determining the precise location of a defective component once the defective stage has been located.

You should be aware that it is possible to make costly mistakes in working with transformerless output stages. A short across the speaker of such stages is effectively an ac short of half of the circuit. If the short is on the transistor side of the speaker coupling capacitor, it is a dc short that may apply full collector supply voltage to one of the output transistors. And the voltage is applied without any series current-limiting resistance. The result is usually the destruction of one power transistor. Then, even if the external short is removed, the second of the two stacked transistors is likely to be destroyed before it is realized that the first was left shorted as a result of its destruction. Thus it may be possible to quickly destroy two expensive power transistors. The moral is clear: Take care to avoid shorting the output terminals of a transformerless audio amplifier.

8.4 MULTIPLEXING

The process by which two or more unique information signals are sent simultaneously over the same communications medium or channel is called *multiplexing*. Many stations in the FM broadcast service transmit simultaneously three information signals: (1) a so-called compatible monaural signal representing the program being produced by the station; (2) a specially coded signal of the same program which permits a stereophonic reproduction of the program by appropriately equipped receivers; and (3) a signal representing a second, completely separate program. Signal elements 1 and 2 are part of the scheme that can provide stereophonic (two-channel) reproduction. Element 3 is the information signal for a service officially called Subsidiary Communications Authorizations (SCA). SCA broadcasts are usually of music without commercials. They are generally used as background music in stores and offices.

Stereo and SCA broadcasting are forms of *frequency-division multiplexing* (FDM). Altogether there are three bands of information signals involved. All three are present at the same time. The three bands taken together form a composite signal carried by a single communications medium—the carrier of the station broadcasting the multiplexed signal. (In reality, all broadcast stations in a particular area are part of a multiplexing scheme, of sorts. Their signals are all present simultaneously; they all use the same communications medium—the earth's atmosphere.) Receivers are designed to permit a user to select one particular signal. Elaborate FDM schemes are used extensively in telephone systems and other communications networks.

Another form of multiplexing is termed *time-division multiplexing* (TDM). In TDM, samples of several different signals are sent one at a time at regular, short intervals over a common medium. In this scheme, the goal is for a given receiver to receive all the information signals. This contrasts with FDM, in which the goal is generally for an individual user to receive only one signal, being uninterested in the other signals. A TDM scheme requires that the receiving equipment "scan" the transmission medium at a rate that is synchronous with the transmission of the separate elements of data. Only thus can the receiver separate the individual information signals.

Stereo Broadcasting

The word *stereo* comes from a Greek root meaning solid or firm, in the sense of three-dimensional. The experience of listening to music reproduced on a "stereo system" is perceived as being more real or lifelike than listening to the same music reproduced by a monophonic system. At the very minimum, a stereo system includes two speakers "with separation."

Separation implies two elements: a physical separation of the speakers, from the listener's point of view; and a separation of the signals driving the speakers. The physical distance between the speakers must be significant compared to the distance of the listener from the speakers. If the listener is, say, 8 ft from the speakers, the speakers should be at least a couple of feet apart.

To create the perception of realism the speakers must be driven by signals picked up at physically separated positions at the location where the music is actually produced. If the stereo reproduction scheme is part of a radio broadcast system, the system must be capable of transmitting the two separate signals simultaneously. The two signals are universally designated the L (for left side) and R (for right side) signals.

The broadcast of signals for stereo reproduction has been a part of the FM broadcast system since the 1950s. A few AM stations are just beginning to broadcast signals for stereo reproduction. In each instance, FM and AM, the FCC has required that proposed new stereo broadcasts be receivable and reproducible by existing monophonic receivers. In a word, the introduction of stereo broadcasting has been required to be *compatible* with the systems as they existed previously. In human terms, this is a simple, reasonable requirement. Technically, the requirement influenced significantly the choice of the form of signals to be broadcast, and thereby, the design of the equipment used to generate, transmit, and receive the signals.

In the FM system, a stereo broadcast consists of the multiplex transmission of two information signals and a pilot carrier. However, the two signals are not simply an L signal and an R signal. Such a broadcast would not be compatible with a monophonic receiver. The two signals are an $L + R$ component and an $L - R$ component. The $L + R$ component is receivable in a normal way by a monophonic receiver. The reproduction of that signal by the monophonic receiver is indistinguishable from a monophonic broadcast. On the other hand, a receiver designed to receive and reproduce stereo broadcasts incorporates a special section that decodes the two signals—$L + R$ and $L - R$—and recombines them so as to produce separate L and R signals. A stereo receiver will have a two-channel audio section—two complete audio amplifiers—which processes the signals and drives two separate speaker systems.

It is not one of the goals of this book to examine in detail the circuits typical of stereo sections of transmitters and receivers. However, a limited goal of investigating the generation and reception of stereo signals in terms of the general functions of block diagrams is intended. This will provide a foundation understanding of the topic which will prepare one to proceed quickly toward an understanding of details when and if that is required.

Let us begin by examining the block diagram of the stereo section of an FM transmitter (see Fig. 8.32). The process starts with the conversion of sound into electrical signals by means of two separate microphones, a left mike and a right mike. The signals from these two are processed by individual audio channels (amplifiers). The $L + R$ signal is produced by adding these separate L and R signals, in phase, in an appropriate linear circuit called an *adder*. This process might well be performed in an identical fashion in a monophonic station. The FM transmitter is frequency modulated by this signal. The result, at this point, is indistinguishable from a monophonic transmission; it is the mono-compatible component of a stereo broadcast.

The R signal is fed to a phase inverter which shifts its phase by 180°, producing a $-R$ signal. The $-R$ signal is combined with the L signal in a second adder to produce an $L - R$ signal.

Simultaneously, a 19-kHz signal, called a *pilot carrier* is being continuously produced by its own oscillator. (It is called a pilot carrier because it is transmitted at a relatively low amplitude and is not modulated directly. It is transmitted to provide a carrier at the receiver for the detection of the $L - R$ signal.) The frequency of the 19-kHz signal is doubled in the 38-kHz doubler and then fed

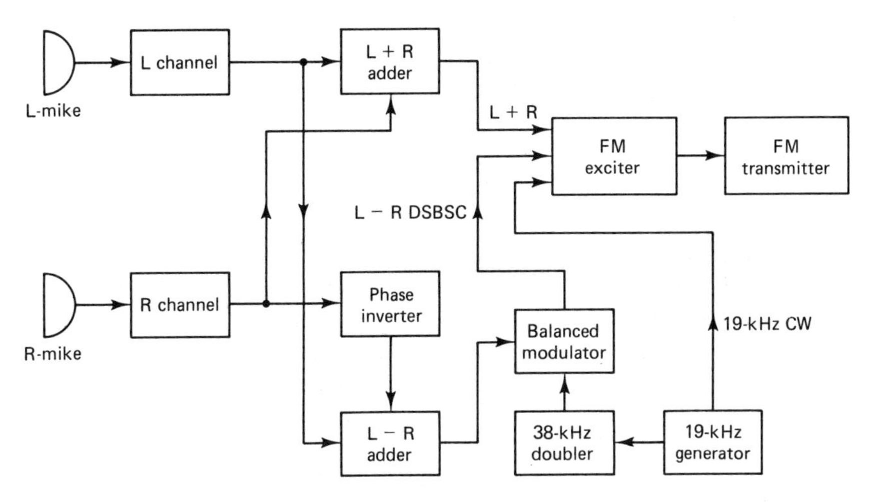

Figure 8.32 Compatible stereo-FM transmitter.

to the balanced modulator. A balanced modulator (see the detailed description in Chap. 9) is a circuit which, in effect, produces amplitude modulation of a carrier but suppresses the carrier in its output as part of the process. It produces two AM-type sidebands with the carrier suppressed. The FM transmitter is modulated by the 19-kHz pilot carrier and the 38-kHz double-sideband, suppressed carrier (DSBSC) signal.

The frequency spectrum of the signals modulating the stereo FM carrier is shown in Fig. 8.33. Remember, the 50-Hz to 15-kHz $L + R$ signal is equivalent to a standard monophonic FM broadcast signal. The 19-kHz pilot carrier is transmitted as an aid in detecting the 38-kHz DSBSC $L - R$ signal at the receiver. The $L - R$ signal will be recombined with the $L + R$ signal in the receiver to produce the separate L and R signals. The concept of frequency-division multiplexing (FDM) is incorporated into the process just outlined in the fact that there are three distinct groupings of frequencies—the 0- to 15-kHz band, the 19-kHz pilot carrier, and the 38-kHz DSBSC band. The FM carrier is modulated simultaneously by all three signals. Each produces its set of side frequencies;

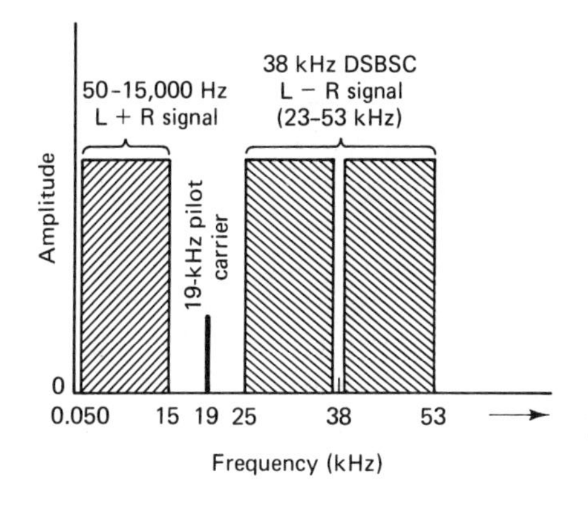

Figure 8.33 Frequency spectrum of signals in FM stereo.

each can be separated from the other at the receiver by appropriate filtering. Let us proceed now to the stereo decoding section of a receiver.

A typical scheme for decoding FM stereo broadcasts is shown in Fig. 8.34. FM stereo and mono receivers are identical up to the output of the FM detector. In a stereo receiver, the output of the detector is passed to the stereo decoder. The first function the decoder performs is to separate the three signal components—$L + R$ signal (0 to 15 kHz), 19-kHz pilot, and $L - R$ signal (23 to 53 kHz). It performs this task by means of appropriate, frequency-selective filters: a low-pass filter is used to separate the 0- to 15-kHz signal from the others; a narrow-band filter selects out the 19-kHz pilot carrier; and a 23- to 53-kHz bandpass filter provides for the separation of the $L - R$ signal.

The 38-kHz DSBSC signal requires a more involved demodulation process than the $L + R$ signal. This processing takes up a very small but definite amount of time. Since the two signals must be recombined in the exact time phase with which they were created, the $L + R$ filter incorporates a time delay to match the delay of the $L - R$ processing.

The decoder section amplifies the 19-kHz pilot carrier and uses it to lock in a 19-kHz local oscillator. The 19-kHz local oscillator, when locked in, is in precise synchronism with the 19-kHz carrier at the transmitter. The frequency of the output of this oscillator is doubled and fed to a balanced demodulator, together with the $L - R$ signal. By this means the $L - R$ signal is recovered.

The recovered $L - R$ signal is recombined with the $L + R$ signal in two separate forms. It is combined in an adder in its normal form to produce a $2L$ signal: $L + R + L - R = 2L$. It is inverted to $-(L - R)$ and combined with $L + R$ in a second adder to produce a $2R$ signal: $L + R - (L - R) = L + R - L + R = 2R$. The now-separate L and R signals are processed further in two separate audio amplifier channels and are used to drive separate L and R speakers, or speaker systems.

SCA Broadcasting

FM stations may be authorized to broadcast a second program simultaneously with the main station program. This is a subsidiary or auxiliary program. It is receivable only on receivers that have been specially equipped with an appropriate decoder. The service is usually a subscription service: the user pays for the decoder and the service. The FCC authorizes selected stations to offer this service through a license called a Subsidiary Communications Authorization (SCA). Regulations pertaining to SCA operations are designed to ensure that this secondary program does not interfere with or degrade the quality of a station's main program. The addition of an SCA program is accomplished by means of frequency-division multiplexing. The process is multiplexing even if the main program is a monophonic program.

A block diagram of one scheme for accomplishing SCA multiplexing is shown in Fig. 8.35. In general, production of the SCA signal which is to modulate the main carrier involves a subcarrier that must be in the frequency range 20 to 70 kHz. Note that the subcarrier frequency in the example of Fig. 8.35 is 67.5

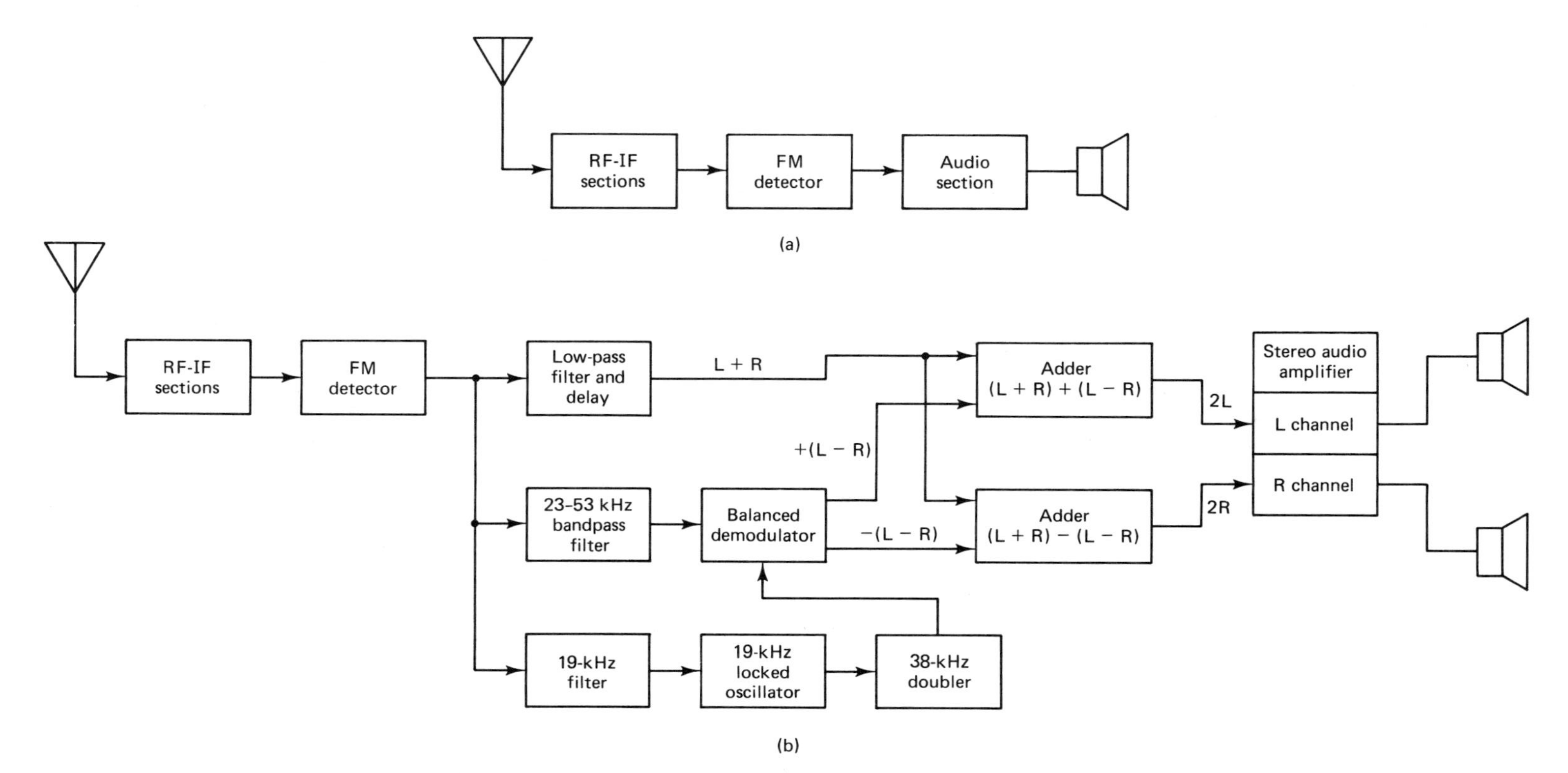

Figure 8.34 (a) Monophonic FM receiver; (b) stereophonic FM receiver.

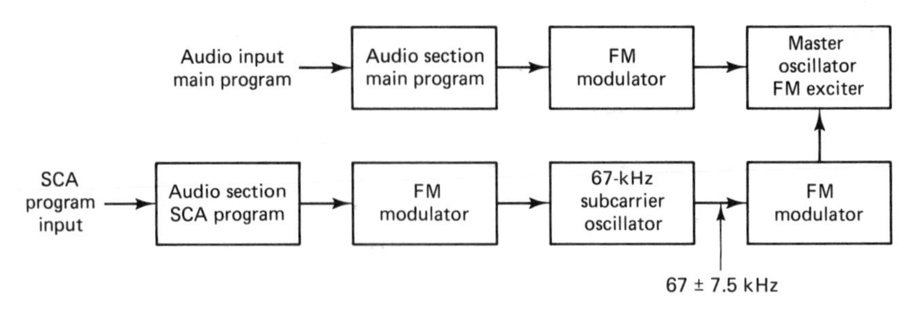

Figure 8.35 SCA multiplexing.

kHz, a value typical of this application. The SCA subcarrier is first frequency modulated by the SCA program signal. Maximum deviation permitted is 7.5 kHz. The modulated SCA subcarrier is then used as an information signal to frequency modulate the main station carrier.

A receiver equipped to receive an SCA transmission passes the signal from the FM detector to a special SCA decoder section. There the SCA signal is separated from other audio signals by means of a 67.5-kHz bandpass filter. The separated signal is passed through a second FM detector, where the audio content of the SCA program is recovered. This audio signal drives the system's speakers after being amplified in the receiver's audio section.

8.5 ALIGNING FM RECEIVERS

In general terms, an FM receiver is aligned in very much the same way as an AM receiver. (You are urged to review the alignment of an AM receiver by looking again at the appropriate section of Chap. 4.) However, there are several important differences between AM and FM receivers in some of the details of their construction and functioning. These differences dictate certain differences in the details of the alignment procedures. For example, for best operation, the IF section of an FM receiver should have a bandwidth of at least 200 kHz. The bandwidth of an IF section in an AM receiver need be only approximately 10 kHz. The detector circuit in an FM receiver requires alignment to center its operation to the operating frequency of the IF section. An AM detector needs no alignment; it has no provision for adjustment.

The steps in the alignment of an FM receiver are as follows: (1) Align the IF section, beginning with the last stage first. (2) Align the FM detector. (3) Check, and if necessary, adjust the tracking of the local oscillator. (4) Check and adjust the tracking of the RF stage. (5) Check, and when indicated, adjust the alignment of the stereo and/or SCA decoder sections. The alignment of an FM IF section can be accomplished by either of two methods—the *steady-frequency method* or the *sweep-frequency method.* The two methods are also applicable to the checking and alignment of the FM detector. We provide information on each method.

Steady-Frequency Method of Alignment

The steady-frequency method is called that because the RF generator used in this technique need be capable of producing only one frequency at a time (the steady frequency). As you will see, the sweep-frequency method requires a special RF generator that "sweeps" its output across a band of frequencies: it is, in effect, frequency modulated. The steady-frequency method is used in the alignment of the IF section of an AM receiver. Since the bandwidth there is comparatively narrow—10 kHz—the IF section is simply aligned for maximum output at the center frequency of the band. The same technique can be used in the alignment of an FM IF section; however, it is inadequate. For best performance of an FM receiver, an IF alignment should ensure that the IF bandwidth is a minimum of 200 kHz. It is necessary that the bandwidth be checked, and adjustments made, by passing frequencies up to 100 kHz on either side of the IF center frequency—10.7 MHz. That is, the IF section should be tested and adjusted to operate with a maximum gain at 10.7 MHz. The output should be down by not more than 3 dB at 10.6 MHz and 10.8 MHz.

The equipment and connections required for performing a steady-frequency alignment on a broadcast FM receiver are shown in Fig. 8.36. As indicated, equipment required includes an RF generator and an instrument for monitoring (i.e., measuring) the output of the IF section of the receiver. The monitoring instrument is used to tell us what effect our adjustments of tuned circuits are producing. We want to know whether our adjustments are producing the desired effect, which is best performance of the receiver. The RF generator should be a typical, laboratory-type generator capable of producing frequencies between 10.5 and 10.9 MHz and 88 and 106 MHz. A frequency scale calibrated so as to permit settings of 10.6, 10.7, and 10.8 MHz is highly desirable. The preferred monitoring instrument is a high-impedance electronic dc voltmeter with an analog scale. It should be possible to electronically "zero" the meter at the *center* of the scale. An instrument with a digital readout can provide the required measurements; however, it is not as convenient to use as an analog meter with a center-zero adjustment. An oscilloscope with a direct-coupled input (for dc voltage measurement) can also be used as a monitoring instrument.

Connecting the generator and monitor to appropriate points in the receiver to be aligned, is, of course, crucial to the success of the procedure. With many receivers it will be appropriate to connect the generator output to the antenna

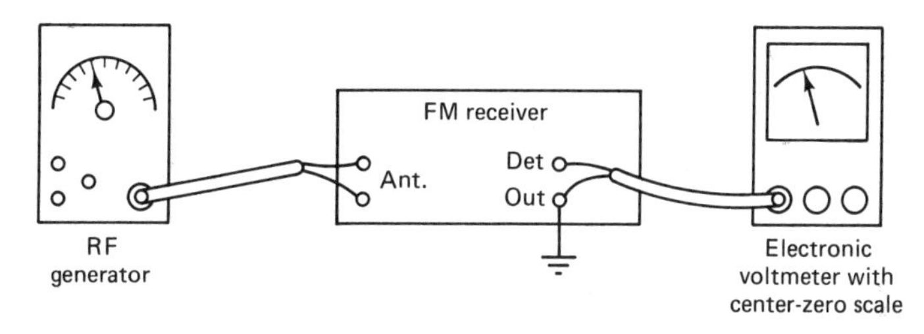

Figure 8.36 Setup for aligning FM receiver, steady-frequency method.

terminals of the receiver. This connection can be used even for the injection of the 10.7-MHz IF signal if the RF and frequency converter sections are not excessively "tight" (narrowly selective).

With some receivers with tight front ends, it may be necessary to inject the IF signal at some point after the frequency converter section. This could be found to be required in order to provide a signal of sufficient amplitude to be observed at the output of the IF section. In fact, when an IF section is severely out of alignment, it may be found necessary to inject the signal at the input of the last IF stage.

The best connection point for the monitor depends in some degree on the type of FM detector the receiver contains. If the detector is a ratio detector, as it is likely to be, an excellent place to monitor the relative amplitude of the IF signal is across the large capacitor always present in a ratio detector. You will recall this capacitor as the one that contributes to the amplitude limiting effect of this detector. The dc voltage across the capacitor which performs the described function is always proportional to the amplitude of the IF signal reaching the detector.

When checking and aligning the detector itself, the monitor should be connected between the point of audio output and ground. This point, in most designs, is at zero volts, dc, when the detector is aligned properly. Tuning-indicator meters on good-quality receivers are connected to monitor the two test points just described.

The alignment of the IF section begins by injecting a signal of 10.7 MHz at an appropriate injection point (see above). The standard intermediate frequency for FM receivers is 10.7 MHz. The amplitude of the injected signal is adjusted for a measurable but not excessive level at the monitor. Manufacturer's instructions, if available, should be consulted for a recommended signal level for alignment. If the section is severely out of alignment the injected amplitude must be reduced as the circuits are brought into alignment. The alignment of the section proceeds with the adjustment of each resonant circuit in the section. Generally, best results are obtained by starting with the last IF stage (the one nearest the detector) first, and then moving stage by stage toward the front end. Each tuned circuit is adjusted for maximum output of the section, as indicated by the monitor. In some FM receivers, IF transformers are *double tuned*. That is, each transformer core contains two ferrite slugs—one for the primary tuned circuit and one for a secondary tuned circuit. Each slug must be adjusted for maximum output. The process of touching up the adjustment of each circuit is repeated several times until no significant improvement in the output of the section is obtainable.

The IF section should now be checked for bandwidth if the RF generator is such as to permit this (it must permit settings at each 100 kHz in the vicinity of 10.7 MHz). For maximum response to the high-frequency content of programs, an FM IF section should have a bandwidth of at least 200 kHz. That is, the output of the section should be down (decreased) by not more than 3 dB when passing frequencies of 10.6 and 10.8 MHz. Thus the frequency of the RF generator is adjusted to these frequencies and the output amplitude measured. A 3-dB reduction corresponds to cutting the measured output voltage by 30%. Hence if the measured output voltage at either 10.6 or 10.8 MHz is equal to less than 70.7% of the value at 10.7 MHz, the selectivity of the section is too narrow.

Increasing the bandwidth involves detuning slightly the previously tuned circuits so that they are not all tuned precisely to the same frequency—10.7 MHz. For example, one might set the RF generator for an injected signal of 10.6 MHz and retune one stage for an increased output at that frequency. Then inject 10.8 MHz and retune a different stage for an increased output at that frequency. Finally, the outputs for frequencies of 10.6, 10.7, and 10.8 MHz should be compared to ensure that the difference between any of them is not greater than 30%.

The task of measuring and adjusting bandwidth with the steady-frequency method is tedious at best. The sweep-frequency method was developed, in part, to facilitate this aspect of the alignment process.

The FM detector is aligned next after the IF section. For this step, the monitor is connected to measure the dc voltage to ground at the audio output point of the detector (a ratio detector is assumed). On a typical ratio detector this point will measure zero volts when the detector is aligned properly. The detector alignment is adjusted by means of the slug in the secondary winding of the IF output transformer. The adjustment is made while a signal of 10.7 MHz is being injected. The slug is adjusted until the voltage on the monitor can be made to vary from one polarity (say, positive), through zero volts to the opposite dc polarity. (This effect is most conveniently observed on a center-zero analog voltmeter or an oscilloscope.) Once it is determined that this effect can be produced, the slug is adjusted and left at the position that produces zero volts on the monitor.

An excellent technique for determining if the best overall alignment of both the IF section and detector has been obtained is as follows (two monitors are required):

1. Connect one monitor (preferably with a center-zero scale) as described in the paragraph above (monitor A).
2. Connect a second monitor to measure the output of the IF section (across the ratio-detector capacitor) (monitor B).
3. Rock the generator frequency dial around 10.7 MHz while observing monitors A and B.

Result: Monitor B should indicate a peak reading at the frequency that causes monitor A to read zero volts. Monitor A should change polarity as the frequency is changed from one side to the other of 10.7 MHz. If these conditions are not observed, further adjustments are needed.

The next steps in the alignment procedure are often called the "front-end alignment." They consist of adjusting the frequency converter (primarily the LO) and/or the tuning dial so that the dial "tracks" the frequencies selected by the frequency converter. Front-end alignment also includes adjusting the tuned circuits in the RF section so that they track the frequency converter.

First, a signal near the low-frequency end of the band is injected at the antenna terminals. A frequency of 90 MHz, for example, would be suitable. The output amplitude of the IF section is monitored in the same way as for IF alignment. The receiver is tuned, in the vicinity of 90 MHz on its dial, for a maximum output on the monitor. The tuned circuit(s) in the RF section is(are) adjusted for maximum output. The dial pointer is adjusted (if possible) so that it indicates 90 MHz. Finally, the receiver is set to 106 MHz; the RF generator is set to inject 106 MHz. The local oscillator trimmer capacitor is adjusted to provide maximum output amplitude, as indicated by the monitor.

This completes the description of a procedure for aligning a broadcast FM receiver using the steady-frequency method.

Sweep-Frequency Method of Alignment

The sweep-frequency method is also sometimes referred to as *visual alignment.* This method displays a plot of the pass band of the IF section and a response curve of the detector on the oscilloscope used to monitor the output of these sections. A diagram showing the equipment and interconnections required is presented in Fig. 8.37. The signal source in this case must be a special sweep-frequency generator. That is, it must be capable of sweeping the output across a band of frequencies. In short, it must be frequency modulated. Commercially available generators of this type generally provide for a selection of sweep-center frequencies and sweep bandwidths. Sweep generators typically are designed with facilities for the sweep alignment of both FM and television receivers.

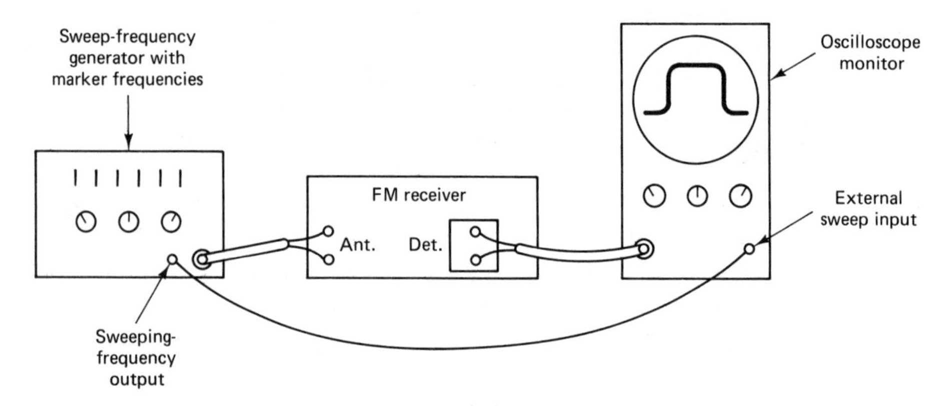

Figure 8.37 Setup for sweep-frequency alignment of FM receiver.

A sweep generator in conjunction with an oscilloscope can produce a plot of output versus frequency of a circuit such as the IF section of a TV or FM receiver. It achieves this result in the following manner. The generator injects a signal that is swept at a constant amplitude across a band of frequencies (set by the user). This frequency-modulated signal passes through the circuit under test and is picked up at its output by a special probe called a DEMOD probe. The DEMOD probe is simply an AM detector circuit—a rectifier with a resistor load and capacitor filter. The output of the DEMOD probe is applied to the vertical input of the monitor oscilloscope. The vertical deflection of the scope, therefore, is at every moment proportional to the envelope of the signal after being processed by the circuit under test. A sample of the signal that sweeps the frequency in the generator is applied as an external horizontal sweep to the oscilloscope. Therefore, the scope trace is swept horizontally in synchronism with the variation in frequency of the test signal. You can think of each point along the horizontal axis of the scope screen as representing a different frequency in the band of frequencies being used to test the circuit. With a proper setup, for example, the left side of the screen might represent 10.5 MHz and the right side, 10.9 MHz. Such a setup could be used to test an FM IF section over a 400-kHz passband. Since the vertical deflection at each point in the trace is proportional to the output for a given frequency in the sweep, the overall display is a plot of output versus frequency.

Sweep generators come equipped with several crystal-controlled, single-frequency oscillators. These are the so-called *marker generators*. When switched on, a marker signal is added to the signal from the DEMOD probe. The marker signal is seen as a disturbance on the scope display. Its horizontal position on the display corresponds to its frequency relative to the band of frequencies included in the sweep. The marker signal therefore "marks" the display with a known frequency reference point. Several marks displayed simultaneously can be used to define a desired bandwidth, or critical frequency points within a band.

When we "sweep" an IF section and display the results on an oscilloscope, we can see immediately the state of the alignment of the section. The display of a section with perfect alignment would be as shown in Fig. 8.38(a): The passband is 200 kHz in width, the amplification is nearly constant across the band, and the "skirts" of the band are steep, indicating good attenuation of signals immediately outside the band. An example of the display for a poorly aligned section is shown in Fig. 8.38(b).

In brief, the alignment of an IF section with such a display consists of adjusting the slugs of the IF transformers until the display is as nearly like that

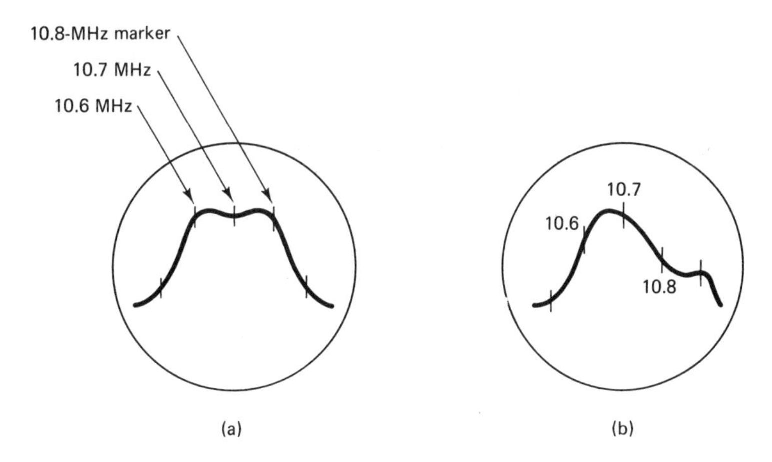

Figure 8.38 Visual alignment displays: (a) display of output of properly aligned FM IF section; (b) display of poorly aligned FM IF section.

of Fig. 8.38(a) as possible. The advantage of the visual or sweep-frequency method is that we can see the effect of an adjustment on the shape of the frequency response curve while we are making the adjustment. Once we have the test instruments connected and operating properly, we proceed to make adjustments following basically the same steps as for the steady-frequency method. We, of course, watch the effects of the adjustments on the display and let the results guide our moves. Becoming proficient at making the right moves requires patience and practice.

The alignment of an FM detector (discriminator or ratio detector) also lends itself to the use of the visual-display method. Examples of correct and unsatisfactory alignments are shown in Fig. 8.39. You can easily guess why the frequency response of an FM detector is commonly called an "S curve." You should be aware of the following points about the detector response: (1) The curve should have a linear portion corresponding to the bandpass of the IF section—generally, 10.6 to 10.8 MHz. (2) The response should cross the zero axis at the center of the bandpass, the intermediate frequency—10.7 MHz. The adjustment of the slug in the detector transformer (the secondary winding of the output IF transformer) serves to center the detector response to the passband, that is, crossover at 10.7 MHz. If a detector response is nonlinear, or the passband inadequate, the cause is usually in the IF section; further alignment of it is needed.

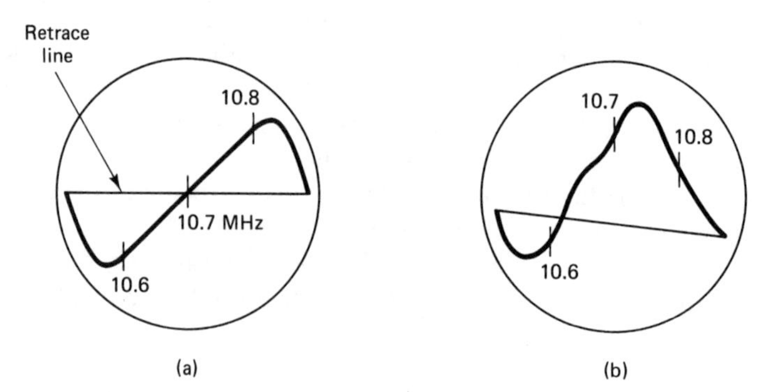

Figure 8.39 Visual displays of FM detector S curves: (a) display of good FM detector alignment; (b) poor detector alignment.

Aligning the Stereo Multiplex Decoder Section

There are many different designs for the decoding of stereo multiplex signals. Consequently, there is not just one, easily stated technique for aligning such sections. Hopefully, if you are confronted with the task of aligning a stereo section, you will have access to the manufacturer's instructions for that procedure. If you do have the manufacturer's instructions, study them carefully and be guided by them.

In general, stereo decoders have adjustments for the following: (1) 19-kHz oscillator operating frequency, (2) maximum output of 38-kHz frequency doubler, (3) threshold level for stereo indicator light, and (4) degree of separation of left and right channels. Signal generators that provide for the stereo modulation of

FM (sweep) signals are available and make the task of aligning a stereo decoder easier and more precise.

8.6 MISCELLANEOUS RECEIVER TOPICS

Combined AM/FM Receivers

Nearly all broadcast receivers and/or tuner components are now manufactured with both an AM and an FM receiver function. When a receiver has only a single function, that function is more likely to be the AM one. It is not too difficult to implement both the AM and FM functions in a single unit. Obviously, the audio section can be the same for both. There must be an RF and frequency converter section for each function since the frequencies involved are vastly different. However, the IF stages are often the same for both functions. A common technique is to use two parallel-resonant circuits in series in the output side of the stages. When operating at 455 kHz, the tank circuit tuned to 10.7 MHz is transparent to the stage because of the relatively low inductive reactance of that tank circuit. When operating at 10.7 MHz, the 455-kHz tank circuit is transparent because of its relatively low capacitive reactance.

The Squelch or Muting Function

The FM system is noted for its comparatively noise-free operation. Noise-free operation refers to the situation when a receiver is processing a carrier whose amplitude at the receiver's antenna terminals is adequate. However, when an FM receiver's tuning is being changed to a different station, the receiver is typically more noisy than an AM receiver. This is true unless some provision has been incorporated for squelching or muting (reducing the amplitude of) the output between stations (i.e., when a carrier is not being processed).

The output noise level of an FM receiver is excessive when it is between stations for two basic reasons: (1) When a carrier is not being processed, the AGC system calls for maximum gain from all affected amplifier sections. All noise signals are amplified significantly more than when a station is being received. (2) The tuned circuits have much greater bandwidth than in AM receivers. Since noise frequencies are random and spread across the entire frequency spectrum, the wider the bandwidth of a circuit, the greater the possibility of the introduction of noise signals.

Interstation noise is perceived by the listener as a hissing sound. This sound is the result of numerous noise frequencies entering the amplifier circuits of the receiver and modulating each other. When these products of intermodulation are in the audio-frequency range, they are demodulated by the detector and amplified by the audio section, finally activating the speaker. The intermodulation process produces both amplitude modulation and the phase modulation which always accompanies amplitude changes. The limiter stages are of little help in eliminating the AM effects because the amplitudes involved are generally too low for limiting to occur.

Circuits that reduce or eliminate the noise present between stations are called *squelch or muting circuits*. "Muting circuits" is usually used in conjunction with broadcast receivers, "squelch circuits" with communications-type equipment (two-way radio). Basically, the muting function consists of interrupting the processing of audio signals by the audio amplifier section. This may be accomplished by shorting the signal or by turning an amplifier stage partially or completely off. The function, then, requires both a means of determining "when to mute" as well as the actual muting itself.

The general idea of muting is represented in block-diagram form in Fig. 8.40(a). A typical approach to muting is to sample the 10.7-MHz IF signal at

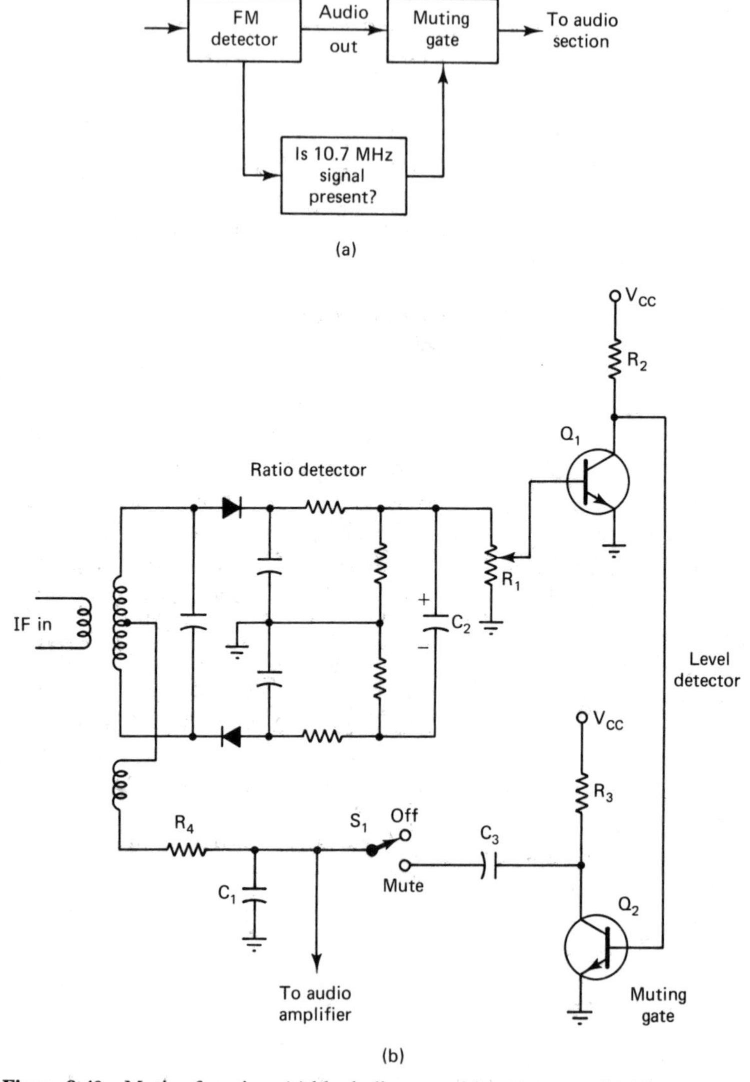

Figure 8.40 Muting function: (a) block diagram of functions required for muting; (b) muting circuit using dc voltage across ratio detector to sense presence of 10.7-MHz IF signal.

the detector circuit. If the 10.7-MHz signal is present, it is converted by rectification to a dc voltage. This dc voltage, in turn, is used to bias a squelch or muting "gate" open so that the audio signal can pass on to the audio section and speaker. If a 10.7-MHz signal is not present, there is inadequate dc voltage to open the muting gate—all audio signals, including noise, are blocked.

A simple circuit demonstrating a possible approach to muting is shown in Fig. 8.40(b). You will observe that transistor Q_2 is connected across the audio output of the ratio detector. The receiver will be muted any time Q_2 is turned on since, when conducting, it shorts the audio signal to ground. Q_2 will be turned on when Q_1 is off. Q_1 is connected to detect the presence of a received carrier of more than minimum amplitude: its base circuit is connected across the ratio detector storage capacitor, C_2. The voltage to ground at that point is typically 1 V when a carrier of adequate amplitude is being processed. Adjustable resistor R_1 is set to provide saturation bias for Q_1 for this condition, or it can be adjusted for any other desired carrier level. Muting is turned on or off by switch S_1.

If you analyze the circuit of Fig. 8.40(b) more than just superficially, you will recognize that its operation may be less than satisfactory under certain conditions. For example, there can be partial muting, even when a carrier is being received, if that carrier is of below normal amplitude. In such an instance Q_1 is not driven to saturation and Q_2 remains partially conducting. If not completely cut off, Q_2 is partially shorting the audio signal. You will find in your work with commercial receivers that many different schemes have been devised to provide various degrees of sophistication of the muting or squelch function.

GLOSSARY OF TERMS

Amplitude limiter A circuit function in an FM receiver that eliminates most of the variations in the amplitude of the IF signal by clipping.

Bessel function In mathematics, the solution of a particular type of differential equation. The function can be used to determine the amplitude of side frequencies in FM.

Capture effect An effect that occurs when two FM carriers of the same frequency and nearly the same amplitude are present in the tuning section of an FM receiver. The stronger carrier appears to take control of the receiver (it "captures" the receiver) and suppresses the second, weaker signal.

Capture ratio The smallest ratio of signal amplitudes at which the capture effect will occur. *See also* Capture effect.

Center frequency The frequency at the center of the frequency swings of a modulated FM carrier; the average frequency of an FM carrier; the carrier frequency of an FM signal.

Channel, FM broadcast A 200-kHz-wide segment of the electromagnetic frequency spectrum assigned for use by an FM station broadcasting program material of interest to the general public.

Deemphasis network A low-pass filter network between the output of an FM detector and an FM receiver's audio section. Its purpose is to correct the distortion of enhanced gain for high frequencies introduced by the preemphasis network in the FM transmitter. The network has an *RC* time constant of 75 μs.

Deviation, frequency In a frequency modulation system, the amount of frequency change of a carrier signal as a result of modulation.

Deviation control A variable-resistance device used to control the amplitude of the modulating signal and, thereby, the frequency or phase deviation of the carrier of an FM transmitter.

Deviation rate In an FM system, the rate at which the frequency of the carrier is changed. Equal to the frequency of the modulating signal.

Double-tuned transformer A type of coupling transformer sometimes used in FM and television IF sections to provide greater bandwidth. It is characterized by having two adjustable controls, usually core slugs—one for adjusting the primary circuit for resonance, and a second for adjusting the secondary circuit for resonance.

Equalization network A circuit found in the audio sections of broadcast receivers for correcting the predistortion introduced during the making of a phonograph record.

Frequency-division multiplexing (FDM) A signal transmission scheme in which two or more signals are transmitted simultaneously over a common transmission medium by using different frequencies for the signals.

Image frequency In signal reception by receivers utilizing the superheterodyne system, the frequency that produces the same intermediate frequency as the desired carrier. In systems in which the local oscillator operates at a frequency higher than the received signal, the image frequency is equal to the desired frequency plus two times the intermediate frequency.

Indirect FM Frequency modulation produced by modulating the phase of a carrier signal downstream from the oscillator, rather than modulating the operating frequency of the oscillator, as in direct FM.

Intermodulation distortion (IM) The distortion of a desired signal by an effect similar to amplitude modulation which occurs when a desired signal and a spurious (or noise) signal are present together in a circuit operating in a nonlinear mode.

Limiter In an FM receiver, a circuit that minimizes the effect of any amplitude modulation of a carrier. Limiting can be achieved by the effect called chopping manifested when an amplifier is overdriven.

Marker frequency The frequency of a signal used to identify a known frequency position on the oscilloscope display produced by a sweep-frequency generator test. *See also* Marker generator.

Marker generator An oscillator-amplifier unit, usually crystal controlled, used to produce a single-frequency signal for identifying a frequency position on the oscilloscope display provided by a sweep-frequency signal generator.

Modulation index In FM systems, the ratio of frequency deviation to the frequency of the modulating signal.

Multiplexing The process of transmitting two or more signals simultaneously over a common transmission medium.

Muting circuit In receivers, a circuit that prevents any speaker output when a carrier is not being received, for example, when the receiver is being tuned between stations.

Phase-locked loop (PLL) A closed-loop system (may be on an integrated-circuit chip) containing a voltage-controlled oscillator (VCO), a phase detector, and a low-pass filter, which operates to provide for a stable operating frequency of the oscillator.

Phase modulation Changing the phase of a carrier signal so as to produce the equivalent of frequency modulation, but without changing the operating frequency of an oscillator.

Pilot carrier A carrier transmitted unmodulated for the purpose of providing a signal required for the demodulation of another transmitted signal.

Preemphasis network In FM systems, a high-pass filter in the audio section of an FM transmitter which boosts the gain of higher audio frequencies so as to enhance the noise-reducing characteristics of the FM system. *See also* Deemphasis network.

Reactance modulator An amplifier circuit that presents itself to an external circuit as a reactance that can be varied by varying the bias level of the amplifier, as, for example, with an audio signal.

SCA broadcasting Broadcasting of a second, subsidiary program by an FM station, authorized by the FCC with a license called a Subsidiary Communications Authorization. The broadcast is simultaneous with the main program and is accomplished through frequency-division multiplexing. It is sometimes referred to as "storecasting," as the programs are typically of background-type music and used in stores and offices.

Separation In stereo-multiplex broadcasting, a term referring to the degree to which the system (pickup, transmitter, receiver, and speakers) is able to reproduce the unique character of sounds coming from the left and right sides of an imaginary listener at the site where a program was originally created.

Stereo decoder A circuit that recovers the left-side and right-side signals of a stereo-multiplex broadcast, enabling an appropriate receiver-speaker system to reproduce the two unique left- and right-side sounds.

Sweep-frequency alignment A method of alignment in which the test signal frequency is swept through the band of frequencies of interest enabling the output to be displayed on an oscilloscope as a visual representation of the bandpass response curve.

Time-division multiplexing (TDM) A method of transmitting, apparently simultaneously, two or more signals over a common communications medium by assigning a portion of each time interval, such as a microsecond, to the transmission of each of the signals to be transmitted.

Varactor (diode) A semiconductor diode that performs as a voltage-variable capacitor by virtue of changes in the width of the so-called "charge depletion region" of the PN junction with changes in junction reverse bias voltage.

REVIEW QUESTIONS: BEST ANSWER

1. In true or direct FM the frequency of the carrier is changed at the: **a.** frequency multiplier. **b.** power amplifier. **c.** buffer amplifier. **d.** oscillator. **e.** none of these.
2. When a carrier is angle modulated at some point after being generated, the result is called: **a.** phase modulation. **b.** frequency modulation. **c.** amplitude modulation. **d.** after modulation. **e.** none of these.
3. In angle modulation, the amount of actual, or apparent, change in the frequency of the carrier is called: **a.** the modulation index. **b.** frequency deviation. **c.** percent modulation. **d.** rate of modulation. **e.** none of these.
4. The rate of deviation, that is, how many times per second a carrier is swung through its pattern of frequency change, in FM represents the: **a.** amplitude of the information signal. **b.** frequency of the information signal. **c.** phase angle of the information signal. **d.** phase angle of the carrier. **e.** none of these.
5. The center or carrier frequency of an FM emission is also its: **a.** minimum fre-

quency. **b.** average frequency. **c.** maximum frequency. **d.** frequency rate. **e.** none of these.

6. During a test, a broadcast FM carrier swings between 101.96 and 101.84 MHz. The carrier frequency is: **a.** 101 MHz. **b.** 101.92 MHz. **c.** 101.88 MHz. **d.** 101.90 MHz. **e.** none of these.
7. The frequency deviation in Question 6 is: **a.** ±60 kHz. **b.** ±96 kHz. **c.** ±40 kHz. **d.** ±101 kHz. **e.** none of these.
8. The carrier in Question 6 is being modulated at a rate of: **a.** 0%. **b.** 50%. **c.** 60%. **d.** 80%. **e.** none of these.
9. In broadcast FM, 100% modulation is defined as a frequency deviation of: **a.** ±110 kHz. **b.** ±100 kHz. **c.** ±75 kHz. **d.** ±25 kHz. **e.** none of these.
10. If the modulating test signal in Question 6 is a 12-kHz audio signal, the rate of deviation is: **a.** 75 kHz. **b.** 60 kHz. **c.** 101.9 MHz. **d.** 12 kHz. **e.** none of these.
11. The modulation index for the situation referred to in Questions 6 and 10 is: **a.** 5. **b.** 12. **c.** 60. **d.** 0.2. **e.** none of these.
12. A 5-kHz audio signal produces 65% modulation of a broadcast FM carrier. The modulation index is: **a.** 5. **b.** 6.25. **c.** 7.85. **d.** 9.75. **e.** none of these.
13. A modulation level of 60% is achieved by a 10-kHz audio signal with an amplitude of 3.5 V. What frequency deviation does a 2.75-V, 7.5-kHz signal produce in the same broadcast FM transmitter? **a.** 7.5 kHz. **b.** 58.93 kHz. **c.** 35.36 kHz. **d.** 45 kHz. **e.** none of these.
14. The effect in which two radio frequencies interact in a receiver circuit, while it is operating in a nonlinear mode, to produce an undesired audible output is called: **a.** mixing. **b.** intermodulation distortion. **c.** nonlinear distortion. **d.** harmonic distortion. **e.** none of these.
15. Although intermodulation distortion (or IM) produces amplitude modulation of the FM carrier, its effect can be heard from an FM receiver because AM also produces: **a.** direct modulation. **b.** SSB. **c.** VCO. **d.** phase modulation. **e.** none of these.
16. Phase modulation produces an effect that is equivalent to the frequency deviation of FM except that the equivalent deviation is: **a.** of opposite polarity. **b.** of the wrong phase. **c.** directly proportional to the modulating frequency. **d.** inversely proportional to amplitude of the modulating signal. **e.** none of these.
17. In a broadcast FM system, a disturbance that produces an AM signal-to-noise ratio of 3.5:1 produces a worst-case FM signal-to-noise ratio of approximately: **a.** 5:1. **b.** 17:1. **c.** 25:1. **d.** 38:1. **e.** none of these.
18. An FM system has a "system modulation index" of 2.5. What is the worst-case S/N if a disturbance produces a maximum phase deviation of 0.225 rad? **a.** 21 dB. **b.** 25 dB. **c.** 33 dB. **d.** 42.5 dB. **e.** none of these.
19. To be standard, the deemphasis circuit in an FM receiver should have an RC time constant of: **a.** 25 ms. **b.** 25 μs. **c.** 75 ms. **d.** 75 μs. **e.** none of these.
20. Direct crystal control of the master oscillator can be used in a/an: **a.** direct FM system. **b.** Crosby system. **c.** indirect FM system. **d.** true FM system. **e.** none of these.
21. A varactor diode functions as a voltage-variable capacitor because varying the reverse junction voltage varies the: **a.** number of minority carriers. **b.** number of holes. **c.** forward current. **d.** depletion-layer width. **e.** none of these.

22. A frequency-modulated oscillator, as used in a Crosby-type direct FM transmitter, is also a VCO (voltage-controlled oscillator) and therefore lends itself to: **a.** AGC. **b.** *S/N*. **c.** AFC. **d.** ANL. **e.** none of these.
23. An automatic frequency control circuit functions as a/an: **a.** phase-locked loop. **b.** amplitude limiter. **c.** AM detector. **d.** noise limiter. **e.** none of these.
24. A circuit that converts a frequency or phase change into a dc voltage is called a/an: **a.** AM converter. **b.** dc phasor. **c.** phase detector. **d.** PLL. **e.** none of these.
25. In addition to a VCO and a phase detector, a PLL includes a/an: **a.** dc phasor. **b.** low-pass filter. **c.** AM phasor. **d.** high-pass filter. **e.** none of these.
26. A deviation or percent modulation control in an FM transmitter adjusts the: **a.** amplitude of the carrier. **b.** frequency of the carrier. **c.** amplitude of the modulating signal. **d.** frequency of the modulating signal. **e.** none of these.
27. When a particular station (or stations) is noisy when received by an FM receiver, it is likely that: **a.** the received carrier amplitude is not sufficient to produce limiting. **b.** there is some kind of interference between the station's antenna and the receiver's antenna. **c.** the station is not filtering its emissions properly. **d.** the receiver is not facing in the proper direction for the particular station. **e.** none of these.
28. Self-limiting is a characteristic of a/an: **a.** synchronous detector. **b.** ratio detector. **c.** F-S discriminator. **d.** quadrature detector. **e.** none of these.
29. Negative ac feedback in an amplifier is effective in: **a.** reducing distortion caused by nonlinear operation. **b.** increasing gain. **c.** minimizing drift of the bias point. **d.** eliminating AM noise. **e.** none of these.
30. If the sound from the left speaker of a stereo system were virtually the same as that from the right speaker, regardless of the program being listened to, the system would be said to suffer from lack of: **a.** discrimination. **b.** multiplexing. **c.** compatibility. **d.** separation. **e.** none of these.

REVIEW QUESTIONS: ESSAY

1. Define *percent modulation* as it is used in frequency modulation. In AM, percent modulation is actually an indication of the relative amplitude of a modulating signal. Compare this relationship with the situation in FM. That is, what characteristic, if any, of the modulating signal in FM is important in determining percent modulation?
2. Define *modulation index* and describe what importance it plays in predicting the characteristics of an FM wave.
3. In AM, the bandwidth required by a modulated carrier is simply equal to twice the highest frequency in the modulating (information) signal. Discuss the factors and relationships that determine the bandwidth of an FM carrier.
4. Why are the higher-frequency products of IM in FM systems of greater amplitude (i.e., louder) than the lower-frequency components of IM noise?
5. Discuss what is meant by *capture effect* and the *capture ratio* of an FM receiver.
6. Describe the most fundamental difference between FM transmitters that produce true or direct FM and those that produce indirect FM.
7. Describe an inherent difficulty in the process used to produce direct FM. Can the difficulty be overcome? How? Describe.

8. Describe briefly the mechanism by which a varactor diode exhibits the characteristics of a variable reactance.
9. Sketch the block diagram of a phase-locked loop as it might be used in an FM transmitter.
10. It is very common for FM transmitters to include frequency multiplier circuits. Why? Give at least two reasons.
11. Discuss the reasons why the power amplifier stages of an FM transmitter are less complicated than those of AM transmitters.
12. For a phase-modulated transmitter to provide the equivalent of true FM, a correcting circuit must be included in its audio section to modify the modulating signal before it reaches the phase modulator. What is the correcting circuit? How does it change the audio? Why is this change necessary?
13. Both frequency multiplication and frequency conversion are used in FM transmitters to obtain the desired frequency deviation at a specific carrier frequency. Compare these two functions, indicating how they differ with respect to their effects on deviation.
14. What is an image frequency? Describe the significant mathematical relationships of the image frequency with other frequencies in a superheterodyne receiver.
15. What is amplitude limiting? Why is it used? How is it accomplished?
16. What is AFC, and why is it used in an FM receiver?
17. Describe the peculiar effect observed when one tries to tune an FM receiver to another station and the AFC is in the "lock" state. How can this situation be avoided?
18. What would probably be the observable effect if the lock range of an AFC circuit were unsuitably matched (too limited, for example) to the operating behavior of the local oscillator of a particular receiver?
19. Describe the significance of the capture range of the AFC function of a particular receiver as it might be observed by a person trying to tune the receiver.
20. Unless a circuit has built-in protective measures, what is likely to be the result of shorting one of the output transistors in a complementary-symmetry audio output stage? Explain the reasoning involved in your answer.
21. What is multiplexing as applied to electronic communications systems? What is the difference between frequency-division multiplexing (FDM) and time-division multiplexing (TDM)? Give an example of each.
22. To achieve the goal of enhanced three-dimensional effect, what technical quality must characterize both the electronic signals and the physical arrangement of a stereo sound reproduction system? Describe.
23. In FM stereo broadcasting, why are there $L + R$ and $L - R$ signals instead of simply L and R signals?
24. What is the reason that a 19-kHz pilot carrier is included in the multiplex signals that make up a stereo broadcast?
25. What is the meaning of the abbreviation "SCA"? Describe briefly the service designated with this abbreviation.
26. Why is sweep-frequency alignment more popular than the steady-frequency method for aligning FM receivers?
27. What is meant by *squelch* or *muting?* Describe, in a general way, how this function can be accomplished.

EXERCISES

1. In a certain FM transmitter an audio (modulating) signal of 3.5 V produces a deviation of ± 27.5 kHz. **(a)** What deviation does a signal of 4.5 V produce? **(b)** What audio signal voltage would be required to produce 100% modulation? Assume that the transmitter is in the broadcast service.
2. On sound peaks a certain FM broadcast program achieves deviation levels of ± 72 kHz. What is the percent modulation of the peaks?
3. An FM broadcast transmitter is being tested with a 12-kHz modulating signal. Determine percent modulation and the modulation index for deviations of: **(a)** ± 36 kHz; **(b)** ± 60 kHz.
4. Estimate the number of side-frequency pairs and bandwidths of the emissions of the tests described in Exercise 3.
5. An FM broadcast transmitter is being deviated to 105.1625 MHz and 105.0375 MHz by a 12.5-kHz test signal. Determine: **(a)** the carrier frequency of the transmitter; **(b)** Δf; **(c)** the modulation index; **(d)** the number of significant side-frequency pairs; **(e)** percent modulation; **(f)** the bandwidth of the modulated carrier.
6. A VHF two-way radio transceiver operates in the business radio-frequency band 148.0 to 174.0 MHz. Maximum permissible deviation is ± 5 kHz. The transmitter is being deviated between 154.5652 and 154.5748 MHz by a 1600-Hz test signal. Determine: **(a)** f_C; **(b)** Δf; **(c)** percent modulation; **(d)** modulation index of the test; **(e)** number of significant side-frequency pairs produced; **(f)** the approximate bandwidth of the carrier.
7. A spurious signal that has a frequency component of 10.71 MHz enters the IF section (f = 10.7 MHz) of an FM receiver. The amplitude of the spurious signal is half that of the IF signal. If the spurious signal causes IM (intermodulation distortion), determine: **(a)** the frequency of the audible noise produced; (the beat frequency); **(b)** the phase deviation of the resultant carrier; **(c)** the equivalent frequency deviation of the interference.
8. Calculate the worst-case signal-to-noise ratios in a broadcast FM system for the following S/N amplitude ratios: **(a)** 2:1; **(b)** 5:1; **(c)** 10:1.
9. Draw the diagram of a simple circuit that could be used as a preemphasis network in an FM transmitter. If a value of 56 kΩ is used for the resistor, what value should be used for the capacitor?
10. An FM transmitter is to have a carrier frequency of 102.7 MHz. **(a)** If the master oscillator will be designed to have an operating frequency in the range 5 to 10 MHz, determine the amount of frequency multiplication that would represent a reasonable design. Remember: Total multiplication is the product of multiple factors of 2 or 3, generally. **(b)** What maximum deviation of the master oscillator should be provided for, given the total multiplication you have chosen in part (a)? Assume that the transmitter is used in FM broadcast service.
11. A broadcast FM transmitter with a carrier frequency of 107.3 MHz (FCC FM channel 297) is to be designed. A combination of frequency conversion and frequency multiplication is to be used to achieve the final frequency while using an oscillator operating below 1 MHz. Phase modulation will be used but is limited to a deviation of about ± 30 Hz. Design a frequency-conversion, frequency-multiplication scheme

that will accomplish the desired result—a 107.3-MHz carrier with a maximum deviation of 75 kHz. Sketch a block diagram of your design and label with all relevant frequencies.

12. Calculate the image-signal frequencies for the following broadcast FM channels: **(a)** 89.9 MHz; **(b)** 100.1 MHz; **(c)** 107.9 MHz. Assume that the LO of a receiver operates at a frequency below the received-signal frequency, that is, that $f_{IF} = f_C - f_{LO}$.

9

TWO-WAY RADIO SYSTEMS

9.1 INTRODUCTION

In terms of sheer numbers, the vast majority of radio transmitters in the world are used in two-way radio systems providing "private" communications for their users. In the United States, for example, there are several thousand radio and television transmitters for the broadcast services (transmitted programs for reception by the general public). By contrast, the number of taxicabs, police and fire vehicles, electric utility vehicles, private vehicles, airplanes, ships and boats, and so on, with installed two-way radio equipment is in the millions. With only a few exceptions, two-way radio systems are used for communication with or between moving vehicles or vessels: automobiles, trucks, airplanes, boats, ships, or spacecraft. Most such applications dictate that the equipment be small, compact, and lightweight. Their antenna systems are simple in comparison to fixed-station installations. Power ratings of transmitters in these applications are low compared to those of transmitters in the broadcast services. It is common for dc power to be supplied by a storage battery.

The most popular form of equipment for two-way radio, by far, is the *transceiver*. A transceiver is a compact equipment package containing both a transmitter and a receiver. Transceivers are available with amplitude modulation (AM), single-sideband amplitude modulation (SSB), and frequency modulation (FM). The form of modulation used in a particular transceiver is determined to a large extent by the service for which it is intended. FCC regulations and laws regulating communications specify forms of modulation, frequency bands, power levels, etc., for each communications service. Services that utilize two-way radio include citizens' band radio, the Private Land Mobile Radio Services, aircraft radio, marine radio, and amateur radio. See Appendix B for information on FCC rules and licensing requirements.

It is our goal in this chapter to gain a knowledge and understanding of the electronics of radiotelephone equipment (transceivers, principally). Circuits used in such equipment generally have critical operating constraints, more critical than those of broadcast AM radio receivers, certainly. We will need to look again and in greater detail at some of the circuits studied previously. Because of unique operating conditions, such as multiple transmitter frequencies, equipment in the radiotelephone services incorporate several circuits, functions, and techniques not found in broadcast equipment. These topics, with titles such as *frequency synthesis* and *single-sideband transmission,* warrant our attention.

9.2 TWO-WAY RADIO: BASIC CONCEPTS

For purposes of studying technical details of construction and operation, the stations used in two-way radio communications are conveniently classified into two types—permanent and mobile. Although the titles of these categories are seemingly self-explanatory, let us look briefly at their meanings. A permanent station is one in which the transmitter/receiver equipment is typically installed in a building of some sort (although it may be a "shack"). The antenna(s) used by the station is(are) installed on land. This last fact has important implications: Land installation provides the potential for more complex antenna systems than mobile stations. The influence of nearby mountains, buildings, water surfaces, etc., on the propagation characteristics of such antenna systems is relatively unchanging in comparison to the antenna systems of mobile stations.

A mobile station, by contrast, is always mounted on a vehicle that has the ability to move about: automobile, truck, boat, ship, airplane, spacecraft, etc. In most cases, the space available for antenna installation is extremely limited; the types of antennas that can be used are limited. The conditions for emission and wave propagation are often changing rapidly as a result of a vehicle's changing position with respect to large buildings or mountain peaks. The physical size of a mobile transmitter is generally severely constrained by available space. The maximum, feasible transmitter power rating is limited thereby.

For purposes of government regulation, the classification of two-way radio stations is somewhat more specific than is indicated in the preceding paragraphs. First, the classifications are *mobile, base,* and *fixed.* A *mobile station* is one which, as indicated above, is associated with a vehicle free to move about. A *base station* is one that is installed in a permanent location and is used principally for communications with one or more mobile stations, and occasionally with other base or fixed stations. A base station is also referred to as a *land station.* A *fixed station* is one installed in a permanent location which is used to communicate with other fixed stations. A system of two or more fixed stations provides what is called *point-to-point communication.* Communication with mobile station(s) is usually not an available option in this form of communication. Point-to-point communication is sometimes used in lieu of land-line telephone communication in remote or developing areas.

Another important method of classifying two-way radio systems is concerned with the method of operation. The basic classifications of this type are illustrated in Fig. 9.1. The most complex of the methods, full duplex, is the easiest to explain. You are using a full-duplex communications system when you are talking to a friend: either of you can speak at any time. You can both speak at the same time (you may not be able to understand each other in that situation, but a third party could hear you both). Your home telephone system is a full-duplex system. A full-duplex two-way radio system requires that each transmitter operate at a different frequency. The frequencies must be sufficiently separated to prevent interference with each other when more than one transmitter is on. Two or more transmitters may be on at the same time. Full-duplex operation is common in point-to-point systems only.

A simplex communications system is one in which all stations involved in a communication use the same frequency. Therefore, only one transmitter can be on at a time if anything useful is to occur. The parties involved generally have an understanding about how the point of transmission is to be transferred from one station to another. Simplex operation is very common in base-to-mobile and/or mobile-to-mobile communications.

Duplex operation (as distinguished from full duplex) is associated with the use of different carrier frequencies between the components of the system. Duplex is different from full duplex, however, in that simultaneous, bidirectional transmission is not permitted or provided for. Usually, full-duplex operation is not possible for one or more technical reasons. Typically, in duplex two-way radio communication, the two frequencies used are sufficiently close together that the same antenna and most of the transmitter/receiver circuitry is the same for both. Duplex operation is common in two-way base/mobile systems: The base station uses one frequency (F_1) for transmission and the mobile stations use a second frequency (F_2). In this arrangement the mobile units can communicate directly only with the base station.

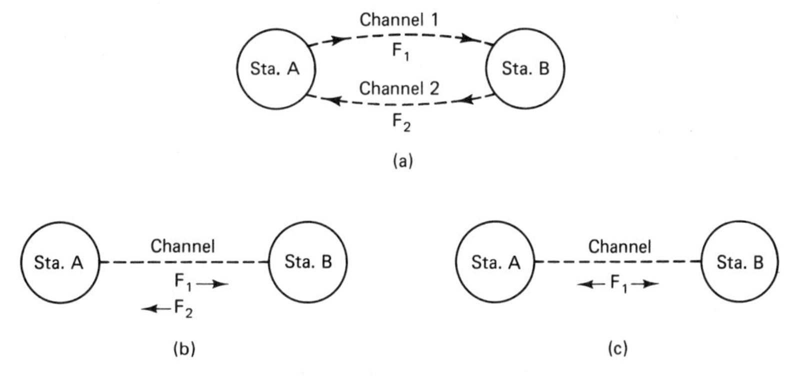

Figure 9.1 Modes of two-way communication: (a) full duplex (or duplex); (b) duplex (or half-duplex); (c) simplex.

Two-way radio systems may also be classified according to the type of modulation used: amplitude modulation (AM), single-sideband (SSB, a special modification of AM), or frequency modulation (FM). All three of these transmission methods are used in radiotelephone systems. Many of the details of the generation and reception of AM and FM transmissions are covered elsewhere in this text. Additional details are provided in this chapter together with a presentation on single-sideband operation.

Finally, two-way radio systems are classified by the FCC according to the kinds of activities which they serve. The four major areas of activities served by radiotelephone systems may be identified as private aviation, small-boat marine, private land vehicle, and all other citizens. Each of these services is allocated various segments of the radio-frequency spectrum. Regulations have been written and are updated from time to time describing requirements for equipment and operating and testing procedures for each "service." Each service is identified with a unique title, for example, Private Land Mobile Radio Services and Citizens' Band Radio Service.

It is not a goal of this chapter (or text as a whole) to present a detailed description of the legal and technical aspects of each of the major radiotelephone services. It is a goal of this text to present basic technical concepts and information about the electronics used in one or more of these services. Fortunately, there are many electronic functions, circuits, and devices which are common, with only minor variations, to the equipment of all the services. Reference to specific radio services is made on occasion.

9.3 AN FM RADIOTELEPHONE TRANSCEIVER

A complete two-way radio communications system requires at least two transmitters and two receivers: Each party to the communications process must have the ability to "send" (transmit) information as well as receive information. Although there are a few exceptions, the transmitter and receiver for one end of a radiotelephone system are packaged in a single unit called a *transceiver*. The transceivers of the various services all serve the same basic functions—to transmit and receive radio waves. However, the details of their design and construction vary significantly.

Let us begin an examination of transceivers by looking at the ULTRACOM 508 VHF-FM transceiver, manufactured by the E.F. Johnson Company. As indicated by its title, the ULTRACOM 508 is an FM unit. It is designed to operate in the frequency range 148 to 174 MHz and is therefore classified as a "VHF" device. We have chosen to look at this particular unit as our first transceiver because it offers a comparatively simple, straightforward design. At the same time, it contains many basic features, the understanding of which will enable us to proceed more efficiently with the study of more complex designs.

A block diagram of the ULTRACOM 508, with minimum detail for a first look, is shown in Fig. 9.2. A glance at the receiver portion of the diagram tells us that it utilizes the superheterodyne principle. In fact, the 508 receiver is called a "double conversion" unit because it contains two heterodyne converter sections and two IF sections. We note that the receiver incorporates an RF amplifier and

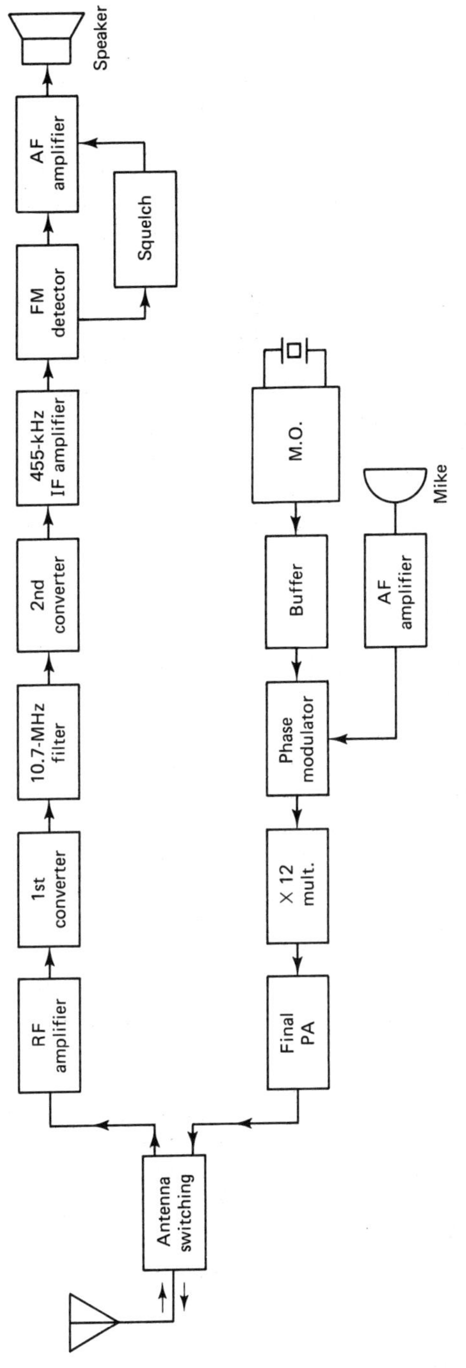

Figure 9.2 Simplified block diagram of ULTRACOM 508 VHF-FM transceiver.

a SQUELCH section. Otherwise, there are no big surprises in this receiver in terms of functions not already familiar to us. Similarly, the transmitter block diagram does not include any unfamiliar functions. It utilizes phase modulation, which permits the use of a crystal master oscillator. Frequency multiplication is used to achieve the desired carrier frequency and frequency deviation.

A block diagram with greater detail of the receive function is shown in Fig. 9.3. One of the items that we may now perceive as new, or at least different, is that the oscillators for the frequency converter sections are crystal oscillators. Crystal oscillators in these positions improve the stability and precision of the receive function.

A second novel element revealed by a diagram of greater detail is the use of a crystal filter in the first IF section. As used here, the crystal filter is a bandpass filter constructed of piezoelectric elements very much like those used in crystal-controlled oscillators. A crystal filter (a more detailed discussion of crystal filters is presented in another section of this chapter) can provide a precisely defined passband. High attentuation of signals just outside the passband is possible. The filter here consists of a set of three separately packaged sections. The packages are provided with leads suitable for soldering to a printed-circuit board.

Receiver Circuitry

Let's start at the antenna and follow the signal all the way through the receiver. Refer to the schematic diagram of the transceiver, Fig. 9.4, as we proceed.

RF energy picked up by the antenna is coupled to the input of the RF amplifier through two tuned circuits: T_1–C_1 and T_2–C_5. These circuits are designed and aligned to provide what is called *broadband* operation: They are designed to function properly across the frequency range 148 to 174 MHz. An individual purchaser may buy crystals with which to operate on only a very narrow segment of the broad range. However, the manufacturer wants the basic design to be suitable for numerous frequency choices. Hence the unit is designed with a broadband front end. The pair of tuned circuits are slightly overcoupled. Overcoupling gives a two-hump response curve and widens the passband. Broadbanding is also aided by tuning the two circuits to somewhat different resonant frequencies. Transformers T_1 and T_2 are tapped to provide impedance matching between the antenna and Q_1.

The RF amplifier, Q_1, employs an MPF 121 dual-insulated-gate MOS field-effect transistor (a MOSFET). It is notable for its low noise and high gain characteristics. The output of this stage also incorporates double-tuned circuits: L_1–C_8 and T_3–C_{15}. These provide broadband coupling to the first mixer (MIXER, Q_2), an MPF 121 MOSFET.

As you will observe on the schematic diagram, the circuit for producing the local oscillator signal input to the mixer has significantly more components than those we have analyzed previously. Included in the LO string are a crystal oscillator, Q_4–Y_1, a buffer amplifier, Q_5, and a frequency tripler, Q_6. The oscillator (RCVR OSC) operates in the frequency range 45 to 54 MHz. The exact frequency

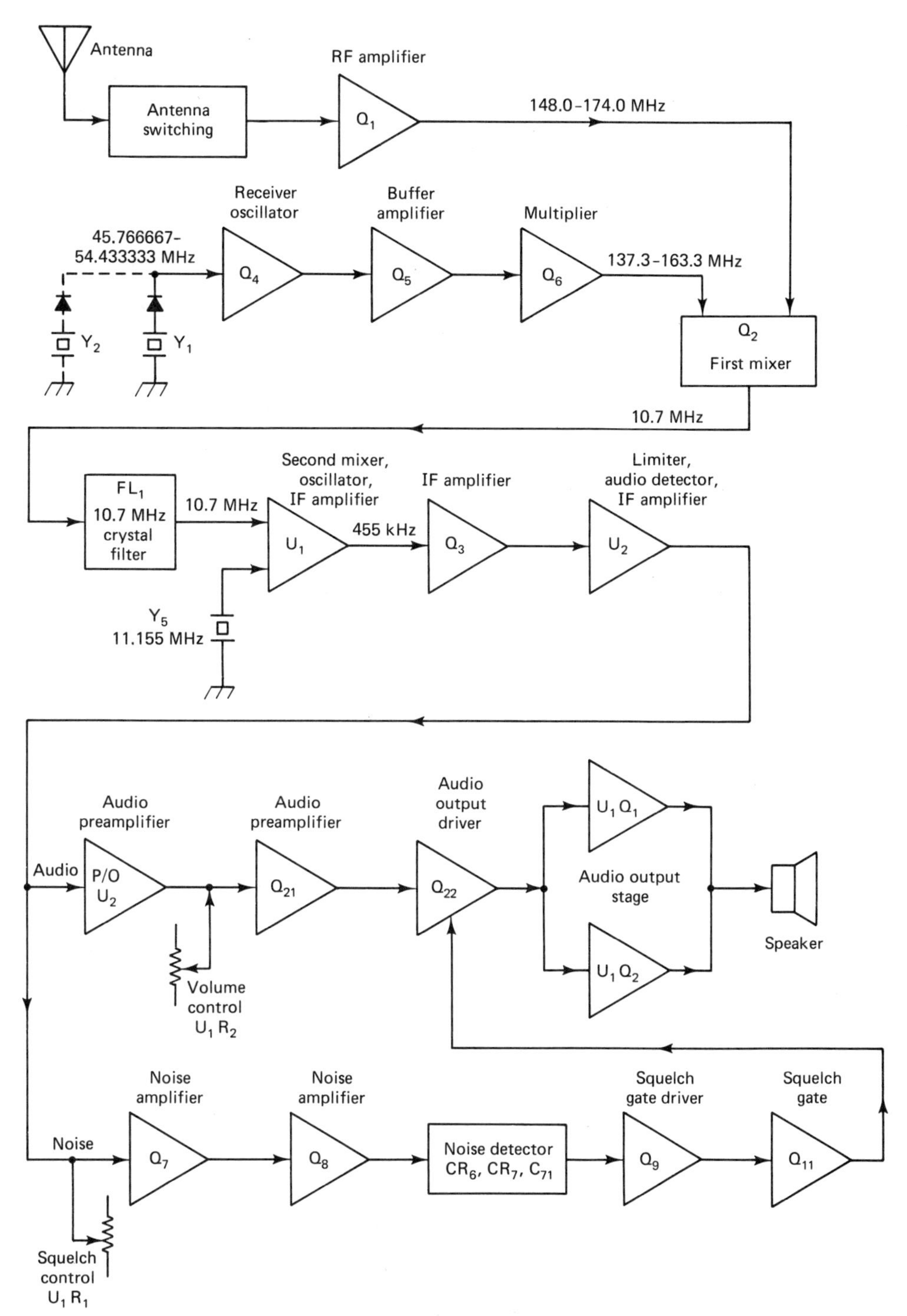

Figure 9.3 Model 508 receiver functional block diagram. (Courtesy of E.F. Johnson Co.)

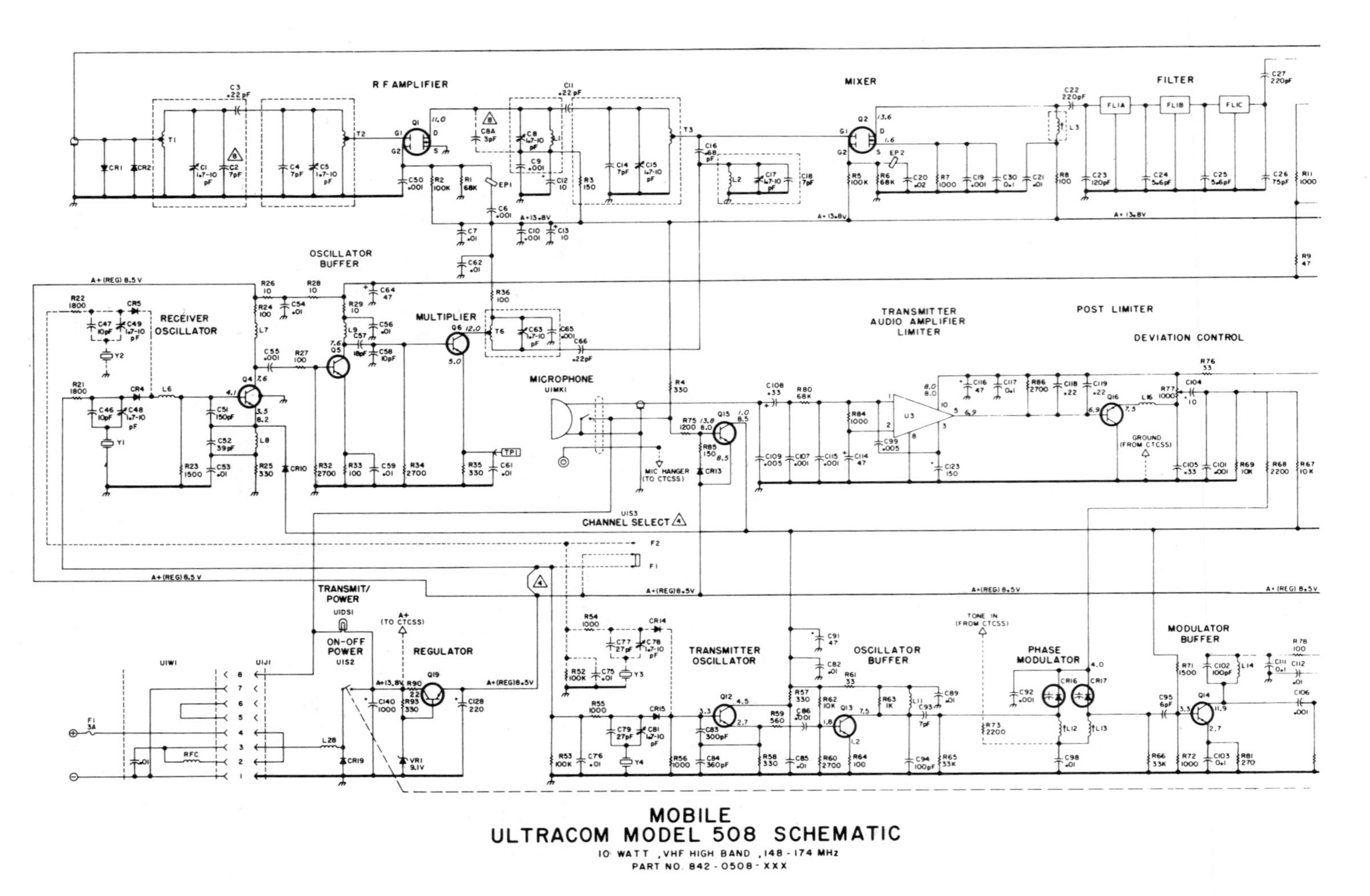

Figure 9.4 Schematic circuit diagram of E.F. Johnson Model 508 VHF-FM transceiver. (Courtesy of E.F. Johnson Co.)

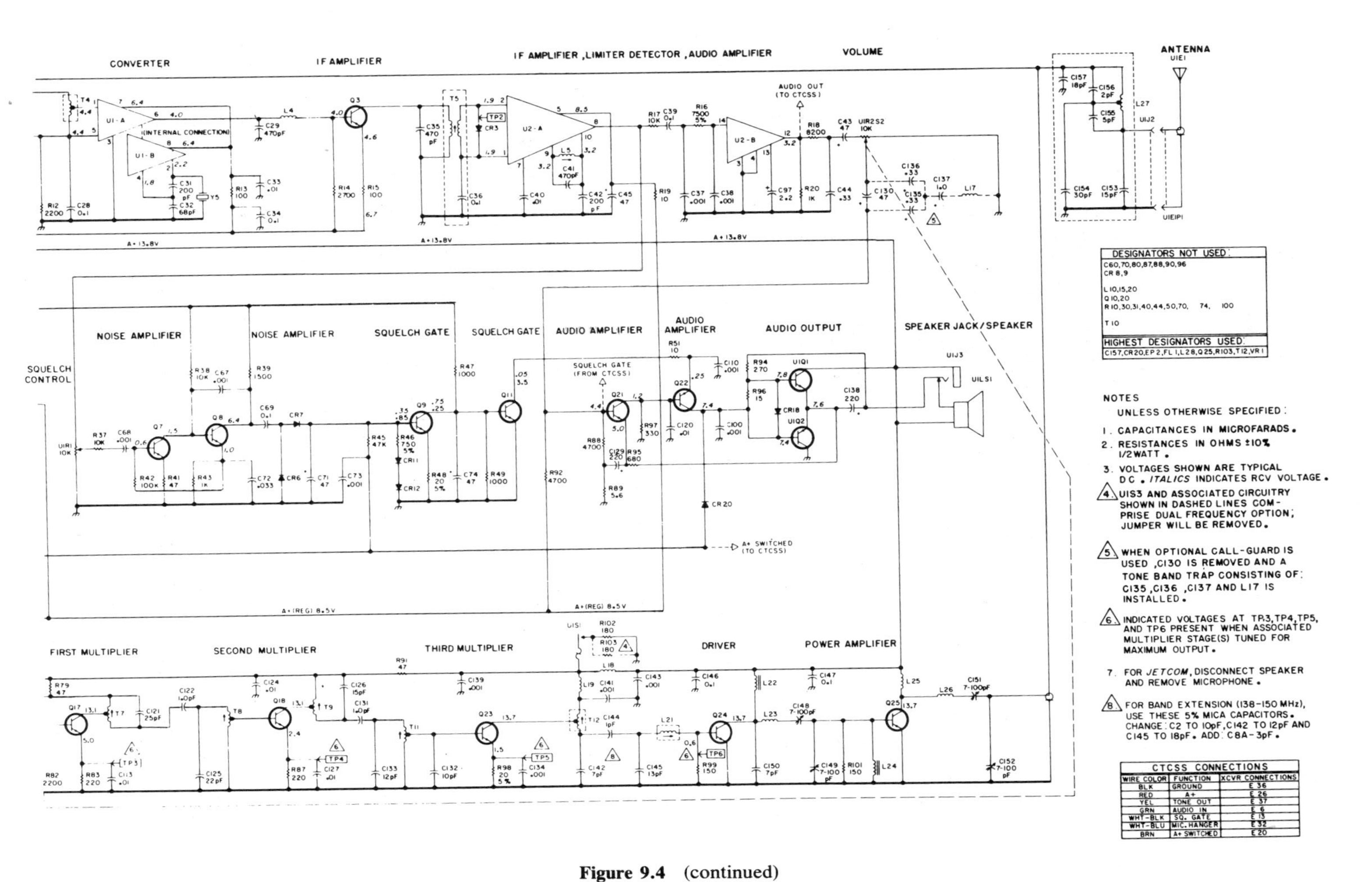

Figure 9.4 (continued)

is determined by the user's choice of frequency for crystal Y_1. The circuit is that of a Colpitts oscillator. The output signal is RC coupled to OSC BUFFER (Q_5). OSC BUFFER provides a signal of relatively constant amplitude to frequency tripler, MULT (Q_6). This buffer stage also serves to isolate the oscillator from the multiplier and minimize frequency pulling. The output of the MULT is tuned to the range 137.3 to 163.3 MHz by circuits T_6–C_{63} and L_2–C_{17}. This three-times signal is capacitively coupled to MIXER through C_{16}.

The first mixer circuit (MIXER) produces a 10.7-MHz IF signal. This is the difference frequency of the 148.0 to 174.0-MHz signal from RF AMP and 137.3- to 163.3-MHz input from MULT. Initial selection of 10.7 MHz is achieved by means of circuit L_3–C_{22}–C_{23}. Two capacitors are used in this circuit to provide a ratio for impedance matching to the crystal filter FL_1.

Crystal filter FL_1 provides a precisely defined, 10.7-MHz bandpass filter. It is specified by Johnson as having a bandwidth of 13 kHz at the 6-dB points, and it has a "six-pole response." This means that the crystal elements provide the equivalent of a complex network of resonant circuits having six resonant frequencies in the passband. The manufacturer indicates that FL_1 provides "the majority of the selectivity and shape factor of the receiver." (Shape factor is defined in a subsequent section in this chapter.) To achieve maximum benefit from the attributes of the filter it must have its impedance carefully matched to the circuit to which it is connected. The circuit containing C_{26}, C_{27}, and T_4 provides this matching to the input of the next element, integrated circuit U_1.

The receiver utilizes two integrated circuits. The first is U_1, an RCA device designated CA3053. It is described by its manufacturer as a differential/cascode amplifier for IF amplifier applications in communications and industrial equipment operating at frequencies from dc to 120 MHz. The second device is U_2, also an RCA IC, designated CA3075. The CA3075 is described by RCA as an FM IF amplifier-limiter, detector, and audio preamplifier. It is suitable for FM IF amplifier applications up to 20 MHz in communications receivers and high-fidelity receivers.

The schematic diagram of the CA3053 chip with pin identification is shown in Fig. 9.5(a). A block diagram of the internal functions of the CA3075 chip is shown in Fig. 9.5(b). The presence of crystal Y_5 at pin 2 of U_1 (CA3053) indicates that U_1 performs the function of a local oscillator for the second mixer. U_1 also mixes the output of that oscillator, 11.155 MHz, with the 10.7-MHz input signal to produce a 455-kHz second IF signal. The CA3053 circuit amplifies the 455-kHz signal and outputs it on pin 6. Inductor L_4 and capacitor C_{29} represent a low-pass filter for the signal, filtering out the 11.155-MHz and 10.7-MHz components.

The 455-kHz signal is amplified further by Q_3 and coupled to the amplifier-limiter of U_2, pins 1 and 2, through transformer T_5 [see the block diagram of the CA3075 in Fig. 9.5(b)]. Integrated circuit U_2 performs four functions: (1) It amplifies the 455-kHz signal, and (2) limits that signal in the amplifier-limiter block. As you will recall, limiting enhances the quality of the signal by sharply rejecting AM and random noise. Limiting also assures a good capture ratio for the receiver. (3) It detects the FM signal using quadrature detection (see Chap. 8). Only a single tuning coil is required outboard of the chip, L_5,

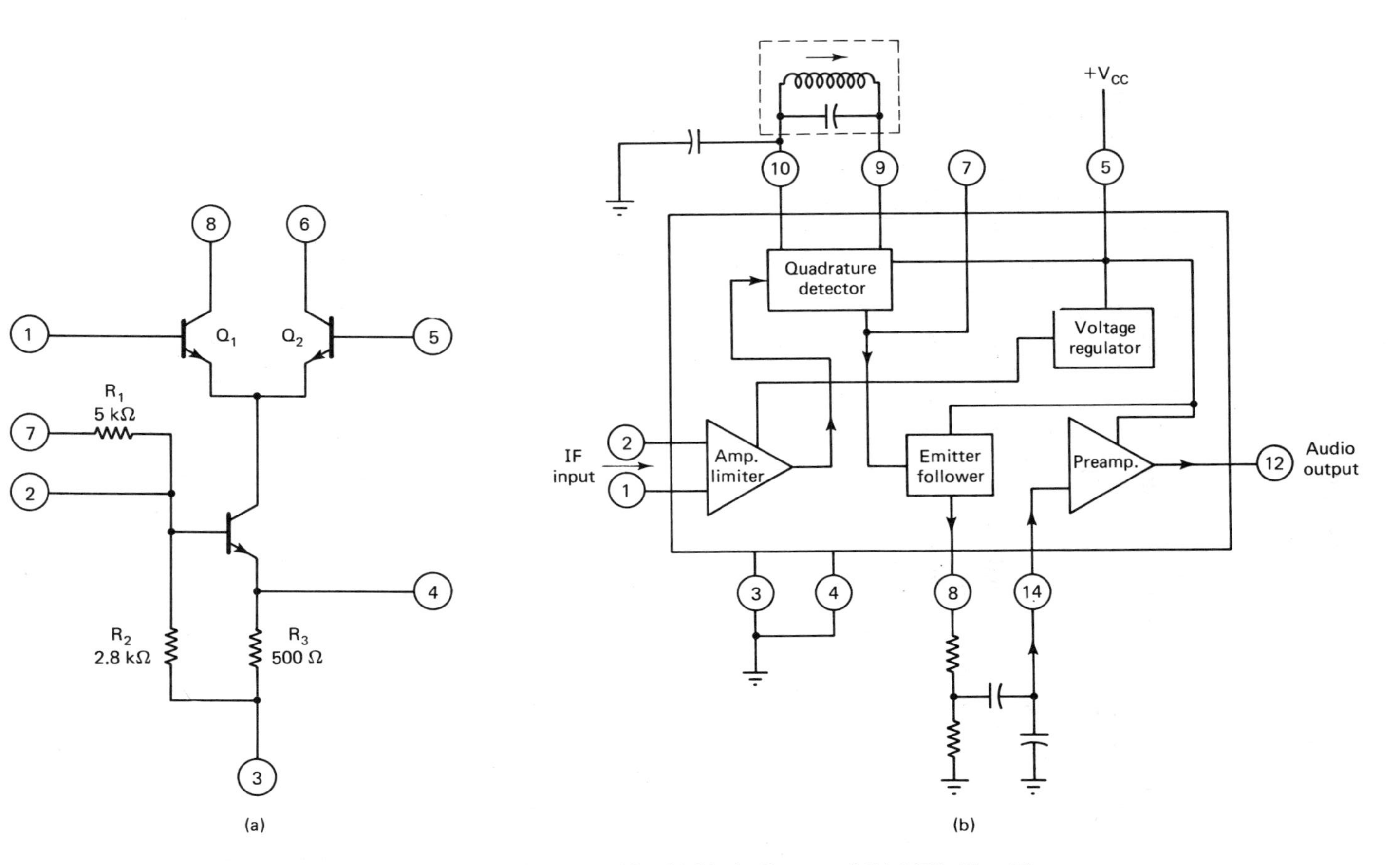

Figure 9.5 (a) Schematic of CA 3053 chip; (b) block diagram of CA 3075 chip. (Courtesy of RCA Corporation, Solid-State Division.)

between pins 9 and 10. (4) It provides preamplification of the audio signal. The audio output from pin 12 is fed through a deemphasis network—R_{18} and C_{44}—and coupled to the input of the volume control, U_1R_2, through C_{43}.

Transceivers like the ULTRACOM 508 are typically installed in land vehicles such as taxicabs, police cars, gas and electric power company service trucks, etc. In these applications a receiver is on all the time the vehicle is occupied. However, message transmissions occur only at random intervals. When a receiver with high sensitivity is not processing a transmitted carrier it is extremely noisy, amplifying noise signals with maximum gain. No one would be willing to keep a receiver turned on for long listening for a message under this condition. Enter the SQUELCH function. SQUELCH (the squelch function) disables the audio section of a receiver when no carrier above a presettable threshold level is being received. On the other hand, an effective squelch function is able to detect a carrier of appropriate strength and enable the audio section so that the operator hears the incoming message.

In FM systems, noise amplitude is proportional to the frequency of the noise. The 508 makes use of this fact in its squelch circuit. A noise signal is obtained from U_2 on pin 8. This is fed to SQUELCH control, U_1R_1, and then to NOISE AMPs Q_7 and Q_8 (see the schematic diagram, Fig. 9.4). This circuit is designed to amplify the high, inaudible noise frequencies and reject audio frequencies. The output of Q_8 goes to a detector-voltage doubler circuit: C_{69}, CR_6, CR_7 and C_{71}. The voltage across C_{71} determines the state of conduction of SQUELCH GATE Q_9. If there is sufficient noise present, Q_9 conducts and Q_{11} is cut off. The audio amplifier of the receiver is disabled when Q_{11} is cut off since its collector–emitter circuit forms the emitter return for audio stage Q_{22}. Of course, when a carrier transmission is being received, the FM process suppresses high-frequency noise; the output of NOISE AMP Q_8 is minimal or nonexistent; Q_9 is cut off and Q_{11} conducts, enabling audio stage Q_{22}. This turns on the entire audio section and any audio content in the received signal is reproduced in the loudspeaker.

The squelch threshold or quieting sensitivity of the receiver is adjustable with U_1R_1, the SQUELCH control on the front panel of the transceiver. This can be set so that a desired carrier level is required to turn on the audio section. Or, for maximum sensitivity, the SQUELCH control is adjusted to its maximum counterclockwise (CCW) position (receiver is "unsquelched") and then adjusted slowly clockwise (CW) until the receiver is just quieted.

The audio section of the receiver consists of predriver stage Q_{21}, driver Q_{22}, and a transformerless, complementary-symmetry push-pull output stage U_1Q_1–U_1Q_2. The designator prefix "U_1" indicates that the output transistors are not located on the printed-circuit board with most of the other components of the transceiver. Instead, they are mounted on the metallic chassis of the transceiver. The chassis thus provides heat sinking for these power transistors. The four transistors of the audio section are direct coupled. This, along with both dc and ac feedback, results in a high-gain amplifier with good thermal stability and minimal distortion. The output stage drives the speaker, U_1LS_1. Audio volume is adjusted by VOLUME control U_1R_2.

You will find located near the lower left corner of the schematic diagram the voltage regulator stage Q_{19}. This is a simple, series-transistor, emitter-follower type of regulator circuit. Its reference voltage is set by the 9.1-V zener diode VR_1. The circuit drops the 13.6-V input from a vehicular battery to a regulated 8.5 V. The regulated voltage is supplied to all circuits in the transceiver whose performance may be affected adversely by a fluctuating supply voltage—primarily, low-power stages. Observe that the points which receive regulated voltage are designated "TO A+ REG."

Transmitter Circuitry

A block diagram showing the functional elements of the transmitter portion of the 508 transceiver appears in Fig. 9.6. We study the diagram to get an overall idea of the design features. As we have already observed, this transmitter utilizes phase modulation and frequency multiplication. As you will recall, phase modulation produces an equivalent to frequency deviation which is proportional to the frequency of the modulating signal. In true FM, however, deviation must depend only on the amplitude of the modulation signal: We may expect to find a circuit in the audio section of the transmitter which compensates for this built-in error of phase modulation. A low-pass filter will, in fact, provide a modulating signal that is inversely proportional to frequency.

The transmitter oscillator (XMTR OSC), Q_{12}, is a crystal-controlled Colpitts oscillator. The output is taken from the emitter circuit and amplified in a buffer amplifier (OSC BUFFER), Q_{13}, before being fed to the PHASE MODULATOR. The transceiver is designed so that it can be purchased as a dual-frequency unit as an option. If purchased as a single-frequency transceiver, only crystals Y_4 in the transmitter and Y_1 in the receiver are connected for operation. If dual-frequency operation is purchased, additional components, such as Y_3, Y_2, C_{77}, R_{54}, etc., are installed by the manufacturer. One of two operating frequencies can then be selected by means of channel selector switch, U_1S_1. You will observe that this switch connects a regulated 8.5 V to one of the two available base circuits for both the receiver and transmitter oscillators.

While we are analyzing the transmitter oscillator let's become familiar with a feature of all transmitters used in two-way communications: a convenient method by which the operator can switch between TRANSMIT and RECEIVE modes. The normal, or standby, mode of a transceiver is always RECEIVE. Switching to TRANSMIT is typically accomplished by depressing a push-to-talk (PTT) switch built into the microphone housing. Refer to the schematic diagram of the 508, Fig. 9.4. Note the lead labeled "PTT" between the microphone assembly, U_1MK_1, and printed-circuit connection point E_{34}. In the nonactuated position, the switch in the microphone housing is open. This, in conjunction with the bias circuit arrangement for Q_{15}, causes Q_{15} to be virtually cut off. Transistor Q_{15} (PTT SWITCH) in this transceiver takes the place of a device which was typically an electromechanical relay in older transceivers. When the PTT switch is depressed at the mike, point E_{34} is grounded, which changes the bias condition for Q_{15} from cutoff to saturation. When saturated, Q_{15} completes the circuit to

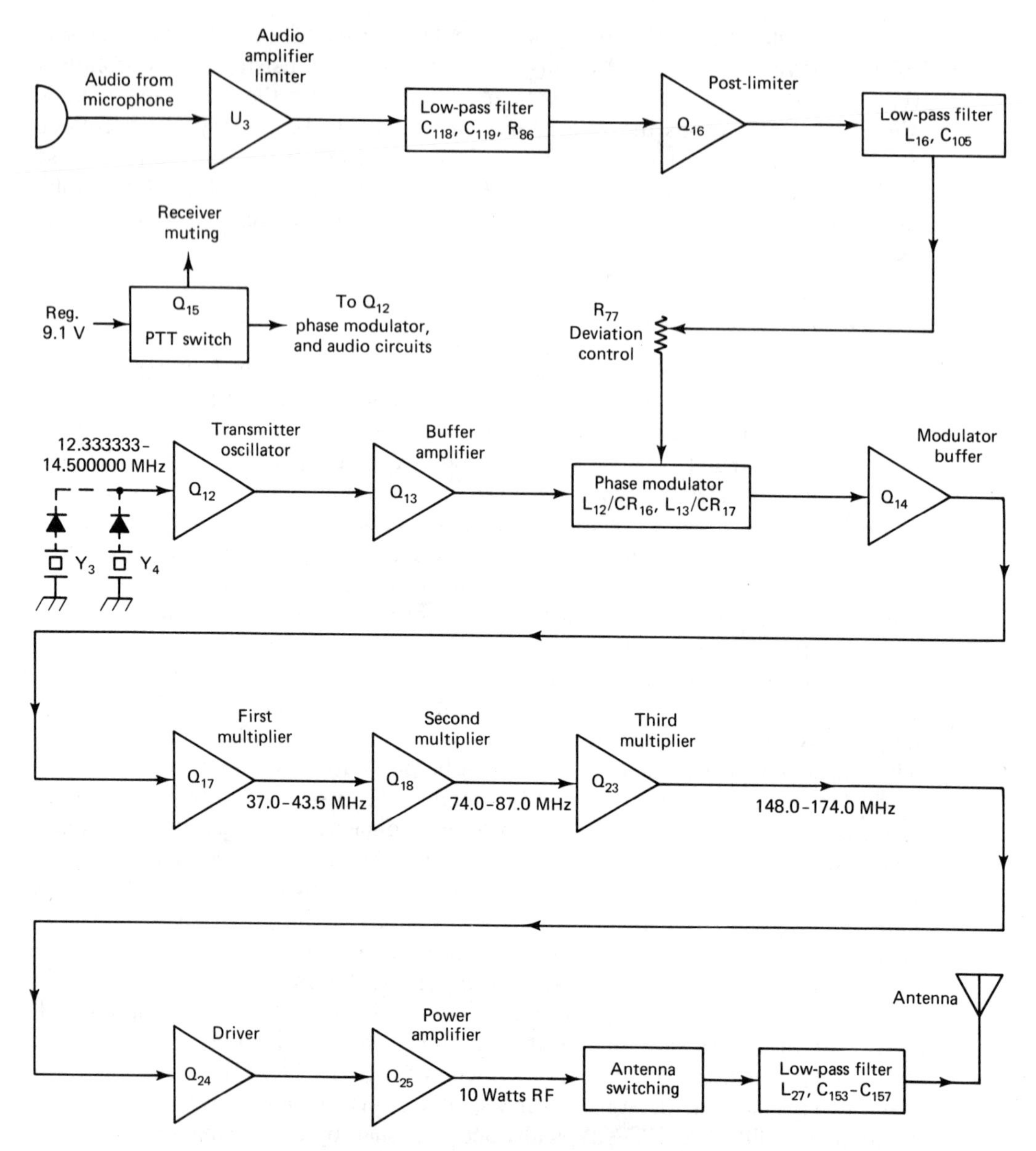

Figure 9.6 Model 508 transmitter functional block diagram. (Courtesy of E.F. Johnson Co.)

provide 8.5 V, regulated, to the transmitter oscillator, oscillator buffer, and the transmitter audio section. A positive voltage is also fed to the SQUELCH GATE base circuit, which disables the receiver audio section. Also, note that when Q_{15} is supplying dc voltage to the transmitter oscillator, the receiver oscillator is cut off by a high dc voltage placed on its emitter by way of diode CR_{10}.

The phase modulator utilizes two varactor diodes, CR_{16} and CR_{17}. These, in conjunction with inductors L_{12} and L_{13}, respectively, provide two tuned circuits from the signal path to ground. A dc voltage varying at an audio rate from the slider of DEVIATION CONTROL R_{77} is connected in parallel to the diodes through R_{68}. The varying voltage on the diodes changes their capacitive reactance and thereby the impedance phase angle of the circuits of which they are a part. Hence the phase angle of the signal at the top of L_{13} varies with respect to the input signal at the top of L_{12}. As we have learned previously, phase displacement of a carrier signal translates into a freqency deviation—thus is the oscillator signal indirectly frequency modulated. The modulated signal is amplified in the MODULATOR BUFFER, Q_{14}, before it is fed to the frequency multiplier stages.

Stages Q_{17}, Q_{18}, and Q_{23} form a $\times 12$ multiplier circuit. Transformer T_7 and C_{121} in the collector circuit of Q_{17}, and T_8/C_{125} in the base circuit of Q_{18} are tuned to three times the frequency of the input signal of Q_{17}: Q_{17} is a tripler. T_9/C_{126} in the collector circuit of Q_{18} and T_{11}/C_{133}–C_{132} in the base circuit of Q_{23} are tuned to twice the output of Q_{17}. The collector circuit of Q_{23}—T_{12}/C_{141}—is tuned to twice its input frequency.

You will note that whereas Q_{14} is provided with dc bias, the multiplier stages Q_{17}, Q_{18}, and Q_{23} are not; they operate in class C mode. The emitters of each of these stages, as well as the base of Q_{24}, have dc voltages that are a function of the amplitude of the signal being processed. These points provide useful test points for alignment and troubleshooting procedures. You will observe that they are identified as "test points" on the schematic, with the designations TP_3, TP_4, TP_5, and TP_6, respectively. A schematic NOTE states that the voltages given for these points correspond to operation at maximum output of the transmitter.

A phase-modulated signal whose frequency is 12 times that of the transmitter oscillator is fed to the DRIVER, Q_{24}. Stage Q_{24} amplifies the signal to a level of 2 W. This signal then drives the transmitter final, Q_{25} (PA), to 10 W. The circuit consisting of L_{26}, C_{151}, and C_{152} on the collector side of Q_{25} is an impedance-matching network which matches the 50-Ω impedance of the antenna to the output impedance of the PA stage.

You will note the circuit consisting of L_{27} and capacitors C_{153} through C_{157} between the output of Q_{25} and the antenna jack, U_1J_2. This circuit is described by the manufacturer as a two-section, elliptic-function low-pass filter. Its purpose is to suppress the harmonics of the carrier, especially the strong second harmonic, and other spurious higher-frequency signals that might originate in the transmitter. The filter is also in series with the signal from the antenna to the receiver.

Since, with a transceiver, one antenna is used for both transmitting and receiving, some method for switching the antenna between the two functions must be provided. It is common practice to use an electromagnetic antenna relay for this purpose. In this unit, however, switching is accomplished by using the special characteristics of a quarter-wavelength transmission line. At connection points E_{45}–E_{46} and E_{47}–E_{48}, the input to the receiver and the output of the transmitter are effectively in parallel, each through a quarter-wavelength of coaxial transmission line—w_3 and w_2, respectively. At the opposite end of the receiver transmission-line stub, diodes CR_1 and CR_2 effectively short the line for the signal from the transmitter. However, the input impedance of a shorted $\lambda/4$ transmission line is infinity (see Chap. 6). Hence the receiver presents an infinite impedance to the transmitter when the unit is operating in TRANSMIT mode. Similarly, looking back into the PA from points E_{45}–E_{46}, the very high impedance of the inoperative Q_{25} stage is transformed to a low impedance at E_{41}–E_{42} by the network, L_{26}, C_{151}, C_{152}. But that low impedance is across the transmission-line stub w_2 and is transformed to a very high impedance at E_{45}–E_{46}—thus is the transmitter output circuit isolated from the receiver input in RECEIVE mode.

The symbol labeled U_1RT_1, which is located near the antenna symbol on the schematic, is for a thermostat (temperature-sensing switch). It controls current to R_{102}, which is a crystal warmer. The thermostat is mounted on the metal chassis of the transceiver so that it senses the internal temperature of the chassis. Together, U_1RT_1 and R_{102} (and R_{103} for dual-frequency transceivers) maintain a more constant temperature for the transmitter crystals. The manufacturer specifies the transmitter frequency variations to be less than $\pm 0.0005\%$ over an environmental temperature range -30 to $+60°$ C.

The audio section of the transmitter consists of a microphone and an audio amplifier for increasing the audio energy to a level sufficient to achieve modulation of the RF carrier in the phase modulator. The audio section must include the preemphasis circuit, a standard requirement for all FM transmitters. In this particular transmitter, since it is phase modulated, the audio section must also include a circuit that will make the audio signal amplitude inversely proportional to the audio frequency. That is, there is required a filter whose output amplitude is proportional to $1/f$ times the input. Remember, phase modulation produces an FM frequency deviation that is proportional to the modulating frequency.

Let us examine again both the block diagram of the transmitter, Fig. 9.6, and the complete schematic of the transceiver, Fig. 9.4. Specifically, let's look at the portions of these diagrams pertaining to the transmitter audio section. We see that an integrated circuit is included—U_3, an RCA CA3011. The CA3011 is a general-purpose, wide-band amplifier with specific limiting characteristics. It is used here in the audio section both for its excellent gain and limiting qualities. Its gain ensures that adequate signal will be available for modulation. Its ability to limit a signal will reduce the possibility of overmodulation.

The circuit between the microphone and the input to U_3 includes components that provide the preemphasis function—a series C_{108} and shunt R_{84}. Between the output of U_3 and the modulator driver, Q_{16} (POST LIMITER), shunt capacitors C_{117}, C_{118}, and C_{119} represent the $1/f$ filter. The POST LIMITER is an emitter-follower stage providing a low-impedance output for driving the phase modulator. Between Q_{16} and the phase modulator, the circuit consisting of L_{16} and C_{105} is a low-pass filter. The filter is designed to have an attenuation or "roll-off" characteristic of -12 dB/octave starting at 3 kHz.

This transceiver is designed to comply with FCC requirements for so-called *narrow-band FM*. A nominal definition of narrow-band FM is that audio frequencies are attenuated at the 12 dB/octave rate above 3 kHz. Also, in narrow-band FM, maximum frequency deviation is limited to ± 5 kHz.

The amount of signal fed to the phase modulator is adjustable by means of R_{77}, DEVIATION CONTROL. You will observe that the audio signal is capacitively coupled to the junction of R_{67} and R_{69}. These two resistors form a voltage-divider circuit for setting the standby (unmodulated) dc voltage level on the varactor diodes of the phase modulator.

The output power rating of a transmitter is an important specification. It is one of the parameters of interest to the FCC. The more RF power radiated by a station, the greater the distance from the antenna the signal is likely to be received. The more the RF power, the more the "talk power." However, the number of individual RF signals that can be accommodated in the RF spectrum is definitely limited. It is logical that a greater number of spectrum users can be accommodated without unacceptable interference to one another if the amount of power for any one station is limited. The manufacturer's power specification for the ULTRACOM 508 is "Power output—10 watts, minimum." Stated in this fashion, the specification means that the manufacturer will consider the unit defective if it does not generate an RF power output of 10 W. However, a typical 508 is not likely to generate significantly more than 10 W. An output of 10 W is within the FCC requirements for a transmitter of this given type (commercial radiotelephone, VHF-FM, 148.0 to 174.0 MHz).

The power output of a transmitter is determined by the design of its output stage. For a solid-state stage, the maximum power is a function of the dc supply voltage of the stage, the voltage across the amplifying device (transistor) when it is saturated, and the equivalent radiation resistance of the antenna, as seen by the output stage. This can be stated in a formula as follows:

$$P_{out} = \frac{(V_{CC} - V_{sat})^2}{2R_{rad}} \tag{9.1}$$

The difference in dc supply voltage and saturation voltage defines the maximum peak-to-peak voltage of the RF signal. In a stage with a tuned output circuit, the RF voltage across that circuit can be as much as twice the net dc voltage difference in the circuit. [Formula (9.1) converts the peak-to-peak value to effective value.]

Equation (9.1) can be solved for the value of antenna radiation resistance,

$$R_{rad} = \frac{(V_{CC} - V_{sat})^2}{2P_{out}} \tag{9.2}$$

The radiation resistance of the quarter-wavelength antenna most generally used with transceivers like the 508 is a nominal 50 Ω. It is evident that if the 508 transmitter produces 10 W of RF power, the impedance-matching network in the PA stage accomplishes approximately a 10:1 impedance transformation ratio.

We have now completed an examination of all the important details of a particular, commercially available FM transceiver. It is highly useful that we reflect on what we have seen and recognize the great similarities between this system and its functions and circuits and those we have seen previously. Electronics has found its way into numerous applications in our modern civilization. The number of applications is growing at virtually an exponential rate. A finite number of basic electronic function blocks can, seemingly, be combined in an infinite number of ways to perform an unlimited number of different tasks. If we maintain any hope or expectation of being able to keep up with even a small segment of electronics applications, we must prepare ourselves to recognize and work with basic building-block circuits and functions: amplifiers, oscillators, modulators, detectors, etc. We can expect new functions to be devised. However, in the interest of personal productivity it is essential that we build and maintain a "mental library" of functions. In performing day-to-day tasks in the world of electronics, there is not time for us to look on each "new-to-us" circuit as a new invention requiring "from the ground up" analysis.

We now want to examine other types of radiotelephone systems. Before looking at specific systems in detail, however, it will be helpful for us to study the electronic details of two basic functions which are being used in an ever-increasing number of radiotelephone systems. These functions are *frequency synthesis* and *single-sideband generation*.

9.4 FREQUENCY SYNTHESIS

An important need of the equipment of most, if not all, of the radiotelephone services is an efficient way of producing two or more carrier frequencies. For example, citizens' band radio (CB) now authorizes 40 channels in the 27-MHz segment of the frequency spectrum. At the same time, FCC (and indeed, practical) requirements relative to carrier frequency stability and precision are such that crystal control of frequency generation is essential. However, special techniques for the efficient use of crystals is called for to ensure that equipment cost is reasonable. In the absence of ingenious techniques for frequency generation, a CB transceiver, for example, could conceivably require 80 or more expensive quartz crystals.

Frequency synthesis is by far the most popular method of generating carrier frequencies in modern communications equipment in 1986. This is a technique that utilizes a combination of crystal control, heterodyne frequency conversion, and frequency division by programmable digital IC counters, all in the context of a phase-locked loop (PLL). This technique is rapidly being improved and expanded with the aid of microprocessor technology, even as this is being written.

Let's look into frequency synthesis by reviewing the basic phase-locked loop and the functions it includes [see Fig. 9.7(a)]. At the heart of a PLL for frequency generation is the voltage-controlled oscillator (VCO). A VCO is an oscillator whose frequency can be controlled/changed by a change of voltage across a varactor diode. Next, there is a means of comparing the frequency of the VCO with an external reference frequency. This function is accomplished in the *phase comparator* or *phase detector*. The phase detector provides an output that is directly related to the *amount* and *direction* of the difference between the VCO frequency and the reference frequency. The detector output is often called the *error signal*. The loop is completed when the error signal is passed through a low-pass filter (LPF) and back to the VCO. In the VCO, the filtered error signal is applied to the control device of the oscillator and changes its frequency to bring it into equality with the reference frequency. This circuit is a closed loop: there are connections from the output of the VCO, through the phase detector and LPF, back to the input of the VCO.

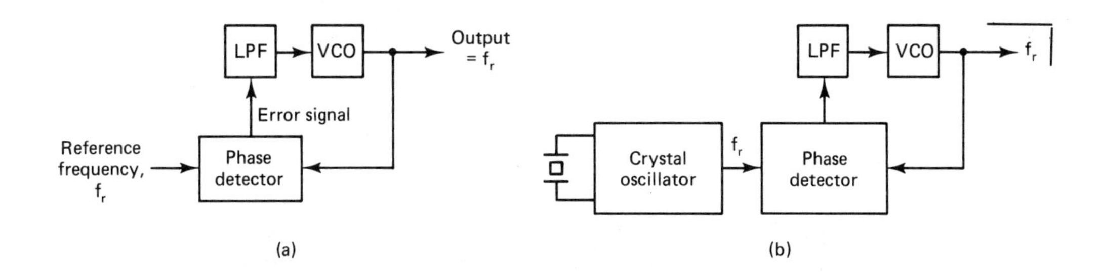

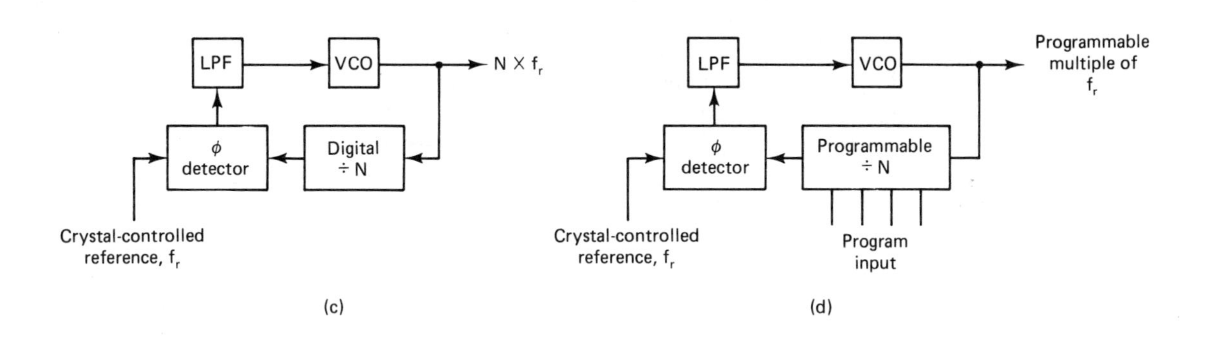

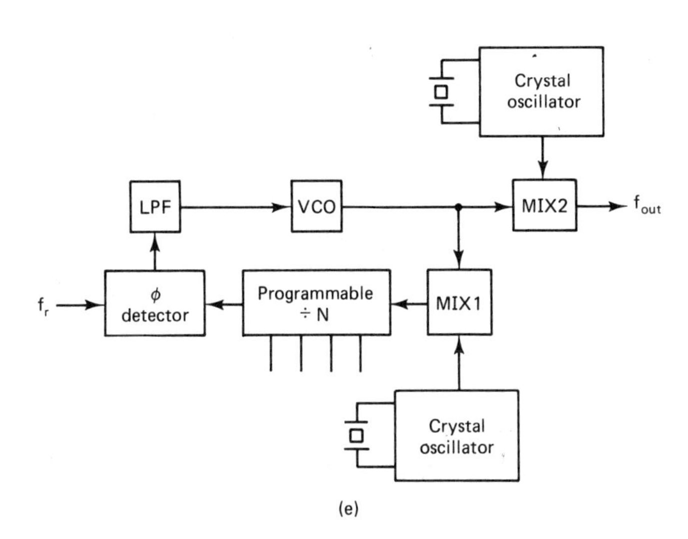

Figure 9.7 Frequency-synthesis schemes: (a) basic phase-locked loop (PLL); (b) PLL with crystal-controlled reference frequency; (c) PLL with digital frequency divider in loop; (d) PLL with programmable digital frequency divider in loop; (e) synthesized frequency source employing PLL and heterodyne frequency conversion.

The basic PLL can easily be modified into a crystal-controlled frequency source by generating the reference frequency for the phase detector with a crystal-controlled oscillator [see Fig. 9.7(b)]. The loop can be tricked into providing a crystal controlled frequency higher than that of the operating frequency of the crystal oscillator by breaking the loop and inserting a binary digital counter-frequency divider between the VCO and the phase detector [see Fig 9.7(c)]. Furthermore, the loop can be made into a multiple-frequency crystal-controlled source by using a programmable digital frequency-divider device [see Fig. 9.7(d)]. Further alterations of the basic loop are, of course, possible. A common one is the insertion of a heterodyne frequency converter in the loop to alter the reference frequency in a way that is not possible by frequency division [see Fig. 9.7(e)].

Let's examine in some detail the way in which a PLL frequency source might be implemented using readily available integrated circuits. A PLL chip especially adaptable to generating carrier signals is the 564. The 564 is available from Signetics Corporation and is designated NE564. It comprises a VCO, phase comparator, amplifier, and low-pass filter. These functions are interconnected as shown in Fig. 9.8(a). The pin configuration for the 16-pin dual-in-line package (DIP) is given in Fig. 9.8(b). An important feature of the 564 device is that the feedback loop is opened between the VCO and the phase comparator. The input and output circuits of these functions are compatible with the most common digital counter ICs (TTL). This enables the user to insert a frequency-modifying network in the loop. In short, a reference frequency radically different from the VCO frequency can be used.

A simple circuit utilizing the 564 to produce an output frequency equal to N times a crystal-controlled reference frequency, f_{ref}, is shown in Fig. 9.9. Study the circuit carefully. Observe especially the fact that the output of the VCO is fed back to the phase comparator through a $\div N$ digital frequency divider. For example a 7490 TTL binary counter chip connected as a divide-by-10 counter with symmetrical (square wave) output could be used in the feedback loop.

You will recognize that by using combinations of digital counters it would be possible to construct a frequency divider with virtually any integer divisor desired. For example, the 7492 TTL chip is a device that can be connected to form a $\div 12$ counter. If a 7490 ($\div 10$) counter and a 7492 ($\div 12$) counter were connected in cascade in the feedback loop of the 564 PLL, an output frequency of $120 f_{ref}$ would be obtained. The number of possible combinations of devices and frequencies is limited only by the imagination of the designer.

Heterodyne frequency converters can also be used to alter the frequency fed back to the phase comparator of a PLL to achieve a desired output frequency. Observe the diagram of Fig. 9.10(a). The 37.660-MHz output of the VCO is being mixed with a 36.38-MHz signal from a crystal oscillator. The difference frequency, 1.28 MHz, is used for comparison and control of loop lock.

Mixer circuits are also available in integrated form. Figure 9.10(b) illustrates the use of the RCA device, CA3053. You will recall that we examined the CA3053 in conjunction with the transceiver of Sec. 9.3. A similar device, the MC1550G, is available from Motorola.

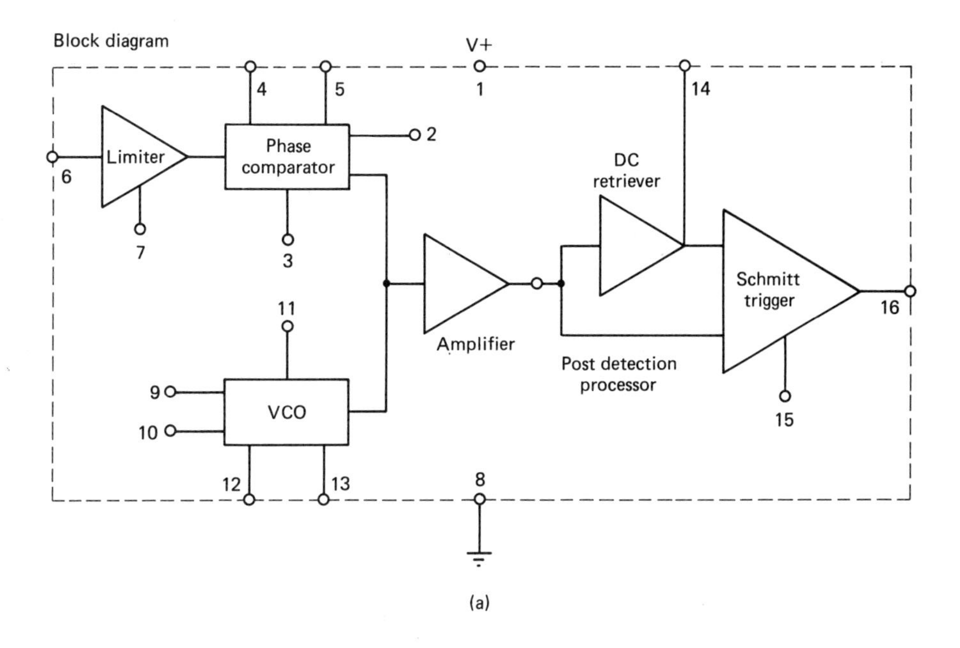

(a)

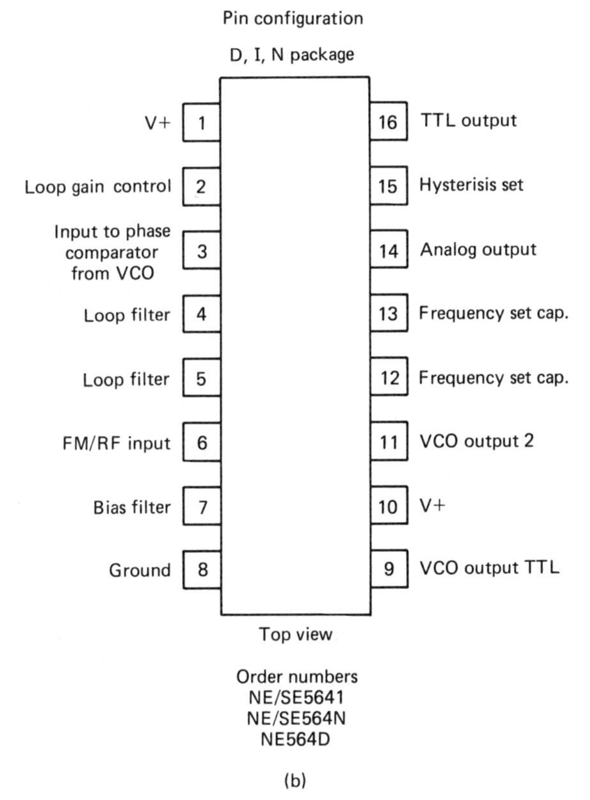

(b)

Figure 9.8 564 phase-locked loop: (a) Block diagram of 564 phase-locked loop; (b) Pinout diagram of 564 (Reprinted with permission of Signetics Corporation.)

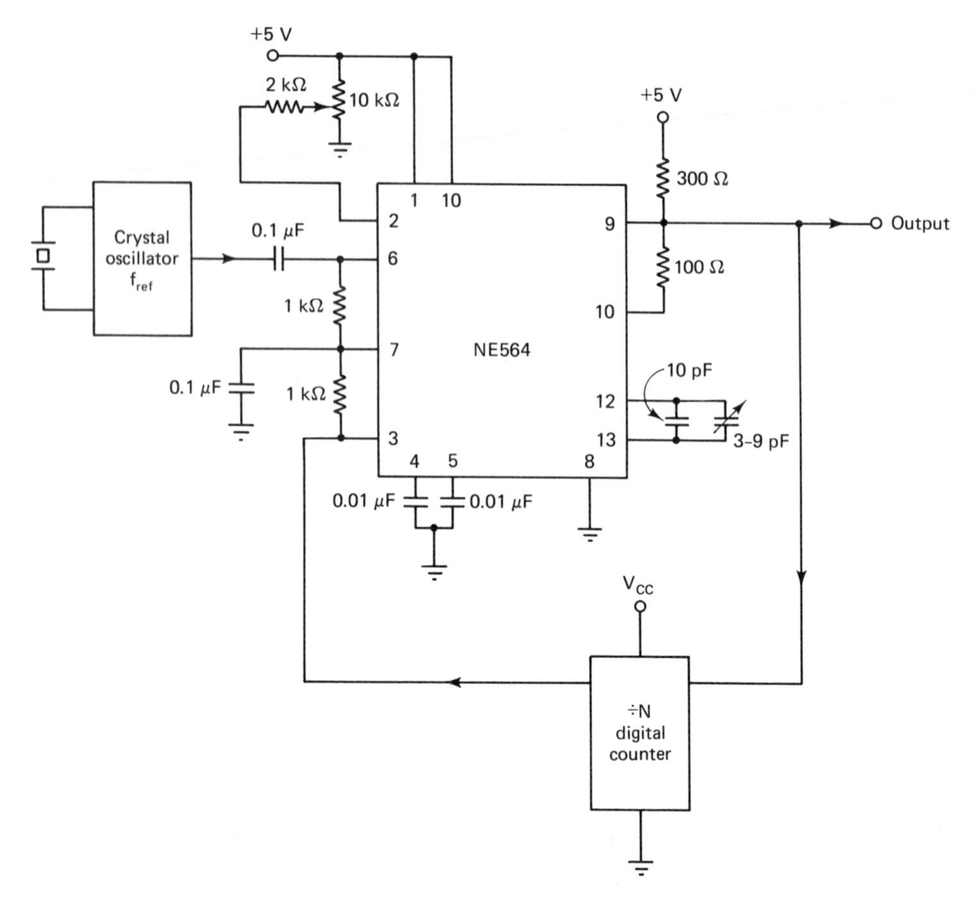

Figure 9.9 Using NE564 PLL and TTL digital counters to produce crystal-controlled frequency. (Reprinted with permission of Signetics Corporation.)

To be useful in many types of communications transceivers, a frequency source must be capable of producing multiple frequencies. Most CB transceivers, for example, can be used on all 40 CB channels. How do we make a PLL synthesizer that can be set to generate a number of different frequencies and still be crystal controlled? One answer is to use a *programmable divider*. The general idea involved is illustrated in Fig. 9.11. The 74192 IC device pictured is a synchronous decade up/down counter. It has two pulse counting inputs, pins 4 and 5. If pin 4, designated CP_D, is held high while pin 5, designated CP_U, is clocked, the device counts up. If pin 5 is held high while pin 4 is clocked, the device counts down. All flip-flops are clocked in parallel, thus providing "synchronous" operation. A block diagram, with pinouts, of the 74192 package (DIP) is shown in Fig. 9.12.

The 74192 is a so-called *presettable counter*. This means that a "count" can be loaded into the counter before counting starts. A desired count is loaded through pins 9, 10, 1, and 15. These pins correspond to binary place values designated D_3, D_2, D_1, and D_0, where D_3 has the decimal-equivalent value of

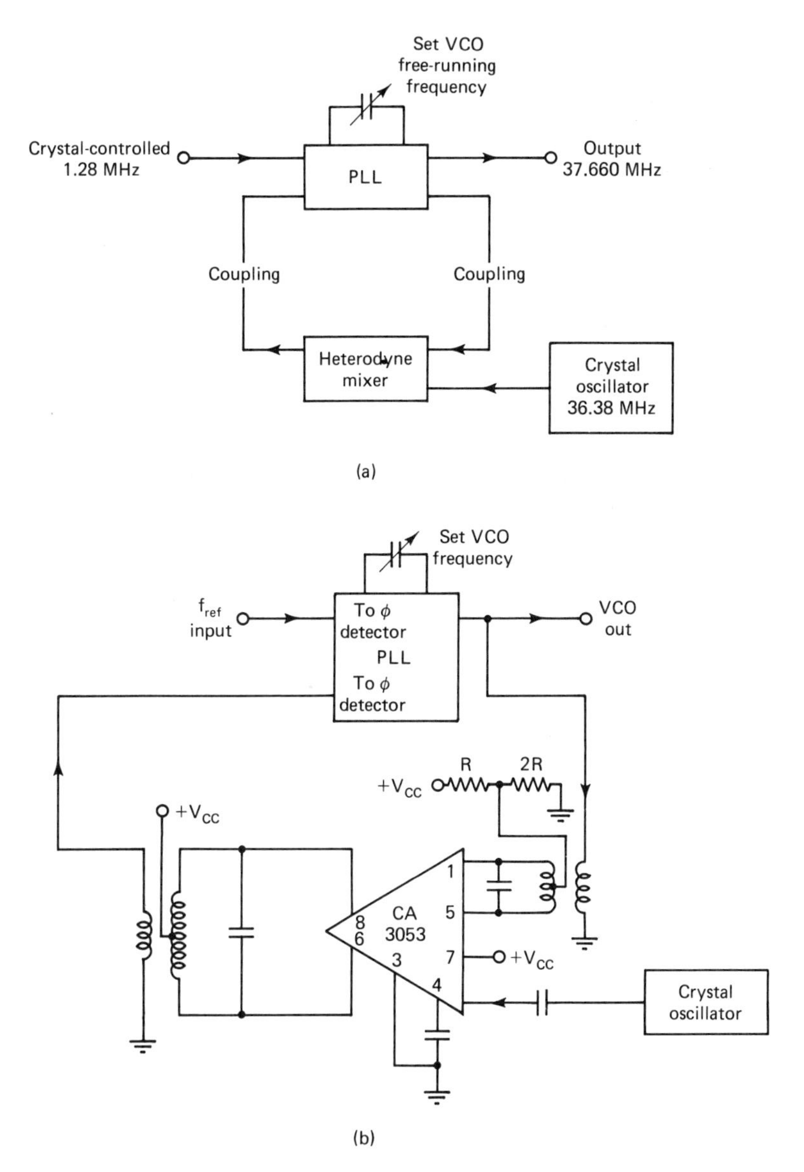

Figure 9.10 Use of heterodyne frequency converter in phase-locked loop: (a) using heterodyne frequency converter between VCO and phase detector to achieve desired frequency; (b) using RCA 3053 as mixer in PLL frequency synthesizer.

8; D_2, 4; D_1, 2; and D_0, 1. Thus a logic "word" of 1001 on pins 9, 10, 1, and 15 would provide an input of 9 to be loaded. As a down counter, the 74192 can be preset with count of up to $15_{10}(1111_2)$. The counter is loaded with the count on the data inputs when pin 11 (*PL*) is pulsed low momentarily.

The device is provided with both "borrow" and "carry" outputs. These are, respectively, pins 13, designated $\overline{TC}_D$, and 12, designated $\overline{TC}_U$. These outputs,

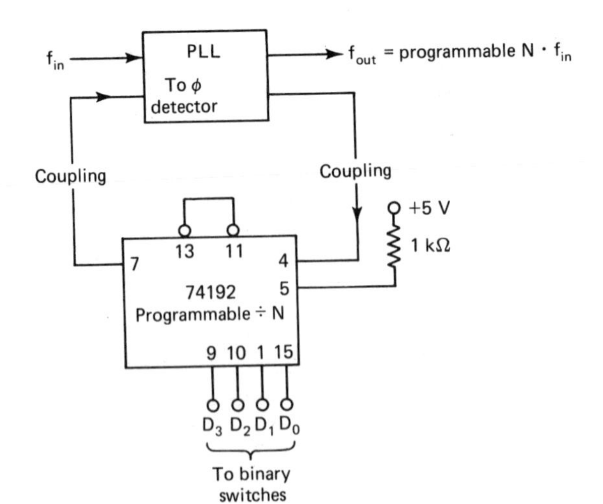

Figure 9.11 Providing adjustable output frequency by using programmable digital divider with PLL.

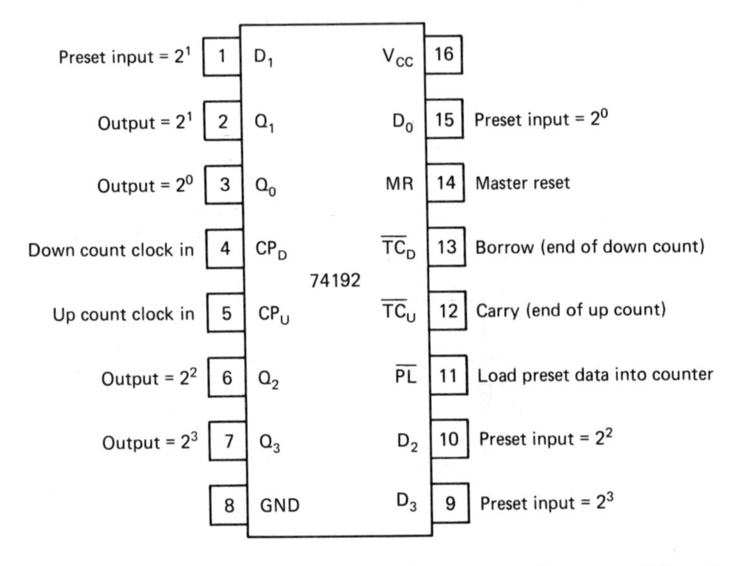

Figure 9.12 Pin functions of 74192 presettable counter. (Courtesy of Signetics Corporation.)

along with the presetting function, make it possible to use the device as a programmable frequency divider. Let's examine the operation of the device when connected as shown in Fig. 9.13. First, the signal to be divided is input on pin 4; and pin 5 is connected high: the counter will count down. Pin 13, the borrow output, will pulse low when the count reaches zero. With pin 13 returned to pin 11, the device will reload the "programmed" data input—pins 9, 10, 1, and 15—each time the count reaches zero. If programmed with a "7," say, the counter will be a modulo-7 (divide-by-7) frequency divider for the following reasons:

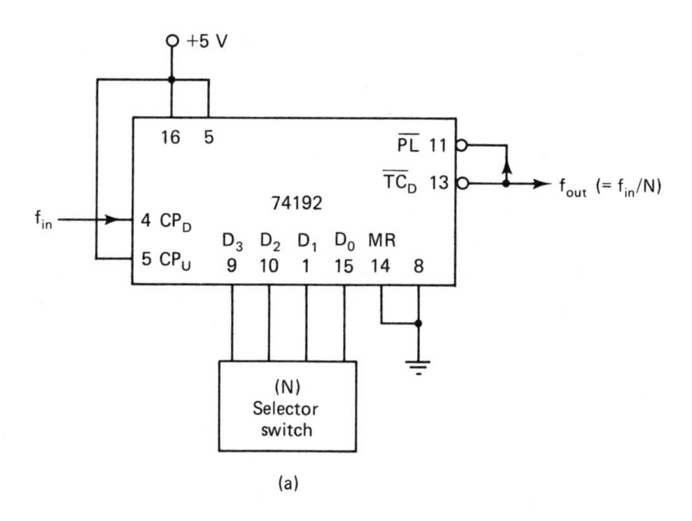

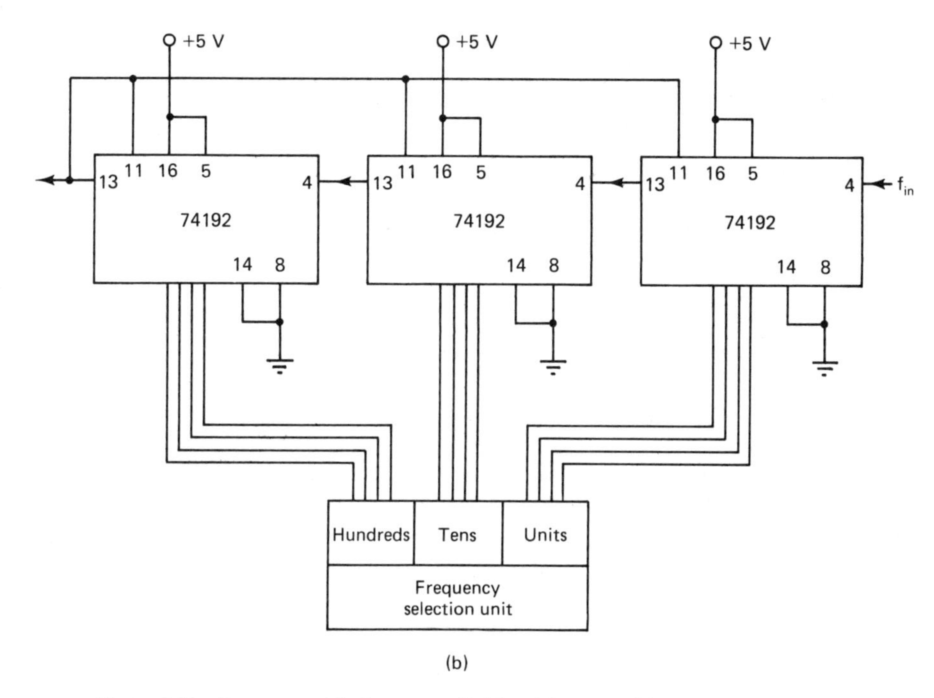

Figure 9.13 Programmable frequency divider: (a) presettable counter 74192 connected as programmable divider (down counter with automatic load of preset data at end of count); (b) cascaded 74192's.

1. The count of the device changes when the signal at pin 4 makes a low-to-high transition.
2. On the seventh low-to-high transition, the count goes to zero (0000) and pin 13, the borrow output, goes low on the next high-to-low transition of the input signal. Simultaneously (on the high-to-low transition), the counter

will reload the count of 7 because the low of pin 13 also appears on pin 11.

3. On the next low-to-high transition the cycle starts to repeat—the count changes from 7 to 6, etc. Pin 13 returns to its high position.

It is apparent that if an arrangement is made to supply a choice of binary input data values from 1 through 10 to the preset inputs of the 74192, it can be programmed to have a modulus of from 1 to 10. On a communications transceiver, for example, the programming device could simply be a manually controlled frequency selector switch.

Of course, 74192 devices can be cascaded to provide even greater choices of frequency selection. Cascaded connections would appear as in Fig. 9.13(b). The borrow output of the lowest-order (units) device is connected to the CP_D input of the second-order (tens) device. The borrow output of the "tens" unit is connected to the count pulse input of the "hundreds" device, etc. Each time one unit completes cycling through its count it borrows "1" from the next-higher-valued unit. This continues until all devices have counted down to zero. Then the cycle starts over. In effect, any modulus from 1 through 999 could be programmed with three 74192 devices. As indicated previously, the programmable divider would require some form of "selector device" which would place the desired logic levels on the four BCD inputs of each chip. These data could also be supplied from a microprocessor controller.

Let us examine how the ideas we have just been discussing could be put together to provide a master frequency source (MFS) for a CB transceiver. Refer to the block diagram of Fig. 9.14. The arrangement depicted does not represent a particular unit. It is intended to illustrate features generally typical of CB transceivers utilizing PLL frequency synthesis. The frequencies noted on the diagram are those required for operation on channel 9 (27.065 MHz). As you will observe, the system utilizes three crystal oscillators:

1. The output of oscillator Y_1 is mixed with the output of the VCO in a heterodyne mixer, TMX_1. The difference frequency, $f_{VCO} - f_{Y1}$, is the input to the programmable digital divider, DD_1. For channel 9, the difference frequency, as indicated, is 1.38 MHz.
2. The output of DD_1 is the input to the loop's phase detector (PHASE DET). The phase detector will compare this signal, generated by the VCO, with the input from crystal oscillator Y_2 and divider DD_2. As indicated, DD_2 has a modulus of 1024; its output frequency is $f_{Y2}/1024 = 10$ kHz.
3. In order for the loop to lock the VCO onto 37.760 MHz, programmable divider DD_1 must be programmed for a modulus of $1.38(10^6)/10{,}000 = 138$. That is, the channel selector switch, CH SW, must be arranged to provide BCD inputs to the three 74192 chips. For channel 9, the BCD inputs to the "hundreds," "tens," and "units" devices would be 0001, 0011, and 1000, respectively.
4. The 37.760-MHz output of the VCO is beat with the 10.695-MHz output

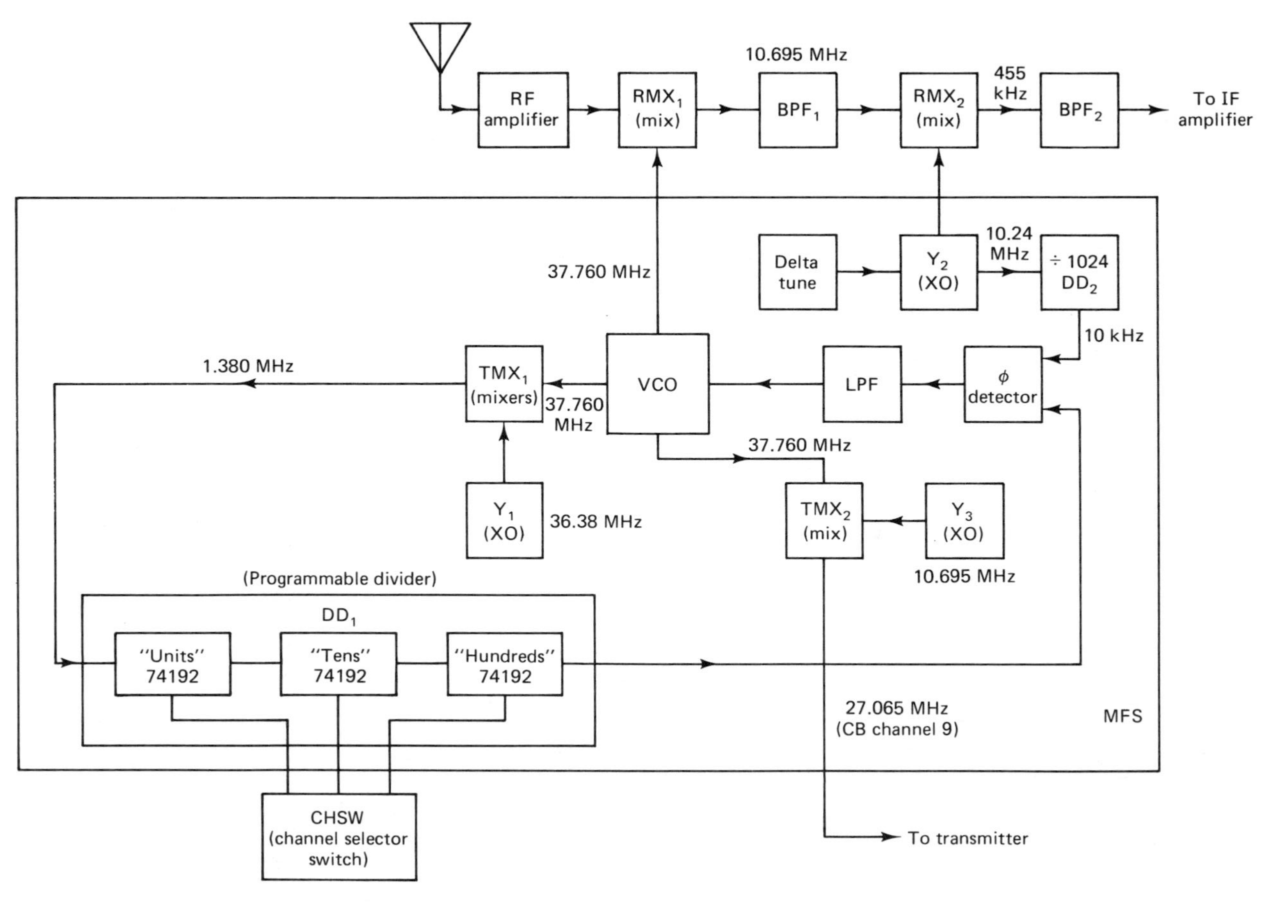

Figure 9.14 Master frequency source (MFS) for 40-channel CB transceiver.

of the third crystal oscillator, Y_3, in mixer TMX_2 to produce the 27.065-MHz carrier frequency required for channel 9.

5. The output of the VCO is the local oscillator input to receiver first mixer, RMX_1. The difference frequency, 10.695 MHz, called the first IF frequency, is fed through a bandpass filter BPF_1 to a second receiver heterodyne mixer, RMX_2. The local oscillator source for RMX_2 is the signal from crystal oscillator Y_2. The output, as indicated, is a conventional IF frequency—455 kHz. This frequency can be modified by the DELTA TUNE function. DELTA TUNE is a simple circuit actuated by an operator-controlled, spring-return toggle switch. While actuated, the switch connects a small amount of capacitance across the crystal, thus changing its operating frequency slightly. The purpose of DELTA TUNE is to permit a receiver to tune in more efficiently a transmission whose carrier signal is slightly off-frequency.

It should be noted that for all 40 CB channels, the output frequency of transmitter mixer TMX_1 is an integer multiple of the reference frequency from DD_2, 10 kHz. As we have seen, for channel 9, the output of TMX_1 is at 1.38 MHz. Let's check other frequencies. The frequency of CB channel 1 is 26.965 MHz. Hence the VCO must operate at 37.660 MHz for this channel. The output of TMX_1 for channel 1 is 1.28 MHz (37.660 − 36.380 MHz), and the required modulus for DD_1 is 128. For channel 33 (f = 27.335 MHz) the figures are: VCO frequency = 27.335 + 10.695 MHz = 38.030 MHz. Output of TMX_1 = 38.030 − 36.380 MHz = 1.65 MHz. A DD_1 modulus of 165 is required.

In summary, the master frequency source (MFS) of Fig. 9.14 provides the frequencies required for the operation of a transceiver on all 40 CB channels. The scheme requires only three crystal oscillators. A VCO provides signals for both the receiver and transmitter functions. The VCO is, in effect, a variable-frequency, crystal-controlled source. The scheme utilizes a phase-locked-loop function. Variable frequency is obtained using a programmable digital divider. A crystal-controlled reference for the loop utilizes a fixed-modulus digital divider. Heterodyne frequency conversion is used to supplement the frequency conversion of the digital frequency dividers.

9.5 SINGLE-SIDEBAND (SSB) TRANSMISSION

As we saw in the preceding section, multiple-frequency operation is a feature of most two-way radio equipment. It is not often found in so-called broadcast operation. (Some international short-wave broadcast stations operate on more than one frequency.) Another operating feature unique to two-way radio is single-sideband operation. You will recall from Chap. 2 that amplitude modulation of a carrier with a single modulating frequency produces two side frequencies. The upper-side frequency is equal to the carrier frequency plus the modulating frequency; the lower-side frequency is the difference of the carrier and modulating frequencies. However, the modulating signal is normally a composite of many frequencies. For example, when someone speaks, the sound contains numerous frequencies. Hence, in normal amplitude modulation, not side frequencies, but bands of side frequencies are produced. There is an upper sideband and a lower sideband.

All of the information of a modulating signal is contained in each of the sidebands, which are the products of amplitude modulation. Both the carrier and one or the other of the sidebands is redundant to a communication transmission: they are not needed. In the interest of saving a significant amount of the limited resource—frequency spectrum—the carrier and one sideband can be eliminated during a communication transmission. Single-sideband transmission also provides for more efficient utilization of the energy available to power a transmitter. Single-sideband (SSB) operation, then, is a method of operating a transmitter such that the carrier and one sideband is eliminated or suppressed. Single-sideband transmission places a special requirement on the receiver of the system. Demodulation of an SSB transmission can occur only in the presence of the carrier. An SSB receiver must therefore be equipped to supply the equivalent of the carrier of an SSB transmission. This is not difficult. A local oscillator, much like the one used in a heterodyne frequency converter, can do the job.

Utilizing SSB transmission in a radiotelephone system simplifies the form of the emission. However, the circuits required for generating the emission are more complex than for amplitude modulation, for example.

The task of generating an SSB emission can be thought of as consisting of two parts: (1) suppressing the carrier, and (2) suppressing one sideband. Carrier suppression is invariably by means of a circuit called a *balanced modulator*. There are at least three methods of suppressing a sideband. The most popular method of sideband suppression is called the *filter method*. It utilizes filters with very narrow passbands to filter out the sideband not to be transmitted. In the *phase-shift method* of sideband suppression, two suppressed-carrier double-sideband (DSB) signals are produced. The carriers of these two signals are identical except for a 90° phase shift between them. The two carriers are modulated by two versions of the audio signal—two versions with a 90° phase shift between them. When the two DSB signals are recombined, one sideband is eliminated as a result of the phasor summation. Let us look in detail, first, at a balanced modulator, then at the filter method of sideband suppression. We will look only briefly at the phase-shift method.

The Balanced Modulator

A balanced modulator can be viewed as a circuit consisting of two parts or branches working together in a certain way. When a carrier signal only is applied to the circuit, the currents in the two branches are equal (i.e., "balanced") but flow in opposite directions in a common branch. Their effects cancel: the carrier is suppressed. When a modulating signal is also applied to the circuit, AM-type side frequencies are produced. The nature of the circuit is such that these currents are not balanced; the side-frequency currents are not suppressed. However, the carrier is still suppressed.

The bipolar-transistor circuit of Fig. 9.15 is an example of a balanced

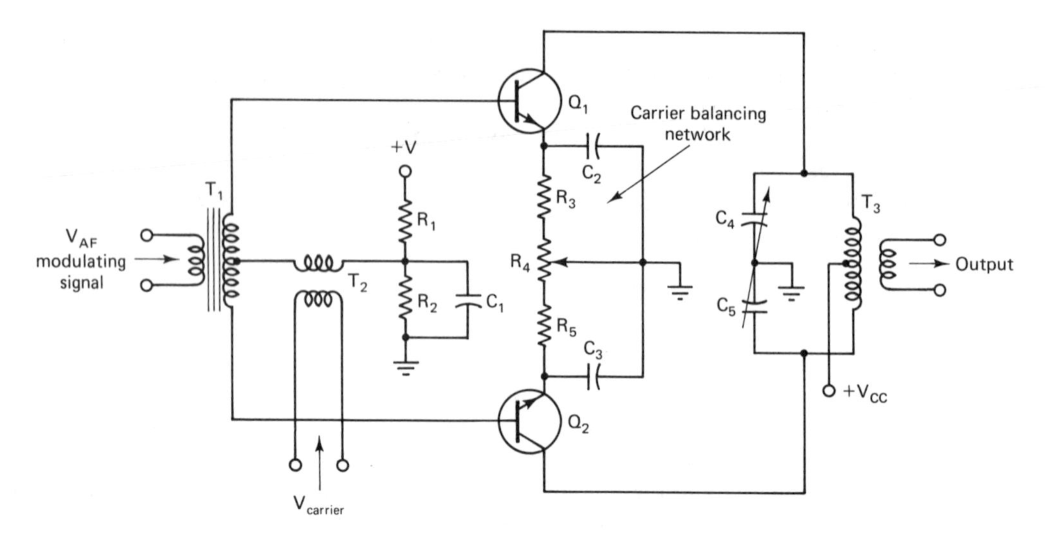

Figure 9.15 Balanced modulator circuit.

modulator. Since it resembles the audio push-pull amplifier that we studied in Chap. 3, it can be useful in learning how a balanced modulator functions. Let us observe these important facts about the circuit:

1. The primary of output transformer T_3 is part of a tank circuit tuned to the carrier frequency.
2. Because transformer T_1 splits the phase of the modulating signal input, the circuit acts in push-pull with respect to the modulating signal.
3. The RF carrier is coupled into the circuit in such a way that the carrier currents in the two halves of the circuit are in phase; they are not in push-pull. If transistors Q_1 and Q_2 are carefully selected for balanced characteristics, the carrier currents will be equal or balanced. Since these currents flow in opposite directions in the two halves of the primary winding of the output transformer, their effects will cancel. The carrier will be suppressed in the output of the circuit.
4. Because the output circuit is tuned to a radio frequency, its impedance at audio frequencies will be insignificant; no significant amount of modulating signal will appear in the output of the modulator.

Let us now consider the operation of the circuit when both an audio and an RF signal are applied to the inputs of the modulator. The audio signal, in effect, shifts the bias of transistors Q_1 and Q_2 in a push-pull fashion. The RF signal has sufficient amplitude to cause the transistors to operate in the nonlinear region of their operating characteristics. With both AF and RF signals present in a nonlinear circuit, typical amplitude modulation action occurs. Upper and lower side-frequency currents are produced in each stage of the circuit. However, because the modulating signal is shifting the bias of the two transistors in a push-pull fashion, the amplitudes of the products of modulation are unequal in the two halves. The side-frequency currents in Q_1 are increasing in amplitude while those in Q_2 are decreasing, and vice versa. The effects of this action in the output transformer are just like those in a push-pull amplifier. An increasing current in one half of the output transformer primary winding is aided by a decreasing current in the other half. The carrier current is suppressed, as described above. The balanced modulator produces the sidebands of amplitude modulation but suppresses the carrier.

A balanced modulator can be constructed using diodes only. A simple two-diode circuit is shown in Fig. 9.16. The arrangement is similar to that of the circuit we have just studied: The RF feed is parallel with respect to the two diodes. The audio feed incorporates a 180° phase shift between the two halves of the circuit. The diodes must be driven sufficiently for their operation to be nonlinear so as to produce modulation products. With no audio input signal present, the conduction of the diodes is balanced and the carrier currents cancel out in the output of circuit. When an audio signal is present, the diodes are biased in opposite directions at an audio rate. The two diodes now conduct at different rates. The side-frequency currents in one half the circuit are increasing while those in the other half are decreasing, etc., producing push-pull action in the output circuit. The output circuit is tuned to the input carrier frequency as in the previous circuit. The output of the circuit is a double-sideband signal with carrier suppressed.

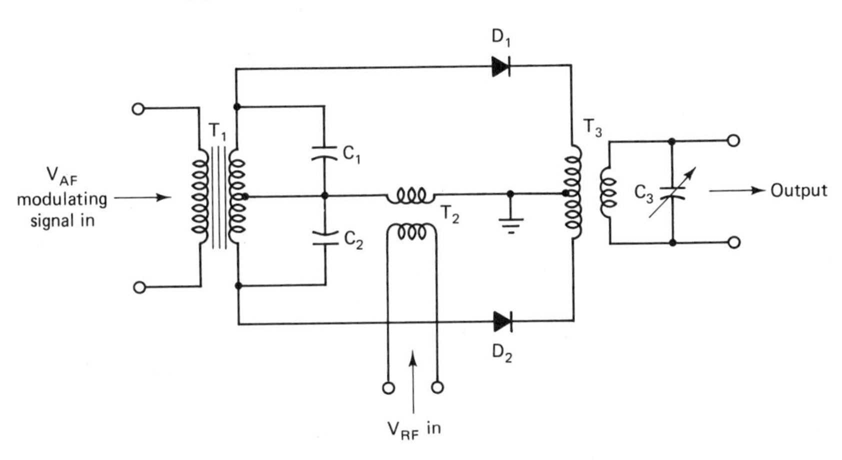

Figure 9.16 Diode-type balanced modulator.

A second, diode, balanced modulator circuit is known as a *ring* or *lattice-type* modulator. A circuit to illustrate this form of the balanced modulator is shown in Fig. 9.17. The particular circuit shown also illustrates the use of *RC* coupling as distinguished from the transformer coupling used in the two previous circuits.

Let's study the action of this circuit. Consider electron flow when an RF signal only is applied to the circuit. The RF input is at the adjustable center of the resistor network connected across the primary winding of the output transformer.

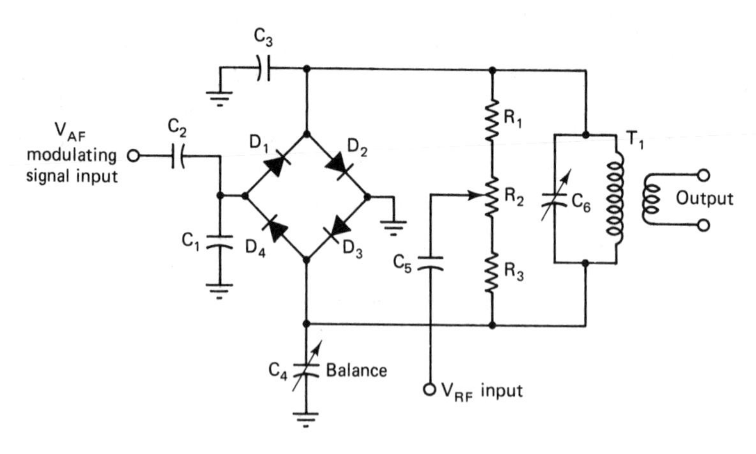

Figure 9.17 Diode ring or lattice-type balanced modulator.

During the positive alternation of the RF signal, electrons will be drawn to this point through two paths: (1) from ground through diode D_2, R_1, and a portion of R_2; (2) from ground through C_1, diode D_4, R_3, and a portion of R_2. At radio frequencies the reactance of C_1 is negligible. The paths are identical or can be adjusted to be so by R_2. The potentials with respect to ground of the two ends of the primary winding of the output transformer are equal: there is no voltage across the output transformer for the carrier alone. During the negative alternation of the carrier signal there are similar identical paths for carrier currents: (1) from the input point through a portion of R_2, through R_3 and diode D_3, to ground; and (2) from the input point through the upper portion of R_2, through R_1, diode D_1, and C_1, to ground. Again, there would be no voltage across the output transformer: the carrier is suppressed.

When an audio modulating signal is applied to this circuit at the junction of diodes D_1 and D_4, the balance of conduction of the diodes will be disturbed. For example, with a positive alternation of audio signal at the junction of diodes D_1 and D_4, the conduction of diodes D_1 and D_2 will be enhanced, that of D_3 and D_4 will be diminished. The drop across R_1 will be increased, that across R_3 decreased. With the products of modulation present there is now a voltage across the output transformer. When the audio signal changes to the opposite alternation, the situation is reversed. The circuit produces outputs of the side frequencies. From experience it is known that for good results, the RF signal amplitude should be six to eight times that of the audio signal. Also, the diodes should be matched for forward-to-reverse resistance values. The circuit in Fig. 9.17 includes adjustments for balancing the circuit. Good circuit balance is required for maximum suppression of the carrier. With matched diodes (good circuit balance), suppression of the order of 40 to 60 dB is obtainable with the diode ring balanced modulator.

Several integrated circuits are available which make for good balanced modulators or heterodyne frequency mixers. One of these, the RCA CA3039, is a diode array. See the schematic diagram in Fig. 9.18. Since the diodes are

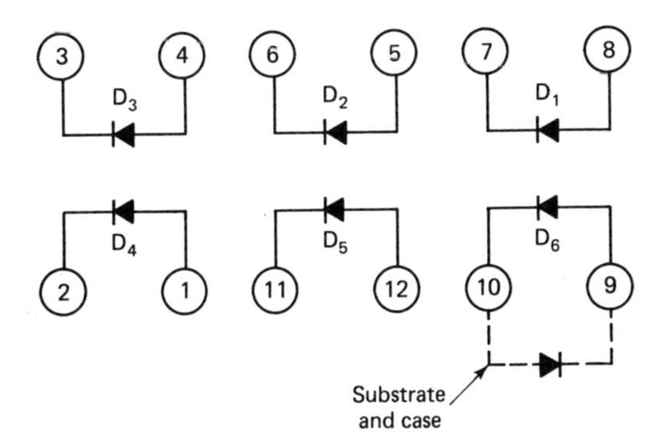

Figure 9.18 Schematic diagram of RCA CA3039 diode array IC.

fabricated on a single substrate (semiconductor wafer) they are especially well matched. The chip is suitable for use in the ring modulator circuit. There is also a chip made just for use as part of a balanced modulator—the Motorola MC1596G or Signetics S5596. A sample circuit is shown in Fig. 9.19. Internally, this chip is an op-amp multiplier: it multiplies the carrier and modulating signals. It can be shown with the appropriate mathematical techniques that the result of multiplying two sinusoidal signals is a composite signal containing only the sum and difference frequencies of the two signals. As indicated, the chip provides for a balance adjustment to achieve maximum carrier suppression.

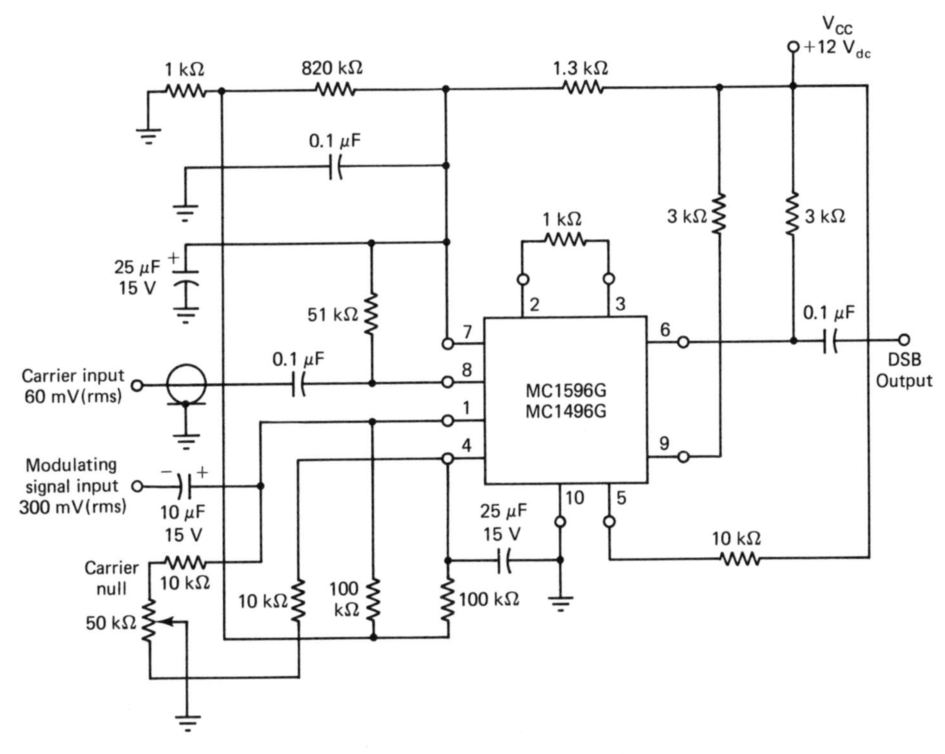

Figure 9.19 Balanced modulator using Motorola MC1596G chip. (Courtesy of Motorola Semiconductor Products, Inc.)

Sideband Suppression: Filter Method

The suppressed-carrier output of a balanced modulator is often refered to as a *double-sideband signal* (DSB). Of course, a true AM signal is a double-sideband signal because it always includes two sidebands. However, the designation "DSB" is generally reserved for the suppressed-carrier signal.

The two sidebands in a DSB signal are separated in frequency only by an amount equal to two times the lowest audio frequency in the modulating signal. This is true, you will recall, because there is a side frequency for each modulating frequency. The lowest upper-side frequency is equal to the carrier frequency plus the lowest audio frequency; the highest lower-side frequency is equal to the carrier minus the lowest audio frequency. For example, if the lowest audio frequency in a modulating signal is 50 Hz, the sidebands will be separated by only 100 Hz.

The narrow separation between the sidebands of a modulated signal makes the task of eliminating one of them difficult. Simple *LC* filters do not have sufficiently sharp cutoff characteristics to do the job. The task can be made somewhat simpler if the lowest audio frequency is limited to about 300 Hz. It has been found that the communications value of voice transmissions are not degraded significantly by this limitation. This means that the response of the ideal filter must change from zero attenuation to maximum attenuation over a range of 600 Hz. The response curve of such a filter is said to have very "steep skirts." The difficulty of achieving this kind of response is also affected by the absolute frequency of the sideband. The higher the operating frequency, the higher must be the Q of the filter circuit. The result is that some form of mechanical or crystal filter system is used for all SSB generating systems operating above about 100 kHz.

Sometimes it is desired that a SSB system afford the user the choice of operating with either the upper sideband (USB) or the lower sideband (LSB). This, of course, complicates system design even further. Before studying filters in detail, let us examine a general scheme for generating SSB signals.

The block diagrams of Fig. 9.20 illustrate two common schemes for producing SSB carrier signals with a choice of using either the USB or LSB. You will observe that in one case, Fig. 9.20(a), the carrier oscillator is switched to obtain sideband selection; only one filter is required. In Fig. 9.20(b), sideband selection is obtained by switch selection of one of two filters. In the first case, to produce an emission using the USB, the carrier oscillator is switched to the lower of two operating frequencies. The center frequency of the USB matches the center frequency of the filter. Selecting the carrier oscillator with the higher of the two operating frequencies produces a LSB that matches the center frequency of the filter.

In Fig. 9.20(b), it is obvious that the choice of using USB or LSB is made by selecting a filter with the appropriate frequency characteristic. In either case, the final carrier frequency is obtained by heterodyne frequency conversion of the output of the SSB generator. You will note that either scheme requires two oscillators, one of them of variable frequency. In a modern transceiver, these

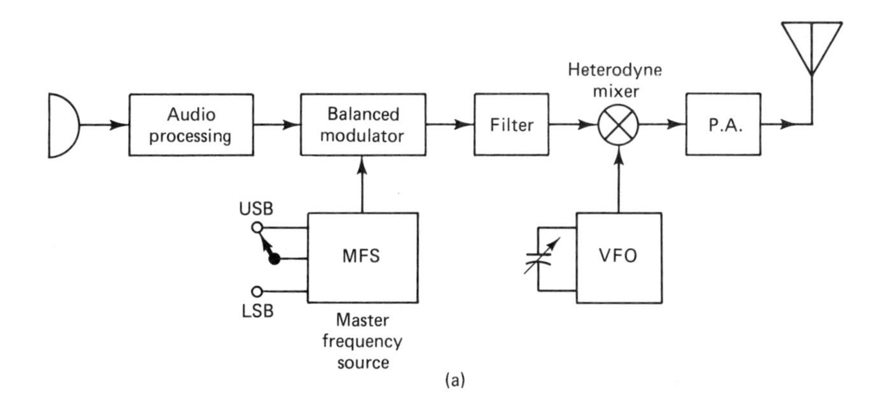

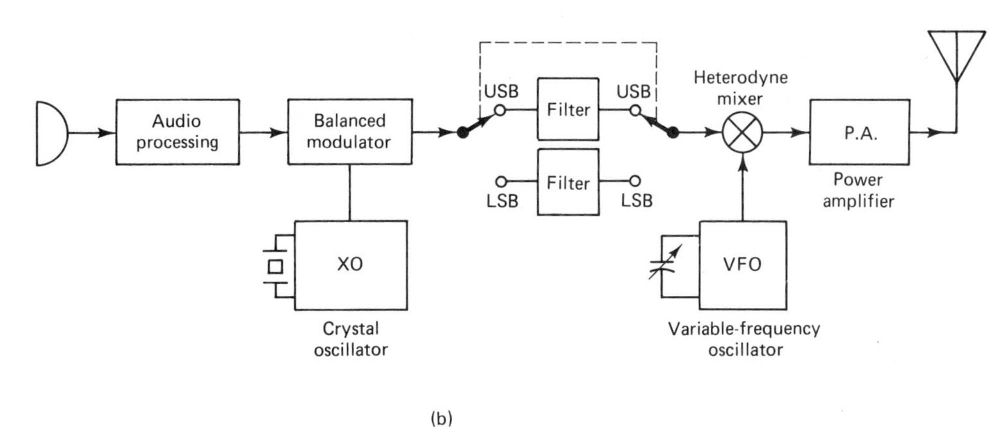

Figure 9.20 Schemes for single-sideband generation: (a) selecting upper or lower sideband by switching carrier oscillator frequency; (b) selecting sideband by switching filters.

oscillators would invariably employ techniques of frequency synthesis similar to those described in the preceding section.

Filters

An electrical filter is, by definition, a circuit or device that allows the energy of certain frequencies to pass freely through itself but impedes the passage of the energy of other frequencies. It is obvious, then, that filters must be composed of elements whose characteristics change with frequency. For example, a very simple filter can be made of a resistor and either a capacitor or inductor. Simple filters, circuits of few parts, have unsophisticated "filtering" characteristics. Filters with high-performance characteristics, steep skirts for example, require many circuit elements and extensive design effort.

The design of filters is one of the most extensively pursued activities in the field of electricity. Entire books have been written on the subject. As one can imagine, the mathematics required is high level. Be that as it may, it is not the goal of this text to present filter design methods. We are interested primarily in being able to cope with a sideband filter when we meet it face to face. We want to know enough about them not to be intimidated when we must be involved with one, in a troubleshooting situation, for example. Therefore, it will be useful to know some of the terminology used in conjunction with filters. We will want to know what are typical characteristics of sideband filters. Certainly, we will want to be exposed to some typical designs.

Filter terminology. There are several terms and phrases that are used by virtually everyone who writes or talks about filters. Learning the meanings of these terms requires some effort. However, the effort is well worthwhile since it makes communication about the subject so much more efficient: If we know the terminology, we can learn a great deal more in a much shorter time.

When we are looking at the performance of a circuit such as a filter, we are usually interested in comparing the output voltage (or current) with the input. The comparison is that of a ratio: output voltage divided by input voltage, $v_{\text{out}}/v_{\text{in}}$. If the ratio is greater than 1, the circuit is said to have a *gain*. If the ratio is less than 1, the circuit is producing a *loss*. Gains and losses in conjunction with filters are almost always expressed in *decibels* (dBs). (An extensive presentation on decibels is included in Appendix A.) A decibel is a logarithmic way of expressing a gain or loss ratio. This method is popular because the human response to sound levels, for example, is logarithmic. Specifically, a voltage gain, or loss, in decibels, is equal to 20 times the common logarithm of the voltage ratio. The formula for the gain, in decibels, of a filter is expressed

$$\text{gain (dB)} = 20 \log_{10} \frac{v_{\text{out}}}{v_{\text{in}}}$$

You can evaluate gains in decibels easily with the aid of your electronic calculator if it has a "LOG" key. When a ratio is less than 1, its logarithm is negative; the gain in decibels is negative. A negative dB gain, then, implies an actual reduction in voltage amplitude. Everything that has been stated here applies equally well to currents.

The band of frequencies in which a filter allows energy to pass relatively freely is called its *passband*. The *stop band* of a filter is the band of frequencies in which a filter is successful in impeding passage. There is never an extremely abrupt change in gain between a passband and a stop band. The definition of the boundary is somewhat arbitrary. The boundary is called the *cutoff frequency*. A cutoff frequency of a filter is that frequency at which its gain is equal to 0.707 times its maximum gain; or the gain is down by 3 dB from the maximum gain, and the gain continues to decrease "beyond" that frequency.

A *bandpass filter* is one that has two cutoff frequencies, a low cutoff at the lower end of the passband and a high cutoff at the upper end of the passband. The band of frequencies between the two cutoff frequencies is called the *bandwidth* of the filter. A filter for obtaining a single sideband from a DSB signal is a bandpass filter. The terminology described here is illustrated in Fig. 9.21.

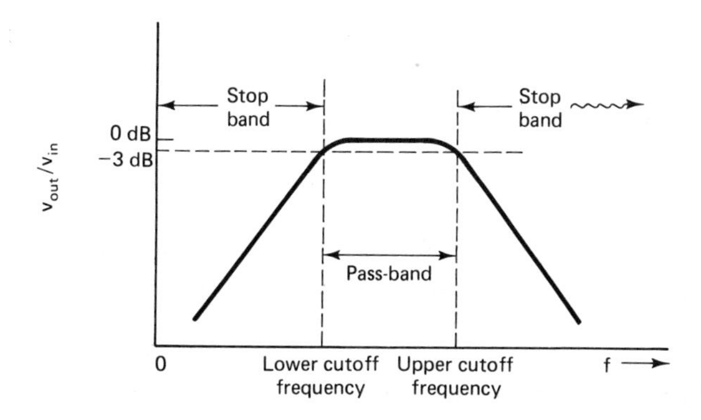

Figure 9.21 Filter performance terminology.

A filter is said to *attenuate* the signal in its stop band. Generally, the attenuation increases with the frequency offset from the cutoff frequency. Let's look at a simple low-pass filter and its response curve to clarify certain terms and concepts (see Fig. 9.22). You will observe that the circuit in Fig. 9.22(a) has only one reactive element, C. The cutoff frequency of such a simple circuit is the frequency at which $R = X_C$. At that frequency the voltage across C will equal 0.7071 times the maximum possible voltage. The cutoff frequency can be determined from the formula

$$f_C = \frac{1}{2\pi RC}$$

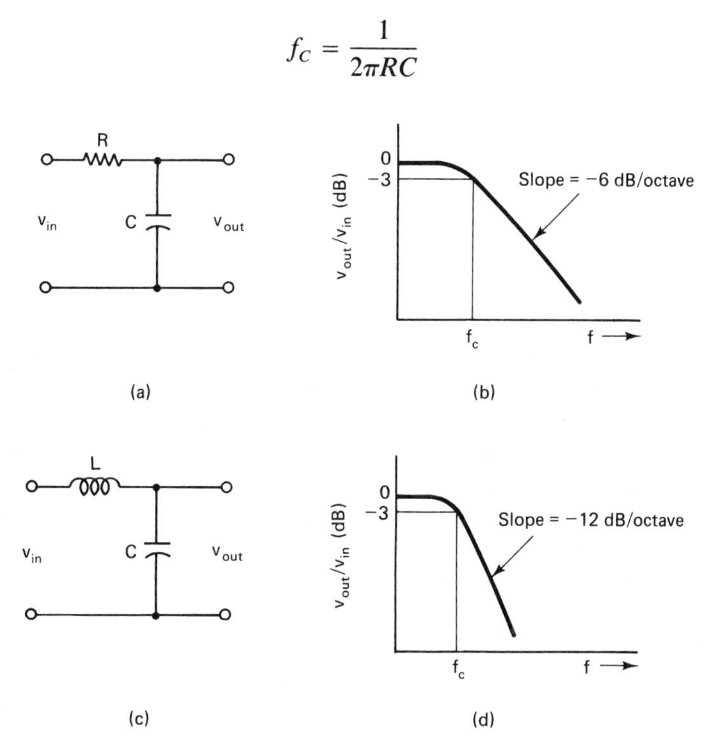

Figure 9.22 Filter concepts: (a) simple first-order low-pass filter; (b) response curve of first-order low-pass filter; (c) simple second-order low-pass filter; (d) response curve of second-order low-pass filter.

At frequencies higher than the cutoff frequency, the attenuation of this filter increases at a constant rate: −6 dB/octave (frequency doubles in one octave). With one reactive element the filter is said to be a *first-order* filter. If an inductor is connected in series with the signal path [see Fig. 9.22(c)], the circuit becomes a *second-order* filter. The attenuation in the stop band increases at twice the rate with respect to frequency: −12 dB/octave. This falloff in gain is proportional to the square of the frequency. A third-order filter would have a falloff proportional to the cube of the frequency. Its attenuation would increase at the rate of −18 dB/octave, and so on.

In general, the order of a filter corresponds to the number of reactances it contains. This is true only if reactances of the same type are not simply in parallel or series with each other. The order also corresponds to the power of the frequency in the equation that states the response of the filter.

A circuit capable of performing a filter function by itself is often called a *filter section*. A section can be first-order, second-order, etc.

Complex filters with sophisticated response curves (steep skirts, for example) can be constructed by *cascading* individual sections (see Fig. 9.23). Sections are cascaded when the output of one becomes the input to the next. There must be no interaction between the sections for true cascading. When filter sections are truly cascaded, each contributes its own effect to the overall response of the combined filter. With careful design of sections and the combining of a number of sections, filters with a variety of desired response characteristics can be constructed.

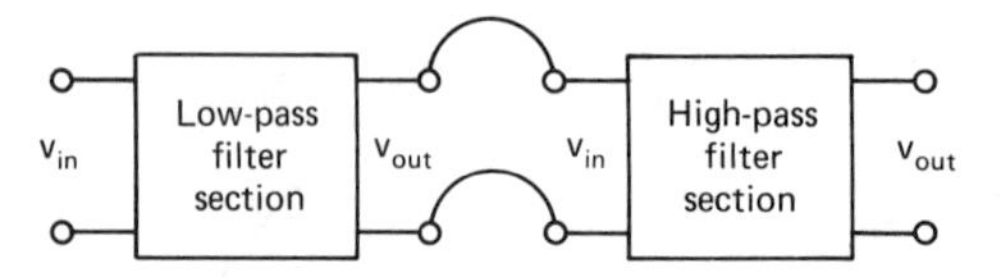

Figure 9.23 Cascading low-pass and high-pass filter sections to produce a bandpass filter.

Often, filters contain elements that can achieve the condition of resonance, either series or parallel (antiresonance). If the resonance effect produces a peaking of the output response, the frequency at which this occurs is called a *pole* of the response. If there is a frequency where there is an abrupt dip in the response of the filter, the point is called a *zero* of the response. If a filter has two peaks in its response, it is said to be a two-pole filter. A three-pole filter would have three peaks in its response curve, etc. In general, the order of a filter is equal to two times the number of poles.

The response curve of some filter designs has peaks and valleys in its pass band (see Fig. 9.24). This effect is referred to as *ripple*. When ripple is expressed in decibels, it refers to the ratio of minimum attenuation in the passband to maximum attenuation in the passband.

In some filters, the minimum attenuation in the passband is not zero. That is, the maximum voltage, say, at the output of the filter is less than the voltage applied to the input. This falloff in voltage can occur because of losses in resistive elements in series with the signal path. The reduction is referred to as *insertion loss* and is commonly expressed in decibels.

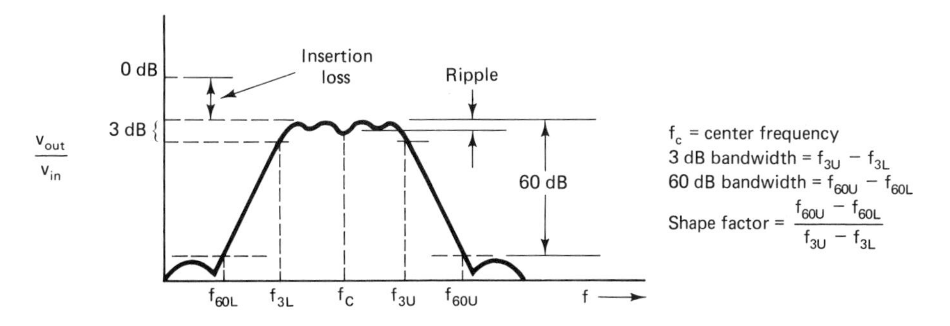

Figure 9.24 Filter response-curve terminology.

Finally, a characteristic often referred to in the description of a filter is the *shape factor.* Shape factor (SF) provides a measure of the steepness of the skirt(s) of a filter. Specifically, shape factor is the ratio of the bandwidth where attenuation is 60 dB to the bandwidth where attenuation is 3 dB (the cutoff frequency). A "perfect" bandpass filter would have vertical skirts; its shape factor would be 1. A perfect filter is not physically realizable; shape factors of real filters are always greater than 1 (see Fig. 9.24).

Practical Sideband-Elimination Filters

Classical filter theory is based on the use of inductors and capacitors for the filter elements. Filters with sophisticated characteristics, such as steep skirts (SF approaching 1), require circuits of high Q. Unfortunately, except for frequencies up to about 100 kHz, LC circuits do not provide Q's of sufficient magnitude for use as filters in sideband suppression applications.

Most commercially available SSB transceivers utilize some form of piezoelectric or metallic-plate filter for sideband suppression. In a sense these are all a form of mechanical filter because their electrical properties are the result of a mechanical vibration of a crystal or metallic plate.

We have already studied the electrical characteristics of quartz crystals (see Chap. 5). You will recall that a mounted crystal has both a series resonance mode and a parallel resonance mode. Crystals may be connected together in various ways to produce filters with extremely sophisticated performance characteristics. We look briefly at crystal filter circuits below.

A metallic-disk filter is a device that utilizes the resonance effect of vibrating metallic disks. A mechanical filter package includes several so-called disk resonators mounted in sandwich fashion between an input and an output transducer. The transducers provide the electrical/mechanical, mechanical/electrical energy conversion required by this kind of filter device. Electrically, the metallic-disk filter is equivalent to six or more parallel resonant circuits connected as the shunt elements in a ladder-type network (see Fig. 9.25). Up to frequencies of about 500 kHz this filter is capable of providing the equivalent of a very high-Q circuit. It has exceptionally high frequency stability. These devices are very reliable and little subject to failure. They are manufactured in packages that can be

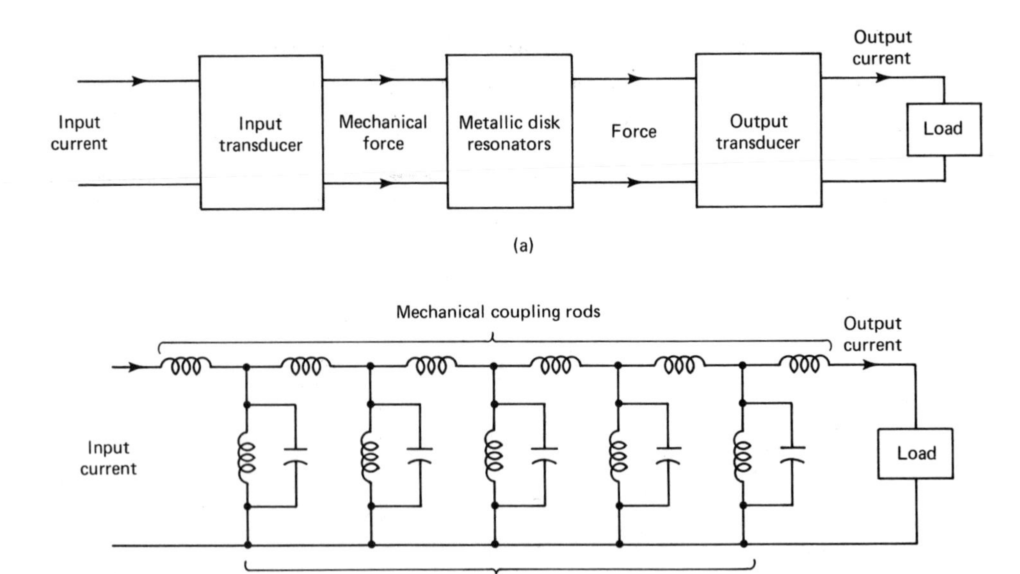

Figure 9.25 (a) Block diagram of metallic-disk filter; (b) circuit analogy of mechanical filter.

mounted on and soldered to printed-circuit boards, somewhat like IF transformers. If failure occurs, the package is simply replaced.

Crystal filters. Crystals with the same or different frequency specifications can be connected together in various ways to produce a variety of filter response characteristics. Because a crystal possesses the characteristics of a circuit that has both a series- and a parallel-resonant mode, it is a *single-pole device*. That is, when considered as a single-element filter, it has both a pole and a zero in its response curve (see Fig. 9.26). These two frequencies are important in the application of crystals as filters. The difference between the two, called the *pole–zero spacing* of the crystal, is also of interest in filter applications.

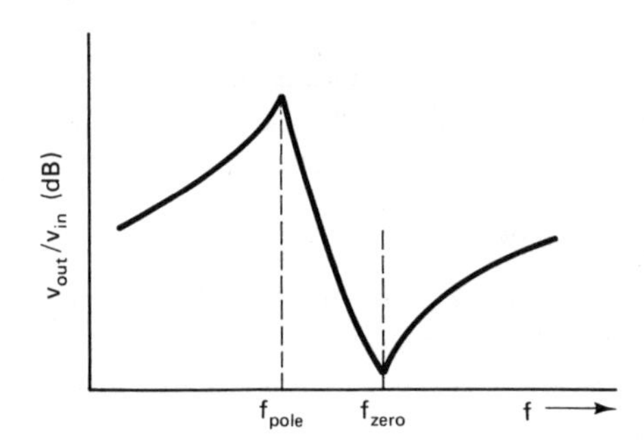

Figure 9.26 Response curve of single crystal as filter.

A classical and still popular crystal filter circuit is the *full lattice* shown in Fig. 9.27. You will observe that crystals Y_1 and Y_2 are in series with the signal path. They are chosen so that their poles (series-resonant frequency) are matched and in the desired passband. They provide a relatively low-impedance path for signal currents at a desired frequency and a narrow band of frequencies to either side of that frequency. Crystals Y_3 and Y_4 are effectively in shunt with the signal path. They are selected with matched zeros (parallel-resonant frequency) which are equal to the poles of Y_1 and Y_2. However, the poles of Y_3 and Y_4 are equal to the zeros of Y_1 and Y_2. Therefore, they shunt little of the signal energy in the desired passband, but effectively short out the signal currents at frequencies outside the desired passband. The high impedance of Y_1 and Y_2 at their zeros greatly attenuate signal currents just beyond the desired passband. Filters with almost ideal shape factors can be obtained if careful attention is given to the selection of crystals with appropriate poles, zeros, and pole–zero spacing.

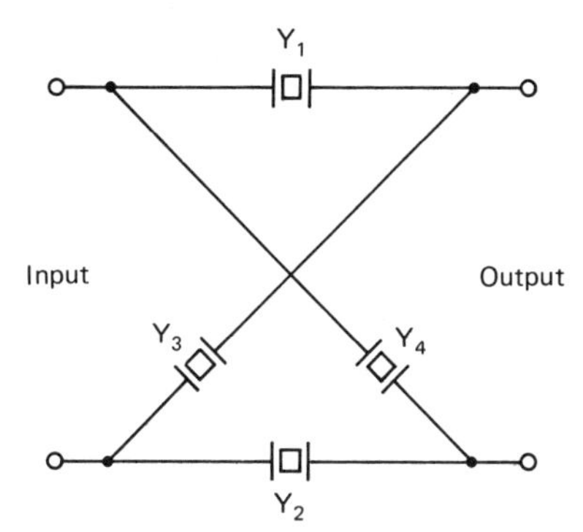

Figure 9.27 Crystal lattice filter.

Other crystal-filter circuits can be and are being used. Complete crystal filters with specified center frequencies and bandwidths are also commercially available as packages ready to mount on printed-circuit boards.

Another type of filter that is a relatively recent development is the *ceramic filter*. You know the word "ceramic" in connection with "clay," "pottery," "earthenware objects," etc. A ceramic material, or clay, is firm, fine-grained earth. It may be composed of a number of different chemical substances. Ordinary clay for pottery making is mostly hydrous aluminum silicate. A ceramic material made of lead zirconate titanate (PZT) is a crystalline compound. When formed and dried as a small disk, it provides piezoelectric properties very much like those of crystal wafers cut from natural quartz crystals. Ceramic disks can be combined in various ways to form effective filter circuits. Packaged ceramic filters are available for immediate installation. They are rugged and relatively less expensive than either metallic-disk or crystal filters. Ceramic filters are produced by a number of manufacturers. Such devices are most commonly available with center frequencies of either 455 kHz or 10.7 MHz. They are available with several different bandwidths: 2 kHz, 3 kHz, 4.5 kHz, and 10 kHz are just some of the bandwidth figures that can be found in advertisements for ceramic filters.

Sideband Suppression: Phasing Method

Because the technology of filter design and construction has advanced very rapidly in recent years, the filter method of sideband suppression has become the most popular method for accomplishing that function. However, the phasing method was once popular and deserves at least a brief look.

The phasing method of eliminating one sideband of a double-sideband signal is accomplished by means of the following individual functions:

1. The audio (modulating) signal is split into two signals of equal amplitude. These two signals are then shifted in phase with respect to each other by 90°.
2. The RF carrier signal is split into two signals of equal amplitude. These two RF signals are also phase shifted by 90°.
3. Each of the phase-shifted RF carriers is modulated by a phase-shifted audio signal in a balanced modulator. This produces two DSB signals.
4. The two DSB signals are recombined in a linear circuit. The resulting signal is the phasor sum of the two DSB signals. The phasing of the sideband components is such that one sideband is eliminated. The other sideband is reinforced.

A block diagram of a scheme for accomplishing the described process is shown in Fig. 9.28. The system can be arranged to provide for selecting either the USB or LSB. All that is required is a switch for reversing connections either between the phase-shifted audio signals or the phase-shifted carrier signals.

The phasing method appears simple in concept. However, practical accomplishment of the process is another matter. Sideband elimination is only as good as the phase shifting. It is not easy to provide circuits that achieve exactly 90° phase shift across the width of a sideband.

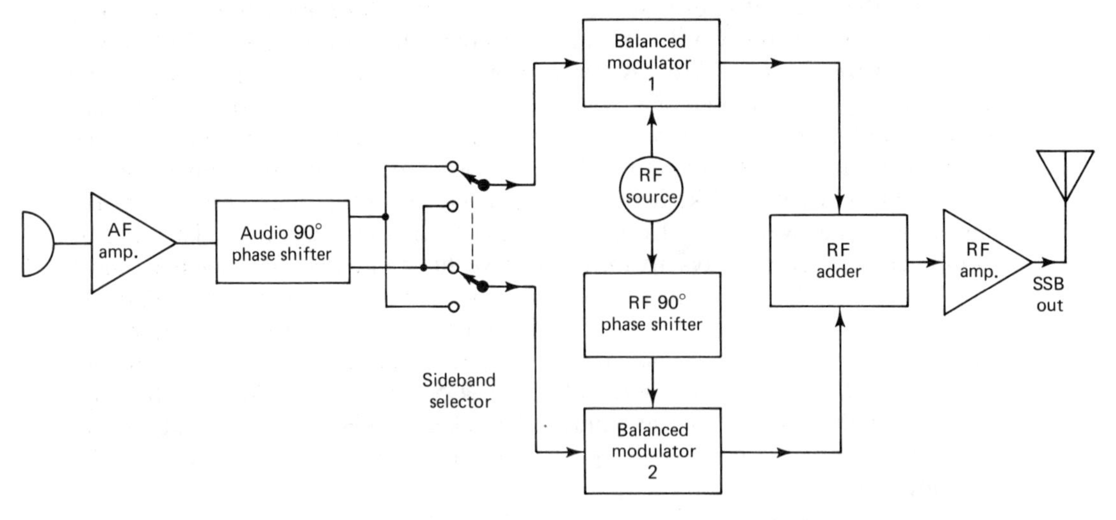

Figure 9.28 SSB by phasing method.

9.6 AN SSB RADIOTELEPHONE SYSTEM

Let us now study the details of the block diagram of a complete SSB radiotelephone system. The diagram of a complete system will illustrate how the various functions with which we are familiar can be fitted together to perform an overall communications function. It will also reveal any functions with which we may not be familiar.

Sideband transmission (SSB transmission is often referred to simply as "sideband" transmission) is used extensively in the range 1.8 to 30 MHz. It has been popular with amateur radio operators for many years. More recently, it is being used to a limited extent in CB radio. SSB operation is now compulsory in the medium-frequency marine band (2 to 3 MHz). It is used by commercial two-way communications services operating in the below-30 MHz range, and so on. There are significant differences in the equipment used by these various services. However, the similarities of the SSB equipments of the various services are sufficient to make it practicable to study one system as a model for the others. This is fortunate since time and space are limited.

We start with the block diagram of Fig. 9.29. You will note immediately that you are familiar with the function titles you find there. One exception is perhaps the FREQUENCY COUNTER DISPLAY unit, which we will study in due course. Your attention is called to the fact that virtually all functions are shared between the TRANSMIT and RECEIVE operations. The block diagram of this figure is useful for indicating the signal flow path for TRANSMIT and RECEIVE functions. Let us go to another level of detail to learn more about how this particular transceiver performs its functions of generating and receiving SSB signals.

Study carefully the diagram of Fig. 9.29. Let's analyze how the system functions when it is in the TRANSMIT mode. We note that generating an SSB signal for emission from the antenna consists of the following processes:

1. A signal from the CAR unit is fed to the AUDIO PROCESSING unit, where it is modulated in a balanced modulator by an audio signal from the microphone.
2. The output of the balanced modulator, a DSB signal, is fed to the IF unit. In the IF unit the signal is amplified and processed by a bandpass crystal filter to eliminate one of the sidebands.
3. The output of the IF unit, now an SSB signal, is passed to the RF unit. In the RF block the SSB signal is converted to its final carrier frequency by mixing with a signal from the output of the VCO of the PLL block. RF amplification increases the signal amplitude in preparation for driving the final stages of the transmitter.
4. The output of the RF unit is fed to a DRIVER amplifier stage in the FINAL block. The POWER AMPLIFIER is driven by the DRIVER to increase the energy content of the SSB signal to its rated level. (The unit depicted is the Kenwood Model TS-130S, HF SSB Transceiver. Its nominal power rating is 200 W.)

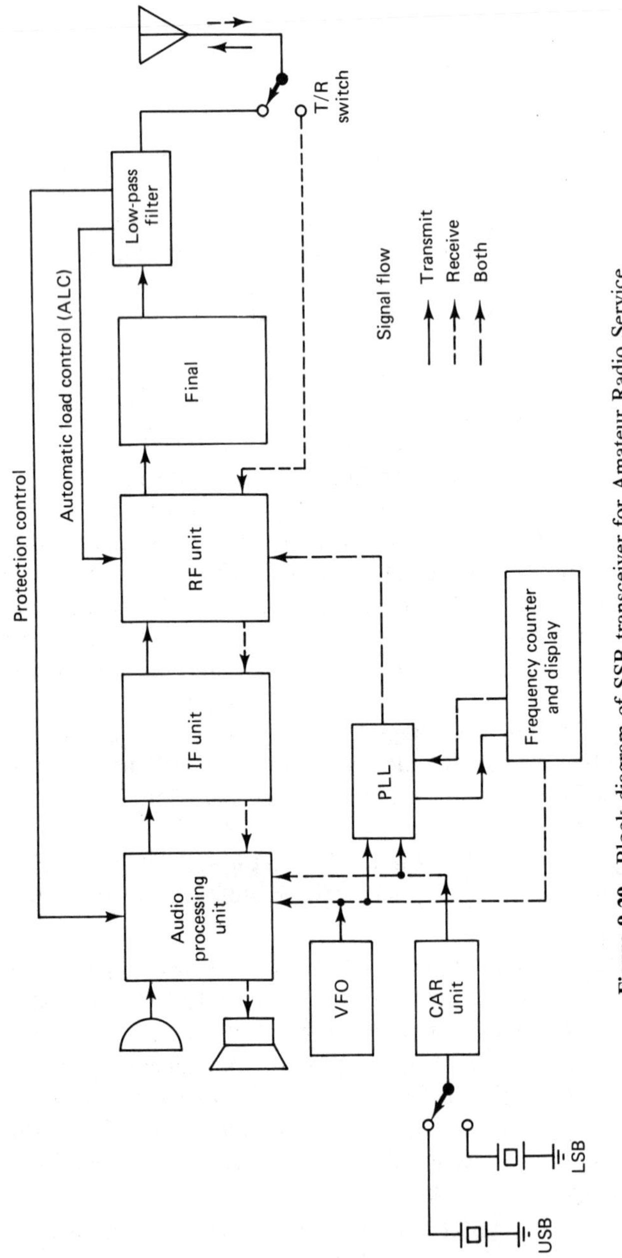

Figure 9.29 Block diagram of SSB transceiver for Amateur Radio Service.

5. The signal is passed through a FILTER unit to remove harmonic and other spurious signals. It is finally passed through the contacts of a TRANSMIT-RECEIVE RELAY to the station antenna.

It will be highly instructional to examine the many interesting details of the functions just enumerated. We will do that in due course. First, however, let's do a first-level analysis of functions for the unit operating in the RECEIVE mode.

Let's examine Fig. 9.29 again. This time we will observe the signal flow path defined by the dashed lines between function blocks. Dashed lines indicate signal flow in RECEIVE mode. Of course, this time we start at the antenna. The process of "receiving" an SSB signal and producing an output through the speaker consists of the following steps:

1. A signal from the antenna passes through the RECEIVE contacts of the TRANSMIT-RECEIVE relay to an input of the RF block. In the RF block the signal receives selection processing in a bandpass filter. After some RF amplification it is converted to the transceiver's intermediate frequency (8.83 MHz) by mixing with an appropriate signal from the PLL block.
2. In the IF block the SSB signal, now converted to the receiver's IF frequency, receives amplification and further filtering. It first passes through a ceramic filter and then through a crystal filter. After further amplification it passes to the AUDIO block.
3. In the AUDIO block the SSB signal is detected and the audio signal is amplified and fed to the receiver's speaker or to headphones. Detection is by means of a circuit called a *balanced demodulator,* about which more later. The balanced demodulator requires an input of an RF signal as well as the SSB signal. The demodulation process can be thought of as a reinsertion of the original carrier which produces a signal equivalent to an amplitude-modulated signal. The demodulator then detects the AM signal.

If you have followed and understood these step-by-step explanations of processes, you have a general understanding of how a particular SSB transceiver performs its tasks of transmitting and receiving single-sideband signals. Circuit functions similar to those used in this particular system have been described in various degrees of detail at other points in this text. However, there is much that we can learn about this system and about systems in general by looking even more closely at the design, construction, and operating details of the function blocks of the transceiver of Fig. 9.29. It is useful to be aware that for this particular transceiver, the function blocks are actually packaged as separate printed-circuit (PC) modules. That is, in the transceiver cabinet there is an RF UNIT PC board, a PLL UNIT PC board, etc.

Let's start by looking at the CAR unit (see Fig. 9.30). The CAR block is the source of what we might call the basic carrier signal for the transceiver. It supplies the RF for both the balanced modulator and demodulator and is a crystal-controlled reference source for the unit's PLL block. The CAR block consists

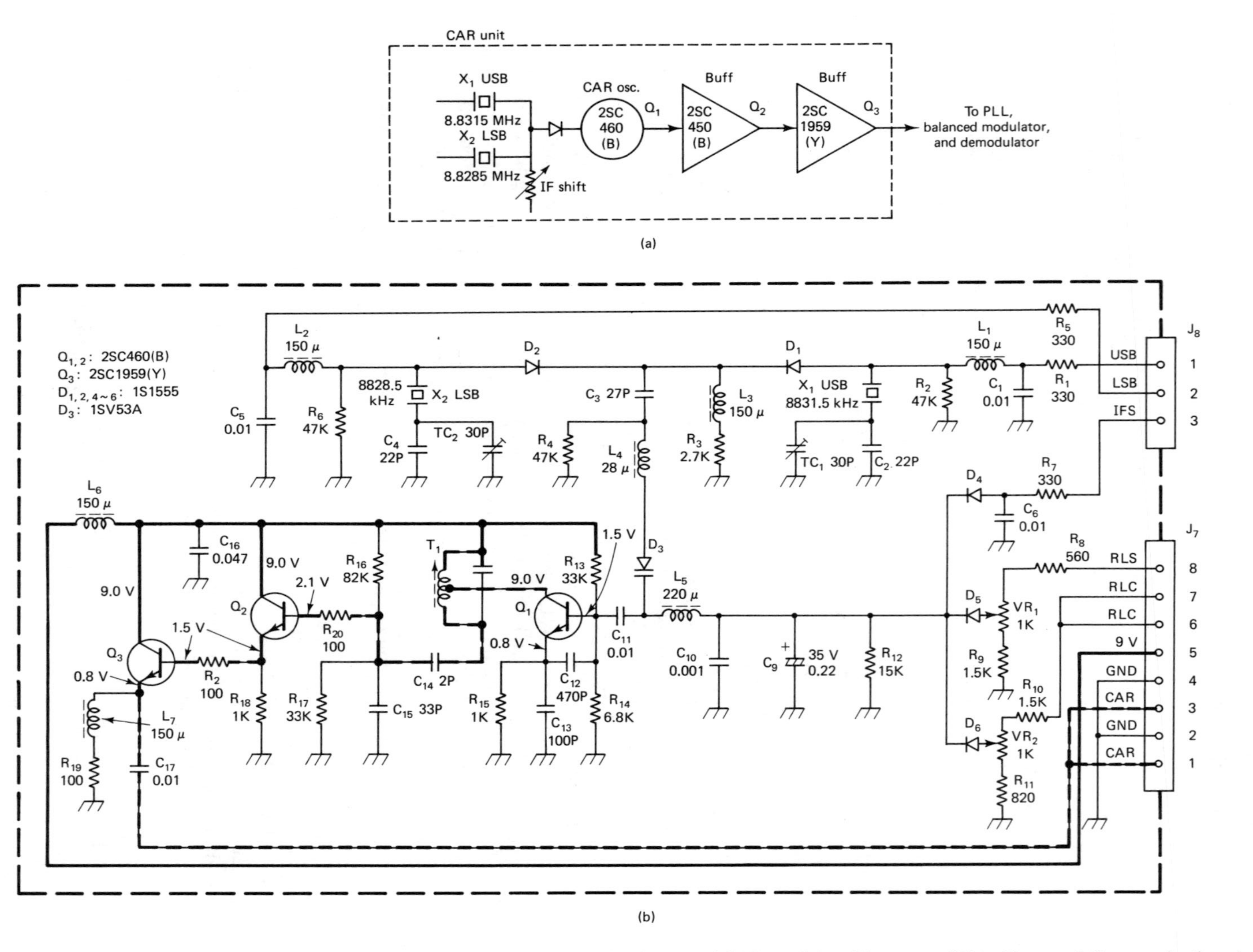

Figure 9.30 CAR function: (a) block diagram of CAR module; (b) schematic diagram of CAR module. (Courtesy of Trio-Kenwood Communications.)

of a variable-frequency crystal oscillator (VXO) and a two-stage buffer amplifier. To provide a choice of operation on either USB or LSB, the CAR circuit incorporates two crystals: X_1, with an operating frequency of 8831.5 kHz, and X_2, with an operating frequency of 8828.5 kHz.

The CAR unit circuit incorporates an idea that has not been presented previously in this book: *diode switching*. The schematic diagram of the oscillator circuit is shown in Fig. 9.30(b). Crystal X_1 (for USB operation) is selected when a positive dc voltage is placed on pin 1, plug J_8. A positive voltage on this point saturates diode D_1, effectively connecting X_1 to the base circuit of Q_1 through C_3 and L_4. A positive dc voltage can be placed on pin 1, plug J_8, by actuating a simple pushbutton switch on the front panel of the transceiver. The switch is labeled "USB," appropriately.

An advantage of diode switching is that the actual switch which controls the dc switching voltage can be located at a point remote to the signal circuit being controlled. The remote location, however, is one convenient to the operator of the equipment. An alternative to diode switching is to connect the signal circuit directly to a switch. This runs the risk of introducing a noise content into the signal, or degrading the signal quality in some other way. To minimize this risk, the switch may be located precisely at the circuit point to be switched. This could represent an inconvenience for the operator or a severe challenge to the designer. Or, if the switch is located remotely, meticulous attention must be paid to conducting the signal from the circuit to the switch and back again to avoid degrading it in any way. Also, diode switching is electronic switching and lends itself to digital logic circuit and/or microcomputer control.

Let's examine further the details of the switching circuit for X_1. The circuit must provide for the desired correct dc voltage on the switching diode. It must also provide for isolating the dc and RF currents. Refer again to the schematic diagram of Fig. 9.30(b). If we want the circuit to operate with crystal X_1, we push a button that puts a positive dc voltage on pin 1, plug J_8. Resistors R_1 and R_2 provide a voltage-divider circuit to set the dc voltage on the anode of D_1. The cathode of D_1 is referenced to ground by resistor R_3; the dc resistance of L_3 is insignificant. However, L_3 is an RF choke which provides a very high impedance to ground for the RF signal at that point: we have separated dc and RF on the cathode side of D_1 by means of L_3 and R_3. Back on the anode side of D_1: Choke L_1 prevents RF signal from getting out on the dc line to the selector switch. Capacitor C_1 bypasses to ground any RF that might have gotten beyond L_1.

Now note that operation with crystal X_2 (LSB) is obtained by actuating a switch that places a positive dc voltage of appropriate amplitude on pin 2, plug J_8. The circuit for X_2 and switching diode D_2 is arranged exactly like that for X_1 and D_1.

It was stated above that the oscillator in the CAR unit is a variable-frequency crystal oscillator. It is a crystal-controlled oscillator, but one in which provision is made for varying the frequency of operation. Generally, we think of a crystal oscillator, especially, as operating at one frequency. However, with the addition of a variable C and/or L to the circuit, we can vary the operating frequency.

The amount of variation is never as much as is possible with an *LC* oscillator. The oscillator circuit of the CAR unit is arranged for some frequency variation. Diode D_3 is a varactor (a voltage-variable capacitor) and is effectively in series with X_1 (or X_2). A variation in the dc voltage (reverse-bias direction) across D_3 will vary its capacitance and thus alter the operating frequency of the oscillator.

Recall that the CAR oscillator is used in both TRANSMIT and RECEIVE modes. It may be expedient during reception to alter the CAR unit frequency slightly as a way of shifting the IF passband for improved reception. For example, when tuned to a desired signal, the receiver may also be processing an interfering signal located near the lower end of the IF passband. By altering the CAR oscillator frequency slightly, the operator can shift the IF passband to a slightly higher frequency, thus attenuating significantly the undesired signal. The basic frequency to which the receiver is tuned is left unchanged. The *IF SHIFT* feature on the CAR board makes it possible to upshift or downshift the IF passband by up to approximately 1 kHz.

IF SHIFT is selected by a pushbutton that places a positive dc voltage on pin 3 (IFS = IF shift), plug J_8. This positive dc voltage forward biases switching diode D_4 and connects the base circuit of Q_1 to a variable resistor off the board. Adjustment of this resistor by the operator varies the voltage on varactor D_3 and thus varies the operating frequency of Q_1. This adjustment is labeled "IF SHIFT." Variable resistor VR_1 is used to calibrate the operating frequency of Q_1 with the IF SHIFT control centered.

A basic building block of the transceiver is the PLL. The output of the VCO of the phase-locked loop determines the channel operating frequency of the transceiver in the following ways:

1. In TRANSMIT mode, the VCO output is beat with the output of the IF unit in the transmitter mixer (TX MIX). The output of the TX MIX sets the carrier frequency of any transmitter emission.
2. In RECEIVE mode, the incoming signal is beat with the VCO signal in the receiver mixer (RX MIX). The output of RX MIX is fed to the IF unit for further filtering and amplification. The VCO frequency therefore determines the frequency of the signal to be received.

As an aid in understanding the PLL, it is important to be aware that this transceiver, as a typical HF amateur unit, is designed to operate on a number of different frequency bands: 11, in fact. An operator selects the desired band with a manual selector switch. The selector switch provides information for coarse frequency selection. This information determines the frequency-division ratio of a programmable digital divider in the PLL. However, in addition to this coarse frequency control, the transceiver design includes a variable-frequency oscillator (VFO) circuit which provides for fine adjustment to a desired operating frequency. This fine adjustment of the VCO is accomplished by various heterodyne mixing arrangements in the phase-detector feedback loop of the PLL.

It is time now to look at the PLL unit in considerable detail. A block diagram is provided in Fig. 9.31. Remember, the basic units of a PLL are (1) a

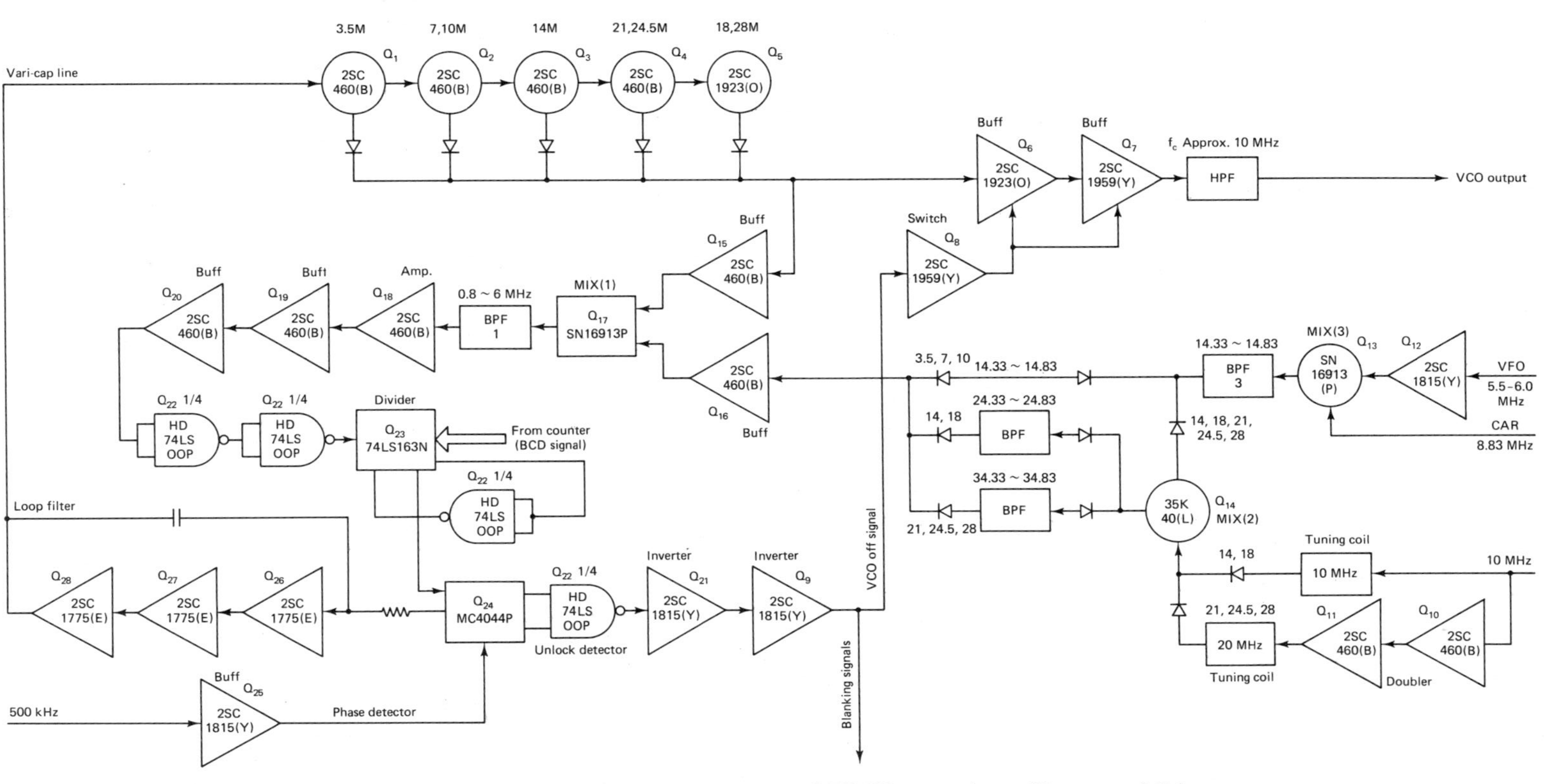

Figure 9.31 Block diagram of PLL of Kenwood TS-130 transceiver. (Courtesy of Trio-Kenwood Communications.)

VCO, (2) a phase detector, and (3) a loop filter. The PLL for the TS-130 transceiver has five VCOs to service the 11 different frequency bands on which it is possible to operate. As indicated on the diagram, diode switching is used to select the desired VCO.

The phase detector is an IC device, a Motorola MC4044P. As you will recall, the phase detector (or comparator) in a PLL compares a version of the VCO signal with a reference signal. Typically, the output of a phase detector is a pulse whose frequency is equal to the frequency of the signals being compared. The amplitude and polarity of the pulse is an indication of the relative phase and frequencies of the two signals being compared. When the two signals are identical in phase and frequency, the output of the detector is virtually zero. If the phase and/or frequencies are off, the detector output pulse amplitude and polarity will reflect the amount and direction of error.

For controlling the VCO, however, a nonpulsing dc voltage is desirable. The amplitude and polarity of this voltage must change to indicate error. But the voltage should not change too rapidly, nor should it contain high-frequency components. The function of a loop filter in a PLL is to integrate the error-signal pulse output of the phase detector and attenuate the higher-frequency components contained in it. The loop filter is an *active* (incorporates amplification) low-pass filter. (A low-pass filter is, in itself, an integrating circuit. An integrating circuit is one that sums or adds up a varying or pulsating signal. The output of an integrating circuit is proportional to the average value of its input; the output may vary, but it does so at a much slower rate than the input.)

The reference signal for the phase detector of the PLL is a 500-kHz signal obtained from a reference oscillator chain in the COUNTER module of the transceiver. (The COUNTER unit contains a 10-MHz crystal-controlled reference signal oscillator. Digital division of a portion of the output of this oscillator produces the 500 kHz for the PLL.) Of course, the signal on the VCO input of the phase detector must be 500 kHz also. Therefore, the PLL module contains various functions with which to derive the 500 kHz from the several VCOs. Contained in the loop between the common output line of the five VCOs and the VCO input of the phase detector is (1) a mixer MIX(1) (Q_{17}) and associated bandpass filter (BPF_1) and amplifier string (Q_{18}–Q_{22}); and (2) a programmable digital divider, DIVIDER (Q_{23}). The inputs to MIX(1) are the VCO signal and a signal whose source depends on the frequency band being used. For the 3.5-, 7-, and 10-MHz bands, the second input is from another mixer, MIX(3), and its filter, BPF_3. The two inputs to MIX(3) are from CAR and VFO. (VFO is the transceiver's *variable-frequency oscillator*. Its frequency is operator adjusted by means of a large knob on the front of the transceiver.) The VFO input permits the operator to fine tune the transceiver to a desired operating frequency. The output of VFO is adjustable between 5.5 and 6.0 MHz. The output of MIX(3) is the heterodyne-sum frequency of VFO and CAR; it is variable between 14.33 and 14.83 MHz.

When the transceiver is set to operate on any of the other bands—14, 18, 21, 24.5, 28, 28.5, 29, or 29.5 MHz—the second input to MIX(1) is from still another mixer, MIX(2). The two inputs to MIX(2) are the output of MIX(3) and either a 10- or 20-MHz signal derived from the COUNTER unit. If the frequency of the band selected is below 20 MHz, the second input to MIX(2) is the 10-MHz signal directly from COUNTER. For the bands above 20 MHz, the 10-MHz signal from COUNTER is doubled in DOUBLER before being input to MIX(2). To summarize: On the 3.5-, 7-, and 10-MHz bands, the non-VCO input to MIX(1) is the 14.33- to 14.83-MHz output from MIX(3). On the 14- and 18-MHz bands, the input to MIX(1) is a 24.33- to 24.83-MHz signal from MIX(2). On the 21-, 24.5-, 28-, 28.5-, 29-, and 29.5-MHz bands the input to MIX(1) is a 34.33- to 34.83-MHz signal from MIX(2).

Before arriving at the loop phase detector, the signal from MIX(1) and BPF_1 is passed through an IC-digital divider, DIVIDER (Q_{23}). DIVIDER is a 74LS163N, TTL, 4-bit, synchronous presettable counter. By presetting a count, the device can be programmed to provide a divisor of between 1 and 16. Four binary inputs (lines whose voltage level is either "high" or "low") to set the divide ratio are obtained from the COUNTER. For example, when the band selector switch is set to the 3.5-MHz position, the output of MIX(1) is a nominal 2.0 MHz. DIVIDER must divide by 4 to produce the 0.5-MHz signal for the phase detector. A binary word of 1100 is generated by a diode encoder in COUNTER from information from the BAND selector switch. The binary word, 1100, is used to preset the counter of the 74163 chip. (The count is 12_{10}.) As a MOD-16 counter, then, it will count "13," "14," "15," and reset to "12." One output pulse, from the highest-order bit position, will be generated for every four input pulses. Other binary words to provide appropriate divisors are generated by information from other positions on the BAND switch.

The output of the PLL module is a signal obtained from one of the five VCOs of the PLL. For a given band, a VCO of appropriate frequency is switched on by connecting a 9-V V_{CC} supply voltage to it through a deck of the BAND selector switch. The VCO signal selected is passed through a buffer amplifier and a high-pass filter (HPF) before exiting the PLL module. Let us remember: the VCO frequency is not the operating frequency of the transceiver. It is beat with the output of the IF unit (8.830 MHz) in the transmitter mixer (TX MIX) in TRANSMIT mode. The difference frequency is the operating carrier frequency. In RECEIVE mode, the VCO output is mixed with the received-carrier signal in RX MIX and produces the 8.830-MHz IF signal as a difference frequency.

Before leaving the PLL it is important that we observe another function provided by this module: an UNLOCK DETECTOR. As we have learned previously, a phase-locked loop can be thought of as having two operating modes: locked and unlocked. It is the nature of the loop that when the two signals being compared by the phase detector approach being very nearly identical, the loop goes into the *lock mode*; or, it locks the VCO frequency onto the reference frequency. Unless powerful external forces operate to change the VCO frequency, the loop will maintain this locked condition very effectively. On the other hand, when the VCO and reference frequencies are relatively far apart, the loop is unable to tune the VCO frequency sufficiently. The loop is in the *unlocked mode*. The PLL will "fall out of lock" if something does not function correctly in the loop, or if an external force is sufficiently strong and persistent to overcome the correcting force of the loop. Obviously, if the loop falls out of lock, the transceiver's operating frequency will not be correct. In TRANSMIT mode, the unit could emit a carrier that would interfere with other stations/services. The MC4044P chip incorporates a function, called *unlock detector,* to detect the unlocked condition. The PLL module utilizes the output of this function to turn off the VCO.

As we have seen, the PLL obtains several of its signals from the COUNTER module. Let's examine the block diagram of the COUNTER next. In addition to generating several signals, a major function of the COUNTER is to count or measure the carrier frequency at which the transceiver is operating and provide a digital display of that information. It is a built-in frequency counter.

The block diagram of COUNTER is shown in Fig. 9.32. The 10-MHz signal used by the COUNTER and PLL is generated in a straightforward, crystal-controlled oscillator, stage Q_1. The crystal circuit is provided with a trimmer capacitor, TC_1. This enables the user to calibrate the frequency of this frequency reference source very precisely against an external standard such as the signal broadcast by National Bureau of Standards station WWV. The 10-MHz signal is split at the output of Q_1: Within the COUNTER the signal is available at the output of buffer amplifier Q_2, an emitter follower. The signal for applications outside the COUNTER module is fed through buffer Q_3.

The output of buffer Q_2 is fed to a divide-by-2, divide-by-5, 7490 TTL, binary counter. The divide-by-2 output is beat with the 5.5- to 6.0-MHz signal from VFO. The difference frequency, 0.5 to 1.0 MHz, is the signal counted by the COUNTER to produce the frequency display. The divide-by-10 output of the 7490 counter is fed to an MC14510BCP IC which is also a $\div 2$, $\div 5$ counter. The $\div 2$ output of this counter is the 500-kHz reference signal used by the PLL phase detector. The MC14510 is connected with its two sections in cascade, producing a $\div 10$ counter with an output of 100 kHz. One application of the 100-kHz signal is to synchronize a free-running multivibrator at a precise 25 kHz. This provides a marker source for calibrating the analog dial of the VFO.

The 100-kHz signal also goes to a $\div 100$ digital divider (using a TC4518 IC). The 1-kHz output of this counter is fed to another $\div 100$ counter. The 10-Hz signal is further divided by 2 and produces a square wave with on and off times of 0.1 s. The positive-going 0.1-s pulse is fed to one input leg of a NAND gate. The 0.5- to 1-MHz signal (recall that it is the heterodyne difference signal derived from the VFO signal) has by now been converted to a square wave. It is fed to the second NAND gate input. This device provides a so-called "0.1-s gate." The signal to be counted is "on" for 0.1 s and off for 0.1 s. The output is fed to a MC14510 counter which counts the pulses during each 0.1-s interval. Finally, the output of this counter is fed to a six-digit (decimal digit) counter. The 10-MHz, 1-MHz, and 100-kHz digit positions of this counter are preset with information from the BAND selector switch. The six-digit counter performs the final counting function for the lower-order digits of the frequency count. Its output is used to drive a seven-segment, six-digit display tube.

In summary, the COUNTER unit contains a 10-MHz master reference frequency source. This source is crystal controlled but can be precision tuned

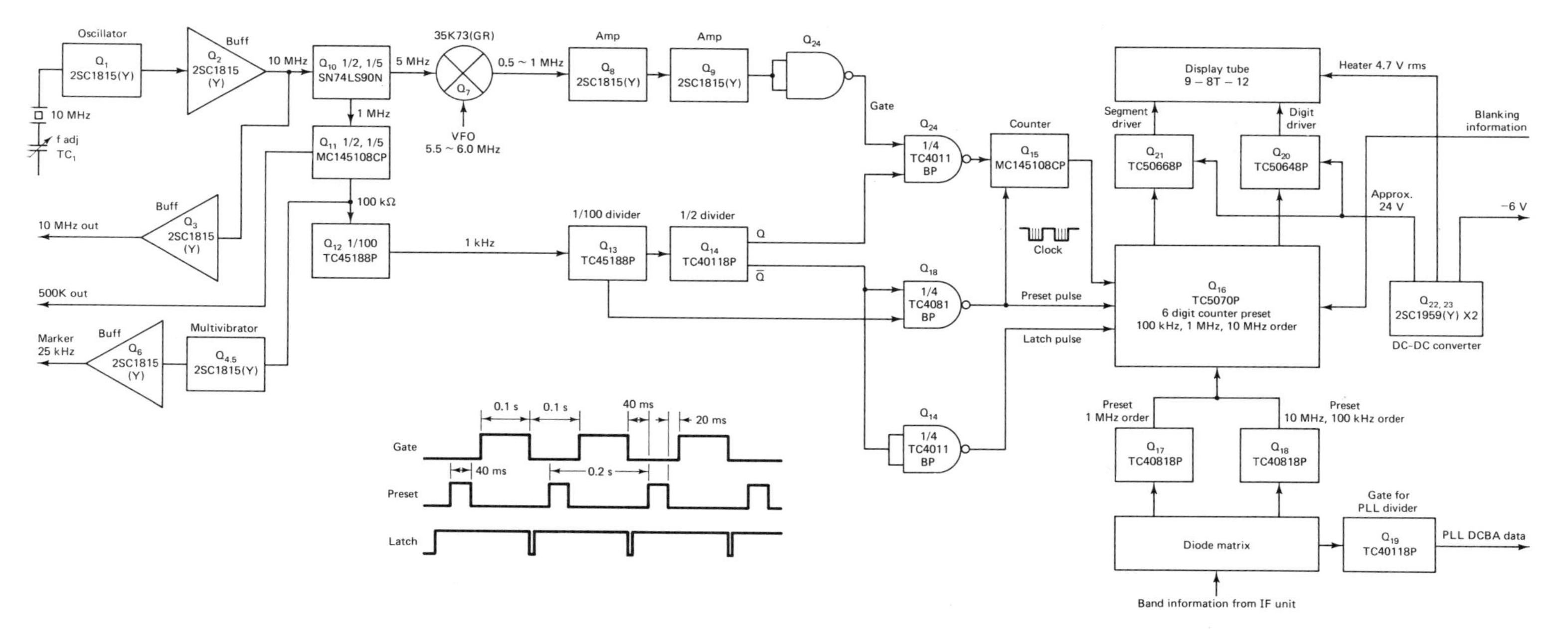

Figure 9.32 Block diagram of COUNTER function of Kenwood TS-130 transceiver. (Courtesy of Trio-Kenwood Communications.)

for calibration to an external standard. The 10-MHz signal is supplied to the PLL module. It is also divided down in a digital divider to 500 kHz, also for use by the PLL. It is further divided and synchronizes a 25-kHz source which can be used as a marker signal for calibrating the manually controlled VFO dial. Further, the COUNTER counts a derivative of the VFO signal. A digital display of the VFO frequency is provided. The three highest-order digits of that display are preset by information from the BAND selector switch. The other digits are produced by the actual frequency-counting process.

The AF-GEN unit is the module on which audio amplification, modulation, and demodulation is achieved. It incorporates an amplifier for the microphone signal, a balanced modulator, a balanced demodulator, an audio amplifier string to drive the speaker, a VOX function, and a sidetone oscillator and amplifier. A second, smaller module, which is used at the operator's discretion in conjunction with the AF-GEN unit, is the PROCESSOR unit. Let's examine these modules and their various functions at the block-diagram level.

Let us start with an analysis of the modulation function. Refer to the block diagram of Fig. 9.33. Observe that the balanced modulator, BM, is a diode ring modulator. The audio input is from the microphone amplifier string; the RF input is the 8.83-MHz signal from the CAR module. After modulation in a buffer amplifier the double-sideband, suppressed-carrier output of the balanced modulator is fed to the IF module.

In cascade with the microphone amplifier string is the PROCESSOR unit, which can be connected in or out of the string by switch. The PROCESSOR unit contains a special integrated circuit: μPC1158. This device provides a function called "speech compression." A speech, or audio, compression amplifier is one that incorporates automatic gain control. The AGC increases the amplitude of low-amplitude audio signals and, conversely, decreases the amplitude of high-amplitude signals. The result is a signal in which there is very little variation in amplitude. The signal is "compressed." This is a signal that is quite different from a "normal" signal from a microphone.

The volume of the voice of most people varies a great deal as they speak; the amplitude of the signal produced by a microphone varies to the same degree. Consider the effect of modulating a radio transmitter with such a signal: If AM or SSB is used, the power level of the emission varies in direct proportion to the level of speech volume. However, the power level of a transmitter is limited to some maximum value. (Both government regulations and heat-dissipating factors are involved in power specifications.) Most speech amplitude levels are significantly less than their maximum a major portion of the time. Therefore, the power output of a transmitter would be less than its maximum rating most of the time if modulated by normal speech. However, if the microphone signal is processed in a speech compressor circuit, the variation in level is significantly reduced: the average power level of a transmitter is increased.

Of at least equal importance to the compression of audio signal variation is the fact that an effective speech processor circuit also automatically limits the maximum level of the audio signal. The chance of exceeding the power rating of the transmitter output stages is reduced significantly. It is safe to operate the

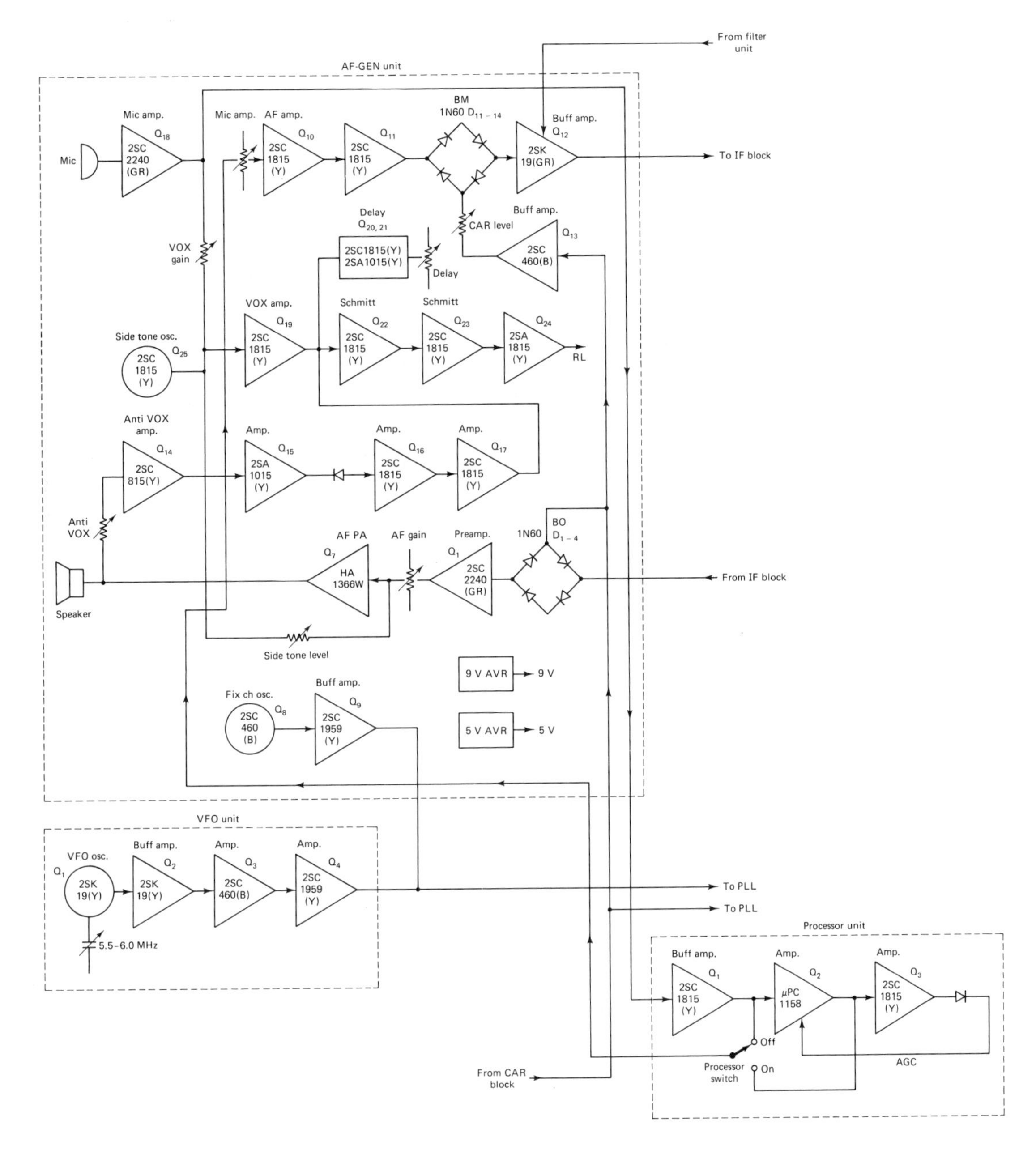

Figure 9.33 Block diagram of AF-GEN module of Kenwood TS-130 transceiver. (Courtesy of Trio-Kenwood Communications.)

transmitter at a higher average power level. Therefore, in this transceiver, turning on the PROCESSOR switch also reduces the time constant of the automatic load control (ALC) circuit in the transmitter FINAL module. This increases the average transmitter power level.

In RECEIVE mode, the received-station signal, after processing by the IF unit, is fed to the AF-GEN module for demodulation. Remember, the signal is an SSB signal; it cannot be demodulated by ordinary AM means. A demodulator circuit must be a nonlinear circuit in which one of the products of the distortion is the audio-frequency signal. If a signal equivalent to the suppressed carrier is combined with the SSB signal in a nonlinear circuit, one of the products will be the audio signal. SSB demodulation, then, is a process involving the reinsertion of the carrier. In typical circuits, the incoming signal and the locally supplied signal are effectively multiplied in the circuit. The circuits are commonly called *product detectors* for that reason.

A diode ring circuit, such as is used in a balanced modulator, is also an effective product detector circuit. In the TS-130 transceiver that we are studying, the balanced detector (BD) utilizes a circuit that is identical to the balanced modulator of the unit. It uses the same type of diodes, 1N60. The circuit that supplies to the product detector the signal to replace the carrier (suppressed at the transmitter) has traditionally been called the *beat-frequency oscillator* (BFO). It is, typically, a separate oscillator used in the demodulation of SSB signals. In amateur radio receivers it is also used to demodulate code or CW signals. In the TS-130, the 8.83-MHz signal from the CAR module is, in effect, the BFO signal. You will observe that it is connected to one of the inputs of the BD.

The output of the balanced detector is fed to an audio amplifier chain. It passes through an audio volume control and then to an audio power amplifier, which drives the speaker.

You will note that the AF-GEN contains a function referred to as "sidetone osc." Using a sidetone oscillator is one method of generating a carrier for code transmission (CW) in a transmitter otherwise designed for SSB use. It is the nature of SSB that a carrier is not generated unless modulation occurs (there is no sideband unless a carrier has been modulated). To produce a carrier (actually, a sideband) for code transmission, then, the sidetone oscillator is keyed on, as well as the RF module. The audio signal from the sidetone oscillator modulates the signal from the CAR oscillator, just as a signal from the microphone would. A sideband created by a single-frequency audio signal is generated. This sideband is emitted as the CW "carrier." To receive this carrier, a receiver would be tuned to a slightly different frequency than if it were receiving an SSB transmission. This difference is referred to as "CW offset." It is common for transceivers to be manufactured so that the basic transmitter carrier frequency (CAR) is offset automatically when CW operation is selected. The amount of the offset would be equal to the sidetone frequency. The sidetone signal is also commonly fed through a station's speaker so that the operator can monitor his/her code transmission.

Another function found in the AF-GEN unit of the TS-130 transceiver is a portion of the so-called *VOX* function. VOX stands for "voice operated."

The alternative to VOX is MOX, manually operated. These terms have to do with "operating" the transmitter (i.e., turning the carrier on for a transmission). Unlike an ordinary telephone which is ready for our transmission all the time that the handset is lifted from its cradle ("off the hook"), a radiotelephone's carrier stands by most of the time. It is turned on only when the user is going to speak. If it is turned on by the operator manually depressing a "push-to-talk" (PTT) switch, it is set for MOX. If speaking into the microphone, without having to push a button, causes the transmitter carrier to be activated, the transmitter is set for VOX. VOX requires electronic circuits that (1) sense the presence of an audio signal of significant amplitude, and (2) turn on the necessary circuits as a result.

In the AF-GEN module of the TS130, the VOX circuit consists of an audio amplifier string followed by a Schmitt trigger. The amplifier increases the audio level from the microphone or sidetone oscillator. The positive alternation is rectified and passed through a Schmitt trigger. The Schmitt trigger converts the signal to a pulse waveform. The output of the Schmitt trigger operates the transmit-receive relay in the IF module. The function is provided with a time-delay circuit which holds the transceiver in transmit mode during short pauses in the operator's speaking. The time delay is adjustable. An anti-VOX function is also included. The anti-VOX circuit monitors the signal driving the speaker. If a signal is present, indicating that a transmission is being received, the anti-VOX circuit has an output. The output of the anti-VOX circuit is in opposition to the signal in the VOX circuit. If the anti-VOX signal is greater, the turning on of the transmitter is prevented.

A fixed channel oscillator (FIX CH OSC) is included on the AF-GEN board. This circuit contains an *LC*-type oscillator. Any one of four preset frequencies are selectable by diode switching. An output of the FIX CH OSC is selected for use in lieu of the output from the VFO by means of the FIX/VFO switch on the front panel of the transceiver. This feature permits the operator to switch quickly to favorite or frequently used frequencies.

Finally, the AF-GEN circuit board contains the circuitry for two dc voltage regulators. These provide 9 V and 5 V regulated for many other parts of the transceiver.

VFO Module

The VFO module, as its designation implies, contains a variable-frequency oscillator. It also contains three stages of buffer amplification (see Fig. 9.34). The oscillator stage uses an *N*-channel, MOSFET active device in a Clapp-type configuration. The operating frequency of the oscillator is continuously variable between 5.50 and 6.00 MHz by means of a variable capacitor. The capacitor rotor is gear driven from a shaft with a large knob and a calibrated analog indicator dial. The gear reduction is such that 20 turns of the frequency-control knob are required for the full range of excursion of the capacitor rotor. This provides for excellent fine-frequency adjustment and resolution. (Remember: The VFO is used to fine tune the transmitter and receiver functions to desired frequencies within a frequency band selected by the BAND selector switch, a coarse selector.)

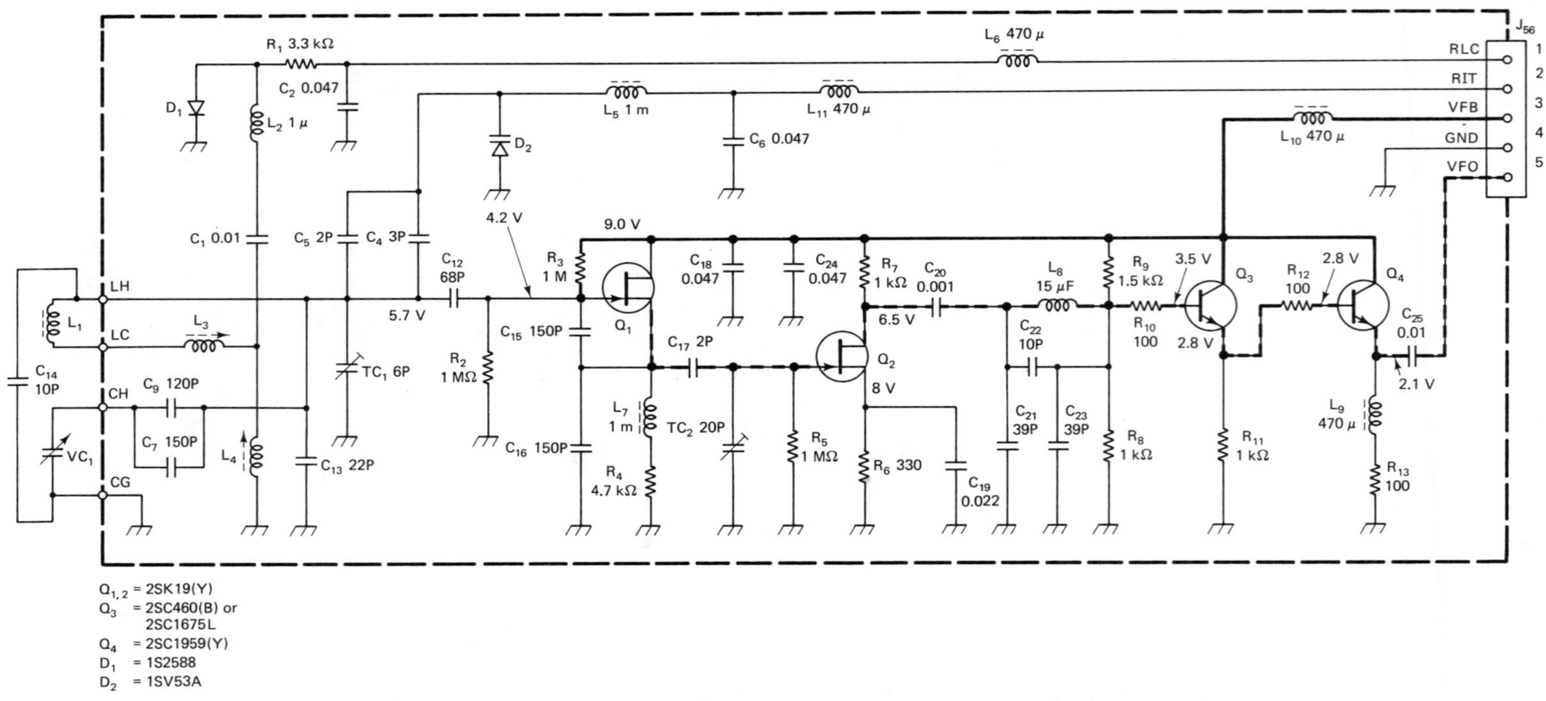

Figure 9.34 VFO circuit of Kenwood TS-130 transceiver. (Courtesy of Trio-Kenwood Communications.)

This VFO assembly includes a varactor diode (D_2) with which to alter the oscillator frequency a slight amount in RECEIVE mode. The diode circuit is activated by means of the *RIT* switch on the front panel. "RIT" stands for "receive incremental tuning." The RIT function provides for off-tuning the receiver without changing the frequency dial. It is used to improve reception when the operating frequency of a received station is slightly off that of the local station.

IF Module

Most of the circuits in the IF module are used in both RECEIVE and TRANSMIT MODE (see Fig. 9.35). It is basically an active filter with filter sections that shape the transceiver's IF passband around the center frequency of 8.830 MHz. We trace signal flow in TRANSMIT mode first:

1. The DSB signal enters the module from the AF-GEN module. This signal is centered on the CAR frequency (USB = 8.8315 MHz, LSB = 8.8285 MHz, and CW = 8.8307 MHz).
2. The DSB is passed through a crystal filter, XF, for the removal of one of the sidebands. When in USB mode, the filter will actually pass the lower sideband and remove the upper. The effect is the contrary of this in LSB mode. (The output of the IF unit will pass through another mixer, TX MIX, before being emitted. Since the difference frequency of TX MIX is utilized, the lowest frequency from IF becomes the highest frequency emitted. Hence the lower sideband in the IF unit becomes the upper sideband at the antenna, and so on.)
3. The signal is amplified in an IF amplifier and fed to the output terminal connected to the RF module.

In RECEIVE mode, the IF section provides more amplification and filtering. It also provides a noise blanking (NB) function. We trace the signal flow path in RECEIVE mode:

1. The received signal enters the IF section from the receiver's first mixer, RX MIX. RX MIX converts the antenna signal to a signal centered on 8.830 MHz.
2. The signal is amplified and fed to a 8.83-MHz ceramic filter.
3. Next in line is a *noise blanking gate* and its associated noise amplifier chain. The noise blanking circuit used in this transceiver is shown in Fig. 9.36. Several schemes for noise reduction in receivers have been devised. The term "blanker" generally indicates a circuit that is designed to detect pulse-type noise energy and literally "blank" the signal path momentarily so that neither noise nor information passes during blanking. This eliminates the noise that can be detected by the scheme. In the circuit of Fig. 9.36, the diodes D_1 through D_4 will be forward biased in the absence of noise, or when NB is switched off. This forward bias is set by resistors R_{70}, R_4, and R_5. When forward biased, the diodes will pass the IF signal being

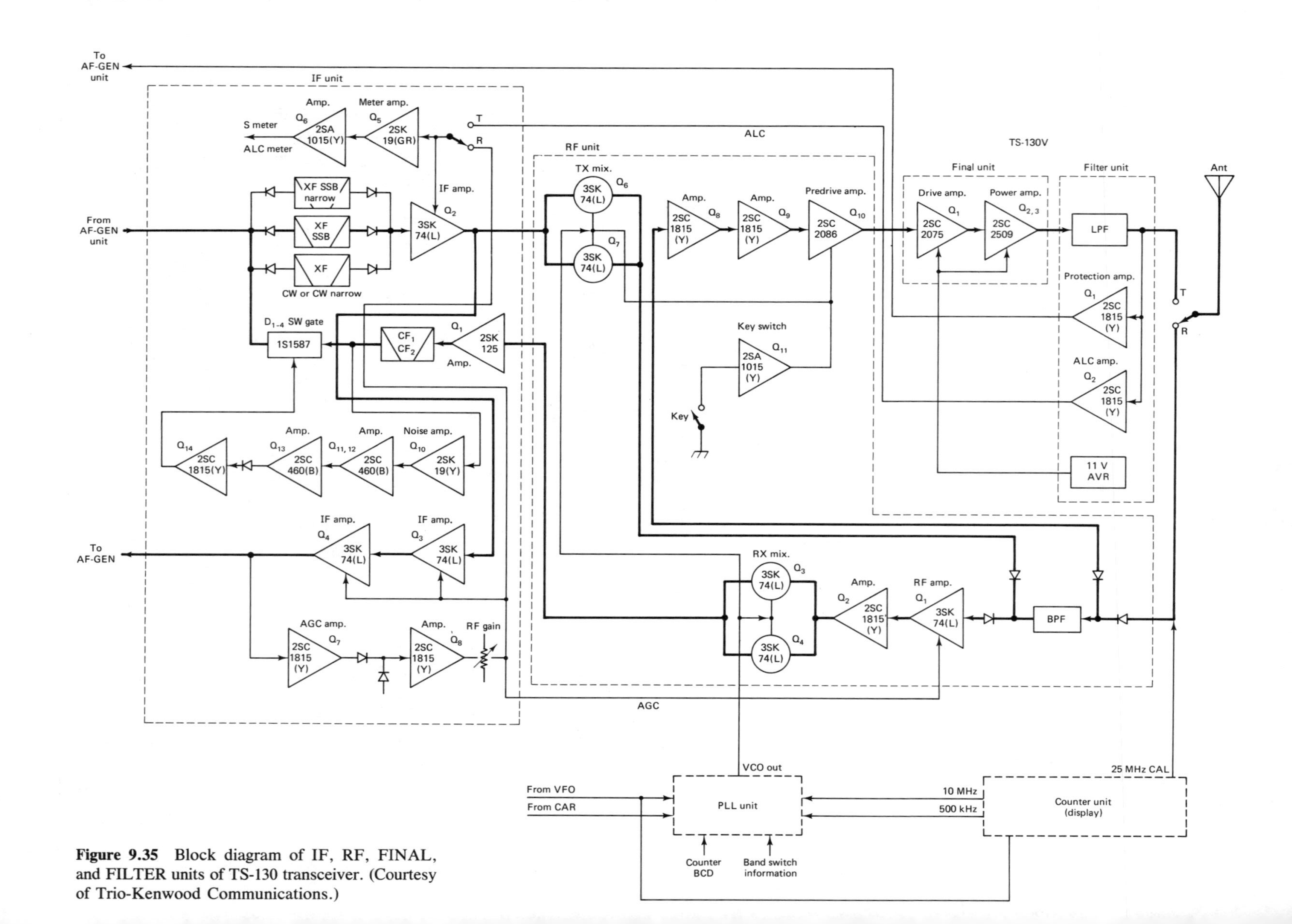

Figure 9.35 Block diagram of IF, RF, FINAL, and FILTER units of TS-130 transceiver. (Courtesy of Trio-Kenwood Communications.)

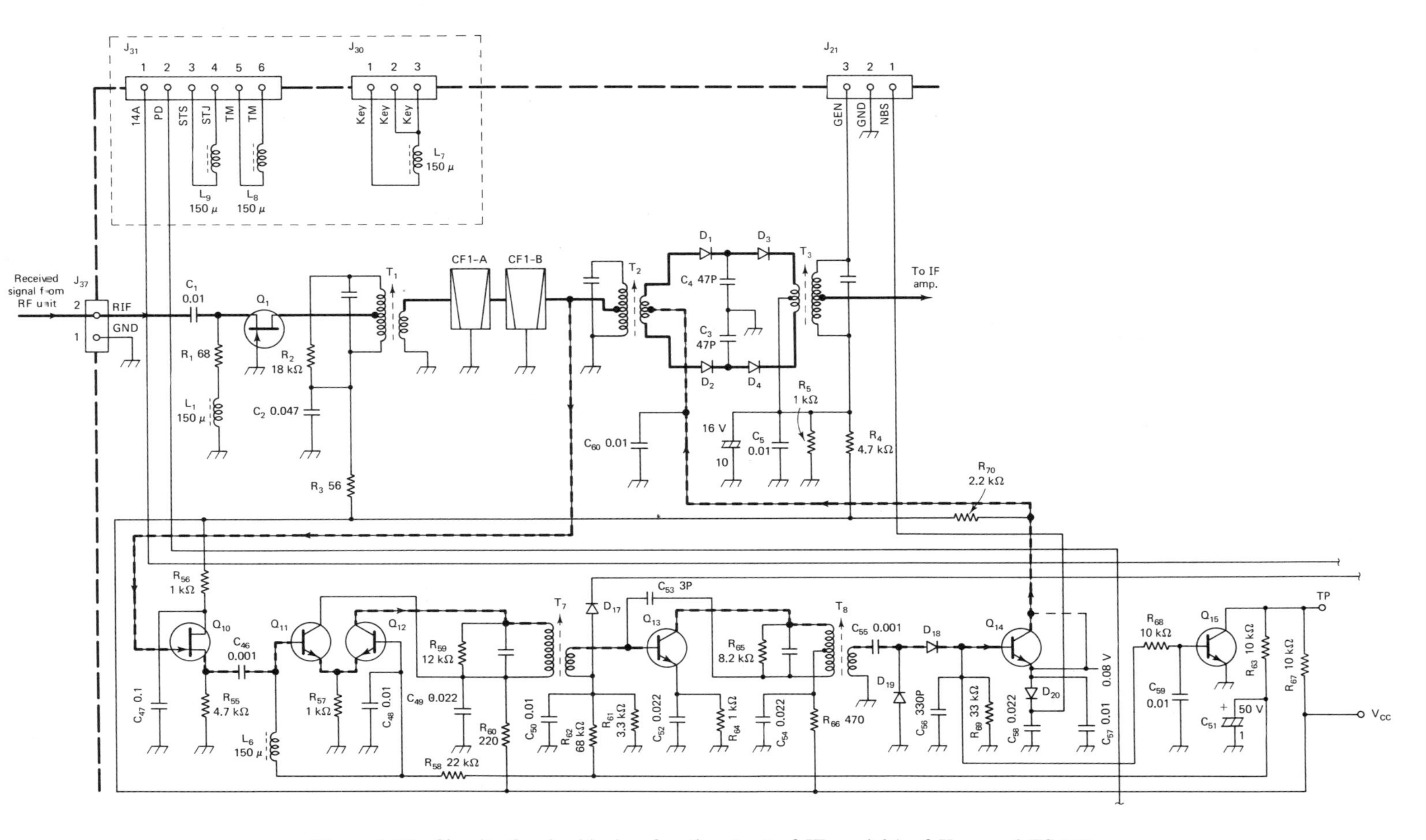

Figure 9.36 Circuit of noise-blanker function (part of IF module) of Kenwood TS-130 transceiver. (Courtesy of Trio-Kenwood Communications.)

processed. The signal at the output of the ceramic filter, CF_1, is fed to the noise blanker chain, which contains a noise amplifier and a transistor gate. When a noise pulse is detected, the blanker chain turns on the transistor gate, effectively grounding the center tap of the secondary winding of T_2. This reverse biases the diodes D_1 through D_4, momentarily cutting off the passage of all signal through the IF section, including the detected noise spike.

4. The signal passes from the noise gate to the crystal filter section (XF). The standard transceiver is shipped with a single filter for sideband suppression. The filter has a center frequency of 8.830 MHz and a −6-dB bandwidth of 2.4 kHz. Two optional filters are available: a CW filter with a −6-dB bandwidth of 500 Hz and a narrow CW filter with a −6-dB bandwidth of 270 Hz. Filter selection is by means of diode switching.
5. After receiving additional amplification in three more IF stages, the signal is passed to the AF-GEN module for demodulation.

The IF module contains an AGC circuit consisting of an IF amplifier, a rectifier, and a dc amplifier. Automatic gain control is applied to two stages of IF amplification used in RECEIVE mode only. It is also applied to one stage of the receiver's RF amplifier. The operator can also adjust RF gain manually by means of a control resistor with knob available on the front of the transceiver.

A two-stage dc amplifier for operating the transceiver's S-meter is located on the IF board. The input to the meter amplifier is the amplified AGC voltage. The AGC voltage amplitude is a measure of the amplitude of the RF signal being processed. The S-meter is literally a *signal-strength meter*. However, since its reading is based on the AGC voltage of the receiver, it is a relative indication. It is not an absolute field strength meter.

A transmit-receive relay that switches several of the transceiver's control functions is mounted on the IF board. This relay does not directly switch the signal path itself.

RF Module

Refer to Fig. 9.35 again. In TRANSMIT mode, the signal from the IF module is fed to a heterodyne mixer, TX MIX (Q_6–Q_7), where it is beat with the VCO output from the PLL module. This stage converts the signal to the frequency at which it is to be radiated. The difference frequency of the heterodyne process is used. The signal is next passed to the bandpass filter (BPF). The BPF filter is used in both TRANSMIT and RECEIVE modes. It is actually eight separate, *LC*, constant-*k* type, π-section, bandpass filters. The separate filters are connected between two decks of the BAND switch; band filter selection is by direct switching. The BPF is switched between the TRANSMIT and RECEIVE signal paths by diode switching.

Continuing with the TRANSMIT mode signal path, the signal passes from the BPF through three broadband RF/predriver amplifier stages. These are called "broadband" amplifiers because, by design, they are not narrowly tuned. They must amplify signals from 3.5 MHz to approximately 30 MHz. The third stage contains impedance-matching elements for coupling the signal to the driver stage on the FINAL module.

A special circuit for keying the transmitter for the transmission of code in CW operation is located on the RF module. When the key is operated, keying stage Q_{11} is cut off. This places a forward bias on the RF predriver stage, Q_{10}, and the RF signal is allowed to pass. You will recall that in order to produce an RF signal for CW in an SSB transmitter it is also necessary to key on the sidetone oscillator (located on the AF-GEN board). TX MIX is also keyed on by the CW key.

In RECEIVE mode, the signal comes into the RF module directly from the antenna. After passing through the BPF it is amplified in a broadband RF amplifier with AGC. It is coupled to the RX MIX through an emitter-follower stage. In RX MIX it is beat with a signal of appropriate frequency generated by the VCO in the PLL module. (For a given operating frequency for the transceiver as a whole, the same VCO frequency is used for both RX MIX and TX MIX.) The output circuit of RX MIX is tuned to the difference frequency—8.830 MHz. The output of RX MIX is, of course, the IF signal and is passed on to the IF module for amplification and filtering, as described above.

FINAL Unit

The FINAL module (see Fig. 9.35) contains a driver stage and two stages of solid-state, push-pull power amplifiers. The power level is increased so that on modulation peaks, in SSB operation, the dc input power to the final stage is 200 W. For CW operation the dc input power to the final stage is 160 W.

The meaning of power-level terminology, for SSB operation, found in equipment specifications and FCC regulations is relatively obscure. It is worthy of further explanation. Levels of power in SSB operation are difficult to define because such levels are very much dependent on the nature of the modulating waveform. For example, in AM, when the modulating signal is zero, the output power of a transmitter is that of the unmodulated carrier which can be described and/or measured relatively easily. In SSB operation, when the modulating signal is zero, the RF power is zero. The waveform of the output of an SSB transmitter, with voice modulation, will resemble that of the sketch in Fig. 9.37(a).

Two definitions are commonly used in connection with SSB waveforms: *Maximum peak amplitude* refers to the greatest amplitude reached by the envelope of the waveform at a given instant. (The *envelope* of a waveform refers to an imaginary line drawn through the tips of individual cycles of variation. There is a positive envelope and a negative envelope.) *Average amplitude* refers to the *average* of all the amplitudes contained in a waveform over a significant time period, for example, the time to say a single syllable.

SSB power specifications are typically in terms of *peak-envelope power* (PEP), the power contained in the transmitter output signal at the maximum peak amplitude. Instances of maximum peak amplitude occur randomly in actual operation where the human voice is producing the modulating signal. The FCC

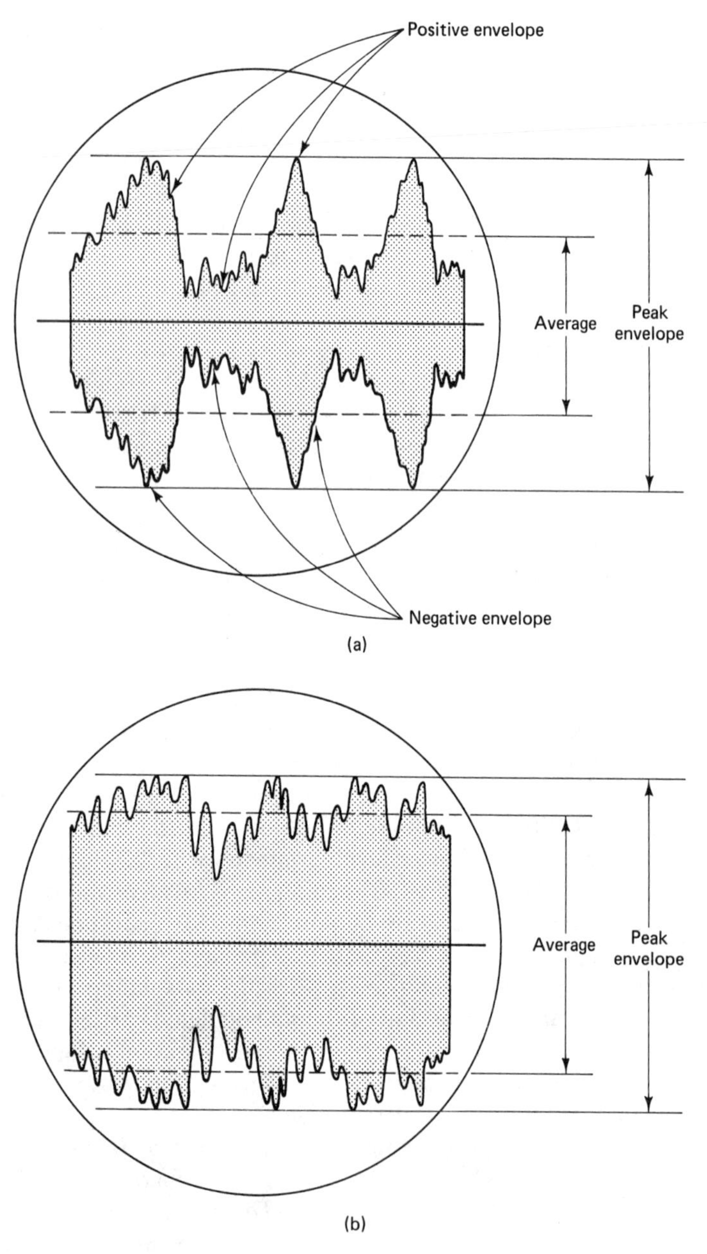

Figure 9.37 (a) Typical SSB waveform with voice modulation; (b) increase in average level of SSB signal after audio processing (compression).

requires that the power rating of SSB transmitter be specified in terms of *dc input power to the final stage at the moment of an envelope peak.* This is called "PEP input" or "peak-envelope dc input."

Since amateur SSB transmitters must be rated for peak power, the ratio of average power to peak-envelope power is an important indicator of the real "talk power" of a transmission. Voice signatures (voice wave patterns) vary from person to person. A *voice processor* (see above) can be used to compress the voice pattern—increase the ratio of average power to peak-envelope power. See Fig. 9.37(b). Compression increases the talk power.

The output power stages of the TS-130 are operated class AB. That is, base-bias circuits are adjusted to provide standby collector currents equal to approximately 1% of maximum peak values. Operation in this manner minimizes the crossover distortion characteristic of push-pull amplifiers operated in true class B. Base voltages are supplied from voltage-regulated sources.

Over-temperature protection is incorporated in the FINAL module by means of temperature-sensing elements mounted on the output transformer core and output transistors' heat sink. When either of these devices detect a temperature exceeding a preselected value, the transceiver is automatically switched to RECEIVE mode. Transmission cannot be resumed until operating temperatures have been reduced to a safe level, resetting the over-temperature protection devices.

FILTER Module

The FILTER unit (again, see Fig. 9.35) contains the low-pass filter (LPF) through which the transmitter signal passes before it is placed on the antenna. The purpose of this filter is to remove harmonics and any other spurious frequencies that may be present in the transmitter signal. The LPF consists of seven separate, multiple-section, constant-*k*-type filter circuits. The circuits are switched directly by two decks on the BAND selector switch. The five bands (24.5 to 29.5 MHz) use a common filter.

Just before being placed on the antenna, the signal passes through a directional energy-sensing element on the FILTER board—a toroidal, voltage-standing-wave ratio bridge. This is the kind of element used in combination RF-power and voltage-standing-wave ratio (VSWR) meters. The output from one side of the bridge circuit is a measure of forward energy; the output from the opposite side of the bridge provides an indication of reflected energy, and indirectly, VSWR. Amplifiers on the FILTER board amplify these two signals. The forward-energy voltage is used in an automatic load control (ALC) function in the transceiver. The ALC voltage is fed back to the IF module and in TRANSMIT is used to control the gain of an IF amplifier: Gain control of this amplifier is switched from AGC to ALC by the TRANSMIT-RECEIVE relay. ALC serves to control the drive to the FINAL, and thus the amount of RF produced. It helps preserve amplifier linearity and reduce distortion due to overdriving.

The ALC circuit contains a so-called *time-constant network*. This is like a low-pass filter in a PLL. It smooths out the ALC voltage, preventing it from responding to peaks of very short duration. With the time-constant network in place, ALC response is effectively based on the average of peak values. When the speech processor circuit (see the discussion of the AF-GEN module above) is switched into the transmitter audio string, the ALC time-constant network is simultaneously switched off. ALC action is now determined by the average transmitter power rather than the average of peak power. The result is that the transmitter operates at a slightly higher power level. Speech compression by the processor circuit tends to eliminate excessive power peaks and thus eliminate the need to monitor primarily for peaks.

The voltage from the VSWR bridge indicating the standing-wave ratio is used in a protective circuit. A transmitter output stage can be damaged by overvoltage if reflected energy becomes excessive. After amplification, the dc voltage from the reflected-energy side of the bridge is fed back to the AF-GEN, where it controls the gain of the buffer amplifier immediately following the balanced modulator. A tendency for reflected energy to increase will result in a reduction of drive.

Power Supply

The TS-130 requires a dc voltage source with a nominal voltage of 13.8 VDC. It can be operated from a "12-V" vehicular battery system. A separate 13.8-V power supply package is available for operation on 120/240 VAC.

Summary

We have now completed a comprehensive and detailed examination of a particular HF, SSB transceiver designed for use in the Amateur Radio Service. We began our examination with a quick overview of the system; we traced the signal flow path in both RECEIVE and TRANSMIT modes. Then we examined each function block of the system in greater detail. Assuming that our efforts were productive, we now have a much better understanding of (1) what each function block consists of, and (2) how the functions work together to accomplish the overall purposes of the system. What is important here is to use the details of a particular transceiver to help us learn more about two-way radio systems in general. It is not our purpose to attempt to become experts on a particular system: Very few, if any, of us will ever have the opportunity to actually work with this particular system, the Kenwood TS-130.

An examination of many two-way radio systems shows that all contain certain basic functions. Many contain more than the basic functions in order to provide enhanced performance, special convenience features, etc. The basic functions, since they provide the same or very similar electronic effects in all systems, are all similar in design. Even so, hardly ever will two identical functions in different systems contain exactly the same circuits. A thorough understanding of basic electronic circuits helps in sorting out the differences that may be found.

Now that we have a basic understanding of the functions of a system of interest and a knowledge of what they contain, a second tracing of signal paths will be more meaningful than the first. Furthermore, it will reinforce our understanding of the functions and aid in remembering that knowledge. Refer to the block diagram of the unit shown in Fig. 9.38. Signal flow with SSB operation selected is detailed in the following steps:

1. Sound picked up by the microphone is converted to electrical energy, which is amplified in the audio string of the AF-GEN block. The audio signal is

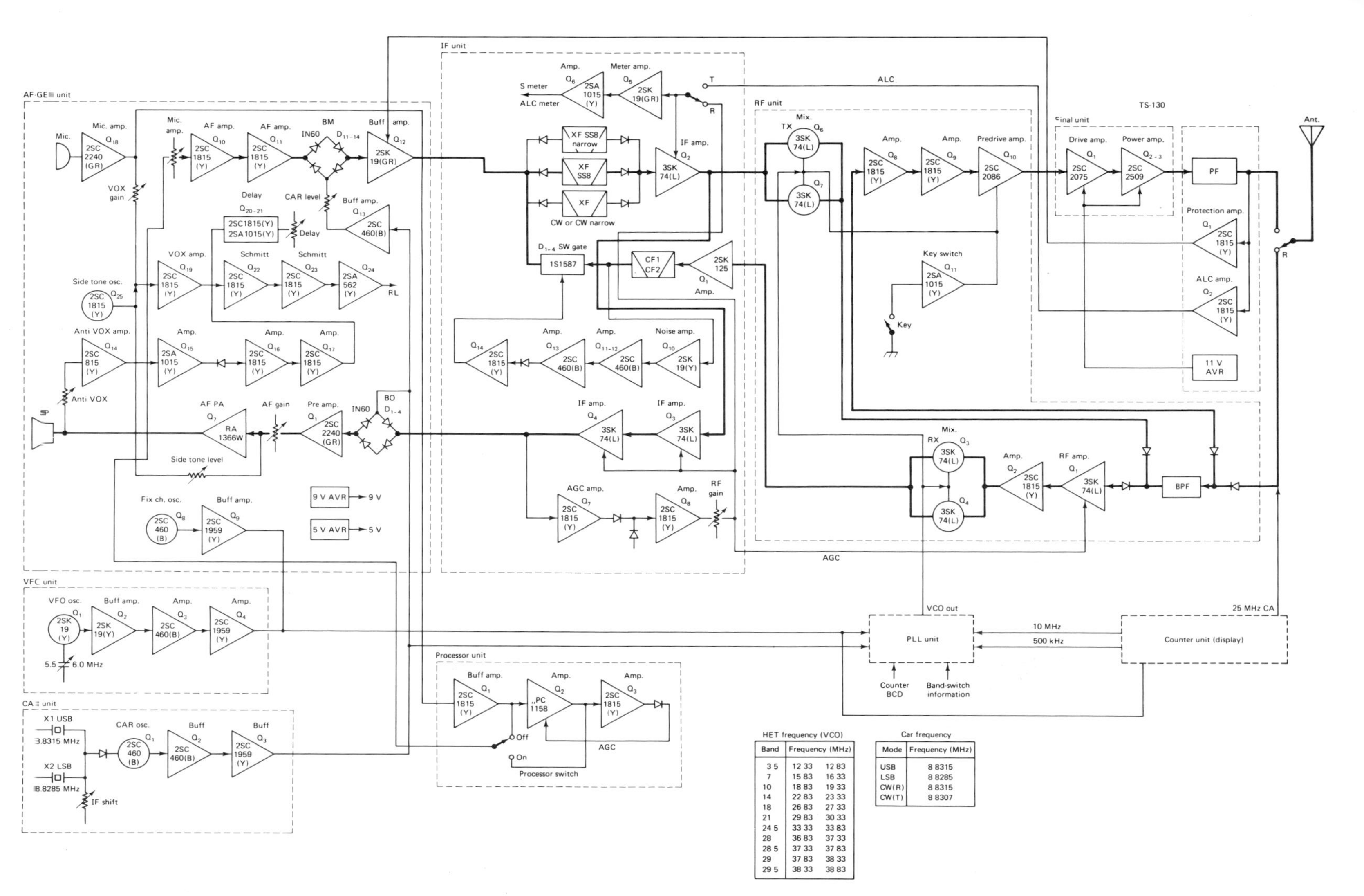

HET frequency (VCO)

Band	Frequency (MHz)	
3.5	12.33	12.83
7	15.83	16.33
10	18.83	19.33
14	22.83	23.33
18	26.83	27.33
21	29.83	30.33
24.5	33.33	33.83
28	36.83	37.33
28.5	37.33	37.83
29	37.83	38.33
29.5	38.33	38.83

Car frequency

Mode	Frequency (MHz)
USB	8.8315
LSB	8.8285
CW(R)	8.8315
CW(T)	8.8307

Figure 9.38 Block diagram of Kenwood TS-130 transceiver. (Courtesy of Trio-Kenwood Communications.)

fed to the balanced modulator, BM, where it modulates an RF signal from the CAR module (8.8315 MHz for USB and 8.8285 MHz for LSB). BM produces a double-sideband signal that is amplified in a buffer amplifier. The buffer amplifier can be disabled by a voltage from the VSWR protection circuit in the FILTER module if energy reflected from the antenna endangers the output stage of the transmitter. The output of the buffer amplifier is fed to the IF module.

2. In the IF module the DSB signal is passed through an 8.830-MHz crystal filter which removes one of the sidebands (the upper SB if in USB mode, the lower SB if in LSB mode). Coming out of the crystal filter, XF, the signal is amplified in an IF amplifier. The gain of the amplifier is controlled by a voltage from the ALC (automatic load control) circuit located on the FILTER module. Three crystal filter packages are available for use in the IF module: a standard SSB crystal shipped as standard with the transceiver, a narrow SSB filter, or a very narrow filter for use during CW transmission. Selection of a particular filter is by means of diode switching.
3. The output of the IF module becomes the input to TX MIX (transmitter mixer) on the RF module. In TX MIX, a balanced mixer, the SSB signal is mixed with an RF signal generated by frequency synthesis in the PLL module. The difference-frequency signal of the heterodyne conversion process becomes the signal to be radiated. That difference frequency is the frequency at which the transceiver is said to be operating. It is displayed on the digital display unit of the COUNTER module. The output of the mixer passes through a bandpass filter, BPF, for removal of all spurious frequencies. It is then amplified in three stages of RF amplification, one of which is designated a predriver amplifier.
4. The output of the predriver is fed to the driver stage of the FINAL module. The driver is designed to increase the energy level of the signal sufficiently to drive the final, output-power stage.
5. The output of the FINAL module is passed through a VSWR sensing element and a low-pass filter (LPF) on the FILTER module. The VSWR element provides signals for VSW protection and automatic limit control (ALC). The LPF, selected by BAND selector switch, filters harmonics and other spurious signals out of the signal to be placed on the antenna.
6. Finally, the signal is fed to the transceiver's antenna through the RF contacts of a transmit-receive antenna switching relay.
7. The transmitter can be switch selected to operate as a CW transmitter for the transmission of code. In this case, when the code key is depressed, a sidetone oscillator on the AF-GEN module is keyed on. Its output is fed to the BM to produce a DSB signal. The frequency of the signal from the CAR module is 8.8307 MHz for CW. Also, the predrive amplifier on the RF module is keyed on and passes the signal generated in the BM and processed by the IF unit.

Signal flow in RECEIVE mode is detailed in the following steps (refer to Fig. 9.38):

1. Air signals picked up by the transceiver's antenna are passed through the "receive" contacts of the transmit-receive relay to the BPF located on the RF module. The BPF is the same one used in TRANSMIT mode. It is switched into the TRANSMIT or RECEIVE string by means of diode switches. Coming out of the BPF the signal is amplified in two stages of RF amplification. The first stage's gain is controlled by the AGC voltage generated in the IF module. Its gain is also controllable by the operator with a control available on the front panel of the transceiver. After amplification, the signal is passed to one input of RX MIX, a balanced heterodyne mixer. The local input to RX MIX is a signal from the PLL module. For a given frequency setting of the transceiver, the frequency of the signal from the PLL is the same for both transmit and receive functions. The output of the mixer—the difference frequency of the mixing process, 8.830 MHz—is fed to the IF module.
2. In the IF module, the 8.830-MHz incoming signal passes first through an IF amplifier and then an 8.830-MHz ceramic filter, $CF_{1/2}$. Out of the ceramic filter, it is fed to a diode noise-blanking gate and then to the crystal filter, XF, which was also used in TRANSMIT mode. From the crystal filter, the signal is amplified by three stages of AGC-controlled IF amplification. After amplification, it is passed to the AF-GEN module.
3. In the AF-GEN module the incoming signal is fed to one input of a balanced product detector (BD). The second input to BD is a signal from the CAR module. The signal from the CAR unit is the same frequency for both send and receive in SSB mode. In CW mode the CAR frequency in send mode is 8.8307 MHz and in receive mode is 8.8315 MHz. The frequency offset is required in CW when an SSB transceiver is used for CW operation. The balanced detector demodulates the received signal and produces an audio-frequency signal at its output. This signal is amplified in the amplifier string of the AF-GEN module and, finally, drives the speaker.

9.7 TRANSCEIVER INSTRUMENTATION

For some time now the need for measuring instruments (current and voltage meters) has been diminishing. A number of years ago, each time a transmitter was powered up, especially one rated 100 W or more, it was necessary for the operator to perform several adjustments of operating parameters. The process was somewhat like performing an alignment on a receiver. In transmitters containing electron tubes, it was especially important for the operator to check and adjust the operating conditions of the power stage: grid current, screen current, plate current, etc. It was essential, therefore, that the transmitter contain instruments with which to measure the quantities that needed regular adjustment for safe and efficient transmitter operation. Today, many transceivers have only one or more indicating lights to inform the user of operating conditions. Some have analog meters, but these are so small and their indications so coarse that their function is little more than that of an indicating light.

Why is the need for transmitter instrumentation and regular adjustment less now than was previously the case? For one thing, more broadband circuits are being used. Another answer is more sophisticated electronics, which has led to more automatic control circuits: automatic gain control, automatic frequency control, automatic limit or load control, etc. Engineers have developed and continue to develop techniques for automatically monitoring and adjusting transmitter operating parameters. These techniques minimize the need for human intervention. In general, automatic controls provide an overall performance which is superior to that obtained with extensive human control. Electronic controls are fast acting, precise, and tireless. Of course, electronic circuits are subject to breakdown. The failure of an automatic control circuit, in some cases, may result in extensive damage to expensive equipment.

Broadcast and amateur transmitters still incorporate measuring instruments: broadcast transmitters because of their high power ratings, amateur transmitters because of flexible operating conditions. Let's examine the details of common instrumentation schemes.

A broadcast transmitter of low or moderate power rating may have three analog-type meters mounted on the front panel of the transmitter cabinet: two for dc voltage and current input to the output-power amplifier and a third, multipurpose meter for RF power and various internal voltages to be used for troubleshooting. A selector switch in conjunction with the third meter enables the transmitter engineer/operator to check quickly any one of several critical operating parameters. Four meters provide for constant indication of RF power as well as dc input power, and a general test meter. Transmitters with very high power ratings—50 kW or more—commonly have a dozen or more panel meters as well as numerous panel indicating lights. A constant indication of operating parameters in many locations in the overall transmitter system is provided.

A high-power-rated transmitter (2 kW or more) almost invariably requires some adjustments when it is powered up as well as any time that changes are made in the loading of the power stage (change in antenna configuration, for example). Adjustments are always made in conjunction with a meter to indicate the effect of the adjustment. High-power-rated transmitters still require the use of electron tubes, at least in the power stage. A typical power-up adjustment with such transmitters is that of the filament voltages of the larger tubes. While observing a panel meter, an operator may set the filament voltages in one or more stages to a specified range by adjusting a knob or knobs on the front panel. The knob may control a motor-driven variac (variable autotransformer) at a location remote from the front panel.

Power amplifier tuning, loading, and power output control are common transmitter adjustments. A PA tuning adjustment involves adjusting the PA output tank circuit to ensure that its resonant frequency is appropriate for best transmitter performance. Generally, best efficiency and minimum distortion is achieved when the PA tank is tuned to a "dip" on the plate (or collector) current meter. Alternatively, best operation may be when this meter is "just off" the dip.

Transmitter loading is a matter of the impedance match between the power output stage and the antenna-transmission line. In day-to-day operation this match should not change unless the antenna configuration is changed. Therefore, transmitter loading, in a broadcast transmitter, is generally not a routine adjustment. For an amateur station, particularly one with several antennas, loading adjustments may be frequent. Loading adjustment is generally performed in conjunction with a VSWR and/or RF power meter. If the station is not fitted with an antenna-matching unit with easily adjustable coils and capacitors (sometimes called a "transmatch"), loading adjustments may involve coil tap changing or even coil changing. In any case, since the possibilities involving transmitter loading adjustments are numerous, it is not practicable to give a detailed step-by-step description. The metering requirement, as indicated, is a means of measuring VSWR and RF power. The meter unit may be an off-panel unit connected in series with the transmission line to the antenna.

Power output, although a function of transmitter loading, is also controlled by the amount of dc input power. Dc input power, in turn, is controlled by dc input voltage. Transmitters of more than modest power rating generally provide a means of power output control by way of dc voltage control. In broadcast transmitters this is generally a front-panel control. For example, it may be a two-step control providing for coarse control at each of two levels: high and low, with a fine control for more precise setting to specified values. The adjustment may involve observing an RF power meter as well as dc input voltage and current meters.

In amateur transceivers, power output control usually involves internal circuit changes. It could involve connecting or disconnecting a completely self-contained outboard power amplifier, a so-called "linear." Dc input current and voltage and RF power measuring functions are generally provided on amateur transceivers. Metering is generally by means of a single, front-panel meter which is switched to measure the various operating parameters. The circuitry required to accomplish the multiple-purpose metering is a study in itself.

Transceivers may incorporate receiver metering. Amateur transceivers invariably include receiver metering. The more expensive CB transceivers include receiver metering. Receiver metering is being provided less often in other types of transceivers: business, marine, and aircraft. With more automatic controls and greater equipment reliability, receiver metering is of no value to operators who are usually busy with other tasks and use radio equipment strictly as a tool. Amateur radio operators, and some CB operators, on the other hand, work with radio equipment as a hobby. They may derive great satisfaction out of monitoring the performance of their equipment and that of others with whom they communicate.

Receiver metering takes the form of the so-called "S-meter." The title is derived, perhaps, from "signal meter" or "signal-strength meter." As already described in the section on an SSB transceiver, an S-meter is typically connected to indicate the AGC voltage in a receiver. From previous study we know that the AGC voltage in a receiver is a function of the amplitude of a received carrier signal in the IF section of the receiver. Therefore, indirectly, the AGC voltage is a measure of the strength of the air signal present in the vicinity of a station's antenna. However, unless an S-meter circuit is carefully designed and calibrated to take into account the gain, or loss, of all the circuits from the antenna (including the antenna) to the measurement point, the meter reading can be only a relative measure of the strength of a particular received signal; and receiver metering circuits are seldom so designed or calibrated. An S-meter is useful in comparisons: comparing the signal strength of the distant station when that station switches to one or more different antennas, for example.

An amateur transceiver S-meter, which is usually the same meter used for the multifunction transmitter measurements, has a unique scale. One arc is marked in decibels. (Another arc may be marked with a simple, 1 to 10 linear scale for reference.) The lower half of the scale is marked 1 through 9, with 9 located at the approximate center of the scale. The scale above 9 is marked in decibels, with the maximum value 40 or more decibels. A typical meter scale is shown in Fig. 9.39.

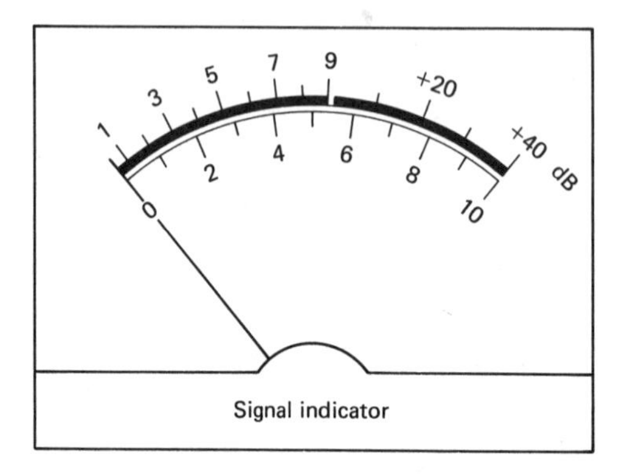

Figure 9.39 Typical scale of receiver S-meter.

The practical meanings of S-meter readings are roughly: S1, barely perceptible; S9, extremely strong. Strength reports are very commonly heard as part of the content of the conversation between two amateur stations, especially at the beginning of the contact. Typical reports might be "your signal is an S7 here," or "you're coming in 10 over 9." The latter means a meter reading at the 10-dB mark above the S9 mark. Practically, it is "10 dB over 'extremely strong.' "

9.8 MICROPHONES, LOUDSPEAKERS, AND HEADPHONES

Every radiotelephone unit must have a microphone and either a loudspeaker or headphones. Technically, these are *transducers:* they convert energy from one form to another. The purpose of a microphone is to convert sound energy into electrical energy. Once converted, the energy can be amplified and used to modulate a transmitter. Loudspeakers and headphones are transducers that convert electrical energy back into sound energy. These are obvious facts with which virtually everyone is familiar simply because of the pervasive influence of our modern mass electronic communications media: broadcast radio and television. Although these devices are essential to radiotelephone systems, they warrant a minimum of time and space. They are little subject to breakdown, and differentiation in technical details between units of different manufacturers is virtually nonexistent.

Microphones may be differentiated on the basis of several characteristics: output level, frequency response, directionality, and output impedance are significant ones. These characteristics, in turn, are related to the method by which sound is converted to an electrical signal. Let us look at each briefly and then examine different microphone types with respect to the energy conversion process.

The output level of a microphone refers to the electrical output for a given sound intensity at its input. When specified formally, output level is usually given in negative decibels. The figure indicates the electrical output in decibels below a reference level of 1 mW for a standard level of sound pressure. The frequency response of a microphone is a measure of the behavior of the output level with respect to the frequencies of sound input components. A "flat" microphone is one that provides a constant output voltage for all input frequencies. For radiotelephone applications a microphone with a flat response over the range 200 to 3500 Hz is appropriate.

Many microphones are omnidirectional, meaning that they respond equally to sound waves from any direction. An omnidirectional microphone is suitable for radiotelephone communication unless the operating environment has a high noise level. In that case, a microphone with limited directionality is more suitable. It requires that the user pay careful attention to positioning with respect to the mouth. But it reduces the amount of pickup of background noise.

For purposes of electrical circuit analysis, a microphone can be thought of as an electrical source. Like all sources, its equivalent circuit includes internal impedance. That equivalent internal impedance is referred to as the microphone's "output impedance." To achieve maximum energy transfer from microphone to external circuit, the impedance of the external circuit must be matched to the output impedance of the microphone.

On the basis of basic materials or conversion process, microphones are classified as carbon, piezoelectric, dynamic, or electret. The *carbon microphone* uses a conversion process which was one of the first in the evolution of electronic communications. A carbon microphone consists of a metal diaphragm in contact with a cup of loosely packed carbon granules. When sound strikes the diaphragm, the carbon particles are alternately compressed and decompressed. The carbon "cup" or button is in series in an electrical circuit. The resistance of the button changes with the flexing of the diaphragm, producing a corresponding change in the current of the circuit. The carbon microphone has a high output and is relatively inexpensive. Although it is still used extensively in telephone handsets, its use has been reduced with radiotelephone applications because of its poor frequency response.

Piezoelectric microphones are also called crystal microphones. They make use of the phenomenon by which mechanical pressure or distortion of certain materials produces a voltage across the material. These microphones are constructed with a diaphragm in contact with a small bar of one of several types of crystalline substances. When sound waves flex the diaphragm, its movement stresses the crystal, producing an output voltage. Crystals that are used include Rochelle salt (after La Rochelle, France, where it was discovered—potassium sodium tartrate) and a barium titanate or lead zirconium titanate ceramic. The output level of piezoelectric microphones is relatively high.

Ceramic microphones, especially, have become popular because they are rugged and unaffected by temperature and humidity. By virtue of their construction they have a capacitive equivalent circuit; their output impedance is high. They are especially suited for use with SSB systems since the unwanted low-frequency response can be attenuated simply by feeding their output to an unmatching, low-input-impedance circuit.

A *dynamic microphone* is an electromagnetic device. Its construction consists of a small, lightweight coil attached to a diaphragm and suspended in the magnetic field of a permanent magnet. When sound waves cause the diaphragm to move, the coil is moved in the magnetic field and an alternating voltage is induced in the coil by electromagnetic induction.

The *electret microphone* is a relatively new development. It is an evolution of the capacitor (or condenser) microphone. In a capacitor microphone, a diaphragm is in close proximity to, but electrically insulated from, a fixed plate. Together, the two plates form a capacitor. As the diaphragm vibrates with impinging sound waves, the spacing between the plates changes. The equivalent capacitance changes along with the spacing. Like the carbon mike, a capacitor microphone does not generate any voltage; it must be part of a circuit with a separate, or bias source. Its changing impedance produces a changing current in the circuit. Now, an electret is an insulator which has, in effect, a permanent static electric charge. If placed between two plates, the electret is like a charged capacitor. A separate bias source is not required. The electret microphone has a high output impedance. An FET impedance-matching circuit is typically used in conjunction with it; its output is higher than that of a dynamic microphone.

Loudspeakers

Loudspeakers, or simply, speakers, are more common than microphones. Although details and quality of construction of speakers may vary, most modern speakers utilize the same basic principle for their operation. They are an electromagnetic device and make use of the "motor" principle of electromagnetism. A speaker consists of a cone to which is attached a coil. Construction design positions the coil in the air gap of a powerful permanent magnet. When the current from an audio power amplifier is caused to flow in the coil, there is an interaction between the magnetic field of the permanent magnet and the changing field produced by the current in the coil. A force is exerted on the coil, causing it and the cone to move. Since the current alternates in direction at an audio rate, the coil and cone move back and forth at an audio rate. This movement of the cone produces sound.

Speakers are comparatively free of failure. Although basically sturdy if mounted properly, a speaker's cone can be damaged easily, especially during handling when the speaker is unmounted. The cone is made of a paper-like material. A speaker's cone can also be damaged by overdriving.

Headphones

Like a loudspeaker, headphones are electromagnetic transducers that change electrical energy into sound energy. Some headphones, in fact, are simply miniature speakers that have been packaged with the usual head band and ear cups. Headphones can be superior to a loudspeaker in situations where the background noise is excessive. They help an operator to shield out some of that noise and concentrate on the message that is being communicated via the radiotelephone system.

9.9 SERVICING RADIOTELEPHONE EQUIPMENT

Servicing radiotelephone (two-way radio) equipment implies adjusting and/or troubleshooting transmitting as well as receiving equipment. Although equipment is available that incorporates only the receive function for one or more of the radiotelephone bands, the vast majority of equipment incorporates both the transmit and receive functions. In most of that equipment, the two functions are included in one package—the transceiver. Separate transmitters and receivers can still be found in the amateur radio field, for example.

In the United States, adjustments or tests during or coincident with the installation, adjustment, servicing, or maintenance of some radio transmitters must be made only by or under the supervision of a person holding an appropriate license. See Appendix B for information about licensing.

There are a number of separately identifiable tasks relating to radiotelephone equipment which must be performed by persons with a technical knowledge of that equipment. Included in a list of such tasks would be: installation, testing, adjustment, alignment, troubleshooting, repair, and preventive maintenance. It is our goal in this section to present basic ideas about two general areas of tasks: adjustment/alignment and troubleshooting/repair. Many professionals would lump these activities together and call them *servicing*.

A presentation of ideas about servicing activities, as they relate to AM receivers, was offered earlier in this text (Chap. 4). You are urged to read that chapter again now. Servicing a sophisticated communications transceiver—troubleshooting it, or performing an alignment—differs from servicing an AM receiver primarily in the details of the actual equipment with which one is involved. All of the basic skills/techniques enumerated in Chap. 4 are still effective.

The component count in communications equipment, generally, is many times that of an AM receiver. This implies that when servicing communications equipment it is even more important to think of a system first, in terms of function blocks and how they work together to perform the overall function. Signal frequencies in communications equipment are almost always higher than for an AM broadcast receiver. Circuits processing higher frequencies require greater attention to techniques that will avoid loading down circuits and/or picking up or injecting noise. Higher frequencies present a need for test equipment with higher frequency ratings than those used for servicing AM broadcast receivers.

Manufacturers' Instruction Manuals

A widely experienced and highly trained technician might well be able to make progress toward aligning or troubleshooting a modern communications transceiver without the aid of a manufacturer's instruction manual on the equipment. However, for most technicians, a good instruction manual is one of the most indispensable of service tools. Manufacturers' instruction or service manuals typically contain the following: complete, formal equipment specifications; operating instructions; system block diagram; schematic and wiring and/or parts layout diagrams; step-by-step alignment instructions; testing and troubleshooting instructions; and a list of required/recommended test equipment.

Many manuals also contain brief but useful "theory of operation" sections. The latter are especially helpful in providing clues about an unusual circuit or novel function whose purpose and design philosophy are not readily perceivable. Theory-of-operation sections are highly useful in helping the inexperienced technician increase his/her basic knowledge of the system of interest. You are urged to make every effort to secure an instruction manual on any equipment you may be assigned to service. You are further urged to develop the useful habit of studying such manuals.

Test Equipment for Servicing Communications Transmitters/Receivers

The test equipment required for servicing communications equipment depends very much on the level of servicing to be performed. The equipment required for emergency field service only is obviously different than that for complete, formal, in-shop alignment and adjustment procedures. The following list starts with the most essential equipment items, those that would be required even for field servicing [designated (f) in the list]. It continues with items that one would expect to have available for use only in a well-equipped shop [designated (s)].

1. *Electronic multimeter* (EMM or DMM) (f). A digital multimeter is not essential, although recommended. However, the meter should be electronic and have ohmmeter and ac/dc voltage and current functions. The low-voltage ohmmeter range for semiconductor in-circuit tests is recommended. Input impedance for voltage ranges should be at least 1 MΩ, preferably 10 MΩ, minimum. Current-measuring capability up to 10 A is recommended.
2. *RF voltmeter* (f). This may be an RF probe for use with the EMM. Voltage ranges from 10 mV to 300 V are desirable.
3. *AF voltmeter* (audio-frequency electronic voltmeter) (f). Recommended frequency range: 50 Hz to 10 kHz (some good-quality DMMs satisfy this requirement). Input resistance: 1 MΩ or greater; voltage range: 10 mV to 30 V.
4. *RF signal generator* (crystal-calibrated stable frequency) (f). Frequency range: 1.8 to 200 MHz; output: 20 dB/0.1 μV to 120 dB/1 V.

5. *AF signal generator* (f). Frequency range: 200 Hz to 5 kHz; output: 1 mV to 1 V, low distortion.
6. *VSWR meter/RF wattmeter* (f). Frequency range: 3 to 200 MHz; power rating: 100 W or more.
7. *Dummy antenna load* (noninductive RF load) (f). Impedance: 50 Ω, 150 Ω; dissipation: 100 W continuous or greater; frequency limits: 1.8 to 100 MHz.
8. *Dummy speaker load* (f). Impedance: 8 Ω; dissipation: 3 W or greater.
9. *Oscilloscope* (s). Calibrated sweep; bandwidth: 0 to 100 MHz.
10. *RF sweep generator* (crystal calibrated) (s). Frequency range: 1.8 to 200 MHz; adjustable center frequency; frequency deviation: ±5 kHz; output: more than 0.1 V.
11. *Frequency counter* (s). Minimum input voltage: 50 mV; frequency range: to 200 MHz minimum.
12. *Spectrum analyzer* (s). Frequency range: 100 kHz to 200 MHz; bandwidth: 1 kHz to 3 MHz.
13. *Miscellaneous dc and RF test leads.*

There are available a number of specialized test equipment units for use with specific types of communications equipment. For example, several manufacturers market CB radio test rigs. These are equipment items that contain several basic test functions: electronic multimeter, signal generators, frequency meter, etc. They include crystal-controlled sources for frequently used CB test frequencies: 10.7 MHz and 455 kHz, for example. Test equipment of this type can be helpful in achieving good productivity in shops that specialize and work only on certain types of equipment.

Adjustments and Alignment

In most electronic systems, there are invariably a few circuits whose operating parameters are extremely critical to the successful performance of the system. The height of the display on a television picture tube is an example. The operating parameters of a circuit—current, voltage, frequency, etc.—depend on many things. Certainly, they depend on the values of components in the circuit. Although a designer may calculate exactly what component values must be to produce a desired result, it is virtually impossible to implement a circuit with those precise values. For one thing, mass-produced components can never be expected to be exactly equal to a specified value. All are permitted to have a tolerance of variation. The smaller the tolerance, the more expensive is the component. Of course, the vast majority of circuits operate satisfactorily with components whose tolerance is quite large—20% even, in some cases.

Some very critical circuits must be fitted with one or more adjustable components: an adjustable resistor (pot) or capacitor, or whatever. Many, perhaps most, such adjustable components can be accessed and adjusted only by removing an equipment panel or cover. These are called *service adjustments*. Some circuits are so critical that they may require component adjustments by the user/operator of the equipment. Such components may be called *variable components,* as distinguished from "adjustable." They are also called *user* or *operator adjustments*. Many of these controls are simply to permit the operator to choose a different operating condition rather than compensating for a slight change in circuit parameters. Adjusting tuning and volume controls on a radio receiver are examples of setting different operating conditions. Adjusting the vertical hold and height controls on a TV receiver, however, are examples of operator intervention to achieve satisfactory operation from a system.

Communications transmitters and receivers typically have several service adjustments which must be made after assembly is complete but before shipment to customers. These include the adjustments of inductors and/or capacitors which "align" the functions with tuned (or resonant) circuits: filters (e.g., IF sections), oscillators, and FM detectors, for example. Adjustment of resistors may also be required to set bias voltages and currents, VCO frequencies, and a number of other operating parameters.

Generally, the more expensive an item of equipment, the more complex it is in terms of numbers of components, circuits, adjustments, etc. A check of the instruction manual of a top-of-the-line amateur transceiver indicates a total of 55 service adjustments and alignment steps to set up the unit for optimum performance. Such an extensive list of initial setup adjustments, generally, do not ever have to be repeated. The exception: some, if not all, will need to be performed after replacement of a failed component. Let's look more closely at receiver and transmitter adjustment and alignment steps: You may be called on to perform them someday.

Receiver adjustment and alignment. First, let's recognize that although all receivers have many similarities, it would be presumptive to imply that a simple 1–2–3 list of service adjustments can be provided that would be appropriate for any receiver with which you might have to work. What we want to do, then, is look at the various types of adjustments, in general, that are made on receivers. Then, when you are faced with the task of setting up an actual receiver, you will be able to use the manufacturer's list of adjustments intelligently. You will be able to proceed with the process with the knowledge that you have a basic idea of what that process is expected to accomplish. You can expect that most receivers that are part of a radiotelephone system will be contained in the combination unit—the transceiver. Since, in many transceiver designs, functions are shared between receiver and transmitter, a number of receiver adjustments will overlap adjustments to be made to the transmitter.

Dc voltage adjustments. Many communications-type receivers, especially those that are part of transceivers, include voltage regulator circuits to provide precisely controlled dc voltages. Regulated voltages may be supplied to all or only a portion of the stages in the unit. In any case, if any of the voltage regulator circuits incorporate provision for adjusting their output voltage, it is essential that the voltage adjustments be made first. The supply voltage level affects the performance of most electronic functions. Changing the supply voltage on a stage after other adjustments have been made would very likely require redoing the other adjustments. Of course, be guided by the manufacturer's recommendations on this matter. Without exception, dc voltage adjustments should be made while monitoring that voltage with a good-quality voltmeter of known accuracy.

Adjustment of tuned circuits. A typical receiver has many tuned circuits. For optimum performance of the receiver, these circuits must be tuned to a specified frequency, a frequency determined by the design process. *If a receiver has been operating satisfactorily, it is not likely that tuned circuits will need to be aligned.* The exceptions to this statement include situations in which failed components have been replaced, or when it is known that a tuned circuit adjustment has been disturbed. The adjustment of a tuned circuit should not be touched unless and until tests have been made on the receiver which indicate that an alignment is necessary. It should never be assumed that an inoperative receiver is caused by a misaligned circuit. Do not touch a tuning adjustment of an inoperative receiver: Get the receiver operating before deciding that an alignment adjustment is called for.

If it has been decided that you are definitely to do an alignment of a receiver, assemble the best equipment available for your use: oscilloscope, RF generator, RF voltmeter, and a frequency counter. If a sweep generator and spectrum analyzer are available, you may be able to use these to advantage. Since alignment of tuned circuits involves adjusting them to operate at a precise frequency, it is essential that you know precisely what frequency you are using for testing. This requires either a generator whose frequencies are known to be stable and can be calibrated and set precisely, or a frequency counter of high precision that can be used in conjunction with a generator. A good-quality oscilloscope and/or RF voltmeter are convenient devices for monitoring the results of your adjustments. Monitoring results is an essential part of the alignment process.

Frequency alignment of circuits that process signals (as distinguished from those that generate signals, such as oscillators) involves (1) *injecting* into the circuit a test signal of appropriate frequency and amplitude, and (2) monitoring the signal at the output of the circuit being aligned to determine if the desired adjustment has been made. Connecting test equipment to a circuit for accomplishing these two steps requires a knowledge of proper techniques and the exercise of considerable skill. Test equipment must be connected to a circuit in a way that will cause the least possible disturbance to that circuit. An alignment or testing procedure is a complete failure if the operating characteristics of the circuit are altered in any significant way by the test equipment attached to it.

In what ways can connecting test equipment to a circuit change that circuit? How can the problem be avoided, or at least minimized? There are several ways that a test device can change a tested circuit:

1. The test device can change the dc voltage operating levels of the circuit under test. If the output circuit of a generator, for example, has low dc resistance

to ground, or if it has its own dc voltage, connecting it to a transistor stage will probably upset the dc voltages in that stage. Avoid this problem by coupling the signal to the test circuit through a capacitor. In some cases, a simple air-core injection coil can be made from a short piece of hookup wire. The receiver manufacturer may describe a very specific method for connecting the signal source to the receiver: Follow those instructions.

2. The test device can "load" the test circuit. If the impedance to ground looking "back into" the test device is of the same order of magnitude as that of the circuit to which the test device is being attached, the test circuit is going to be loaded excessively. Its operating characteristics will be changed. The test results may be meaningless. Worse, you may be lead to "adjust" the circuit to operate with the test device attached. When the test device is disconnected, the circuit may not work at all with its normal signal.

It is usually possible to find a circuit point or a method for injecting a signal or for monitoring its passage without loading the circuit being tested. The impedance of the antenna input circuit of receivers is usually low, of the order of 50 Ω. If this input can be used to get a signal where you want it to be, it is an excellent location for injection. The base side of a transistor stage is usually much more susceptible to loading than the collector side: Inject on the collector side, rather than the base side, if possible.

3. An injected test signal can simply be of the wrong amplitude. If the amplitude of a test signal is excessive, (a) it may overdrive one or more amplifier stages, producing distortion and accompanying spurious signals; (b) it may activate an AGC function which then causes an entirely misleading interpretation of results; or (c) the injected signal may be so strong as to be coupled across a stage rather than being processed by the stage. The result, again, is a completely misleading observation of cause and effect. When injecting test signals, it is absolutely essential that you (a) inform yourself about the injection point (What is the amplitude of the signal at that point when the receiver is working normally and processing an "air" signal?), and (b) go about ensuring that the signal you inject is that same order of magnitude. For example, do not inject a 1-V signal at a point that normally processes a 10-mV signal, etc. Of course, if the signal you inject is too small in amplitude, you will probably not be able to observe an output of any significance.

In conclusion, know what you are about. An item of test equipment is only a tool. The results you obtain with various types of equipment can be no better than the knowledge, skill, technique, and thought with which you use them.

The operating frequencies of the various tuned circuits in a receiver are generally related to each other, even if they are not the same frequencies. Furthermore, there is often a degree of interaction between two or more tuned circuits. For these reasons, there may be an optimum or "best" procedure for accomplishing an overall alignment, with best results, with minimum steps, in a minimum amount of time. A manufacturer's technical staff will have determined this best procedure. It is the one outlined in the instruction manual: *Follow it.* If you do not have the instruction manual, get it.

Someday you may have to do an alignment without access to an instruction manual. If you find yourself in that situation, try to find and talk to someone who has already done an alignment on the type of receiver with which you are confronted. If there is simply no procedure list available from any source, you need some suggestions. Start with these:

1. Determine suitable circuit points for injecting a signal and for monitoring the output of sections to be aligned. Remember and apply precautions about loading, dc isolation, signal amplitude, etc.
2. Inject an appropriate signal and verify that you can measure the resulting output signal on your chosen monitoring device: oscilloscope, RF voltmeter, or other.
3. Align filter circuits (e.g., IF sections) first, starting with sections nearest the output. Working toward the front end of the receiver, align all signal processing sections that tests indicate need alignment. Assure yourself that you are using the correct frequency(ies) and amplitude(s). In the typical situation it is necessary to reduce an injected-signal amplitude as sections are brought into alignment. To ''tune out'' the effects of interacting circuits, repeat all adjustments until no further improvements can be produced.
4. Proceed to the adjustment of signal-generating circuits (PLLs, oscillators, etc.) after all signal-processing circuits have been aligned. If available, use a frequency counter to measure the operating frequency of source circuits.
5. After you complete an alignment, assess the performance of the receiver while receiving an air signal. If the result is not satisfactory, analyze your procedure, modify it, and repeat, to improve performance. If you have previously performed alignments using manufacturer's instructions, recall and use the techniques you learned from such instructions.

Remember, the purpose of receiver adjustments is to set those circuit parameters which are adjustable to values that will produce optimum performance of the receiver, all things considered. The manufacturer knows best how to achieve that result. Follow his instructions. Receiver adjustments generally include more than just the alignment of high-frequency circuits.

Transmitter adjustments and alignment. When working on a transmitter it is important to remember that anytime you key it on when it is connected to an antenna, RF energy is being radiated. To minimize interference to others using the spectrum, during testing and alignment procedures, replace the antenna with a *dummy antenna,* an RF load that will dissipate the RF energy instead of radiating it. Some air testing will inevitably be desirable. Reconnect the antenna for final testing after you have completed prescribed adjustments and have tested for optimum performance using a dummy antenna. Overvoltage is likely to destroy an output stage if a transmitter is tested without either an antenna or a dummy load.

Although the goals of transmitter adjustments and alignment are similar to those of a receiver, there are significant differences. Since the purpose of a transmitter is to generate RF energy (actually, to convert dc energy to RF), its optimum performance coincides with maximum production of RF energy. This occurs when each stage in the RF string is operating to pass along the maximum possible amount of energy. In some transmitters, each stage in the RF string will have adjustable *L* and/or *C* elements, very much like the IF section of a receiver. A transmitter alignment includes, among other things, the adjustment of these circuits for best performance. In the case of a transmitter, the signal from its own master oscillator (or frequency source, whatever it may be) is used for alignment. An RF wattmeter connected to measure actual RF power into a load (dummy antenna, preferably) is the usual monitor for this alignment.

Inasmuch as the operating parameters of transmitters are regulated by the FCC, a significant portion of transmitter testing and adjustment relates to ensuring that a transmitter operates as prescribed. This is particularly true of the testing and adjustment procedures that are performed by the manufacturer before a transmitter is put into operation by a purchaser/user. The various broadcasting services—AM, FM, and TV—typically use high-power-rated transmitters: 1000 W and up. Stations in these services are required to make routine periodic transmitter tests to ensure that transmitters are operating correctly. These tests include measurement of RF power, operating frequency, and modulation level. Stations in these services may receive on-site visits from FCC personnel at the time of completion of a transmitter installation, or later, to verify that a station transmitter is operating according to regulations.

In contrast with this, the several millions of transmitters which are part of the compact transceivers used in two-way radio installations receive FCC *type acceptance*. That is, a manufacturer with a new transceiver to be placed on the market submits a complete set of specifications and design details to the FCC for consideration. If the FCC technical staff agrees that the design will perform within regulations, as claimed, the design is type accepted by the FCC. The manufacturer has the responsibility of ensuring that the transmitters which it produces perform within regulations, as claimed. The following is a brief listing and description of the types of tests and/or adjustments that may be performed on transmitters to ensure operation within regulation. The tests described might be performed at the factory before a unit is shipped. Or they might be performed in the field on a transmitter that has been in use but which has had one or more components replaced after failure. You might someday be involved in performing tests such as these:

1. Testing the RF string for optimum alignment and production of specified power. Actual power should not exceed prescribed power, but it should not be significantly less than specified. Alignment consists of processing an RF signal, from the transmitter's own frequency source, through the RF string and adjusting any available *L* and/or *C* circuits for a performance that indicates correct frequency tuning. This is typically either a maximum voltage at a designated test point, or a minimum voltage, a dip in a meter reading, at the test point.

2. Measuring and setting the transmitter's operating frequency(ies). One of the most stringent FCC regulations regarding radio transmitters relates to their operating frequency. Except for amateur radio equipment, which has its own

special regulations, radio transmitters are required to operate at fixed frequencies. Depending on the service, a transmitter may be permitted to operate on a number of fixed frequencies. Each frequency, however, must be switch selected and crystal controlled. This is to ensure that operating frequencies are precise and stable.

Consider as an example the Johnson ULTRACOM 508 VHF/FM transceiver unit described in Sec. 9.3. The FCC frequency tolerance for transmitters of this type is ±0.002% of the assigned carrier frequency. Required frequency stability (freedom from drift) is 0.0005% of the carrier frequency. In adjusting a transmitter of this type for correct operation, the master crystal oscillator frequency must be adjusted by means of a trimmer capacitor. This adjustment is made while measuring the final carrier frequency with a frequency counter. The counter must have an accuracy greater than the frequency tolerance of the transmitter (e.g., ±0.00025% or better).

3. Testing modulation process and adjusting for correct operation. In FM transmitters, modulation testing involves modulating the transmitter with an audio test signal. The test signal is of accurately known frequency and amplitude. The transmitter's output is measured with a deviation meter. Deviation must be adjustable within prescribed limits. Maximum deviation for two-way radio applications is generally ±5.0 kHz. If maximum deviation is greater than prescribed, the circuitry for limiting deviation must be investigated for faulty operation.

On the other hand, a transmitter must be capable of achieving a maximum deviation of at least 80% of prescribed maximum deviation. Modulation, by regulation, must be achieved with not more than a prescribed maximum of distortion.

A *spectrum analyzer* can be used to monitor the modulated output of a transmitter. The spectrum analyzer is an oscilloscope-based instrument. However, it presents a frequency-based display. A conventional oscilloscope display is time based. The spectrum analyzer indicates the frequency content of a signal. Thus it will reveal harmonics and other spurious frequency components. If a signal is distorted, harmonic content, as revealed by the analyzer, will be excessive.

In AM transmitters, greater than 100% modulation is undesirable for two reasons: it produces distortion of the information signal, and it produces spurious RF signals which may cause interference (called "splatter") with other frequency channels. In testing, the transmitter is modulated with an accurately known audio signal. The transmitter's ability to achieve maximum or near-maximum modulation is checked. Necessary adjustments, if any, are made. The transmitter is checked for automatic modulation limiting and distortion.

The modulation testing of SSB transmitters is moderately more complex than for either FM or AM transmitters. If an SSB transmitter is modulated with a single-frequency audio test signal, a single-frequency RF signal is produced since the side frequency is the signal, unlike either FM or AM. Unless the RF stages are driven excessively hard, there is little chance for distortion to occur. Hence modulation testing of SSB transmitters usually involves the so-called "two-tone" test. That is, the transmitter is modulated with an audio frequency input signal containing two different frequencies. With this test, an oscilloscope display of the modulated output of an SSB transmitter looks very much like an AM-modulated carrier with 100% modulation. The envelope of the display approaches the waveform of a perfect sine curve. The presence of any distortion in the RF signal also presents itself as distortion of the envelope of the two-tone test display. A spectrum analyzer is effective in revealing distortion in a two-tone modulation test. A complete test involves observing distortion levels versus modulation levels.

Generally, SSB transmitters utilize some form of automatic level or load control circuitry (ALC). Adjusting the operating levels of this circuitry is normally part of the adjustment procedure. The purpose of such a circuit is to automatically adjust circuit operation to maximize RF power output while preventing exceeding acceptable distortion levels.

Associated with modulation testing in SSB transmitters are two other facets of the SSB generation process: sideband and carrier suppression. Transmitter operation must be tested for correct achievement of this aspect of the SSB process. The spectrum analyzer, again, is helpful in this test.

4. Testing the transmitter output for the suppression of harmonics and other spurious signals. The radiation of frequencies other than those within a prescribed frequency channel causes interference with other channels. A spectrum analyzer can detect the presence of excessive levels of undesired frequencies.

In summary, transmitter testing, adjustment, and alignment procedures have several goals: (1) to achieve high, efficient RF output, but output within FCC regulation limits; (2) to ensure that operating frequency(ies) are correct and stable within FCC tolerances; (3) to ensure that the modulation process is being achieved adequately and within FCC limitations; and (4) to ensure that desired distortion limits and spurious frequency suppression goals are being achieved. Performing typical transmitter tests requires that you have a knowledge of and skill in using the following: precision multimeter, oscilloscope, RF generator, frequency counter, spectrum analyzer, RF wattmeter, and AF and RF dummy loads.

Although there are similarities in the procedures for testing, adjusting, and aligning all transmitters, there is usually a preferred procedure for each transmitter. Transmitter manufacturers evolve such procedures and describe them in instruction manuals. You are urged to obtain the instruction manual and follow the procedure outlined there when you have the task of performing work on a transmitter.

Troubleshooting Two-Way Radio Equipment

The second part of servicing two-way radio equipment involves diagnosing the defect(s) of inoperative or improperly or poorly functioning equipment—troubleshooting. We add that repairing the defect is an implied part of troubleshooting. Troubleshooting communications equipment—transceivers primarily—is very much like troubleshooting a relatively simple, AM broadcastband receiver. A detailed description of that task is provided in Chap. 4. You are urged to reread and review Sec. 4.1 now.

There are also differences between troubleshooting an AM receiver and a communications transceiver:

1. The transceiver is more complex in the sense of having more functions and a higher component count. A transceiver is more likely to have state-of-the-art integrated circuits.
2. A transceiver (or transmitter) may have circuits which handle more power than any circuits found in an AM receiver. Operating such circuits incorrectly can cause the destruction of expensive components.
3. Operating a transmitter is an action that can produce interference with the communications operations of other users of the radio-frequency spectrum. Operating a transmitter is an activity regulated by law. It is your responsibility to know and observe the applicable regulations. Whenever possible, use a dummy antenna load on a transmitter when keying it on for testing.
4. When replacing a defective component, it is possible to disturb the performance of a transmitter in a way that will cause its operation to fail to be in accordance with regulations—to be illegal. It is your responsibility to avoid this result.

To troubleshoot a two-way radio system, follow the basic procedure outlined in the flowchart of Fig. 4.1. It is very important to know the block diagram of the system you are going to troubleshoot, or to have the diagram at hand. The first step is to determine which subsystem is inoperative. In a two-way radio system, there are at least two transmitters and two receivers: we consider each receiver and each transmitter a subsystem.

In a typical situation, you will be called because a particular transceiver is "dead." Hence, is the problem in the transmitter or receiver of that particular unit? Sometimes the answer will be obvious: If the unit receives but an attempted transmission cannot be heard by another station with which communication is know to be possible, the area of uncertainty is narrowed to the transmitter. If a unit neither receives nor, apparently, transmits, is the transceiver defective? Or is the station simply out of range or in a propagation shadow with respect to all other stations on the same operating frequency? The method for finding the answer to this question depends very much on the situation. The point is that you must use logical thinking and your knowledge of how a two-way radio system works. Devise and make operational tests that will provide information leading to the answer. It is important to make all the obvious observations and checks before assuming that the problem is a difficult technical one: Is the unit getting an appropriate dc voltage? Is the antenna connected? Is the cable from antenna to unit in good shape, not cut or broken?

Once you have reached a conclusion about which subsystem contains the problem, the next step is to determine which *section* of that subsystem is affected. Inform yourself of the signal flow path. Think of the unit in terms of the various functions that process the signal. In many typical situations, transceivers are mounted in vehicles—cars, trucks, boats, airplanes, etc. You may be asked to "look" at a problem unit in a vehicle in the "field," that is, away from a service shop. You may not have available your favorite test equipment: RF generator, oscilloscope, etc. You may be limited to a multimeter with RF probe and an

RF wattmeter. What tests might you make that would enable you to pin the problem down to a particular section, given the test equipment available? Perhaps you have a similar transceiver in your own vehicle. You could use it to call another station and get their cooperation in making a transmission with which you could test the functioning of the receiver subsystem of the defective transceiver.

In any case, your knowledge of the subsystem and a touch of ingenuity are essential factors in your ability to make maximum use of the resources that you have available. In many cases it will not be possible to find the problem without the use of a greater variety of test equipment. The transceiver must either be removed from the vehicle and taken to a shop, or the vehicle moved to a location where other test equipment can be used. If a transceiver can be taken to a fully equipped shop, the task of locating the defective section is a normal one. With appropriate signal-generating and signal-monitoring equipment, signal tracing and/or signal injection are effective techniques for quickly narrowing the problem area to a section.

As we learned in Chap. 4, after reaching a conclusion about the problem section, an effective next step is to locate the defective stage. The same techniques used in locating a defective section can be used to pinpoint the defective stage: signal tracing and/or signal injection. An analysis of the dc voltages of the stages of a section will reveal a defective stage in many, though not all, types of faults. This technique is useful in situations where the kinds of test equipment available is limited.

If the equipment is not very old, it is inevitable that you will find integrated circuits and/or modular circuit boards in communications-type equipment. In many designs, modularization corresponds relatively closely to the function-block diagram. That is, each function section is implemented on its own printed-circuit "card" or module. The result is that the division of a system into separate sections (function blocks) is literally physical, not just conceptual. Although this technique can have disadvantages, it can aid the troubleshooting and repair process. For example, if the IF module has a "signal in" and a proper "signal out," the problem is not in the IF section; it is elsewhere. On the other hand, if there is a signal in but none out of a particular module, that module has the problem. Or does it? Some care must be exercised in reaching that conclusion. It will probably be correct most of the time. It could be wrong, for example, if the module were not receiving the correct dc voltage. In any event, if a spare module known to be good is available, substitution of it for the suspected module might be the method your employer would prefer that you use.

One of the hoped-for advantages of modular construction when it first began to be used was an increase in productivity in the service area (troubleshooting and repair). Whether increased productivity is realized depends on a number of factors. It is not one of the goals of this text to analyze the pros and cons of modular construction, however. As a prospective service technician, it is important that you are aware that modules exist. You should know that performing troubleshooting tasks in which you replace a defective module with a new or repaired module requires a specialization or honing of limited skills. In effect, your progress through the several steps leading to the ultimate diagnosis of a defective component always stops at the level of diagnosing the defective section. This is good in the sense that, in general, you will spend less time on any one problem. You have bypassed the more tedious steps of finding the defective component. You may be able to gain experience with a greater variety of problems. On the other hand, in not performing such steps you will tend to lose your "touch" in finding defective components.

Integrated circuits, to a lesser extent than modules, combine the functions of two or more discrete-component stages into a single package. Here, too, it is more obvious than with normal circuitry that there must be a signal in and a signal out. At least this is true of linear ICs. If a linear IC is getting the proper dc voltage, is properly grounded, and otherwise correctly "in" the circuit (pins soldered properly, no solder bridges, or making good contact in a socket, etc.) and has a "good" signal in and no signal out, it is obviously defective. It must be replaced. (This does assume that the output circuit, external to the IC, is not defective and loading down the IC's output.)

Let us assume that we have to troubleshoot to the component level, and so, let us continue reviewing the procedure outlined in Chap. 4 as it applies to communications equipment. Once the defective stage has been identified, dc-voltage analysis (see Chap. 4) is an effective method of identifying the defective component. This is true for most, but not all, cases. For example, open coupling or bypass capacitors will not be revealed by this technique. The manufacturer's instruction manual is an important resource to have available and to know how to use, especially in conjunction with this phase of troubleshooting. It is the rare manufacturer that cannot or will not provide schematic diagram(s) with all important dc voltages marked on the diagram. Transceiver schematics typically have two sets of voltages: one for a specified receive mode, and one for a transmit mode. It is essential that you read all notes on a schematic so that you know under what conditions voltages were measured. You must duplicate those conditions if you are to take voltage measurements which are useful for comparison with those specified. Review Chap. 4 carefully to refresh your memory on how to use dc-voltage analysis to pinpoint defective components.

When replacing a soldered component, it is important to observe good soldering techniques, especially on printed-circuit boards. Use a vacuum-type solder remover when unsoldering a component. Exercise care that you do not "lift" foil from the board. When soldering the new component, use the proper-size soldering iron so that the heat is adequate but not excessive. One of the most common causes of electronic equipment failure is the damage done to one or more components during a soldering procedure.

As part of the process of replacing a component, you must analyze its effect on the performance of the equipment in the likely event that its electrical parameter is not exactly the same as the component replaced. That is, can the component replaced change the power output, frequency of operation, modulation level, amount of spurious signal, etc., of the transmitter? If the answer is yes, then, after the repair is completed, the transmitter must be tested, its performance checked, and any necessary adjustments made to bring the operation into line with regulations. You (or your supervisor, if you are not certified) have a legal responsibility to ensure that such steps are taken.

Troubleshooting Logic Circuits

Before leaving troubleshooting, let's focus briefly on troubleshooting logic circuits, that is, circuits containing digital ICs. Since virtually all modern communications devices contain some logic circuits (the digital dividers in PLLs, for example) it is desirable to have a working knowledge of useful techniques in that area. The skill of troubleshooting linear circuits is of great help with digital circuits, but it needs some augmentation.

In working with the analog circuits of radio receivers and transmitters, we get used to the idea of a signal "flowing" through the system. We think of each function as contributing its share to an overall purpose by "processing" the signal in its unique way: changing its frequency, as in a mixer; increasing the signal amplitude, as in an amplifier; recovering the information signal, as in a detector; etc. An important part of this concept is our interpretation that the signal is present at every moment in every part of the system. The form of the signal changes, to a degree, as the information changes. With an appropriate instrument, we can monitor the signal at any time and at any point in the system. Of course, there are functions, like local oscillators, which are not processing the basic signal directly, but contribute to the process. But their signal is just as continuous (always present), if not more so, than the main signal.

In working with logic circuits, the concept of a common signal being continuously present, in some form, in every part of the system, is no longer appropriate. Logic circuits, quite literally, are switching circuits. You will recall that we call the basic building-block elements, *gates*. There are AND and OR, NAND and NOR gates. We use these gates as switches to define paths over which an information signal, or more commonly, control signals are to flow. The magic of logic gates is that they can be opened and closed electronically, and thus rapidly. At one instant a signal path is along one circuit route; 10 ns later it is along another route, etc.

In troubleshooting logic circuits, the analog technique of signal tracing becomes one more of tracing logic levels. The basic element is the gate, rather than a stage, as in linear circuits. A logic circuit fails when a gate fails to "open" and "close" the way it is designed to do. The symptom of a failed gate is that it "hangs up": its output is either always "high" or "low," regardless of the conditions on the input legs. The goal of the troubleshooting process is to find that one gate which is hung up. Logic troubleshooting tools are the oscilloscope and a special device called a *logic probe*. The purpose of each of these devices is to let us "see" whether or not a particular logic circuit point is changing levels the way it is supposed to, given the conditions of the circuit.

A logic probe is, itself, a logic circuit. In its simplest form it is just an LED (light-emitting diode). With one side of the LED grounded, the other side can be touched to different points in the circuit. If the LED goes on, the state of the point is high; if the light stays off, the point is at a low state. One step of sophistication, using voltage comparators, provides a two-light indication: red for a high state and green for a low state. More sophisticated probes are available which provide information about points where pulses are present. For example, a one-shot multivibrator can be used as a "pulse stretcher." Its output will indicate a high even though the point tested receives only an occasional short high pulse. An oscilloscope can provide much more information. It can be used to assess the waveshape of a pulse as well as its simple presence. A dual- or multitrace scope permits a time-occurrence comparison of two or more pulse waveforms.

Effective logic troubleshooting requires an organized "search" procedure just as in analog troubleshooting. Logic systems can be broken down into subunits, as with analog systems. The digital circuitry of communications equipment, generally, will not be extensive. Nevertheless, subdivision in terms of functions will usually be possible. Again, the manufacturer's block and schematic diagrams, now showing block and logic diagrams of the digital sections, are a must.

You are using an organized procedure when you work to reduce the area of uncertainty, first, to large subdivisions, then, to smaller and smaller units of circuit. You monitor the operation of a subdivision by checking logic levels at key points while the unit is being exercised. Ultimately, if you are successful, you find a single gate that hangs up. (This ignores the possibility of very ordinary failures such as broken circuit board, defective resistor, capacitor, etc.) A gate is normally part of an IC that has several gates. You replace the entire package. If ICs in the system are socket mounted, as they sometimes are, you merely unplug the defective IC and plug in a new one. If the IC is soldered to the circuit board, you must unsolder it and solder in a new one.

To troubleshoot logic you must know what it is supposed to do. You get that information by studying and interpreting the manufacturer's diagrams. Knowing what a logic circuit does usually involves "when," not just "what." That is, relative timing of events is an important part of logic functions. Timing is often indicated by a *synchrogram.* A synchrogram is an oscilloscope display of two or more waveforms, or a line drawing or sketch of such a display, real or imagined. The purpose of the synchrogram is to show the time relationships of the waveforms that it includes. When you use an oscilloscope to check time relationships, you must take care to set up the scope so that the traces do, in fact, have identical time bases. For example, dual-trace scopes generally have two methods or modes of producing dual traces: In ALTERNATE mode, the beam completes one waveform and then starts again on the left side of the screen to produce the second trace. The two traces do *not* have a common time base. In CHOP mode, the beam switches back and forth between two traces while it is completing one pass from left to right. The two traces, in this case, have a common time base.

Let's examine a simple logic circuit diagram for the purpose of getting a feeling for synchrograms and the processes of determining what a logic circuit does and how to troubleshoot it. Study the diagram of Fig. 9.40.

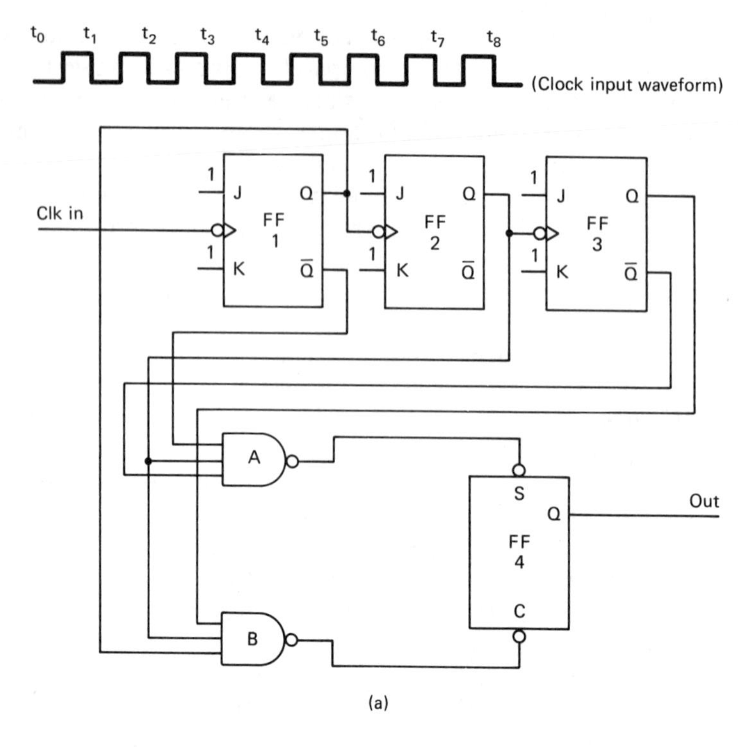

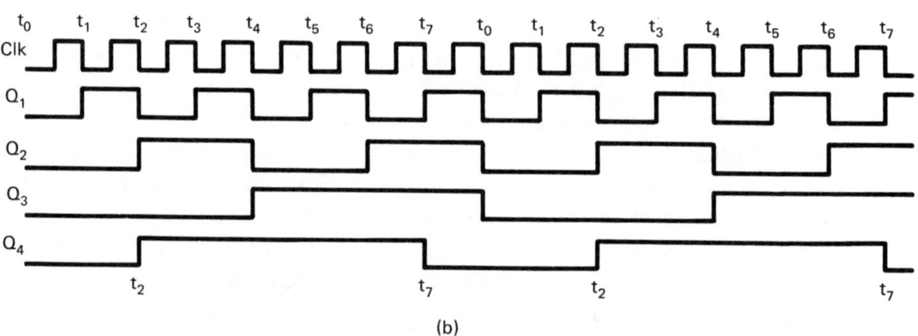

Figure 9.40 (a) Logic circuit and waveform for Example 1; (b) synchrogram for circuit of (a).

Example 1

Assuming that the logic circuit shown in Fig. 9.40(a) represents a "function" of a larger system, what is its purpose? The input to the function is the clock signal, "clk," and the output is taken from the Q terminal of FF_4. Explain the operation of the circuit. Add the output waveform to the synchrogram of the circuit.

Solution. The purpose of the function of Fig. 9.40(a) is to generate a pulse waveform with the following specifications: Using the notations on the clock waveform, a positive-going pulse at time t_2 with a duration (on-time) of five clock periods. The circuit consists of a MOD-8 ripple counter and two NAND gates to decode the count of 2_{10} (010_2) and 7_{10} (111_2). When "2" is decoded, the output of NAND gate A goes low, setting FF_4. The Q output of FF_4 goes high to begin the desired pulse. When the count of 7 is reached, the output of NAND gate B goes low, clearing FF_4. The pulse is terminated after five clock periods. The output waveform, coordinated with the clock waveform, is shown in Fig. 9.40(b). The cycle repeats itself every eight clock periods.

Example 2

The function driven by the output of the circuit of Fig. 9.40(a) fails to operate correctly. A check with a logic probe indicates that the output of FF_4 is always high. What conclusion would you reach concerning the cause of the failure of this circuit?

Solution. FF_4 is not being reset. Either FF_4 is defective or it is not getting a low input from NAND gate B. Check the output of B. If the output of B is always high, check its inputs. If the inputs are correct—all high at time t_7—then B is defective. (This assumes that the chip of which B is a part has the correct dc voltage and a proper ground connection.)

9.10 MICROPROCESSOR-CONTROLLED TRANSCEIVERS

Many radiotelephone services provide users with the choice of operation on many channels. For example, CB radio provides for 40 channels in the range 26.965 to 27.405 MHz. The number of CB channels was increased to 40 from 23 in 1976 (in the United States). Amateur radio operators not only have numerous "bands" in which to operate, they are permitted to use equipment that has continuously variable operating frequencies within the various bands. Other services, marine and aircraft, for example, also have a number of channels in which to communicate.

Transceivers that can operate on a number of frequencies are considerably more complex than those that operate on a single frequency. In the evolution of such equipment, some of the first techniques involved using multideck, multiposition selector switches. These were wired so that each operating channel had its own master oscillator. RF amplifiers shared active elements—electron tubes—but tank circuits were switched so that the amplifiers were correctly tuned for each channel. Early CB transceivers, for example, sometimes had 46 or more crystals (two for each channel).

Equipment manufacturers began searching for ways to produce transceivers at less cost. They recognized the need to reduce the component count. Fewer components could mean lower labor costs, for one thing. Digital ICs were being found useful in many applications other than just computers. One of the major developments in communications equipment was the marriage of the phase-locked loop and digital ICs. So-called frequency synthesis was the result. Employing PLL frequency-synthesis techniques, 40-channel CB transceivers using only three crystals began to be common. The concept of broadband design of RF circuits was also developed and used in the design of multifrequency units, with a resulting reduction in component count.

Another, more recent electronic development which is having a significant impact on the design of multifrequency communications equipment is the microprocessor controller. In simplest terms, a microprocessor is almost a "computer on a chip." We say "almost" because most microprocessor chips require one or more support chips to be an effective computer.

Another way to look at a microprocessor is as a huge quantity of highly organized logic circuitry. And logic circuits are simply very highly organized, highly flexible switching circuits. Switching circuits, you say; but switching is exactly the source of our problem in multifrequency transceivers. That is, there is so much switching required that it makes the task of accomplishing it complex, and therefore expensive. At this stage of design evolution, the application of microprocessors in communications equipment is basically one of automating elaborate switching arrangements. The result is, so far, not really lower-cost equipment, but equipment that has many more, and more desirable, operating features, as well as improved and more precise performance.

As microprocessor technology and its applications in communications electronics advance, it is inevitable that the cost of communications equipment will be reduced. It is likely that some form of microprocessor control will find its way into virtually all forms of communications equipment, even AM broadcast receivers. It is highly important, then, that you know something about this development.

Gaining a thorough understanding of "how" microprocessors and microcomputers do what they do requires a comprehensive background in computer logic. Having an elementary knowledge of computer programming is also, typically, a characteristic of persons doing an intensive study of microprocessor technology. It is not feasible to include extensive coverage of computer logic and programming in this book. Therefore, be warned that this section does not pretend to provide an explanation of how microprocessor control works. However, there are two areas of knowledge about microprocessors that we can profitably venture into without being fully qualified with the normal prerequisites: terminology (or vocabulary) and the "what" of microprocessor control. In this section, then, we examine the most relevant terms of the jargon of microprocessor technology as well as "what" microprocessor control "is" or "does" in a radiotelephone transceiver. If you are using this text as a student in an electronics curriculum in a community college or other institution providing technical education, you may already have studied computer programming and logic. If you have not studied these topics, you are urged to do so.

The microprocessors used in transceivers are, generally, not the CPUs used at the heart of microcomputers. They are, rather, chips designed especially for the purpose of controlling transceivers. That is, they are *custom* CPUs. Let us first examine a list of functions which are typical of those performed by microprocessors in transceivers. Then we will examine the items on the list in some detail. By this process we will be working at achieving our goal of learning "what is microprocessor control"? Learning the terminology associated with microprocessors occurs automatically as we encounter new terms and are provided with explanations of their meanings. It is essential in all of this that we utilize our knowledge of the function blocks of transceivers and view the application of a microprocessor as that of taking over the control of certain of these functions. As we will see, this taking over of control relieves the operator of some control tasks by providing automated, but at the same time, more flexible control. The payoff, for the operator, is more operating options. In terms of equipment construction, the payoff is a lower component count and smaller unit size.

Typical tasks performed by microprocessor control circuits in transceivers include, but are not limited to, the following:

1. Polling the front-panel switches for the desired operating conditions: band, mode (CW, AM, FM, or SSB, etc.), transmit, receive, etc.
2. Programming digital dividers, etc., in PLL circuits to generate the required frequencies for selections made by the operator through band switch and VFO setting. Frequency selection input may also be through the means of a calculator-type keypad.
3. Providing information for the front-panel visual display unit.
4. Storing in either internal memory or external RAM (random-access memory) chips, information on the display and operating frequencies, even when transceiver power is off (if a backup battery is used).
5. Supervising transmit mode: allows transceiver to go into the transmit mode only if the PLL function(s) is locked and the selected frequency is within preset limits.
6. Performing diagnostic tests on itself when powered up after the backup power has been off.

Now, let's examine each of these tasks in greater detail to increase our understanding of the part a microprocessor control function plays in the operation of a transceiver:

1. The microprocessor "polls" the front-panel switches. "To poll" in this context means "to ask the position of." The microprocessor chip is provided with input pins for each of several switches. For example, an amateur transceiver would be provided with switches for selection of CW or SSB, USB or LSB, etc. The microprocessor, under the control of a clock timing pulse, goes to each switch and asks "What is your position?" The switch answers, giving a logic level of high or low for "on" or "off," respectively. The microprocessor stores each answer in its memory section. The stored answer is used to determine exactly how the transceiver will operate. The microprocessor comes back to each switch on a periodic basis, say every 10 μs, and repeats the question and stores the answer.

In the case of a band switch (frequency selector switch), the answer to

"What is your position?" is more complex. The band switch has numerous positions rather than simply "this" or "that." These positions are indicated by a binary code. For example, four conductors, each of which can be high (1) or low (0), can indicate 16 different positions of a switch: 0000, 0001, 0010, 0011, . . . , 1111. The microprocessor accepts this kind of information from the band switch and stores it in its memory to use in programming the digital dividers in the PLLs of the system.

The information received from the VFO frequency control function is more complex than the band switch. Since the microprocessor can use only binary data, the input from the VFO control knob must also be binary. Users of variable-frequency oscillators (ham radio operators, mostly) are accustomed to the continuously variable (as distinguished from stepped) feature of older, analog-type controls. (Turn a knob; the feeling is smooth, without notches.) The control knob shafts of digital VFOs are, typically, provided with an arrangement called a "shaft encoder" which provides binary digital data to the microprocessor while maintaining the "feel" of a continuously variable analog control. The shaft encoder provides binary information which indicates the direction of rotation of the VFO shaft and the speed at which it is being turned. This feat is accomplished by means of two small light-source, photoelectric pickup units which are positioned to detect slots in a disk mounted on the VFO shaft. The microprocessor contains circuitry that converts the relative position and width information of the pulses produced by the photoelectric pickups into binary data. The microprocessor uses these binary data to determine the amount and direction of the change it makes in the operating frequency of the transceiver. A separate "variable-frequency oscillator" circuit is not required.

2. Programs digital dividers in PLL circuits to generate the required frequency(ies) as determined by the operator's selection. The microprocessor uses the information it has stored from all inputs: band switch, "VFO" control, CW/SSB switch, etc., and generates a binary code. This binary code is sent to the appropriate PLL circuit. (Typical microprocessor-controlled transceivers use two or more PLL circuits.) The binary code is of the form to correctly program digital frequency dividers in a loop to produce the selected frequency.

Let's learn some new, specialized terminology in connection with this item. The CPU and the phase-locked loops are connected together by a set of conductors called a *bus*. Since the bus transmits data, it is called a *data bus*. In this application, the data bus, typically, consists of eight conductors. With eight conductors, 256 unique codes can be sent: Each conductor can be made either high or low, 1 or 0.

If there is more than one PLL, as is typical, each is connected to the same set of eight conductors. Hence a means must be provided for the CPU to "address" a particular PLL. "To address" means "to select." In some systems, one of the bus lines is used as an "address line." In other systems, a separate "address bus" may be used. Getting back to the data bus, the conductors on the printed-circuit board representing the bus are, of course, connected to appropriate pins on the CPU chip. This set of pins, as a group, is referred to as a *port*.

To send data to a PLL, then, a CPU places 1's and 0's on its "data port" pins. The corresponding bus lines are energized, high or low, by the electrical potentials on the data port pins. Simultaneously, the input pins of all PLL digital divider devices are energized by the potentials on the bus conductors. However, only one PLL device will actually "take in" the data. Which one? The one that is addressed by the CPU. How and when does the divider device actually receive the data? The device is said to *latch* the data on a *strobe* from the CPU. A latch is a simple flip-flop which takes the state of the bus conductor to which it is connected, but only at the instant when it receives a second, time-significant signal called a *strobe signal* or, usually, simply *strobe*. Again, the typical description is that the "divider IC latches the data on the bus, on the strobe from the CPU." After strobing, since the data are now stored in the latches in the divider IC, the bus is available for other purposes, and the CPU can go on to other tasks.

The digital divider devices in the PLLs of this description are the equivalent of *output devices* in a conventional computer system. On the other hand, the band-selector switch is equivalent to an *input device* of a conventional computer system. Typical computer systems, to provide flexibility, are equipped to receive data from a number of input devices, and to output data to a number of output devices. (The combination term "input/output" is abbreviated I/O.)

With several I/O devices, the CPU must have a means of addressing (selecting) particular ones. In some computer systems, I/O devices are addressed using the same instructions as are used for addressing memory (selecting memory locations). That is, I/O devices are treated as if they were simply another memory location. In other systems, there are unique instructions for input and output, and I/O devices have their own unique address codes. The first method of I/O addressing is called *memory-mapped I/O;* the second method is called *direct I/O*. You will meet these terms in connection with microprocessor-controlled transceivers, since both techniques are used in the custom CPUs found in these applications.

3. Provides information for the front-panel display unit. Microprocessor-controlled transceivers invariably incorporate a digital-display unit for indicating operating frequency. Other operating parameters, such as USB/LSB, TRANSMIT, etc., may also be displayed on the display panel. Since the CPU is the clearinghouse, as it were, for such information (it needs the information in order to correctly control operations), it is not difficult to arrange matters so that the information can be displayed on an electronic display panel. The set of pins that connect to the conductors going to the display panel represent a second output port for the CPU.

The display device may contain elements for displaying as many as eight decimal digits, as well as other operation indicator display elements. The decimal displays use the universal seven-segment arrangement for each digit. Each segment is driven separately. The logic circuitry for determining "digit" and "segment" to be driven, in accordance with the actual operating frequency, is contained in the CPU. The display is *multiplexed*. That is, only one digit at a time is actually energized. All digits appear to be lighted all of the time to the human eye. This is a physiological phenomenon known as *persistence of vision*. The multiplexing operation is such that the display is *refreshed* many times per second: Each digit receives current for a brief time, many times per second.

4. Stores frequency information in memory. This enables an operator to save an "operating frequency" and recall that frequency quickly. When using a variable-frequency transceiver, an operator must adjust dial(s) until he has brought all adjustments to the required settings for a particular frequency. This, of course, requires some time. With memory provision, an operator, by depressing a single button, can store that frequency in memory and recall it for use at a later time. With a battery backup, to provide the very minimal amount of power required to keep the memory energized, frequency information can be retained even when the transceiver's main power is off.

5. Supervises transmit mode. The CPU is programmed to monitor the operating frequency of the transmitter. If the operator selects a frequency that is outside predefined limits, or if a PLL is not in lock (and thus not producing the selected frequency) the CPU prevents the transmitter from being energized. Also, if a PLL goes out of lock during a transmission, the CPU senses this and forces the unit into receive mode.

6. Performs diagnostic tests on itself when powered up after backup power has been off. When backup power goes off, data required for correct operation are lost. Diagnostic tests, programmed routines that exercise the various functions of the CPU, reveal any malfunction in the CPU. This is indicated to the operator. The operator is thus informed as to whether or not it is feasible to proceed to reestablish operation.

9.11 BROADBAND DESIGN

We have used the term "broadband" a number of times in this chapter. So far, we have not defined it carefully. You will find it occurring frequently in technical literature relating to communications circuits—usually without adequate definition or explanation of its meaning. To an extent, it is self-explanatory. However, let's examine its meaning in some detail.

In general, and in simplest terms, the description "broadband" may be applied to a circuit whose transfer characteristic changes with frequency. If the transfer characteristic changes very little over a frequency range that is "large" compared with the center frequency of the range, the circuit has a broadband characteristic. If the frequency range of little change is "small" compared to the center frequency, the circuit is a narrow-band circuit.

"Large" and "small" are very relative. Can we be more specific? Take the IF section of a typical AM broadcast band receiver. The passband may be 10 kHz wide and the center frequency is 455 kHz. What is the passband as a percentage of the center frequency? Answer: $10 \times 100/455 = 2.2\%$. The passband is approximately 2% of the center frequency. What about an FM IF section? The passband is typically 200 kHz; the center frequency is 10.7 MHz. $0.2 \times 100/10.7 = 1.9\%$. Again, the passband is approximately 2% of the center frequency.

Now consider the power stages of an amateur HF transmitter. The section is expected to amplify signals over the range 3.5 to 30 MHz. What is the "passband" as a percent of center frequency? 26.5 × 100/17 = 156%. That's "broadband"!

The terms "RF amplifier" and "tuned amplifier" are often considered to be synonymous. And, of course, a tuned amplifier, of which IF amplifiers are examples, is normally a narrow-band amplifier. Designing RF amplifiers in multifrequency transmitters to have broadband characteristics is a relatively modern approach. In transmitters using narrow-band amplifiers, tuned circuits must be switched to provide operation on more than one frequency. Switching the tuned circuits of RF amplifiers often leads to difficulties. When using non-broadband transmitters, ham operators, typically, find it expedient to perform a short alignment procedure when they change from one frequency band to another. Broadband design provides good performance over a wide frequency range without readjustment of operating circuits.

Broadband design is now used in the transmitters of all radio services that employ multifrequency operation, that is, operation on numerous frequency channels. Broadband amplifier operating characteristics are achieved by the use of coupling/impedance matching transformers with high-permeability cores and negative-feedback techniques. Coupling transformers in broadband design, typically, use ferrite, toroidal (doughnut-shaped) cores. A transformer with a high-permeability core has characteristics which approach those of the ideal transformer: The transformation ratio is virtually independent of frequency because leakage reactance is negligible with the high-permeability core. Negative feedback in an amplifier reduces those characteristics that vary with frequency. If an amplifier has more gain at 20 MHz than at 10 MHz, with negative feedback there will be more signal fed back at 20 MHz: the gain will be reduced more. The result is a flattening of the gain versus frequency characteristic of the amplifier. Broadband design has emerged partially because of the development of new materials (ferrite cores and semiconductor devices) and partially because of the evolution of our understanding of electronic circuits.

9.12 CONTINUOUS TONE-CONTROLLED SQUELCH SYSTEMS (CTCSS)

One of the disadvantages of two-way radio systems, in which there are a number of stations operating on the same frequency, is lack of privacy. For example, everyone who has a receiver that is turned on and tuned to the frequency will receive all transmissions on the frequency. If many of those transmissions are of no concern to the listener, the experience can be fatiguing. An example of the situation is a commercial business with a number of vehicles equipped with two-way radio, all operating on the same frequency. All operators can receive all messages.

A scheme that avoids the problem described and provides operation resembling the private-line (PL) operation of ordinary telephone systems is called a *continuous tone-controlled squelch system,* abbreviated CTCSS. This system requires two functions: one at each transmitter and one at each receiver. A tone-generating circuit is provided at each transmitter. The tone, a very low-frequency subaudible signal, is added to the signal from the microphone. The result is that the transmitter, when activated, is continuously modulated by the CTCSS tone, as well as by the voice signal. At each receiver, a circuit is provided that keeps the audio section of the receiver squelched unless the CTCSS tone is being processed by the receiver. A block diagram illustrating the basic features of a CTCSS system is shown in Fig. 9.41.

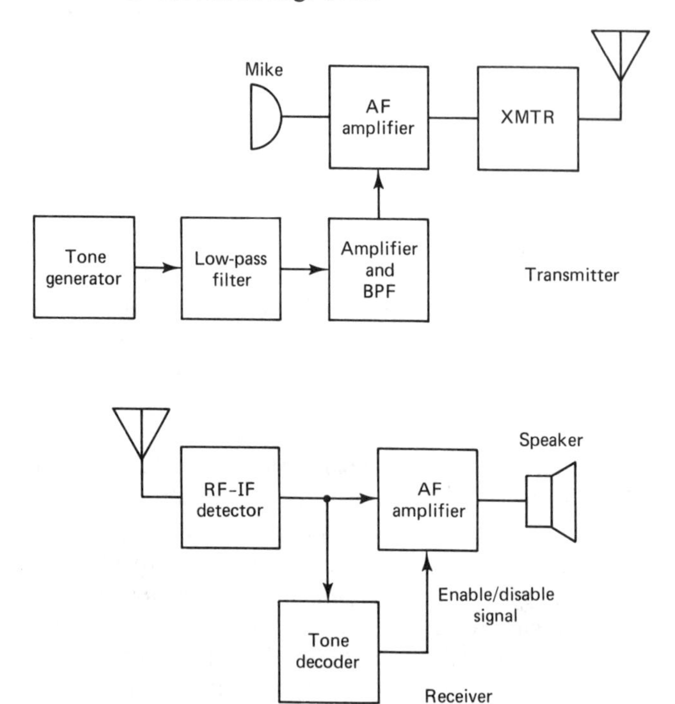

Figure 9.41 Principles of continuous tone-controlled squelch system (CTCSS).

9.13 ANTENNAS FOR RADIOTELEPHONE SYSTEMS

The vast majority of nonamateur radiotelephone stations are mounted in vehicles—automobiles, trucks, boats, and airplanes. As a result, the most common type of antenna for such stations is the $\lambda/4$ vertical antenna, often called the vertical whip antenna. Its advantages are convenience of mounting and a radiation pattern, horizontally omnidirectional, well suited to the service for which it is intended. Because, on most vehicles, the antenna is mounted relatively close to the ground, the propagation range is limited. This is also appropriate since the frequencies assigned to radiotelephone service are heavily used in most areas. Limited range ensures that interference between stations is limited to a reasonable level. The input impedance of a $\lambda/4$ vertical whip is, typically, a nominal 50 Ω.

The antenna systems of base stations, because they do not suffer the physical mounting limitations of mobile stations, are often more elaborate. However, in some services, CB radio for example, the configuration and mounting height of base station antennas may be limited by FCC regulation. The purpose of such regulations is to limit interference between stations. The antennas of amateur stations are some of the most elaborate in the world. See the chapters on antennas for information about antenna arrays and other special antenna configurations.

9.14 RADIOTELEPHONE REPEATER STATIONS

The basic idea of a repeater station is one of receiving an incoming transmission, demodulating it, etc., and then completely retransmitting it. The idea has been used for a long time in long land-line telephone services. It is also used extensively in two-way radio systems. The goal of repeater-station use is extended range.

The basic functions of a repeater-station system are illustrated in Fig. 9.42. A basic characteristic of two-way radio repeater-station systems is that the repeater station is located at a higher elevation than the surrounding land area. Typical locations are on the top of a mountain peak as near as possible to the center of the area to be served. A repeater station can also be mounted on the top of a tall building in an urban area. The ultimate mounting is on an earth-satellite space vehicle.

There are numerous schemes in use concerning operating frequencies and control of the repeater station. To be feasible, control of a repeater station must be automatic. That is, it must contain control functions that cause it to retransmit signals of appropriate quality and characteristics, without human intervention. Figure 9.42 depicts an automatic repeater using split-frequency operation: the station receives on frequency A and retransmits on frequency B. This means, of course, that the stations communicating with each other, through the repeater, must also operate split frequency: transmit on frequency A and receive on frequency B.

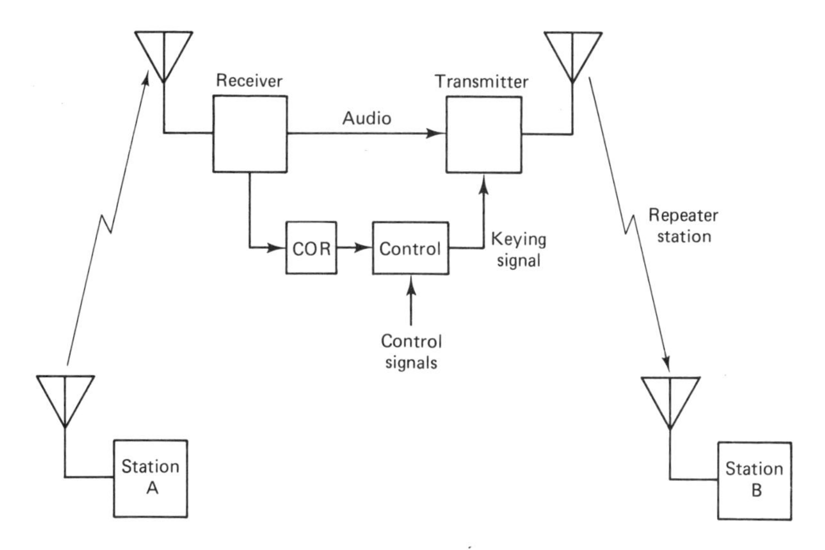

Figure 9.42 Principles of radio-repeater station.

An essential function in a repeater station is a device called a *carrier-operated relay* (COR). The COR is a circuit that (1) detects when a signal of adequate strength is being received, and (2) turns on the repeater transmitter. The audio output of the receiver must be connected to the audio input of the transmitter. The COR circuit generally operates off the squelch circuitry of the receiver. Recall that a squelch circuit detects the presence of a carrier signal of adequate strength. Normally, the squelch circuit controls the audio section of the receiver, enabling it when a carrier is being received and disabling it at other times.

A highly sophisticated form of radiotelephone service, somewhat related to repeater service, is coming into being just as this is being written. It is called *cellular radiotelephone service, mobile phone,* or *cellular service.* In a cellular telephone system, a densely populated area is divided into smaller geographical areas or *cells.* Each cell is served by a computer-controlled base station with features similar to those of a repeater station. Persons in vehicles with cellular equipment installed can communicate, through the automatic base station, and then conventional telephone systems, with anyone, at any location in the world. The signal strength of the received signal at a base station is monitored under the control of a computer. When a vehicle moves out of one cell into another cell, the operation is switched to the base station of the new cell. Of course, all of the behind-the-scenes monitoring and switching of operation is transparent to the user. The cellular mode of operation reduces interference and thus increases the number of users which can be accommodated. It is claimed that the number of users which a system can ultimately handle is nearly unlimited. The cellular system provides a two-way radio communications system of high reliability.

9.15 THE CORDLESS TELEPHONE

Another application of two-way radio with which you may be familiar is the cordless telephone. The cordless telephone permits a person to communicate in an apparently normal manner over a conventional telephone system while unencumbered by a cord connecting the handset to the base instrument. With a cordless phone, a person can wander around within a range of approximately 500 ft of the base instrument and still carry on a telephone conversation. This is possible because of a two-way radio link between the handset and the base instrument. Both the handset and the base unit contain transceivers. The audio output of the receiver in the base unit is connected as the input to the telephone line; the incoming audio signal on the telephone line is used to modulate the base transmitter.

GLOSSARY OF TERMS

Balanced demodulator A circuit used for demodulating an SSB signal. It is termed "balanced" because the arrangement of components provides balanced paths for current flow.

Balanced modulator A circuit, which can take many forms, that produces AM-like sidebands but suppresses the carrier.

Bandwidth Conceptually, the group of frequencies which a circuit, filter or otherwise, allows to pass with little or no attenuation. Mathematically, the difference between the highest frequency passed and the lowest passed.

Base station A radio station, part of a two-way radio communications system, which is fixed in location.

BFO Beat-frequency oscillator, the circuit that generates the signal in a receiver for detecting an SSB or CW signal.

Broadband Characteristic of a circuit whose bandwidth is greater than is deemed normal or typical for circuits of the type. The term is frequently used to describe circuits in radio transmitters which are not narrowly tuned for a single carrier, but process equally well, carriers of a wide range of frequencies.

Carrier-operated relay (COR) A circuit that activates the transmitter portion of a radio repeater when it detects that the station is receiving a signal (carrier) of adequate strength.

Cellular telephone A mobile telephone system in which a city's area is divided into smaller areas called cells. Telephones (radiotelephone transceivers, actually) are mounted in vehicles and connect to conventional telephone facilities through a network of radiotelephone stations, one for each cell. As a vehicle moves from cell to cell, the radio transmission link is switched from the station of one cell to another under computer control.

CTCSS Abbreviation for "continuous tone-controlled squelch system." A scheme, used in radiotelephone systems, in which a transmitter, when activated, is continuously modulated by a subaudible tone. Only the target receiver contains a circuit that will enable (desquelch) its audio section when the tone is received and demodulated. Provides operation similar to the private-line operation of conventional telephone systems.

Cutoff frequency The frequency(ies) at the edge(s) of the passband of a circuit (usually applied to filters) at which the output of the circuit has been attenuated by 3 dB.

Diode switch A circuit consisting of a diode and a dc voltage source which can provide either a low-impedance path or an extremely high-impedance path to a signal by virtue of a change in the level of dc voltage supplied to the diode.

Direct I/O An arrangement involving a central processing unit (CPU) and input/output equipment in which that equipment is addressed with unique I/O address codes, as distinguished from the arrangement in which I/O equipment is addressed as if part of memory. *See also* Memory-mapped I/O.

Dummy antenna A resistor, usually noninductive, used to provide a load for a transmitter and to avoid radiating RF energy.

Duplex A term used to describe a method of operating a communications system in which two frequencies are used—one for each direction of the communication. Does not imply that both frequencies or directions are used simultaneously. The term "half duplex" is used with this same meaning.

First-order filter A filter whose equivalent circuit includes only one reactive element.

Fixed station A radio station, part of a two-way radio system, which is always located at one geographical position.

Frequency synthesis Literally, the generation of a signal by putting together two or more component signals of different frequencies. Usually refers to method of producing signals for transceivers by means of a phase-locked-loop circuit.

Full duplex A term indicating that a two-way communications system is provided with the facilities to permit communications in both directions simultaneously. *See also* Duplex.

Insertion loss The unavoidable attenuation of a signal in the passband of a filter.

Lattice circuit A circuit in which there are elements that cross from side to side as well as run along each side.

Memory-mapped I/O Reference to a method of addressing the I/O devices of a computer, that is, as if each were another memory location.

Mobile station A radiotelephone station mounted in a vehicle, thus one that does not have a fixed geographical position.

Narrow-band FM Description applied to the operating conditions of most FM two-way radio systems to indicate that frequency deviation and maximum audio frequency are limited so that the bandwidth of an operating channel does not exceed 16 kHz (approximately).

Noise blanker A circuit for reducing pulse-type noise by amplifying and rectifying the signal containing noise and using the resulting negative-going pulses to cut off an amplifier section during a noise pulse.

PEP Peak-envelope power, a term used in connection with SSB transmission to describe the power contained in the signal at the maximum peak amplitude. Used in the power rating specification of SSB transmitters.

Pole (of response) A term describing a peaking of the response of a filter and corresponds, in frequency, to the resonant frequency of a series-resonant, series branch of the filter, or to the resonant frequency of a parallel-resonant, shunt branch circuit.

Port Name given to the circuitry of a central processing unit, which, conceptually, represents the point of entry or exit of data logic signals.

Product detector Circuit for demodulating a single-sideband signal utilizing the principle of demodulation by product of SSB signal and locally supplied carrier.

Repeater station Radiotelephone station, consisting of receiver and transmitter, used to extend range of two-way communications by automatically retransmitting all signals received of adequate strength.

Ring modulator One of several forms of balanced modulators, so-called because circuit configuration resembles a ring.

Second-order filter Filter whose equivalent circuit includes two types of reactive elements.

Shape factor (SF) Mathematical ratio of bandwidths of filter at two attenuation levels: -60 dB and -3 dB (i.e., SF = -60 dB bandwidth/-3 dB bandwidth).

Simplex Term applied to two-way communications system in which same frequency is used for both directions of the communication.

S-meter Signal strength meter commonly found on radiotelephone transceivers to give indication of relative strength of received signal.

Spectrum analyzer Oscilloscope-like instrument providing frequency-domain display as distinguished from time-domain display of conventional oscilloscopes; useful in evaluating frequency components of a complex waveform.

Squelch Circuit function commonly found in radiotelephone transceivers to suppress audio output unless a channel signal of adequate strength is being received.

SSB Abbreviation for "single sideband"; commonly used to designate a radiotelephone communications system which transmits only one of the two AM sidebands.

SSBSC Abbreviation for "single-sideband, suppressed carrier."

Transmatch Name of equipment item containing adjustable inductors and capacitors for matching output impedance of transmitter to input impedance of transmission line/antenna of radiotelephone station.

Variable-frequency oscillator (VFO) In transceivers, source of signal to provide variable-frequency operation of transmitter and receiver and normally used in conjunction with, not in place of, a crystal-controlled master oscillator.

VOX Abbreviation for "voice operated," meaning, literally, "voice-operated transmitter-receiver switching control"; a circuit function of a radiotelephone station (transceiver, typically) which switches the station from RECEIVE mode to TRANSMIT mode when an audio (voice) signal at the microphone input has an adequate amplitude level; to be distinguished from mode switching controlled by manual closing of a switch [i.e., from MOX ("manually operated")].

Zero (of response) Point in frequency spectrum at which attenuation of a filter is a maximum; corresponds to resonant frequency of a parallel-resonant circuit connected in series with signal path, or resonant frequency of series-resonant circuit connected as shunt branch.

REVIEW QUESTIONS: BEST ANSWER

1. The term used for radiotelephone stations that are free to be moved about is: **a.** variable. **b.** vehicle. **c.** mobile. **d.** duplex. **e.** none of these.
2. The significant difference between "base station" and "fixed station" is that a/an: **a.** fixed station communicates primarily with mobile stations. **b.** base station communicates primarily with mobile stations. **c.** fixed station communicates primarily with other fixed stations. **d.** base station communicates only with other base stations. **e.** both b and c.
3. Two frequencies, one for each direction of communication, are used in: **a.** duplex operation. **b.** simplex operation. **c.** full-duplex operation. **d.** none of these. **e.** both a and c.
4. Provision for simultaneous communication in both directions is a characteristic of: **a.** simplex operation. **b.** duplex operation. **c.** full-duplex operation. **d.** both b and c. **e.** none of these.
5. The ULTRACOM 508 receiver input circuit—T_1–C_1 and T_2–C_5—is described by the unit's manufacturer as a broadband front end. The 508 is designed to be operable over the frequency range 148 to 174 MHz. The bandwidth of the front end, as a percentage of the center frequency of the operating range, is approximately: **a.** 16%. **b.** 24%. **c.** 36%. **d.** 73%. **e.** none of these.
6. The 508 receiver is described as a dual-conversion superheterodyne design. Because of this, it requires two: **a.** RF amplifiers. **b.** detectors. **c.** local oscillators. **d.** heterodyne mixer circuits. **e.** both c and d.
7. The frequency of the LO for the 508's first mixer is controlled by crystal Y_1. The signal from the LO is tripled in stage Q_6 before being mixed with an incoming carrier signal in mixer Q_2. If installed crystals provide operation on a channel frequency of 154.47 MHz, the operating frequency of crystal Y_1 is: **a.** 154.47 MHz. **b.** 165.17 MHz. **c.** 143.77 MHz. **d.** 47.92333 MHz. **e.** none of these.
8. Recall that the image signal frequency for a superheterodyne receiver in which $f_{LO} < f_{carrier}$ is equal to received-carrier frequency minus two times IF. The image

frequency for the 154.47-MHz channel is: **a.** 165.17 MHz. **b.** 175.87 MHz. **c.** 143.77 MHz. **d.** 133.07 MHz. **e.** none of these.

9. Assume that a channel adjacent to the 154.47-MHz channel has a frequency of 160.94 MHz. The shape factor of crystal filter FL_1 is more important for: **a.** adjacent channel rejection. **b.** image rejection. **c.** modulation rejection. **d.** harmonic rejection. **e.** none of these.

10. The SQUELCH function of the 508 is designed to turn off the audio section when high-frequency (superaudio) noise is being processed. Using high-frequency rather than low-frequency noise is an effective technique because an FM system: **a.** attenuates high-frequency noise more than low-frequency noise. **b.** amplifies low-frequency noise more than high-frequency noise. **c.** attenuates low-frequency noise more than high-frequency noise. **d.** most people prefer to hear high frequencies. **e.** none of these.

11. One of the effects of closing the PTT (push-to-talk) switch on the 508 is to: **a.** disconnect power to the receiver. **b.** energize the transmitter oscillator. **c.** switch the antenna relay from RECEIVE to TRANSMIT. **d.** cut off receiver oscillator stage Q_4. **e.** both b and d.

12. The 1.5 V specified for test point TP_5 (emitter of Q_{23}, third MULT): **a.** can be checked with the transceiver in receive mode. **b.** is an ac voltage. **c.** is present only when transmitter is operating and tuned for maximum output. **d.** is best measured with an RF voltmeter. **e.** none of these.

13. To achieve its prescribed equivalent maximum deviation of ± 5 kHz, the phase modulator of the 508 must produce an equivalent deviation of approximately: **a.** 417 Hz. **b.** 765 Hz. **c.** 2.5 kHz. **d.** 5 kHz. **e.** none of these.

14. Refer to Eq. (9.2). Assuming that the collector supply voltage for ouput stage Q_{25} is 13.7 V, V_{sat} is 1.0 V and power out is the specified 10 W, the radiation resistance of the antenna, as seen by Q_{25}, is approximately: **a.** 50 Ω. **b.** 25 Ω. **c.** 15 Ω. **d.** 8 Ω. **e.** none of these.

15. If stage Q_{25} operates at 78% efficiency when producing 10 W of RF power, the dc input power and current to the final stage are approximately (assume $V_{CC} = 13.7$ V): **a.** 7.8 W, 0.5693 A. **b.** 12.82 W, 0.9358 A. **c.** 9.8 W, 0.7852 A. **d.** 20.55 W, 1.5 A. **e.** none of these.

16. A frequency-synthesizer loop contains a divide-by-18 digital IC and the reference frequency is a crystal-controlled 2.7364 MHz. When in lock, the VCO frequency will be approximately: **a.** 49.2552 MHz. **b.** 152.02 kHz. **c.** 23.6937 MHz. **d.** 1.3486 MHz. **e.** none of these.

17. Three 74192 chips connected in cascade to form a three-decade programmable divider (they operate as binary down counters) are programmed with the following binary words from a channel selector switch: 0001-0111-1001. If this divider is the only frequency-modifying device in a PLL circuit, and the reference frequency is 387.562 kHz, the VCO-output frequency should be: **a.** 0.387562 MHz. **b.** 2.165 kHz. **c.** 376.323 MHz. **d.** 69.3736 MHz. **e.** none of these.

18. Implementing a circuit that suppresses one of the sidebands in a SSB system is more difficult than suppressing the carrier because of: **a.** the narrow separation between the USB and LSB. **b.** FCC regulations. **c.** high audio frequencies. **d.** inherent noise. **e.** none of these.

19. A filter with a response curve described as having "steep skirts" corresponds to a shape factor: **a.** of infinity. **b.** of zero. **c.** approaching 1. **d.** approaching zero. **e.** none of these.

20. An advantage of diode switching is that the conductors from the circuit to be switched to the mechanical switch device carry only: **a.** audio signals. **b.** RF signals. **c.** IF signals. **d.** dc current. **e.** none of these.

21. In the Kenwood TS-130S transceiver, the VFO varies the channel operating frequency by: **a.** changing the presetting of a programmable divider. **b.** changing the intraloop frequency through heterodyne mixing with the CAR frequency. **c.** adding to or subtracting from the frequency set by the band selector switch. **d.** fine tuning the master crystal oscillator. **e.** none of these.

22. The UNLOCK DETECTOR function provided by some IC phase-locked-loop devices is a valuable feature when used in transmitter channel-frequency-determining applications because it: **a.** prevents inadvertent off-frequency operation. **b.** protects very expensive equipment from theft. **c.** prevents transmission of incorrectly coded messages. **d.** provides more flexible operating procedures. **e.** none of these.

23. Speech processor circuits that compress the audio modulating signal find popular application in SSB transmitters because the net effect of compression is: **a.** lower emission of spurious signals. **b.** higher peak RF power. **c.** higher average RF power. **d.** lower audio power. **e.** none of these.

24. The power output of an SSB transmitter when the modulating signal is zero is: **a.** equal to the carrier power. **b.** the peak-envelope power. **c.** average power. **d.** zero. **e.** none of these.

25. For SSB transmitters, peak-envelope power (PEP) corresponds to: **a.** the average of the carrier peaks. **b.** the output power at the maximum peak of the output signal waveform. **c.** the dc input power at maximum envelope peak. **d.** the output power at the peak of the modulating signal. **e.** none of these.

26. For purposes relating to FCC regulations, SSB transmitter power ratings are in terms of: **a.** peak-envelope dc input. **b.** PEP. **c.** average envelope power. **d.** maximum RF carrier power. **e.** none of these.

27. Aside from modulation level and transmission line/antenna loading, the most important determinate of transmitter RF output power is: **a.** RF current. **b.** dc input voltage to final stage. **c.** propagation conditions. **d.** operating frequency. **e.** none of these.

28. Something that a person servicing electronic equipment must constantly be at pains to avoid, if desirable results are to be obtained, is that of: **a.** violation of FCC regulations. **b.** overloading dc voltmeters. **c.** excessive loading of signal circuits by connected instruments. **d.** overuse of instruments. **e.** none of these.

29. A logic probe indicates that the output of a NAND gate in the digital circuitry of an inoperative transceiver is always high. A synchrogram for the circuit shows the point should have a pulse waveform on it. A good next troubleshooting step would be to: **a.** get a new logic probe. **b.** replace the NAND-gate chip. **c.** call the manufacturer about the synchrogram. **d.** check the inputs to the NAND gate. **e.** none of these.

30. In microprocessor-controlled transceivers, an area where the functions of a microprocessor are used extensively is in: **a.** automatic gain control. **b.** generating the numerous frequencies of multiple-channel systems. **c.** filtering IF signals. **d.** regulating power supply voltages. **e.** none of these.

REVIEW QUESTIONS: ESSAY

1. Describe the difference(s) between *fixed* and *base* stations as these terms are used in governmental regulations.
2. Describe briefly the three methods of operating two-way communications systems: full duplex, duplex, and simplex.
3. What is the meaning of the term *double-conversion receiver*?
4. What is broadband design? When and how is it used?
5. Why is the shape factor of the IF section of a receiver important to the selectivity of the receiver?
6. Describe the two functions performed by the CA3053 integrated circuit in the Johnson 508 transceiver.
7. List three functions that the CA3075 IC performs in the Johnson 508 receiver.
8. What is the purpose of the squelch function in a communications receiver? In general, how does a squelch section accomplish its purpose?
9. What does "PTT" stand for? Describe the function represented by this abbreviation.
10. What is narrow-band FM? Give details.
11. Give the formula for calculating maximum RF output power for a transmitter.
12. What is frequency synthesis? How can it be used to generate numerous frequencies, all controlled by a single crystal?
13. What is a programmable frequency divider? How is a presettable digital counter used to perform this function? How is a frequency divider used to actually provide a frequency that is a multiple of a reference frequency?
14. Compare the uses of heterodyne frequency converters and programmable frequency dividers in PLLs. Why are they both sometimes used in the same loop?
15. What does "SSB" stand for? Describe similarities and differences of SSB and AM. Upon what theoretical principle is SSB based (i.e., why is it possible to communicate using a single sideband)?
16. What basic operating characteristic of a balanced modulator makes it useful as a function in an SSB system?
17. After processing in a balanced modulator, what remains to be done to a signal in preparation for transmission as an SSB signal? Describe a method for accomplishing the additional step(s).
18. What characteristic of a DSB signal makes eliminating one sideband difficult? Describe the characteristic(s) a filter must have in order for it to be effective in suppressing a sideband.
19. Give the formula for shape factor (SF) of a filter. Describe in your own words the practical significance of shape factor.
20. Give the meaning of the following terms: cutoff frequency, bandwidth, insertion loss, stop band, and passband.
21. What is meant by the *order* of a filter? What are *poles* and *zeros* of response?
22. What is *diode switching*? Give an example of how it is used in a transceiver. Discuss advantages/disadvantages of diode switching.
23. What is the meaning of *VFO*? Describe how a VFO is used in an amateur transceiver.
24. What is speech compression? How is it used in SSB transmission? Why is this feature popular among SSB transmitter users?

25. What is a *BFO*? What is it used for?
26. What does the abbreviation "VOX" stand for? Discuss the concept that it represents.
27. Explain, in your own words, what a noise blanking gate is, what it does, and how it does it.
28. Describe a signal strength meter. Where and how is it used?
29. Give the definitions of the following terms and discuss the concepts which they represent in connection with SSB operation: maximum peak amplitude, envelope, average amplitude, and peak-envelope power.
30. What is automatic limiting control? What is its purpose? How is it accomplished, in general?
31. Why is the loading of a transmitter, represented by the impedance match between transmitter and transmission line/antenna, important to a transmitter operator?
32. Why do circuits that process higher frequencies require greater care during testing and servicing procedures than those that process lower frequencies?
33. Why is a government concerned about and involved in regulating the alignment of transmitters but is, seemingly, unconcerned about the alignment of receivers?
34. List and discuss at least four important goals of transmitter testing and alignment procedures.
35. Discuss the significance to a troubleshooting operation of the notes on a manufacturer's schematic diagram, particularly the notes about operating conditions for dc voltage measurements.
36. Describe a responsibility of a person replacing a defective component in a transmitter, a responsibility not usually associated with a similar task in many other types of equipment.
37. Describe in your own words what microprocessor control is as applied to communications transceivers.
38. What does *to poll* mean in the context of microprocessor control?
39. What is a *bus* as the term is used in computer technology?
40. What is meant by the term *to address*? Explain in detail; do not simply give a synonym.
41. What is a data port?
42. What is meant by the abbreviation "I/O"? What is memory-mapped I/O? What is direct I/O?

EXERCISES

1. Calculate the bandwidth (148 to 174 MHz) of the front end of the ULTRACOM 508 as a percentage of the center frequency of the band. Assume that the front ends of an AM broadcast band receiver and an FM broadcast band receiver are broadbanded. Calculate their bandwidths as a percentage of band center frequencies.
2. If an ULTRACOM 508 transceiver were set up to operate on a channel frequency of 167.71 MHz, what would be the LO frequency from MULT stage Q_6 to MIXER stage Q_2? What would be the operating frequency of a crystal used for Y_1 of RCVR OSC stage Q_4? for Y_5?
3. If an FL_1 filter in an ULTRACOM 508 receiver has a shape factor of 1.1 and a −3-

dB bandwidth of ±6.5 kHz, what is the −60-dB bandwidth? (See the definition of shape factor under "Filters" in Sec. 9.5.)

4. The ULTRACOM 508's manufacturer specifies that the receiver attenuates a signal spaced 30 kHz from a desired signal by 70 dB. Assume that the signal level of a 167.71-MHz signal at the input to stage Q_1 is 0.75 μV and that its level at the collector of Q_3 is 1.0 V. What would be the input signal level of a 167.74-MHz signal if it were producing a 0.5-μV signal at the collector of Q_3?

5. Refer to stage Q_{24} on the schematic diagram of the 508 (Fig. 9.4). Assume that the collector voltage of Q_{24} swings between 13.7 and 1.0 V when the stage is being driven. The manufacturer indicates that this stage produces 2 W in driving stage Q_{25}. Calculate the approximate load resistance seen by Q_{24}.

6. The manufacturer of the 508 transceiver specifies that the frequency drift of the transmitter is less than ±0.0005% over a wide range of environmental temperatures. If the transmitter is operating on the 167.74-MHz channel, what is the maximum expected frequency drift, in hertz?

7. The reference frequency for a PLL is a crystal-controlled 2.793654 MHz. The loop incorporates a ÷28 digital frequency divider. Calculate the operating frequency of the VCO.

8. Refer to Exercise 7. Two changes are made in the design: (1) The reference frequency for the loop is obtained by mixing the 2.793654-MHz signal with a 3.932843-MHz signal and using the difference frequency. (2) The frequency-divider ratio is changed from 28 to 134. Calculate the VCO frequency for the changed design.

9. Refer to the discussion of an imaginary master frequency source (MFS) for a CB transceiver in Sec. 9.4 (see also Fig. 9.14). For each of channels 7 (27.035 MHz), 17 (27.165 MHz), and 38 (27.385 MHz), calculate: **(a)** VCO operating frequency; **(b)** output frequency of TMX_1; **(c)** modulus for DD_1; **(d)** binary input words for each of the decade counters of DD_1. (*Hint:* Convert each digit of the modulus into its BCD equivalent.)

10. The following voltage ratios represent output/input voltages for filter circuits. Express each ratio as a pure decimal number and as a decibel value (remember, the ratios must be of voltage values, not just numbers): **(a)** 2.754 V/123.7 mV; **(b)** 275.6 mV/378.9 μV; **(c)** 0.06453 V/0.1739 V; **(d)** 42.75 mv/0.1864 V; **(e)** 392.4 μV/87.51 mV.

11. The response of a filter is 3 dB down at the frequencies of 545.7 kHz and 557.7 kHz. What the bandwidth of the filter? What is its center frequency? What is the bandwidth as a percentage of center frequency?

12. The output of a filter, in its passband, varies between 1.378 and 1.836 V when the input is a steady 2.0 V. Calculate: **(a)** the passband ripple; **(b)** maximum insertion loss of the filter. Express both answers in decibels.

13. A filter's output is 3 dB down at 10.70675 MHz and 10.69325 MHz, and 60 dB down at 10.708438 MHz and 10.691563 MHz. **(a)** Calculate the shape factor of the filter. **(b)** Assuming that the −3-dB frequencies remain unchanged, what would be the −60-dB frequencies and the −60-dB bandwidth if the shape factor were changed to 1.15?

14. Ham radio operators generally refer to the various frequency bands on which they are authorized to operate in terms of wavelength rather than frequency. For example, in the HF band (3 to 30 MHz) they are authorized operation in the following meter bands: 80, 40, 20, 15, and 10. They also use a 6-m and a 2-m band. When specified by frequency, ham bands include the following: 144 to 148 MHz, 50 to 54 MHz, 28 to 29.7 MHz, 21 to 21.45 MHz, 14 to 14.35 MHz, 7 to 7.3 MHz, 3.5 to 4.1 MHz, and 1.8 to 2.0 MHz. Make a list of the frequencies given; opposite each frequency indicate the name of the band, using an appropriate nominal wavelength. (*Hint:* Refer to Chap. 6 for a formula for calculating wavelength, given the frequency.)

15. A station that operates at 165 MHz is permitted a frequency tolerance of $\pm 0.0005\%$ by FCC regulations. By how much could the station's frequency be off and still be within regulation?

16. The frequency meter used to measure the frequency of the station in Exercise 15 has an accuracy of $\pm 0.00005\%$. What should the measured frequency error be limited to in order to assure compliance with regulation?

17. The dc current to the final stage of an SSB transmitter is measured while the transmitter is being modulated with a two-tone audio test signal. The amplitude of the modulating signal is equal to the maximum rated amplitude specified by the manufacturer. The dc current measures 20.86 A and the dc supply voltage to the stage is 13.7 V. Calculate the PEP dc input power.

18. While making the test described in Exercise 17, the RF voltage is measured across a 50-Ω dummy antenna. The RF voltage measures 100 V (rms). **(a)** Calculate the RF power being delivered to the load. **(b)** What is the efficiency of the output stage?

19. Refer to Fig. 9.40 and the associated examples. The function driven by the output of the circuit is not operating correctly. A check with a logic probe indicates that the output of FF_4 is always low. **(a)** Describe other tests you would want to make on the circuit. Give your reasons for wanting to make the tests. **(b)** Assume various results from your tests. What do your assumed results indicate about the circuit? **(c)** Describe possible causes for the malfunction.

20. Find an advertisement for a microprocessor-controlled transceiver. (Advertisements for ham transceivers can be found in ham magazines, for marine transceivers in sailing or yachting magazines, and for CB transceivers in magazines such as *Time, Newsweek,* etc.) Study the ad and summarize the microprocessor-controlled features of the unit.

10

TRANSMISSION and PROPAGATION: ADVANCED TOPICS

10.1 TRANSMISSION-LINE EQUATIONS

Although we have already examined transmission lines in some detail (Chap. 6), we did so primarily in a qualitative way. That is, we looked at certain properties and concepts of transmission lines without making any extensive use of mathematics. However, on occasion it is desirable to predict the performance of a transmission-line-load condition to a comparatively precise degree. This requires a mathematical approach.

In this section we look at the most fundamental and important equations for transmission lines. Then in Sec. 10.2 we go on to a graphical tool, the Smith chart, which is based on these equations. We will look at a few of the ways it can be used to provide answers about the expected performance of transmission lines. To help prepare yourself to understand this more mathematical approach to transmission lines, look over again, right now, the portions of Chap. 6 that have to do with transmission lines.

Distributed Parameters

In Chap. 6 we were introduced to the idea that it is useful to think of a transmission line as consisting of virtually an infinite number of tiny sections containing discrete components: series R and L and shunt C and R. This concept is illustrated again in Fig. 10.1(a). Recall that the sections and components are considered

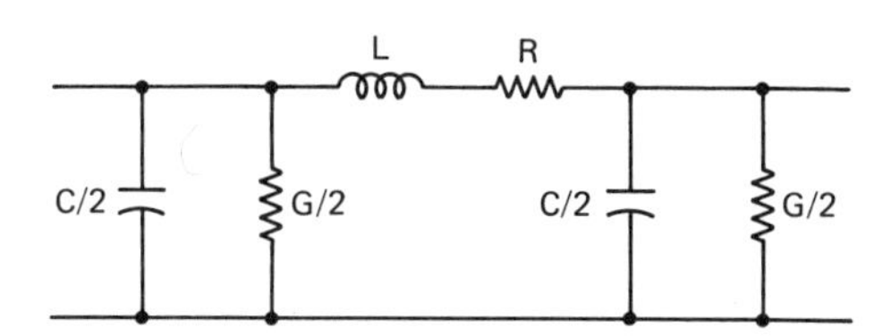

Figure 10.1 Distributed parameters for per-unit length of transmission line.

to be distributed evenly throughout the line. They are called *distributed parameters*. You will recall that G represents conductance, the reciprocal of resistance: $G = 1/R$. The symbols shown on the diagram—R, L, C, and G—represent the resistance, inductance, capacitance, and conductance per unit length of the line. Unit length can be any convenient length—inch, foot, centimeter, meter, and so on.

Characteristic Impedance: Z_0

For a line of infinite length, excited by a sinusoidal signal, the input impedance is called its characteristic impedance, Z_0. The input impedance, as could be expected, is related to the distributed parameters:

$$Z_0 = \sqrt{\frac{R + j\omega L}{G + j\omega C}} \quad \text{ohms} \tag{10.1}$$

When the series R and shunt G of a line are significant, the line is said to be a lossy line, and Eq. (10.1) holds. If the line is a lossless line, and many practical lines can be so considered, R and G are insignificant and drop out of Eq. (10.1). The formula for characteristic impedance then reduces to

$$Z_0 = \sqrt{\frac{L}{C}} \quad \text{ohms} \tag{10.2}$$

Propagation Constant: γ

As we learned in Chap. 6, of considerable value in analyzing transmission lines is the concept of waves of voltage and current traveling down the line. These waves are produced when a source is connected to and "excites" the line. At any given point on a transmission line, voltage and current amplitudes and phase vary as the waves travel or "propagate" down the line. It is possible to describe variations along the line by means of a complex (i.e., containing real and "imaginary" parts) mathematical concept called the *propagation constant*. The propagation constant is conventionally represented with the Greek lowercase letter γ (pronounced "gamma"). It is related to a unit length of line, like the values of R, L, etc. The formula for the propagation constant is

$$\gamma = \alpha + j\beta = \sqrt{(R + j\omega L)(G + j\omega C)} \quad \text{per unit length} \tag{10.3}$$

(α is the Greek lowercase letter "alpha," β is the Greek lowercase letter "beta.") For a lossless line, R and G are insignificant, and

$$\gamma = 0 + j\omega\sqrt{LC} \qquad (\alpha = 0, \beta = \omega\sqrt{LC}) \tag{10.4}$$

The propagation constant is the exponent in an exponential expression.

The expression describes mathematically how voltage or current vary with distance along a line (an infinite line, or one with no reflections):

$$V_x = V_1 e^{-\gamma x} = V_1 e^{-\alpha x - j\beta x} = V_1 e^{-\alpha x} e^{-j\beta x} \tag{10.5}$$

$$I_x = I_1 e^{-\gamma x} = I_1 e^{-\alpha x - j\beta x} = I_1 e^{-\alpha x} e^{-j\beta x} \tag{10.6}$$

The x in the exponent represents the distance between the point on the line where the voltage is V_1 or the current is I_1, and the point where the values are V_x or I_x (see Fig. 10.2).

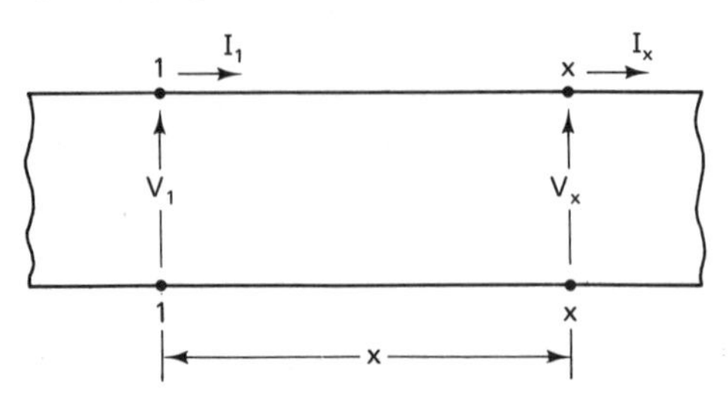

Figure 10.2 Voltages and currents at different points on a transmission line.

Because there is no j in the first factor of the expression, $e^{-\alpha x}$, it is a real quantity: it expresses amplitude. Further, since the sign of the exponent is negative, the factor *decreases* with distance, x. This factor, $e^{-\alpha x}$, is therefore called the *attenuation factor* because it reveals that the voltage or current is attenuated with distance along the line. Component α of the propagation constant is called the *attenuation constant.*

If we separate the amplitude factor from the remainder of the expression of Eq. (10.5) and rearrange, we obtain

$$e^{-\alpha x} = \frac{V_x}{V_1}$$

Then

$$\alpha x = -\ln \frac{V_x}{V_1}$$

and if x is a unit length,

$$\alpha = -\ln \frac{V_x}{V_1} \tag{10.7}$$

The form of the expression of Eq. (10.7) is remindful of the definition of the *bel*. The difference is that common (base 10) logarithms are used in the definition of bels or decibels. Natural or Naperian logarithms are used with α. The unit of α is therefore the *neper*. The unit is named after Napier, the man who first defined natural logarithms. (It is said that his name was originally spelled "Neper.") The value of α obtained from an evaluation of Eq. (10.7) is usually expressed as "nepers per unit length." It can also be expressed in decibels. To express this value in decibels per unit length, multiply nepers by 8.686: 1 Np = 8.686 dB.

The presence of j in the second factor, $e^{-j\beta x}$, means that this portion of the propagation constant is associated with the *phase* of the propagating waves with respect to distance along the line. In fact, β is commonly called the *phase constant*. Mathematically, the expression $e^{-j\beta x}$ is another way to write a complex number since $e^{j\theta} = \cos\theta + j\sin\theta$. Therefore, another way of expressing the voltage at point x (see Fig. 10.2) in terms of the voltage at point 1, is

$$V_x = V_1 e^{-\alpha x}(\cos\beta x - j\sin\beta x) \tag{10.8}$$

It is apparent that the product βx is an angle. The angle is in radians. Hence the constant β represents the number of radians by which the phase of a voltage or current wave changes in traveling a *unit length* along a transmission line. Or, the unit of β is radians per unit length. Furthermore, the negative sign in the exponent indicates that the phase of the voltage at x lags the voltage at point 1. The voltage wave is traveling from point 1 toward point x, so it is logical that a point in the pattern of variation occurs later (lags) at x.

Let's look at β in another way. In one period, T (the time of one cycle), of a waveform, the corresponding wave on a transmission line advances by one wavelength, λ. This is by definition. The corresponding phase change is 2π radians. Thus we write

$$\beta\lambda = 2\pi \tag{10.9}$$

or

$$\beta = \frac{2\pi}{\lambda} \tag{10.10}$$

or

$$\lambda = \frac{2\pi}{\beta} \tag{10.11}$$

The time, t, for an object to move a distance, x, at a velocity, v, is

$$t = \frac{x}{v} \tag{10.12}$$

Similarly, the time, T (the period), for a point on a wave to move a distance, λ (a wavelength), is

$$T = \frac{\lambda}{v} = \frac{1}{f} \tag{10.13}$$

Hence

$$v = f\lambda = f\frac{2\pi}{\beta} \tag{10.14}$$

but

$$\beta = 2\pi f\sqrt{LC} \qquad \text{(for a lossless line)} \tag{10.15}$$

Therefore,

$$v = \frac{2\pi f}{2\pi f\sqrt{LC}} = \frac{1}{\sqrt{LC}} \qquad \text{(for a lossless line)} \tag{10.16}$$

The quantity v in Eq. (10.16) is called the *velocity of propagation*. It is important to note that it is a function of the distributed parameters of the transmission line. This contradicts our previous notion (see Chap. 6) that the velocity of propagation on a transmission line is the same as the speed of light: approximately 300,000,000 m/s. For transmission lines with air dielectric, and, in fact, for many practical twin-lead or coaxial lines, the velocity of propagation is indeed the same as the speed of light. There are transmission media, however, especially in the microwave region, in which the velocity of propagation must be considered as less than the speed of light.

Example 1

An approximation for the attenuation constant of many transmission lines is $\alpha = R/2Z_0$, where R is the resistance per unit length (includes both conductors). This approximation applies when the loss in the dielectric between the conductors is insignificant in comparison to the I^2R loss of the conductors. A transmission line of this type is constructed of No. 18 copper wire; resistance is 6.510 Ω per 1000 ft of wire. What is the attenuation factor per 100 ft of this two-wire line if the characteristic impedance is 300 Ω? What is the total attenuation for 2 mi of the line?

Solution. The total resistance per 100 ft of line is

$$2 \times 100 \times \frac{6.510}{1000} = 1.302\ \Omega$$

("2 ×" because the transmission line is two-wire). Thus

$$\alpha = \frac{1.302}{2 \times 300} = 0.002170\ \text{Np}/100\ \text{ft}$$

or

$$\alpha = 8.686 \times 0.00217 = 0.01885\ \text{dB}/100\ \text{ft}$$

Total attenuation for 2 mi:

$$2 \times 0.002170 \times \frac{5280}{100} = 0.2292\ \text{Np}$$

$$2 \times 0.01885 \times \frac{5280}{100} = 1.991\ \text{dB}$$

Example 2

The propagation constant at 300 MHz for a certain transmission line is $2.5 \times 10^{-5} + j0.6667\pi$ per meter. If the 300-MHz input signal at a given instant is $100 \angle 0^\circ$ V, calculate the voltage at the 1247-m point for the same instant. Determine amplitude and phase angle.

Solution. The amplitude is determined by using the attenuation expression

$$V_x = V_{\text{in}}e^{-\alpha x}$$

$$V_{1247\text{m}} = 100e^{-2.5(10^{-5})1247}$$

$$= 100(0.9693) = 96.93\ \text{V}$$

The relative phase angle at 1247 m is obtained by evaluating the phase expression exponent, βx,

$$\beta x = 0.6667\pi \times 1247$$
$$= 831.37\pi \text{ radians}$$

Since each cycle is 2π radians, the wave goes through

$$\frac{831.37\pi}{2\pi} = 415.685 \text{ cycles}$$

over the 1247-m segment of line. Hence its relative phase angle is

$$0.685 \times 360° = 246.6°, \text{ lagging}$$

Let's refer again to the velocity of propagation: $v = 1/\sqrt{LC}$. Be aware that for a line with air dielectric (all or most of the space between the conductors is filled with air), v evaluates to the speed of light. However, when the space between the conductors is filled with an insulating material whose dielectric constant is significantly different from that of air, the value of C in the formula is changed. Consequently, the value of v is changed; the wavelength is also changed, because it is a function of v as well as of f. Several common dielectric materials have a dielectric constant, k, in the range 2 to 8 ($k_{air} = 1$). The velocity of propagation is changed by the reciprocal of the square root of the dielectric constant:

$$\text{velocity of propagation (for } k) = \frac{300{,}000{,}000}{\sqrt{k}} \quad \text{m/s}$$

Example 3

A 300-MHz signal is to be transmitted on a coaxial cable that is filled with an insulating material whose dielectric constant is 1.89. Calculate the velocity of propagation and the wavelength of the signal.

Solution.

$$v = \frac{300(10^6)}{\sqrt{k}} = \frac{300(10^6)}{\sqrt{1.89}}$$
$$= 218.2(10^6) \text{ m/s}$$
$$\lambda = \frac{v}{f} = \frac{218.2(10^6)}{300(10^6)} = 0.7274 \text{ m}$$

Reflection Coefficient: Γ

We learned in Chap. 6 that when a transmission line of infinite length is excited by a voltage source, voltage and current waves travel down the line without reflection. Waves that travel from the source and toward the load end of a transmission line are conventionally called *incident waves*. We identify such voltage and current waves with the symbols V_i and I_i.

We also learned that if the line is terminated with a load impedance Z_L such that $Z_L = Z_0$, there will be no reflections. However, if these conditions are not met, some percentage of the energy will be reflected. The reflection will be in the form of a pair of voltage and current waves traveling back toward the source and away from the condition causing the reflection. We will call the reflected waves V_r and I_r.

The ratio of reflected voltage to incident voltage is called the voltage reflection coefficient, often designated Γ_v. (The symbol "Γ" is the Greek capital letter gamma.) Expressing this mathematically, we have

$$\Gamma_v = \frac{V_r}{V_i}$$

There is an analogous relationship for the current reflection coefficient:

$$\Gamma_i = \frac{I_r}{I_i}$$

The reflection coefficients are a function of the relationship between the characteristic impedance of the line, Z_0, and the terminating or load impedance, Z_L. We state without proof or derivation the equations for the reflection coefficients in terms of impedances:

$$\Gamma_v = \frac{Z_L - Z_0}{Z_L + Z_0} \tag{10.17}$$

$$\Gamma_i = \frac{Z_0 - Z_L}{Z_0 + Z_L} = -\Gamma_v \tag{10.18}$$

Standing-Wave Ratio

A reflected wave may reach the source that is providing the excitation of a line. If the internal impedance of the source, Z_S, is not equal to Z_0, a portion of the reflected wave will be reflected from the source: there will be a rereflection. The reflection coefficient applies here also, with Z_S substituted for Z_L in the formulas above.

The rereflected wave propagates down the line as a secondary incident wave. It is reflected at the unmatched load. And so on. This process of the reflection of reflections repeats itself numerous times. The amplitudes of voltage, or current, along the line are resultants. That is, the amplitude at each point along the line, at a given instant, is determined by the phasor summation of all the components on the line at the point at the given instant. Ultimately, a steady-state condition is achieved on the line. ("Ultimately" here is an extremely short time, however.) The steady-state condition on a line with reflections is the pattern of *standing waves* with which we became familiar in Chap. 6.

You will recall that the standing-wave pattern (of voltage or current) contains points of minimum (or zero) amplitude called *nodes*. A quarter-wavelength away from the nodes are points of maximum amplitude (which vary in a sinusoidal pattern) called *loops* or *antinodes*. The ratio of the amplitude at a loop or antinode to the amplitude at a node is called the *standing-wave ratio* (SWR). The voltage-standing-wave ratio (VSWR) is the standing-wave ratio for the voltage wave; the current-standing-wave ratio is the ratio for the current wave. The two are equal at a given point on the line. A common designation for the ratio is ρ (Greek lowercase letter rho). It is usually expressed in terms of voltage,

$$\text{VSWR} = \rho = \frac{V_{\text{max}}}{V_{\text{min}}} = \frac{V_i + V_r}{V_i - V_r} = \frac{1 + \Gamma}{1 - \Gamma} \tag{10.19}$$

An expression for the reflection coefficient in terms of the VSWR can be found by manipulation of the second part of the formula,

$$\Gamma = \frac{\rho - 1}{\rho + 1} \tag{10.20}$$

Example 4

What is the VSWR for a shorted, lossless line? for an open lossless line? for a line terminated with $Z_L = Z_0$?

Solution. For a shorted, lossless line, $V_r = V_i$; hence

$$\text{VSWR} = \frac{V_i + V_i}{V_i - V_i} = \frac{2V_i}{0} = \text{infinity}$$

For an open, lossless line it is also true that $V_r = V_i$, hence VSWR = infinity. For a line with the load impedance matched to the characteristic impedance (i.e., with $Z_L = Z_0$), $V_r = 0$, hence

$$\text{VSWR} = \frac{V_i + 0}{V_i - 0} = \frac{V_i}{V_i} = 1$$

Summary: The VSWR on an open, or shorted, lossless line is infinite. The VSWR is unity on a line terminated with its own characteristic impedance (or on a line of infinite length).

Input Impedance of a Transmission Line: Z_{in}

One of the performance characteristics of a transmission line of frequent concern is that of *input impedance*, Z_{in}. By definition, the input impedance of a line is equal to the ratio of the voltage to the current at its input when excited. That is,

$$Z_{\text{in}} = \frac{V_i}{I_i} \tag{10.21}$$

As we saw in Chap. 6, for certain line lengths and certain terminations it is relatively easy to estimate what the input impedance is. For example, the input impedance of an odd quarter-wavelength of a shorted lossless line is infinity; the input impedance of an odd quarter-wavelength of an open lossless line is zero; etc. These are special cases; we examined them in Chap. 6 because they required little or no mathematics for evaluation. Let's look at the general formula for the input impedance of a lossless line of length x, excited by a signal whose wavelength is λ and which is terminated in Z_L:

$$Z_{in} = Z_0 \frac{Z_L \cos(2\pi/\lambda)x + jZ_0 \sin(2\pi/\lambda)x}{Z_0 \cos(2\pi/\lambda)x + jZ_L \sin(2\pi/\lambda)x} \quad (10.22)$$

The ratio $2\pi/\lambda$ is equal to β, the phase constant of the propagation constant:

$$\beta = \frac{2\pi}{\lambda}$$

The equation for input impedance is sometimes seen expressed in terms of β:

$$Z_{in} = Z_0 \frac{Z_L \cos \beta x + jZ_0 \sin \beta x}{Z_0 \cos \beta x + jZ_L \sin \beta x} \quad (10.23)$$

Let us not forget that even Eqs. (10.22) and (10.23) are for the special case of a lossless line. For the sake of completeness we give the completely general form of the equation,

$$Z_{in} = Z_0 \frac{(Z_L + Z_0)e^{\gamma x} + (Z_L - Z_0)e^{-\gamma x}}{(Z_L + Z_0)e^{\gamma x} - (Z_L - Z_0)e^{-\gamma x}} \quad (10.24)$$

or

$$Z_{in} = Z_0 \frac{Z_L \cosh \gamma x + Z_0 \sinh \gamma x}{Z_0 \cosh \gamma x + Z_L \sinh \gamma x} \quad (10.25)$$

The expressions "sinh γx" and "cosh γx" mean "the hyperbolic sin of γx" and the "hyperbolic cosine of γx," respectively. They are functions found not uncommonly in advanced topics in mathematics. They have the following values in these equations:

$$\sinh \gamma x = \frac{e^{\gamma x} - e^{-\gamma x}}{2}$$

$$\cosh \gamma x = \frac{e^{\gamma x} + e^{-\gamma x}}{2}$$

As you can observe, the computation of the input impedance involves several steps. It is not surprising that a timesaving, graphical technique was developed for performing such computations: the Smith chart. We look at this popular tool in the next section.

10.2 THE SMITH CHART

A sample Smith chart is shown in Fig. 10.3. At first glance it appears to be a complex maze of circles and arcs of circles: that it is. However, the circles and arcs have meanings. Altogether, the chart is an imaginative application of a mathematical concept: the *locus*. (The Smith chart, developed in the late 1930s, is a contribution of an American engineer, Phillip H. Smith.) A locus is a system of lines, circles, etc., which satisfy a particular condition.

The concept of locus can be demonstrated with a simple series circuit. Consider the circuit of Fig. 10.4(a): a series R of fixed value and a variable X_L supplied by a constant voltage source V. The impedance of the circuit is $R + jX_L$. The current in the circuit, $I = I_x + jI_y$, is equal to $(V_x + jV_y)/(R + jX_L)$.

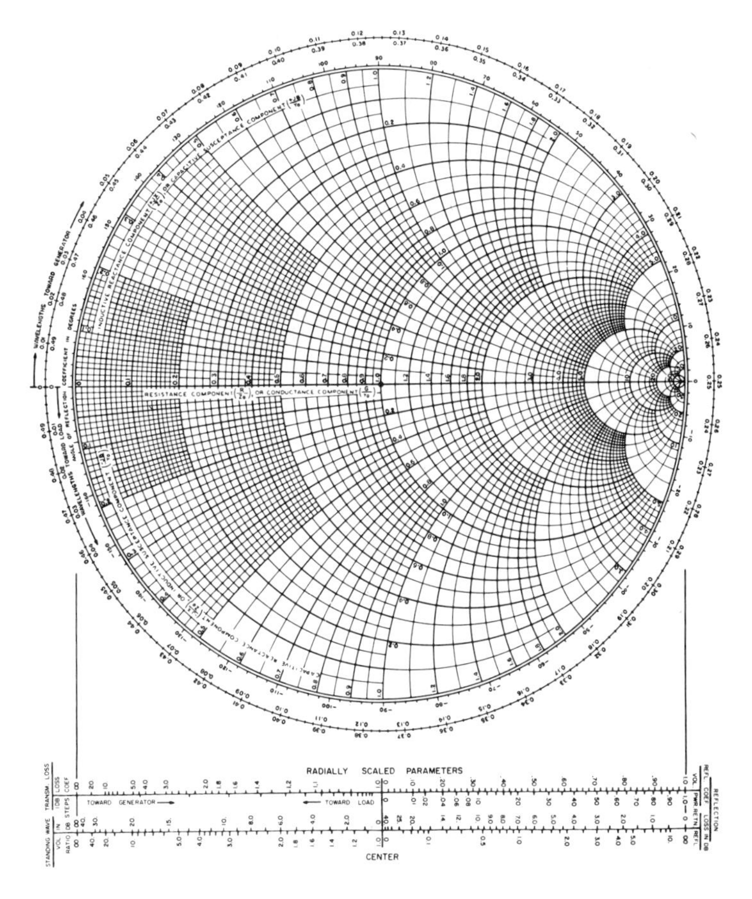

Figure 10.3 Smith chart. Permission to use representation of Smith chart throughout this book has been granted by Phillip H. Smith, Murray Hill, N.J., under his renewal copyright issued in 1976.

It can be shown that no matter what value X_L may have, the locus of the values of I is a semicircle [see Fig. 10.4(b)]. The diameter of the circle is V/R; the center of the circle is at $V/2R$ on the horizontal axis.

It can also be shown that the semicircle of Fig. 10.4(b) is the locus of the admittance Y of the cicuit. Remember: Admittance is the reciprocal of impedance. Hence

$$Y = \frac{I}{V}$$

Admittance is directly proportional to I; it follows the same locus.

The Smith chart is usefully thought of as a set of impedance coordinates to represent the R and X values of transmission-line impedance. You are familiar with the conventional representation of impedance on rectangular coordinates: R along the horizontal or real axis and X along the vertical j or imaginary axis (see Fig. 10.5). On the Smith chart, the R "coordinates" are a set of circles all

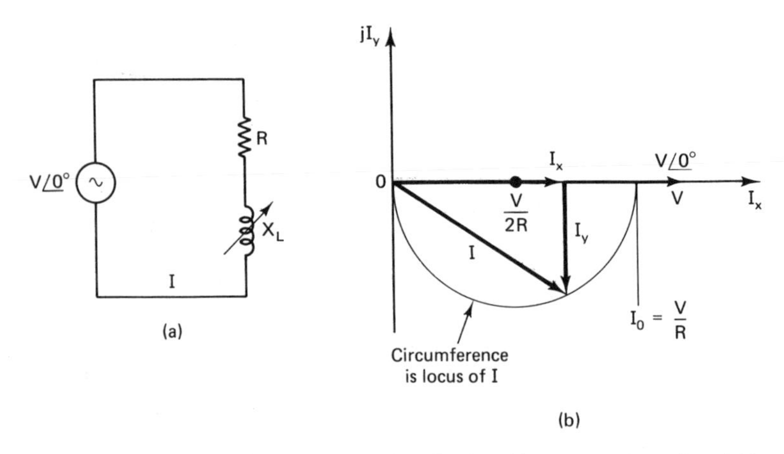

Figure 10.4 Concept of locus: (a) series circuit with constant R and variable X_L; (b) current locus.

tangent at the right end of their horizontal diameters. Study Fig. 10.6(a) and compare the diagram to the Smith chart in Fig. 10.3. Each circle represents a constant R value. Values not shown explicitly can be located by interpolation using values shown.

Reactance values are represented by a set of arcs of circles. See Fig. 10.6(b) and compare with the Smith chart in Fig. 10.3. Observe, from Fig. 10.6(b), how the center and radius of an arc are determined.

A very important aspect of the Smith chart is that its values are *unit* values. Unit values are also called *normalized* values. These terms, in this instance, mean that the values are "per unit of Z_0." Normalization, or the use of unit

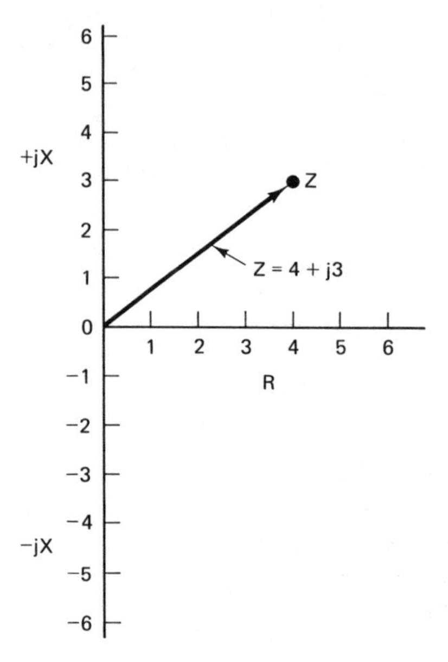

Figure 10.5 Conventional plot of impedance on complex plane.

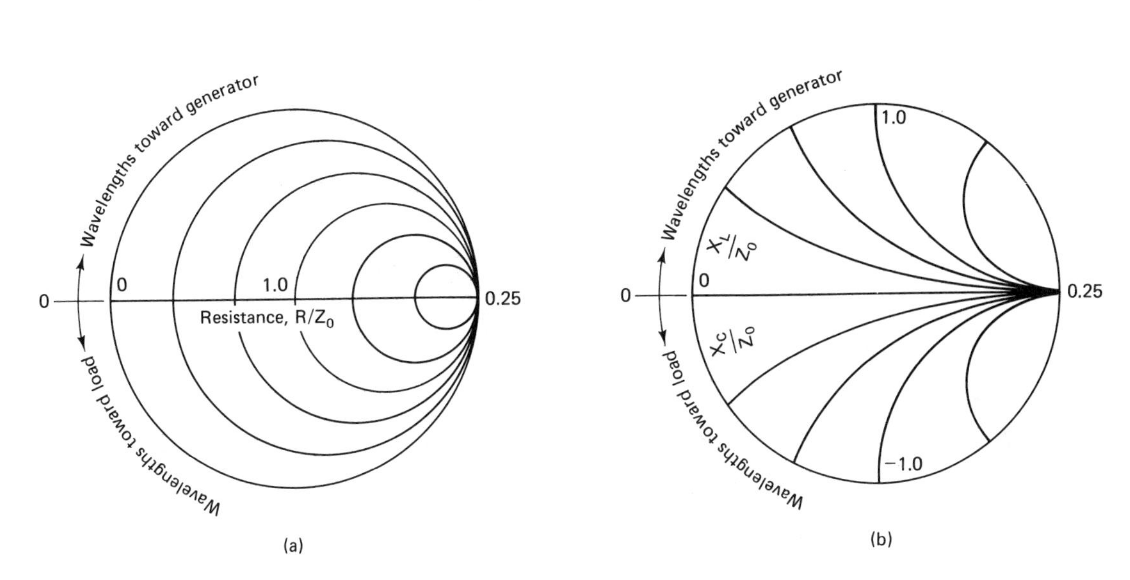

Figure 10.6 Coordinate elements of Smith chart: (a) circles of constant resistance, R/Z_0; (b) arcs of constant reactance, X_L/Z_0 and X_C/Z_0.

values, makes the chart universal: It can be used with transmission lines of any value of Z_0. After unit values are obtained from the chart, they are converted to explicit values for a line of a particular Z_0 simply by multiplying the values by that Z_0.

One of the many advantages of the Smith chart is that one can learn to use it effectively without knowing and/or understanding its mathematical derivation. In fact, studying the derivation does not teach one how to use the chart. That is best achieved by the guided working of examples in its use.

Let's get started using the Smith chart by learning to read off the values associated with explicit points on the chart. We will also interpret the values as parameters of transmission lines of various characteristic impedances. The following facts about the mechanical arrangement of the coordinate lines of the chart will assist us in learning to read off the values of identified points:

1. The largest circle (the circle defining the chart) represents a constant value of zero ohms of unit R. The value "0" is marked at the left end of the horizontal diameter. Circles of smaller diameter represent larger values of unit R. Values of unit R are marked on the horizontal diameter of the chart. A chart with the circles of constant unit R values of 0, 0.4, 1.0, 2.0, 5.0, and 20 emphasized and identified is shown in Fig. 10.7.
2. The horizontal diameter corresponds to a value of zero ohms of unit reactance. The upper half of the chart contains coordinates for positive values of unit reactance (i.e., inductive reactance). The lower half of the chart contains coordinates for negative (capacitive) unit reactance. Reactance values are marked along the 0-ΩR circle and the 1.0-ΩR circle. A chart with emphasized arcs representing the unit reactance values of $+0.2$, $+0.6$, $+1.0$, $+2.0$, $+10.0$, -0.2, -0.6, -1.0, -2.0, and $-10.0\ \Omega$ is shown in Fig. 10.8.

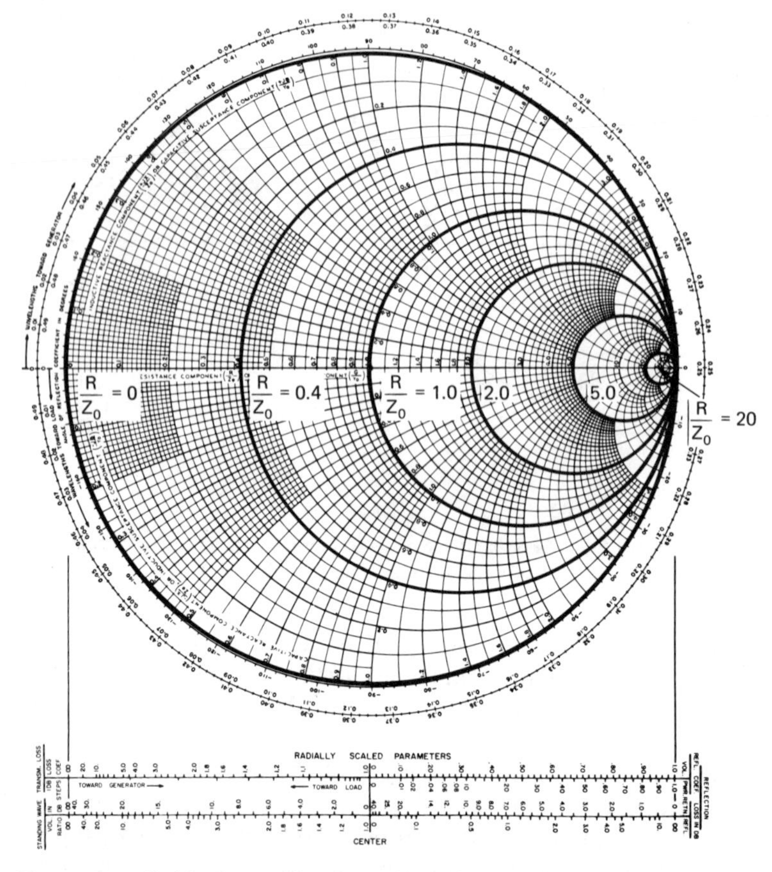

Figure 10.7 Smith chart with R/Z_0 values of 0, 0.4, 1.0, 2.0, 5.0, and 20 Ω emphasized.

Example 5

Refer to the chart of Fig. 10.9. (a) Read off the values of unit impedances associated with the points on the chart marked A, B, C, and D. Convert the unit impedances to the actual impedances of lines with Z_0's of (b) 50 Ω and (c) 300 Ω.

Solution. (a) A: $0.2 + j0.9\ \Omega$

B: $1.25 + j0\ \Omega$

C: $0 - j1.1\ \Omega$

D: $0.175 - j0.625\ \Omega$

(b) A: $50(0.2 + j0.9) = 10 + j45\ \Omega$

B: $50(1.25 + j0) = 62.5 + j0\ \Omega$

C: $50(0 - j1.1) = 0 - j55\ \Omega$

D: $50(0.175 - j0.625) = 8.75 - j31.25\ \Omega$

(c) A: $300(0.2 + j0.9) = 60 + j270\ \Omega$

B: $300(1.25 + j0) = 375 + j0\ \Omega$

C: $300(0 - j1.1) = 0 - j330\ \Omega$

D: $300(0.175 - j0.625) = 52.5 - j187.5\ \Omega$

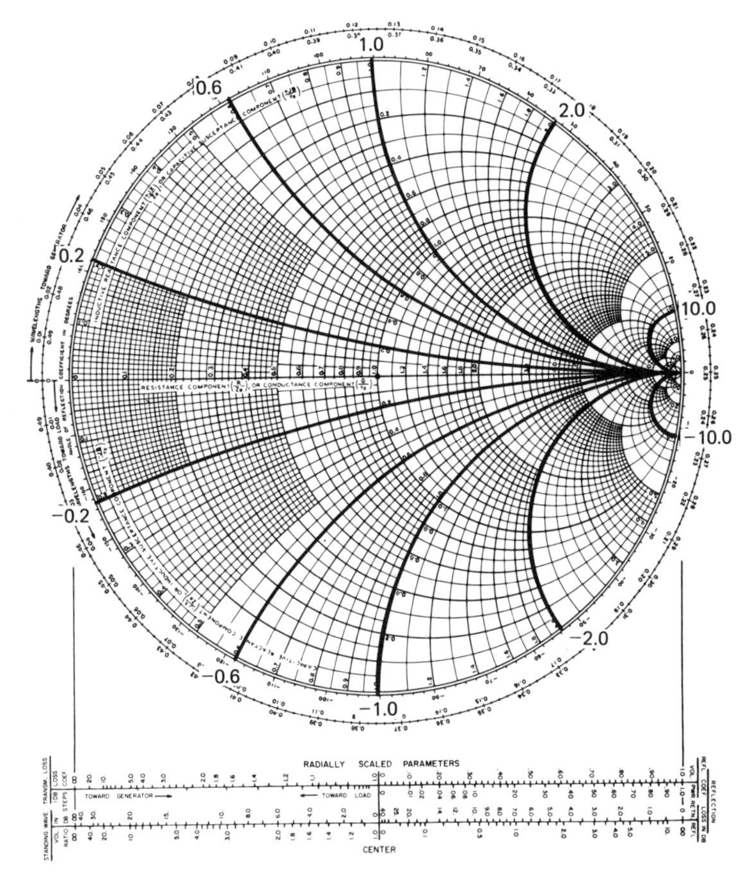

Figure 10.8 Smith chart with emphasized X/Z_0 reactance values of ± 0.2, 0.6, 1.0, 2.0, and 10.0 Ω.

In Example 5 we interpreted the chart as representing unit impedance values. The chart may also be interpreted as representing admittance values. Please recall that admittance is the reciprocal of impedance,

$$Y = G + jB = \frac{1}{Z} = \frac{1}{R + jX}$$

and

$$G = \frac{R}{R^2 + X^2} \qquad \text{and} \qquad B = \frac{X}{R^2 + X^2}$$

The values of G and B on the chart are unit values, just as R and X are unit values. In this case the relationships with Z_0 are as follows: Values read off the chart are equal to GZ_0 and jBZ_0. Or, it is also correct to say that the values on the chart are normalized by division by Y_0. That is, the values are G/Y_0 and jB/Y_0. To obtain actual values of G and B for a particular line, divide the normalized values read from the chart by the Z_0 of the line or multiply by the Y_0 of the line. The unit of conductance (G), susceptance (B), and admittance

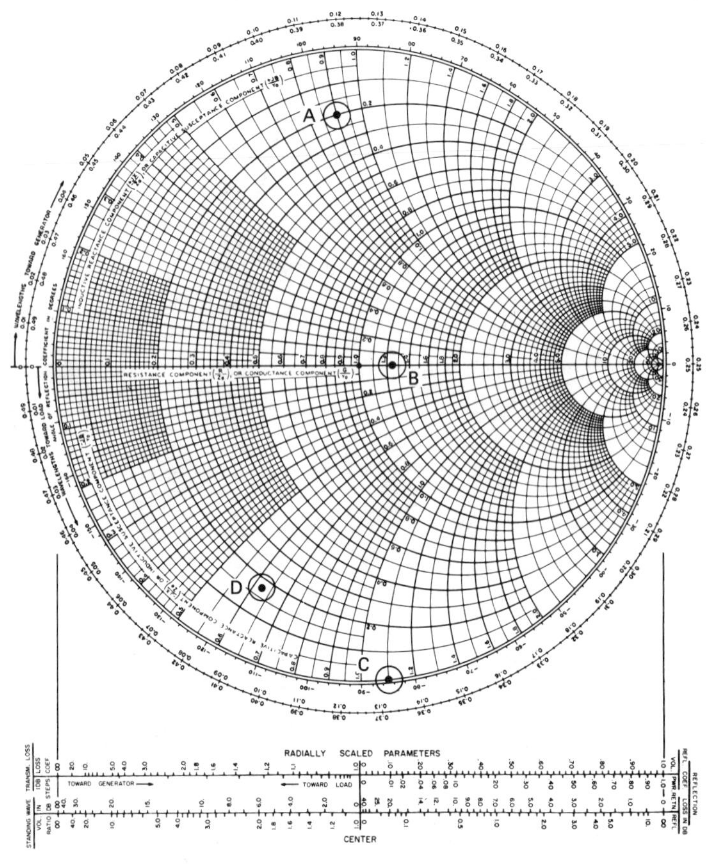

Figure 10.9 Impedance values plotted on Smith chart.

(Y) is the *siemens*, abbreviated S. The susceptance values on the upper half of the chart are for capacitive (positive) susceptance; susceptance is negative on the lower half of the chart, for inductive susceptance.

Example 6

Refer to Example 5 and the points marked A, B, C, and D on the chart of Fig. 10.9. Interpret the points as representing admittance. (a) Give the value of admittance of each of the points. Convert the normalized (unit) values to actual values of admittance for transmission lines with the following characteristic impedances: (b) 50 Ω and (c) 300 Ω.

Solution. (a) The points are "read" as in Example 5. However, the interpreted values now have the unit S (siemens).

(b) We divide the unit values by the Z_0's (or multiply by the Y_0's).

$$A: 0.02(0.2 + j0.9) = 0.004 + j0.018 \text{ S}$$

$$B: 0.02(1.25 + j0) = 0.025 + j0 \text{ S}$$

$$C: 0.02(0 - j1.1) = 0 - j0.0220 \text{ S}$$

$$D: 0.02(0.175 - j0.625) = 0.0035 - j0.0125 \text{ S}$$

(c) A: $0.003333(0.2 + j0.9) = 0.6667 + j3.0$ mS
B: $0.003333(1.25 + j0) = 4.167 + j0$ mS
C: $0.003333(0 - j1.1) = 0 - j3.667$ mS
D: $0.003333(0.175 - j0.625) = 0.5833 - j2.083$ mS

To perform transmission-line evaluations with the chart, it is often necessary to "enter" the chart with a known value of impedance and identify that impedance as a point on the chart. The process consists of converting actual impedances to unit impedances and then locating the unit values on the coordinates. Unit (or normalized) values are obtained by dividing values by the Z_0 of the line being analyzed.

Example 7

Assume that the following impedance values represent actual impedances at particular points on the specified transmission lines. Plot the impedances on a Smith chart. (a) $12.35 + j17.59\ \Omega$ on 50-Ω line (b) $38.42 - j27.86\ \Omega$ on 50-Ω line (c) $235.6 + j127.8\ \Omega$ on 300-Ω line (d) $189.7 - j276.4\ \Omega$ on 300-Ω line.

Solution. In preparation for plotting the values we normalize them by dividing by the characteristic impedance of the particular line:

(a) $\dfrac{12.35 + j17.59}{50} = 0.2470 + j0.3518\ \Omega$

The value is plotted as the point labeled (a) on Fig. 10.10.

(b) $\dfrac{38.42 - j27.86}{50} = 0.7684 - j0.5572\ \Omega$ [see point (b) on Fig. 10.10].

(c) $\dfrac{235.6 + j127.8}{300} = 0.7853 + j0.4260\ \Omega$ [point (c) on Fig. 10.10].

(d) $\dfrac{189.7 - j276.4}{300} = 0.6323 - j0.9213$ [point (d) on Fig. 10.10].

Phase and Electrical Distance

One of the major convenience features of the Smith chart is that it enables one to determine the impedance at other points on a transmission line once the impedance is known at a given point. Study, again, the sample chart in Fig. 10.3. Observe that the scale along the outside of the outermost circle of the chart is labeled "wavelengths toward generator." The scale increases going in a clockwise direction. Its zero is at the left end of the horizontal diameter.

Starting at the same point and going counterclockwise on the inside of the same circle is a scale marked "wavelengths toward load." These scales are marked in hundredths of a wavelength. They run from 0 to 0.50λ.

Also, along the outside circle of the chart itself is a scale labeled "angle of reflection coefficient in degrees." This scale runs from 0 at the right end of the horizontal diameter to +180° at the left of that diameter, around the upper half of the chart; it runs to −180° around the lower half of the chart.

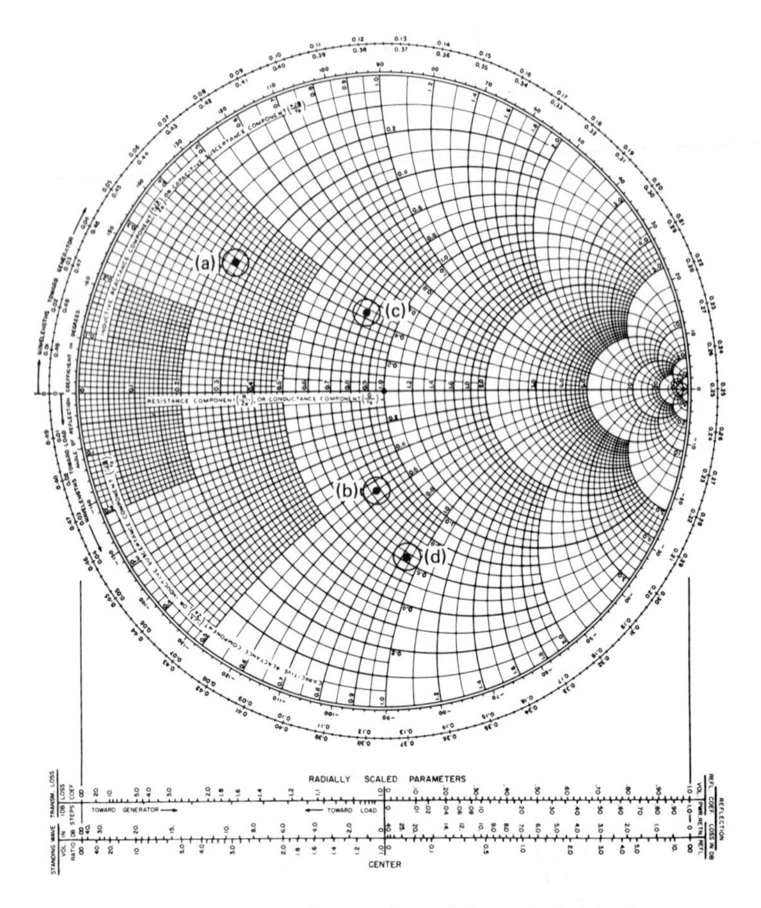

Figure 10.10 Impedance values of Example 7 plotted.

There is an important, basic reason why the chart represents a half-wavelength or 180 electrical degrees along a transmission line: The impedance along a lossless transmission line, which is not terminated in its own characteristic impedance, is cyclical over a half-wavelength. Or, very simply, the variation of impedance values along the line repeat themselves every half-wavelength. These facts, plus the design of the chart, have three important implications for its use. Learn these and be able to apply them:

1. The impedance along any lossless, uniform line with $Z_L \neq Z_0$ is represented by a circle plotted around the center of the chart, the point $1.0 + j0$.
2. The radius of the circle represents the normalized impedance of the line at every point along the line.
3. Movement along the circle is equivalent to movement along the line. Distance can be measured with the scales around the circumference by projecting radii to the scales.

Example 8

A 50-Ω coaxial transmission line is terminated with an impedance of 5.0 − j20.0 Ω. Determine the impedance of the line at a distance of 0.2 wavelength from the load.

Solution.

1. We normalize the load impedance:

$$\frac{5.0 - j20.0}{50} = 0.10 - j0.40\ \Omega$$

2. Next, enter the chart and plot the impedance, the impedance at the load end of the line. See the point labeled "Z_L" on the chart of Fig. 10.11.
3. Using the point 1.0 + j0 as a center, draw an arc of a circle clockwise from the point 0.10 − j0.40. We want to find the impedance at a distance of 0.2λ "from the load." Of course, "from the load" must be "toward the generator," or clockwise around the chart. Observe the arc on Fig. 10.11.
4. Draw a radius through the point representing the impedance at the end of the line: 0.10 − j0.40. Extend the radius until it crosses the outermost circle. Note the point of crossing: approximately 0.439λ. Locate a point at a distance of 0.2λ from this point clockwise along the "wavelengths toward generator" scale. Draw a radial line through the point just located [i.e., at 0.139λ (0.139 = 0.439 + 0.2 − 0.50)].
5. The impedance at a distance of 0.2λ from the load is the impedance at the intersection of the arc just constructed and the second radial line. The unit impedance at the intersection (see Fig. 10.11) is 0.20 + j1.16 Ω. The actual impedance of a 50-Ω line at 0.2λ from a 5.0 − j20-Ω load is

$$50(0.20 + j1.16) = 10.0 + j58.0\ \Omega$$

Standing-Wave Ratio (SWR) and the Smith Chart

Another useful feature of the Smith chart is that the radius of a circle drawn on the chart with the point 1.0 + j0 as the center corresponds to the VSWR on a lossless transmission line. (Please recall that we said just above that the radius of such a circle also represents the impedance of the line.) The VSWR is read off the chart at the point where the described circle crosses the horizontal centerline to the right of the center of the chart. For example, the circle representing the impedance of the line of Example 8 crosses the horizontal centerline (see Fig. 10.11) at a reading of approximately 12. The VSWR on the line is about 12.

The voltage-standing-wave ratio (VSWR) is called a *radially scaled parameter.* This means that its value, read from the chart, is proportional to the radius of a circle concentric with the chart. Scales for several radially scaled parameters are provided just below the Smith chart (see Fig. 10.11). Please observe that one of the scales is the standing-wave-ratio scale—the left half of the lower scale. VSWR can be read from this scale. The process requires two steps:

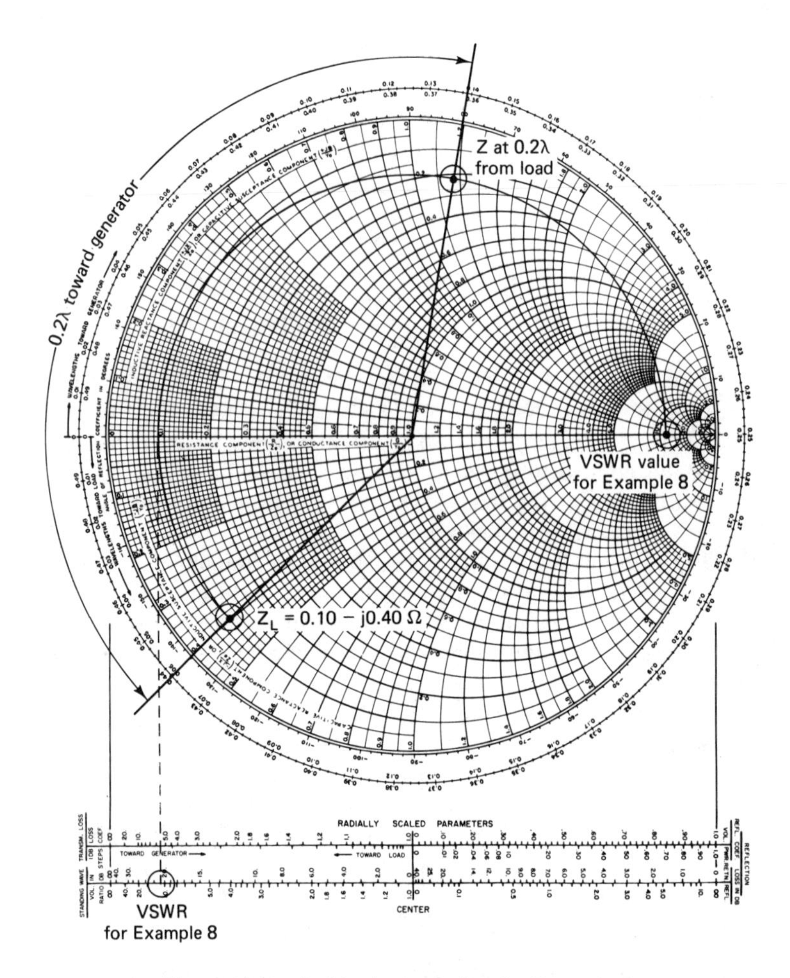

Figure 10.11 Smith chart solution for Example 8.

1. Continue any circle representing a VSWR until it crosses the left half of the centerline of the chart.
2. Project a line perpendicular to the centerline (and to the scales below the chart) from the point of intersection of the VSWR circle with the centerline, to the VSWR scale below. Read the VSWR from the scale at the point the projected line crosses the scale.

This method of scaling the VSWR for Example 8 is also shown on the chart of Fig. 10.11.

The value of a known load impedance can be used in conjunction with a Smith chart to estimate the VSWR and the input impedance of a line of a particular electrical length.

Example 9

A load circuit connected to a 300-Ω transmission line has an impedance of 450 + $j275$ Ω when the line is excited by a 100-MHz signal. The line is 78 cm in length. Assuming that the line is lossless, determine the input impedance and the VSWR.

Solution.

1. We first normalize the load impedance by dividing by Z_0:

$$\frac{450 + j275}{300} = 1.500 + j0.9167\ \Omega$$

2. Next, we locate this unit impedance on the chart. See point L on the chart of Fig. 10.12. We draw a circle through point L with C $(1.0 + j0)$ as the center.

3. The radius, CL, of the circle is equivalent to the VSWR on the line. We read off VSWR on the horizontal centerline to the right of point C,

$$\text{VSWR} = 2.3$$

(*Note:* VSWR can also be determined by laying off the radius, CL, on the VSWR scale at the bottom of the chart, starting at the SWR value of 1.)

4. Extending the radius CL to the outer rim of the chart, we read the reference wavelength from the wavelengths-toward-generator scale: 0.1922λ. To find the input impedance of the 78-cm stub, we must first convert to electrical length.

$$\lambda \text{ (for 100 MHz)} = \frac{300}{100} = 3 \text{ m}$$

$$l(\text{electrical}) = \frac{\text{physical length}}{\lambda} = \frac{0.78}{3} = 0.26\lambda$$

The input of the line is 0.26λ toward the generator from the load. The input is located at

$$0.1922 + 0.26 = 0.4522\lambda$$

5. We go around the chart to the point 0.4522λ toward the generator. Construct a line from the 0.4522λ point to the center of the chart. The point of intersection, I, with the constant VSWR circle identifies the unit input impedance, which we read off the chart:

$$Z_i = 0.47 - j0.245\lambda$$

The actual impedance is obtained by multiplying by Z_0:

$$Z_i = 300(0.47 - j0.245) = 141 - j73.5\lambda$$

Indirect Measurement of Impedance: Slotted-Line Technique

The fact that a circle centered on the point $1.0 + j0$ represents the constant VSWR of a lossless line, as well as its normalized impedance, is interesting. Is

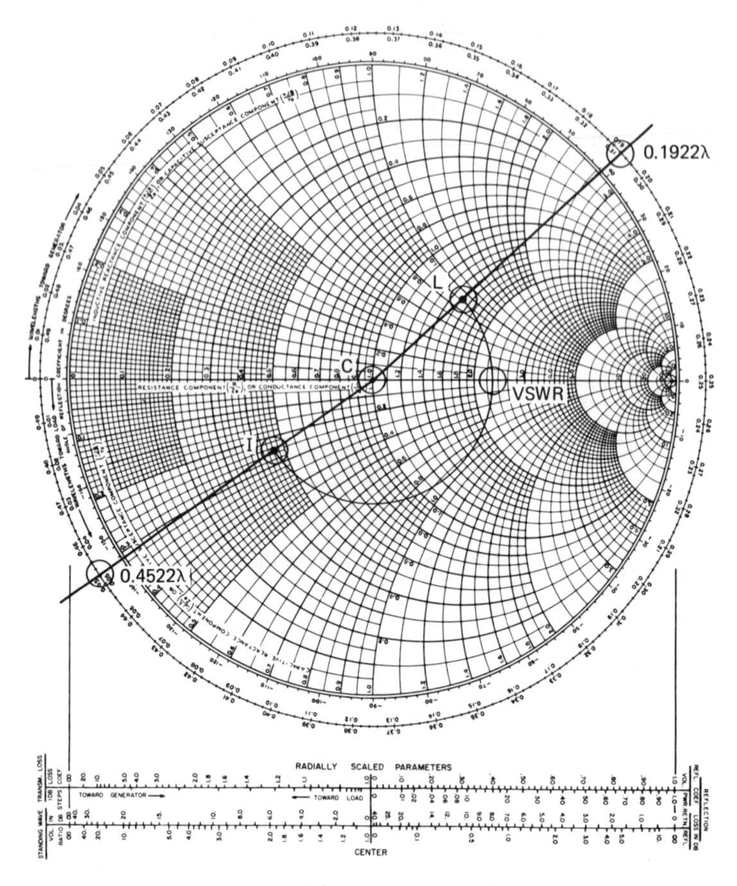

Figure 10.12 Smith chart solution for Example 9.

there not some way of obtaining impedance from a measured VSWR and the Smith chart? The answer is "yes," of course. This is a common application of the chart. Measuring the VSWR on a line is a relatively straightforward process. (A discussion of SWR meters was presented in Chap. 6.) Measuring impedance directly is more difficult.

Consider the situation of an impedance of unknown value connected as a load to a lossless transmission line. The line can be excited and the standing-wave ratio measured. The VSWR value can be entered on a Smith chart and the circle of constant VSWR constructed. We know that this circle can tell us something about the impedance along the line, including the impedance at the end, the load impedance. However, without specific information about where to start on the circle, we are stuck. Information that relates impedance and distance from the load is required.

Let's look at a well-known technique for obtaining the required impedance/distance information. It makes use of the fact that points of minimum voltage (nodes) in the standing-wave pattern shift with changes in the load impedance. The amount and direction of the shift of voltage minima contains the information needed to relate impedance and distance.

Examine the diagram of Fig. 10.13 and confirm these observations:

1. Voltage minima of the standing-wave pattern on a shorted line are located at the short and at multiples of $\lambda/2$ away from the short.
2. Voltage minima on a line terminated by an impedance not equal to the characteristic impedance of the line are shifted away from those of the shorted line. The first voltage minimum is located at a distance of less than $\lambda/2$ from the load; all others are spaced at multiples of $\lambda/2$ from the first.
3. The value of the standing wave on the unmatched, lossless line at voltage minima points for the shorted line is the same as that at the plane of the load.

The observations above are the basis for the fact that the impedance of the line (with unknown Z_L), at points of voltage minima for the shorted line, is equal to the unknown load impedance. The distance between the points of voltage minima for the two conditions—unknown impedance and a short—provides the information we need for using the Smith chart to read off the unknown impedance from the circle of constant VSWR. First, we must make use of the following information: (1) Points of voltage minima on the standing-wave pattern correspond to points of minimum line impedance. (2) The impedance of such points is purely resistive. In other words, the impedance of points of voltage minima can always be plotted to the left of center on the centerline of the Smith chart. (The circle of constant VSWR passes through this point.) Furthermore, measuring the distance and direction of shift of the points of voltage minima

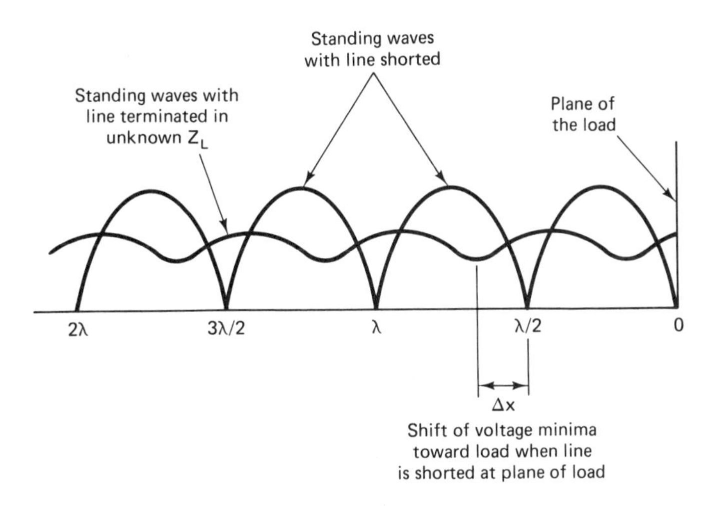

Figure 10.13 Shift in location of points of voltage minima of standing waves when line termination is changed from unknown Z_L ($Z_L \neq Z_0$) to short.

when a line is shorted provides the distance and direction to go around the VSWR circle to determine the unknown impedance.

It is physically possible and practicable to determine an unknown impedance using a technique based on the general ideas we have just examined. The technique requires the measurement of standing-wave ratio and also the location of points of voltage minima in the standing-wave pattern on a line. The latter requires that the line be an "open" line as distinguished from a "closed" line such as a coaxial line.

Twin-lead transmission lines are a natural for application of the technique. Specially manufactured apparatus for use with coaxial lines and waveguides are available. The devices are called *slotted lines*. One form of slotted line consists of a segment of a cylindrical coaxial line in which the outer conductor (cylinder) is slotted to permit the insertion of a voltmeter probe. A photograph of a slotted-line unit is shown in Fig. 10.14. The voltmeter-probe assembly is mounted so that it can be moved along the line with the probe in the slot to determine the position of voltage minima.

In actual application, the process of indirect measurement of an unknown impedance with a slotted line consists of the following steps:

1. Connect the slotted line to the impedance to be measured.
2. Excite the combination with a source of known frequency.
3. Probe the slot for VSWR and the location of a voltage minimum.
4. Short the line at the plane of the unknown impedance and determine the

Figure 10.14 Hewlett-Packard Model 816A slotted-line section in Model 809C. Universal probe carriage. (Photo courtesy of Hewlett-Packard Company.)

location of a voltage minimum at the *minimum* distance from the minimum located with the unknown impedance connected. (The shift in the location of the minimum is never more than $\pm\lambda/4$. There are minima in each direction from the first measured one: the *nearest* minima must be located.)

5. Carefully measure the *amount* and *direction* of the shift of the voltage minimum.
6. Plot the VSWR value on a Smith chart (to the right of the $1 + j0$ point, on the horizontal centerline). Construct a circle through the VSWR point with the center at $1 + j0$.

From this point on the process of indirect impedance measurement becomes that of interpreting the Smith chart. The point where the VSWR circle intersects the horizontal centerline to the left of the center of the chart represents a point of minimum impedance and voltage. The unknown impedance is also located on the circle. Its exact location is determined by the direction and amount of voltage minimum shift measured during the test. The following instructions are used in locating the point from which to read the value of the unknown impedance using the plotted circle:

1. Convert the physical distance of the measured shift of the voltage minimum to an electrical distance in fractional wavelengths for the signal frequency used to excite the test setup.
2. Proceed from the short-circuit impedance point (where the VSWR circle intersects the zero reactance line to the left of the chart center) around the circle to the location of the unknown impedance. This is accomplished by laying off a distance along the "toward generator" or "toward load" scales starting at the zero point on the left side of the chart. The distance to be laid off is the electrical-length equivalent of the shift distance measured on the slotted line. The direction to proceed is exactly the same as that determined during the test measurements. If the voltage minimum shifted toward the generator when the line was shorted, proceed clockwise around the Smith chart, and so on.
3. Mark a point on the appropriate scale at the location ascertained in step 2. Construct a line from this point through the center of the chart, intersecting both sides of the circle.
4. Read the value of the unknown impedance at the point where the line of step 3 intersects the near side of the circle (i.e., the side of the circle between the point on the outer scale and the center of the chart). The admittance of the unknown impedance can be read from the point of line/circle intersection opposite the center.

The following facts are useful in verifying that this procedure has been followed correctly:

1. A shift of the voltage minimum toward the load when a line is shorted is indicative of a net capacitive component in the unknown impedance.

2. A shift toward the generator is indicative of net inductance in the unknown impedance.
3. If shorting the load causes no shift in the voltage minimum, the impedance is purely resistive and equal to Z_0/VSWR.
4. If shorting the load causes a shift of exactly $\lambda/4$, the impedance is purely resistive and equal to $Z_0 \times$ VSWR.
5. A shift measurement is correct only when the shift is not more than $\lambda/4$.

The slotted-line technique is easier than it sounds. Let's follow it in application to a specific case.

Example 10

An unknown impedance is tested with a slotted-line setup in which the generator frequency is 1500 MHz. Before the line is shorted, the VSWR is measured as 3.9 and a voltage minimum is located at a reading of 26.5 cm on the slotted-line scale. When the line is shorted at the load, the voltage minimum (nearest) shifts to a reading of 28.9 cm on the slotted-line scale. The shift is toward the generator. Determine the normalized value of the unknown impedance. What is the actual value of the unknown impedance if the slotted line has a Z_0 of 50 Ω?

Solution.

1. We determine the wavelength of the exciting signal. The dielectric in the slotted line is air; we can use the speed of light as the velocity of propagation:

$$\lambda = \frac{v}{f} = \frac{300}{1500} = 0.20 \text{ m} = 20 \text{ cm}$$

2. The amount of shift in fractional wavelengths is

$$\Delta x = \frac{28.9 - 26.5}{20} = 0.12\lambda$$

3. We enter the VSWR on a Smith chart (see point V on the chart of Fig. 10.15) and construct the VSWR circle. Next, we measure the distance $\Delta x = 0.12\lambda$ along the "toward generator" scale from 0 on the scale. Identify the point to which the minima shifted as X. Construct a line from X through the center of the chart. The normalized impedance is located at the point marked Z. Its value is $0.46 + j0.825$. If the slotted line has a characteristic impedance of 50 Ω, the actual value of the unknown impedance is $50(0.46 + j0.825) = 23.0 + j41.25$ Ω.

A useful characteristic of the Smith chart is that the equivalent admittance of a given impedance can be determined easily. The admittance value of an impedance is located diametrically opposite the impedance. Thus, for the impedance of Example 10, we can read the admittance at the point on the VSWR circle marked A. The normalized value of admittance of the unknown impedance is $0.51 - j0.92$.

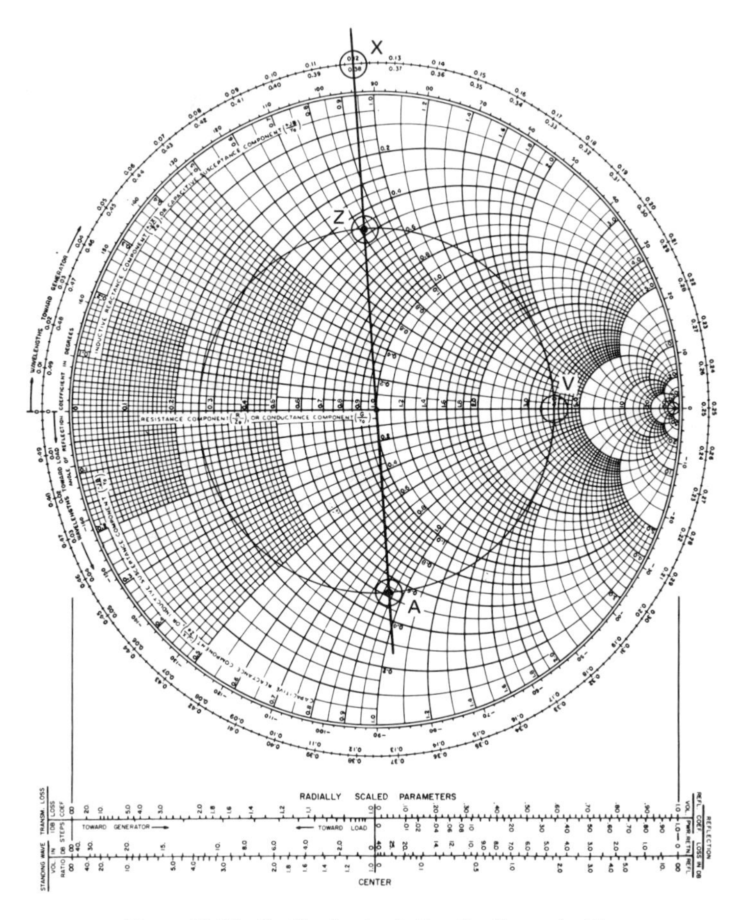

Figure 10.15 Smith chart solution for Example 10.

Correcting Mismatch Conditions with the Aid of the Smith Chart

The primary purpose of transmission lines is to transmit energy from one location to another. When this is the purpose it is usually important that it be accomplished as effectively as possible. Effective transmission is accomplished when the impedance of a load attached to a line is matched to the characteristic impedance of the line.

Since the Z_0 of a line is usually a pure resistance, the impedance of the load, *as seen by the line*, should be a resistance of the same value. If such a match does not attain, energy will be reflected from the load. Standing waves will be set up. Energy transfer effectiveness will be less than optimum.

If the actual impedance of a load does not match a transmission line, all is not lost, however. Remember the phrase just above, "as seen by the line." If the load and some auxiliary circuit or component taken together can be made to "appear" to match the impedance of the line, energy will not be reflected. Transmission will be relatively effective.

A common technique for making a load match a transmission line is to connect a transmission-line stub in parallel with the load at or near the plane of the load. A Smith chart can be used to simplify the calculations for determining the exact nature of this fix. Let's look at an example to observe the application of the technique.

Example 11

An antenna presents a load of $225 - j175\ \Omega$ to a 300-Ω twinlead transmission line. See the circuit diagram of Fig. 10.16. The operating frequency of the system is 500 MHz. It is desired to improve the line-antenna match by connecting a shorted transmission-line stub across the line near the antenna. Estimate l, the length of stub required, and d, the distance away from the antenna at which it should be connected. Assume that the dielectric of the line is equivalent to that of air.

Solution.

1. We normalize the antenna impedance,

$$Z_L = \frac{225 - j175}{300} = 0.75 - j0.5833\ \Omega$$

and plot it on a Smith chart. See point L on the chart of Fig. 10.17. Next, we construct the circle of constant VSWR through the point L.

2. In matching problems where the matching circuit is to be connected in shunt, it is convenient to work with admittance. We read the normalized admittance of the antenna at the point M,

$$Y_{\text{antenna}} = 0.84 + j0.64\ \text{S}$$

3. Next, we move around the constant-VSWR circle to its intersection with the constant-conductance circle of 1 (i.e., to the circle for $G/Y_0 = 1$, which passes through the center of the chart). See point B on Fig. 10.17. At point B, the admittance of the line looking toward the antenna is $1.0 + j0.72$ approximately. If we attach a negative susceptance of 0.72 (i.e., $-j0.72$) at this point, the $+j0.72$-S susceptance of the antenna, as seen from this point, will be canceled out. What is the location of this point? How do we get the $-j0.72$ S of normalized susceptance?

4. The location of the matching stub is the same as the chart distance between points M and B: Point M is at 0.131λ toward the generator; point B is at 0.153λ toward the generator. The stub location, d, is $0.153 - 0.131 = 0.022\lambda$ toward the generator. To determine the physical location, we calculate the value of λ for the system,

$$\lambda = \frac{v}{f} = \frac{300}{500} = 0.60\ \text{m} = 60\ \text{cm}$$

$$d = (0.022)(60) = 1.32\ \text{cm}$$

5. The shorted stub to be attached at the 0.022λ point must look like the $-j0.72$ S of susceptance. We start at point P: $0 - j0.72$, on the rim of the chart. The length of the stub is determined by going around the circle of constant VSWR of infinity (i.e., the outside rim of the chart) to the first point corresponding

to a short circuit (i.e., infinite conductance). We have identified the point as S. We moved in the "toward load" direction for the obvious reason. The stub length corresponds to the "electrical distance" on the chart between points P and S. That distance is $0.25 - 0.099 = 0.151\lambda$. A stub length of $0.151 \times 60 = 9.06$ cm is required.

Answer: A shorted stub, 9.06 cm in length, connected across the line at a distance of 1.22 cm from the antenna will make the antenna "look as if" its impedance were $300 + j0\ \Omega$.

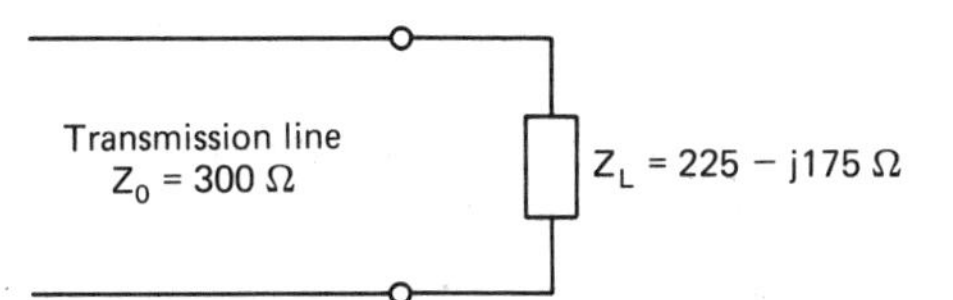

Figure 10.16 Transmission line with $Z_L \neq Z_0$.

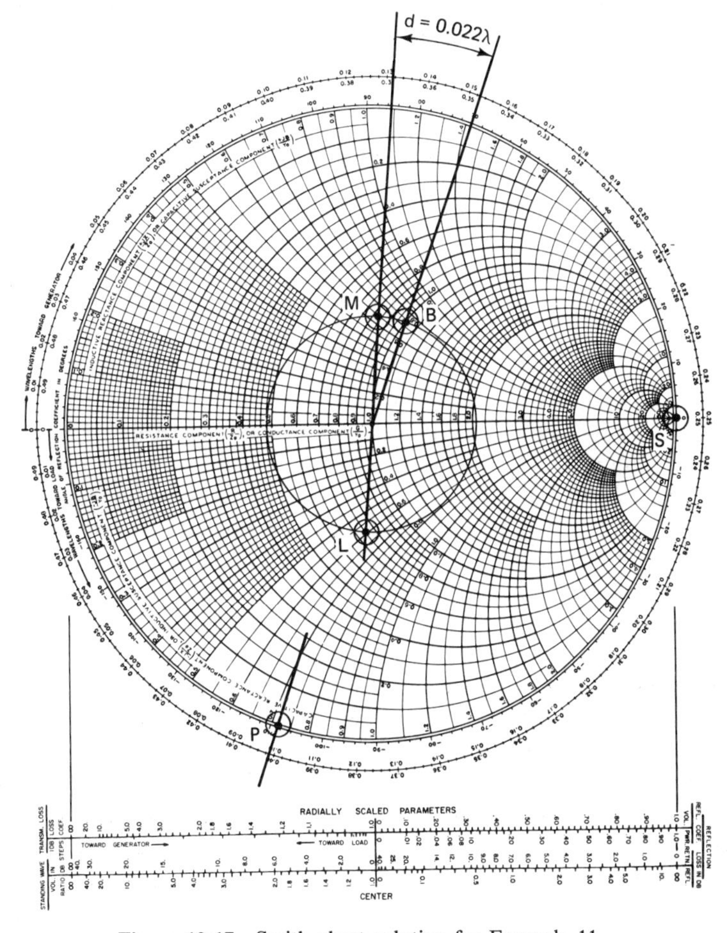

Figure 10.17 Smith chart solution for Example 11.

When a single segment of transmission line is used to match the impedance of a load to a power transmitting line, the device is commonly called a *single-stub transformer*. The term "transformer" is used because it is remindful of transformers used to match impedances in other contexts: audio systems, for example. The single-stub transformer suffers the disadvantage that it must be located at a precise location on the main transmission line. This is often difficult, if not, in fact, impossible to accomplish in practical situations. For example, it is impossible with coaxial lines, except with specially manufactured sections of line. A two-stub matching transformer can be used in a manner similar to that of the single-stub. However, locating the two stubs allows for much more flexibility. The two-stub transformer is typically used instead of the single-stub in practical applications.

The application of the Smith chart is not limited to any "official" list. Application is limited only by the knowledge, skill, and ingenuity of the user. Many applications have been devised. However, because of the nature and objectives of this text, space can be allocated only to the very brief sampling of applications provided above. Persons interested in learning more about the Smith chart and its applications are referred to the Bibliography at the back of the book.

10.3 MICROWAVES AND WAVEGUIDES

As you are already aware, wavelength is inversely proportional to frequency. For example, at 1.0 GHz (1000 MHz), λ is 30 cm (about 1 ft). At 3 GHz, the beginning of the SHF (super high frequency) range, λ is 10 cm; at 30 GHz, the lowest frequency in the EHF (extra high frequency) range, λ is only 1 cm. Signals with wavelengths of this order of magnitude are referred to as *microwaves*. Transmission lines used for transmitting energy in these very high frequency ranges often take the form of completely hollow cylindrical or rectangular tubes called *waveguides*. In concept, any transmission line could correctly be called a waveguide since it "guides" the "waves" of voltage and current along its length. However, there are special reasons why the title "waveguide" seems to be more appropriate for the hollow tubes used for transmitting short-wavelength energy.

In this section we examine a few of the most basic concepts associated with the transmission of microwaves in waveguides. Like many other subjects we have touched on, microwaves and waveguides are broad topics. Thorough and comprehensive coverage would require more than one volume the size of this book for these subjects alone.

A common approach for leading the uninitiated into the world of waveguides is to evolve the "metallic insulator" of Fig. 6.14. Look at the drawing of that device again now, and imagine adding another shorted quarter-wavelength stub opposite the one shown, then adding numerous other shorted quarter-wavelength stubs along the length of the primary transmission line (see Fig. 10.18). Since the input impedance of a quarter-wavelength stub is infinite, the addition of one, or many, has no effect on the operation of the primary line. However, the

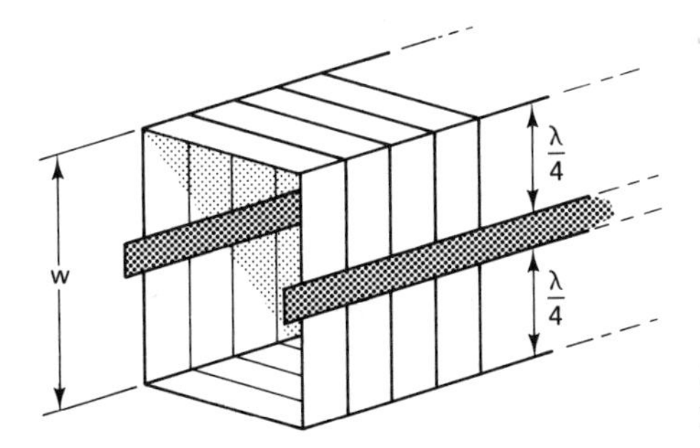

Figure 10.18 Evolution of waveguide from metallic insulator concept.

process can be extended until the effect is that of a completely enclosed, rectangularly shaped tube—a waveguide. Of course, the enclosed transmission line is usable only at the frequency that makes one of its dimensions slightly more than a half-wavelength. Waveguides are feasible transmission lines only at the super high frequencies that use guides of reasonable dimensions.

Although the waveguide we have just evolved seemingly has no effect on the operation of the original primary line, this is not entirely true. For one thing, waveguides have a number of advantages over parallel-wire lines especially and over coaxial lines to some extent as well. Explanation of their theory of operation involves several new concepts not required for more conventional types of transmission lines.

Advantages of Waveguides

A major problem with twin-lead transmission lines at higher frequencies is that the amount of direct radiation from such lines increases with the frequency of the signal being transmitted. The result is that a twin-lead transmission line radiates virtually all of the energy it is carrying, and transmits little, if any, to a load at frequencies above several hundred megahertz. The problem of energy loss due to radiation is almost totally eliminated with coaxial lines and waveguides because these forms of transmission lines "enclose" the signal and prevent its radiation.

A second source of energy loss for both parallel-lead and coaxial transmission lines is in the dielectric that supports the separation of the conductors. This is called *dielectric loss*. Although, theoretically, there is no current flow in an insulator, a dielectric, there is some current flow in actual, practical dielectrics, and there is dissipation. Of course, this dissipation is extremely small. However, it increases with frequency. Again, at very high frequencies, an energy loss becomes consequential. Because waveguides are completely hollow and in most cases filled with air, dielectric loss is virtually nil.

A third form of energy loss in transmission lines is in the I^2R heating of the conductors of the line. Heating or "copper" loss is directly proportional to the resistance of a conductor, for a given current. And the resistance of conductors of RF energy increases with frequency! This is the result of the phenomenon called "skin effect." As the frequency of a current increases, it "travels" more and more on the surface of a conductor. The penetration of the disturbance of electron movement becomes shallower. This means that a smaller cross section of a conductor is utilized for current flow. And the consequence of that, in turn, is an increase in the resistance of the conductor since resistance is inversely proportional to cross-sectional area. Coaxial lines represent some improvement over parallel-wire lines in the matter of heating loss since the conduction area of the outer conductor is significantly larger than that of the inner conductor. However, a waveguide has a major advantage in this regard: the inner (or one) conductor is completely eliminated; and the conduction area of the inner surface of the guide is significantly larger than that of the coaxial line.

The use of waveguides is not all gravy. In comparison with other forms of transmission lines, they are difficult and expensive to install. The skills required for installation are more like those of a plumber than of an electronics technician, or even of an electrician. Waveguides, in most instances, are rigid devices. Their routing must be carefully planned. Joints or connection points must be carefully made to avoid discontinuities in the inner, reflecting surfaces and the consequent creation of standing waves. Other forms of transmission lines are relatively flexible and can simply be unrolled and positioned to conform with almost any surface contour.

Waveguides are more expensive to manufacture. They must be precision made. Inner surfaces must conform to precise dimensions and be free of burrs, unevenness, etc., which could disturb the reflection patterns of the guided waves. Several waveguide sections of various shapes used to accommodate a variety of routing situations are shown in Fig. 10.19.

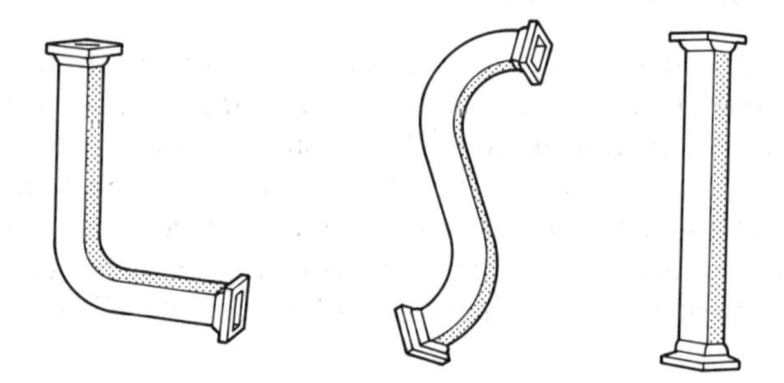

Figure 10.19 Miscellaneous waveguide sections.

Theory of Operation

In many respects, waveguides can be dealt with in ways not unlike those used for other transmission lines. Matters of characteristic impedance, the need for impedance matching, etc., are not significantly different for waveguides. However, learning a few new ideas is required if one wishes to gain an understanding of how a waveguide transmits energy. This understanding is useful as a basis for a working knowledge of some of the operating peculiarities and limitations of waveguides. Conventional explanations of waveguide theory utilize the concepts of electric and magnetic fields extensively. Be forewarned of the necessity for getting involved with these concepts.

It is useful to think of a waveguide as a special environment for the propagation of electromagnetic (EM) waves. The ideas involved are not significantly different from those we examined in connection with the propagation of such waves from antennas (see Chap. 6). In fact, waveguides are energized or excited by a probe which acts very much like an antenna. Energy is removed from a waveguide by an antenna-like probe (see Fig. 10.20).

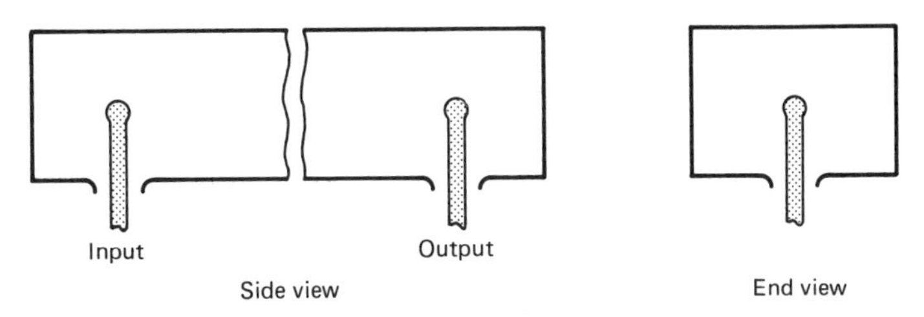

Figure 10.20 Input–output coupling for waveguide operation.

For electromagnetic waves, the "special environment" which is a waveguide is like a tunnel. However, the waves are not able to simply streak straight down the tunnel like a train going through a tunnel. Rather, the waves are guided in their motion through the tunnel by bouncing from side to side of the tunnel. Indeed, although the propagation velocity of the waves along their zigzag journey is the same as that in free space (the speed of light), the velocity along the axis of the waveguide is less than the speed of light. This, of course, is the result of the actual distance traveled being greater than the length of the tunnel.

You will recall that electromagnetic waves consist of two inseparable components: an electric field component designated E, and a magnetic field component designated H. These components are vector quantities since they have both magnitude and direction in space. The vectors are always at right angles to each other. Together, the two vectors define a plane. The direction of travel (of propagation) of an EM wave is always perpendicular to the plane of the vectors.

When radiation is emitted from a point source—a small, simple antenna—the electromagnetic energy travels (is propagated) away from that source in all directions. Since the source of energy varies in amplitude at a radio-frequency rate, the intensity of the energy being propagated varies at the same rate. The result is that in the space surrounding the antenna, the energy intensity varies in a wave-like pattern. The pattern travels out from the source. The leading edge of this energy disturbance is called, appropriately, a *wavefront*. In the first instant of emission, the wavefront is like a small sphere surrounding the source. With time the wavefront travels away from the source, expanding the sphere. The action is like that of a spherical balloon being inflated—the wavefront corresponds to the surface of the balloon. With each succeeding alternation of the source, a new wavefront is generated, and so on.

A wavefront is an *equiphase* surface. That is, the surface represents all points of equal intensity of the electric and magnetic fields. Between successive wavefronts, fields vary in a sinusoidal pattern. As the sphere of the wavefront gets larger, with greater distance from the source, the wavefront approaches the nature of a flat plane. A wavefront, or simply, a wave, can never be perfectly flat. However, in examining the theory of wave propagation in waveguides it is convenient to think in terms of waves with perfectly flat fronts. Such a wave is called a *uniform plane wave*.

Imagine, now, RF energy injected into a waveguide by means of a small antenna. Imagine, further, a uniform plane wave traveling away from the antenna and striking the side of the waveguide at an angle θ (see Fig. 10.21). In the diagram, the electric field of the wavefront is represented with small circles with dots at their centers. Think of these as the end view of field lines pointing out of the paper. The direction of travel is indicated with the heavy dashed line labeled "ray." The velocity of the wavefront in the direction of the ray is the free-space velocity—the speed of light. The angle of the ray with a line perpendicular to the side of the waveguide—the reflecting surface—is also θ. We know from a previous presentation in Chap. 6, that for radio waves, like light, the angle of reflection is equal to the angle of incidence. Therefore, the ray leaves the side of the waveguide at the angle θ with respect to the normal to the surface.

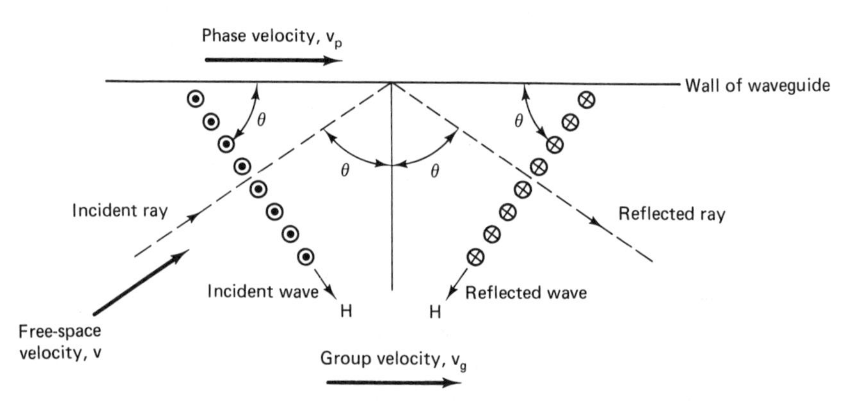

Figure 10.21 Reflection of electromagnetic wave from wall of waveguide.

A very important question is: What happens to the wavefront upon reflection? In answering this question we fall back on our knowledge of more conventional electricity: *The wave must be reflected so that there is no electrical potential at the point of reflection.* The reflection occurs at the surface of a very good conductor. *No difference of potential can exist along the surface of a perfect conductor.* (*IR* drops occur along conductors because they have significant resistance; they are far from perfect.) How can the condition of zero potential at the reflection "boundary" be satisfied? It is satisfied only by the wave being reversed 180° without change in amplitude upon reflection at the boundary. In formal terminology, it is said that the wave "satisfies the *boundary condition* of a zero electric field component in parallel with the surface of the conductor." The situation is analogous to that of a transmission line terminated in a short. The reflected wave is indicated with small circles with "x's" to represent field lines entering the paper. The direction of the magnetic field, *H*, is also indicated on the diagram of Fig. 10.21.

Let's not forget the following facts about the uniform plane waves we are using to explain propagation inside a waveguide: (1) Between successive wavefronts, field intensities vary in a sinusoidal pattern. (2) The distance between successive wavefronts, the wavelength, is the same as that in free space. (3) The wavefront extends to infinity (or to the other "side" of the waveguide, and to the top and bottom of the waveguide). Reflected waves, then, interact with incident waves. Standing waves are produced. The standing-wave pattern exists *across* the waveguide, that is, from side to side. This phenomenon is depicted in Fig. 10.22. Note that, in Fig. 10.22, there are points of maximum electric field and minimum electric field. These lie along straight lines parallel to the reflecting surface (the side of the guide); they are labeled E_{max} and E_{min}. Between these lines, the fields vary in a sinusoidal pattern.

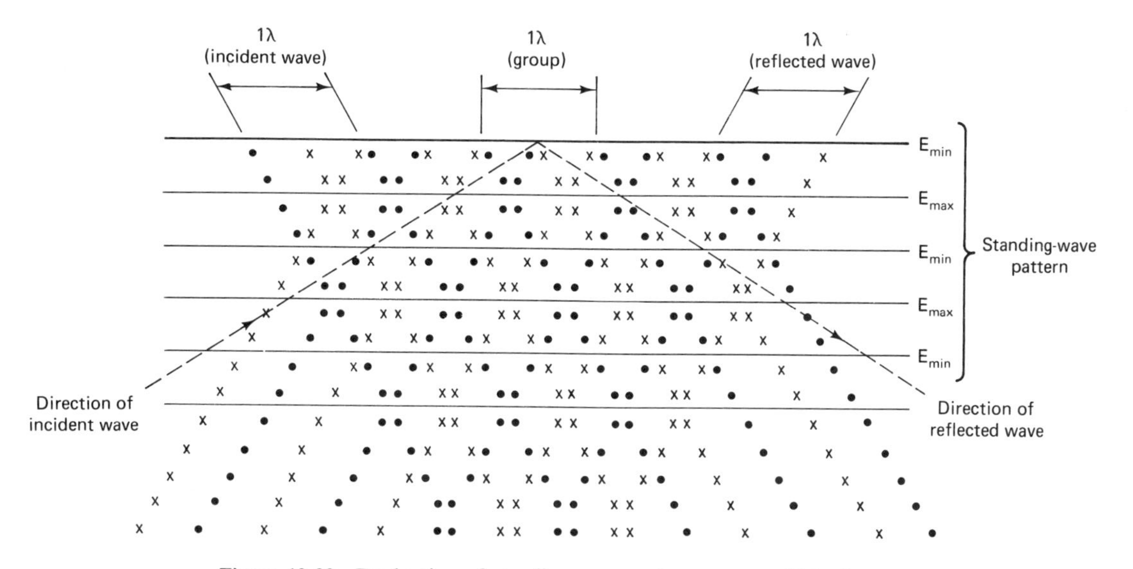

Figure 10.22 Production of standing-wave pattern across width of waveguide.

Again, the standing-wave pattern exists *across* the waveguide. However, it is simultaneously moving down the waveguide: The energy of the wave is being propagated down the waveguide. Details of a simple or primary pattern of reflection and standing waves are depicted in the drawings of Fig. 10.23. Figure 10.23(a) represents the top view of a waveguide in which plane waves are reflected from side to side and reinforce each other only once: at the center of the guide. A similar view is contained in Fig. 10.23(b) except that the guide is wider; notice that this affects θ, the angle between the directions of propagation at the center of the guide. (Guide width, of course, also affects angles of incidence and reflection.) The variations of the intensities of the fields moving down the waveguides are depicted in Fig. 10.23(c) and (d). Please observe carefully that the wavelength of the propagating energy varies with the angle θ! This wavelength can be called the group wavelength and designated λ_g. The velocity of propagation of the energy in the waveguide, the group velocity, v_g, also varies with the angle

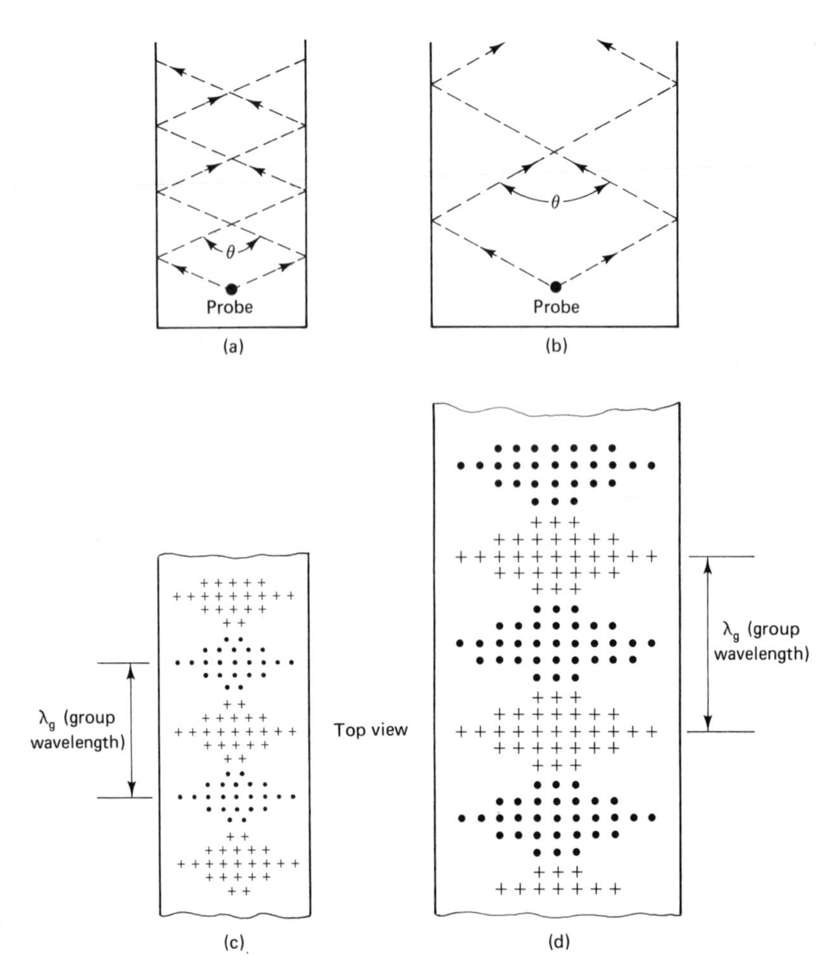

Figure 10.23 (a) and (b) Effect of wavelength and waveguide width on propagation patterns; (c) and (d) patterns of variation of field intensity of waves propagating in waveguide.

θ. The angle θ is determined by the physical width of the waveguide and the free-space wavelength of the propagating energy.

In our examination of waveguide phenomena in the last few paragraphs we have made, but not stated, an important assumption: The effects and results we have described are true for the condition of the electric field vector being perpendicular to the top and bottom of the waveguide. This is a common way or *mode* of operating waveguides. It is referred to as the *transverse electric* or *TE mode*. It is not, however, the only way in which a waveguide can be excited for successful propagation.

In the last several paragraphs we have also pretty much ignored, if not forgotten, the magnetic field component of the electromagnetic wave. But it is still there. It, too, must satisfy boundary conditions: The magnetic field must exist in closed loops parallel to the surface of conductors and be perpendicular to the electric field. Fortunately, if the boundary conditions for the electric field can be met, those for the magnetic field can also be met. The magnetic field patterns, represented by dashed lines and corresponding to the two electric field configurations of Fig. 10.23, are shown in Fig. 10.24.

This completes the presentation of a brief and basic view of the theory of propagation of radio waves in waveguides. Let's help fix this view in our minds by reviewing it and condensing it to an even briefer summary. Then we will proceed to a more quantitative and formal examination of the operating characteristics and limitations of waveguides. First, then, the summary:

1. When injected into a waveguide, an electromagnetic wave will, if possible, be reflected from the sides of the guide in a fashion that will satisfy the so-called boundary conditions, that is, the electric field will be either perpendicular to conductor surfaces or will be reflected in a way that produces a zero-field component parallel to the surface of a conductor.
2. Reflected waves combine with incident waves in the waveguide to produce standing waves across the width of the waveguide. The standing-wave pattern must have voltage zero points at the sides of the guide (to satisfy boundary conditions).
3. If boundary conditions are satisfied, a wave of electromagnetic energy propagates down the guide. The propagation velocity and wavelength of the energy wave are different from those it would have in free space. These characteristics depend on the nature of the reflection pattern set up in the

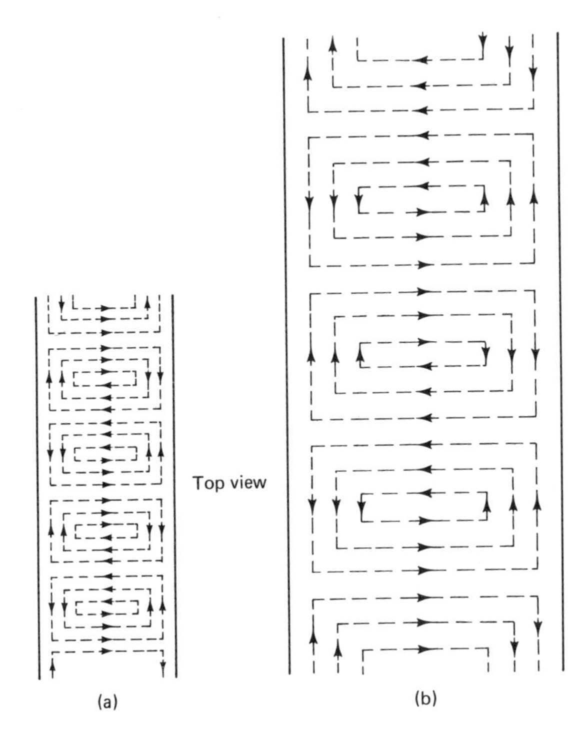

Figure 10.24 (a) Magnetic field pattern for corresponding electric field pattern of Fig. 10.23(c); (b) this magnetic field pattern goes with the electric field pattern of Fig. 10.23(d).

waveguide. The reflection pattern, in turn, is a function of the free-space wavelength (or frequency) of the energy and the dimensions of the waveguide. If boundary conditions cannot be satisfied, energy will not be propagated, or, at best, will be severely attenuated.

Cutoff Frequency

When the width, w, of a waveguide is exactly equal to half of the free-space wavelength of a wave (i.e., when $\lambda = 2w$), the wave will bounce from side to side in the waveguide and will make no progress down the guide. The angle of incidence or reflection is zero. The frequency that makes this condition true is called the *cutoff frequency* for the guide. The associated wavelength is called the *cutoff wavelength*. Furthermore, the waveguide will severely attenuate all frequencies lower than the cutoff frequency (or all wavelengths longer than the cutoff wavelength). A waveguide is like a high-pass filter. Its attenuation is like that of a resonant wave trap: it does not dissipate the energy in the heating of a conductor, it simply blocks the passage of the energy.

Group and Phase Velocities

We have already noted that the velocity at which energy actually moves down a waveguide is less than the speed of light. The actual velocity of propagation is the velocity at which the crosswise standing wave moves down the guide. This is called the *group velocity*. We previously stated that it is designated v_g. The group velocity is mathematically related to the propagation velocity of the wave in free space, v, and the sine of the angle of incidence (see Fig. 10.21). Specifically,

$$v_g = v \sin \theta$$

We have now described two velocities associated with the movement of a wave in a waveguide: (1) the velocity, v, of the wave along its zigzag path in the guide, and (2) the group velocity, v_g, the velocity of actual advancement of the phenomenon along the straight-line journey down the guide. There is a third velocity of interest in this matter—the *phase velocity*, v_p.

The phase velocity is the speed at which the wavefront advances along the side of the guide where it is being reflected. A simply analogy is that of a long ocean wave striking a beach at an angle instead of head on. The wave approaches the shore at a certain velocity; this velocity is analogous to v, the basic propagation velocity of the EM wave along its zigzag course. If you stood watching the ocean wave, you would observe it racing along the beach from the point at which it first contacted the beach. The speed at which its point of contact moves along the beach is observably greater than its actual approach velocity. This velocity of the point of contact is analogous to the phase velocity of a propagating wave in a waveguide. It, also, is related to v and the angle of incidence,

$$v_p = \frac{v}{\sin \theta}$$

The phase velocity may be greater than the speed of light. Indeed, if the angle of incidence is zero, the phase velocity is infinite.

Modes of Operation

In the introduction of the basic principles of waveguide propagation (see above) we examined the phenomena associated with a *transverse electric wave*. That is, we studied the behavior of a wave in which the electric field lines were perceived to extend between the bottom and top of the waveguide, and to be perpendicular to those surfaces. Furthermore, we said that the lines were all in the plane perpendicular to the direction of travel of the wave. Formally, a transverse electric wave is one in which all components of the electric field lie in a plane *transverse*, that is, perpendicular, to the direction of propagation. A transverse electric wave is conventionally designated with the letters TE.

There are numerous crosswise standing-wave patterns of TE waves that can satisfy boundary conditions (i.e., that will propagate). We have already studied the simplest one (see Fig. 10.23). The pattern of Fig. 10.23 contains a single voltage maximum at the center of the guide and nodes at the sides (as required to satisfy boundary conditions). The pattern is that of a half sine wave. It is called a *half-sine* configuration. Other common configurations are *full sine* and *one-half sine*. These are diagrammed in Fig. 10.25. With such configurations the waveguide is said to be operating in the "TE mode," or the wave is propagating

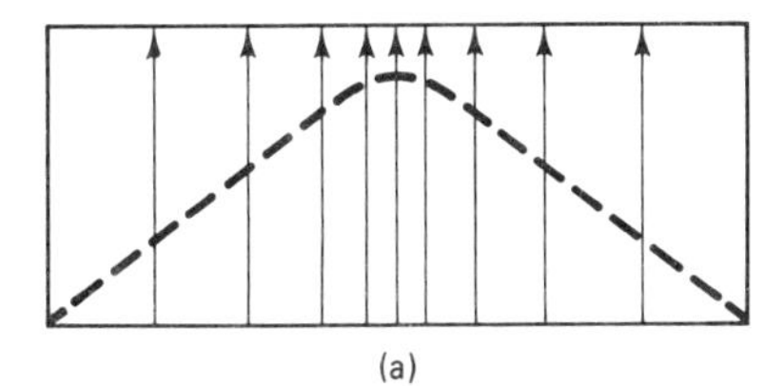

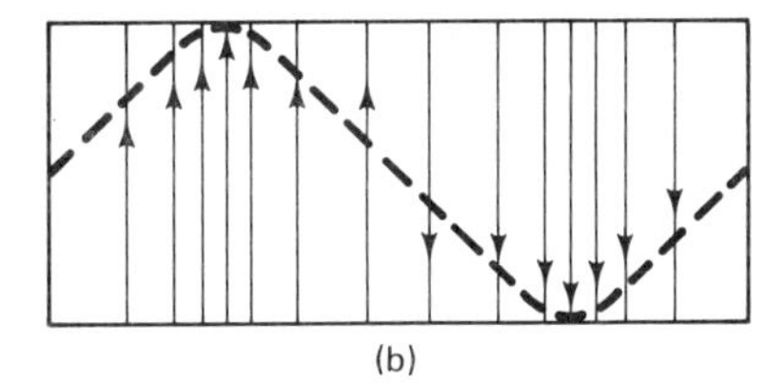

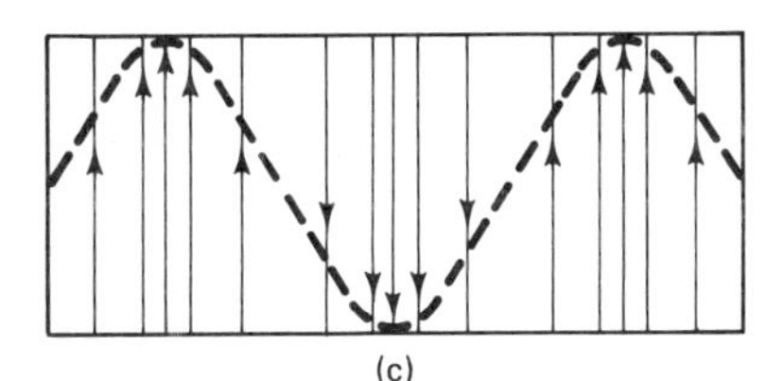

Figure 10.25 Standing-wave patterns across width of waveguide. Line density represents field intensity: (a) half-sine configuration; (b) full-sine configuration; (c) one-half-sine configuration.

in a "TE mode." Please be aware that in the TE mode, the magnetic field component of the wave has lines of force in both the transverse plane and along the direction of propagation.

It is possible for a wave to propagate when the lines of force of the magnetic field component lie entirely in the plane transverse to the direction of propagation. In this case the propagation is referred to as a "TM mode," that is, a transverse magnetic mode. In this configuration, the electric field has components both in the transverse plane and along the direction of propagation.

A subscript convention is used to identify the mode of operation of a waveguide, or the mode of propagation of a wave. The convention consists of a double subscript, used as in TE_{mn} or TM_{mn}. The first subscript, m in the examples, specifies the number of half-sine patterns in the transverse plane along the width of the guide. For the half-sine pattern of Fig. 10.23, m is 1. The second subscript, n in the examples, designates the number of half-sine variations in field intensity along the narrow dimension of the waveguide. In the examples provided so far (see Figs. 10.23 and 10.25) there has been no variation in intensity in this direction: The n subscript for the configurations of Figs. 10.23 and 10.25 is 0. The complete designations for the configurations of Fig. 10.25 are TE_{10}, TE_{20}, and TE_{30}, respectively. Configurations of several types, including TM modes, identified with their mode designations, are diagrammed in Fig. 10.26.

The TE_{10} mode is sometimes referred to as the *dominant mode* of a waveguide of rectangular cross section. It is a mode which corresponds to the configuration that would propagate the lowest possible frequency for a waveguide of given dimensions. More complex modes are referred to as *higher-order modes.*

Waveguide shapes are not limited to those of rectangular cross section which we have been studying. Cylindrical guides are also used extensively.

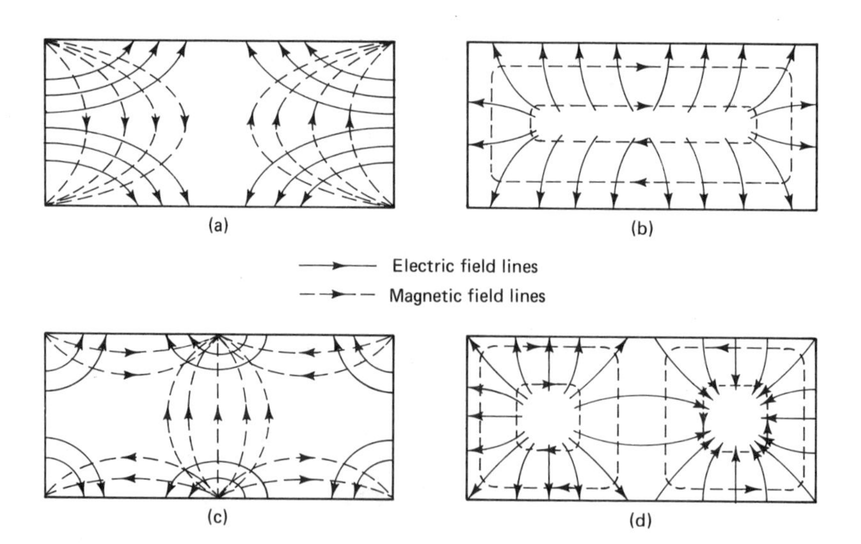

Figure 10.26 Waveguide propagation modes: (a) end view, TE_{11} mode; (b) end view, TM_{11} mode; (c) TE_{21} mode; (d) TM_{21} mode.

Modes of propagation in cylindrical guides are similar to but not exactly the same as in guides of rectangular cross section.

Methods of Excitation

As we have seen, waveguides are literally "guides for electromagnetic waves." Waveguides must be excited by a generator in such a way that waves capable of being propagated will be produced at the point of excitation. Unlike methods of exciting twin-lead and coaxial lines, it is not enough simply to connect a generator to two points on the guide.

Waveguides can be excited by connecting a generator to either a probe or loop inserted in the guide, or through a window usually called an iris or slot. Diagrams of several forms of these coupling methods are shown in Fig. 10.27.

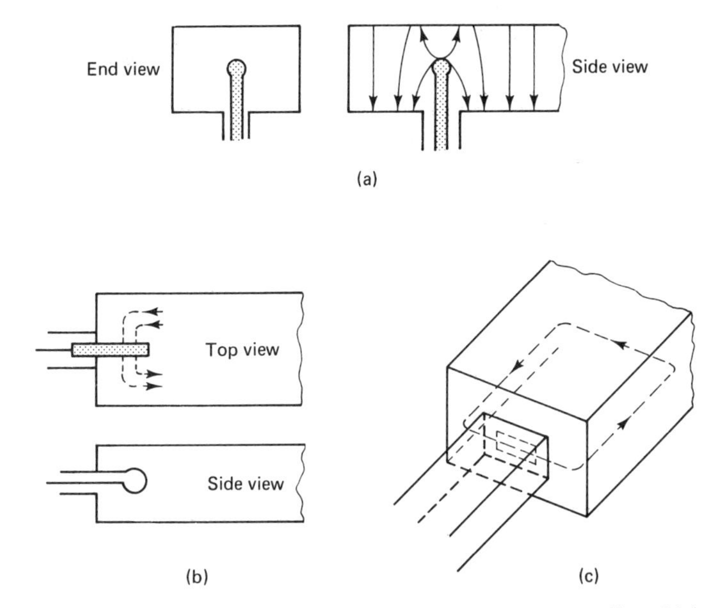

Figure 10.27 Waveguide coupling methods: (a) solid probe coupling; (b) loop coupling; (c) window or slot coupling.

The loop probe provides coupling through the magnetic field component; the straight probe is a means of electric field coupling. The removal of energy from a waveguide—output coupling—is accomplished by means of the same methods as input coupling. In window coupling, electromagnetic energy effectively "leaks" into or out of the waveguide.

Characteristic Impedance

When dealing with waveguides we think in terms of electromagnetic waves, not separate waves of voltage and current. This requires a few adjustments to our thinking about characteristic impedance as well.

The characteristic impedance of a waveguide can be thought of as the ratio of the amplitude of the electric component to the magnetic component. The amplitudes are those at right angles to the direction of energy propagation. Of course, these components vary with frequency, dimensions of the guide, modes of propagation, and so on. The characteristic impedance also changes with these factors. Unlike twin-lead and coaxial transmission lines, the characteristic impedance of a waveguide is not simply a function of its physical dimensions.

It is beyond the intended scope of this discussion to examine the details of characteristic impedance of waveguides. However, it is important to note that basic concepts of impedance matching are observed in couplings to waveguides. Measures similar to those of other types of transmission lines are used to improve matching.

Various techniques can be used to adjust the impedance of a waveguide as seen from a point of coupling. For example, when a metallic cylinder or strip is inserted in a waveguide and extends across the full depth of the guide, it alters the wave pattern to produce a result exactly like a shunt inductive reactance on conventional transmission lines [see Fig. 10.28(a)]. The device is called an *inductive post*.

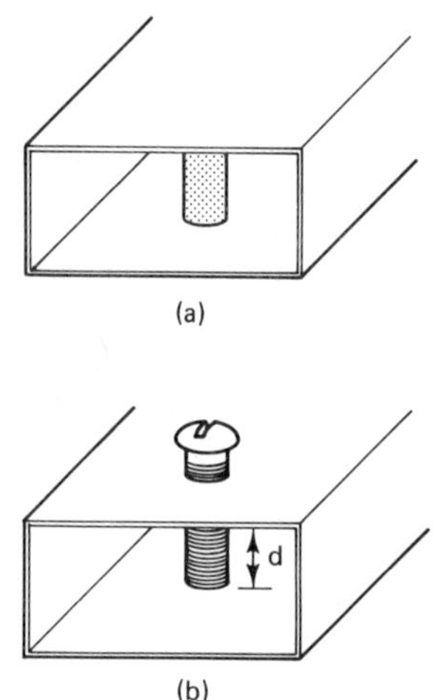

Figure 10.28 Methods of altering impedance of waveguide: (a) inductive post; (b) adjustable reactance post.

An adjustable screw may be inserted in the center of a waveguide to provide adjustable reactance [see Fig. 10.28(b)]. If the depth of the screw, d, is slight, the disturbance of the wave pattern will be seen as a capacitive reactance. If d is exactly equal to a quarter-wavelength, the effect is like a resonant circuit: there is no net reactance. When d is greater than a quarter-wavelength, the height of the guide permitting, the effect is inductive.

Strips placed inside waveguides also provide impedance altering. Examples of symmetrical inductive, capacitive, and resonant windows or iris diaphragms are shown in Fig. 10.29.

Waveguide stubs may also be used to perform impedance-matching functions. Stubs, open or shorted, of appropriate lengths may be connected in series or in parallel with a mainline waveguide transmission line. See Fig. 10.30 for examples

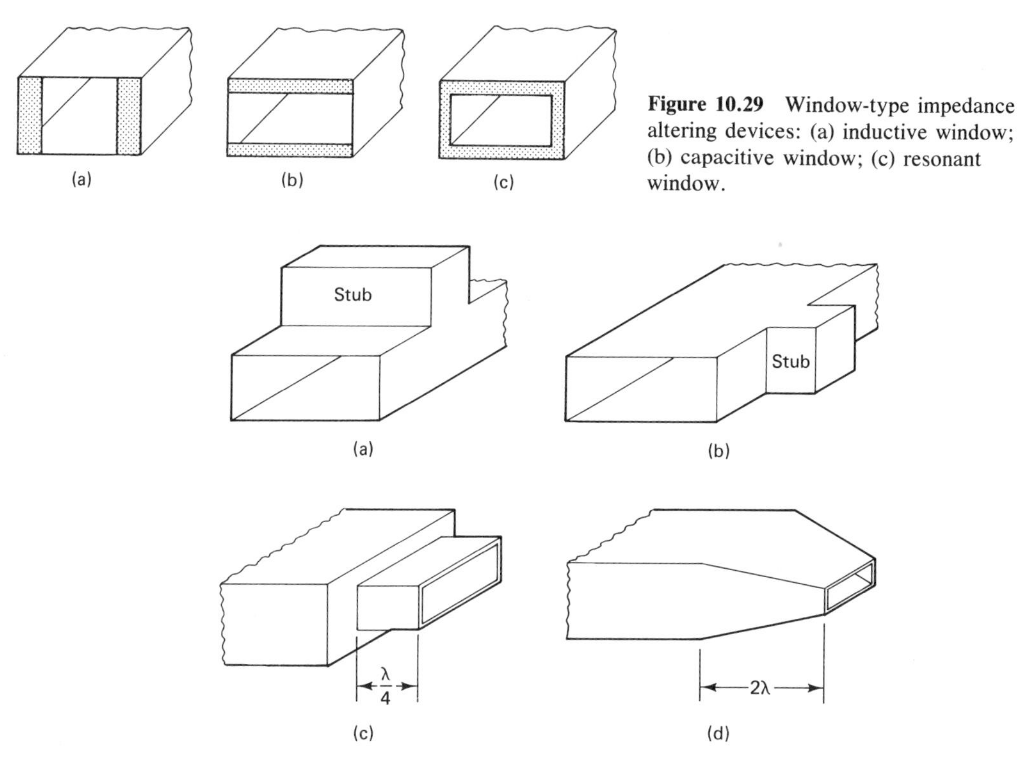

Figure 10.29 Window-type impedance altering devices: (a) inductive window; (b) capacitive window; (c) resonant window.

Figure 10.30 Waveguide stubs for impedance matching: (a) series stub impedance matching; (b) shunt stub impedance matching; (c) $\lambda/4$ stub transformer; (d) tapered section impedance transformer.

of these applications. Note the tapered section shown, which is typically used as an *impedance transformer*.

The Resonant Cavity

It is possible to construct a waveguide section so that it behaves like the parallel-resonant tank circuit of more conventional circuits. Such a section is called a *resonant cavity*. The resonant cavity, as its name implies, is a hollow box. It can have a number of different shapes. Some typical shapes are shown in Fig. 10.31.

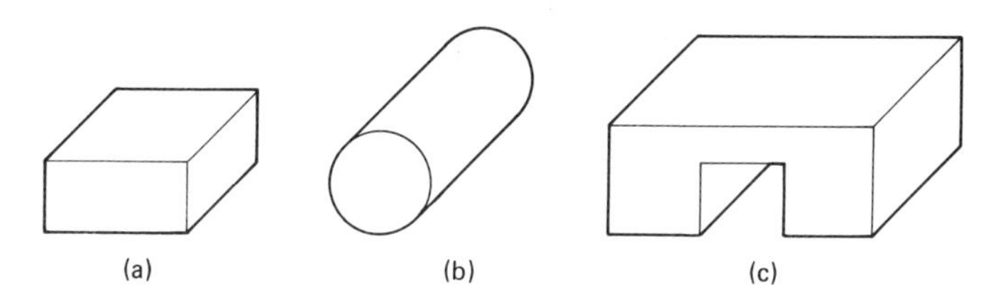

Figure 10.31 Resonant cavity shapes: (a) rectangular; (b) cylindrical; (c) reentrant.

As could be expected, a cavity is resonant only when it exhibits very special behavior under excitation by a microwave generator. The required behavior (behavior like that of a resonant circuit) occurs when energy is reflected in the cavity and produces a standing wave. For example, a resonant cavity can be made from rectangular waveguide by closing both ends of a section a half-wavelength long. When excited by a signal of the appropriate frequency, the energy is reflected back and forth between the ends of the section (the cavity) and a standing-wave pattern will be set up as shown in Fig. 10.32.

The applications of resonant cavities are similar to those of resonant circuits made of discrete L and C components. Of course, the applications are in circuits operating at microwave frequencies. A common application is as the tank circuit of a microwave oscillator. Cavity resonators are used as filters—bandpass and band elimination.

A very important application of resonant cavities is as part of a microwave frequency-measuring device. The device is called a *wavemeter.* Wavemeters are common measuring devices at nonmicrowave frequencies also. The device, in general, consists of a tunable resonant circuit and a means of measuring the signal level of the energy in the circuit. The tuning mechanism is calibrated to indicate the frequency to which the circuit (or cavity) is tuned. The frequency-measuring process consists of the following steps:

1. Excite the resonant circuit or cavity with the signal whose frequency it is desired to be determined.
2. Tune the circuit or cavity until it resonates at the signal frequency. (Resonance is observed as a peaking of measured signal level in wavemeters which are designed as so-called *transmission* or *absorption* devices; a dip in measured signal level occurs at resonance in a *reaction* wavemeter.)
3. Read the unknown frequency off the calibration chart of the wavemeter opposite the setting of the tuning mechanism.

From the foregoing it is obvious that cavity resonators are tunable. Methods of tuning include an adjustable screw or post inserted in the cavity (see Fig.

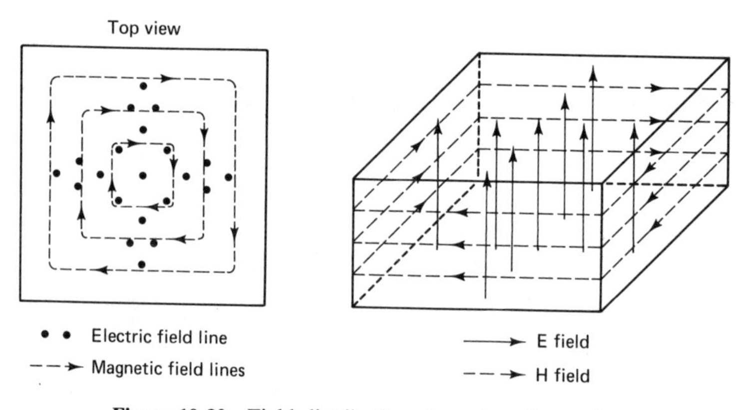

Figure 10.32 Field distributions in rectangular cavity.

10.28), or an adjustable shorting plunger. The latter is a common tuning technique in commercially available cavity wavemeters. The diagram of Fig. 10.33 illustrates the adjustable shorting plunger and a method of input and output coupling.

Like common *LC* resonant circuits, cavity resonators can be described in terms of the performance characteristic *Q*. Please recall that *Q* is a measure of the ratio of stored energy to dissipated energy in a resonant circuit. Energy is dissipated in the walls of a cavity as the energy wave is reflected back and forth between opposing walls. The *Q* of a cavity is lowered if energy is removed and fed to a load.

Another cavity of special interest is the *reentrant cavity*. The adjective "reentrant" is associated with the concept of pointing inward. A reentrant angle in a multisided figure, for example, is one that points inward instead of outward as is true of the typical angle of such figures. Cross sections of two common reentrant cavity shapes are shown in Fig. 10.34. The complete cavities are hollow devices produced by rotating the shapes around the line marked "axis" on the diagrams. The cavity of Fig. 10.34(a), for example, is toroidal (doughnut-like). Reentrant cavities are most commonly found as part of microwave electron devices (vacuum tubes), notably the *reflex klystron*.

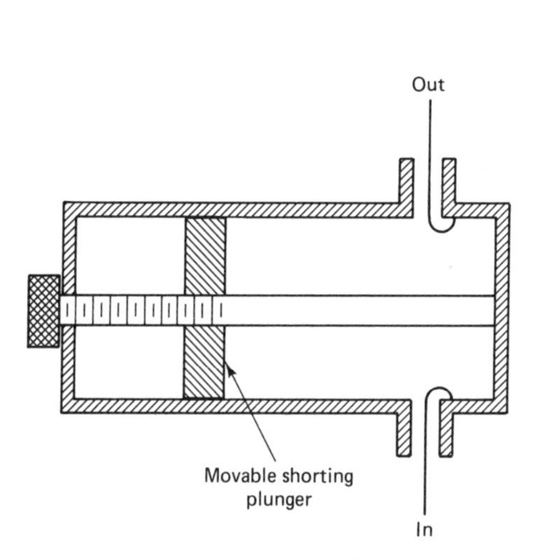

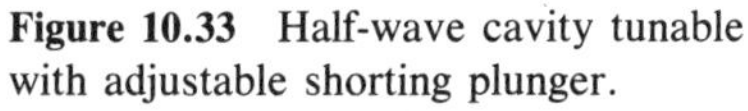
Figure 10.33 Half-wave cavity tunable with adjustable shorting plunger.

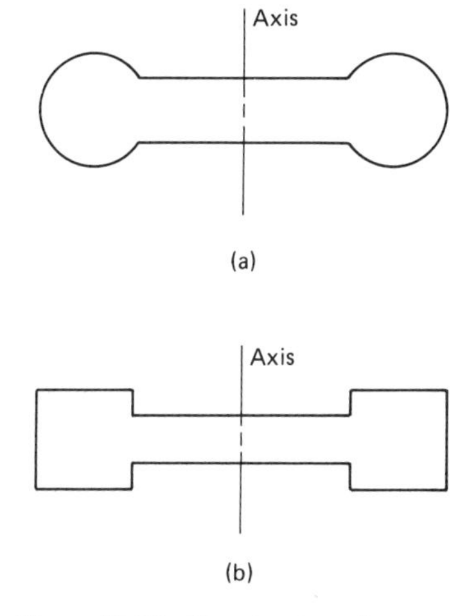

Figure 10.34 Reentrant cavity cross sections: (a) toroidal reentrant cavity; (b) center-gap reentrant cavity.

Waveguide Attenuators

As with other energy-handling devices, there are situations in waveguide applications in which it is desirable to be able to adjust the level of energy being transmitted by the system. In lower-frequency circuits, amplitude control devices typically are designed to adjust the level down from some maximum level and are called attenuators, appropriately. Common devices for controlling the amount of transmitted energy in a waveguide are called waveguide attenuators.

Attenuators, at whatever frequency, may take one of two forms, or a combination of these two: they may reflect all or a portion of incident energy, or they may absorb and dissipate energy by conversion to heat. Also, attenuators generally are designed so that their impedance matches that of the system in which they are installed. This is so as to disturb as little as possible the energy that is to be allowed to pass.

A drawing depicting a waveguide *flap attenuator* is shown in Fig. 10.35. The "flap," whose depth into the guide is adjustable, is covered with a resistive material—it dissipates energy. The shape of the flap is designed to provide impedance matching between the attenuator and the waveguide.

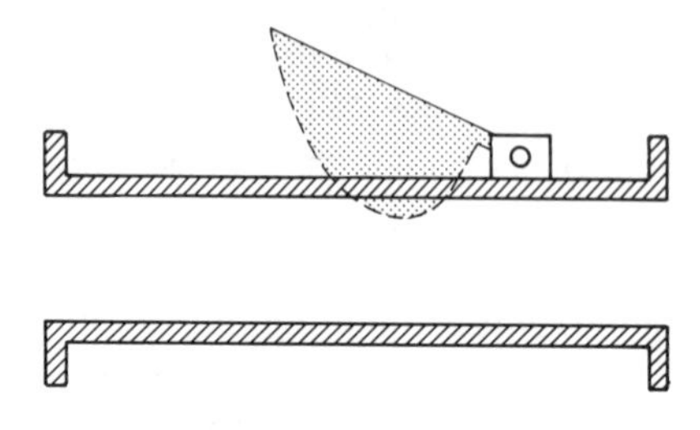

Figure 10.35 Cross section of waveguide section with flap attenuator.

10.4 MICROWAVE ANTENNAS

Because of their short wavelengths (measured in centimeters) the behavior of radio microwave emissions is easily compared with that of light. The comparison is more frequently made than for lower-frequency radio emissions. The propagation of microwaves in space, for example, is strictly line-of-sight, like light. The result is that microwaves find common application in point-to-point communications systems; microwaves are seldom "broadcast." Microwave links between earth stations and satellite space stations now provide thousands of communications channels between all the major cities of the world.

As we learned in Chap. 6, antenna lengths are of the order of a half-wavelength of the signal being transmitted. Microwave antennas, accordingly, have dimensions measured in centimeters. Their small size permits their easy combination in microwave antenna arrays. Antenna arrays provide for the concentration—focusing or "beaming"—of microwave emissions.

A broadside array containing eight full-wavelength antenna elements is shown in Fig. 10.36 along with its radiation pattern. Please note the narrow, beam-like nature of the radiation pattern. If used with a reflector, the array has the radiation pattern shown in Fig. 10.36(c). It is readily apparent why a system of this type is so adaptable to point-to-point communications.

A microwave antenna that especially utilizes the light-like properties of microwaves is the parabolic or "dish" antenna. Some of these can even be seen sprouting in the backyards of American homes in the mid-1980s. Let's examine the theory of the parabola.

A parabola is a mathematical figure or curve. The curve represents all points equidistant from a fixed point called the *focus* and a straight line called the *directrix* (see Fig. 10.37). If this curve is rotated about a line through the focus and perpendicular to the directrix, a parabolic surface—a dish—is formed. If a light source is placed at the focus, the parabolic surface will reflect the light

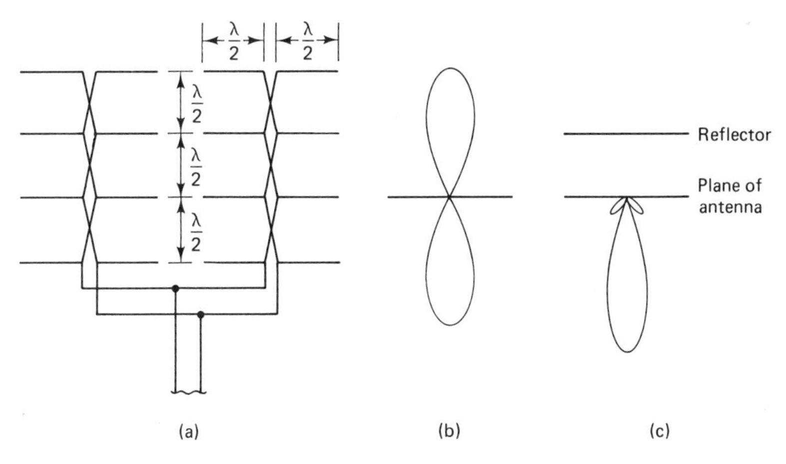

Figure 10.36 (a) Eight-element broadside microwave antenna array; (b) radiation pattern of broadside array; (c) radiation pattern of boradside array with reflector.

to form a beam of parallel light rays perpendicular to the directrix. The reflector surface of automobile headlight elements is generally parabolic in shape, for example.

As mentioned briefly above, parabolic reflector, or dish-type units are very popular in microwave applications. Reflection of light waves is most effective with polished, or at least, smooth surfaces. Microwaves require a metal surface. The surface need not be continuous. Some reflectors are, in fact, made of a screen-type material with a very coarse weave.

A parabolic reflector is typically used in conjunction with a dipole, half-wave antenna located at the focus. A simple reflector element is used to reflect radiation back to the parabolic surface (see Fig. 10.38). This reflector improves the operation of the dish, not only by getting more energy back onto the dish, but also by attenuating the broad-beam, direct radiation in the forward direction from the dipole. The active antenna element is fed energy from a microwave transmitter through a waveguide transmission-line system.

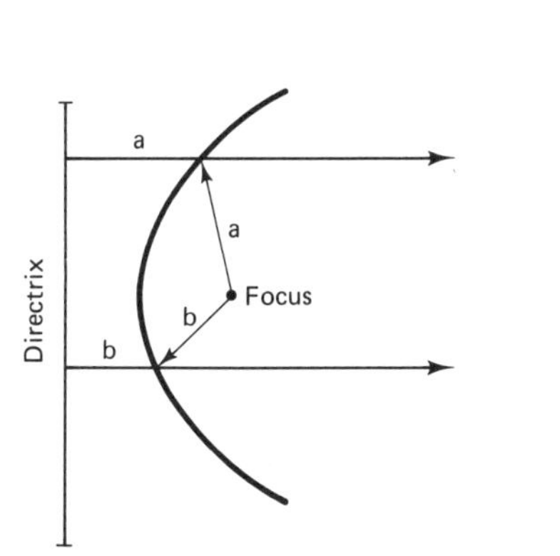

Figure 10.37 Parabola.

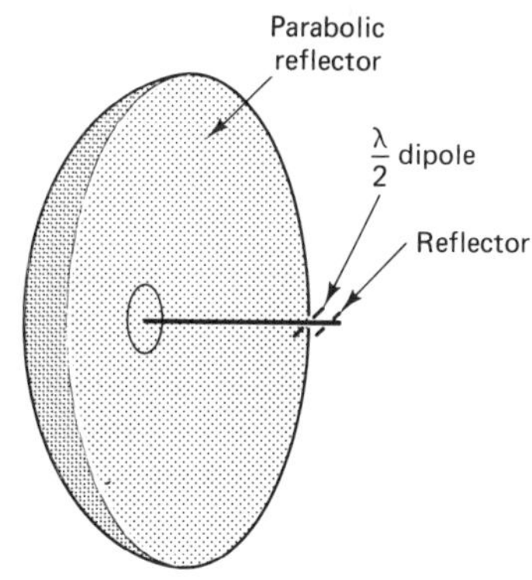

Figure 10.38 Dish reflector microwave antenna assembly.

GLOSSARY OF TERMS

Absorption wavemeter A frequency-measuring device utilizing a resonant circuit or resonant cavity. The resonant cavity absorption wavemeter is characterized by having a single entry port. Maximum energy amplitude is observed at the frequency at which the circuit or cavity is resonant.

Attenuation constant, α The real portion of the complex propagation constant for transmission lines; may be used to calculate the reduction in signal amplitude over distance; has the unit neper per unit length.

Cutoff frequency (of a waveguide) The lowest frequency for which successful propagation can be achieved in a waveguide of particular dimensions; the frequency for which the width of a waveguide is exactly one-half wavelength.

Directrix *See* Parabola.

Dominant mode The mode of operation of a waveguide for which the longest wavelength can be propagated, usually TE_{10} mode for a rectangular waveguide.

Equiphase surface (of an electromagnetic wave) The surface of the imaginary sphere representing propagating electromagnetic energy over which the amplitude and orientation of the electric and magnetic field components are the same.

Focus *See* Parabola.

Full-sine configuration A mode of propagation of an electromagnetic wave in a waveguide such that the standing-wave pattern across the width of the guide is a full sine wave with the electric field at zero at the side walls.

Group velocity The velocity of propagation of electromagnetic energy along the axis of a waveguide, as distinguished from the velocity of the wavefront along its zigzag path in the waveguide; the velocity at which the crosswise standing-wave pattern (the group) in a waveguide moves down the guide.

Half-sine configuration A mode of propagation of an electromagnetic wave in a waveguide such that the standing-wave pattern across the width of the guide is a half sine wave with the electric field at zero at the side walls.

Neper The natural logarithm (base *e*) of a voltage or current ratio.

Normalized impedance The value of an impedance (often associated with a transmission line, e.g., characteristic impedance, load impedance, etc.) after it has been divided by a reference impedance value, usually the characteristic impedance of the line of interest.

One-half-sine configuration A mode of propagation of an electromagnetic wave in a waveguide such that the standing-wave pattern across the width of the guide is one and one-half sine waves, with the electric field zero at the side walls.

Parabola A curve which represents all points equidistant from a fixed point called the focus and a straight line called the directrix.

Phase constant, β The imaginary component of the propagation constant of a transmission line used to determine the phase shift over distance of a voltage or current wave; has the unit radian per unit length.

Phase velocity The velocity with which an equiphase (see above) plane of an electromagnetic wave advances along the side of a waveguide.

Propagation constant, γ A complex expression (contains real and imaginary parts) used to determine the amplitude and phase change, over distance, of a voltage or current wave on a transmission line.

Reaction wavemeter A resonant circuit or cavity frequency-measuring device that absorbs energy from the source being measured and produces an energy minimum indication at the frequency at which the wavemeter is resonant.

Reentrant cavity A resonant cavity shape produced by returning one or more surfaces inward.

Reflection coefficient The ratio of reflected voltage (or current) to incident voltage (or current) on a transmission line.

Resonant cavity A section of waveguide, shorted at both ends, whose length is equivalent to exactly a half-wavelength for the frequency of interest; any of a number of hollow shapes that produce a standing wave of a half-wavelength at the frequency for which it is called resonant.

TE mode One of the basic methods of exciting a waveguide such that an electromagnetic wave propagates in the guide with all components of the electric field in a plane transverse (at right angles) to the axis of the guide (the direction of propagation).

TM mode One of the basic methods of exciting a waveguide such that an electromagnetic wave propagates in the guide with all components of the magnetic field in a plane transverse (at right angles) to the axis of the guide (the direction of propagation).

Uniform plane wave An electromagnetic wave in which it is assumed that equiphase surfaces are plane as distinguished from spherical.

Unit length Any uniform length of transmission line (e.g., centimeter, inch, foot, meter, etc.) chosen for use with "per unit length" parameters.

Velocity of propagation The velocity with which equiphase points of an electromagnetic wave travel through or along a medium: air, space, transmission line, etc.

Wavemeter An adjustable resonant circuit or cavity used for measuring the frequency of a source (of unknown frequency) of electromagnetic energy.

REVIEW QUESTIONS: BEST ANSWER

1. The distributed parameters of transmission lines are for uniform lengths of a: **a.** foot. **b.** yard. **c.** centimeter. **d.** meter. **e.** any of these.
2. When the distributed parameters of R and G are so small as to be insignificant, a transmission line is said to be: **a.** matched. **b.** lossy. **c.** lossless. **d.** characteristic. **e.** none of these.
3. The voltage at a distance of 1000 m from the input of a transmission line is down to 90% of its original value. The attenuation constant is: **a.** -0.1054 Np/km. **b.** 0.1054 Np/km. **c.** 0.1054 Np/m. **d.** 0.4576 Np/yd. **e.** none of these.
4. When the unit length being used for transmission-line calculations is equal to λ, the phase constant, β, has a value of: **a.** 180°. **b.** π radians. **c.** 2π radians. **d.** $3\pi/2$ radians. **e.** none of these.
5. A 75-Ω transmission line has a resistance (total) of 0.05144 Ω/m. The dielectric loss in the line is negligible. The total attenuation, in nepers, for 1000 m is approximately: **a.** 51.44. **b.** 0.3429. **c.** 150. **d.** 0.2783. **e.** none of these.
6. The attenuation constant of a certain transmission line is $1.365(10^{-5})$ Np/m. The

amplitude of the voltage at a point 786.4 m down the line, when the input voltage is 35 V, is: **a.** 34.98 V. **b.** 33.69 V. **c.** 34.63 V. **d.** 32.49 V. **e.** none of these.

7. Refer to the transmission line of Question 6. If the phase constant of the line is 2.356 rad/m, what is the phase angle of the signal at the 786.4 m point relative to the input signal? **a.** −5.502 rad. **b.** −23.89 rad. **c.** −278.94°. **d.** −315.25°. **e.** either a or d. **f.** none of these.
8. The insulation of a certain transmission line has a dielectric constant of 1.395. The velocity of propagation in the line, in meters per second, is approximately: **a.** $267.8(10^6)$. **b.** $278.8(10^6)$. **c.** $248.6(10^6)$. **d.** $254.0(10^6)$. **e.** none of these.
9. If the transmission line of Question 8 is carrying a 450-MHz signal, the wavelength of the signal while on the line is: **a.** 0.6554 m. **b.** 0.5644 m. **c.** 1.772 m. **d.** 0.6667 m. **e.** none of these.
10. The reflected wave on a transmission line has an amplitude of 87.5 V when the incident wave is 97.89 V; the reflection coefficient is: **a.** 0.8939. **b.** 0.8856. **c.** 0.8743. **d.** 0.9052. **e.** none of these.
11. A 50-Ω coaxial line is terminated in a 300-Ω resistance. The line will have a reflection coefficient of approximately: **a.** 1.962. **b.** 0.9834. **c.** 0.8749. **d.** 0.7143. **e.** none of these.
12. The reflection coefficient for a transmission line with a VSWR of 7 is: **a.** 0.6667. **b.** 0.75. **c.** 0.5. **d.** 0.90. **e.** none of these.
13. When looking at a Smith chart, the circles that are all tangent at a common point (the right end of the horizontal centerline of the chart) represent values of: **a.** X. **b.** B. **c.** Z. **d.** R. **e.** none of these.
14. The outside circle of the Smith chart represents an R value of: **a.** infinite ohms. **b.** 0 Ω. **c.** 50 Ω. **d.** 1 Ω. **e.** none of these.
15. The horizontal centerline of the Smith chart represents an X value of: **a.** infinite ohms. **b.** 0 Ω. **c.** 50 Ω. **d.** 1 Ω. **e.** none of these.
16. The normalized value of the impedance $78 + j35$ Ω, when considered in conjunction with a 50-Ω transmission line is: **a.** $1.5 + j0.75$ Ω. **b.** $1.34 + j0.85$ Ω. **c.** $1.56 + j0.70$ Ω. **d.** $1.75 + j2.5$ Ω. **e.** none of these.
17. The impedance at a certain point on an unmatched 75-Ω transmission line is read off a Smith chart as $0.67 - j0.38$ Ω. The actual value of the impedance is: **a.** $50.25 - j28.50$ Ω. **b.** $111.9 - j197.4$ Ω. **c.** $8.933 - j5.067$ Ω. **d.** 75 Ω. **e.** none of these.
18. The normalized impedance $0.8 - j1.3$ Ω is located on a Smith chart in the: **a.** upper right quadrant. **b.** upper left quadrant. **c.** lower left quadrant. **d.** lower right quadrant. **e.** none of these.
19. Refer to Question 18. Assume that the impedance given is the load impedance on a lossless transmission line. If excited, the line will have a VSWR of approximately: **a.** 3.9. **b.** 6. **c.** 8. **d.** 4.1. **e.** none of these.
20. If the line of Question 19 were a 300-Ω line a quarter-wavelength long, its input impedance for the condition stated would be approximately: **a.** $240 + j390$ Ω. **b.** 0 Ω. **c.** infinite ohms. **d.** $105 + j165$ Ω. **e.** none of the above.
21. The distance, in wavelengths, from the load to the first point of voltage minimum (the point where the impedance is a pure resistance and a minimum) for the line of Question 19 is approximately: **a.** 0.338. **b.** 0.162. **c.** 0.09. **d.** 0.25. **e.** none of these.

22. In a slotted-line impedance-measuring setup, the exciting frequency is 1800 MHz. A voltage null (minimum point) is observed to shift 3.5 cm toward the load when the line is shorted. In electrical distance, the null shift is: **a.** 0.543λ. **b.** 0.139λ. **c.** 0.21λ. **d.** 0.3334λ. **e.** none of these.
23. Refer to Question 22. If the VSWR with unknown impedance connected is 4.2, the normalized value of the unknown impedance is approximately: **a.** $0.21 + j0.46\ \Omega$. **b.** $2.1 - j2.0\ \Omega$. **c.** $1.9 + j2.6\ \Omega$. **d.** $0.44 - j0.68\ \Omega$. **e.** none of these.
24. A uniform plane wave is one in which equiphase surfaces are thought of as: **a.** spherical. **b.** flat. **c.** rectangular. **d.** cubical. **e.** none of these.
25. The nature of reflections of EM waves in waveguides is determined in part by the fact that "no difference of potential can exist along the surface of a perfect conductor." This is called: **a.** Ohm's law. **b.** Maxwell's law. **c.** impedance matching. **d.** a boundary condition. **e.** none of these.
26. For successful propagation in a waveguide, one dimension of the guide's cross section, usually referred to as the width of the guide, must be at least equal to: **a.** $\frac{1}{8}\lambda$. **b.** $\frac{1}{4}\lambda$. **c.** $\frac{1}{3}\lambda$. **d.** $\frac{1}{2}\lambda$. **e.** none of these.
27. When a waveguide is excited so that the vector of the electric field component is perpendicular to the direction of propagation, it is operating in the: **a.** FM mode. **b.** TE mode. **c.** half-sine mode. **d.** TM mode. **e.** none of these.
28. The cutoff frequency for a 5-cm waveguide is: **a.** 150 MHz. **b.** 300 MHz. **c.** 1.0 Ghz. **d.** 3.0 GHz. **e.** none of these.
29. Of the velocities associated with waveguide propagation, the one that can be greater than the speed of light is ______________ velocity. **a.** free-space **b.** phase **c.** sonic **d.** group **e.** none of these.
30. The mode of waveguide operation that permits the propagation of the lowest possible frequency for given waveguide dimensions is called the: **a.** TE mode. **b.** TM mode. **c.** dominant mode. **d.** TEM mode. **e.** none of these.
31. A waveguide section that has many of the electrical characteristics of a resonant circuit is called a/an: **a.** resonant cavity. **b.** hollow box. **c.** inductive post. **d.** reflex klystron. **e.** none of these.

REVIEW QUESTIONS: ESSAY

1. What are the two components of the propagation constant of a transmission line? Describe how the propagation constant can be used in predicting the behavior of a particular transmission line.
2. Is the velocity of propagation of electromagnetic energy on a transmission line always equal to the speed of light? If not, when is it different? Is it more or less than the speed of light?
3. Why does the distance-parameter scale on the Smith chart (the outside scale) run from zero to exactly one-half wavelength?
4. Where is VSWR read from or entered on the chart?
5. What is meant by *normalized impedance*? How do you convert from regular to normalized values? from normalized to regular values?
6. Describe the relationship, on a Smith chart, between an impedance and its corresponding admittance value. For example, given one value, how do you find the other?

7. What is a microwave? How is it distinguished from other electromagnetic waves?
8. What is the connection between microwaves and waveguides? Is "Waveguides are practicable only for microwaves" a true statement? If yes, why?
9. What major problem with the transmission of extremely high frequency energy over open, twin-lead transmission lines is solved with the use of waveguide transmission lines?
10. How is a waveguide like a tunnel? Discuss the theory of propagation in a waveguide using an analogy that includes the concept of a tunnel.
11. Describe the boundary condition for the electric field component for successful propagation of an electromagnetic wave in a waveguide.
12. Name and describe two modes of propagation in a waveguide.
13. Define *cutoff frequency* and *cutoff wavelength.*
14. Describe the subscript convention for identifying waveguide propagation modes. Give examples to illustrate the convention.
15. Describe methods of coupling energy into and out of waveguides.
16. Describe methods of achieving relatively narrow beams of microwave radiation.

EXERCISES

1. A 300-Ω lossless line (R and G are negligible) has an L of 2.5 μH/100 m. What is the value of C per 100 m?
2. What is the L/C ratio for a 50-Ω coaxial line? a 75-Ω coaxial line? a 150-Ω twin-lead line? Assume that the lines are lossless.
3. The total resistance of a certain coaxial transmission line is 5.669 Ω/100 m. If the characteristic impedance of the line is 75 Ω, estimate the attenuation factor per 100 m (see Example 1).
4. The phase constant of the line of Exercise 3 is 2.365 rad/100 m at 1.0 MHz. What is the phase constant of the line, per meter, at 500 MHz?
5. A 75-Ω coaxial line feeds a load whose impedance is $50\,\angle 10^\circ$ Ω at 100 MHz. Calculate the voltage reflection coefficient for the condition described. [*Hint:* See Eq. (10.17).]
6. Refer to the line-load condition described in Exercise 5. Use the Smith chart to determine the input impedance of the line, loaded as described, if it is 50 m in length. Assume that it is lossless. (Remember: For a lossless line, the Smith chart repeats itself every $\lambda/2$.)
7. The line of Exercise 5 is loaded with an impedance of $150\,\angle 75^\circ$ Ω (at 90 MHz) and is 45 m long. Use the Smith chart to determine its input impedance when excited by a 90-MHz signal.
8. Convert, using a Smith chart, the following impedance values to their equivalent admittances: **(a)** $45 + j75$ Ω (50-Ω line). **(b)** $90 - j58$ Ω (75-Ω line); **(c)** $250 + j50$ Ω (300-Ω line).
9. Convert, using a Smith chart, the following admittance values to their equivalent impedance values: **(a)** $6.0 - j10.0$ mS (50-Ω line); **(b)** $12 + j15$ mS (75-Ω line); **(c)** $2.78 + j1.84$ mS (300-Ω line).
10. Assume that the impedance values given in Exercise 8 are the impedances of the specified lines at points which are 0.18λ from their respective loads. For each case,

determine what the impedance of the line would be at a point 0.15λ closer to the generator.

11. Assume that the following normalized impedances exist at some point on energized, lossless transmission lines; determine the VSWR for each case: **(a)** $1.0 + j0\ \Omega$; **(b)** $0.5 - j1.3\ \Omega$; **(c)** $1.5 + j2.5\ \Omega$; **(d)** $0 - j0.8\ \Omega$.
12. The antenna for a 158-MHz two-way radio base station has an impedance of $50\ \angle 18^\circ\ \Omega$. A 30-m length of 50-Ω coaxial cable connects the antenna to the transmitter. Use a Smith chart to determine the impedance seen by the transmitter at the input to the transmission line. Estimate what the VSWR on the line might be during a transmission.
13. An antenna is tested with a slotted-line test rig at a frequency of 150 MHz. The VSWR with the impedance connected is 5.4. Upon the shorting of the line at the plane of the load, a voltage minimum shifts 18 cm toward the load. Determine the input impedance of the antenna. The Z_0 for the slotted line is 50 Ω.
14. Design a single-stub transformer that will make the antenna of Exercise 12 appear to be a proper match for the transmission line feeding it. That is, determine the length of the stub and the distance it should be located away from the antenna for the desired match.

11
TELEVISION SYSTEMS

"Television" means, literally, "seeing at a distance." Since "showing" and "seeing" are forms of human communication, electronic television systems are part of the endeavor that we call electronic communication. It is the purpose of this book to present technical information about the "electronics" of electronic communication, and this chapter is devoted to a presentation on the electronics of television systems. However, television is a vast subject. In terms of allocated resources—money, time, talent etc.—television is an endeavor of greater consequence, perhaps, than both broadcast and two-way radio combined. At most, a single chapter can be but a brief introduction to television. As we shall see, however, a television system is, in a very real sense, only a specialized "radio" system. There are many books available which contain comprehensive and detailed information on the many aspects of television for those who have a need and/or desire to know more about it.

11.1 BASIC PRINCIPLES

Let's approach a television system using the technique we have used with other systems. Although we probably all have a reasonably good idea of what a television system is "supposed to do," let's examine that question a little more formally. Then let's look briefly and in general at "how it does what it is supposed to do." Then we will look at a television system a third time in a somewhat more detailed, technical way. Of course, the amount of detail we can examine in the space of one chapter is limited.

What a television system is supposed to do for human beings is to enable them to see at a distance. It continuously reproduces at a remote location a reasonable replica of what may be a "live" scene—a view of a visible "happening" as it happens. The "product" or "result" of a television system, then, is an electronically created visual image which is a replica of a visible object or panoramic scene at a location which may be, and generally is, remote from the actual object or scene.

"How" does a television system produce a replica of a remote visible image? A relatively general, nontechnical explanation of the "how" of television is as follows. A television system (1) continuously converts the light energy of a visible object or scene into an electronic signal, (2) it transmits the signal to a remote location, and (3) reconverts the signal to a visible image.

At this level of explanation, then, the electronics of a television system performs three basic functions. The first and third functions above are completely new in the sense that they have not been touched on previously in this book. The second function, however, is essentially "old hat" to us. If the signal representing the visual image (i.e., the video signal) is transmitted by means of a broadcast RF carrier signal, the process is virtually identical to that of transmitting an audio signal. A vast majority of video transmissions are by this means—RF carrier. Some are not. In some special-purpose systems the signal is carried by coaxial cable. Such systems are called closed-circuit television systems (CCTVs).

A very simple, basic block diagram of a broadcast television system could be identical to that of a broadcast radio system (see Fig. 11.1). It consists of a block representing a transmitter with an antenna and a receiver and antenna. The RF carrier generated by the transmitter is modulated by the video signal; the RF carrier is propagated through the space between the transmitting and receiving antennas; the receiving antenna picks up the RF carrier and passes it to the receiver; the receiver processes the modulated signal, ultimately demodulating it and using the recovered video signal to reproduce the image on a cathode ray tube.

The diagram of Fig. 11.1 and the brief description of it are very general. They provide no information whatsoever concerning the generation of the video signal. To learn more about television systems let's proceed to the examination of a block diagram with more details. See Fig. 11.2, a block diagram of a television station.

One of the important facts to be gleaned from Fig. 11.2 is that a television broadcast station incorporates the equivalent of two RF transmitters—one for the video signal and one for the audio content of a TV program. The signal for the latter is referred to as the *aural signal* in the conventional terminology of the video world. Frequency modulation is utilized in the aural transmitter; a modified form of amplitude-modulated transmission called *vestigial sideband* transmission (A5C emission) is utilized in the video transmitter. We give more information about these details later.

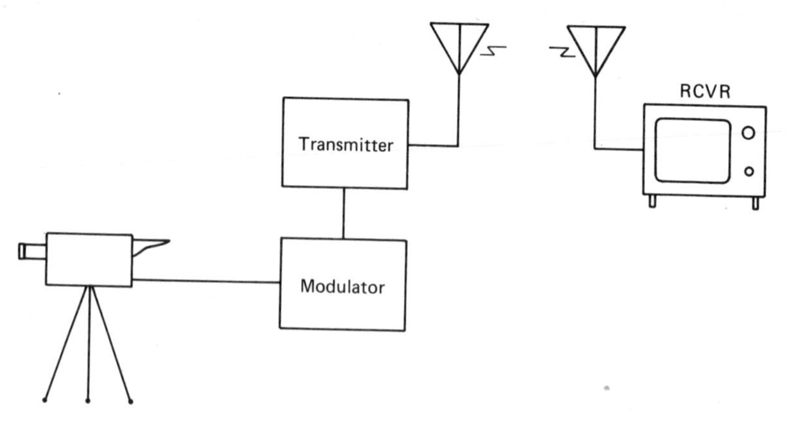

Figure 11.1 Basic television system.

The blocks labeled "scanning timing" and "deflection generators" in Fig. 11.2 provide clues to the process of converting a visual image into an electronic signal. Similarly, the process of the subsequent reconversion of the electronic signal to a visual image is implied by these functions.

A very fundamental description of the generation of a video signal is as follows:

1. The light from a scene or object to be televised is focused by means of an appropriate optical lens to form an image on a special surface in an electronic device called a *camera tube*.

2. The image on the special surface is methodically and repeatedly scanned by an electron beam in the camera tube. Scanning by the electron beam is accomplished by magnetic deflection forces, vertical and horizontal. The deflection forces are the result of currents from vertical and horizontal deflection generators flowing in coils positioned appropriately in close proximity to the camera tube. The scanning process is controlled very precisely by the scanning timing functions. By virtue of the scanning process the

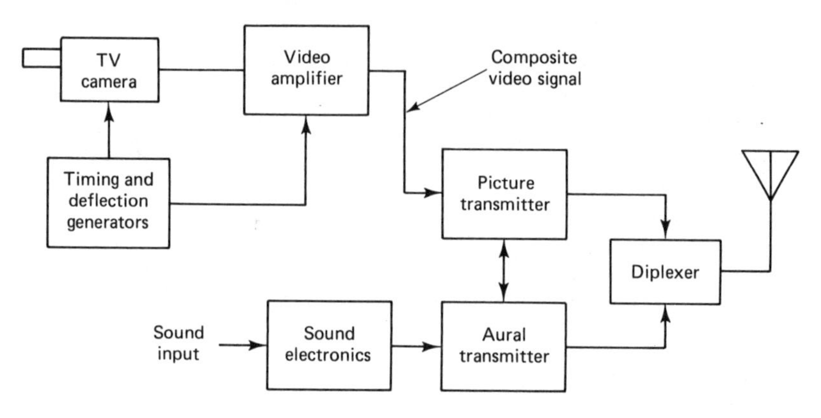

Figure 11.2 Basic functions of television station.

total image is effectively divided into extremely small areas called "picture elements."

3. The interaction of the electron beam with the scanned image alters the intensity of the electron beam; the interaction modulates the electron beam, producing a varying beam current. The variations in the beam current correspond to variations in the light intensity of the scanned image. The resulting modulated current flows through a resistor producing a voltage. This voltage is the video signal since it contains information about the light intensity variations of the visual image. When coordinated with the scanning timing, the video signal variations represent light-intensity information about each of the individual picture elements of the total image.

The reproduction of a televised image on the picture tube of a television receiver is, to a degree at least, the reverse of the process just described. Consider the following fundamental description:

1. The picture tube contains a surface (the screen) coated with a material called a *phosphor* which emits light when struck by an electron beam. The intensity of the light emitted is proportional to the intensity of the electron beam (i.e., the amplitude of the beam current) and the velocity of the electrons in the beam.
2. The viewing surface of the picture tube is methodically and repeatedly scanned by an electron beam. The process produces a set of visible lines called a *raster*. Scanning of the electron beam is accomplished by magnetic deflection: Currents, produced by horizontal and vertical deflection circuits, flow in deflection coils located on the neck of the picture tube.
3. When a television broadcast signal is processed by the receiver, information is contained in the broadcast signal which the receiver uses to precisely coordinate the scanning of the phosphor of the picture tube with the scanning of the image in the camera tube.
4. Simultaneously, the received video signal is used by the receiver to modulate the intensity of the electron beam producing a variation in light intensity on the picture-tube screen. The result: a "light" picture is "painted," picture element by picture element, on the picture-tube screen.

This very fundamental description of the television process is useful in providing a first round of understanding of the system. It is time now, however, for us to look at the system in much greater detail. There are several broadcast systems in use in the world. We confine our attention to the system used in the United States. The discussion in this section will focus on black-and-white (monochrome) systems. Details of color television will be presented in a subsequent section.

The Composite Video Signal

Let's think of a television broadcasting station for the moment as a factory, a factory producing two products. The consumers of the products are all the television receivers that are set to receive the particular station. The products manufactured by the station are a *composite video signal* and a related *aural signal*. The essence of these products is information—the information required to recreate at each receiver the images and sound of the TV program being broadcast. Let's focus on the composite video signal first. Let's examine the signal so as to become familiar with its several components. Then we will look at how it is generated at the station and, finally, how it is used by a receiver in recreating images.

Figure 11.3 is headed "Television Synchronizing Waveform for Monochrome Transmission Only." It is the technical specification by the Federal Communications Commission of the composite video signal for black-and-white (monochrome) television. It is extracted from Part 73 of the FCC Rules and Regulations. The waveform shown is the model that all monochrome television stations are

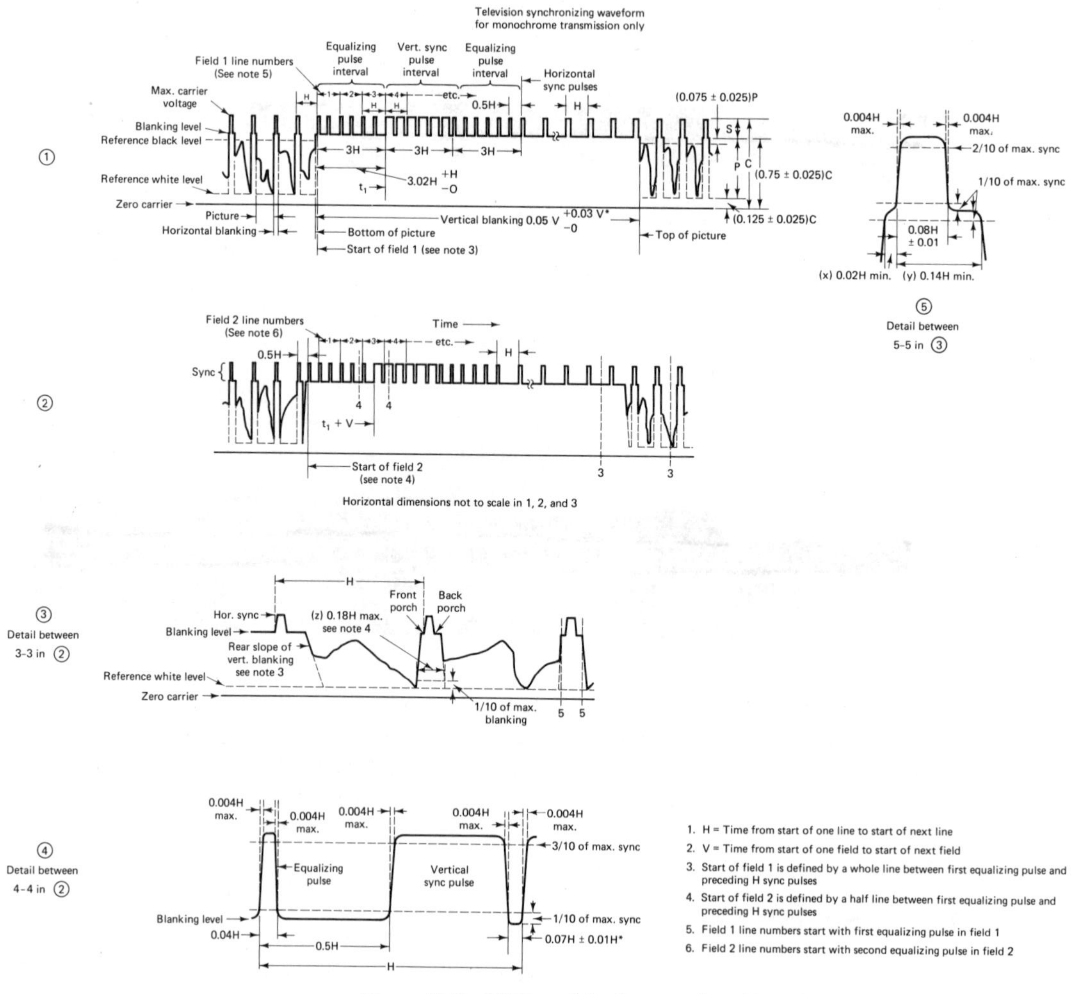

Figure 11.3 FCC model of composite video signal.

expected to emulate in their operation. The specification devotes itself almost completely to the task of *synchronizing* the two scanning processes: the scanning of the projected image in the camera tube and the scanning of the picture-tube phosphor.

A careful analysis of the model composite video signal indicates that it is made up of the following components: (1) pulses for synchronizing horizontal scanning; (2) pulses for synchronizing the vertical scanning process; (3) signal levels for blanking the screen when the electron beam is retracing either between horizontal or vertical scans; (4) "equalizing" pulses for achieving interlaced scanning; and finally, (5) signal levels providing light-intensity information about picture elements (i.e., video information). Do not be dismayed if there is little that you understand from a reading of the preceding statement. You could not possibly understand it unless you already have a knowledge of the details of the television scanning process. Let's turn to those details now and then return to the composite video signal.

As has been indicated previously, a video signal is produced by scanning an image in a camera tube. That signal is used to produce an image by scanning the phosphorescent screen of the picture tube. What constitutes one complete "image?" How many scanning lines are there in one complete image? How many images are there per second? The answer to these questions and a few related questions provide the information we need to know about television scanning.

A *scanning line* is the phenomenon of the electron beam being driven by the deflection mechanism across the face of a tube (either the camera or picture tube) once. When the image on a television receiver is viewed from the normal position, scanning is from the left side to the right side of the screen. Scanning is considered to start at the top left corner of a viewed image and proceed downward. In the U.S. broadcast television system, a complete image incorporates 525 scanning lines. A complete image is called a *frame*. The system broadcasts information for 30 frames each second. However, to improve the quality of the reproduced image, the system utilizes a technique called *interlaced scanning*. In this technique, the 525 lines are divided into two groups of $262\frac{1}{2}$ lines each. The images produced by the two groups are called *fields*. Each group of $262\frac{1}{2}$ lines scans the image from top to bottom, so that each field is in effect a full image, albeit a coarsely scanned one. Figure 11.4 illustrates the concepts of interlaced scanning.

In the context of the picture tube, as soon as the scanning beam reaches the right side of the screen, the system drives it back to the left side. The *retrace*, as it is called, is, of course, at a much greater velocity than the original trace. However, the picture will be better if the beam is turned off, or *blanked*, during the retrace (otherwise, the beam will produce an annoying visible "white" line along the retrace path). A signal equivalent to a blacker-than-black video signal is used for this purpose. The station provides this horizontal blanking level in its transmitted composite video signal. Look at Fig. 11.3 again and find this component labeled on the diagram as *blanking level* and *horizontal blanking*.

Similarly, when the electron beam completes its downward pass for one

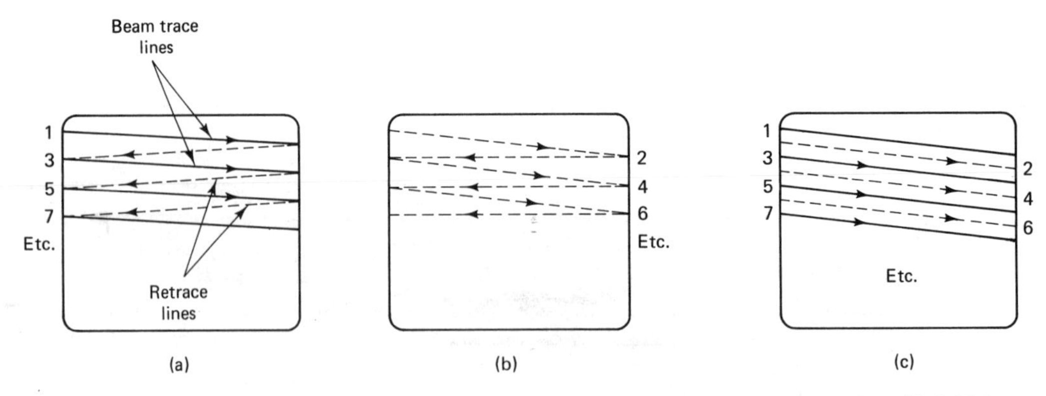

Figure 11.4 Principle of interlaced scanning: (a) field 1, odd-line scanning; (b) field 2, even-line scanning; (c) scanning of complete frame: field 1 plus field 2.

field, it must return to the top of the viewed screen. The vertical retrace "uses" up several horizontal lines in its completion. The beam, if on, actually zigzags from side to side of the screen during vertical retrace. Again, a visible retrace is annoying to the viewer. The composite video signal incorporates a blanking level as part of the *vertical interval* component which can be used by the receiver to blank the vertical retrace. (The vertical retrace, especially, is sometimes seen in TV sets that are not working properly.) Study the diagram of Fig. 11.3 and find the portion of the waveform labeled *(e) vertical blanking*, etc. in the part of the diagram designated ①.

The process of scanning the image in the camera tube and the screen in the picture tube must be coordinated precisely: Corresponding scanning lines in both functions must begin at exactly the same time and relative position and be completed at the same time. Achieving this result is called *synchronization*. A television system achieves synchronization through the use of *horizontal* and *vertical sync pulses*. Sync pulses are rectangular pulses superimposed on top of the blanking pulses—horizontal and vertical—of the composite video signal. Look at Fig. 11.3 another time and find the portions of the waveform with labels such as "sync," "horizontal sync," or "vert. sync pulse interval." Observe the very detailed specification of the horizontal sync pulse in the portion of the diagram designated ⑤. Also, note that in ① the vertical sync pulse has a duration of $3H$. The designation H represents the time from the start of one line to the start of the next line—the time for one complete horizontal trace/retrace. Sync pulses are produced by a pulse generator at the television transmitter. They are used by both the camera unit and the receiver to synchronize the scanning process.

Thus far we have examined these two components of the composite video signal: blanking pulses and sync pulses. Let's next determine what the composite signal provides to assist the system in achieving interlaced scanning. The component that is involved in this feature is a series of special pulses called *equalizing pulses*. In the vertical blanking segment of the portion of the figure labeled ① you will find two equalizing pulse intervals—one before the vertical sync pulse interval and one after it. Each equalizing pulse interval in diagram ① has a duration of $3H$ (three horizontal periods). The pulses are spaced at $0.5H$ (half-line) intervals. Further, the first equalizing pulse occurs at exactly $1H$ after the last horizontal sync pulse to its left on the diagram (see ①). Alternate equalizing pulses are also horizontal sync pulses: they keep the horizontal deflection circuitry "in sync" during the vertical retrace time. You will note that the diagram of portion ① of Fig. 11.3 is labeled "Start of field 1."

Now study the diagram labeled ②. Note that the separation between the last horizontal sync pulse and the first equalizing pulse is only $0.5H$ in this case. Examine the diagram carefully and you will discover that the vertical sync pulse now begins at $3.5H$ after the last horizontal sync pulse, instead of at $3.0H$. The effect of this is to terminate the last line of field 1 at the center of the scan and to begin the first line of field 2 at the top center of the screen. The lines of field 2 interlace or trace between the lines of field 1 (see Fig. 11.4).

Although this system produces only 30 completely scanned images or frames per second, there are 60 fields per second. If there were only 30 "pictures" per second, there would be a noticeable "flicker" to the viewed image on a television receiver. By breaking the 30 images up into 60 fields using interlaced scanning, flicker is effectively eliminated. The fact that there is no apparent flicker with 60 fields per second is the result of a characteristic of human vision. The characteristic is called *persistence of vision*. In effect, our personal viewing systems retain an image for something over $\frac{1}{60}$ of a second. When we view 60 separate images of a given scene per second, our sighting mechanism gives it to us as one persistent image. If the image rate is only of the order of 30 per second or less, the sighting mechanism perceives separate images and the subjective effect is flicker.

The spaces between the horizontal blanking levels in the waveforms of Fig. 11.3 contain the "payload" of the composite video signal. These spaces contain the actual video signal—information about the light levels of the viewed image along each scanned line. Black is represented by a maximum signal amplitude; white is represented by a minimum amplitude. Recall that the blanking level—referred to as "blacker than black"—is at a greater amplitude than any portion of the video signal. Verify that for yourself right now by looking again at Fig. 11.3. Of course, sync pulses also represent blacker-than-black video levels since their amplitudes are greater even than the blanking level. The television receiver utilizes the video signal to modulate the electron beam of the picture tube and thus to paint the "light pictures" which we know as the TV image.

Let's review the description of the composite video signal. A review will help us see the relationships of the various components more clearly and thus help fix the knowledge in the memory bank of our mind:

1. The composite video signal for monochrome television broadcasting in the United States contains five components: blanking levels, horizontal and vertical sync pulses, equalizing pulses, and video signal.
2. Blanking levels, or pulses, occur during horizontal and vertical retrace intervals. They may be used by camera and receiver equipment to blank

(turn off) the scanning beams during the time when these beams are returning to the left side of the image after a horizontal scan, or the top of the image after a vertical scan.

3. Horizontal sync pulses are inserted in the waveform during horizontal retrace intervals. They are used to synchronize horizontal scanning at the camera tube and picture tube. Similarly, vertical sync pulses are inserted in the vertical retrace interval to synchronize the vertical scanning process.
4. Equalizing pulses are inserted in the composite video signal in the vertical retrace interval to coordinate the process of interlaced scanning used in conjunction with the production of 60 fields per second from 30 frames (complete images) per second.
5. The actual video signal is the portion of the composite video signal between horizontal blanking levels. The video signal represents light-level information about the original viewed scene or object. It is produced by the scanning of the projected image in the camera tube. The video signal is used in the television receiver to modulate the electron beam of the picture tube.

Broadcast Television

In broadcast television, the composite video signal is used to modulate a radio-frequency transmitter. The modulated carrier is placed on an antenna and radiated for use by anyone with a television receiver within propagation range of the station's antenna. The form of modulation used is basically amplitude modulation. However, it is a special form of AM. The upper sideband is transmitted complete; the lower sideband is partially eliminated. The part remaining is only a trace or *vestige* of the original sideband. Accordingly, the process is referred to as *vestigial sideband transmission*. The sound for a program is represented in what is referred to as the aural signal. The aural signal is transmitted by a separate frequency-modulated carrier. A full television broadcast signal, then, includes two separate but precisely related carriers—the picture carrier and the sound carrier.

In television broadcasting in the United States, each television station is allocated a band of frequencies called a *channel*. A standard channel is 6 MHz wide. A channel includes both the picture carrier and sound carrier and their side frequencies. A graphical representation of a standard television channel is shown in Fig. 11.5. In the diagram, P represents the picture carrier and S represents the sound carrier. The diagram also includes an indication of the relative location of the *color subcarrier*, C, used in color television broadcasts.

Television channels have been allocated by the FCC in two basic frequency groupings. A listing of the frequencies of the 82 channels allocated is shown in Table 11.1. Channels 2 through 13 are the original television channels. They were allocated when TV broadcasting was just beginning. For several years they were the only frequency assignments available for TV broadcasting. As you can see from the table, the highest frequency in this group is 216 MHz. The entire group is in the range of frequencies designated VHF and, as a result, these channels are referred to as the "VHF channels." Channels 14 through 83

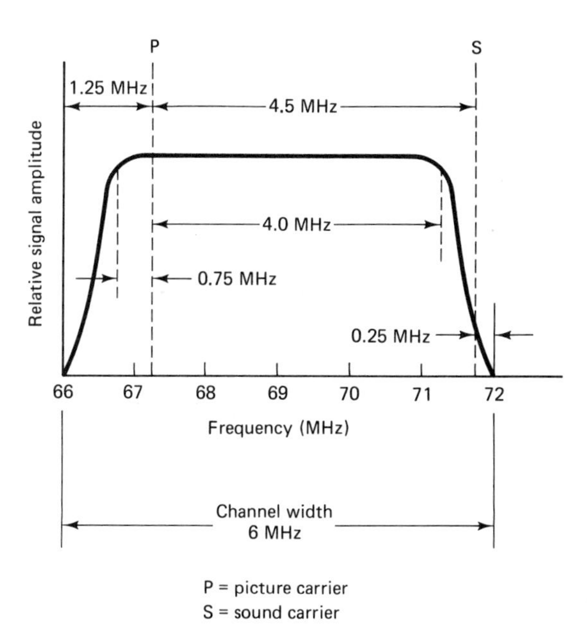

Figure 11.5 Standard broadcast television channel (channel 4 illustrated). P, Picture carrier; S, sound carrier.

have been allocated more recently. Their frequencies are in the UHF range and they are referred to as the "UHF channels."

Let's examine channel 4 as an example to see exactly how the picture and sound carriers fit into an assigned channel. The frequency range for channel 4 is 66 to 72 MHz. Figure 11.5 indicates that the picture carrier is located 1.25 MHz above the lower channel limit. The picture carrier frequency for channel 4, therefore, is $66 + 1.25 = 67.25$ MHz. The sound carrier is 4.5 MHz above the picture carrier, or 0.25 MHz below the upper channel limit. The sound carrier frequency for channel 4 is $67.25 + 4.5 = 72 - 0.25 = 71.75$ MHz.

A carrier, amplitude modulated by a composite video signal, is depicted in Fig. 11.6. The drawing is a representation of what might be viewed on an oscilloscope when the scope's time base is set at a frequency that displays the details of a signal corresponding to only one or a few horizontal scanning lines. Such a "horizontal scanning rate" can be estimated using the details of television scanning:

1. 525 lines per frame $\times$ 30 frames per second $=$ 15,750 lines/s
2. Period for one line $= 1/15{,}750 = 63.49\ \mu s$
3. Total time for viewing three lines $= 3 \times 63.49 = 190.5\ \mu s$
4. A scope sweep-time setting of 20 μs/div gives a total display time of 200 μs, enough time for slightly more than three TV scanning lines.

The amplitude modulation process used by television transmitters is sometimes referred to as "down modulation." This particular description is appropriate because of the fact that when there is no image to be transmitted (i.e., when the image is totally black), the modulation produces maximum carrier amplitude.

TABLE 11-1 FCC TELEVISION CHANNEL ALLOCATIONS[a]

	Freq. (MHz)	Ch. No.	limits of channel (MHz)
			54
P	55.25	2	
S	59.75		
			60
P	61.26	3	
S	65.76		
			66
P	77.25	4	
S	81.75		
			72
			76
P	77.25	5	
S	81.75		
			82
P	83.25	6	
S	87.75		
			88
			174
P	175.25	7	
S	179.75		
			180
P	181.25	8	
S	185.75		
			186
P	187.25	9	
S	191.75		
			192
P	193.25	10	
S	197.75		
			198
P	199.25	11	
S	203.75		
			204
P	205.25	12	
S	209.75		
			210
P	211.25	13	
S	215.75		
			216
			470
P	471.25	14	
S	475.75		
			476
P	477.25	15	
S	481.75		
			482
P	483.25	16	
S	487.75		

	Freq. (MHz)	Ch. No.	limits of channel (MHz)
			488
P	489.25	17	
S	493.75		
			494
P	495.25	18	
S	499.75		
			500
P	501.25	19	
S	505.75		
			506
P	507.25	20	
S	511.75		
			512
P	513.25	21	
S	517.75		
			518
P	519.25	22	
S	523.75		
			524
P	525.25	23	
S	529.75		
			530
P	531.25	24	
S	535.75		
			536
P	537.25	25	
S	541.75		
			542
P	543.25	26	
S	547.75		
			548
P	549.25	27	
S	553.75		
			554
P	555.25	28	
S	559.75		
			560
P	561.25	29	
S	565.75		
			566
P	567.25	30	
S	571.75		
			572
P	573.25	31	
S	577.75		
			578
P	579.25	32	
S	583.75		
			584
P	585.25	33	
S	589.75		
			590
P	591.25	34	
S	595.75		

	Freq. (MHz)	Ch. No.	limits of channel (MHz)
			596
P	597.25	35	
S	601.75		
			602
P	503.25	36	
S	607.75		
			608
P	609.25	37	
S	613.75		
			614
P	615.25	38	
S	619.75		
			620
P	621.25	39	
S	625.75		
			626
P	627.25	40	
S	631.75		
			632
P	633.25	41	
S	637.75		
			638
P	639.25	42	
S	643.75		
			644
P	645.25	43	
S	649.75		
			650
P	651.25	44	
S	655.75		
			656
P	657.25	45	
S	661.75		
			662
P	663.25	46	
S	667.75		
			668
P	669.25	47	
S	673.75		
			674
P	675.25	48	
S	679.75		
			680
P	681.25	49	
S	685.75		
			686
P	687.25	50	
S	691.75		
			692
P	693.25	51	
S	697.75		
			698
P	699.25	52	
S	703.75		

	Freq.	Ch. No.	limits of channel (MHz)
			704
P	705.25	53	
S	709.75		
			710
P	711.25	54	
S	715.75		
			716
P	717.25	55	
S	721.75		
			722
P	723.25	56	
S	727.75		
			728
P	729.25	57	
S	733.75		
			734
P	735.25	58	
S	739.75		
			740
P	741.25	59	
S	745.75		
			746
P	747.25	60	
S	751.75		
			752
P	753.25	61	
S	757.75		
			758
P	759.25	62	
S	763.75		
			764
P	765.25	63	
S	769.75		

	Freq.	Ch. No.	limits of channel (MHz)
			770
P	771.25	64	
S	775.75		
			776
P	777.25	65	
S	781.75		
			782
P	783.25	66	
S	787.75		
			788
P	789.25	67	
S	793.75		
			794
P	795.25	68	
S	799.75		
			800
P	801.25	69	
S	805.75		
			806
P	807.25	70	
S	811.75		
			812
P	813.25	71	
S	817.75		
			818
P	819.25	72	
S	823.75		
			824
P	825.25	73	
S	829.75		
			830
P	831.25	74	
S	835.75		

	Freq.	Ch. No.	limits of channel (MHz)
			836
P	837.25	75	
S	841.75		
			842
P	843.25	76	
S	847.75		
			848
P	849.25	77	
S	853.75		
			854
P	855.25	78	
S	859.75		
			860
P	861.25	79	
S	865.75		
			866
P	867.25	80	
S	871.75		
			872
P	873.25	81	
S	877.75		
			878
P	879.25	82	
S	883.75		
			884
P	885.25	83	
S	889.75		
			890

[a] P, picture carrier frequency (MHz); S, sound carrier frequency (MHz).

The presence of light in the viewed image, on the other hand, results in a reduction of the carrier amplitude—the brighter the picture element, the greater the carrier reduction, hence the descriptive phrase "down modulation."

The Television Sound Signal

Television viewing would be less than satisfactory if we could not hear the sounds that normally accompany the events televised. The signal that a TV receiver uses to reproduce such sounds is provided by a subsystem which is virtually a separate radio station. At a television station the sound or aural signal is obtained from "live" microphones or some recorded source. The audio signal thus obtained is used to frequency modulate the aural transmitter. The frequency of the aural transmitter is always 4.5 MHz higher than the frequency of the video transmitter.

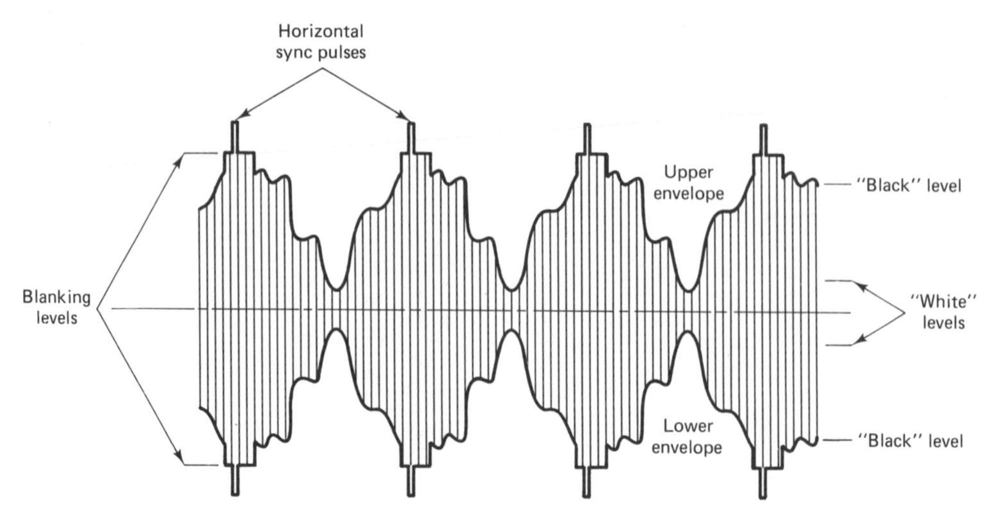

Figure 11.6 Television carrier amplitude modulated by composite video signal.

In TV sound broadcasting, 100% modulation is specified as a deviation of ±25 kHz instead of the ±75 kHz for stations in the broadcast FM service. The audio chain of the aural transmitter is expected to process audio signals of frequencies up to 15 kHz.

The Television Station Antenna

A common antenna is used to radiate both the picture and sound carriers. Coupling to the antenna from the two separate transmitters is by means of a special balanced bridge circuit called a *diplexer*. The function of the diplexer is to permit each signal to excite the antenna but, at the same time, prevent each signal from feeding back into the output circuitry of the other transmitter. The TV station antenna must be designed to be broadband in its characteristics in order to radiate effectively the broadband energy of the combined picture and sound carriers. Furthermore, the TV station antenna in most locations will be omnidirectional in order to serve the population areas that the station desires to reach. The radiated TV signal is horizontally polarized as witnessed by an abundance of horizontally polarized TV receiving antennas in most areas of the world.

11.2 TELEVISION STATION EQUIPMENT

In the preceding section we have had a general, fundamental look at the most essential concepts of a broadcast television system. Let's now look at the transmitting end of the system in greater detail. Please recall our analogy: A TV station is a factory that manufactures a composite video signal and a sound signal. The composite video signal contains (1) information about the light intensity of individual picture elements, and (2) information with which to synchronize the scanning process. Let's look first at the equipment used to produce the picture information: the camera equipment. Practical considerations of space and objectives limit our look to the most fundamental and essential aspects of such equipment.

Essentials of Television Camera Equipment

As could be expected, several different schemes for the conversion of light images into an electronic signal have resulted in several competing forms of equipment. Camera tubes have evolved from the *iconoscope* to the *image orthicon* to the most commonly used tubes today: the *vidicon* and *plumbicon*. However, it is not correct to infer from this statement that each successive tube is simply an improved version of the previously listed tube. The principles of operation of the vidicon, for example, are quite different from those of the image orthicon.

Despite some sharp differences in operating principles, all camera tubes have several features in common. Since each tube utilizes a scanning electron beam, all tubes incorporate a component called an *electron gun* and require an associated deflection mechanism for deflecting the electron beam. The same functions are required by the picture tube of a television receiver. (They are also used by the cathode ray tube of an oscilloscope.)

The purpose of an electron gun is to produce an electron beam—a narrow stream of electrons moving at relatively high speed through a vacuum. An electron gun requires (1) a source of free electrons, (2) a means of accelerating the electrons to the desired speed, and (3) a means of focusing the moving electrons into a narrow beam.

Figure 11.7 is a generalized representation of an electron gun. Free electrons are provided by the *cathode* that is heated by the *filament* to a temperature sufficient to liberate electrons from the metal. Acceleration of the electrons is provided by a high positive potential on an *accelerating grid* between the cathode and the anode. Simultaneously, this stream of electrons is focused into a narrow beam as a result of the electric field produced by the *focusing grid*. The process is called *electrostatic focusing*. The effect is analogous to that of focusing light with optical lenses.

The scanning process as used in television systems, in both the camera and picture tubes, is produced by magnetic deflection of the electron beam.

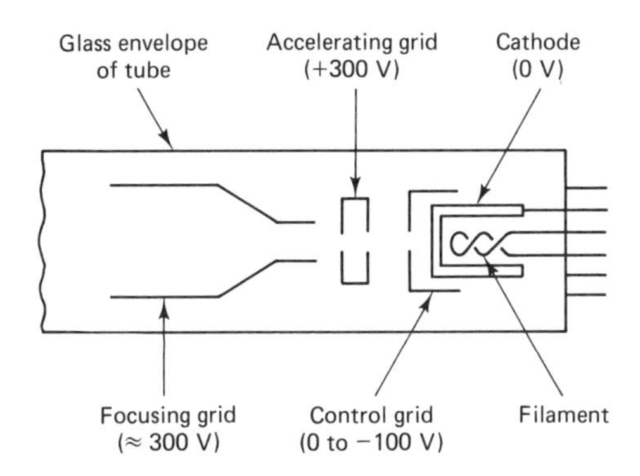

Figure 11.7 Electron-gun fundamentals.

Recall the so-called "motor principle" or motor action which you learned in your study of basic electronics (electricity and magnetism). The idea of motor action is that when a current-carrying conductor is placed in a magnetic field, a force is exerted between the conductor and the field. The force is the result of the interaction between the magnetic field produced by the movement of electrons in the conductor and the existing magnetic field. Electric motors are a practical application of this principle. Loudspeakers also utilize the principle. If an electron beam in a vacuum passes through a magnetic field with at least a component of the motion of the electrons perpendicular to the field, the beam will be deflected. The deflection is the result of the force of motor action.

Electron-beam scanning in either a TV camera tube or picture tube is the process of deflecting the beam from side to side of the image while deflecting it at a slower rate from top to bottom of the image. We examined the exact parameters of scanning in Sec. 11.1: 525 lines per frame, 30 frames per second, etc. A magnetic deflection scheme for television, then, requires two sets of coils placed so as to produce two magnetic fields in the space through which the electron scanning beam passes. A pair of coils placed above and below the tube (camera or picture) would, when supplied by an appropriate current, provide horizontal deflection of the beam. Similarly, a pair of coils on either side of a tube can produce vertical deflection if energized with an appropriate current. A simplified diagram to indicate the basic concepts of magnetic deflection is shown in Fig. 11.8.

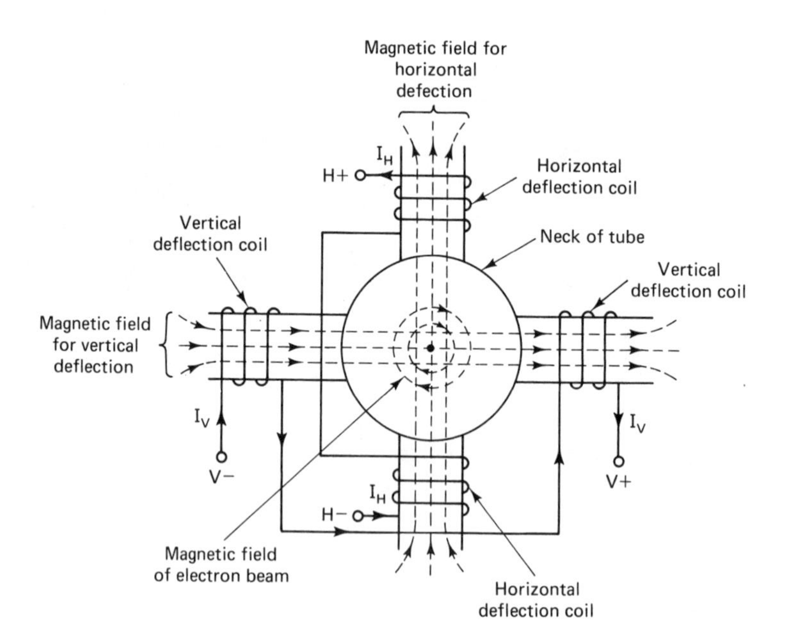

Figure 11.8 Concept of magnetic deflection of electron beam. Pictured are conditions required for deflecting beam down and to the right when electrons are travelling toward the observer.

To produce an undistorted image of a televised object or scene, it is essential that the electron beams in camera and picture tubes be deflected in the scanning process at a constant rate. That is, when scanning horizontally, for example, the beam should proceed across the image at a constant speed. A given displacement of the point where the beam lands on its target requires a corresponding change in the angle of direction of the beam. The amount of force required for a given angular displacement is proportional to the displacement. As the beam moves across the image, the displacement angle increases at a linear rate with respect to time: the displacement force must increase at a linear rate. A linearly increasing magnetic force is produced by a linearly increasing current through a coil. The waveform for the current in a magnetic deflection system, then, is the sawtooth of Fig. 11.9. The basic waveform is the same for either vertical or horizontal deflection in TV systems. Of course, the frequency or repetition rate is different for the two directions: 60 Hz for vertical deflection and 15,750 Hz for horizontal.

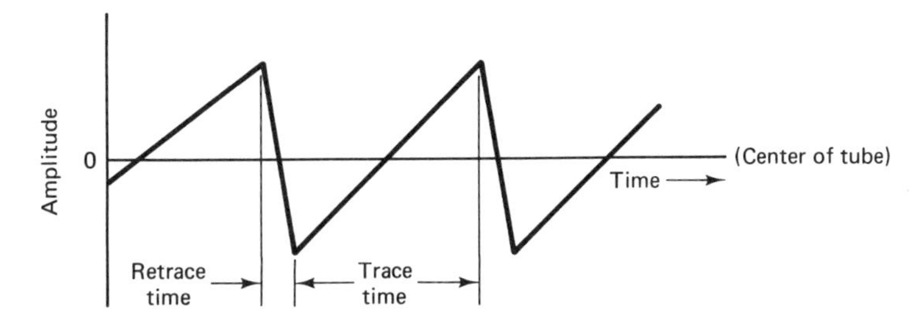

Figure 11.9 Sawtooth waveform of force (current) required for deflection of electron beam in television camera or picture tubes. (Repetition rate = 60 Hz for vertical deflection, 15,750 Hz for horizontal deflection.)

The electronic circuits used to provide the sawtooth currents for camera-tube deflection coils are conventionally named *horizontal* and *vertical deflection generators*. Historically, deflection generators have been forms of circuits called relaxation oscillators or astable multivibrators. Integrated-circuit packages that fulfill the same purpose are referred to as "function generators." Mathematically speaking, a function is a statement of the dependence of one quantity on another. The trigonometric functions—sine, cosine, and tangent—are classic illustrations of the concept of function. A function generator, then, is a circuit that can generate the curves or waveforms of several common mathematical functions: sine wave, square wave, sawtooth, etc.

A sawtooth generator circuit typically makes use of a short segment of the curve of the voltage across a charging capacitor. You will recall the formula for this voltage as

$$v_c = V_{\text{applied}}(1 - e^{-t/RC})$$

The curve for v_c is shown in Fig. 11.10. You will observe that the beginning 10% of the curve, approximately, is quite linear. This portion of the voltage curve can be used effectively as a signal to control the generation of the current to drive deflection coils.

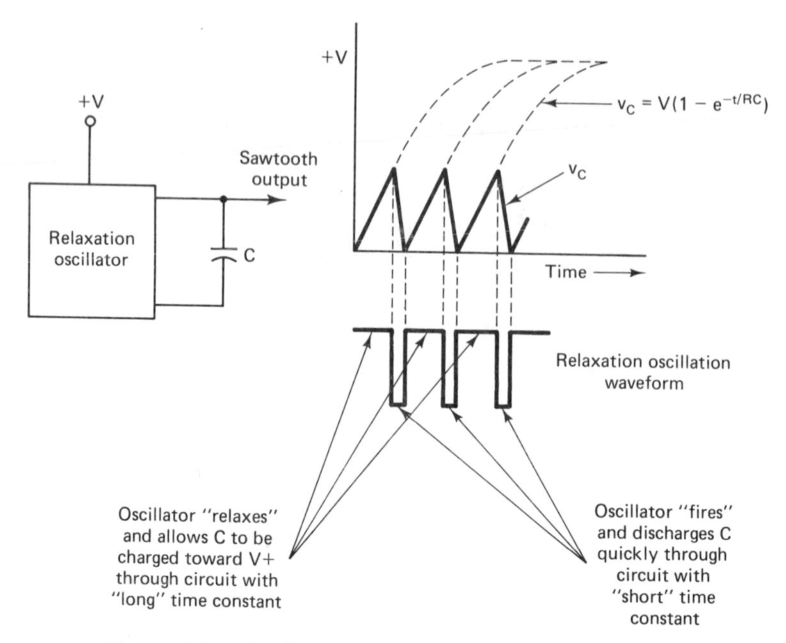

Figure 11.10 Fundamentals of sawtooth waveform generator.

Let's turn our attention now to the process by which a picture signal is generated. The picture or video signal in a monochrome TV system contains information about the relative brightness of each elemental picture area. An elemental picture area can be considered to be equivalent to the cross-sectional area of the scanning beam. A simplified diagram of the essential features of a vidicon camera tube is shown in Fig. 11.11. Observe the presence of the following elements identified on the diagram: electron gun, deflection coils, optical lens, image conversion plate. The electron gun and deflection coils produce the scanning of the image conversion plate by an electron beam, as described above. The image conversion plate converts the light image focused on it by the optical lens to an electrical signal. Details of the construction of the image conversion plate and of the conversion process include the following (see Fig. 11.11):

1. The image conversion plate consists of a sandwich of three very thin films or layers of special materials coated on the inside surface of the end of the glass envelope of the tube. The first layer, the layer on which the light image is focused, is a transparent conductive film. The middle layer is a film of semiconductor photoresistive material. The third layer is a photoconductive material. However, it is a mosaic: It consists of tiny islands or areas of photoconductive material developed on the second layer by a photographic process. Each mosaic island, in conjunction with a corresponding area on the front conductive layer, and the photoresistive material between them, is like a tiny capacitor.
2. The resistivity of the middle layer is very high—of the order of 50 MΩ when there is no light striking it. The conductivity of this second layer

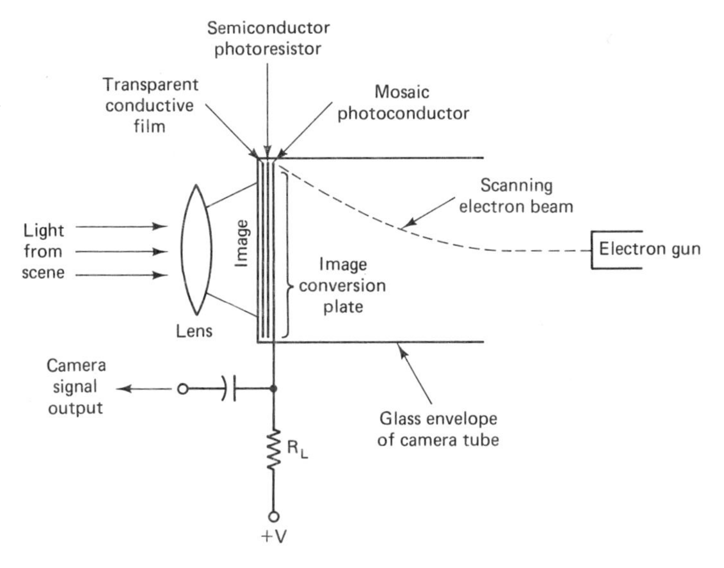

Figure 11.11 Converting a light image into an electrical signal.

increases, however, as it is struck by light from the image on the first layer. Its effect, in the sandwich of layers, is like a dielectric whose leakage characteristic is variable with the light striking it.

3. The scanning process causes the electron beam to "land on" each mosaic island (of the third layer) in turn. The amount of current "leaked" by the middle layer while the beam is on a mosaic island is proportional to the amount of light striking the corresponding area of the first layer. The leakage current flows through the external resistor connected to the first layer and produces a voltage whose amplitude reflects the light intensity of the particular picture element being scanned at that exact instant. The voltage waveform produced by this process is the *video signal.*
4. The electron beam completes the series circuit consisting of the external power supply, the video load resistor R_v, the image conversion plate, and the vidicon cathode.

Television Transmitter Block Diagram

Let's now look at a more detailed block diagram of a monochrome television transmitter (see Fig. 11.12). Observe that the output of the SYNC GENERATOR is both directed to the camera tube and added to the composite video signal for transmission. Sync pulses generated at the transmitter are used by the system to make the scanning precise: Scanning electron beams in the camera tube and picture tubes are positioned at identical locations, relative to the televised image, at every instant. Similarly, blanking information is both used by the camera tube and transmitted. The video signal is amplified for use by the camera monitor and for modulating the transmitter. Deflection generators provide the signals for electron beam scanning in the camera tube.

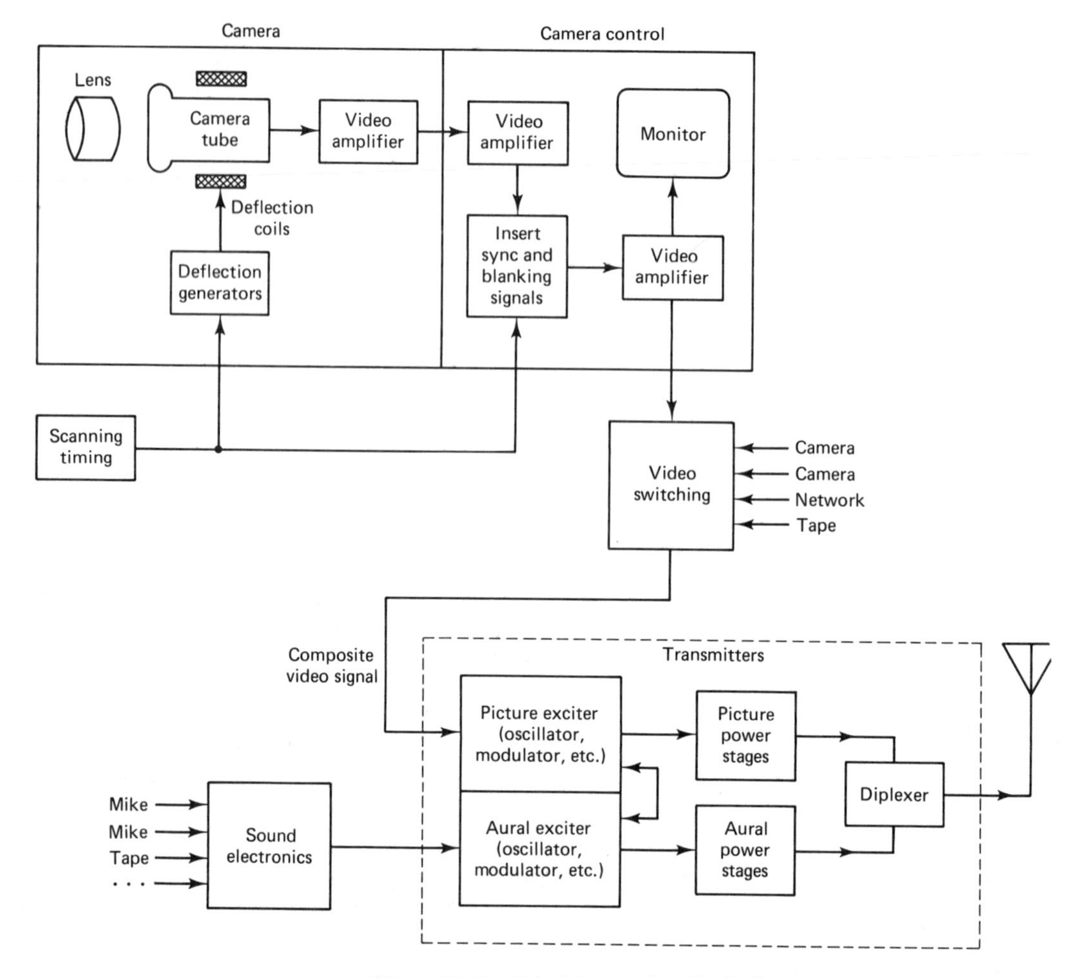

Figure 11.12 Television station block diagram.

The sound transmitter for a television station is a more-or-less conventional FM transmitter. The major difference from a conventional FM broadcast transmitter is that the television sound channel is effectively 50 kHz wide as compared to the 200 kHz of a broadcast FM channel. Also, for television sound a frequency swing of ±25 kHz corresponds to 100% modulation (as opposed to ±75 kHz for broadcast FM).

11.3 THE TELEVISION RECEIVER

The second major subsystem of a broadcast television system is the television receiver. A look at a block diagram showing only major functions of a TV receiver (see Fig. 11.13) reveals that it includes the following: a signal-receiving function, a video-signal-processing function, an audio section, a deflection section, a power supply section, and, of course, the picture tube.

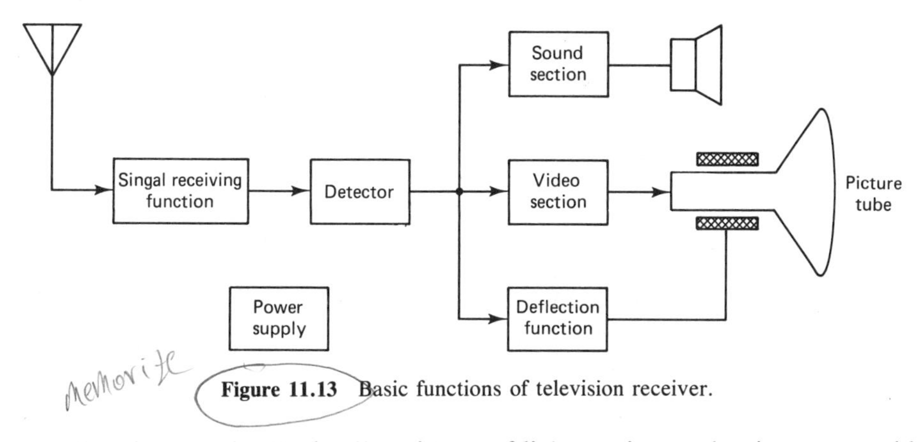

memorite

Figure 11.13 Basic functions of television receiver.

The picture tube "paints" a picture of light, a picture that is a reasonable replica of the original scene or object televised. This picture is painted by scanning the "screen" of the tube with an electron beam. The scanning is synchronized with the scanning of the image on the camera tube. The screen of the picture tube is a coating on the inside of the glass envelope of the tube. The coating is a material called a *phosphor*. The phosphor produces light when struck by electrons traveling at high speed. The amount of light produced at a given spot on the screen is proportional to the intensity of the electron beam (i.e., the beam current) if the speed of the electrons is held constant. The "video" portion of the composite video signal transmitted by the television transmitter is used by the picture tube to control the beam current and thus the intensity of light produced at each elemental picture location. The function of a large portion of the circuitry of a television receiver is to get the video signal to the picture tube and to achieve the synchronized scanning of the electron beam in the tube. Let's see how the various functions of a receiver work together to accomplish this result.

Signal-Receiving Function

The signal-receiving function of a television receiver is very similar to the RF-IF sections of an AM or FM radio receiver. Refer to the block diagram in Fig. 11.14. Television receivers utilize the superheterodyne design for achieving selectivity and sensitivity. As indicated in Fig. 11.14, the RF amplifier and frequency converter sections are packaged in a separate shielded container and called the *TV tuner*. Circuits for processing the VHF channel signals (2 through 13) are packaged as the VHF tuner; circuits for processing the signals of all other channels are packaged as the UHF tuner.

In the VHF tuner, signals in the range 54 to 216 MHz of the 30- to 300-MHz VHF band are first processed by the RF amplifier. The RF amplifier as well as the local oscillator are *gang tuned*. This means that all the relevant circuits for tuning in a particular channel are selected by a switching arrangement that provides a discrete position for each channel.

Numerous ganged switching schemes have been devised. Typically, they

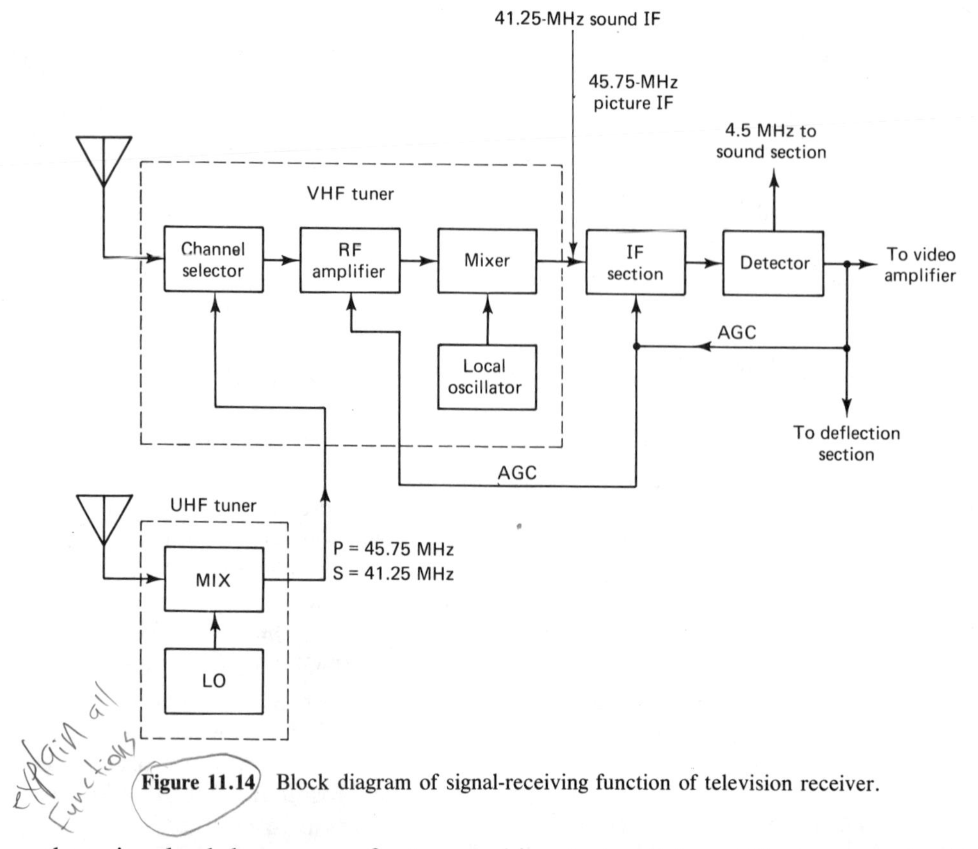

Figure 11.14 Block diagram of signal-receiving function of television receiver.

have involved drum- or wafer-type multicontact switches with the several coils required connected between the appropriate contact points. In other words, the inductances for the resonant circuits for the RF amplifier and the local oscillator for each channel were changed by the operator in making a selection with the channel selector switch. A more recent innovation in channel selection technology makes use of varactor diodes. Separate diodes and appropriate dc voltage circuits are provided for each channel. Channel selection in this case is accomplished by switching dc voltages to the varactor diodes of the desired channel. With the proper voltage, of course, a varactor diode will have the capacitive reactance to resonate a circuit in the RF amplifier or local oscillator at the frequency required to select a desired channel. Switching dc voltages is much less subject to difficulty than is switching RF circuits directly.

The local oscillator of the VHF tuner is designed to operate at a frequency above the desired channel frequency. The output of the LO is combined with the picture and sound carriers in the mixer. The result is a picture IF signal of 45.75 MHz and a sound IF signal of 41.25 MHz. (Note that the sound IF is lower than the picture IF. The sound carrier is always 4.5 MHz higher than the picture carrier. Why is the sound IF the lower of the two IFs?) The output of the frequency converter—the IF signal—is fed to the IF section of the receiver.

The fine-tuning control is an adjustable inductance that can alter the operating frequency of the LO. The amount of frequency change possible is minimal but usually sufficient to achieve optimum tuning of the receiver to the desired channel.

You will note from Fig. 11.14 that the UHF tuner is only a frequency converter. It converts signals in the range 470 to 890 MHz of the 300- to 3000-MHz UHF band to the 45.75 and 41.25-MHz IF frequencies used by the receiver. These are fed to the channel 1 position of the VHF tuner. (Although channel 1 was at one time assigned to TV service, it has since been removed.) At this position the VHF tuner serves as an IF amplifier. The RF amplifier and mixer circuit parameters are changed so as to amplify IF. The VHF local oscillator is turned off.

The IF Section

Like the IF sections of AM and FM radio receivers, the IF section of a television receiver provides most of the selectivity and sensitivity of the receiver. However, the shaping of the passband of the TV IF section is much more critical than that of a radio receiver. The TV IF section is not simply a filter with a 6-MHz passband. However, even achieving a 6-MHz passband would require somewhat different measures than are sufficient for the comparatively narrow passbands of radio receivers. Let's look first at the shape of the ideal passband of a TV IF section and then briefly at how it can be achieved.

A diagram showing the relative gain levels in the IF passband of a TV receiver is provided in Fig. 11.15. You will observe that the 45.75-MHz picture IF signal is not at the center of the passband. Rather, it is at a 50% of maximum gain level on the higher-frequency skirt of the response curve. To understand more easily the reason for this seemingly strange characteristic of the passband, recall that the TV picture carrier is transmitted in a vestigial sideband mode (see Fig. 11.5). Both upper and lower side frequencies up to about 0.75 MHz on

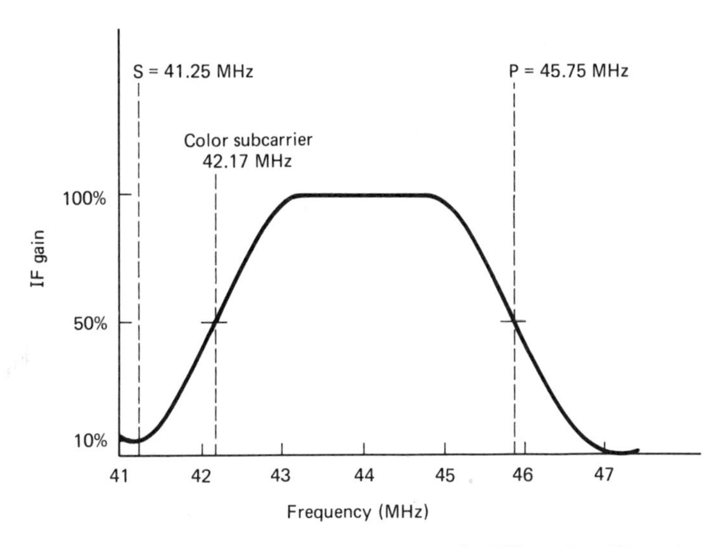

Figure 11.15 Ideal frequency response curve for TV receiver IF section.

either side of the picture carrier are transmitted. The energy levels of these frequencies in the transmitted signal is greater, then, than the energy levels of side frequencies further removed from the carrier. The result: frequencies up to 0.75 MHz on either side of the picture IF signal in the TV receiver need less amplification than other side frequencies.

Observe the 50% gain level marked on the lower-frequency skirt of the IF response curve of Fig. 11.15. It is labeled "color subcarrier—42.17 MHz." In color TV broadcasting (about which, more later) the information needed by a receiver to reconstruct a color picture is "carried" by a subcarrier. The color subcarrier is removed from the picture carrier by 3.58 MHz. You will recall the principle of a subcarrier from the presentation on the SCA subcarrier in FM broadcasting (Chap. 8). The location of the color subcarrier was determined by careful system design.

The sound IF signal is also processed by the receiver's IF section. However, because the sound signal is less vulnerable to the effects of spurious noise signals, the IF gain requirement for the sound signal is less. IF gain for the sound frequency (41.25 MHz) is typically about 10% of the maximum gain of the IF section. When a TV receiver is designed to process the sound carrier through the same IF section as the picture signal, as just described, it is said to use *intercarrier sound.* Intercarrier sound processing in a receiver is possible because of the precise 4.5-MHz frequency difference (intercarrier separation) between the transmitted picture and sound carrier frequencies.

Achieving a passband response curve approaching that of the ideal of Fig. 11.15 requires numerous tuned circuits. The use of a number of adjustable circuits make it possible, with careful alignment, to spread the gain of an IF section over a comparatively wide passband. The concept is called *stagger tuning*: Each of a number of resonant circuits is adjusted to resonance at each of several frequencies within the desired band. The overall result—a summing of the individual effects—is a response curve of the desired shape. Wave traps—resonant circuits that attenuate the signal at certain frequencies—are also used to help shape the skirts of the response curve.

The process of manually adjusting the several tuned circuits in the IF section of a TV receiver is called IF alignment. It is a difficult task to accomplish without the aid of special equipment. The most common and effective approach utilizes a *sweep generator.* A sweep generator is a signal generator whose output frequency can be frequency modulated or "swept" over a range of frequencies. For TV IF alignment, a sweep generator with a basic operating frequency in the range 40 to 50 MHz and a sweep width of 7 to 10 MHz is required.

A diagram illustrating the test equipment setup for sweep alignment is shown in Fig. 11.16. The swept-frequency signal is applied to the input of an operating TV IF section. The output of the IF section is AM detected, either by the receiver's video detector or a separate diode demodulator probe. The detected signal is applied to the vertical input of an oscilloscope. The signal that sweeps the frequency of the signal generator is applied to the horizontal input of the scope and is used for its horizontal sweep drive. The result is a visual display of the response curve of the IF section: The amount of vertical deflection at

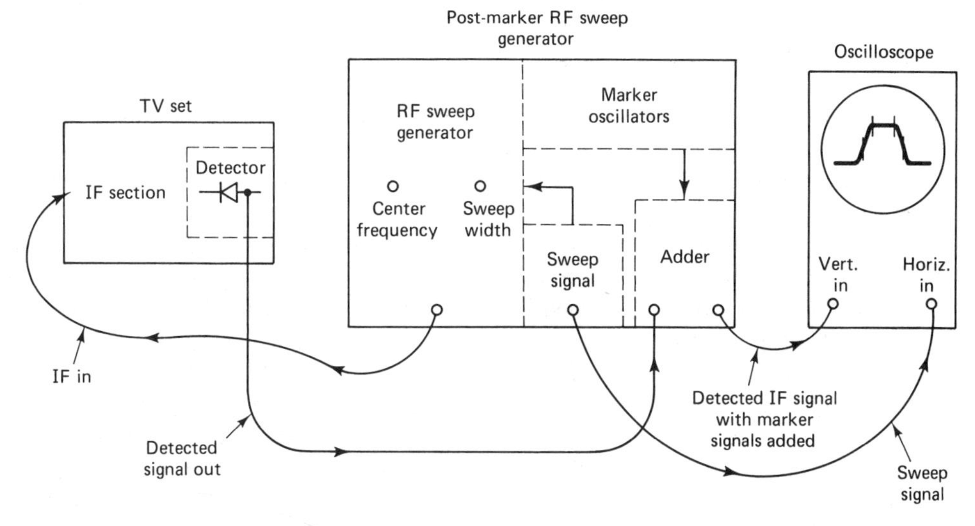

Figure 11.16 Test equipment setup for visual alignment of TV receiver tuned circuits.

any given point on the display correlates with the frequency of the signal generator at that particular instant. Anyone adjusting the circuits of an IF section can see immediately the result of any adjustment. The passband can be "shaped" to resemble the model shape. This technique is often called *visual alignment*.

An essential part of the "front end" of any TV receiver is an automatic gain control (AGC) function. Generally, both the RF amplifier and the IF section receive automatic gain control. In principle, the AGC function in a television receiver is exactly analogous and similar to the AGC functions in radio receivers—AM and FM. However, because the need for AGC in a TV receiver is even more critical to acceptable performance than in a radio, the AGC system is typically more sophisticated. The AGC voltage is often amplified; and the AGC system may utilize a technique called "delayed AGC." In delayed AGC, a reduction in gain is applied to the RF section only if a reduction of gain in the IF section is not sufficient to reduce a received signal to a predetermined amplitude. Delayed AGC improves the performance of a TV receiver by holding noise to a minimum. AGC voltage is usually obtained from the output of the video detector. This voltage, as obtained, is a function of the blanking level of the composite video signal; its amplitude is not influenced by the variations of the actual picture signal.

The Video Detector

As we have observed previously, the picture signal in broadcast television systems is transmitted by the amplitude modulation of an RF carrier. Recovering that signal requires an AM detector. Thus video detector circuits typically utilize a diode for the same purpose as for the diode in an AM radio receiver detector.

The polarity of the diode in a video detector is of critical significance. You will recall that the composite video signal "manufactured" and transmitted by the television station contains deflection sync signals and blanking levels. It is essential that the blanking level of the video signal reach the control grid of the picture tube with a negative polarity. A strongly negative blanking level on the control grid cuts off the electron beam during the blanking periods. The polarity of the video detector diode and the number of stages in the video amplifier section must be coordinated to achieve a negative blanking level at the picture tube.

In monochrome television receivers the video detector is typically used in the process of separating the sound signal from the video signal. Since the detector circuit is a nonlinear circuit, sum and difference frequencies are produced in its output. The difference between the picture IF signal and the sound intercarrier IF is 4.5 MHz—45.75 MHz − 41.25 MHz. This 4.5-MHz signal is one of the products of detection. It is tapped off from the video detector output and fed to the sound IF section, which is designed to operate at 4.5 MHz.

Since the output of the video detector is the recovered composite video signal, it contains sync signal information as well as the picture signal and the 4.5-MHz sound carrier. A sample of the video detector output is sent to the deflection section, where a *sync separator* function will retrieve the sync signals.

The Video Section

The video section of a TV receiver contains a special-purpose amplifier—a *video amplifier*—to increase the energy level of the picture signal, and, of course, the picture tube itself. See the block diagram of Fig. 11.17. The term "video amplifier" is often used to describe a class of amplifiers. In this sense the most telling characteristic of a video amplifier is its extreme bandwidth—extreme in comparison to other classes of amplifiers. For example, the amplifiers used to process the signal for the vertical deflection of an oscilloscope can have a bandwidth of 0 to 100 MHz. These are obviously "video amplifiers." A high-quality video amplifier in a TV receiver may have a bandwidth specified as 0 to 4 MHz.

When an amplifier's low-frequency cutoff is 0, the amplifier must be direct coupled. It must be capable of preserving/amplifying dc levels as well as amplifying

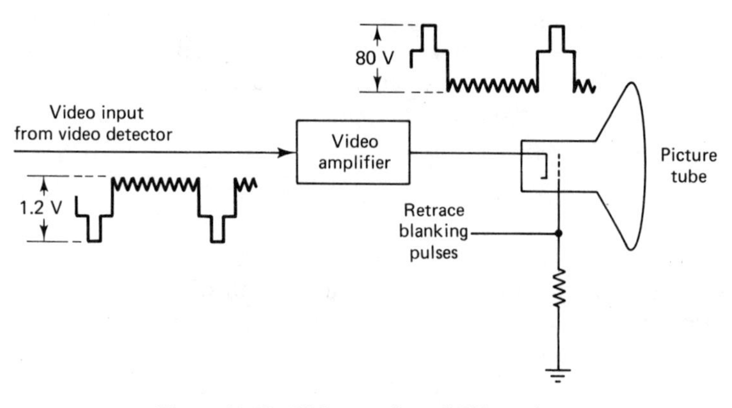

Figure 11.17 Video section of TV receiver.

ac signals. Therefore, there must be no capacitance in series with the signal path. Many TV receiver designs do use direct-coupled video amplification. An advantage of direct coupling is that the receiver can capture and pass along the exact dc levels of the picture signal that were generated and transmitted by the TV station. The dc levels of a picture signal are involved in determining the overall brightness level of the final picture.

A disadvantage of direct-coupled amplifiers is that their design is more sophisticated. And more sophisticated design generally equates with more expensive design. More engineering design time is required. More components are required.

Video amplifiers may and some designs do use ac coupling which does not preserve the transmitted dc levels. A special function called a *dc restorer* circuit is required in such designs to provide a necessary dc level for the picture signal at the picture tube control grid.

As already indicated, video amplifiers are characterized by a wide passband. The low-frequency response limit can be extended, to 0 Hz if desired, by the use of direct coupling. Extending the high-frequency limit of a basic amplifier is also required to make it a video amplifier. As a natural phenomenon, the gain of an amplifier inevitably begins to decrease with higher frequencies. This effect is the result of the decrease of the capacitive reactance which shunts the signal path in any amplifier. The capacitance in question is the inherent capacitance between signal-carrying conductors and ground, for example.

In general, increasing or at least preserving the gain of an amplifier at higher frequencies can be accomplished by increasing the inductive reactance component of the load circuit of an amplifier stage. If chosen properly, the series inductive reactance can balance out the effect of the lowered shunt capacitive reactance. The inductive reactance, of course, increases with frequency while the capacitive reactance decreases. Inductances (coils) used in this fashion are called *peaking coils* because they "peak" the gain at higher frequencies.

The video section brings the broadcast picture signal to the control grid of the picture tube. This element, like the control grid of any electron tube, controls the quantity of electrons that will be allowed to leave the cathode of the tube and reach its plate. The equivalent "plate" of a picture tube is its phosphor-coated viewing surface. By scanning this surface with an electron beam, using the picture signal to control the intensity of the beam, the picture tube paints a light picture on its viewing surface. Let's postpone looking at this process in detail until after we have examined the deflection section of the receiver.

Deflection Section

The deflection section of a television receiver generates the voltages/currents which are required to deflect the electron beam of the picture tube. See the block diagram of Fig. 11.18. Precise deflection—both vertical and horizontal—achieves the scanning of the viewing area by the tube's electron beam. When this scanning is synchronized, through the medium of the composite video signal, with the scanning of the image on the camera tube, a replica of that image is produced on the picture tube's screen. Thus the deflection section also processes

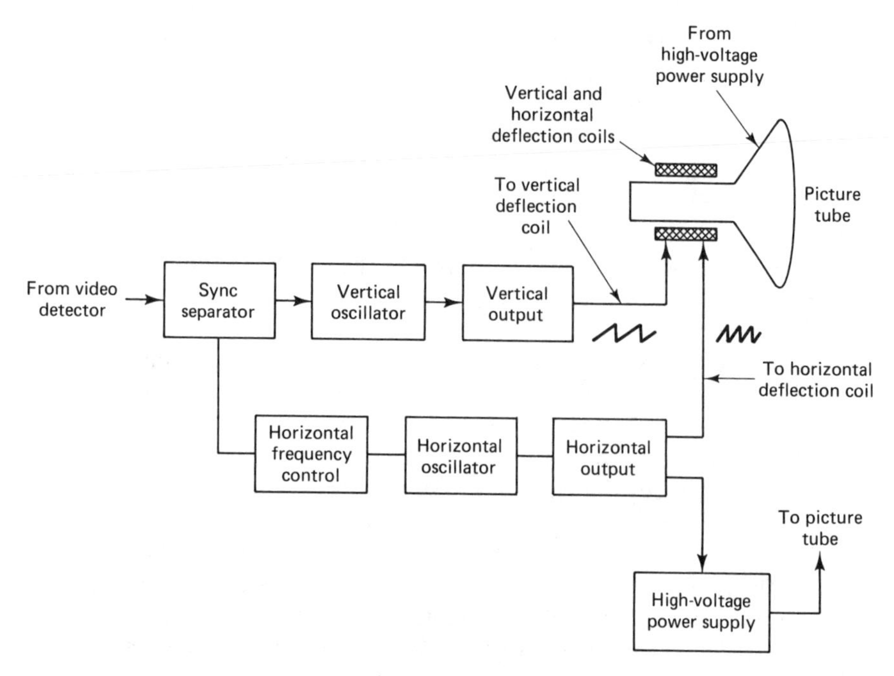

Figure 11.18 Deflection section of TV receiver.

the sync signals recovered from the transmitted composite video signal, and it uses them to synchronize the scanning process at the picture tube.

Deflection of the electron beam in a picture tube is different from the deflection of the beam in a camera-tube primary only in the amount of deflection required. For example, the diameter of a vidicon tube is of the order of 2.5 cm (1 in.). Television picture tubes with a diagonal measurement of 27 in. are very common. Magnetic deflection is used with picture tubes as well as with camera tubes. Greater deflection requires stronger magnetic fields. Stronger magnetic fields imply either more turns per deflection coil or more deflection current, or both.

The deflection force, and therefore the deflection current, must have the sawtooth pattern described in connection with camera-tube deflection. The repetition rate (pulses per second) for vertical deflection is 60 Hz. This corresponds to the number of fields per second. The horizontal repetition rate is 15,750 Hz, corresponding to the number of lines per second. Again, as with the camera-tube deflection generators, the generating circuits can be any of a number of different types of relaxation oscillators. These circuits are required to have asymmetrical operating characteristics: a short on-time, corresponding to the retrace time of the scanning beam, and a comparatively long off-time or relaxation time, corresponding to beam trace time.

When a TV receiver is operating but not receiving a transmitted signal, it produces a simple pattern or grid of bright trace lines on the viewing screen of the picture tube. The pattern is called a *raster*. The deflection of the beam to produce the raster is a result of the deflection generators operating in a *free-*

running mode. That is, their repetition rates are determined by their design: the selection of components that determines operating frequency. Generally, the free-running frequencies are slightly lower than the frequencies that will attain when the generators are synced with the transmitter deflection generators. As soon as the receiver is adjusted to reproduce a televised picture, the deflection generators must be controlled to operate in precise synchronism with the scanning process at the TV station. The result of the lack of synchronism is a picture that "rolls" vertically or from side to side. It may be worse: it may exhibit "tearing," be a series of dark and white diagonal bands, etc.

Deflection synchronism is achieved through the use of the sync signals transmitted by a TV station as part of the composite video signal. A sample of the detected composite video signal is passed, usually after amplification, to a set of *sync separators*. The essential part of the vertical sync separator is a low-pass filter, generally the equivalent of a series R and a shunt C. This configuration is also called an *integrating circuit*. The integrating circuit integrates (literally, adds up) the energy in the serrated pulse components of the vertical sync pulses (study Fig. 11.3 again at this point). This integrated pulse is fed to the vertical deflection oscillator, where it "triggers" that oscillator into synchronism with the signal from the station.

Horizontal sync pulses are recovered from the composite video signal by the horizontal sync separator. Horizontal sync pulses are much shorter in duration and occur at a much higher repetition rate than the vertical pulses. The horizontal sync separator uses a technique that is functionally the inverse of the vertical separator. Its essential circuit is a *differentiating circuit*—a circuit that detects the edges of the pulses in the composite waveform. It consists of a series C and shunt R. At a pulse edge, the capacitor attempts to charge, producing a voltage spike across R. The horizontal deflection section is typically a phase-locked loop (see Chap. 8). The voltage spike from the horizontal sync separator is used as the external reference signal in the loop's phase detector. When operating correctly, the horizontal deflection frequency and phase is "locked" precisely to the phase and frequency of the horizontal deflection generator at the TV station.

The output stages of the vertical and horizontal deflection sections drive the associated deflection coils. That is, the output stages force currents with sawtooth waveforms through the coils of the deflection yoke, which is mounted on the neck of the picture tube. The amount of energy required to achieve the desired deflection is significant. Horizontal deflection requires the greater amount of energy because much greater acceleration of the electron beam, and, therefore, a higher current is required to achieve the 15,750 lines per second versus the 60 vertical fields per second. Electronically speaking, the deflection output stages are power stages.

The Picture Tube

We have looked briefly at the functions of deflecting the electron beam in the picture tube and of getting the picture signal to that final element in the picture chain. It is time now to summarize the process of producing the image on the

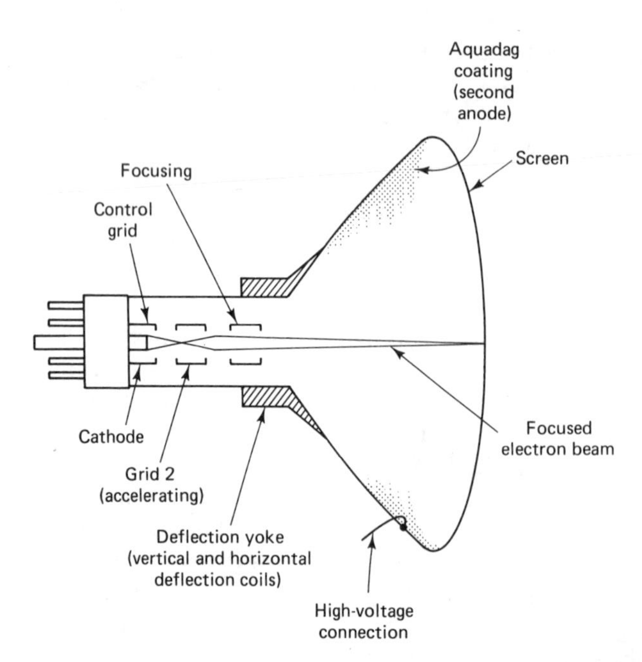

Figure 11.19 Essential details of monochrome picture tube.

viewing screen. Figure 11.19 shows the essential details of a monochrome picture tube.

The picture tube produces an electron beam—a stream of electrons confined to a narrow cross section—by means of an electron gun in conjunction with a high-voltage accelerating potential. The beam is deflected from top to bottom and from side to side by a deflection system. The deflection system consists of horizontal and vertical deflection generators which drive current through the horizontal and vertical deflection coils. These coils make up the deflection yoke mounted on the neck of the picture tube. In the absence of a television signal, the electron beam/deflection system produces a set of white horizontal lines on the viewing screen: the raster. When the receiver system is processing a television signal from a station, the video or picture component of the signal is applied to the control grid of the picture tube. The signal on the control grid modulates the intensity of the electron beam, thereby producing elemental picture areas on the viewing screen which are white, black, or intermediate levels of gray. The overall system—transmitter and receiver—is synchronized with respect to deflection and the video signal. The result is that the light-level-modulated picture elements on the picture-tube screen are precise replicas of corresponding picture elements on the camera-tube input screen. The relationship between camera- and picture-tube picture elements is correct both with respect to locations on the screens and relative light intensities: The system produces an image on the picture tube which is a replica of the object or scene viewed by the camera tube.

The Sound Section

The sound section of a television receiver has many of the same functions as an FM radio receiver (see Fig. 11.20). You will recall that transmitted picture

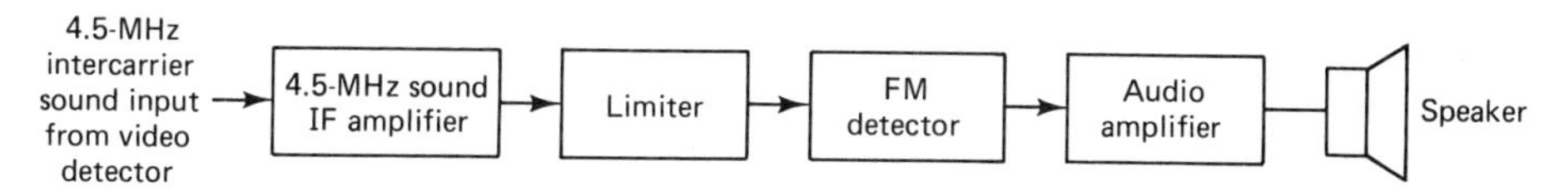

Figure 11.20 Sound section, TV receiver.

and sound carriers have an intercarrier spacing of 4.5 MHz. With both carriers present in and processed by the receiver's IF section, it is not difficult to obtain a 4.5-MHz intercarrier sound signal. In monochrome receivers, a portion of the output of the video detector output is fed to the sound section. The 4.5-MHz intercarrier sound signal is one of the components of the output of the video detector. It is produced in the detection process. The sound and picture carriers heterodyne in the nonlinear detector circuit, producing the difference frequency—4.5 MHz—among other components.

The sound sections of TV receivers generally incorporate an integrated-circuit chip containing functions for amplification and detection. In fact, chips are available that contain both TV sound IF amplification and detection and audio amplification. For example, TDA1190 is a general industry designation for a chip of this type. A block diagram of a CA1190Q, the RCA version of the chip, with typical external circuit connections is shown in Fig. 11.21. You will note that the chip contains IF amplification, limiting and filtering, FM detection, and audio power amplification. The FM detector is described as a differential peak detector and requires only one tuned coil.

The Receiver Power Supply

A television receiver has both typical and unique requirements for dc voltages. Most of a modern TV receiver's circuits are solid state. That is, they are either in the form of integrated circuits or contain discrete transistors. Dc voltage requirements for these circuits are modest: 12 V, 24 V, and 48 V are typical.

A TV receiver's unique power supply requirement is the high dc voltage required as the accelerating potential for the picture tube's electron beam. This potential is of the order of 12,000 V for the typical monochrome set and up to 34 kV for large-screen color receivers. Every TV receiver incorporates what is termed a "high-voltage power supply."

Let's examine in a general way the operation of a typical TV high-voltage power supply. The operation includes several interesting features. As a beginning, a couple of general statements are helpful: (1) A typical TV high-voltage power supply is an effective energy recycler; it recycles the energy stored in the magnetic field of the horizontal deflection coil. (2) It is a form of switching power supply. The combination block-schematic diagram of Fig. 11.22 depicts the salient features of this circuit.

At the heart of a high-voltage power supply is the horizontal output transformer. This transformer interfaces (connects, with appropriate impedance matching) the horizontal deflection coil to the horizontal output stage. This transformer is also referred to as the *flyback transformer*. We shall see why in

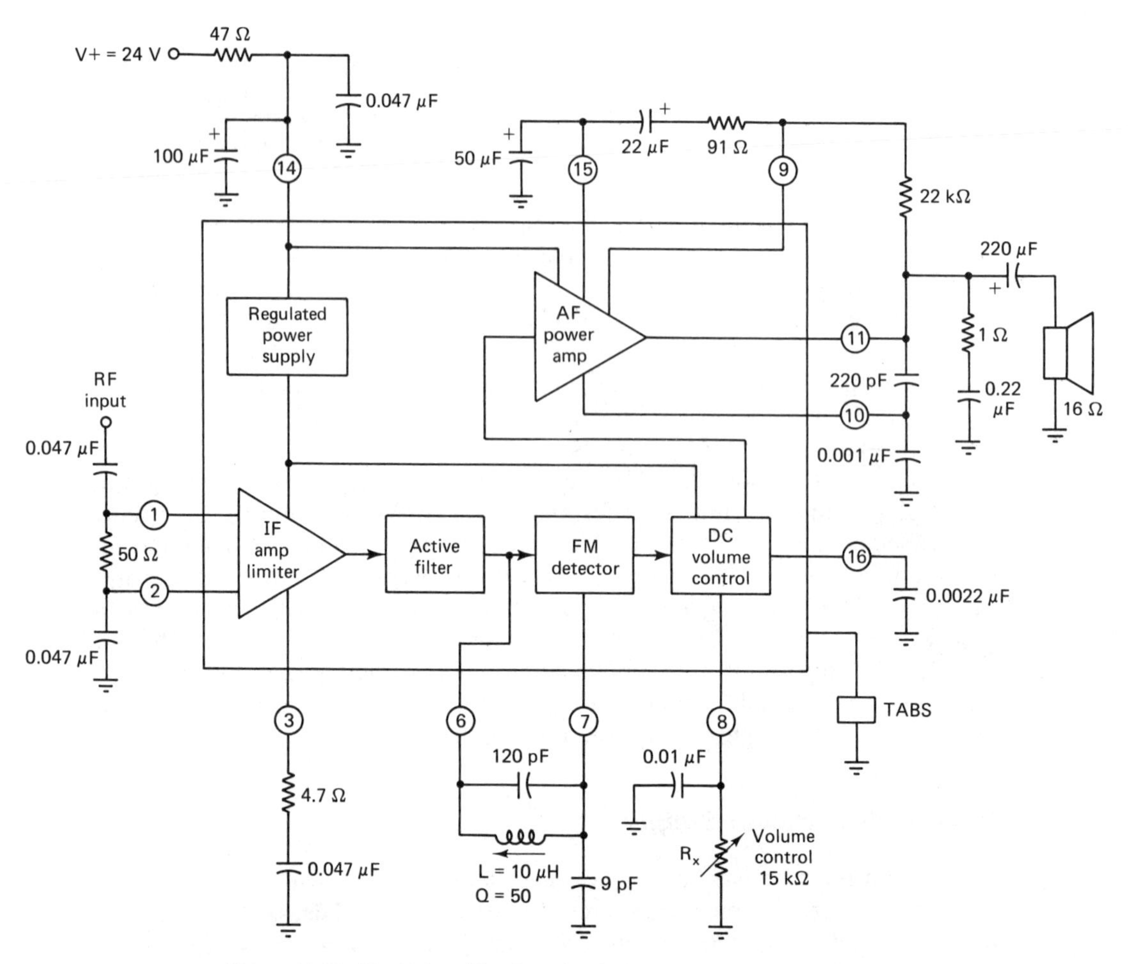

Figure 11.21 Typical application circuit for CA1190Q TV sound IF and audio output subsystem chip. (Courtesy of RCA Corporation.)

a moment. Let's consider the sequence of events in this process of the horizontal deflection of the picture-tube electron beam:

1. The current driven through the horizontal deflection coil by the horizontal output stage gradually increases until it is at a maximum when the beam reaches the limit of its excursion—the right side of the screen. The intensity of the magnetic field of the deflection coil is also at its maximum and represents a very large quantity of stored energy.
2. The beam starts its rapid return (flyback) to the opposite side of the screen at the instant the horizontal output stage is cut off, stopping the increase of deflection current. The flyback phenomenon is actually a half-cycle of damped oscillation of a resonant circuit composed of the horizontal deflection coil and transformer and the capacitance in that circuit—stray and added. The resonant frequency of that circuit is designed to give a half-period

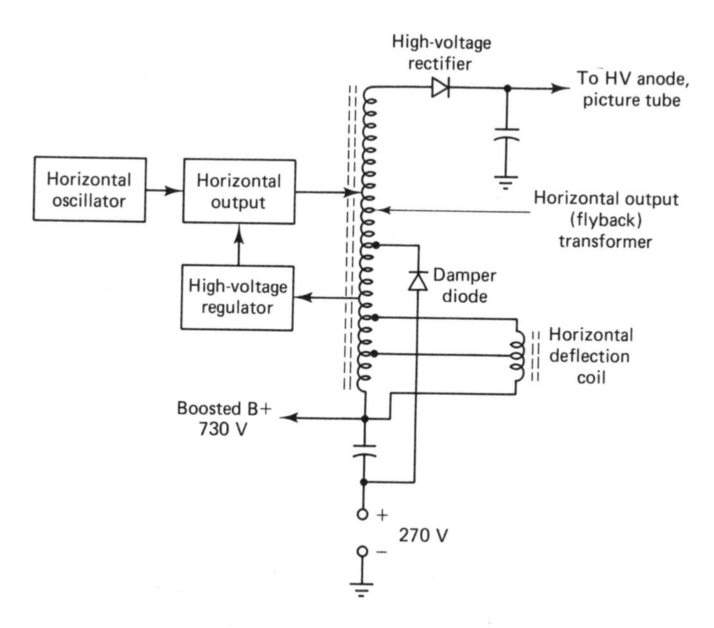

Figure 11.22 High-voltage power supply—TV receiver.

approximating the desired flyback time (horizontal retrace time). The rate of change of the current in the flyback transformer is very high during flyback time. A high voltage is induced in the high-voltage winding.

3. Once the electron beam has reached the left side of the screen, the oscillation phenomenon of the deflection circuitry is controlled by the damping diode. The effect of the damper is to provide linear deflection of the electron beam, using recycled energy, to approximately the center of the picture-tube screen. The horizontal circuit picks up the process at this point and completes the deflection to the right side of the screen.

The flyback process, then, results in a high voltage being induced in the flyback or horizontal output transformer. The high-voltage power supply includes this high-voltage winding, a high-voltage rectifier, and a high-voltage regulator function. A well-regulated high voltage for the picture tube is essential for the production of a good-quality television image. The brightness of the image is a function of the accelerating potential as well as the intensity of the electron beam. A fluctuating high voltage can cause a fluctuation in image brightness. It can also cause image size and focus problems. (The so-called "blooming" of a television image is the result of a weak high voltage. The weak high voltage is, in turn, typically caused by an overload on the high-voltage supply.)

The typical high-voltage power supply in television receivers can legitimately be termed a switching power supply for the following reasons: (1) It receives dc energy from a more conventional power supply in the receiver. (2) It produces high voltage in a transformer winding by virtue of a high current change rate controlled by an electronic switching operation.

This concludes our consideration of the monochrome television receiver. We look at color television in the next section. We will be examining modifications in the television system required for the reproduction of images in color.

11.4 COLOR TELEVISION

A full-color television image on a color picture tube is, in a very real sense, the result of superimposing three monochrome images of red, blue, and green. The task of a color television system is to provide information in electronic form to a color picture tube for generating these three images. In short, a color television system is a modified monochrome TV system that produces images in triplicate.

The Color Television Camera

A television camera for color television is one that supplies three video signals. These signals represent light-intensity information of three basic colors in the televised image: red, blue, and green. One method of generating this information is to use three identical monochrome camera tubes. The tubes are mounted in close proximity to each other. And the camera as a unit is fitted with special mirrors which "split" the light image of the televised scene according to color information. Mirrors with this ability are called *dichroic* mirrors. A set of mirrors that splits the light from the scene to be televised into components representing red, blue, and green information in the scene is used. This concept, of dichroic mirrors and three camera tubes, is illustrated in Fig. 11.23. There are also color

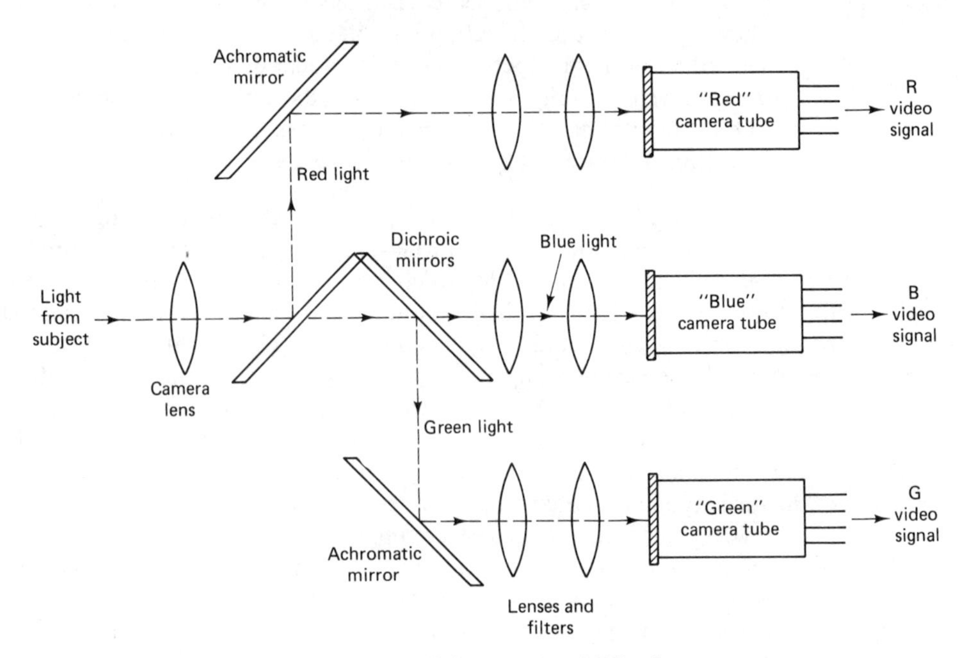

Figure 11.23 Beam-splitting concept of TV color camera.

camera tubes available which have the capability of supplying tricolor information as a single tube.

The tubes in three-tube color cameras may be vidicons (see above). However, they are more likely to be *plumbicons*. A plumbicon is similar in its operation to a vidicon. The image conversion plate of a plumbicon is an equivalent PIN diode. (PIN stands for "P-type–Intrinsic–N-type"). The title "plumbicon" is derived from the symbol for lead—Pb. Lead oxide, PbO, a semiconductor, is used in the construction of the image plate of plumbicons.

Of course, each of the tubes requires deflection coils. The deflection coils are driven by common deflection generators which are part of a television station's complement of equipment.

The Color Television Transmitter

Color television broadcasting was initiated in the United States several years after the introduction of monochrome television broadcasting. As a result, the government required that the color system be compatible with the existing monochrome system. That is, it was required that the new system produce a signal that could be received successfully by the existing millions of black-and-white receivers. This requirement led to the adoption of certain features found in the color broadcasting system now in use.

The most obvious of the features of the color broadcasting system which resulted from the compatibility requirement have to do with the nature of the video information provided by the system. One component of the color video signal represents basic black-and-white image light-intensity information. This component is called the *luminance signal.* The luminance signal is the basic compatibility link between the monochrome and color broadcasting systems. It can be used by a monochrome receiver to produce a normal monochrome image. The second component of the color video signal contains color information and is called the *chrominance signal.* The chrominance signal is not used by a monochrome receiver at all.

Figure 11.24 illustrates the functions and their relationships used in producing the video signal of a color television station. As stated previously, the luminance signal contains the light-intensity information which can be used by a receiver to produce a black-and-white image. It is the signal that would be generated by a single camera tube receiving all of the lens-gathered light from a scene. In typical television station practice it is generated by the addition of signals from the red (R), blue (B), and green (G) camera tubes. However, as shown in Fig. 11.24, Y, the luminance signal, is not the simple sum of these three separate signals. Because of the way in which colors are perceived by human vision, the color signals are combined as 30% of the red signal, 11% of the blue, and 59% of the green. This formula produces a black-and-white image from signals generated by a color system which is the equivalent of the image of a monochrome system.

Study the block diagram of Fig. 11.24 again. Pay special attention to the signal paths that lead to the C (chrominance) signal. You will note that color information is not simply transmitted as the three signals from the red, blue,

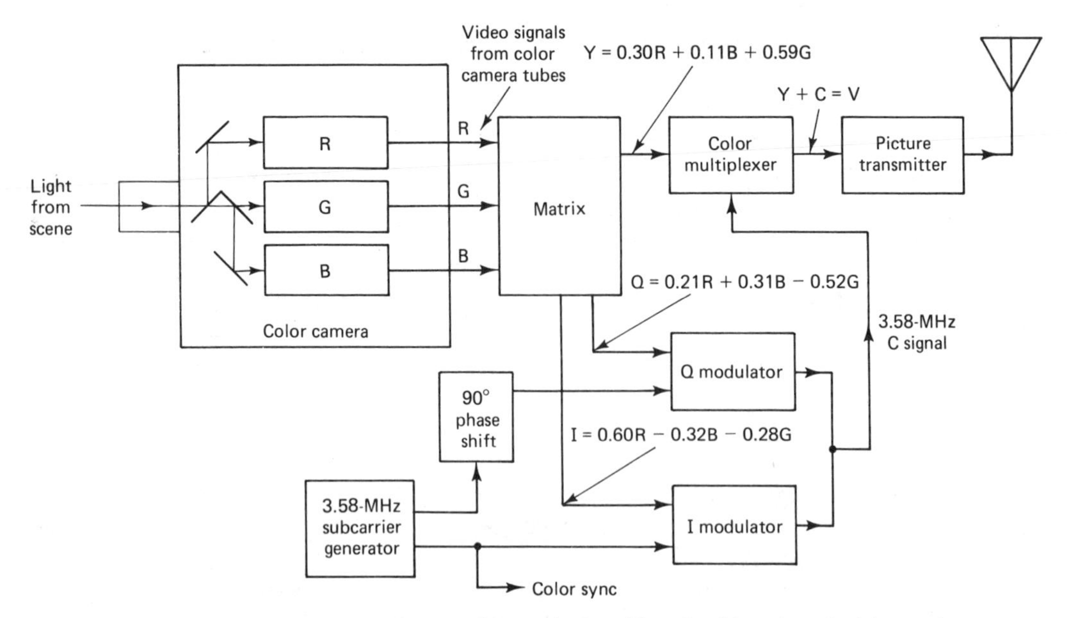

Figure 11.24 Functions used in producing video signal in color television station.

and green outputs of the color camera. Preparing the video signal(s) to modulate the station transmitter involves several steps:

1. The three signals from a color camera, R, B and G, are fed to a special function called a *matrix*. The matrix is a network that combines the signal voltages in desired proportions. As already noted, the proportions of the Y signal, one of the outputs of the matrix, are 30% of the R video, 11% of the B video, and 59% of the G video. As a formula, $Y = 0.30R + 0.11B + 0.59G$.
2. A second output of the matrix is a color signal designated the I signal. The formula in this case is $I = 0.60R - 0.32B - 0.28G$. The negative signs in the formula indicate that a signal is inverted from its normal electrical polarity before being combined with the other components in the formula.
3. The third output of the matrix is also a color signal and is designated Q. The designation Q is for "quadrature," meaning a 90° phase shift. As we shall see, the Q signal modulates the 3.58-MHz color subcarrier 90° out of phase from the I signal. The proportions for this signal are $Q = 0.21R + 0.31B - 0.52G$.

The use of the I and Q signals to transmit color information is the result of decisions made during the process of designing the color transmission system. The representatives of many companies worked on the design of the system. There were competing designs. Lengthy consideration was given to many technical factors during the design process. Some of the more important factors considered included (1) how to make the most effective use of the limited bandwidth of a television channel—6 MHz, and (2) how to achieve good color images without complex, expensive receiver design. Ultimately, a design decision was made by the FCC. This commission now requires that all color television stations use the *Y*, *I*, and *Q* signals.

As indicated by Fig. 11.24, transmitting color information requires the use of a subcarrier. A subcarrier is a signal that is modulated in some fashion and then serves, itself, as a modulating signal to modulate a main station transmitter. [You will recall that the concept of a subcarrier was introduced in connection with the SCA service of FM broadcasting (see Chap. 8).] The color subcarrier frequency is 3.58 MHz. The *I* signal is produced with a bandwidth of 1.5 MHz. It amplitude modulates a component of the color subcarrier directly. The *Q* signal is limited to 0.5 MHz of bandwidth. It amplitude modulates a component of the subcarrier that has been phase shifted by 90°. The introduction of the 90° phase shift serves to preserve the individual character of the two signals: *I* and *Q*.

The *I*- and *Q*-modulated components are combined and become the 3.58-MHz modulated chrominance signal, *C*. The *C* and *Y* signals are combined in an adder network, sometimes called a color multiplexer, to produce a total video signal, *V*. The *V* signal is used to modulate the picture transmitter of the color television station in the same fashion as the monochrome video signal modulates the picture transmitter of a monochrome television station.

The *I*- and *Q*-modulated 3.58-MHz components that make up the *C* signal are actually modulation side frequencies only. The 3.58-MHz color subcarrier is modulated in a suppressed-carrier mode of amplitude modulation. The demodulation of chrominance signal in the receiver requires the presence of the equivalent of a 3.58-MHz carrier. A 3.58-MHz oscillator for this purpose can be provided in the color television receiver easily enough. However, good performance requires that both its frequency and phase be precise. The phase of the chrominance signal literally determines the color reproduced.

The color television system design includes provision for the transmission of a station-generated 3.58-MHz synchronizing signal. This signal is called a *color sync burst* signal. The word "burst" in the descriptive title indicates that it is not transmitted continuously. A short series of 8 to 11 cycles, a burst, of the 3.58-MHz color subcarrier is transmitted during the interval between each horizontal scanning line. Specifically, the 3.58-MHz color sync burst is transmitted on the "back porch" of each horizontal blanking pulse, the portion of the blanking pulse following the horizontal sync pulse. A drawing showing the details of a horizontal blanking pulse with deflection pulse and color sync burst for the color television system is provided in Fig. 11.25.

The details of other aspects of a color television broadcasting station are similar to those for a monochrome station: generation of the composite video waveform with horizontal and vertical blanking and sync signals, etc. The video portion of the waveform for a color transmission, of course, is made up of the *Y* and *C* signals. The process of transmitting and receiving sound in a color system is the same as that for a monochrome system.

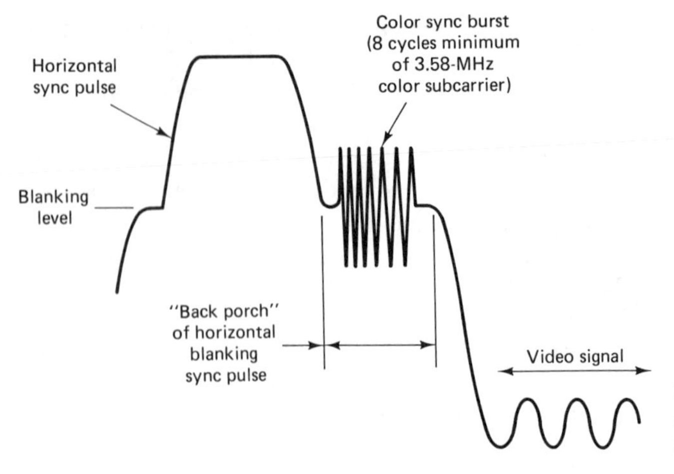

Figure 11.25 Horizontal blanking and sync pulse with color sync burst for color television.

The Color Television Receiver

A color television receiver differs from a monochrome receiver primarily by its containing (1) a color picture tube, and (2) a color video section for decoding the color information of the chrominance signal. However, a color receiver may differ from typical black-and-white (B/W) recievers in other less critical ways. For example, a color receiver will generally have an IF section with a broader pass band than that of most B/W receivers. This enables it to pass the broader range of frequencies required for good color rendition. The 4.5-MHz intercarrier sound IF signal is typically tapped off before the video detector in a color receiver, instead of at the output of the detector as in the typical monochrome TV set.

Let's examine, first, the details of one of several available types of color picture tubes, and, second, the concepts of a color video section. Study the color picture tube in Fig. 11.26. A color picture tube creates color images of many different colors or *hues* by creating, effectively, three superimposed images of the three primary colors: red, blue, and green. You will observe that Fig. 11.26 indicates the presence of three electron guns labeled "*R* gun," "*B* gun," and "*G* gun." These are like the electron guns we have examined previously; they contain focusing elements, and when a high positive accelerating voltage (of the order of 30 kV) is present, they create electron beams. The guns and the beams they produce are labeled red, blue, and green, of course, not because the electrons are of these colors but because of the colors produced by their respective phosphor targets on the viewing screen.

The viewing screen of a color picture tube is coated with several hundred thousand phosphor dots (or, in some tube designs, stripes). The phosphors used are of three types: one type fluoresces red when struck by an electron beam, the second fluoresces blue, and the third fluoresces green. The dots are coated in groups of three called *triads*.

Between the screen and the electron guns, in close proximity to the screen, is a steel plate with tiny holes. There is a hole for each phosphor triad; each

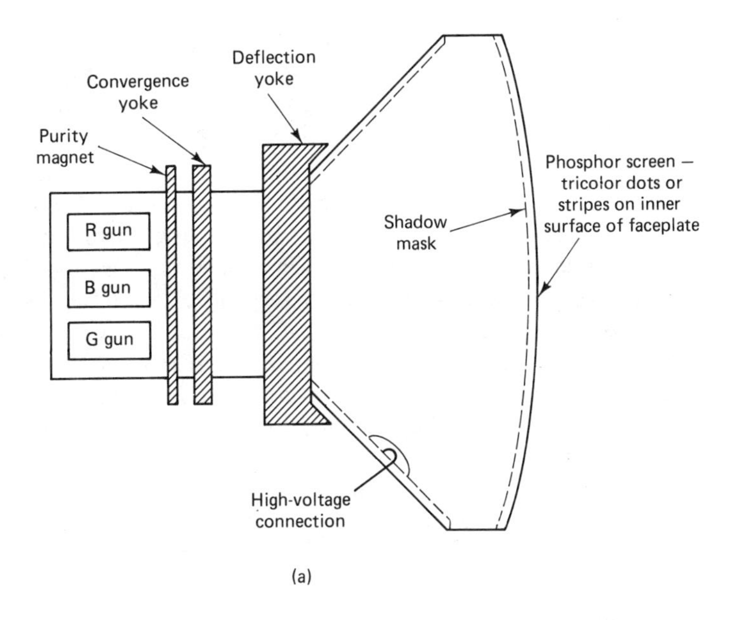

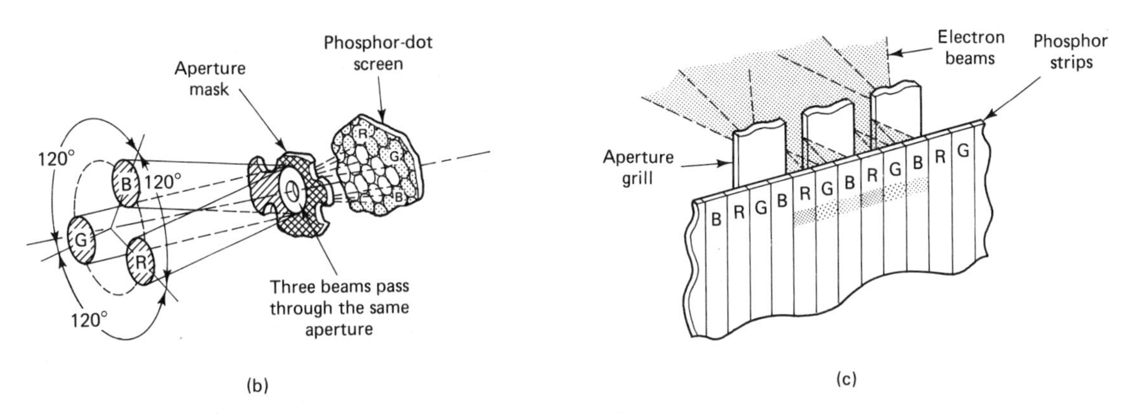

Figure 11.26 Color picture tube: (a) basic concepts of color picture tube; (b) detail of aperture mask-phosphor dot construction; (c) detail of aperture grill-phosphor strip construction. (b), (c) Courtesy of Clyde N. Herrick, *Color Television: Theory and Service,* 2nd Ed., © 1977, pp. 63, 64. (a Reston Publication) Reprinted by permission of Prentice-Hall, Englewood Cliffs, N.J.

hole is centered on its triad. In operation the beams are focused and aimed to pass through the holes in the steel plate and strike phosphor areas of a particular type—a red beam strikes red phosphors, etc. The phosphor areas are "shadowed" from the beams of the other "colors" by the steel plate, which is called, appropriately, a *shadow mask.*

One of the tasks of the deflection functions associated with a color picture tube is to ensure that the three electron beams *converge* at each shadow mask hole, that is, that the three beams pass through the same hole at each hole location. If this convergence does not occur, the three color images will not be superimposed; images of separated colors will be perceived by the viewer.

A lack of "convergence" is not an uncommon problem in color TV sets. It usually appears only in certain areas of an image: part of the screen will be mostly red, or green, etc., at all times. Because of the geometry of a picture tube—a flat instead of spherical viewing screen, for example—convergence requires specially designed circuits to alter the normal deflection currents in the deflection coils. These circuits typically have nine or more interactive adjustments. When a picture tube is "out of convergence," a readjustment of the convergence controls is required. "Converging" a picture tube (adjusting the convergence controls) is usually required when a replacement tube is installed. It may be required as the result of lack of convergence due to aging of components, inadvertent changing of the controls by unauthorized or inexperienced persons, etc. Converging a color picture tube is a skill that requires a sustained, patient effort to develop.

Each of the electron guns has its own control and screen grid. This permits the intensity of each beam to be controlled separately. The perceived color of each phosphor triad is determined by the relative intensity of each of the primary colors. A wide variation of hues can be obtained by varying the relative intensities of the three colors. (A comprehensive study of color television includes a detailed study of the physics of color.) A color picture tube "paints" a color image by creating three superimposed monochrome images: a red image and a blue and a green image. Each image is created by the scanning process, not unlike that of a B/W set. Each image is created by a variation in electron-beam intensity in conjunction with the scanning process. The perceived color of an image at any point in the overall image is determined by the relative intensities of the three primary images at that point. The relative intensities of the three images at any given point is determined by the electronic signals applied to the electron guns. The electronic signals are controlled by the information encoded in the video signal transmitted by the television station.

Let's examine the concepts involved in recovering from a broadcast video signal the signals required to control the separate electron guns of a color picture tube.

Color Section of a Color Television Receiver

Figure 11.27 shows the basic functions of a typical color section of a color television receiver. It also illustrates the relationship of the color signals to the luminance (B/W) signal with respect to the overall control of the electron beams in the picture tube. Let's examine the process of controlling the picture tube in a 1–2–3 fashion. Refer to Fig. 11.27 as you follow the description below.

1. The composite video signal from the video detector is passed to the video preamplifier for amplification and tapping off of signals to be passed to various functions: (a) A composite video signal of appropriate amplitude and phase is passed to a *delay line* and the *Y* amplifier. This is the luminance signal—the equivalent of a B/W signal. (b) A sample is passed to the deflection and

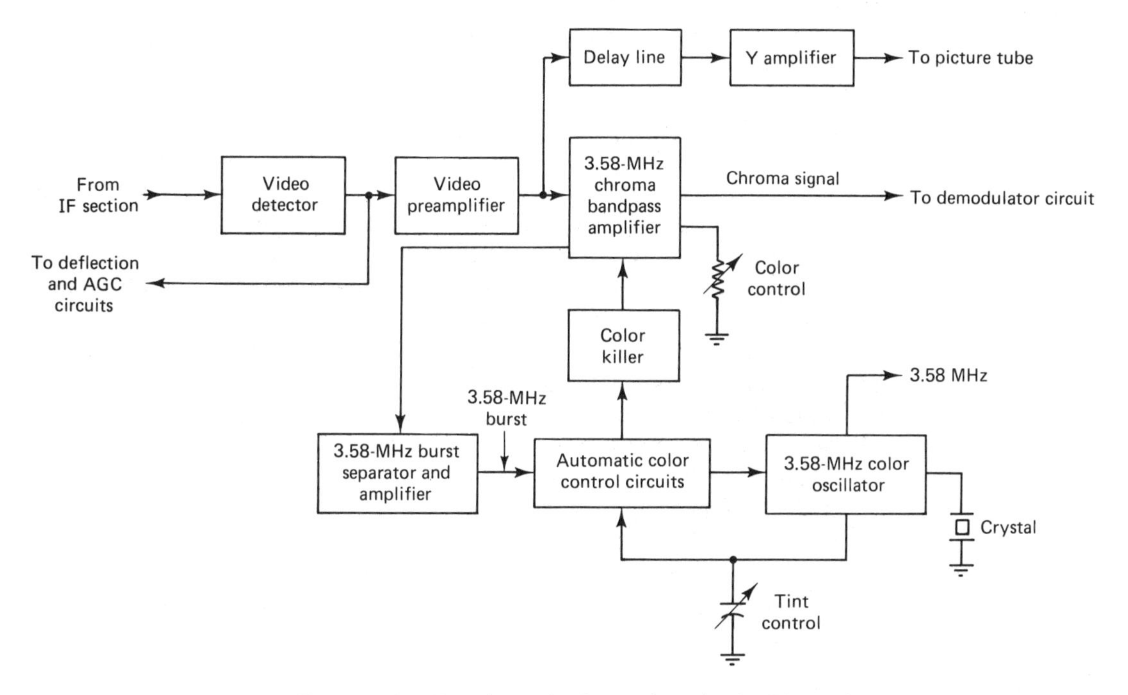

Figure 11.27 Functions of color section of color TV receiver.

AGC sections for synchronizing scanning and AGC control. (c) A third output of the amplifier goes to the chroma (color) section of the receiver.

2. The luminance signal receives a time delay in the delay line; it is then amplified in a video amplifier and finally, applied to the picture tube. This signal is often, although not necessarily, applied to the cathodes of all three electron guns. This permits a common control for the brightness of the screen display. The luminance signal, in conjunction with "neutral" signals from the chroma section, for example, can create a normal black-and-white picture.

The time delay provided by the delay line is required to make the time of arrival of the luminance signal at the picture tube coincide with that of the chroma signal. The chroma signal arrives later, inherently, because it passes through more electronic processing circuits on its way to the picture tube grids.

The delay line is a synthetic transmission line. It is a coil usually about 8 in. long which is coated with an insulating material and then wrapped with foil. The construction effectively provides distributed inductance and capacitance as in a transmission line. Its electrical length is much greater than might be expected for its physical length because of the way it is constructed.

3. The color or chroma section of a color television receiver must perform several functions as its part in creating color images. The ultimate effect of the output of this section is to provide signals to the three electron guns of the picture tube. These signals, in combination with the *Y* signal, control the electron beams in a way that produces color images. A description, at a conceptual level,

of the functions and steps required to produce the ultimate color signals is as follows:

a. The composite video signal from the preamplifier is passed through an IF-like function called the *chroma bandpass amplifier*, or simply, the *chroma amplifier*. The chroma amplifier incorporates tuned circuits with which it is tuned to a center frequency of 3.58 MHz, the color subcarrier frequency. (You will recall that the 3.58-MHz color subcarrier itself is missing since it is suppressed at the transmitter.) The chroma amplifier generally has a pass band of 1 MHz—from 3.08 to 4.08 MHz.

   The chroma signal, the output of the chroma amplifier, is passed to the color demodulators. If provided with a gain control for adjusting the amplitude of the chroma signal, that control is called the "color" or "saturation" control. The saturation of a color refers to the depth of the color (e.g., whether it is "deep red" or just "pink").

   The chroma section generally includes the function referred to as "color killer." This is a small circuit with the task of determining whether or not the composite video signal is that of a color broadcast. It accomplishes this by looking for the presence or absence of the color burst, which is transmitted only during horizontal blanking time. If the color burst is not present, the color killer disables the chroma amplifier to prevent distracting color streaks in a black-and-white picture. A chroma section typically utilizes an integrated circuit to perform the major portion of its functions.

b. In order to demodulate the chroma signal and recover information with which to create color images, the receiver must provide a locally generated signal to serve as the 3.58-MHz subcarrier. Accordingly, a 3.58-MHz oscillator and the circuitry to precisely synchronize its frequency and phase with that of the transmitter is generally considered part of the chroma section. Although a 3.58-MHz oscillator is virtually always crystal controlled, provision is always made for slight frequency correction and for phase synchronization. The synchronizing signal is the color burst signal transmitted by the television station on the back porch of every horizontal blanking pulse. A sample of the composite video signal is passed through the *burst amplifier*, where horizontal pulses are stripped from the signal and the color burst signal is amplified. The amplified burst signal is used, generally in a phase-locked-loop type of circuit, to synchronize the frequency and phase of the 3.58-MHz oscillator.

c. The color television receiver combines the amplified chroma signal with the synchronized, locally generated 3.58-MHz subcarrier in a demodulator function to recover color information. You will recall that the color information from the color cameras was encoded into signals designated I and Q, and the I and Q signals were used to modulate the 3.58-MHz subcarrier. The combined output of this process was the chrominance or C signal. Television receivers are virtually never designed to recover I and Q signals as such. A very popular demodulator design produces what may be referred to as *color-difference* signals. A block diagram of a demodulator of this type is shown in Fig. 11.28.

The $R - Y$ and $B - Y$ demodulators of Fig. 11.28 utilize *synchronous detectors*. Typical circuits, which will not be examined here, include integrated circuits for the detector function. Of essence in this type of detector is that the precise phase of the reinserted 3.58-MHz CW (unmodulated) signal determines the amplitude of the color-difference signal. And of course, ultimately, the amplitude of a color-difference signal determines the color of an image area on the picture-tube screen.

You will note that the phase of the 3.58-MHz CW signal at the input to the $R - Y$ demodulator is $-90°$. This phase angle is with respect to the phase of the burst signal received from the transmitter. Similarly, the $B - Y$ demodulator receives a "180°" CW signal. Combining these CW signals with a common chroma signal, the detectors produce $-(R - Y)$ and $-(B - Y)$ signals. The reason for such results is revealed only by a detailed phasor analysis of the complete chrominance signal.

After amplification, the output signals of the demodulators are fed in precise proportion to a $G - Y$ amplifier. This is a circuit that generates a $G - Y$ signal from the $R - Y$ and $B - Y$ signals.

d. The three color-difference signals from the color-difference amplifiers are added to, or *matrixed* with the Y signal to reproduce, finally, the R, B, and G signals generated originally by the color camera. The matrix circuit may be a separate circuit. Or, as is often the case, the picture tube is used to matrix the signals. For example, if the Y signal is applied to the cathodes

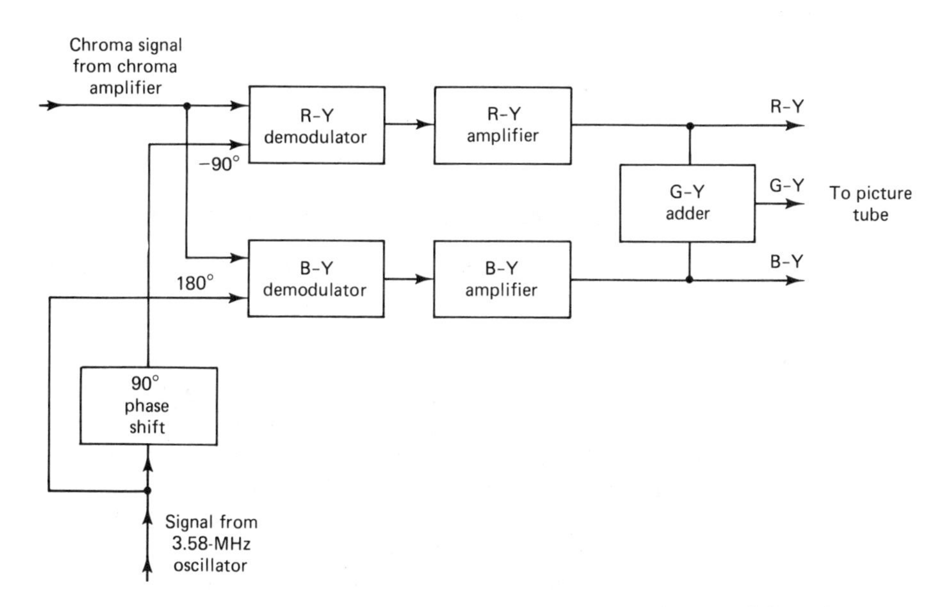

Figure 11.28 One of several possible schemes for recovering color information.

of the R, B, and G guns and the difference signals applied to the control grids of the three guns, the following algebraic results are obtained:

$$(R - Y) - (-Y) = R$$
$$(B - Y) - (-Y) = B$$
$$(G - Y) = (-Y) = G$$

The sign of the Y signal is negative because the effect of a signal on the cathode of a tube is opposite that of a signal on the grid.

The regenerated R, B, and G signals are used by the color picture tube to regenerate the color image. The process utilizes the concept of a scanning electron beam just as in a monochrome receiver. The difference is that in a color picture tube there are three beams.

We have now traced a color image from the original scene or image at the color television camera to the reproduction of the image on the screen of a color television picture tube. The process includes the generation of electronic signals representing the images produced by the red, blue, and green components of the light from the scene being televised. These three signals are combined in one way to produce a luminance signal that can be used by a monochrome receiver to produce an image compatible with monochrome television broadcasts. The three signals are processed in a different way to produce the I and G signals, which are side frequencies of a 3.58-MHz subcarrier. Combined, these are the C or chrominance signal. The luminance or Y signal and the C signal are used to modulate the color television transmitter.

The color television receiver receives the color broadcast in a manner virtually identical to that for monochrome receivers. However, the color receiver has additional functions that process and decode the chrominance signal. The decoded color information signals are combined with the luminance signal to produce color images.

11.5 ALTERNATIVE TELEVISION TRANSMISSION SYSTEMS

In the preceding sections of this chapter, reference to television broadcasting has been made in several instances. From its beginning, the term "television broadcasting" referred to a communication system very similar to a radio broadcasting system. A television broadcasting system consisted of picture and aural transmitters driving a broadcasting antenna, on the one hand. On the other hand were numerous individual receiving stations—individual receivers and associated antennas. Television broadcasting still includes such communications systems, but now also includes other means of getting television broadcast signals to individual users of those signals.

Let us examine, first, some of the important details of antenna-to-antenna transmission of television signals. Then we will look briefly at two alternative methods by which television broadcast signals are transmitted to individual users.

Antenna-to-Antenna Transmission

A typical television broadcasting system consists of a broadcast television station and numerous individual television receivers. The station is owned by an individual, a group of individuals, or a company. The receivers are owned by members of the public at large. The station includes a picture transmitter and an aural transmitter. These are connected to drive a common transmitting antenna through a special network called a diplexer.

Because of the frequencies used for television broadcasting—54 to 890 MHz—propagation of such signals is limited to line-of-sight transmission paths. To cover the greatest possible service area and reach a maximum number of prospective viewers, television stations typically employ omnidirectional transmitting antennas; and the antennas are generally mounted on the tops of tall buildings or on the peaks of available, strategically located hills.

A number of types of antennas are used in television broadcasting (telecasting). All are characterized as having significant power gain. The antenna gain is taken into consideration in a typical specification of a television station: the *effective radiated power* (ERP). Effective radiated power is equal to the peak power of the picture transmitter times the power gain of the antenna. As a formula,

$$\text{ERP} = \text{power (peak, picture transmitter)} \times \text{power gain of antenna}$$

For example, if a picture transmitter has a peak power of 50 kW and drives an antenna with a power gain of 10, ERP $= 50 \times 10 = 500$ kW.

The peak power of a picture transmitter is defined by FCC regulations as being equal to 1.68 times its average power. The average power rating of a picture transmitter is determined while the transmitter is being operated into a nonreactive dummy load. During the test the transmitter is to be modulated between sync pulses with a constant-amplitude signal equivalent to the blanking level—75% of peak amplitude. The load is to have a resistance equal to the characteristic impedance of the transmission line normally used to feed the antenna. The power is to be determined by direct measurement—by RF wattmeter, for example. The transmitter final amplifier, when driving the antenna, must operate with the same dc voltage and current value produced during the power measurement test.

During the course of operation the peak power of a picture transmitter must be held in a range of 80 to 105% of the rated peak power. The aural transmitter is required to be operated in a power range of 10 to 20% of the picture transmitter power. The power variability of the aural transmitter has the same limits as for the picture transmitter—80 to 105%.

Telecast signals are propagated as horizontally polarized waves. This is evidenced by the orientation of the "arms" of the ubiquitous home television antenna. Persons located within short range—10 to 15 mi—of a television transmitting antenna are able to use simple, half-wave dipole antennas. This is provided, of course, there are no intervening obstructions—hills or tall buildings—between the transmitting and receiving antennas. When the distance for a transmitting antenna is greater, reception is improved by the use of directional antennas—

antennas that incorporate a reflector element and a few to many director elements (see Chap. 6). Such antennas provide gain in a particular direction. When properly aimed at the transmitting antenna, they increase the received signal amplitude of desired telecast signals and attenuate ghost or noise signals.

In some large metropolitan areas that support many television stations, all stations have located their transmitting antennas relatively close together in a strategic location. This permits directional receiving antennas to be aimed in a single, fixed direction for good reception. In areas where this is not feasible or not the practice, directional antennas are often mounted on remote-controlled direction-changing mechanisms. In situations of this type the television viewer may be obliged to "tune" the direction of the antenna each time a different channel is selected.

The maximum range over which telecast signals may be received, as determined by the line-of-sight limitation, is typically of the order of 50 to 60 mi. The actual distance is determined by the heights of the transmitting and receiving antennas and/or any intervening obstructions. Some persons, living in relatively isolated areas adjacent to large metropolitan regions, seemingly defy the laws of physics in order to participate in the joy of television viewing. For example, television receiving antennas have been seen at homes on the deserts of southern California at distances up to 100 mi or more from Los Angeles, the location of the nearest television transmitting antenna. Of course, the antennas are placed atop unusually tall mounting structures.

Cable Television Systems

As hinted in the preceding discussion, there are limitations to the reception of broadcast television signals by all those who desire to participate in the process. Distances exceeding the range of useful reception and intervening obstructions are two obvious limitations. In the early years of television broadcasting, attempts to overcome these limitations were very largely individual. In some cases, a group of householders living in the propagation "shadow" of a hill between their location and telecasting antennas would organize a cooperative effort (called a *community antenna system*) to solve their TV reception problem: They would go together and mount an antenna on the offending hill and split the signal and feed it to each participant's home over ordinary television twin-lead lead-in wire. Occasionally, a system achieved a degree of sophistication by the use of coaxial-cable transmission line. Thus was born an alternative method of distributing telecast signals to viewers' receivers—*cable television systems*, or more commonly, simply *cable television*.

A common, early application of cable television systems was in large apartment buildings. These systems typically included a single high-gain antenna mounted on the roof of the apartment building. In larger installations, a signal amplifier was installed at the antenna. The signal was then split with an appropriate network circuit and fed over coaxial cable to individual apartments. Or, in large buildings with many apartments, the signal was split and fed into branches and subbranches with line amplifiers applied at strategic locations to maintain adequate signal levels to all apartment units. Limiting the electrical noise that could find its way into the system was also an important objective of such systems.

In modern cable television systems the concepts just described are expanded to include entire cities or urban subregions. The application has been referred to as the "wired city." Early cable systems were subscribed to by users because the users did not have good reception of all of the desired, available telecast channels. Reception was poor because of either one or the other or both of the limitations described above. The problem of inadequate reception was aggravated with the advent of color telecasting: Good color reproduction requires a stronger received RF signal than monochrome reproduction. Early commercial cable systems were, in effect, a taking over, by individual entrepreneurs or companies, of the tasks of supplying "community" telecast signals.

In its basic form, a cable television system consists of (1) a single high-gain antenna mounted and aimed so as to receive television signals from local TV stations, and (2) an antenna signal fed through a network of coaxial cables, line amplifiers and signal dividers to individual paying subscribers. The interconnecting cables are typically mounted on electric power utility poles. A television cable appears to be simply another telephone cable. RF amplifiers—line amplifiers—are used to boost the levels of a cable's signals at appropriate points along the line. Transmission-line signals must be reamplified before line attenuation has become excessive. If signal amplitudes are not maintained above certain minimal levels, serious degradation of the signal occurs. The inherent noise of the system becomes comparable to the signal level. And the noise is amplified along with the signal. In minimal systems, the signals supplied to subscribers by the cable company are those broadcast by local television stations and picked up by the system's master antenna.

Many commercial cable TV systems today, however, are anything but minimal systems. The advent of telecommunications satellites has expanded the horizons of television viewers and cable system operators tremendously. Many cable systems include earth stations that receive television signals from communications satellites. Satellite signals include retransmitted broadcast television signals from all parts of the country (or, indeed, the world). Cable systems also generally provide special programming, such as first-run movies shown without commercial interruption.

The signals transmitted over coaxial cable by a cable company are normal television-channel carrier signals. In some instances, the cable company's service cable—a length of cable that runs from a junction box on a power pole into the subscriber's premises—is connected directly to a coaxial antenna fitting on the subscriber's TV set. The subscriber selects desired channels in the regular way by changing the set's tuner(s). In other installations, especially those in which many signal channels are supplied, the cable is connected to a frequency-converter "box" supplied by the cable company. The converter is connected to the antenna input of the TV set with a short length of cable. In this application, all signals are converted to the frequency of a single channel—typically, channel 3. The viewer sets the tuner on his/her TV set to the prescribed channel and selects desired programs by a selector on the cable company's converter box.

So-called "movie channels" are typically extra-charge services. Their signals are specially encoded—scrambled—before transmission. Special descrambling converters are provided susbscribers who contract to pay the extra charge. Extra-charge signals may be received by all subscribers, but no satisfactorily viewable image is produced if the signal has not been decoded properly.

Television Signals via Satellite

As this is being written (1985) there are numerous telecommunications satellites in orbit around the earth. Of these, almost 20 retransmit broadcast TV signals which are receivable in most locations in the continental United States. The signals are intended only for reception by cable companies and/or member stations of television networks. Cable companies, for example, pay large fees for the signals provided. However, the signals can be and are being received by individuals. What is required for such reception is an installation called a *TVRO*. The designation TVRO comes from the phrase "*t*ele*v*ision *r*eceive *o*nly." A limited presentation on satellite communications is provided in Chap. 12.

Among other components, a TVRO installation requires the installation of a microwave "dish" for the reception of signals from satellites. A dish is a parabolic reflector (see Chap. 10) and is used in conjunction with a microwave antenna located at its focus. Typical dishes have diameters of 6 to 20 ft. Because of their large size they are not suitable for installation in many congested city locations. TVRO installations cost several thousand dollars. Besides, of course, the signals that can be received are not intended to be free. The companies that produce the retransmitted signals are searching for ways to eliminate the loss of income which noncompensated reception represents.

It is not likely that satellite television will expand as the "free" reception of provided-for-a-fee signals. However, direct-to-home satellite TV service is beginning to be developed as a subscriber service. It is referred to as *DBS* (*d*irect *b*roadcast *s*atellite) service. A complete DBS service includes providing, installing, and maintaining, at each subscriber's home, all of the equipment necessary for the reception and processing of satellite TV signals. A major advantage of DBS service is that it includes a more powerful retransmitted satellite signal. This permits the use of much smaller dishes. Dishes 2 to 4 ft in diameter are being used successfully. Installations of this type are feasible in most locations. It is reasonable to expect that DBS service will succeed as an alternative method for the distribution of telecast signals. It is likely that there will be significant expansion of satellite TV in this direction in the coming years.

GLOSSARY OF TERMS

Aural signal Of a television broadcast station, the RF wave produced by a transmitter (called the aural transmitter) frequency modulated by the sound (audio signal) of a television program.

Blanking The temporary turning off, usually during retrace, of the electron beam of a device such as a television camera or picture tube which utilizes a scanning electron beam to perform its function.

Chrominance signal The component of a composite video signal which carries information about the color of the image being televised.

Color sync burst In color television broadcasting, a short interval of 8 to 11 cycles of the 3.58-MHz color subcarrier broadcast during the interval between successive horizontal scanning lines and used to synchronize (sync) the 3.58-MHz oscillator located in each television receiver.

Deflection Of an electron beam, the moving of the beam from side to side and up and down to accomplish the methodical scanning of an area by the beam.

Delay line Of a television receiver, a circuit possessing the characteristics of a transmission line and used to literally postpone temporarily the arrival of the monochrome picture information so as to match the arrival time of color picture information.

Diplexer A network that allows a television broadcasting antenna to be excited simultaneously by both the picture and aural transmitters without feedback of signal from one transmitter to the other.

Effective radiated power (ERP) A power figure for a television station equal to the product of average power and the power gain of the transmitting antenna.

Field In television broadcasting, one-half of a complete image frame produced by the scanning of 262.5 horizontal lines. *See also* Frame.

Frame One complete television image, of which there are 30 per second, consisting of 525 horizontal lines (in the system used in the United States).

Horizontal interval Time between conclusion of one horizontal scan line and beginning of next scan lines in composite video signal, the horizontal interval is occupied by the horizontal blanking pulse, horizontal sync pulse, and in color broadcasting, the color sync burst.

Hue Color (i.e., red versus orange or blue, etc.).

Intercarrier sound In a television receiver, the sound IF signal produced as a result of the mixing of the picture carrier and the aural carrier whose intercarrier frequency difference is 4.5 MHz.

Luminance signal In color telecasting, the monochrome-compatible component of a picture signal, which contains information about image light levels only (as distinguished from information about image colors).

Phosphor The material used to coat the inside surface of the glass envelope at the viewing end of a television picture tube to produce the viewing screen and which fluoresces (gives off light) when struck by electrons traveling at high speed.

Raster The grid of white (light) lines produced on the viewing surface of a picture tube by its scanning beam when no video signal is being processed.

Retrace The phenomenon of an electron scanning beam returning to its starting position, usually at a much higher speed than during the trace motion.

Retrace time The time required for an electron scanning beam to complete a retrace motion. *See also* Retrace.

Shadow mask A steel plate containing approximately 300,000 holes mounted just behind the viewing screen of a color picture tube and used to ensure that each of three electron beams strikes a phosphor dot of one color only at each location of approximately 300,000 phosphor dot triads.

Sync pulse A pulse transmitted by a television station to be used for synchronizing the scanning of television picture tubes with the scanning of the image in the camera tube. *See also* Synchronize.

Sync separator A circuit used to recover horizontal or vertical synchronizing pulses from the composite video signal.

Synchronize To make one event occur simultaneously with another event.

TVRO (*t*ele*v*ision *r*eceive *o*nly) A term applied to an equipment combination designed to receive only (as distinguished from transmit and receive) broadcast television signals from a telecommunications satellite.

Vertical interval A time span in the composite video signal corresponding to the vertical retrace time and used for the transmission of vertical blanking pulses, vertical sync pulse, and equalizing pulses. *See also* Retrace.

Vestigial sideband transmission A method of AM transmission, used in television broadcasting in the United States, in which one full sideband and only a part (vestige) of the second sideband is transmitted; designated by FCC as A5C emission.

Visual alignment A procedure for adjusting a bandpass filter or amplifier for a desired passband shape accomplished in conjunction with a display of the passband on an oscilloscope.

REVIEW QUESTIONS: BEST ANSWER

1. The first step in the television process is: **a.** turning on the TV set. **b.** converting a light image into an electrical signal. **c.** putting up the TV antenna. **d.** hiring TV performers. **e.** none of these.
2. A broadcast television system is similar to a broadcast radio system because: **a.** it transmits information signals by means of an RF carrier. **b.** it converts a light image into an electrical signal. **c.** it converts an electrical signal into a light image. **d.** all of these. **e.** none of these.
3. Electron-beam scanning is performed in: **a.** the camera tube only. **b.** the picture tube only. **c.** both the camera and picture tubes. **d.** neither the camera nor the picture tube. **e.** none of these.
4. In a video signal, a voltage level represents a/an: **a.** sound level. **b.** light level in the televised image. **c.** blanking level. **d.** synchronizing pulse. **e.** none of these.
5. Images on the screen of a TV receiver are the result of the: **a.** projection of the electron beam. **b.** projection of the electron gun from behind the screen. **c.** screen's phosphor material fluorescing when struck by the scanning electron beam. **d.** video signal turning image cells on and off. **e.** none of these.
6. The average overall brightness of a TV screen can be increased if the electron-beam current is: **a.** decreased. **b.** increased. **c.** phase shifted. **d.** slowed down. **e.** none of these.
7. The blanking levels or pulses in the composite video signal are present during: **a.** both vertical and horizontal (retrace) intervals. **b.** vertical intervals only. **c.** horizontal intervals only. **d.** between fields 1 and 2 only. **e.** none of these.
8. In broadcast television in the United States there are: **a.** 30 frames per second. **b.** 60 fields per second. **c.** 525 lines per frame. **d.** 15,750 lines per second. **e.** all of these.

9. Since there is one horizontal sync pulse for each horizontal line in a TV signal, the horizontal sync pulse repetition rate is: **a.** 525 Hz. **b.** 15,750 Hz. **c.** 60 Hz. **d.** 31,500 Hz. **e.** none of these.
10. When a transmitter is amplitude modulated but transmits all of one sideband and only a part of the other sideband, the emission is called: **a.** single sideband. **b.** upper sideband. **c.** partial sideband. **d.** vestigial sideband. **e.** none of these.
11. The frequency width of a standard TV broadcasting channel is: **a.** 4.5 MHz. **b.** 6 MHz. **c.** 200 kHz. **d.** 3.58 MHz. **e.** none of these.
12. The width of the vestigial sideband of a television broadcast signal is approximately: **a.** 1.25 MHz. **b.** 0.25 MHz. **c.** 3.58 MHz. **d.** 4.5 MHz. **e.** none of these.
13. An electron-tube component that includes a source of free electrons and a means of focusing the electrons into a narrow beam when they are accelerated is a/an: **a.** cathode. **b.** electron gun. **c.** phosphor. **d.** focusing grid. **e.** none of these.
14. The currents in the deflection coils of TV camera and picture tubes have a sawtooth waveform so as to produce: **a.** linear motion of the electron beam during the trace and a high-speed retrace. **b.** equal trace and retrace times. **c.** simultaneous vertical and horizontal motion. **d.** vertical movement of the electron beam only during horizontal retrace. **e.** none of these.
15. If a TV receiver is "varactor tuned," changing the channel selector switches: **a.** inductors. **b.** capacitors. **c.** crystals. **d.** dc voltages. **e.** none of these.
16. The frequency response characteristic of the IF section of a TV receiver, for best results, is carefully shaped to compensate for: **a.** UHF interference. **b.** adjacent FM channel interference. **c.** low video frequencies. **d.** vestigial sideband transmission. **e.** none of these.
17. When the 41.25-MHz sound IF and 45.75-MHz picture IF are mixed in a nonlinear circuit such as a detector, a 4.5-MHz sound signal is one of the available signals. Obtained in this fashion, the 4.5-MHz signal is referred to as the: **a.** chrominance signal. **b.** luminance signal. **c.** intercarrier sound. **d.** video signal. **e.** none of these.
18. To achieve the relatively broad bandwidth required of a TV IF section, a special tuning technique may be used. It is called: **a.** stagger tuning. **b.** broad tuning. **c.** varactor tuning. **d.** peak tuning. **e.** none of these.
19. One of the defining characteristics of a video amplifier is its: **a.** high frequency response. **b.** ac coupling. **c.** rising frequency response at higher frequencies. **d.** relatively broad bandwidth, starting at dc. **e.** none of these.
20. The light pattern on the screen of a TV picture tube produced when the deflection circuits are free running and not responding to the synchronizing signals of a TV broadcast is called: **a.** brightness. **b.** contrast. **c.** interlaced scanning. **d.** the raster. **e.** none of these.
21. High dc voltages—12,000 V and up—are used in TV receivers to: **a.** impart high speed to the electrons in the scanning beam. **b.** excite the phosphor coating of the picture-tube screen. **c.** deflect the scanning beam. **d.** return the beam electrons to the electron gun. **e.** none of these.
22. A device that can split a light beam and provide light intensities related to the color sources of the light is called a/an _____________ mirror. **a.** zoom **b.** antiastigmatic **c.** dichroic **d.** trifocal **e.** none of these.
23. After the 3.58-MHz subcarrier is modulated by the signal from a tricolor camera

chain it is called the ______________ signal. **a.** chrominance **b.** *I* **c.** *Q* **d.** luminance **e.** none of these.

24. The composite video signal for color television contains an additional synchronizing signal. It is called the: **a.** back porch. **b.** color sync burst. **c.** chrominance signal. **d.** subcarrier. **e.** none of these.

25. A delay line is required in the path of the luminance signal of color television because processing: **a.** delays the luminance signal. **b.** speeds up the luminance signal. **c.** speeds up the chrominance signal. **d.** delays the chrominance signal. **e.** none of these.

26. A 3.58-MHz oscillator is required in a color TV receiver for: **a.** synchronizing the chrominance signal. **b.** reinserting the carrier for the chrominance signal as required for demodulation. **c.** producing the intercarrier sound signal for color telecasts. **d.** all of these. **e.** none of these.

REVIEW QUESTIONS: ESSAY

1. Describe at least three ways in which a broadcast television system is like a radio broadcasting system.
2. Why does a TV station have two transmitters? What is each used for? What type of modulation is used by each?
3. List and describe briefly the components of the composite video signal as used in broadcast television in the United States.
4. Using all of the following terms, give the makeup of a television picture: frame, field, scanning line, and interlaced scanning.
5. It is said that the scanning electron beam of a TV picture tube "paints" a picture of light on the screen of the tube. In monochrome TV, picture "colors" are light, dark, and shades of gray between these two extremes. Given that a video signal varies between 0 and 100 V, what would be the "color" of the portion of an image when the video signal is: **(a)** 90 V; **(b)** 2 V; **(c)** 45 V; **(d)** 70 V; **(e)** 120 V?
6. Describe a standard television broadcast channel. That is, give the bandwidth and the relative locations of the picture carrier, aural carrier, and color subcarrier.
7. What is the repetition rate of horizontal sync pulses? of vertical sync pulses?
8. Why is it important that the deflection of electron beams in camera and picture tubes be linear? What would be the effect if the rate of vertical deflection in a receiver were less over the top half of the screen than over the bottom half? (What would be the effect on the appearance of people on the screen, for example?)
9. Describe the waveform of the force that is needed to properly deflect the electron beam in a television tube. Explain the function of the two different components of the waveform.
10. What is the relationship between beam current and image brightness in a TV picture tube?
11. List the typical picture and sound intermediate frequencies for TV receivers.
12. Define *stagger tuning*. Why is stagger tuning used?
13. Discuss the characteristics that are important in classifying an amplifier as a video amplifier.

14. Describe briefly a sync separator. How does this function differ for vertical and horizontal sync pulses?
15. What is a dichroic mirror? Where are dichroic mirrors used in television systems?
16. Describe the difference(s) between the luminance signal and the chrominance signal.
17. What is meant by the *red gun*, the *blue gun*, and the *green gun* when referring to a color picture tube?
18. Describe three methods for getting a TV broadcast signal from a TV station to a viewer's TV receiver.

EXERCISES

1. Sketch and label the channel diagram (see Fig. 11.5) for the following TV channels (see Table 11.1): **(a)** 5; **(b)** 13; **(c)** 28.
2. A television station has an average power of 50,000 W. Its transmitting antenna has a power gain of 20. Calculate the effective radiated power for the station.
3. A television station has an average power of 35,000 W. Its transmitting antenna has a field (voltage) gain of 4. Calculate the effective radiated power for the station. (Power gain is equal to the square of voltage gain.)

12

DIGITAL COMMUNICATIONS SYSTEMS

12.1 INTRODUCTION

A pair of the hottest buzzwords in the world of communications in the mid-1980s make up the term *digital communications*. This topic is hot because it appears as if the techniques of digital communications provide solutions to many of the problems and limitations of conventional or *analog communications*. Knowledgeable experts in electronic communications have already predicted that by the mid-1990s virtually all new communications systems and additions to existing systems will incorporate digital techniques. At the present time, the expansion of the facilities and usage of communications systems is nothing short of a revolution. The techniques of digital communications are perceived as being applicable with great advantage to everything from data communication between computers to television broadcasting and the high-fidelity recording of sound.

What is digital communications? Why is such a great future being forecast for it? How is it different from analog communications? What are some of the things—terminology, concepts, etc.—that one needs to know in order to have a chance at participating in the opportunities that such an exciting and rapidly expanding field would be presumed to offer? It is the purpose of this chapter to provide some of the answers to these and other questions about digital communications. It is surely obvious, however, that in a single chapter we can hardly do more than scratch the surface of this subject.

Even a casual reading of any form of written material on a topic in this field leads one to conclude that digital communications is nothing if not a prolific generator of new and exotic terms, abbreviations, and acronyms. Hence little, if any, progress can be made in learning the essentials of digital communications without a working knowledge of the most important and most frequently used of these terms and expressions. A major objective of this chapter, then, is to provide the information with which you can develop such a working knowledge. The goal is to help you develop a knowledge of the concepts that the terms represent, not simply word definitions of the terms. If we—you and I—are successful in this endeavor, you will be ready to proceed to a more detailed study of the field and qualified to seek entry-level employment in it (assuming that you already have a knowledge of analog communications). Let's get started with "What is digital communications?"

12.2 WHAT IS DIGITAL COMMUNICATIONS?

Virtually all computers these days are digital computers. (Yes, there have been and still are analog computers. Their numbers are very small.) Digital computers handle information in digital form. There are more and more computers coming into the world. Computers are "talking" to each other more and more. So digital communications has to do with communications between computers, right? Wrong!! That is, the statement is at least partially wrong. Digital communications includes much more than the communications between computers. *Digital communications is the transmission of information in digital form.* The information may be of any type that can be converted into digital form. That includes all information that can be put into or is normally in the form of analog electronic signals: sound, vision, physiologic, etc.

The information from a computer (i.e., computer data) can be said to be a *natural* form of digital information: It comes from the computer in digital form. The transmission of computer data is termed *data transmission.* Data transmission can be, has been, and still is to some degree, accomplished as an analog transmission. Analog transmission requires that the digital information be converted to a form of analog signal. For example, bursts of two different-frequency signals have been used to represent the binary ones and zeros of typical computer data. These analog signals are then transmitted over conventional analog transmissions systems: the facilities of public or private telephone systems, microwave radio links, etc. Let's reinforce the concept: Digital communication is not synonymous with (digital) data transmission; a data transmission may actually be an analog transmission.

Although data transmission and digital communications are distinctly different ideas, data transmission, computer technology, and digital communications are intimately entwined. As we shall see, there are advantages to digital communications over analog communications. Improved accuracy of the transmission of the information, for example, is one advantage. Computer data is natural digital information and must be transmitted with virtually 100% accuracy. In recent years there has been an explosion in the numbers of computers in use and a similar increase in data transmission. It is not surprising, then, that there is an intense interest in the development of techniques for transmitting digital information in digital form: digital communications. Furthermore, the developments of computer technology have made the techniques of digital communication more and more feasible and attractive and not just possible.

But digital communications facilities can be and will be used to transmit all forms of information now transmitted by analog communications facilities: telephone voice conversations, radio and television broadcast signals, etc. More important, digital communications makes possible and encourages the development of other forms of communications: electronic mail, magazines, and money; cellular telephony; and many others not now even imagined.

Before going any further let's look at a puzzling situation before it has a chance to cause us any real discomfort. The puzzle has to do with the meaning of the terms *communications* and *telecommunications*. Do they mean the same thing? If not, what is the difference in their meanings? A study of technical literature pertaining to "communications by electronic means" indicates that the two words have the same or very similar meanings. This is true especially if "communications" is used with the modifier "electronic," or if used in a context in which it is obvious that "electronic communications" is implied. Further, it appears that "telecommunications" has a European flavor and just plain "communications" has an American flavor. That is, authors whose birth, education, and/or work experience prompts the label "European" appear to prefer the term "telecommunications." American authors appear to prefer using the term "communications" to mean the same thing. In any case, "telecommunications" (and therefore "communications") appears to be a generalization of the concept implied by the terms "telephone," "telegraph," "radiotelephone," etc. Of course, the meanings of these terms, especially "telephone," have expanded tremendously since their introduction into the language. In conclusion, whatever the word used—communications or telecommunications—"transmission of information by electronic means" appears to be a useful working definition.

We will soon be leaving this general discussion of what digital communications is and moving on to some of its technical details. Before we do, however, let's look at one related item which is the cause of some confusion about the meaning of digital communications. The topic is *satellite communications*. Satellite and digital communications are often presented together in technical literature: textbooks, technical journal articles, etc. Again, however, these two are not synonymous. Digital communications techniques can be and are being used with virtually all transmission systems: radio, coaxial cable, waveguide, and land-based microwave link. Digital communication is associated with the study of satellite transmission systems because it is the technique of choice of new satellite systems. Indeed, it is becoming the only transmission method chosen for use in new satellite systems. And most new communications systems involve the use of transmission by satellite. We examine some of the aspects of satellite transmission in this chapter.

12.3 A BASIC DIGITAL COMMUNICATIONS SYSTEM

It is time to "jump in" and start learning something about how a digital communications system functions. As with other systems, an excellent way to get started is to study a block diagram. Refer now to the block diagram of Fig. 12.1. The diagram is intended to illustrate general concepts only; it does not represent any particular, actual system. The diagram illustrates three important characteristics of digital communications systems:

1. Such systems utilize relatively conventional communications techniques and apparatus in the transmission segment of the system: radio-frequency transmitters and receivers, coaxial transmission lines, line-of-sight transmission by land-based microwave facilities, and transmission via satellite microwave links.
2. Analog-to-digital coders and digital-to-analog decoders are essential functions of digital communications systems which process information in analog form—voice and video signals, for example.
3. A great advantage of digital communications systems is their adaptability to the transmission of many channels of information. Commercial systems invariably provide many communications channels. They are capable of receiving information from many sources, and for this reason are called *multiple-access* systems. The process of using a common transmission link for the transmission of multiple signals is called *multiplexing*. A corresponding process is required for separating signals after transmission. The separating

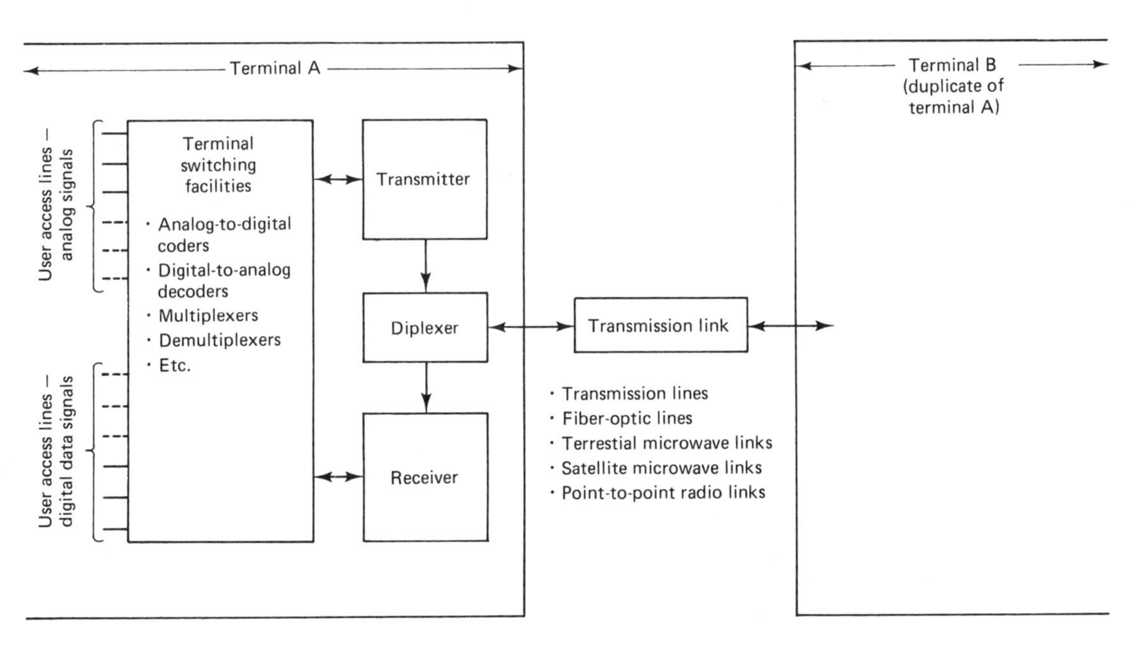

Figure 12.1 Block diagram of digital communications system.

process is called *demultiplexing*. The respective electronic functions are called a *multiplexer* (MUX) and a *demultiplexer* (DEMUX). You will note that these functions are included on the system block diagram in Fig. 12.1.

A short description of the operation of the digital system illustrated in Fig. 12.1 is as follows. Information signals, in both analog and digital form are received at the transmitting facility. Analog signals are converted to digital form by an analog-to-digital (A-to-D) encoder (usually called simply a "coder"). Converted digital and natural digital signals are selected by a multiplexer and used to modulate the facility's transmitter. The modulated carrier is placed on the transmission medium: coaxial transmission line, microwave radio link, etc. At the receiving facility, a receiver processes the received carrier, demodulating it and passing it to a demultiplexer for distribution of the multiple information signals to the desired destinations. Analog-derived signals are reconverted in digital-to-analog (D-to-A) decoders for delivery in analog form.

We have looked at multiplexing and demultiplexing briefly in other chapters. However, A-to-D and D-to-A conversion have not been mentioned previously in this book. These functions are basic to digital communications systems. Gaining a working knowledge of them is an essential step in the study of digital communications systems. The most widely used technique for converting analog signals to digital form is called PCM, an abbreviation for *pulse-code modulation*. Let's have a look at it.

PCM

Before looking at any of the details of the "how" of PCM, let's examine what is meant by "digital" and "analog" as these terms are used in the context of information signal processing. Refer to the waveforms of Fig. 12.2.

An analog signal might be as simple as a sine wave, as in Fig. 12.2(a). A complex analog waveform is shown in Fig. 12.2(b); it resembles the waveform of speech. You are aware that in an analog waveform, the shape of the wave from moment to moment is the characteristic that represents information. In the case of sound reproduction, for example, the motion of a speaker diaphragm is determined by the waveform of the signal driving the speaker (causing the diaphragm motion). If the waveform is changed even slightly, the character of the sound heard by a perceptive listener is also changed. When noise is heard over desired sound from a speaker, the noise is a result of an unintended alteration of the driving waveform.

Study the waveform of Fig. 12.2(c). Observe that it appears to be only a series of irregularly spaced pulses of irregular width. In fact, if analyzed against a time grid as in Fig. 12.2(d), this waveform can be said to contain the binary information, 101100101101. If analyzed in still greater detail, the waveform can be said to represent 1011, 0010, and 1101, or 11, 2, and 13. We are obviously examining a *digital waveform* in which a binary "1" is represented by the high level of the waveform and a "0" by its low level.

Let's digress for just a moment from our goal of examining the "how" of PCM. Now that we know the difference between analog and digital signals (if

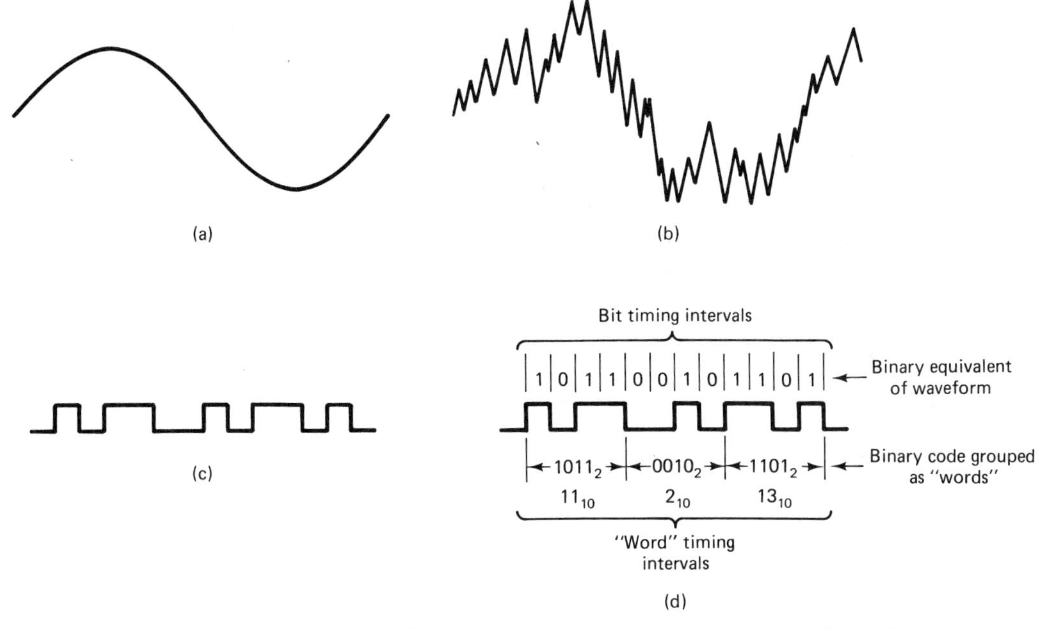

Figure 12.2 Analog and digital waveforms: (a) sine-wave analog waveform; (b) complex analog waveform; (c) pulse stream (digital) waveform; (d) pulse stream decoded.

you don't feel you know the difference yet, spend some more time right now studying Fig. 12.2 and the paragraphs relating to it), we can appreciate the reason for one of the major advantages of digital transmission: its inherent ability to overcome the hazards (to information) of noisy transmission environments. Consider the following:

1. All transmissions, digital and analog, are degraded by the transmission process. Amplitudes are reduced by the loss of energy to the transmission medium. Wave shapes are changed as a result of the addition of noise energy to the information signal.
2. In the case of analog signals, noise energy cannot effectively be eliminated from the signal once it has been introduced. A "damaged" waveform can never be repaired (completely restored to its original form). The effect of noise can be minimized best by repeatedly reamplifying the information signal at frequent intervals along a transmission path. This procedure aims at maintaining the signal amplitude well above that of the noise level of the system and thus holding down the effect of noise.
3. A digital signal, on the other hand, if reprocessed before it is degraded excessively, can be completely regenerated as a fresh, new waveform. The process can be repeated as many times as necessary along a transmission path. Added noise is completely eliminated at each regeneration. All that is required for regeneration is for the degraded signal to be detected. The detected information is used to control the process of generating a clean, new signal containing exactly the same information as the original signal.

Well, effectively immunizing a signal from noise is an extremely important advantage. PCM is a process by which an analog signal, such as is shown in Fig. 12.2(b), can be converted to a digital signal like that of Fig. 12.2(c). After conversion into digital form, the signal can, of course, be processed as a digital signal and the advantages of digital transmission realized. Now back to the "how" of PCM.

The conversion process of PCM involves three elements or operations which have commonly accepted, descriptive titles. The operations are called *sampling*, *quantizing*, and *coding*. In general, the process consists of (1) electronically "looking at" (sampling, determining the amplitude of) an analog signal at regular time intervals, (2) converting the "quantity" of the sampled amplitude into one of a fixed set of explicit values (quantizing), and (3) converting the explicit values into a series of pulses (coding). The application of these three operations to the conversion of a portion of a pure sine wave is illustrated in Fig. 12.3. The analog waveform is shown being sampled and quantized in Fig. 12.3(a). The "quantity" of the amplitude at each sampling moment is indicated as a decimal number on the diagram of the waveform. A table showing the binary conversion values of the amplitude "quantities" is given in Fig. 12.3(b). Finally, the result of encoding the binary "quantity" values into a digital waveform is illustrated in Fig. 12.3(c).

I hope you now have the feeling that you "know" PCM in this very rudimentary way in which it has been presented in the paragraphs above. Spend some more time with the paragraphs if you do not feel that you are quite "there" yet. From this point on we shall be looking at the basic operations in greater detail. You will be making a mistake if you read on without having a grasp of the process of PCM in general terms. Do not let the rather forbidding title "pulse-code modulation" put you off. Remember, we "look" at an analog signal at regular intervals, "quantize" the amplitude at each "sampling" moment, and convert the amplitude "quantity," expressed as a binary number, into a digital waveform.

You may be puzzled as to why the word "modulation" is included in the title of the process. Your puzzlement is not unreasonable. The process, indeed, does not involve an operation similar to the amplitude or frequency modulation of RF carriers which we have examined in connection with radio communication systems. In fact, as we shall see, digital waveforms derived through the process of PCM are used to modulate RF carriers prior to transmission. The title is what it is and is widely accepted and used. Let us think of the process of "changing" a pure dc voltage "waveform" into the digital waveform containing binary-coded information as the "modulation" referred to in PCM.

If you are concerned because you don't yet know "how" these things can be accomplished, that's good. You will be more receptive to an explanation of "how." But you cannot expect yourself to know the "how" at this point because we have not begun yet to explore techniques.

I am reluctant to state it, but, as usual, there are a number of different ways of accomplishing the "how" of PCM. Each has advantages and disadvantages. A great deal of engineering effort has been expended in examining the

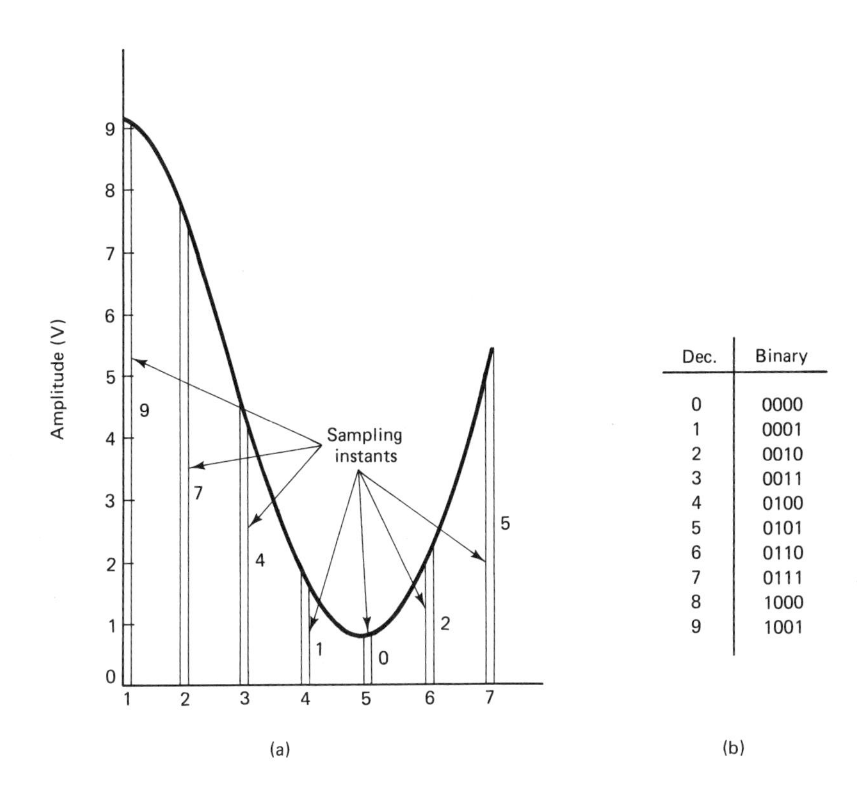

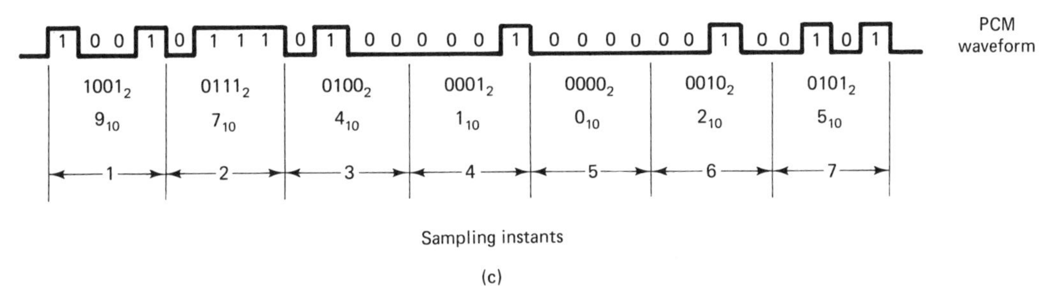

Figure 12.3 Analog-to-digital PCM conversion process: (a) analog signal being sampled and quantized; (b) decimal/binary conversion table; (c) amplitude values at sampling instants converted to PCM waveform.

relative pros and cons of various PCM techniques. If you are interested in this aspect of the topic, you will be pleased to know that numerous books have been published on the theoretical aspects of PCM. The titles of some of these are listed in the Bibliography at the back of the book. Here, however, we shall be happy if we can achieve a useful, if limited, working knowledge. We want to learn the most important ideas of digital communications as related to the practical application of this form of communication. We shall examine briefly only the most widely used techniques of the "how" of PCM.

12.4 SAMPLING

As an idea, sampling an electrical signal is very simple. It can be represented schematically by the circuit of Fig. 12.4(a). You will note that the circuit here incorporates a mechanical device—a rotating cam—to open and close a switch. The switch simply connects the output of the sampler to its input during the moments when it is closed. A mechanical device is useful in conveying an idea because its function is "visible." But a modern high-speed digital communications system is no place for a mechanical switcher. In an actual communications system, switching would be by electronic means—an "electronic switch." The electronic switch could be controlled by a waveform like that labeled "sampling waveform" in Fig. 12.4(b). What is important to grasp at this point is that the sampler circuit enables its output circuit to "look" at a signal on a regular time schedule and provide information about the amplitude of that signal at the moment of each "look." These "look moments" are referred to as *sampling instants*.

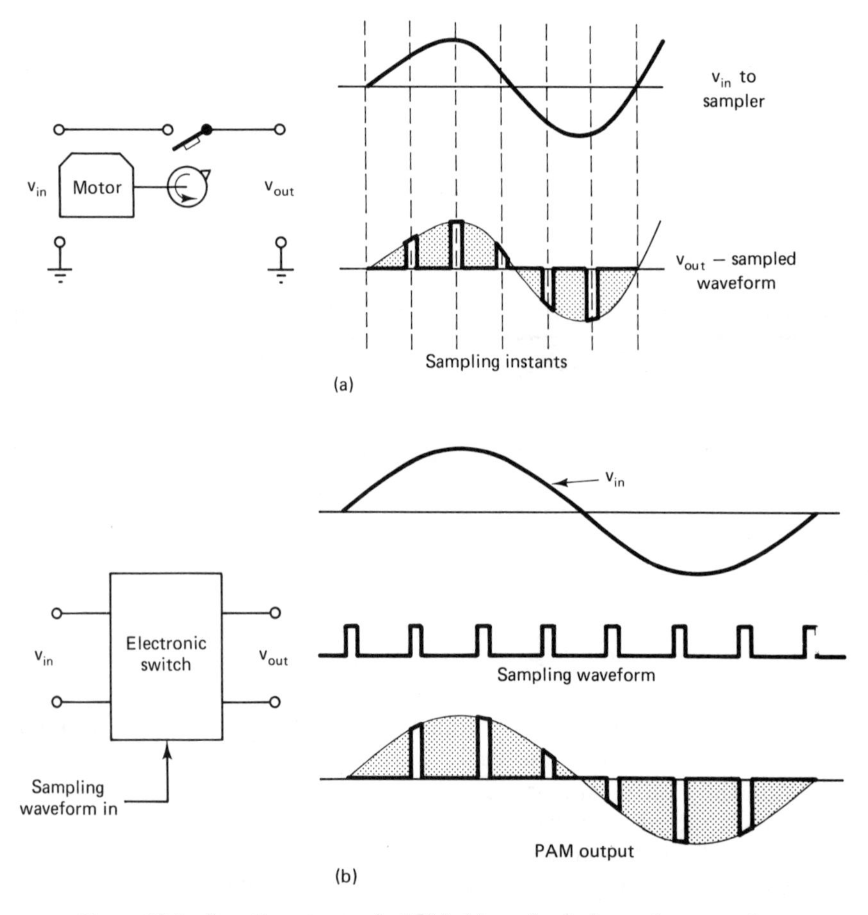

Figure 12.4 Sampling process in PCM: (a) mechanical waveform sampler; (b) sampling with an electronic switch.

Observe that the output of the "sampler" function is a series of pulses with varying amplitudes [see Fig. 12.4(b)]. This type of sampling produces what is referred to as *pulse-amplitude modulation* or PAM (another item for our growing list of abbreviations and acronyms). A PAM signal is an analog signal since the amplitudes of these pulses still contain essential information about the original signal. PAM is used as an intermediate step in the conversion of analog signals to PCM signals.

There are other methods of sampling which present alternatives to the simple on-off concept that we have just examined. One of these makes use of a triangular or sawtooth waveform to control the sampling operation (see Fig. 12.5). With a sawtooth sampling waveform it is possible and feasible to achieve two other commonly identified forms of "modulation." These are *pulse-position modulation* (PPM) and *pulse-width modulation* (PWM) (more ingredients for our growing alphabet soup!). Pulse-width modulation is also referred to as PDM—*pulse-duration modulation*. For illustrations of these modulation forms, refer to Fig. 12.5. You will note that the intersection of the rising portion of the sampling waveform with the signal waveform is used to determine the *positions* of pulses of constant amplitude in the PPM output. On the other hand, in PWM (or PDM) the *time period* over which the sampling pulse is greater than the signal is used to determine the *width* or *duration* of pulses of constant amplitude. Of course, the vertical component of the sawtooth waveform could be used to control a

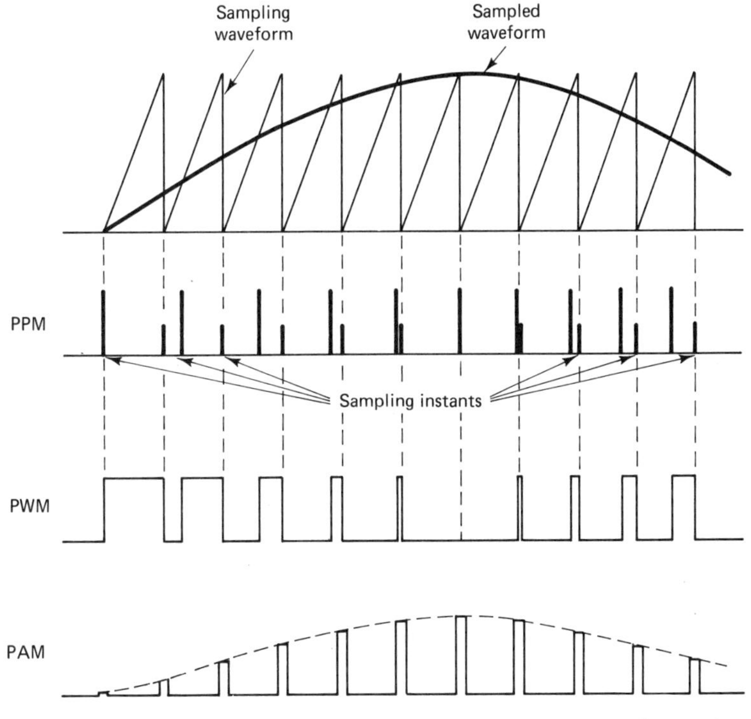

Figure 12.5 Three forms of modulation resulting from sampling of an analog waveform.

sampling circuit to produce PAM. A PAM output waveform is provided for comparison with the other modulation formats in Fig. 12.5.

Sampling Rate

The number of times per second a sampling circuit "looks at" an information signal is called the *sampling rate*. Common sense tells us that if a signal is not sampled often enough (i.e., if the sampling rate is too low), there will not be enough information in the modulated waveform. It would be impossible to reconstruct an acceptable facsimile of the original analog signal at a destination somewhere down the line. On the other hand, if we sample too often—if the sampling rate is too high—we throw away some of the advantages of the whole process.

What, then, is the minimum sampling rate that must be used? If we think about the situation for a moment, we will not be surprised to learn that the minimum sampling rate is different for different types of information signals. If a signal waveform is such that changes in amplitude take place relatively slowly, then, obviously, a low sampling rate will be adequate to capture the needed information. If a signal has rapidly changing amplitudes, like a video signal, for example, a high sampling rate is required.

This entire matter of sampling rate has been examined thoroughly and scientifically with the aid of mathematics. The result is widely accepted and often stated as the sampling theorem: *An information signal whose highest frequency is equal to f_m may be completely represented by a sampling technique in which the sampling rate is at least $2f_m$.*

A very specific terminology is commonly used in conjunction with discussions of the present type. It is important that you be familiar with it. The phrase *baseband signal* is used to signify an unaltered information signal. For example, in voice communication the baseband signal is the audio signal as it comes from a microphone. The binary-coded data signal from a computer is considered a baseband signal if *the electrical signal is a low-frequency signal and extends to approximately 0 Hz or dc. Band limited* is a term used to identify a signal, usually a baseband signal, in which there are no frequency components above a specific limit, say, f_m. The minimum sampling rate, then, for a band-limited baseband signal is equal to $2f_m$, where f_m is the highest frequency present in the band-limited signal. Many "natural" information signals, for example speech communications signals, are not naturally sharply band limited. In digital communications systems, such signals are generally processed through sharp-cutoff, low-pass filters to achieve a definite band limit.

In summary, sampling is a process of "looking at" an analog information signal at regular intervals and converting amplitude information into a signal of a form different from that of the original. Pulse amplitude modulation (PAM), pulse width or pulse duration modulation (PWM or PDM), and pulse position modulation (PPM) are three common types of signals used to represent sampled information. A sampling communications system may transmit signals in one or another of these forms. However, generally, these signal types are used only as temporary, intermediate forms in the process of converting an analog signal into a true digital signal. They do not provide the advantages, in transmission, of true digital signals.

12.5 QUANTIZATION AND CODING

We are looking at the task of preparing analog signals for transmission in digital communications systems. The first of three elements of the process is sampling, as described in the foregoing paragraphs. The second element, *quantization*, is one of converting the sampled information into digital information in preparation for *coding* as a digital signal—the third concept of the process. The two operations—quantization and coding—are closely related. Indeed, in real systems these two steps of the digitalization process are invariably accomplished in a single hardware function. The function is called a *coder*.

Remember, the information to be converted to digital form is amplitude information about an analog signal. By definition, "digital" implies that in a given system, there will be a finite number of amplitude levels represented. *Therefore, the quantization process always involves an approximation or "rounding off" of level information.* This is true because the amplitude of an analog signal changes continuously: It has an infinite number of amplitude levels. Approximating a value always results in some error. The approximation error introduced by quantization is called *quantization error* or *quantization noise*.

In this section we examine three basic schemes for quantization of analog signals: *uniform quantization, differential quantization* and *delta modulation*.

Uniform Quantization

As an electronic process, quantization consists of *comparing* an analog signal to a reference signal of x discrete levels. The value of x, the number of levels in the reference signal, is dictated by the size (i.e., width—number of bits) of the binary code with which the system is designed to operate. For example, if a 3-bit code is to be used, the reference signal will have eight levels (counting the zero level). Sixteen levels would be appropriate for a 4-bit code, and so on. When the amount of change—the "height" of the "steps"—of the reference voltage is constant or *uniform*, say Δv, then the process will be one of *uniform quantization*. The waveform, called a *ramp*, for a reference voltage for uniform quantization is illustrated in Figure 12.6. The ramp shown contains eight levels and is therefore suitable for use with a 3-bit binary code.

It is useful to think of quantizing as a process of (1) converting an analog (continuously varying) signal into a signal that changes in discrete steps, and (2) determining which step or level is present at each sampling instant. An example of the result of the application of part 1 of this process to a sine wave is illustrated in the drawing of Fig. 12.7(a).

The error of this rounding process, as introduced at each sampling instant, is depicted in Fig. 12.7(b). Recall: The error is called quantization error or quantization noise.

The overall process of quantization (coding) can be accomplished in a

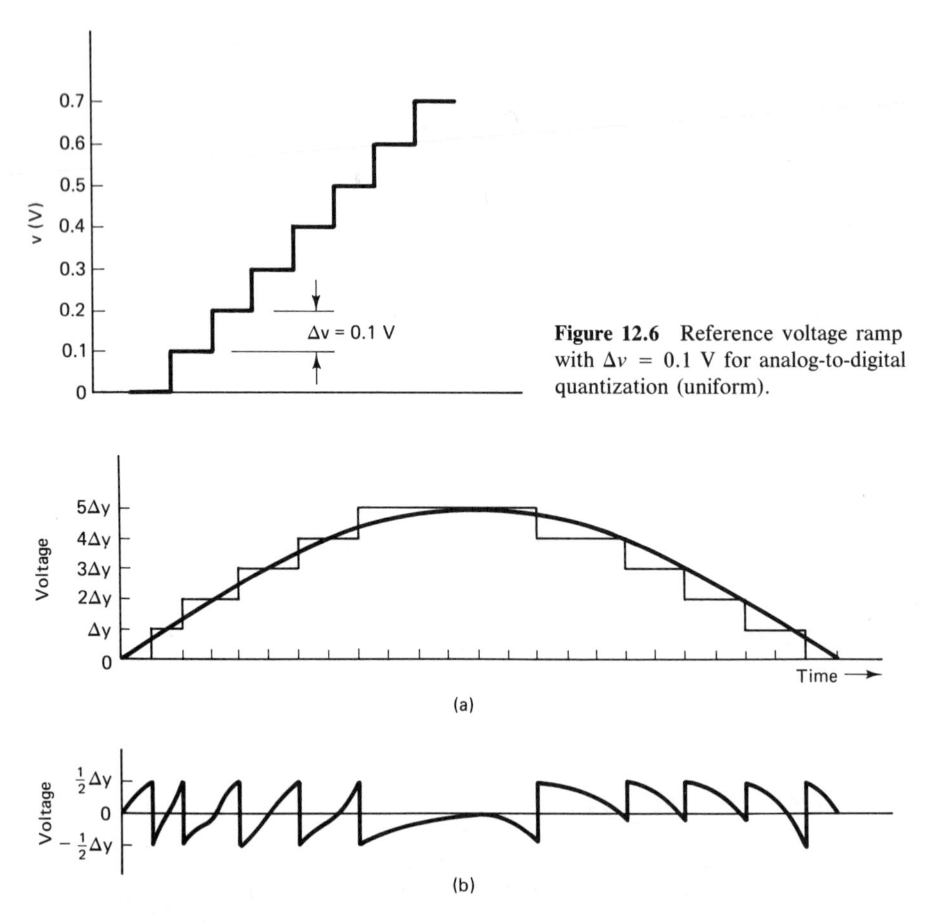

Figure 12.6 Reference voltage ramp with $\Delta v = 0.1$ V for analog-to-digital quantization (uniform).

Figure 12.7 Quantization and quantization error: (a) quantization of analog waveform; (b) quantization error due to rounding.

number of ways. The schemes can be classified as *parallel coding* or *serial coding*. The terminology refers to the manner is which the binary data are produced: all bits simultaneously as in parallel coding, or bit by bit as in serial coding. Let's examine a parallel coding scheme first.

Parallel coding. Figure 12.8 illustrates an eight-step, parallel PCM coder. The output of this function is a three-bit binary "word," a unique word for each input condition. In the scheme depicted, each of seven *voltage comparators* compares the sampled voltage from an analog signal to one of seven, nonzero levels which, together, make up the "reference signal." A comparator output is high only if both inputs are equal to or above a predetermined threshold level. The *sample and hold* function "holds" a sample of the analog signal long enough at each sampling instant for the comparator outputs to be determined. The outputs of the comparators are fed to a digital integrated-circuit (IC) function called a *priority encoder* where the information is changed to a 3-bit binary code.

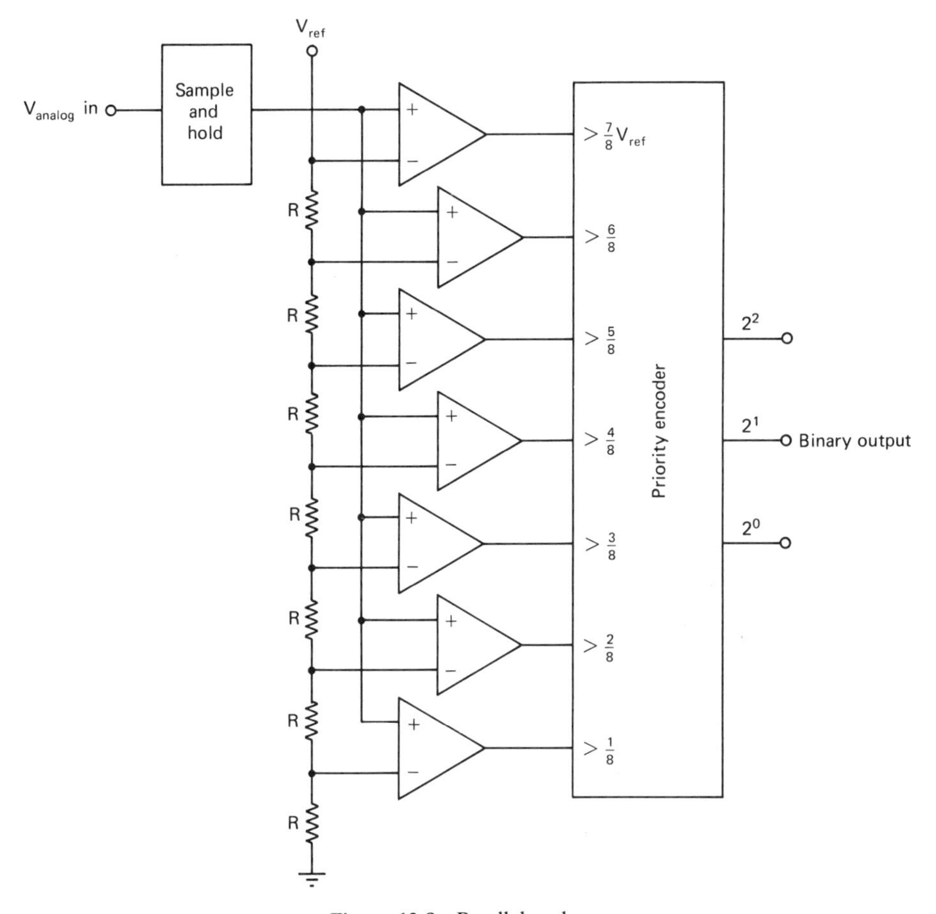

Figure 12.8 Parallel coder.

For example, if the outputs of all seven comparators are high, indicating the analog signal is present at the highest amplitude level, the binary output code is 111. If comparators ">1/8" through ">5/8" have high outputs, the binary output will be 101, etc.

Let us digress briefly to take note of an important detail of the priority encoder. Digital encoders are very common functions in the world of digital equipment. For example, when a particular key on a calculator or computer input terminal is pressed, the information of that event is sent to the system as a binary code. The function which converts the simple on/off signal from a single switch to a coded signal is an encoder—a *keyboard encoder*. In the case of the keyboard operation, the encoder recognizes only a single input at a time—the input which became active first if more than one key is pressed at one time. In the priority encoder, two or more or all of the inputs may be active. The internal logic "assigns" a priority to each input. When more than one input is active, the logic recognizes the one assigned the highest priority. In the case of the scheme of Fig. 12.8, the encoder is connected so that higher-priority inputs correspond to higher analog amplitude levels. An example of a commercially available priority encoder is the TTL package designated 74148 with eight inputs and a 3-bit binary output.

The parallel coder scheme of Figure 12.8 is very fast. The output is available virtually instantaneously upon the application of an input. The only delay for a given analog signal sample is the propagation delay through the logic functions of the encoder. The input to the sample-and-hold function should be a PAM (pulse amplitude modulation) signal.

Although the parallel coder is fast its hardware requirements increase directly with the number of sampling levels. For example, a 16-level coder would require 15 comparators. The parallel coding scheme is used primarily for wideband signals such as video.

Serial coding. In serial coding, each sample (of the analog signal) is converted bit by bit to a code word. The process is obviously slower than parallel coding. However, the amount and complexity of hardware is less than for parallel coding. The process is entirely adequate for narrow bandwidth signals such as that of speech. Serial coding is generally popular in speech transmission systems.

Several schemes for achieving serial coding have been devised. We examine one of the most popular schemes: a *counter-controlled analog-to-digital converter* (ADC). The counter-controlled ADC is also sometimes called a digital-ramp or staircase ADC. An interesting feature of most serial analog-to-digital (A/D) coders is that they employ *digital-to-analog converters* (DACs) as subfunctions. We look at digital-to-analog conversion in the next section. It is an essential function of digital communications systems: DACs or *decoders*, as they are also called, are needed to regenerate an analog signal from the transmitted digital signal.

A counter-controlled A/D converter is depicted in Figure 12.9. In this scheme, a binary counter is restarted from zero at each sampling instant. The ramp reference voltage also starts from zero at the beginning of each sampling interval and is fed to one input of a voltage comparator. The other input of the comparator is the sample of the analog voltage. When the ramp voltage reaches the level of the sampled analog voltage, the output of the comparator goes high. This high level is used to stop the counter. The changing of levels in the ramp waveform is coordinated with the "counts" of the counter by means of the "clock" pulse stream: Each count of the counter corresponds to a level of the ramp reference voltage. The count on which the counter stops indicates the "step" of the ramp waveform at which it became equal to or greater than the analog voltage sample. Hence, the output of the counter indicates, in binary-coded form, the level of the analog voltage at each sampling instant. As we see in the next section, the ramp voltage can be produced by a digital-to-analog converter: the ramp generator is a DAC.

Comparator. The *comparator*, *voltage comparator* or *analog comparator* (all are used as titles of the same function) is an important function in analog-

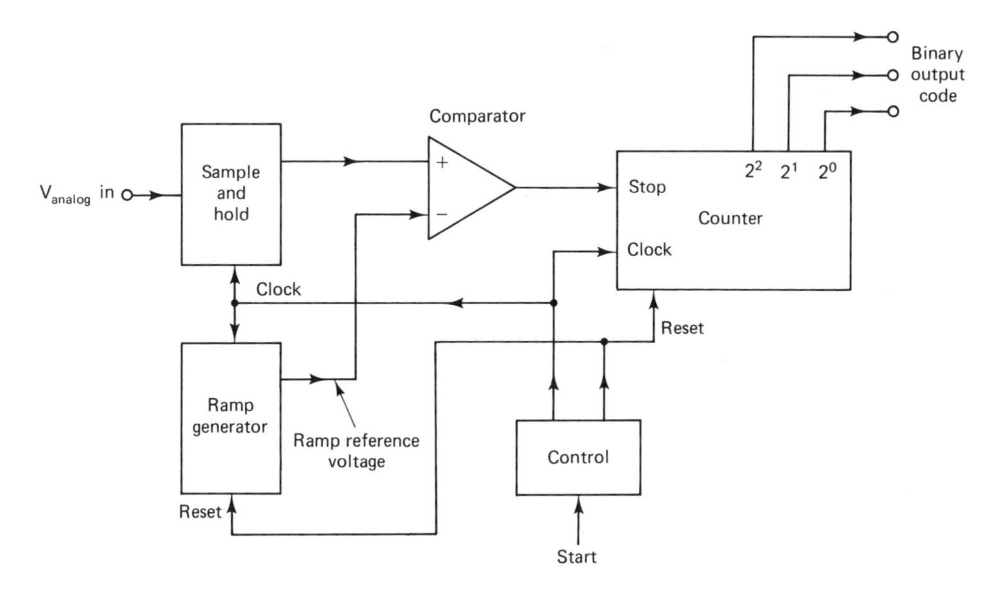

Figure 12.9 Counter-controlled coder.

to-digital conversion schemes. We have observed its use in both the parallel and serial coders. Let's examine its operation in some detail.

An analog comparator for use in A/D converters is basically an operational amplifier (op-amp) which has been modified to produce an output compatible with digital logic circuits. For example, if the comparator is to be used with TTL logic circuits, its output would be approximately 5 V to represent a logical 1 and approximately 0 V to represent the logical 0. A wide variety of voltage comparators for use in A/D converters is now available in IC chip form. Examples are the Motorola LM111 and LM139 devices, as well as many others. Comparable RCA devices are numbered CA311 and CA339.

A very much simplified schematic diagram of an IC voltage comparator is shown in Figure 12.10. Transistors Q_1 and Q_2 form a simple differential amplifier. The operation of Q_3 is controlled by the relative conduction of Q_1 and Q_2; Q_4 is controlled by Q_3. If the voltage at A is greater than the voltage at B, $I_{c1} > I_{c2}$ and Q_3 is biased off. The voltage at the base of Q_4 (collector of Q_3) is high positive and Q_4 is biased off. The voltage at the point labelled OUTPUT is 0 V or logic 0. The answer to the logic question Is $B > A$? is "no." If $B_{in} > A_{in}$, then $I_{c2} > I_{c1}$ and Q_3 conducts turning on Q_4 and producing a voltage at OUTPUT approximately equal to the supply voltage, V_{CC}. V_{CC} is chosen to correspond to logic 1 for the system in use. A 1 at OUTPUT indicates that $B > A$.

Summary: Uniform Quantization

Quantization or coding, in the context of this chapter, is the process of converting a sampled, communications, analog signal into a digital signal. The digital signal obtained is a stream of binary words—groups of pulses representing 1's and 0's.

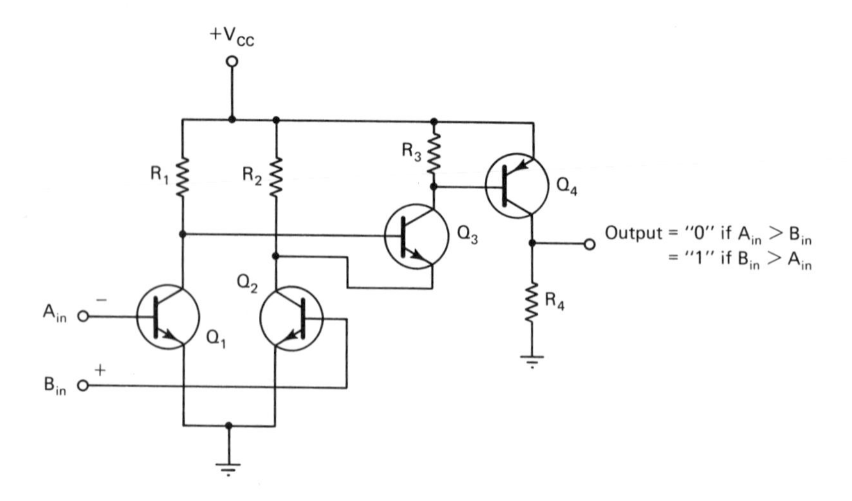

Figure 12.10 Simplified circuit of voltage comparator.

The set of electronic functions, which, acting together, accomplish the result of analog-to-digital conversion is called a coder.

We have now examined, at a block-diagram level, the details of two coding schemes (i.e., coders) for uniform quantization. The highest speed in coding is accomplished by a parallel coder—a scheme that generates all bits of the output word simultaneously (in parallel). This coder is referred to as a *flash coder*, especially by those who manufacture it as an integrated circuit. A counter-controlled (or digital ramp) coder converts an analog signal to digital by generating the output word one bit step at a time. It is a relatively simple scheme but also relatively slow. It is a form of serial coder. (A faster serial A/D converter is the successive-approximation coder.) The serial coder we examined requires the function of a digital-to-analog converter and one voltage comparator. A parallel coder requires several comparators. The number is equal to the number of sampling levels minus 1.

Nonuniform Quantization

When an analog signal is quantized by comparison with a reference voltage that changes in discrete steps of unequal or *nonuniform* value, the process is called *nonuniform quantization.* Uniform quantization, quantization using discrete steps of uniform value, seems to be very straightforward. If it works, why would anyone want to fool around with a changing (nonuniform) voltage step size? The answer is that there are several advantages to be gained by using a variable step size. To "appreciate" what these advantages are and why they arise, let us examine some additional technical aspects of uniform quantization.

As we have already observed, quantization carries with it some inherent error called quantization error or quantization noise. The significance of this error is that when an analog signal is reconstructed from the information contained in a derived digital signal, the reconstructed signal will not be exactly the same as the original analog signal. The reconstructed signal will differ from the original by the amount of the error.

Quantization error can be reduced to a desired value by the use of smaller and, consequently, a greater number of steps in the quantization process. However, more steps mean more digital information must be transmitted: Each step requires more bits to represent it; more words—groups of bits—are required in a given time period, and so on. The effect is that the message requires a greater bandwidth. A given transmission medium—coaxial cable, microwave link, etc.—is limited in the number of bits of digital data it can carry per second. Although the limits of typical transmission media are high, all transmission links in commercial systems are expected to carry numerous messages simultaneously. Increasing the number of bits required for satisfactory transmission of each message, to reduce error, simply reduces the number of messages that can be transmitted in a given time span. Such a measure is economically undesirable. Technical alternatives which have the potential of getting around the problem are attractive. One such alternative is nonuniform quantization.

In a given transmission system, with uniform quantization steps, the quantization error is greatest for signals of small amplitude changes and least for signals of large amplitude changes. As an aid in visualizing this effect, consider the following. A uniform quantization scheme utilizes eight steps of 0.5 V each. A sine wave with a peak-to-peak amplitude of 4 V (−2 V to +2 V) requires all eight steps for representation. A sine wave with a peak-to-peak value of 1 V is represented with only two steps. Clearly, if the system used more and smaller steps for signals of low amplitudes and fewer and larger steps for signals of high amplitudes, the conversion process would be more precise.

It is widely recognized that the distribution of amplitudes in typical information signals—speech signals, for example—is not uniform. The probability of the occurrence of low amplitudes is greater than that of high amplitudes. It is quite reasonable, then, to imagine a system that utilizes smaller steps for signals of low amplitude and larger steps for high-amplitude signals. Most important, this is accomplished without an increase, on the average, of the total number of steps required (as compared to a system with uniform steps).

Well, nonuniform quantization sounds great, doesn't it? Why don't we go for it? Unfortunately, implementing such a system directly is far easier said than done.

The most obvious method of achieving nonuniform quantization would be to modify one of the techniques we have already examined, for example, the parallel quantizer (see Fig. 12.8). The step differences between the reference voltage inputs for each of the comparators would be modified. The steps between the first few comparators (>1/8, >2/8, etc.) would be decreased; the steps between the comparators of high amplitudes would be increased. The difficulty in this scheme is in providing the precision, drift-free reference voltages required.

Another scheme that has been used successfully first quantizes a signal using a large number of small steps. Then, after the information is encoded in digital form it is processed with digital circuit functions that translate it into a new form. The new form is equivalent to a representation of the original analog signal using nonuniform steps.

The most popular scheme used in commercial systems for achieving the equivalent of nonuniform quantization is a technique called *companding*. "Compand" is a manufactured word: it is made from *com*press-ex*pand*. The process utilizes a function called a *compressor* at the transmitting end of a system and an *expandor* in the receiver. The compressor reduces (compresses) the range of amplitudes in the information signal. The compression is nonlinear. When processed by a compressor, an analog signal is deliberately distorted: signal components of high amplitude are attenuated more than components of low amplitude. The output of the compressor is processed in a quantizer using uniform steps. The result is identical to nonuniform quantization. At the receiver, after decoding (conversion from a digital to an analog signal), the reconstructed analog signal is processed by the expandor, which restores the normal amplitude relationships. A block diagram illustrating the basic functions of a compandor system is provided in Fig. 12.11.

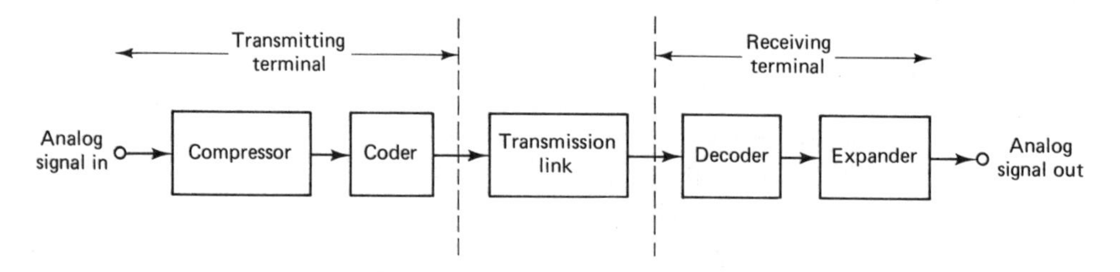

Figure 12.11 Block diagram of compandor system.

Differential Quantizing: Delta Modulation

In the methods of quantizing that we have considered thus far, each word of digital data (i.e., each group of bits) represents amplitude information about an analog signal. That information is for a particular sampling instant. Furthermore, the amplitude information is absolute. That is, the system measures the amplitude of a point on a waveform with respect to an absolute or fixed reference: usually ground or 0 V.

Another way of communicating amplitude information is to transmit information about a *change* from an immediately preceding level. For example, a digital 1 could be used to indicate a positive change from the last level and a 0 to indicate a negative change. As a general concept, this technique is termed *differential quantizing*. When the measure of a change is a fixed amount, for example $\Delta v = 0.1$ V, the system is called *delta modulation* (DM).

An example of an information signal quantized by delta modulation is shown in Fig. 12.12. Remember: If the analog waveform changes by an amount equal to or greater than Δv, a 1 is transmitted (the delta-modulated equivalent waveform increases one step). If the analog waveform amplitude change is less than Δv, a 0 is transmitted (the delta waveform decreases). The digital signal resulting

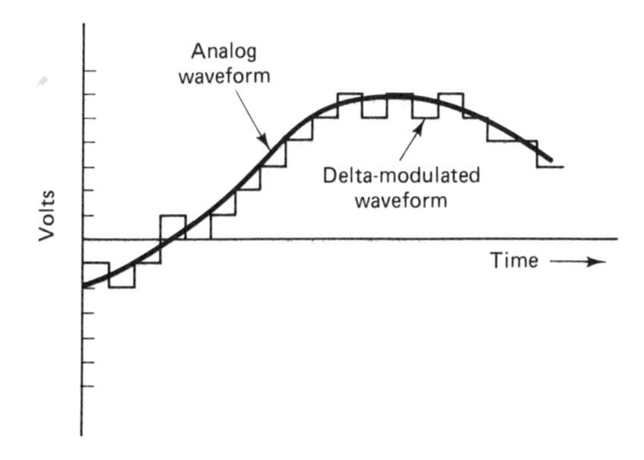

Figure 12.12 Delta modulation.

from delta modulation is simply a stream of 1's and 0's representing positive or negative changes in the analog signal.

Implementation of delta modulation requires some form of memory function: a circuit which "remembers" an amplitude level that can be changed up or down upon receipt of a 1 or 0 signal. One such circuit capable of performing this function is called a *staircase generator.* A staircase generator produces a delta-modulated waveform (see Fig. 12.13). Figure 12.13 shows a block diagram illustrating the implementation of delta modulation using a staircase generator.

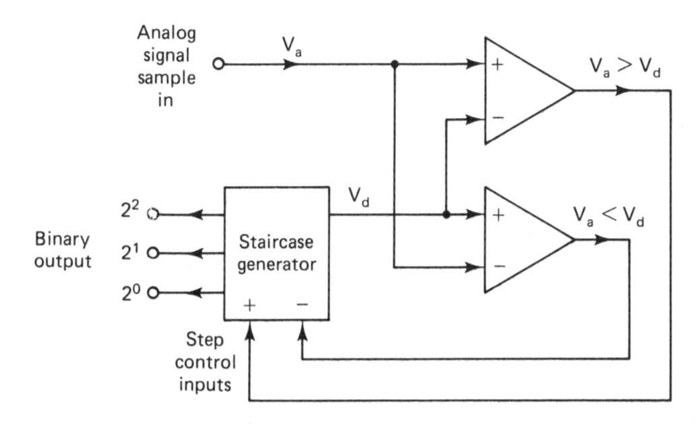

Figure 12.13 Delta modulation using staircase generator.

As with other forms of quantizing, DM is subject to error or quantizing noise. The process of delta modulation is sometimes referred to as "tracking the analog signal" (*to track* means to trace or follow along the path of). When plotted against the original waveform, the errors of a DM waveform become obvious because the DM waveform does not "track" the information signal well. An example of this is shown in Fig. 12.14(a). The result is called *slope clipping*, for an obvious reason. The problem is identified as *overloading*. Overloading is an appropriate word, as the process is not able to keep up with the task required of it. The quantized waveform does not change rapidly enough. The remedies are (1) to limit signals to relatively slowly changing ones, (2) to increase the step size, or (3) to increase the sampling rate. If the step size is too great, however, the result is overquantization, as illustrated in Fig. 12.14(b). If the

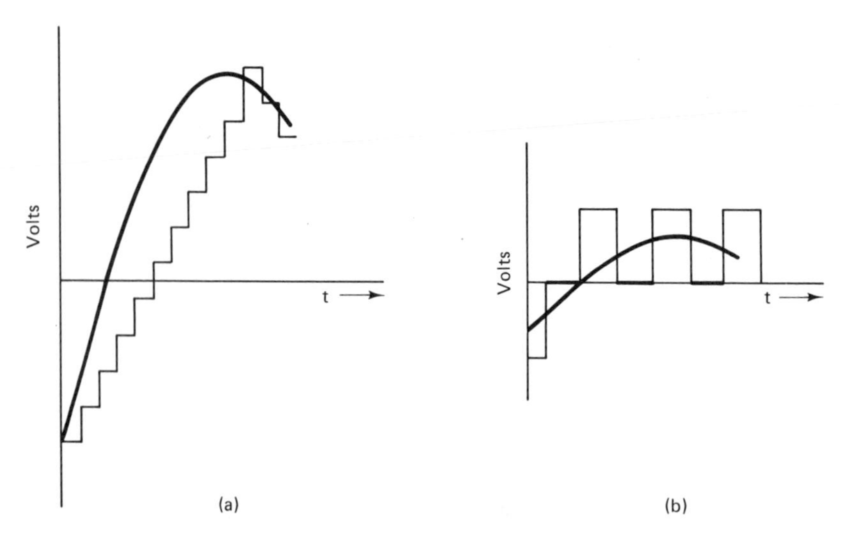

Figure 12.14 (a) Slope clipping in delta modulation; (b) overquantization in delta modulation.

sampling rate is increased significantly, the transmission requires an excessive amount of channel space.

Summary: Quantization

The process by which successive samples of an analog information signal is converted into digital information is called quantization. There are numerous schemes to accomplish the process. We have examined in a general way uniform quantization, nonuniform quantization and delta modulation, a form of differential quantization. All methods have inherent error called quantization error or quantization noise because all methods involve approximation or rounding off.

12.6 DECODING

A digital communications system consists of many functions. Coding is only one of several functions performed at the transmitting end of the system. Before looking at these other functions, however, let's look at the other side of coding, that is, *decoding*. When an analog signal, which is being transmitted by digital means, reaches its destination, it must be changed back into analog form before it can operate a speaker, produce a TV picture, etc. In a word, it must be decoded.

A digital communications decoder is, fundamentally, a digital-to-analog converter. Digital-to-analog conversion is the process of converting a set of specific digital values into a continuously varying (analog) voltage or current. The amplitude of the analog voltage or current, at a given instant, is in accordance with a predetermined formula or code. That is, the circuit produces a given amplitude for each input digital value. Generally the digital value is represented as a binary code.

The general concept of a DAC is represented as a simple block diagram in Figure 12.15. Please observe, from the table included, the relationship between the 4-bit input to the device and V_{out}. DACs are widely available as IC devices. It is not an efficient use of our time (and text space) to examine in detail the nuances of different circuit designs. However, a knowledge of the basic ideas used in D/A conversion circuits is useful as a background to a study of their application.

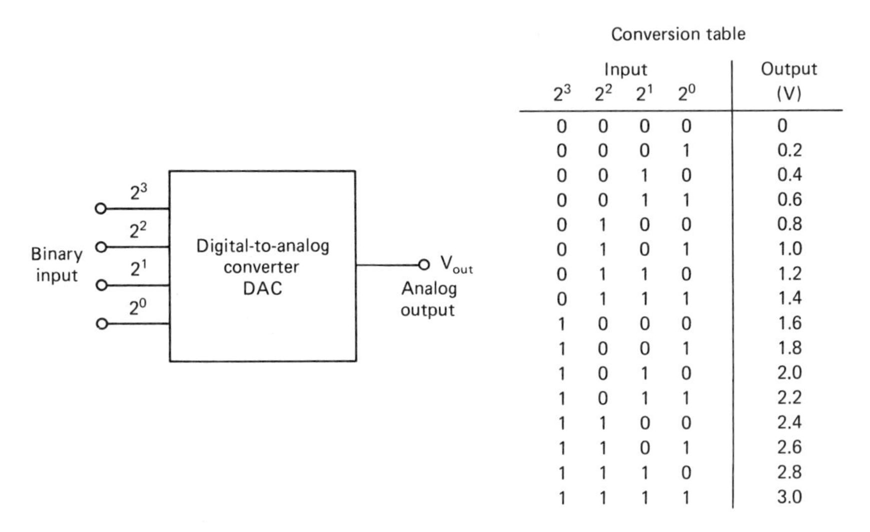

Conversion table

Input 2^3	2^2	2^1	2^0	Output (V)
0	0	0	0	0
0	0	0	1	0.2
0	0	1	0	0.4
0	0	1	1	0.6
0	1	0	0	0.8
0	1	0	1	1.0
0	1	1	0	1.2
0	1	1	1	1.4
1	0	0	0	1.6
1	0	0	1	1.8
1	0	1	0	2.0
1	0	1	1	2.2
1	1	0	0	2.4
1	1	0	1	2.6
1	1	1	0	2.8
1	1	1	1	3.0

Figure 12.15 General concept of digital-to-analog conversion.

One technique for converting a binary-coded input to an analog output is by means of a simple, resistor network. The schematic of such a circuit is shown in Figure 12.16(a). In this circuit, single-pole, double-throw toggle switches are used to represent a 4-bit binary input. The network is supplied by a voltage source V_s; the analog output is V_{out}. You will observe that the network requires only two resistor values: R and 2R.

The output of a D/A conversion network is predicted by analyzing the circuit produced by a given switching condition. Before analyzing specific examples of input conditions, let's observe the general effect of placing a switch in the "0" position. First of all, when a switch is in the binary 0 position it grounds the circuit at that point. For example, with all switches at 1 except the 2^0 switch, the resistance to ground for point C is R and for point B, 2R. Study the diagram of Figure 12.16(b). For the switching condition 1100, the resistance to ground of point *B* is *R* and for *C* it is 1.2*R*, and so on. With four switches there are 16 possible analog voltage output levels, counting 0 V as a level. With all four switches in their 0 positions, the resistor network is totally grounded, is not connected to V_s, and $V_{out} = 0$.

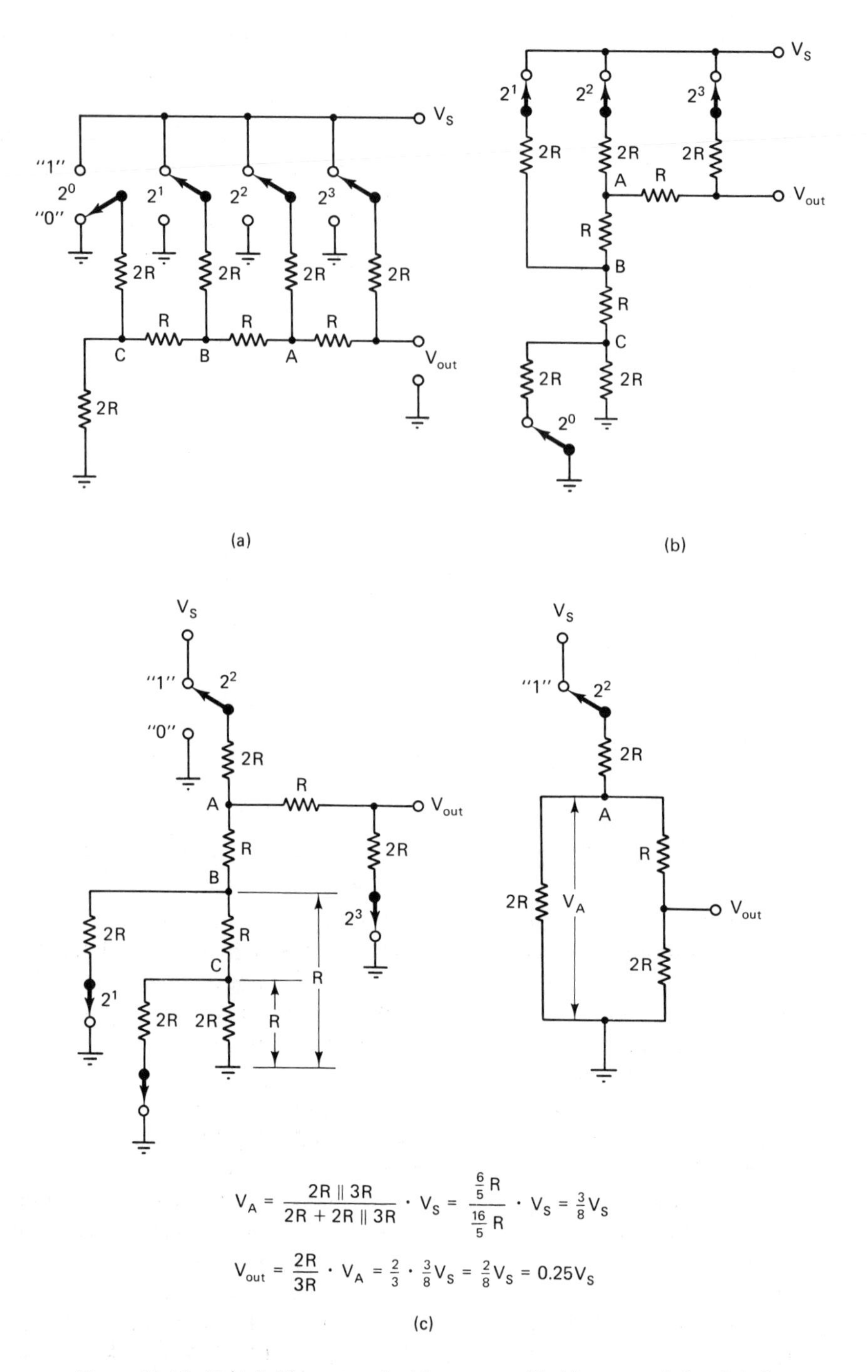

Figure 12.16 D/A ladder network: (a) concept of ladder network for digital-to-analog conversion; (b) equivalent circuit with binary input of 1110; (c) analysis of circuit with 2^2 = "1" all others = "0" (0100).

Let's analyze the network for the input condition 0100 [see Fig. 12.16(c)]:

$$V_A = \frac{2R\|3R}{2R + (2R\|3R)} \times V_s = \frac{1.2R}{3.2R} \times V_s = 0.3750V_s$$

$$V_{out} = \frac{2R}{3R} \times V_A = \frac{2}{3} \times 0.375V_s = 0.25V_s$$

Careful analysis will show that each additional binary value of 1 produces an additional analog value of $V_s/16$. That is, V_{out} for the binary input 0001 is $V_s/16$; for 0010, $V_s/8$; for 0011, $3V_s/16$; etc. Analyze additional switching conditions using the technique demonstrated to verify for yourself that the circuit functions as stated. Correct operation of the circuit assumes that the input impedance of any circuit driven by V_{out} will be very much larger than the impedance seen looking back into the resistor network.

In practical D/A converters the network must respond to binary logic levels. This can be accomplished by transistor switching circuits in place of the manual switches shown in Fig. 12.16. A circuit to illustrate the idea of transistor switches capable of responding to logic level input is shown in Fig. 12.17.

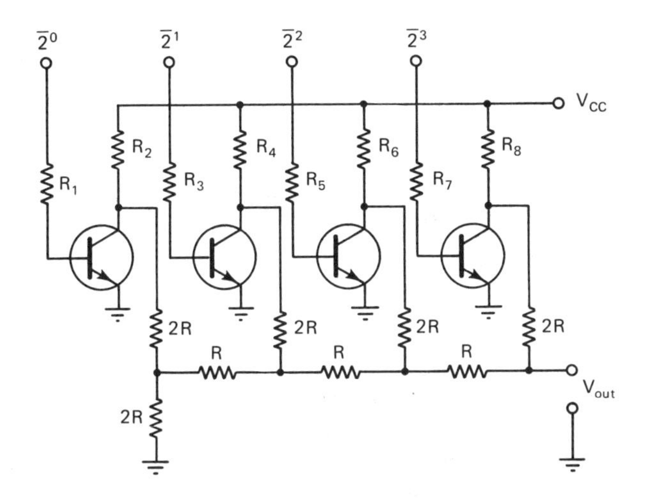

Figure 12.17 D/A ladder network with transistor switches.

Several integrated-circuit packages that implement digital-to-analog conversion are available. For example, the Motorola AD562 IC is a 12-bit high-speed D/A converter. It functions in a manner similar to that of the circuit of Fig. 12.17. The AD562 can be adapted for use with either TTL logic or CMOS logic. It is one of a family of chips referred to as data conversion devices.

Telecommunications systems are virtually always bidirectional. Those constructed to provide telephone-type services always provide for two-way transmission of messages (i.e., full-duplex operation). Consequently, system terminals—equipment sites where messages are transmitted and received—require both coders and decoders. These are physically and logically related. The combination is given the title *codec* (*co*der-*dec*oder) in the terminology of the communications industry.

Single codecs may be used to process many simultaneous transmissions in a sharing procedure called *frequency division multiplexing* (FDM). Or multiple, so-called *single-channel codecs* may be used in a multiple-message transmission scheme called *time-division multiplexing* (TDM). We look at multiplexing and other system details next.

12.7 A BASIC DIGITAL COMMUNICATIONS SYSTEM: A SECOND LOOK

We began this chapter by looking at a generalized block diagram depicting the function blocks of a digital communications system (see Figure 12.1). We discovered that the diagram contained a few functions with which we have some familiarity: transmitter, receiver, and transmission media such as a coaxial cable.

However, the diagram indicated a need for several functions that we had not encountered previously in this text: coders, decoders, multiplexers, and demultiplexers. Subsequently, we looked into coders and decoders. If that effort was successful, we now have at least a basic working knowledge of those functions. It is time to take a second look at the system. We want to fill in a few more gaps in our knowledge about how the system operates. And there are other subfunctions of the system with which we are unfamiliar. We will examine these in a manner similar to our look at coders/decoders.

In digital communications systems there are two other areas of major technical interest in addition to the area that we have just examined: coding/decoding. One of these areas is multiplexing/demultiplexing and the other is the transmission of digital signals. We shall look at each of these in the following pages.

Multiplexing/Demultiplexing

Many of the advances that have been made in telecommunications have been the result of attempts to keep up with, or better anticipate the growing demand for communications facilities. A major portion of these advances are concerned with schemes for the sharing of facilities. It was never feasible to run a pair of wires between each telephone user and every other user. In the early days of telephony, "party lines" were a commonplace. Telephone company customers could listen in on their neighbors' phone conversations. Today, telephone customers generally are connected to a local switching station with their own private line (a twisted-pair transmission line). From that point on, however, telephone users share facilities. Sharing of long-distance facilities, especially, is an economic necessity. Transmission media, such as a single coaxial cable or a single microwave radio channel, carry numerous simultaneous telephone conversations in normal operation. The technical name for this method of operation is *multiplexing*.

Prior to the rapid growth of digital communications techniques, a relatively recent phenomenon, multiplexing generally meant *frequency-division multiplexing* (FDM). One of the advantages of digital communications, and one of the reasons for its rapidly growing popularity, is that it utilizes very effectively a different multiplexing technique—*time-division multiplexing* (TDM). Time-division multiplexing, as we shall see, can be implemented relatively easily with low-cost digital ICs. This gives it an economic advantage over the more complex FDM and adds an attractive attribute to digital communications systems.

A brief look at FDM is provided in Chap. 8 in connection with the stereo multiplexing of FM broadcasting. It is not an objective of this section to extend significantly a coverage of FDM. However, a quick review of it and a brief look at its application in telecommunications systems will provide a useful base for an examination of TDM.

Frequency-division multiplexing (FDM). In Chap. 8 multiplexing was defined as a process by which two or more unique information signals are sent simultaneously over the same communications medium or master channel. When the process is accomplished through the technique of transmitting separately modulated subcarriers, it is called frequency-division multiplexing. The rudimentary ideas of a simple telecommunications system using FDM are illustrated in the block diagram of Fig. 12.18(a). You will observe that each information channel requires its own subcarrier generator and modulator and a bandpass filter at the transmitting end of the system. At the receiving end, each channel requires a demodulator, a circuit for the regeneration of the subcarrier, and a bandpass filter.

The use of subcarriers enables the system to transmit multiple information signals simultaneously; it spreads the signals over a segment of the radio-frequency spectrum. The idea is depicted in the spectrum diagram of Fig. 12.18(b). In systems that include a radio link, the subcarriers modulate a master carrier. FDM is a highly developed technique. Systems are in use which enable several thousand simultaneous telephone conversations, for example, to be combined to form a single signal transmitted over a single transmission link.

Time-division multiplexing (TDM). In time-division multiplexing two or more signals are sent over a shared transmission system in a scheme in which each signal is allocated utilization time—a time slot—on a regular, recurring basis. A straightforward analogy of the scheme is illustrated with the motor-driven rotary selector switch depicted in Fig. 12.19. Observe that a 10-channel system is represented. If the motor drives the switch at a speed of one revolution per second (1 rps), each channel is enabled to transmit over the single transmission link for a period of approximately 0.1 s, once each second. In a very narrow, technical view of the scheme, transmission of messages is not truly simultaneous. Indeed, the method is not appropriate for the transmission of unconverted analog signals. However, the technique is virtually a natural for the transmission of pulse-code-modulated analog signals. This is a result of the fact that in PCM, a signal is already time-divided by sampling, an integral step in the analog-to-digital conversion process.

Time-division multiplexing is not a function exclusive to large digital communications systems. In fact, TDM is used in much of the digital equipment

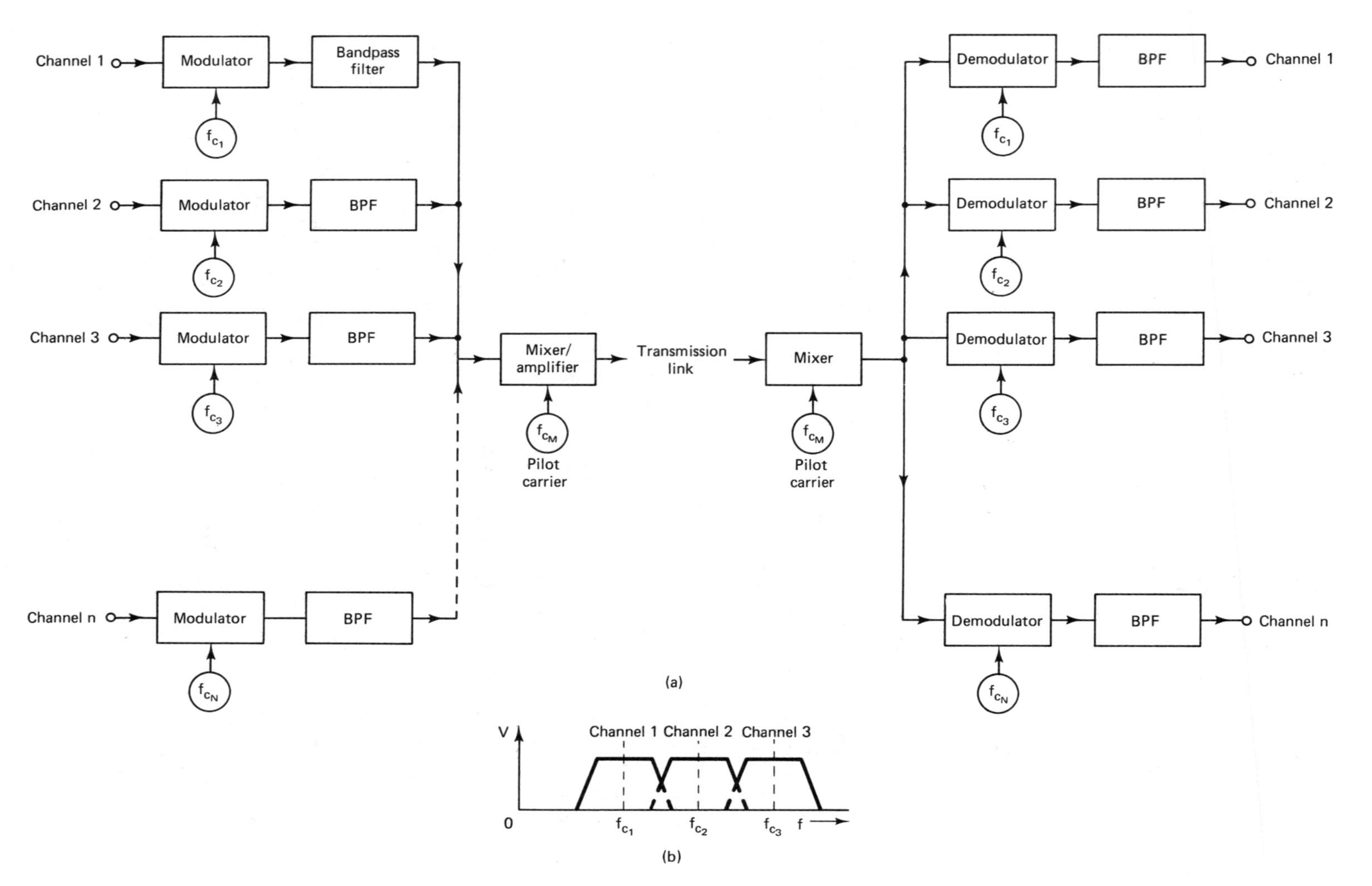

Figure 12.18 Frequency-division multiplexing: (a) rudimentary frequency-division-multiplexed system; (b) frequency spectrum of FDM.

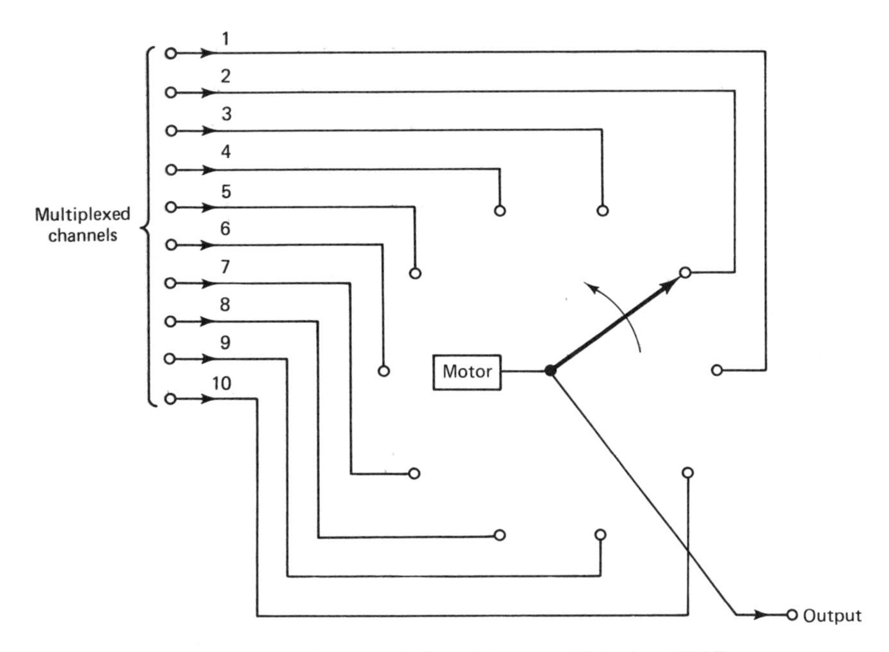

Figure 12.19 Concept of time-division multiplexing (TDM).

with which you are already familiar. For example, hardware sharing by means of TDM is a feature of the design of most equipment with digital readouts: hand-held calculators, digital multimeters, frequency counters, digital watches, and many more. High-speed IC devices with switching rates in the megahertz range are readily available.

A simple TDM system. Let's look now at some basic details of a rudimentary, multichannel TDM communications system. Refer to Fig. 12.20. The block diagram shown is a model of a *single-channel codec system*: It is a system in which each information channel has its own coders and decoders. (There are systems in which all channels share a set of common coders/decoders—shared codec systems.) The system shown is an 8-channel system. The number of channels is an arbitrary choice, in this case. For one thing, it illustrates easy implementation of the multiplexing/demultiplexing functions with standard, readily available ICs. For example, a 74151 chip, an 8-line-to-1-line multiplexer, could be used along with a 74138 chip, a 1-line-to-8-line demultiplexer. You are urged to obtain a manufacturer's TTL data manual and look up these devices. Studying a data page on a device will help make it more real for you.

A multiplexer (mux) is logically equivalent to a multiposition selector switch. Its positions are determined electronically rather than manually. Electronic selection is determined by the logic levels on the *select inputs* of the device. For example, an 8-line mux requires three select inputs—S_2, S_1, and S_0. A binary word 000 on these inputs would select input line 0, or communications channel 1, in the sample system of Fig. 12.20. A binary word of 101 would select input 5 or channel 6, and so on.

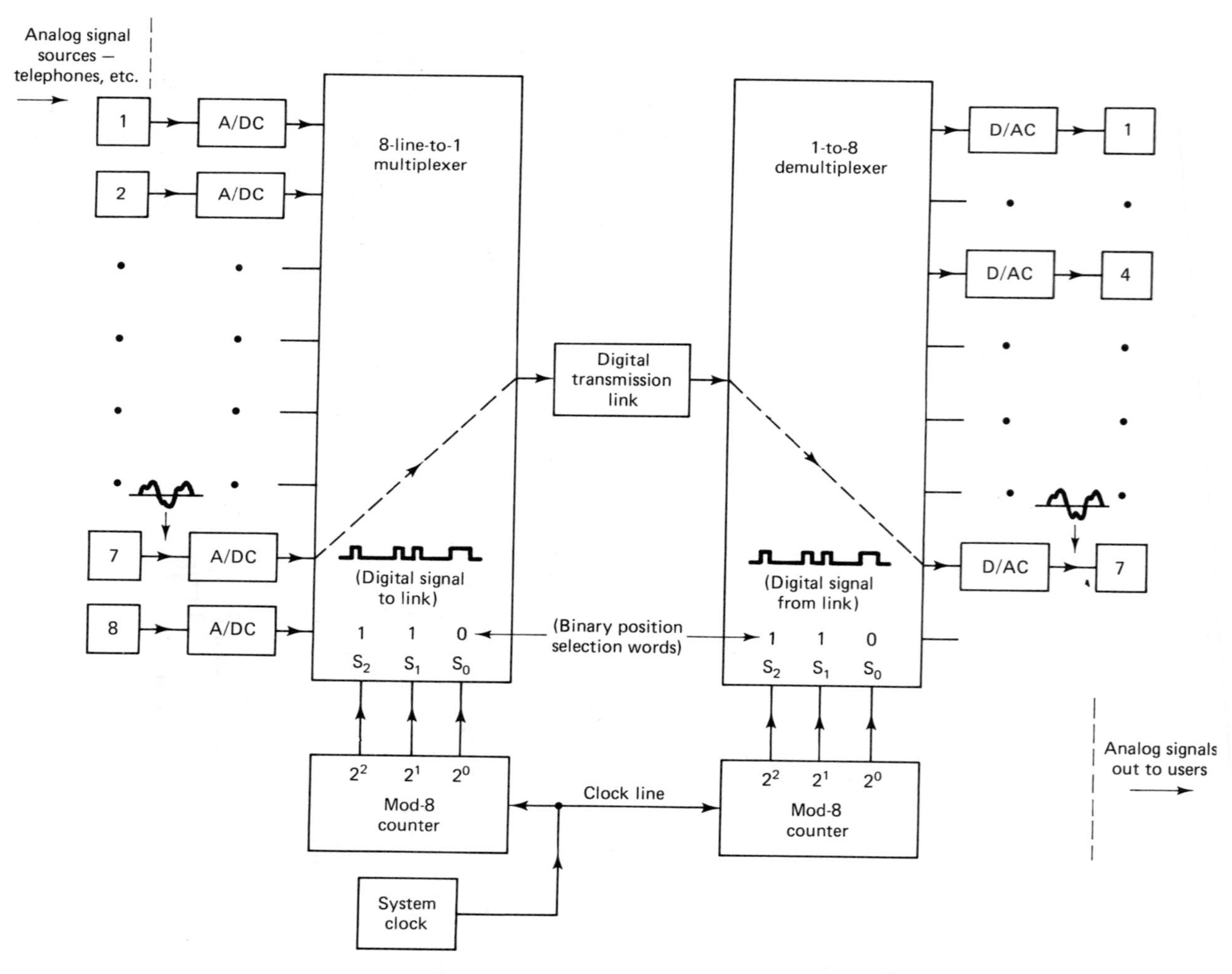

Figure 12.20 Rudiments of TDM digital telephone communications system (left-to-right transmission half of system only, shown).

A demultiplexer (demux) is logically equivalent to a multiposition distributor switch. It "steers" the signal on an input line to a selected output line—one of eight possible output lines in the case of the 74138 chip, for example. Selection of the desired output line is, again, by means of a logic word on the select inputs of the device.

In a simple TDM system the mux and demux functions are cycled through all of their positions (eight in the case of the system of Fig. 12.20) on a regular basis. This procedure allocates a time slot to each channel. As an example, if the system cycles through the eight positions once a second, each channel would be "on line" for approximately $\frac{1}{8}$ s in every second.

The cycling function for the system of Fig. 12.20 is easily implemented by means of 3-bit binary counters. The counters are connected to count pulses from a system clock. The output of the counters are 3-bit words that are fed to the select inputs of the mux and demux. As the counters count from 000 to 111, the mux and demux cycle through their eight positions, connecting first one channel and then another to the transmission link of the system. On the first clock pulse after the count of 111 is reached, the counters reset to 000 and the process repeats itself. The pulse repetition rate (PRR) of the clock signal would be required to be 8 Hz.

Obviously, one of the functions which the system must provide is a means of ensuring that the multiplexer and demultiplexer are interconnecting the correct channels. (When cross connections occur, we end up talking to strangers!) One method of accomplishing this is to provide for using the identical clock signal at both ends of the system. Then, if the counters are reset by a common reset pulse and work correctly all the time, they will count in step and maintain the correct selections of mux and demux. Such a system is called a *synchronous* time-division multiplex. The system of Fig. 12.20 illustrates this feature by means of a clock line running between the two ends of the system.

It is important that you be aware that the functions shown in Fig. 12.20 are only half of a complete two-way communications system. A complete system would duplicate the functions shown, with the transmit functions of the second half on the right and the receive functions on the left. We are not examining the coder/decoder functions. These could be any one of the several types presented in the preceding section.

More Details of TDM Systems

The TDM system presented in Fig. 12.20 is very basic. It leaves out some of the details of actual systems in the interest of simplicity. The diagram was made simple because learning is a step-by-step process. We have taken one small step along the road toward a working knowledge of TDM systems. It is time now to take another step toward the complexity of real TDM systems.

In actual PCM communications systems, each sample of the analog information signal is coded into a code word of a specified number of bits (binary digits). One of the decisions the designers of the system must make is that of the number of bits to be used per sample word. For example, in one of the PCM transmission systems used for voice telephone communication by the American Telephone and Telegraph Company in the United States, the Bell T1 TDM system, an 8-bit word is used. The T1 system is also a 24-channel system; it transmits 24 telephone conversations simultaneously on a single, twisted-wire-pair transmission line.

One of the details ignored in the system of Fig. 12.20 is the fact that the output of the A/D coder is available in parallel form. That is, in an 8-bit system such as the Bell T1, for example, the code word for each sample is available at the outputs of the eight flip-flops of the register or counter of the coder. During transmission, however, a word must appear simply as a string of pulses representing the 1's and 0's of the word. In short, a word is formed "in parallel"; it is transmitted "in serial." A parallel-to-serial hardware function is required. An 8-to-1 mux is also easily adapted to this purpose. The block diagram of a scheme for a *parallel-to-serial converter*, or *serializer*, is given in Fig. 12.21. You will observe that the multiplixer is cycled to select, in turn, the output of each of the eight FFs of the coder by means of a bit counter. The counter cycles by counting a system clock signal (pulse waveform).

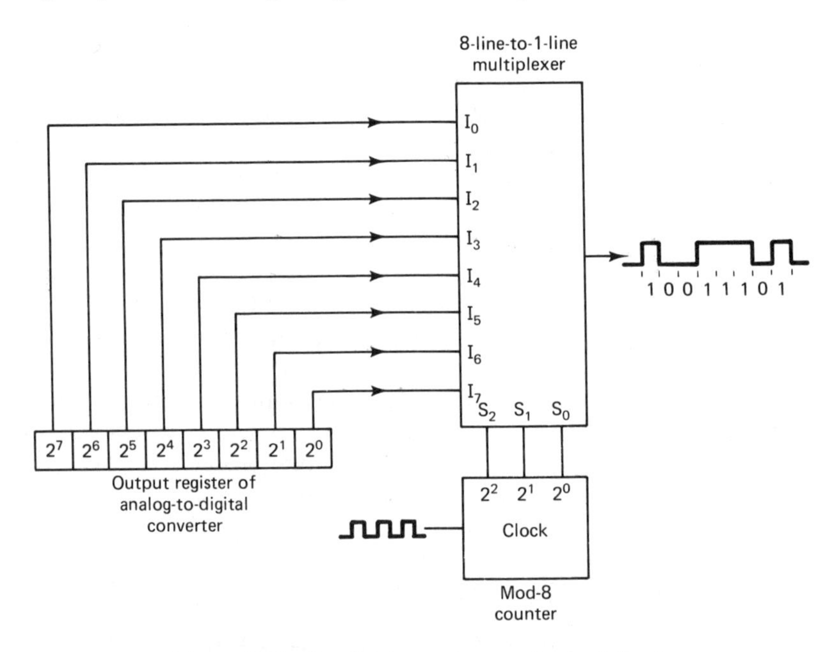

Figure 12.21 Parallel-to-serial converter (serializer).

Interleaving. In TDM, as we have seen, the system scans, on a regular time basis, the various input channels that it services. During the time slot for a channel, it places a coded signal from that channel on the common transmission system. This process is called *interleaving*. If a complete *word* (e.g., an 8-bit word in the T1 system) is accepted from each channel during each pass, the process is called *word interleaving*. If a time slot is equivalent to the period of a single bit, the process is called *bit interleaving*. That is, in bit interleaving, one binary digit per time slot is accepted from each of the channels serviced by the system. These concepts are illustrated in Fig. 12.22.

The actual form of interleaving used is determined by the sequencing of the channel-select multiplexer at the transmitting end. The sequencing of the mux is a function of the channel-select counter (i.e., the speed at which it counts). The speed of counting of the channel-select counter is, of course, determined by the frequency of the clock signal fed to it. The clock signal is typically produced by a very stable crystal oscillator. The frequency of the clock signal fed to the channel-select counter is chosen to provide the desired form of interleaving. The timing of the serializer must be coordinated with that of the channel-select multiplexer.

Of course, as we observed in the case of the model system of Fig. 12.20, the demultiplexer at the receiving end of the system must step in sequence with the channel multiplexer. Some means of synchronizing the stepping of these two functions so that the proper channels at each end of the system are aligned must be provided by the system. An alternative scheme to the separate line for synchronizing the mux and demux of Fig. 12.20 is generally used. One alternative technique involves the transmission of a special-purpose aligning signal on the common line. Such a signal is called a *frame aligning word* (FAW). The title is derived from the idea of a *frame* of data. A frame of data is a sequence of consecutive data bit time slots. The number of slots in the sequence is predetermined. Consequently, the position of each slot with respect to a FAW is predictable.

Various schemes relating to the transmission of frame aligning words (FAWs) are employed. A common technique employs two FAWs per frame. Both *bunched*

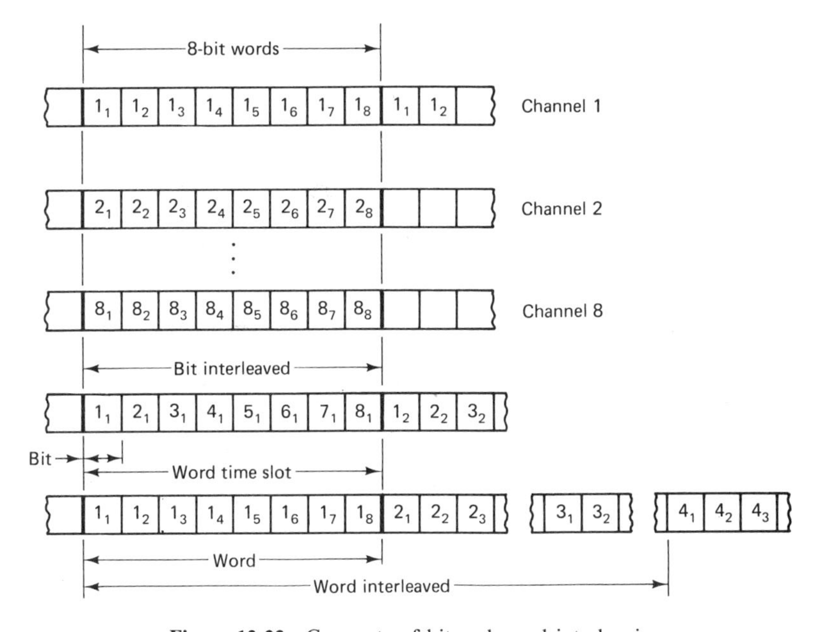

Figure 12.22 Concepts of bit and word interleaving.

framing and *distributed framing* can be used. In a bunched, two-FAW system, both FAWs are transmitted in consecutive time slots. In distributed framing, the FAWS are separated in a frame. These ideas are illustrated in Fig. 12.23. The FAW is used at the receiving end of the system to resynchronize the separate clock at that end.

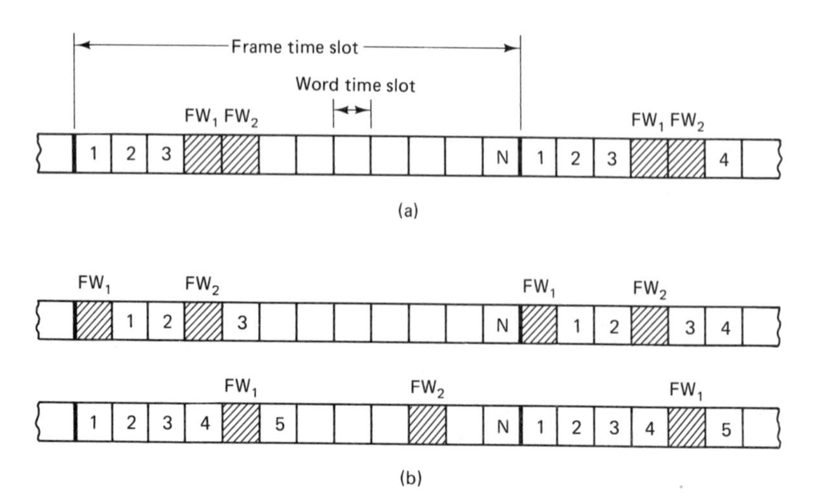

Figure 12.23 Concept of frame aligning words: (a) bunched frame words; (b) distributed frame words.

Let's see how the ideas of framing are worked into our sample system—the Bell T1 system. We have already observed that it is a 24-channel, 8-bit system. The channels operate with a design bandwidth of 4.0 kHz. Sampling is done at a rate of 8 kHz. In the T1 system a frame is defined as 24 words—one 8-bit word from each of the 24 channels. The FAW consists of a single bit added to each frame. A diagram showing the details of a T1 frame is shown in Fig. 12.24. The special, nonsequential order of the interleaving is used to make the system compatible with existing communications systems. The FAW transmission alternates between a 1 and a 0. A complete frame consists of 193 bit slots $(24 \times 8) + 1$. Since each channel is sampled 8000 times per second, there are 8000 frames per second. (Remember: One frame contains one sample from each of 24 channels.)

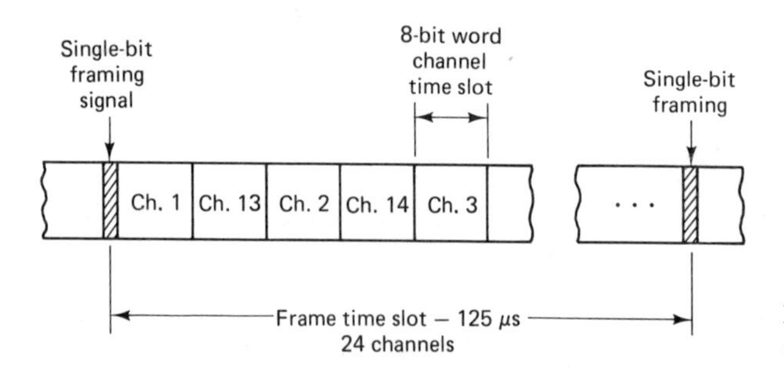

Figure 12.24 Framing for Bell T1 system.

The transmission rate of the T1 system is 1,544,000 bits per second (bps), calculated from 8000 frames per second times 193 bits per frame. Since there must be a clock pulse for each sampling pulse, the master system clock frequency is 1.5440 MHz.

Start-Stop Digital Transmission

In the preceding paragraphs we have examined several aspects of TDM which can be grouped under the heading *synchronous digital transmission*. Let's look very briefly at another type of digital transmission—*start-stop* or *asynchronous transmission* (meaning "not synchronous"). We cannot spare the resources—time and page space—for more than a short look. However, some of the terminology used in start-stop transmission appears so frequently that it would be unwise to skip over the subject completely.

The application of techniques of start-stop transmission predate the development of most electronic devices, and certainly, modern digital transmission systems, by many years. These early digital communications systems utilized electromechanical devices—motors and relays—mostly in the form of *teleprinters*. These were slow-speed devices with transmission capabilities of approximately 12 to 15 characters per second. Synchronous transmission, as described above, was out of the question at that time. The operating speeds of motors simply could not be controlled precisely enough to permit synchronous transmission.

Successful transmission of information with these early digital systems was achieved by the technique of *transmitting synchronizing information with the code for each character transmitted*. The synchronizing information is referred to as *start* and *stop bits*. Start and stop bits are still used today in many transmission applications. For example, they are used in the transmission of character data, in serial-bit form, between microcomputers and some types of printers.

In the early systems, it was found that the terminal equipments remained in synchronism to an acceptable degree over the time required to transmit *one character*, but not much longer. Hence the system, in effect, initialized the terminal equipment at the start of transmission of each character, by means of a start bit. Stop bits—one or two were used according to the requirements of the system—signaled the end of a character. The stop bit(s) also gave the equipment time to complete a character cycle and settle to a starting position for the next character. The code used was a 5-bit code called the Baudot code. An example of code containing start and stop bits is given in Fig. 12.25.

The use of more than 5 bits per code word was found to increase the

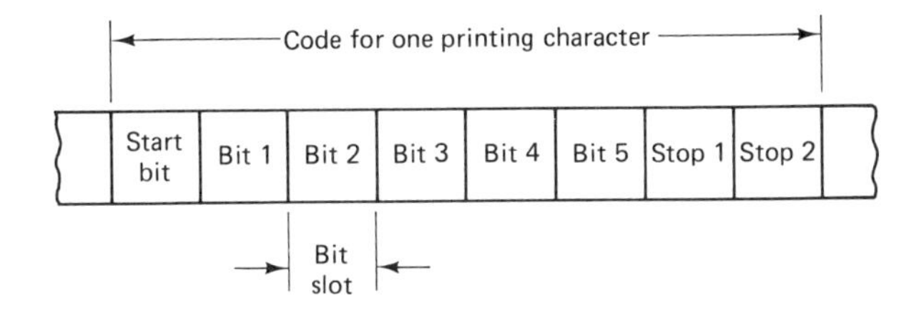

Figure 12.25 Illustrating 5-bit code for teleprinting with 1 start and 2 stop bits.

chances of error through loss of synchronism of the electromechanical equipment. Of course, these early systems were not sampling systems for the transmission of analog signals. They are more correctly characterized as printing telegraph systems—a code word represented a single printing character.

The use of start and stop bits in a transmission system produces a form of internal or self-synchronization—"internal" or "self" with respect to each code word. The technique can be and is used to good effect in systems with much higher bit rates (bits per second) than the electromechanical systems described above. However, these extra bits sent along with each code word represent a significant "overhead." The overhead reduces the quantity of information that can be transmitted per second over a transmission link of a given bit-per-second capacity. The use of start and stop bits reduces the utilization efficiency of the system.

Conclusion

One of several advantages of pulse-code modulation (PCM) is that it is easily adaptable to time-division multiplexing (TDM). TDM is an attractive alternative to frequency-division multiplexing (FDM). It is attractive because it permits the use of many of the numerous digital integrated-circuit functions developed primarily as part of the computer technology revolution. These devices are mass produced and are therefore inexpensive and readily available.

In this section we have examined the general ideas and basic details of TDM communications systems. We have looked at the functions of interleaving and synchronization which are performed at the terminals of TDM systems. There remains to be examined an absolutely essential function: the transmission process. By transmission we mean getting the signals from one geographical location to another. As we shall see, there are several types of transmission links used in digital communications systems. All of these have been and are being used in "conventional" communications systems, that is, systems that process analog signals as analog signals. In other chapters in this book we examined some of the basic details of systems for the transmission of analog signals. In the next section we look at basic ideas and details of the transmission links of digital communications systems.

12.8 TRANSMISSION LINKS FOR DIGITAL COMMUNICATIONS SYSTEMS

Block diagrams of digital communications systems we have used thus far include Figs. 12.1 and 12.20. On these diagrams we see the portion of the system between sending and receiving terminals designated as the *transmission link*. We define "transmission link" as that portion of a communications system actually involved in getting information signals—in whatever form, analog or digital—from one geographical location to another. A transmission link may incorporate any one or a combination of several processes and transmission media: land line (twisted-wire pair or coaxial line), fiber-optic cable, point-to-point radio link, land-based microwave link, or satellite microwave link. Transmission links are really complex electronic systems in and of themselves. We can simplify the process of learning about such systems by categorizing them and breaking them down into their various parts.

There are several ways of categorizing transmission systems. They can be categorized according to the medium used. In this case, there would be a category for systems that use copper conductors as the medium and another category for those that use space as the medium. Examples of the first category are telephone systems using twisted-wire-pair-cable or coaxial-cable transmission lines. Fiber-optic links would fit in this category. Examples of the second category are point-to-point radio systems, land-based microwave systems, and satellite systems.

Transmission systems can also be categorized according to the modulation method used: no-modulation or baseband systems, AM systems, FM systems, etc. Let's start our examination of transmission systems by looking at the simplest type: a baseband system. Some of the things we will learn about a baseband system will be useful in learning about other systems.

Baseband Transmission

In a baseband transmission system for analog signals, the signal is transmitted without modification. For example, when you talk by telephone to a friend living nearby, the telephone facilities you use are most likely a baseband system. The audio-frequency signal produced when you talk into the microphone element of the telephone handset is transmitted, as is, to the earphone element of your friend's handset. On the other hand, if you call your grandmother or another relative or friend on the other side of the continent, you will use a system that transmits the signals representing your voice by modulating a carrier signal. This system, for long-distance telephony, is not a baseband system.

As we have seen, in a PCM digital communications system analog signals are converted to binary code. The electrical signal of a person's voice talking on the telephone, for example, is converted to a stream of pulses representing 1's and 0's. If the transmission link in a PCM system is not too long (i.e., is not more than about 20 mi), the stream of pulses may be placed directly on the line. Such an operation could be called a digital baseband operation. No use is made of a carrier.

However, the pulse stream used in actual digital baseband systems is virtually never an unmodified version of the simple pulse waveform in which a 1 is represented by say +5 V and a 0 is represented by 0 V. Such a waveform is a *unipolar* waveform. An example is shown in Fig. 12.26(a). The simple unipolar waveform output of an A/D codec has several disadvantages as a line code.

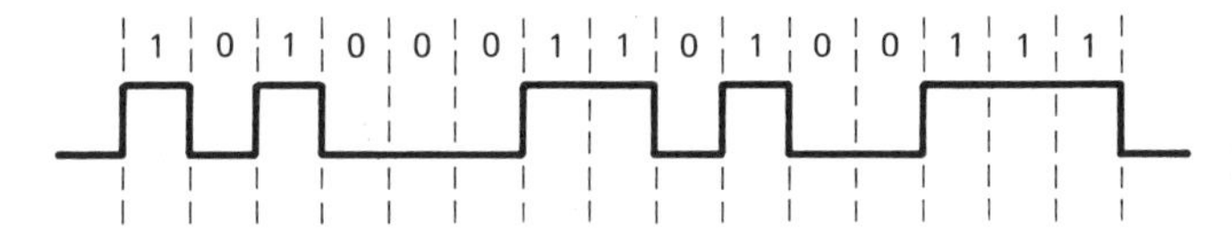

Figure 12.26 Unipolar line code pulse stream with bit-time slots indicated.

(*Line code* is the term used to identify a pulse waveform used especially for PCM transmission on a telephone line.) Most of the disadvantages of a unipolar line code can be overcome, or at least minimized, by slight modifications of it. Several such modifications are in use.

One of the disadvantages of the unipolar line code is that it contains a significant dc level. That is, its average value is not zero but is, for example, a positive dc value in the case of a unipolar waveform that goes to the positive side. This average value will change if there is a long sequence heavy with 1's or 0's. A changing dc level tends to obscure momentarily the difference in pulse levels and thus to cause confusion to the terminal circuits that must interpret such differences as 1's and 0's. The result is an increased probability of error in the interpretation of transmitted signals. The effect of a changing dc voltage level, sometimes called *dc wander*, is illustrated in Fig. 12.27.

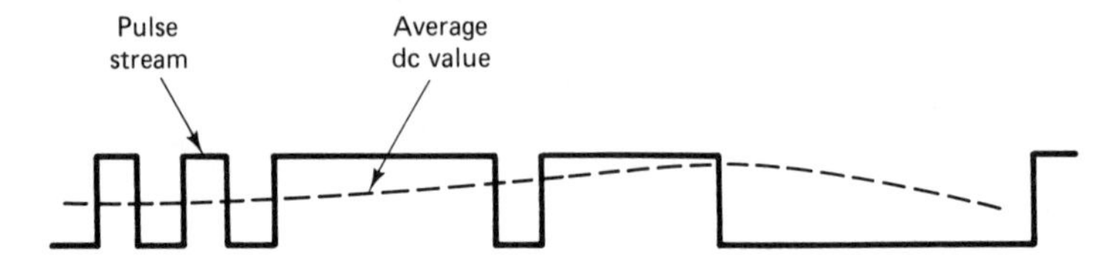

Figure 12.27 Dc wander caused by long strips of 1's or 0's in unipolar line code.

Besides a low dc content, there are other characteristics that make one line code a better choice than another. The meaning and significance relative to performance of other characteristics are much less apparent than those of dc level.

A second, desirable characteristic of a satisfactory line code is that the most energetic (powerful) of the frequency components of the waveform match the pass band characteristic of the transmission link. For example, the energy in a unipolar PCM waveform is greatest at very low frequencies—near 0 Hz. On the other hand, most practical transmission lines used for baseband operation transmit better at somewhat higher frequencies. Modification of the unipolar line code to shift the concentration of energy to slightly higher frequencies is thus desirable.

A third requirement of a good line code is that it must contain a certain minimum of timing information. The access, switching, and repeater stations along a transmission system typically do not contain their own local, independent timing circuits. Rather, they have circuits that might best be called "parasitic" timing circuits. That is, they have circuits which provide timing information for decoding the received digital signals all right. But these circuits must receive periodic information from the information signal itself in order to remain in step with the timing of the system. (The process is called "extracting the timing information.") To be usable, the transmission code must contain timing information at sufficiently short intervals to meet the requirements of the "local" timing facilities. Timing information usually means simply a crossing of the zero-voltage axis at the edge of a system time slot by the information signal.

How, then, can the idea of a simple unipolar code be modified to achieve a more useful code? Let's examine a couple of codes that have been and are being used.

One of the simplest codes which is an improvement over the transmission qualities of the basic unipolar code is called *alternate mark inversion.* It is abbreviated AMI; it is a *bipolar code.* The basic PCM code can be converted to AMI with an appropriate digital logic circuit. In the conversion process, a binary "1" of the PCM code is referred to as a *mark.* And, *alternate marks are inverted* (hence alternate mark inversion). The "0" values of the original code remain unchanged. An example of the conversion of a simple binary signal to AMI is shown in Fig. 12.28.

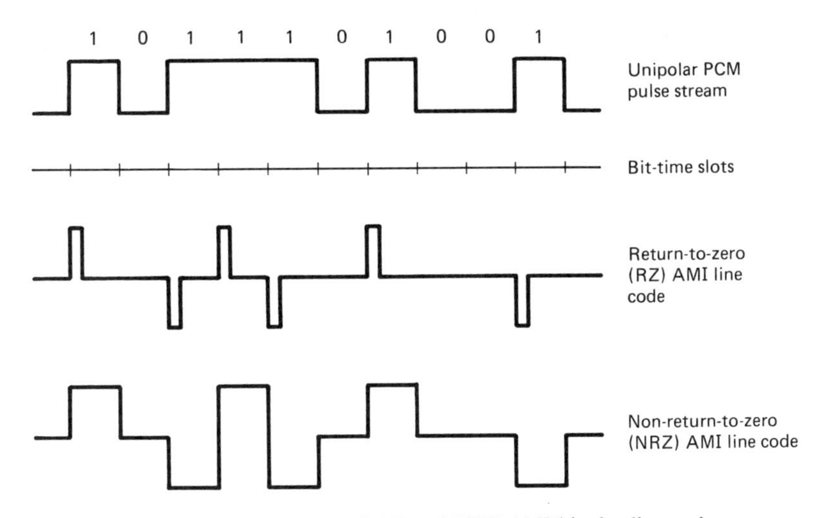

Figure 12.28 Unipolar, and RZ and NRZ AMI bipolar line codes.

You will note from Fig. 12.28 that there are two forms of the AMI conversion: a return-to-zero form (RZ) and a non-return-to-zero form (NRZ). In the RZ form, a mark (remember, a binary 1) occupies only a fraction of the mark time slot and then returns to zero; and, of course, alternate marks are of opposite polarity (inverted). In the NRZ format, the mark occupies the full time slot of the binary 1 of the original binary code; and alternate marks are inverted. Either of these forms of the AMI has an equal number of positive and negative pulses. The result is that an AMI waveform has zero dc content. Careful measurements have also demonstrated that an AMI signal has its power concentration shifted to a higher frequency than that of a simple unipolar binary signal—another sought-for characteristic.

The timing circuit of a local terminal in a transmission system is usefully compared to an RF frequency multiplier (see Chap. 5). You will recall that a frequency multiplier as used in an AM or FM transmitter is a class C amplifier. A class C amplifier utilizes the "flywheel" effect of an *LC* resonant circuit. The current pulses of the amplifier keep the "flywheel going" by injecting energy, in the form of short-duration spikes, into the resonant circuit. Of course, the polarity and timing of the spikes must be just right to match the activity (damped oscillation) of the resonant circuit (the "tank circuit"). In a frequency multiplier, the output tank circuit is tuned to a multiple of the input signal frequency: The output tank circuit gets an energizing spike only on every second ($\times 2$) or every third ($\times 3$) cycle. In a nonmultiplying class C amplifier, the output tank gets a boost in every cycle.

The timing circuit of a local terminal in a PCM system is basically a resonant circuit which gets a boost only when the incoming signal crosses the zero line. In AMI, the signal crosses the zero line every time a mark (binary 1) is transmitted. However, in actual transmissions, who knows how often 1's will be transmitted? The transmission of 1's is determined by the information being transmitted. It is possible for there to be times when there is a long (several milliseconds, say) dry spell with no 1's. Local timing circuits do not receive their updating when this occurs; the "flywheel" quickly starts to run down or get off speed. Timing is upset. Unless special conditions are placed on the coding (to ensure a sufficient number of 1's), AMI cannot guarantee the transmission of adequate timing information. This is, therefore, a serious disadvantage of AMI and limits its application.

A solution to the problem of lack of sufficient timing information in a line code is to use a modification scheme that arbitrarily increases the number of 1's in the transmitted signal. A number of such schemes have been conceived and put into practice. Implementing a scheme of this type involves comparatively complex logic circuits. This is not the place to delve into the design of such circuits. However, it is useful for you to have a working knowledge of the basic ideas of such schemes. Let's take a brief look at a particular scheme—the so-called *binary 6-zero substitution* (B6ZS) code. The B6ZS is a special code used in the United States by the Bell Telephone System on certain types of transmission links.

The title "binary 6-zero substitution" is very descriptive of the way in which this special code works. Remember, its purpose is to increase the number of 1's in a binary transmission. More to the point, its purpose is to eliminate the occurrence of segments of a binary transmission in which there is a long string of 0's (six 0's, to be specific). Very simply the B6ZS code (1) detects, at the time of transmission, segments of a signal in which six 0's occur in sequence; and (2) substitutes segments of special replacement code. The replacement code contains more 1's, obviously. The replacement code is an incorrect representation of the information being transmitted; it must be detected at the receiver and changed back to the correct code. The presence of replacement code is detected by a process based on an idea called *code violation*. Let's examine, first, the meaning of code violation and then look in more detail at B6ZS code.

As we observed above, the essence of AMI code is that alternate 1's of a binary message have opposite polarities. The result is that two pulses of the same polarity (a zero is a nonpulse) never occur in sequence. We can demonstrate the effect of this "rule" of AMI by representing a positive-going pulse with a "+" and a negative-going pulse with a "−". For example, the pulse train for the binary code 1 0 0 1 1 0 1 0 1 1 1 could be described with the notation + 0 0 − + 0 − 0 + − +. If we align these two representations in a table, the result would be as in Table 12.1.

The effect of "violating" the rule of AMI is self-evident in the three cases of violation illustrated in Table 12.1. A violation occurs and is easily detected any time two successive polarity signs are the same, regardless of the position and number of 0's in a code sequence.

TABLE 12.1

Binary code:	1 0 0 1 1 0 1 0 1 1 1
Bipolar AMI code:	+ 0 0 − + 0 − 0 + − +
Violation 1	+ 0 0 − − 0 − 0 + − +
Violation 2	+ 0 0 − + 0 − 0 + + +
Violation 3	+ 0 0 + + 0 − 0 + − +

The B6ZS system has two substitution patterns for replacing any segment of code containing six 0's in succession: The pattern used is determined by the polarity of the 1 pulse just preceding the segment of six 0's. The two patterns are illustrated in Table 12.2.

TABLE 12.2 SUBSTITUTION SEGMENTS FOR B6ZS

Preceding one-pulse polarity	Substitution code segment
+	0 + − 0 − +
−	0 − + 0 + −

Study the substitution segments for B6ZS carefully in Table 12.2. You will observe that violations are built into the code. When the preceding one-pulse polarity is +, that + and the second bit of the code produce the violating sequence + 0 +. The violation, − 0 −, is also present in the third through fifth bits. The corresponding violations for a preceding 1 pulse with − polarity are − 0 − and + 0 +. Illustrations of the substitution of the two types of segments are provided in the following examples.

Example 1

Substitution of B6ZS code when polarity of preceding 1 pulse is +:

Binary code: 1 0 1 0 0 1 0 0 0 0 0 0 1 0 1 0 0 0 0 0 0 1

Modified code (B6ZS): + 0 − 0 0 + (0 + − 0 − +) − 0 + (0 + − 0 − +) −

(substituted segments)

Example 2

Substitution of B6ZS code when polarity of preceding 1 pulse is −:

Binary code: 1 0 1 1 0 1 0 0 0 0 0 0 1 0 1 0 0 0 0 0 0 1 0

Modified code (B6ZS): + 0 − + 0 − 0 − + 0 + − + 0 − 0 − + 0 + − + 0

substituted segments

Regeneration versus Repeater Stations

Signal attenuation with distance along a transmission link is an ever-present problem in any communications system. An acceptable system incorporates measures to meet this problem. For example, in an analog baseband system utilizing a twisted-wire-pair transmission line, the system is equipped with electronic amplifiers located at strategic points along the line. The amplifiers are typically called *repeaters*. Their purpose is to boost the amplitude of the transmitted signal and to do this before the signal is attenuated significantly. If the signal amplitude is not maintained above a certain critical level, the ever-present noise of any system will creep in to an excessive degree. The signal-to-noise ratio will become unsatisfactory to users of the system. Repeater station locations are determined by the characteristics of the system. They are located to achieve a desired result in terms of system signal-to-noise ratio, cost, etc.

In a system transmitting analog signals the degradation of the signal by distortion and the unwanted injection of noise is inevitable. The effect of quality-destroying forces is cumulative along the transmission link. Even with the best of repeaters and other facilities, analog signals suffer with distance transmitted.

Once an analog signal is "damaged" (i.e., suffers from distortion or the addition of noise content) it is virtually impossible to "repair" it. The situation is quite different with digital signals. *If a digital signal is not degraded so badly that it cannot be decoded, it can be regenerated as a like-new signal.*

Transmission links for digital communications systems incorporate equipment facilities located along the line which "regenerate" the signals before they have become severely damaged. As with repeaters, regenerators must be located strategically. A strategic location is determined by the characteristics of the system, particularly the attenuation and noise characteristics of the line.

Digital signals become attenuated, distorted, and infected with noise just as analog signals do. However, if this degradation is not allowed to occur to an excessive degree, the signal can be detected and remanufactured, as it were, into a new signal. A *regenerator*, then, must include the facilities of both a digital receiving and transmitting facility. It must incorporate the circuits with which to decode the incoming digital signal; and it must have circuits with which to generate a new coded signal. Such circuits sound more complex than the amplifiers of a repeater station for an analog system. However, the effects of regeneration are far more beneficial to the communications process than repeated amplification.

Systems Testing: The "Eye" Diagram

If you pursue a career as a technician in the field of electronic communication, there is an excellent chance that your career will lead to work with digital communications systems. If that should turn out to be the case, you very probably would some day be involved in testing the performance of a transmission link for digital signals. Let's look at a popular technique for evaluating digital transmission. The technique involves displaying a digital signal on an oscilloscope. The resulting display, when the scope is properly adjusted, is usually called an "eye diagram" because it reminds one of an eye [see Fig. 12.29(d)].

The signal of a digital transmission system is a stream of randomly occurring positive, negative, and zero pulse levels, as in Fig. 12.29(a). However, after traveling a short distance on a transmission link, the pulses lose their sharp

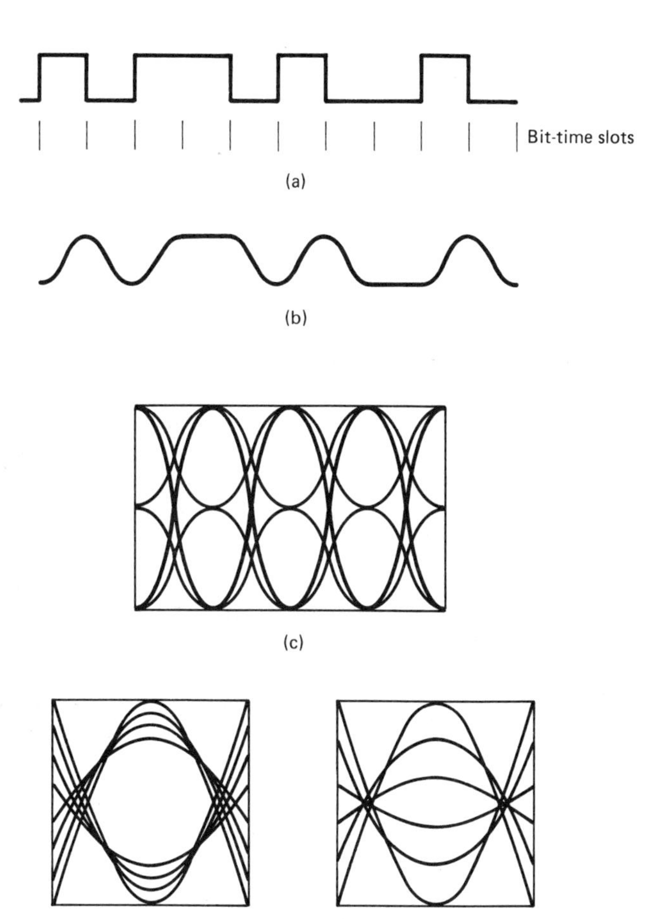

Figure 12.29 Signal degradation and the eye diagram: (a) "ideal" random pulse stream; (b) degraded pulse stream; (c) typical oscilloscope display of random pulse stream; (d) the "eye" diagram test pattern illustrating minimum distortion; (e) severe closing of eye diagram illustrating significant distortion.

corners and become deformed to a degree by spreading (an increase in pulse width). The effect is illustrated in Fig. 12.29(b). This deformation makes the task of determining which part of a waveform is a 1 and which a 0 much more difficult. In short, the chances of an error in information transmission increases with the level of disturbance of the pulse waveform. It is often desirable, therefore, to evaluate the performance of a transmission link by observing the transmitted signal on an oscilloscope. But the task is more than what it might seem.

When a simple signal such as a sine wave or even a rectangular pulse waveform is displayed on an oscilloscope screen, it can often be adjusted so that it appears as a stationary, single-trace waveform. It appears this way only because each succeeding trace is identical to the previous one. When each succeeding trace is not identical to all preceding traces, the display appears to "move" or to be "smeared." When an actual digital signal is displayed on an oscilloscope, at first glance it appears to be a meaningless jumble [see Fig. 12.29(c)]. Or it appears as if the scope has not been synchronized to the signal. This is especially true for someone accustomed to looking at sine waves. However, because of the random nature of the pulses, it is not possible to obtain what one would normally call a "synced" display.

Nevertheless, the display of Fig. 12.29(c) can be used to assess in an approximate way the degree of deformation of the digital signal. Let's focus our attention on one "eye" of the diagram of Fig. 12.29(c), as in Fig. 12.29(d). The "opening" of the eye proves to be a fair measure of the degree of deformation of the signal: (1) If the center portion of the eye is comparatively large and clear, the signal is being transmitted with minimum deformation. (2) If the clear area at the center of the eye is decreased, that is, if signal traces close down the eye, as in Fig. 12.29(e), deformation is present to a significant extent.

Transmission Links Using Modulated Carriers

When a transmission link in a communications system is to carry signals—either analog or digital—for more than a few miles, it typically uses one or more modulated carrier signals. This fact is often the result of an economic decision. The use of carriers makes it possible to use frequency-division multiplexing as well as time-division multiplexing. In short, by using carriers we can make a basic system carry many times as many simultaneous information signals than would otherwise be possible. This last fact is very important when the transmission distance is long. The cost of numerous transmission lines is traded for the lower cost of somewhat more complex terminal facilities.

The use of carriers to achieve frequency-division multiplexing (FDM) implies the process of modulation. If you have studied the earlier chapters in this book, you are familiar with three basic schemes for placing information signals on carrier signals: amplitude modulation (AM), frequency modulation (FM), and phase modulation (PM). We have examined in previous chapters each of these techniques as they are used with analog signals. The same three basic ideas are also used in the modulation of carrier signals by information signals in digital form. The techniques of implementation are modified somewhat to better match the nature of digital signals. As a result we have a new set of abbreviations. In this section we look at ASK (*amplitude-shift keying*), FSK (*frequency-shift keying*), and PSK (*phase-shift keying*). Let's start by looking at amplitude-shift keying, the digital version of amplitude modulation.

Amplitude-shift keying (ASK). If a radio-frequency carrier signal were amplitude modulated by a bipolar digital signal, a simple AMI signal, for example, the result would be like that of Fig. 12.30. The operation could be accomplished by borrowing a conventional modulator from an analog system. In effect, we are simply using a digital baseband signal to amplitude modulate a carrier in a conventional, "analog" way. The digital baseband signal can be recovered in a conventional AM receiver by conventional AM demodulation techniques (see Chap. 3). The recovered baseband signal must be decoded in a fashion typical of baseband systems.

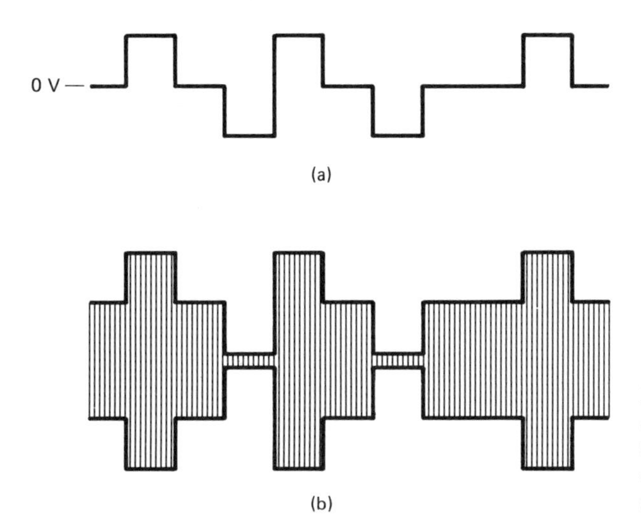

Figure 12.30 Amplitude modulation of analog signal with digital signal: (a) PCM signal (AMI); (b) RF signal amplitude modulated with PCM signal.

There is a more direct, alternative way by which the equivalent of amplitude modulation by digital signal can be achieved. In rudimentary form it simply involves turning a transmitter on for a binary 1 of the digital information signal, and turning it off for a 0. The result would be as shown in Fig. 12.31(b). The "modulator" in this scheme is a circuit to "key" the transmitter on and off under the control of the output from a digital logic circuit [see Fig. 12.31(a)]. The "keying" operation "shifts" the transmitter output between two amplitude levels. One of the levels can be zero amplitude, but it need not be; the 0 amplitude need simply be different from the 1 amplitude [see Fig. 12.31(c)]. In any case the process is one of "amplitude-shift keying." And we have an explanation for the title given to this type of modulation. The title is used to identify both of these modulation methods: "analog-like" modulation or the more direct "on-off" modulation.

A demodulator for the direct on-off modulation process performs the decoding function simultaneously with the demodulation function. It need simply be a

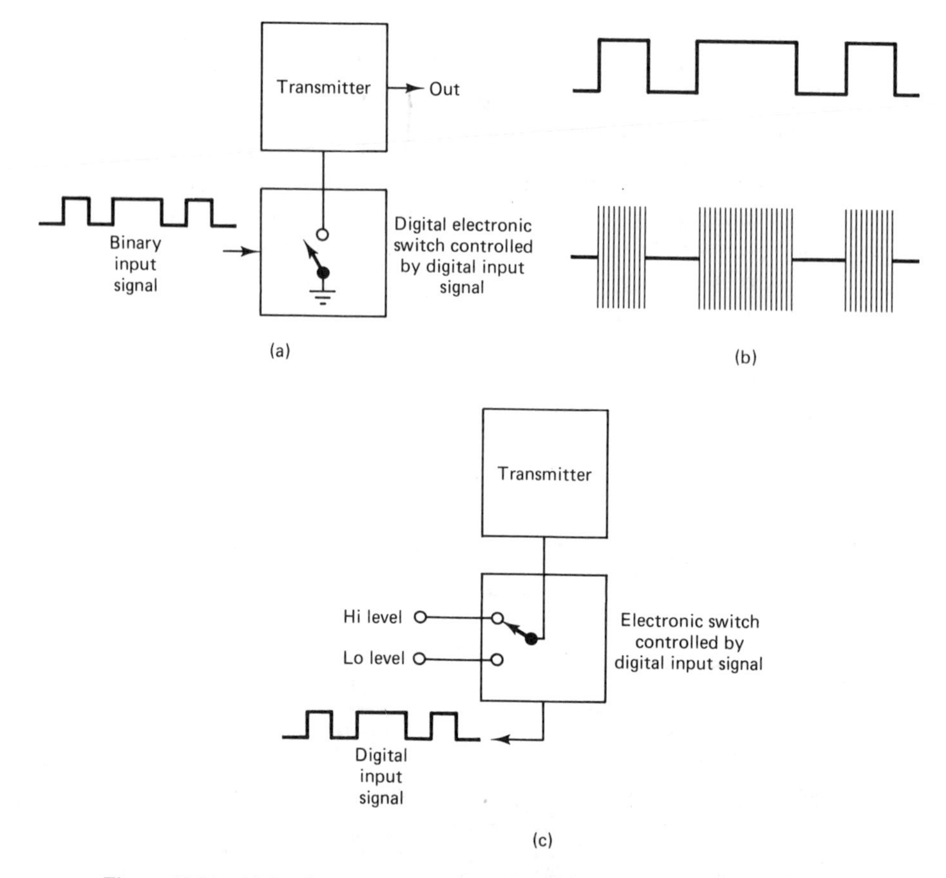

Figure 12.31 (a) Rudiments of on/off keying for digital data transmission; (b) waveform produced by on/off keying; (c) rudiments of amplitude-shift-keying (ASK) modulation.

circuit that recognizes each of two different types (amplitudes) of transmitted signals. It produces a binary digital output based on the level of each signal segment received.

We say again that an important goal of ours for this chapter is to provide a working knowledge of the most widely used *terminology* in the field of digital communications. If we study technical literature written for the benefit of those persons involved in designing a communications system, we find two words popping up frequently in connection with demodulation schemes: *coherent* and *incoherent*. It is desirable that we have a fundamental knowledge of these terms and the processes they represent.

A beginning statement about a coherent demodulator is that it is a function which requires information about the *phase* of a carrier signal as well as its amplitude. It is said to require information of the complete signal shape. An incoherent demodulator, by contrast, uses only information about the amplitude of a carrier signal to recover the information signal. These are very general statements and leave much to be desired by way of a "working knowledge."

Before proceeding to more useful descriptions, however, let's exercise the intuitive side of our brain a bit: The incoherent demodulator appears to be a "cheap and dirty" (read "unsophisticated") way of recovering a signal, one with some relative disadvantages, perhaps. The coherent demodulator sounds like something that would be more difficult to implement but might offer advantages over the simpler one.

The terms "coherent" and "incoherent" are approximately equivalent to "synchronous" and "nonsynchronous." A coherent or synchronous demodulator is one that recovers an information signal "in synchronism" with the original carrier signal. The original carrier signal may be transmitted or may be generated locally by the receiver. You will recall that in normal AM, the carrier is transmitted as part of the complete signal. The carrier is available, therefore, at the receiver for synchronous or coherent demodulation of the transmitted signal. On the other hand, in single-sideband transmission (SSB; see Chap. 9), the carrier is not transmitted. In the transmission of analog signals by SSB, the carrier must be regenerated at the receiver in order for the analog signal to be recovered by demodulation. Both of these situations are examples of synchronous or coherent demodulation.

What, then, is incoherent demodulation? In incoherent demodulation, the demodulator circuit is designed to recover information only about the amplitude of the transmitted signal (i.e., information in the envelope only). An example: A circuit looks at an SSB transmission modulated with digital information. The received transmission is not mixed with a local carrier. The circuit simply rectifies the signal and indicates its relative amplitudes at appropriate sampling moments. The process can work since with digital transmissions we are ultimately interested only in whether a signal is representing a 1 or a 0 at any given moment. If a system can provide this information without the aid of a locally regenerated carrier, good enough. Unfortunately, the incoherent demodulator is much more susceptible to the damaging effects of noise, noise added to the signal along the transmission path. Noise can easily distort amplitude-only information and thus confuse the decoding circuitry.

We have now looked at the basic details of amplitude-shift keying as a means of placing digital information on an RF carrier. A baseband digital signal can be used like an analog information signal to amplitude modulate the carrier. Or, the digital signal can be used to turn a transmitter on and off with two levels of output amplitude possible. In a receiver, demodulation of a digitally modulated signal may or may not occur in the presence of the carrier (usually, a locally regenerated version of the carrier). If a carrier signal is used in demodulation, the process is called *coherent demodulation*. Coherent modulation is potentially a superior method since it recovers phase as well as amplitude information. If the carrier is not used in demodulation, the process is termed *incoherent*.

You are hereby warned that there are variations on the theme of amplitude-shift keying. If you proceed with a study of digital communications systems beyond what is provided in this book, you are likely to encounter terms such as quadrature-amplitude-modulated (QAM) and others. Let's turn our attention now to frequency modulation as applied to the transmission of digital information.

Frequency-shift keying (FSK). Recall our study of frequency modulation (FM) in relation to the transmission of analog signals (see Chap. 8). In FM, the *frequency* of a carrier signal is changed in order to incorporate information from the modulating information signal. The amount of the frequency change is used to represent the amplitude of the information signal; the rate of the change represents the frequency of the information signal.

The "information" of a baseband digital signal is much simpler than that of an analog signal. As we know, a baseband digital signal can be as simple as a stream of pulses of only two amplitudes representing 1's and 0's. This binary information can be transmitted over an RF carrier link by switching a transmitter between two basic operating frequencies: a low frequency, say, representing a 0 and a higher frequency representing a 1 of the digital information. Such a scheme would be a form of frequency modulation. The switching operation is often called "keying" and the keying "shifts" the transmitter between the two frequency sources. The process is called *frequency-shift keying* (FSK). The diagram of Fig. 12.32 illustrates the concepts of FSK.

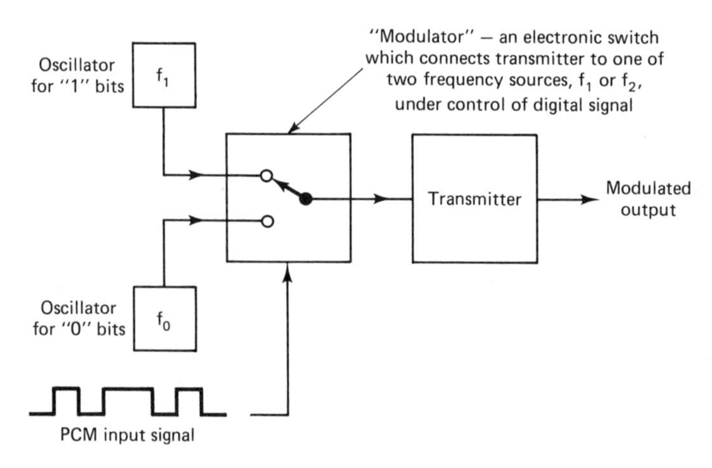

Figure 12.32 Concept of frequency-shift-keying modulation (FSK).

The frequency modulation of a carrier can also be accomplished by means of an analog-type FM modulation scheme. The digital signal is fed to the modulator as a continuous waveform. The modulator circuit responds to the signal as if it were simply a peculiarly shaped audio signal.

Just as with ASK, there are coherent and incoherent demodulators for frequency-shift-keying systems. If the carrier signal is present in the receiver and available for the demodulation process, it is possible to extract phase as well as frequency and amplitude information from the received signal; the process is coherent demodulation. A scheme may consist of passing the incoming FSK signal through each of two FM detectors (see Fig. 12.33). Each of the detectors is tuned to one of the two "shift" frequencies of the system. The outputs of the detectors are designed to give opposite polarities and are fed to a summing

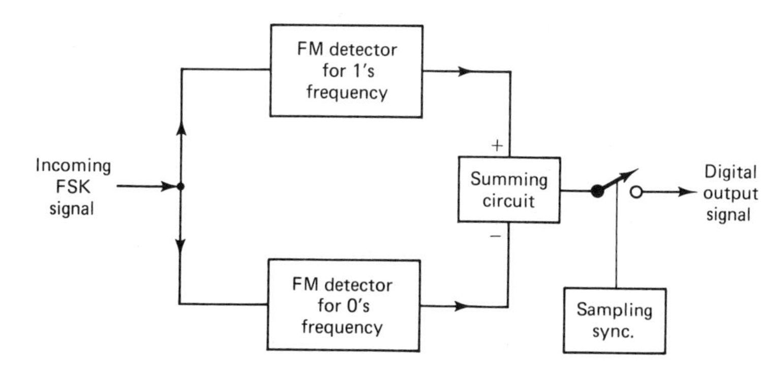

Figure 12.33 Elements of coherent demodulator.

circuit. The output of the summing circuit indicates at a given sampling moment whether the received information is a 1 or a 0.

A simpler scheme is termed an incoherent detector. The incoming FSK signal is split and passed through two bandpass filters (BPF). The filters are tuned to the two "shift" frequencies of the system. The outputs of the filters are fed to envelope detectors (basically, rectifier circuits) and then to a summing circuit. Again, outputs of the detectors are normally opposites. The "net" output of the summing circuit indicates, by its polarity, whether a 1 or a 0 was detected at a given sampling moment. Figure 12.34 illustrates the concept.

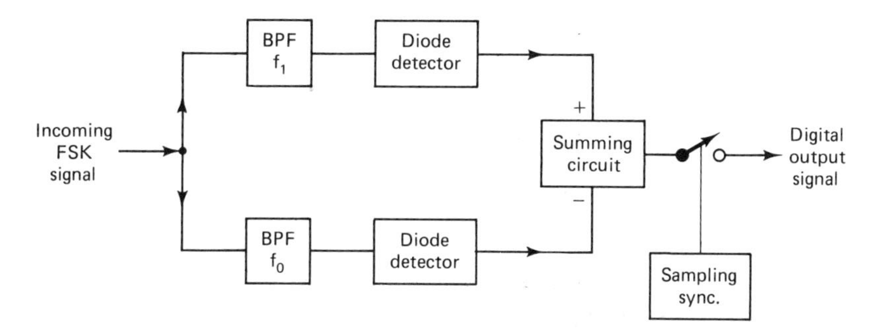

Figure 12.34 Elements of incoherent demodulator.

The Phase-Locked-Loop Detector. Extensive operating details of the phase-locked-loop circuit are presented in Chaps 8 and 9. You will recall that a PLL includes a voltage-controlled oscillator (VCO), a phase detector, and a low-pass filter (LPF). The phase detector has two inputs: one from the VCO and a second from any other desired source. The phase detector thus compares the phase (and frequency) of the VCO with that of the other signal. The output of the phase detector is a bipolar dc signal. It contains information about the amount and direction of the phase/frequency offset between its two input signals. This "error" signal is passed through the LPF and back to the VCO to correct its phase/frequency. This input to the VCO is, in effect, a frequency demodulation of the second input to the phase detector. If the second input is an FSK signal containing digital information, that information is available at that point.

Phase-shift keying (PSK). Phase-shift keying is a form of phase modulation (see Chap. 8). You will recall that in phase modulation in analog communications systems, *phase changes* in the carrier signal are used to represent the information to be transmitted. A simple scheme for a digital system is to key shift a transmitter between two sources of the same frequency but different phase. The idea is illustrated in Fig. 12.35. Of course, the two phases represent the two binary digits. Also, as with FSK, an analog-type phase-modulator scheme can be used to transfer the digital information to a carrier.

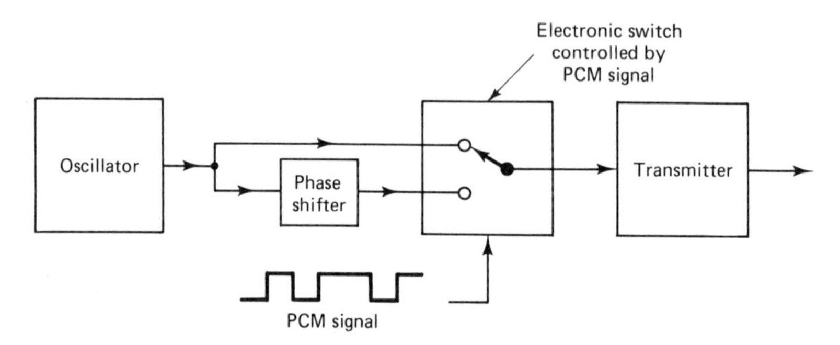

Figure 12.35 Elements of phase-shift-keying (PSK) modulation scheme.

Demodulation of a PSK signal can be accomplished in a coherent demodulator virtually identical to that described above for FSK. However, the incoherent demodulator described for FSK is not able to detect phase differences. You will recall that it detects frequency changes without any knowledge of phase changes. It cannot be used for PSK. For this and other reasons a modified form of PSK—*differential phase-shift keying* (DPSK)—is a popular alternative form of phase-shift keying.

Differential phase-shift keying (DPSK). In one DPSK scheme, the phase of a carrier is shifted only when a 0 is transmitted. The shift is not to an absolute phase but simply to the "other" of two possible phases. Thus information is not transmitted by an absolute phase value but by a *change* in a phase value. If, at one sampling moment, there is no phase change from the preceding moment, a 1 is being transmitted; if there is a change, a 0 is being transmitted. A demodulator need only detect changes in phase, not absolute phase values. The presence of the carrier is not required for detection. Incoherent detection is possible. Let's look at a simple but very practical detector of this type.

You will observe that in Fig. 12.36 a DPSK signal is beat with a sample of the same signal delayed in time by an amount equivalent to a sampling interval. That is, successive signal samples are mixed in a heterodyne mixer. The result is that if the two samples are of the same phase, a large amplitude output will be produced, indicating a 1. If the two signals are of different phase, for example different by 180°, the output will be different. In fact, the circuit can be designed

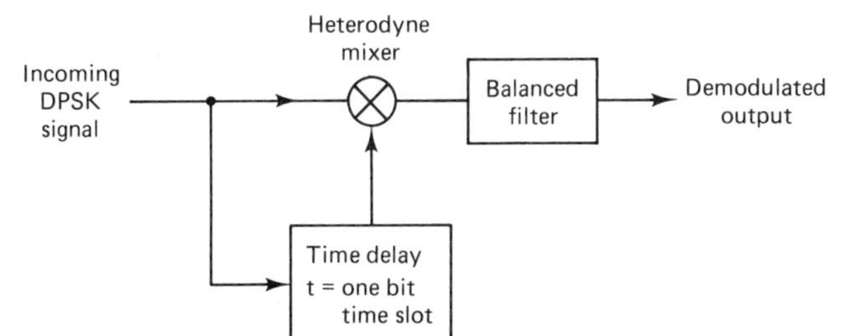

Figure 12.36 Simple demodulation scheme for DPSK signal.

so that the outputs for the two cases are of opposite polarity. The circuit detects "change" or "no change" in phase. It thereby detects a 0 or a 1.

Transmission of digital data by means of PSK is a popular technique. The performance of PSK systems in the presence of noise is on a par with FSK systems. Since a single frequency is used, PSK has the additional advantage of using less bandwidth. In fact, because of its nature, PSK makes possible its own special kind of multiplexing. This might be called *phase multiplexing*. In this scheme two separate information signals are used to achieve DPSK modulation of a carrier. Each information signal uses a 180° differential phase-shift scheme. However, the displacements between corresponding phases of the two signals is 90°. Multiplexing is achieved by interleaving the phases. In ac theory a phase displacement of 90° is termed a *quadrature* displacement. The described form of PSK is called *quadrature phase-shift keying* (QPSK).

Conclusion

This section is about "transmission links for digital communications systems." We started off by looking at the ideas involved in a baseband system operating over twisted-wire-pair transmission lines. Next we looked at several ways of placing digital information signals on carrier signals (i.e., modulation). This digression, if we must call it that, was necessary since the use of carrier signals is an essential element in the design and operation of other types of transmission links. Possessing a rudimentary knowledge of modulation techniques, we can now proceed to a look at the basic ideas of other types of links.

Transmission links that utilize signal carriers can be classified into two groups. Group A includes systems that utilize physical, land-based, transmission lines. Group B includes systems that utilize electromagnetic radiation over the major portion of the transmission distance. Transmission-line media used in group A systems include cables made of numerous twisted-wire pairs, coaxial cables, waveguides, and optical-fiber cables that use light as the carrier vehicle. Examples of group B transmission links include two-way, point-to-point radio-telephone systems, microwave transmission systems using land-based repeater towers, and satellite systems.

All of the transmission links enumerated in the preceding paragraph have been and still are used for the transmission of analog information signals. For most of them, frequency-division multiplexing is used to provide literally thousands of channels for simultaneous, two-way transmissions on a single system. A portion of the facilities of most of these systems are gradually being converted for the transmission of information in digital form. Of all these systems, however, satellite transmission links are of greatest interest in the context of digital communications systems. As this is being written, communications facilities are being expanded at a very high rate in the United States, and, indeed, virtually all over the world. Such expansion is an attempt to satisfy the rapidly increasing demand for communications services around the world. Virtually all new or expansion communications systems utilize satellites for their long-distance transmission links; and these links utilize the techniques of digital communications. Let's examine the rudiments of a satellite system.

12.9 SATELLITE COMMUNICATIONS SYSTEMS

A first, not-too-detailed look at an elementary satellite communications system reveals that it contains the functions pictured in Fig. 12.37. As shown, the system consists of two or more "earth stations" and the satellite itself. The satellite is a "space vehicle" placed in orbit around the earth by means of a rocket. (Since 1982, satellites are being launched by means of the "space shuttle," a sort of combination space rocket and airplane.)

Most new communications satellites are placed in *geosynchronous orbits*. This means that the speed and position of the satellite is such that it appears to be stationary to an observer on the earth's surface, a ground station antenna, for example. A geosynchronous orbit is one approximately 22,000 mi or 35,400 km above the earth's surface and in a plane that includes the equator. At such a height a satellite's "natural" speed is just right to keep it in synchronism with the rotation of the earth.

If a satellite is not in geosynchronous orbit, communicating earth antennas must be equipped with sophisticated, expensive tracking equipment to keep them aimed at the satellite. Even with geosynchronous satellites, slight adjustments in the aim of earth station antennas are necessary from time to time: a satellite may "wander" in its orbit to a slight degree, over time.

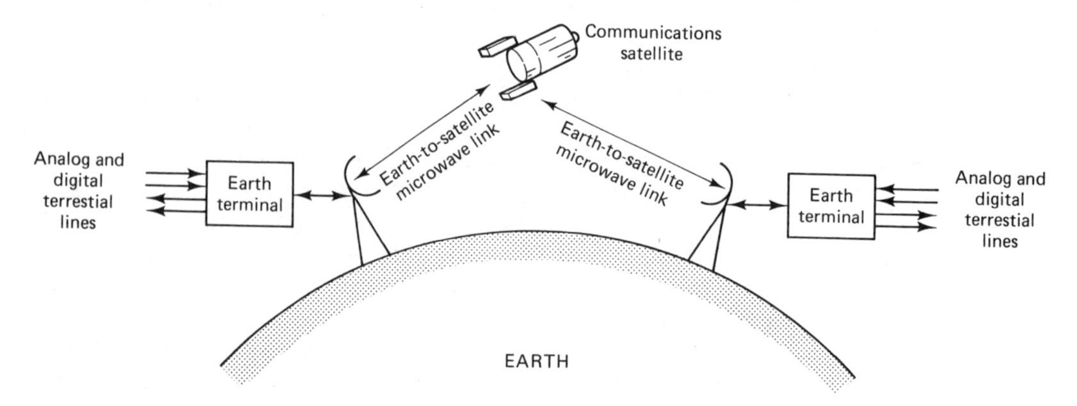

Figure 12.37 Elements of a satellite communications (satcom) system.

A satellite in a satellite communications system is not simply a passive antenna-type reflector. The "vehicle" has installed in it equipment with which it receives signals from a transmitting earth station. Further, it has transmitter-type equipment with which it retransmits information signals back toward the earth. In imagining the details of even the simplest of systems we would include the idea that it provides for simultaneous, independent, two-way communication between two earth stations (i.e., true duplex operation). Therefore, an equipment complement would include at least two transmitters and two receivers, a pair of each for each communications direction. The combination—a receiver and its related transmitter—is generally called a *transponder*. As a matter of interest, the Intelsat 5, a commercial U.S. communications satellite launched in 1980, carries 27 transponders and provides the equivalent of 12,500 two-way voice channels.

A satellite vehicle must provide a source of energy to operate the communications equipment that it carries. At present such energy is provided by means of solar cells mounted on large panels attached to the vehicle. The construction of the vehicle and the solar panels of modern satellites is such that the entire surface of the panels face the sun at all times. Energy levels of the order of several kilowatts are now possible. By contrast, units launched in 1965 could provide only 40 W to their communications equipment.

The actual transmission links between a communications satellite and its several earth stations utilize the medium of narrow-beam microwave electromagnetic radiation. The transmission from an earth station up to a satellite is called an *uplink*. And—you guessed it—the transmission down from satellite to earth is over a *downlink*. Some satellites operate with so-called C-band links: 6-GHz uplinks and 4-GHz downlinks. The Ku-band is another satellite assignment: 14-GHz uplink and 12-GHz downlink. In the K-band, uplink frequencies are 29 to 30 GHz, and downlink, 19 to 20 GHz.

Satellite communications systems are extremely expensive as total systems. The launch costs—the cost of the launching rocket, fuel, launching facilities, highly skilled personnel, etc.—are a significant part of the total cost of a satellite system. Ground terminal facilities are sophisticated and costly. The result is that in order to be economically feasible, a satcom system (satellite communications system) must provide a very large number of equivalent individual communications channels. Early satcom systems used only frequency-division multiplexing (FDM) to achieve more intensive utilization of these expensive facilities. New systems are using TDM to increase the information-carrying capacity of equipment. The development of digital communications techniques has made the utilization of TDM possible. Let's look more closely at the major details of a typical TDM satcom system.

Time-Division-Multiplexed Earth Terminal

A block diagram showing only the most basic of details of an earth terminal for a TDM digital satellite communications system is shown in Fig. 12.38. The diagram is for a system utilizing C-band links: 6-GHz uplink and 4-GHz downlink.

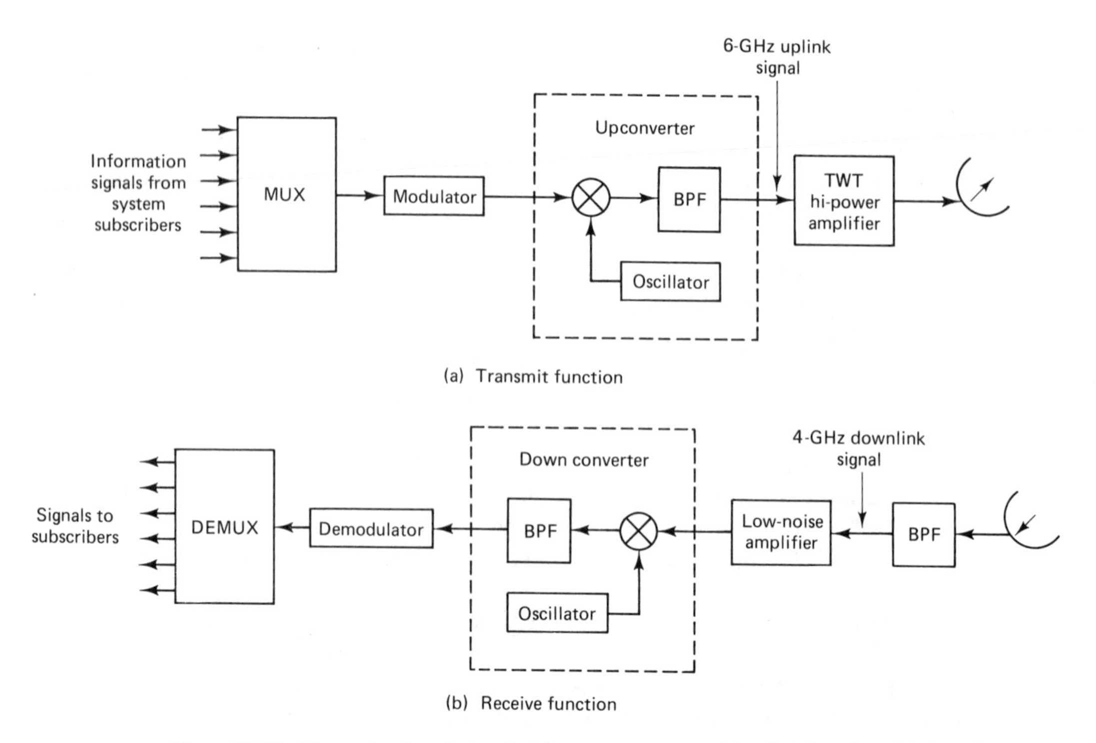

Figure 12.38 Elements of earth terminal for satcom system: (a) uplink function; (b) downlink function.

You will observe that the "sending" side of the terminal includes a multiplexer for selecting, in turn, each of the incoming information signals. For purposes of this diagram it is assumed that all incoming information signals have already been converted to some form of PCM (i.e., to digital form). These signals would typically arrive over twisted-wire pair or coaxial transmission lines from various subscribers located in the vicinity of the earth terminal.

Each incoming signal is allocated a time slot on the uplink carrier. This allocation is the result of the action of the multiplexer. An RF carrier is modulated by the digital signals. Modulation is by one of the techniques discussed in the preceding section—ASK, FSK, PSK, etc. After modulation, the carrier frequency is shifted (converted) to that of the uplink—6 GHz—by means of a heterodyne-type frequency converter. It is referred to as the *upconverter*. The output of the upconverter is passed through a bandpass filter (BPF) to ensure that its bandwidth is properly limited. The signal is then amplified to increase its energy level to that sufficient for transmission over the radio link between the earth terminal antenna and the satellite antenna. Amplification is typically by means of a special electronic device for microwave frequencies. The device is called a *traveling-wave tube*, abbreviated TWT. We will take a brief look at the operating principles of traveling-wave tubes in a subsequent section.

The antenna for the microwave frequency of the 6-GHz uplink signal is of the parabolic reflector (or "dish") type described in Chap. 10. This antenna confines the radiation to a relatively narrow beam. A narrow-beam radiation pattern has at least two significant advantages:

1. It concentrates the radiation so as to increase the ERP (effective radiated power) in the desired direction. This effect is especially important for the downlink since the amount of power available to operate a transmitter in the satellite vehicle is extremely limited.
2. The narrow beam reduces the potential for signals intended for one satellite from interfering with other, nearby satellites. (Yes, the number of satellites in use is approaching a figure where this kind of interference must be taken into consideration.)

A photograph of a typical dish antenna for a satcom earth terminal is shown in Fig. 12.39.

Figure 12.39 Typical standard earth station antenna (10 to 15 m in diameter). Photograph reprinted with permission of Dr. Elizabeth L. Young, Comstat General Corporation, Washington, D.C.

Refer again to the block diagram of Fig. 12.38. Study the portion of the diagram that refers to the "receive" function of the terminal. You will observe that from the antenna the signal first passes through a BPF and then a *low-noise amplifier* (LNA). These two items are typically incorporated as part of the antenna assembly. It is important to the success of the system that the downlink signal receive low-noise amplification immediately off the antenna. This measure helps to ensure that the amplitude of what may be a relatively weak signal is boosted before the noise content has become excessive. From the LNA the signal is transmitted, typically through a section of waveguide, to the downlink receiver. The downlink receiver is a form of superheterodyne receiver. The 4-GHz signal is converted in the *down converter* to a lower intermediate frequency (IF), amplified, and finally, demodulated. Remember, the carrier is transporting several information signals by means of time-division multiplexing. These information signals are now separated in a demultiplexer and sent on their way over terrestrial (land-based) facilities to their ultimate destinations.

The satellite transponder portion of a satcom system is represented in Fig. 12.40. In brief, the transponder has a receiver section for receiving the modulated

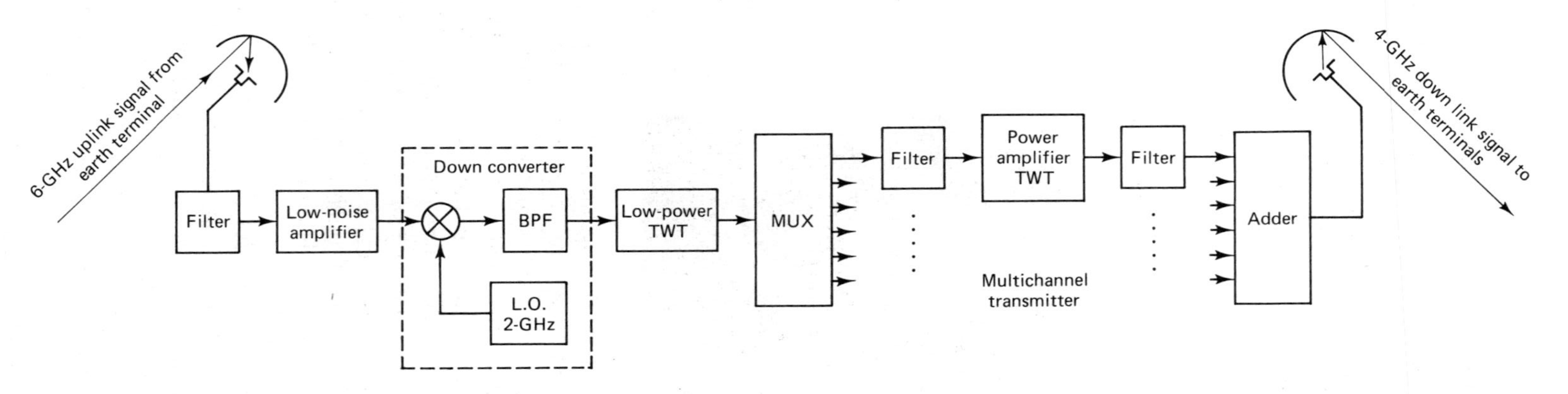

Figure 12.40 Elements of communications satellite transponder.

uplink signal (a 6-GHz signal in the example being illustrated here). The receiver converts this signal to the downlink frequency (4 GHz in this example) in the function labeled "down converter" in the diagram. The signal is filtered and receives some preamplification in a TWT device in the receiver portion of the transponder. It is then fed to the transmitter section of the transponder. In the transmitter section it receives further filtering; its energy level is amplified to the desired level for retransmission. Finally, the downlink signal is fed to a dish-type antenna which is aimed at a particular area of the earth's surface.

This description, of course, provides only the barest of details of a hypothetical satellite communications system. Because of their great cost, satcom systems must be designed and operated so that intense utilization of their facilities is achieved. Frequency-division multiplexing, as well as TDM, is used to increase the message-carrying ability of systems. This increases the complexity of a system significantly. However, just on the basis of this minimum of information, it is not difficult to recognize that satcom systems have a great deal in common with the other communications systems we have studied in this book. In fact, the utilization of a digital code for the transmission of all types of information, analog as well as digital, is perhaps the most radically different aspect of the satcom system described here. Familiar electronic building block functions such as amplifiers, oscillators, etc., may be associated in different arrangements than we have seen before. The design features of such functions may differ significantly from those we have examined previously, this because of the extremely high frequencies used in satcom systems. What should be heartening, however, is that information about older, more conventional communications systems is useful as a basis for the study of such a new, seemingly exotic, type of system.

The Traveling-Wave Tube (TWT)

An electronic device that has achieved considerable popularity in the field of satellite communications is the *traveling-wave tube*. The vast majority of electron devices—electron tubes and semiconductor devices—used at subgigahertz frequencies do not operate satisfactorily at the frequencies used in satcom systems. The primary problem for electron devices at extremely high frequencies is tied to an idea called *transit time*, meaning the transit time of electrons (or current carriers) across (or through) the device. Very simply, at extremely high frequencies, electrons in "normal" or "low-frequency" devices just do not have time to travel from one electrode to another during the period of a signal cycle. Electron transit time is "long" compared to the period of a cycle. The result is that "low-frequency" devices do not amplify when signal frequency increases to the point that makes transit time significant in comparison to the time for one cycle.

Several devices have been invented to get around the problem of excessive transit time. A list of names of these so-called "microwave" electron devices includes *magnetron*, *klystron* and traveling-wave tube. These are all electron-tube-type devices. Several semiconductor devices have also been developed which are functional at extremely high frequencies.

The traveling-wave tube has found a place in satcom technology because of several unique characteristics. First, it has a broadband capability. TWTs have been utilized in designs that amplify RF signals over a 5-to-1 bandwidth. This feature is very useful in satcom systems in that it makes possible the processing of several channels in a frequency-division-multiplexed mode. TWTs have also proven to be capable of operating continuously over a long time span. The value of this characteristic for a device installed in a satellite is obvious. Additional advantageous features of TWTs include high gain, high power-handling capability, high efficiency, and good linearity and stability. Let's look at the basic construction details and principles of operation of a traveling-wave tube.

The key to an understanding of how a TWT works is in its title: "traveling wave." Of course, as in all electronic amplifying devices, the device must make it possible to convert "raw energy" (i.e., dc power from a power supply) into signal energy. In essence, in a traveling-wave tube, electrons accelerated (given energy) by a dc source give up some of that energy to a traveling wave (the signal) and thus increase the energy of the signal wave.

A simplified representation of the physical details of a TWT is given in Fig. 12.41. The entire assembly is built around a relatively long glass envelope evacuated to a high vacuum. The internal elements of the tube include a source of free electrons, a means of accelerating these electrons with a dc voltage, and a helical coil along which the signal wave travels. The electron source is a more-or-less conventional, indirectly heated cathode (as in a TV picture tube, for example). Near the cathode is an accelerating anode, an electrode with a large hole in it that receives a positive dc potential from a dc power supply. This anode produces a beam of electrons traveling toward the opposite end of the tube at relatively high speed (and high energy level). The electrons are "collected" at the other end of the tube and returned to the power supply circuit by means of the electrode labeled "collector" in Fig. 12.41.

The path of the electron beam is through the center of the helical coil. This coil is excited near the cathode end of the tube by the signal to be amplified. Being of a very high frequency, the excitation produces a traveling wave (of

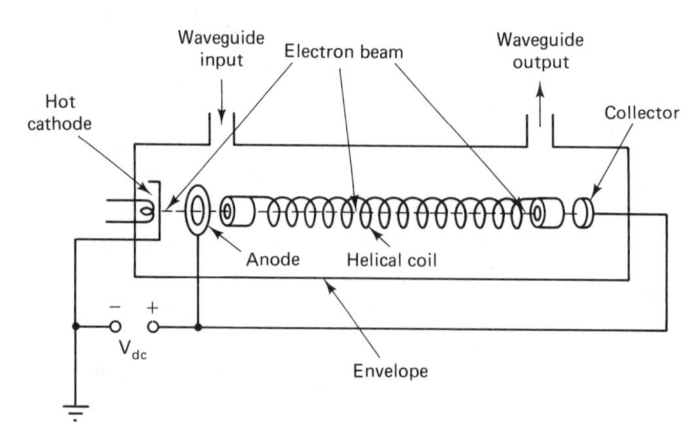

Figure 12.41 Traveling-wave tube.

voltage and current) along the helix. Although the speed of the wave along the helix conductor is virtually the speed of light, the speed of the wave down the tube is significantly less than that. This is so, of course, because the wave travels a much greater distance in following the conductor around the helix.

In fact, the speed of travel of the wave along the axis of the tube is adjusted to be just slightly less than that of the electrons in the tube's electron beam. The electrons and the RF field of the wave interact. Some electrons are slowed by the interaction; others are speeded up. A bunching of electrons is produced. The electron bunches interact with the RF wave. The important, net result is that there is a transfer of dc-derived energy from the electron beam to the RF wave. The wave arrives at the opposite end of the tube and is taken off significantly increased in energy level. Dc energy has been converted to RF energy; amplification has been produced. A gain of 30 dB is a common achievement for commercial TWT amplifiers.

In addition to the features described, a TWT device must include a function to maintain the electron stream as a relatively narrow beam. This function is typically achieved by some form of electrostatic or magnetic focusing.

SUMMARY

1. A digital communications system is one that transmits information signals, both analog and digital, in digital form. Digital transmissions are not confined to the digital data of computers.
2. One advantage of the technique of transmitting information in digital form is that it permits more intensive use of communications facilities. This is true because it is feasible to employ TDM (time-division multiplexing), as well as FDM (frequency-division multiplexing) when information is in digital form.
3. In TDM, numerous signals can be transmitted over a single channel by each being allocated a time slot in a transmission "cycle" that repeats itself many thousands of times per second.
4. Another advantage of communications involving signals in digital form is that a pulse waveform can be regenerated to a like-new waveform if it is reprocessed before excessive damage is done to it by the addition of noise. Digital information signals are more immune to the hazards of noisy communications facilities than are analog signals.
5. Transmission links of communications systems are the facilities common to thousands of individual users of the systems. Transmission links include twisted-wire-pair and coaxial transmission lines, terrestrial (land-based) microwave transmission systems, fiber-optic light-transmitting cables, point-to-point HF, VHF, or UHF radiotelephone systems, and satellite microwave systems. Transmission links include terminal facilities such as transmitters and receivers as well as the transmission media.
6. Satellite communications systems are closely associated with digital communications systems. This is primarily because most new communications systems, or major expansions, utilize satellite facilities for long range transmission. And these new satellite systems utilize digital transmission techniques almost exclusively.

GLOSSARY OF TERMS

Alternate mark inversion (AMI) The technique of inverting alternate 1's (marks) in a transmission code containing only 1's and 0's, to make a bipolar waveform out of a unipolar waveform.

Baseband transmission A communications transmission in which an information signal of relatively limited bandwidth is transmitted "as is," that is, without utilizing the process of modulating a higher-frequency signal to "carry" the information signal.

Binary 6-zero substitution (B6ZS) A technique of inserting additional 1's into an AMI code transmission anytime the original code contains six consecutive 0's, to increase "zero crossings" and therefore the timing information content of the coded signal.

Bipolar code A signal code waveform in which pulses vary in both a negative and positive direction from a zero-voltage baseline.

Bit interleaving In time-division multiplexing, the scheme in which single bits from each of several sources are accepted during the time slot allocated to each source by the multiplexing process.

Code violation In binary 6-zero substitution coding, the difference from normal code (easily detected by receiving equipment) introduced by the code substituted for six consecutive 0's in the original code.

Code word In PCM, a group of binary digits representing a single amplitude "sample" in the analog-to-digital conversion process.

Codec In a digital communications system, a combination COder-DECoder.

Coding In PCM, the portion of the process, of converting an analog signal into one in digital form, in which sampled amplitude information is finally represented as a binary code.

Companding A scheme of alternately COMpressing and then exPANDING amplitude information in the PCM coding/decoding process in order to reduce the quantizing error introduced by quantization with uniform quantizing levels. *See also* Quantization.

Comparator *See* Voltage comparator.

Dc wander The variation in "average" dc level of a unipolar pulse waveform of transmitted code caused by the random occurrence of pulses representing the two binary quantities—1's and 0's.

Delta modulation A form of PCM quantization in which a binary 1 is produced (transmitted) if the amplitude of a "sample" of an analog signal is greater by a "delta" amount than at the preceding sampling and a 0 is produced if the amplitude is not greater by the "delta" amount.

Frame A set of time slots in a time-division-multiplexed communications system and in which the time slot for each information source has a fixed time/position relationship to a frame aligning signal.

Frame aligning word (FAW) One or more pulses transmitted as part of each "frame" of data and used by a TDM system in synchronizing sending and receiving functions—multiplexer and demultiplexer.

Interleaving The process of sequencing the time slots allocated to the various signal sources in a time-division-multiplexed communications system.

Line code The digital code (stream of pulses) used to actually transmit information in digital form in a baseband transmission system.

Mark The pulse representing a binary 1 in a digital communications code.

Non-return-to-zero form (NRZ) The description applied to a digital waveform in which a pulse does not return to a zero-amplitude level until the end of its normal time slot.

Quantization One phase in the overall process of converting a signal in analog form to digital form, the phase in which the sampled amplitude information of the analog waveform is converted into a binary quantity.

Regenerator An electronic function, located at regular intervals along the transmission line of a digital communications system, which extracts information from a degraded digital waveform and produces and retransmits a like-new waveform containing the extracted information.

Return-to-zero form (RZ) A form of line code in which a pulse returns to the zero-amplitude level before the end of its time slot.

Sampling One of the phases of the process of converting an analog signal into a digital signal, the phase in which an electronic circuit ''looks at'' the amplitude of the analog signal at periodic intervals.

Sampling instant The instant at which a measuring circuit ''looks at'' an analog waveform and assesses its amplitude. *See also* Sampling.

Sampling rate The number of sampling instants per second. *See also* Sampling *and* Sampling instant.

Satcom Abbreviation for ''satellite communications.''

Slope clipping The effect of cutting across the slope of an analog waveform as a result of the delta step in delta modulation being too small for the coded result to track the analog waveform.

Time-division multiplexing (TDM) A scheme to permit the sharing of hardware in the simultaneous transmission of two or more signals by allocating periodic time slots to each of the signals to be transmitted.

Transit time Of an electron device, the time required for current carriers (electrons, etc.) to travel from one device electrode to another.

Transponder The combination of electronic functions mounted in a communications satellite which receive the uplink signal, convert it to the lower-frequency, downlink signal, amplify the downlink signal, and feed it to the downlink antenna for reradiation toward the earth.

Unipolar code A PCM line code in which pulse excursion is in only one direction, positive or negative, from the zero reference level.

Voltage comparator An electronic function, typically a linear integrated circuit, which compares voltage amplitudes at its two input terminals and gives either a high- or a low-level output, compatible with integrated logic functions, to indicate whether or not the compared amplitudes are equal.

Word interleaving A scheme of allocating time slots in a TDM system such that the time slot for each source is sufficiently long to permit the transmission of a group of bits—a code group or word—as distinguished from time-slot allocation for single bits.

REVIEW QUESTIONS: BEST ANSWER

1. The techniques of digital communications can be used for: **a.** computer data only. **b.** speech signals only. **c.** video signals only. **d.** virtually any analog or digital information signal. **e.** none of these.

2. One of the advantages of transmission of information signals in digital form is that digital transmission: **a.** is noise free. **b.** lends itself to TDM as well as FDM. **c.** lends itself to FDM better than analog transmission. **d.** uses less power. **e.** none of these.
3. Another advantage of digital over analog transmission is that a digital signal, if not excessively degraded by noise, can be regenerated. This means that: **a.** positive feedback can be used. **b.** detection is simpler. **c.** amplifiers are not required. **d.** a like-new signal can be produced for further transmission. **e.** none of these.
4. Sampling, quantizing, and coding are essential elements in the process of: **a.** pulse-code modulation. **b.** pulse-width modulation. **c.** pulse-position modulation. **d.** demultiplexing. **e.** none of these.
5. The minimum satisfactory sampling rate for an information signal in which the highest frequency is 4.8 kHz is: **a.** 2.4 kHz. **b.** 4.8 kHz. **c.** 6.4 kHz. **d.** 9.6 kHz. **e.** none of these.
6. The fact that there must be only a limited number of reference amplitude levels in the quantization of an analog signal means that: **a.** digital signals are noisier than analog signals. **b.** there are inherent rounding-off errors in the process. **c.** important information should not be transmitted over digital systems. **d.** digitally transmitted audio signals have very low quality. **e.** none of these.
7. Companding is a special technique used to achieve: **a.** uniform quantization. **b.** delta modulation. **c.** nonuniform quantization. **d.** differential quantization. **e.** none of these.
8. If a PCM system makes full use of 5-bit code words in the coding process it effectively provides for ______________ quantizing levels. **a.** 5 **b.** 8 **c.** 16 **d.** 32 **e.** none of these.
9. A priority encoder differs from a simple digital encoder in that: **a.** it has fewer inputs. **b.** it permits more than one input to be active at a time. **c.** it has more inputs than a basic digital encoder. **d.** its output is a decimal digit. **e.** none of these.
10. A linear integrated circuit that can give a high or low logic level output to indicate whether or not the amplitudes of its two input signals are equal is called a/an: **a.** voltage comparator. **b.** D/A converter. **c.** parallel coder. **d.** differential analyzer. **e.** none of these.
11. A certain ADC has a step size of 1.5 V and a maximum amplitude rating of 12 V. Its resolution is: **a.** 1.5%. **b.** 5%. **c.** 8%. **d.** 12.5%. **e.** none of these.
12. The function used to recover the analog form of an information signal transmitted as a digital code is called a/an: **a.** decoder. **b.** transponder. **c.** down converter. **d.** up converter. **e.** none of these.
13. TDM is especially adaptable to digital communications because both PCM and TDM: **a.** involve the idea of sampling or time division. **b.** are low-frequency phenomena. **c.** are limited-amplitude techniques. **d.** are most efficient at high frequencies. **e.** none of these.
14. A device that is ideal for sequencing MUXERS and DEMUXERS because its output is in binary and it increments its count on each clock pulse is the: **a.** shift register. **b.** synchronizer. **c.** codec. **d.** binary up counter. **e.** none of these.
15. In TDM, if all the bits from a PCM "word" are accepted for transmission at one time the process is called: **a.** bit interleaving. **b.** word interleaving. **c.** serializing. **d.** distributed framing. **e.** none of these.

16. In asynchronous transmission systems, start and stop bits are used for the self-synchronization of a/an: **a.** single PCM sample. **b.** single code word. **c.** single frame. **d.** line of data. **e.** none of these.
17. The term associated with the transmission mode which does not involve the modulation of a carrier signal by an information signal is: **a.** TDM. **b.** FDM. **c.** radiotelegraph. **d.** baseband. **e.** none of these.
18. There are two forms of AMI: the one in which a "mark" pulse does not use its entire time allocation is designated: **a.** RZ. **b.** NRZ. **c.** ASK. **d.** PWM. **e.** none of these.
19. The binary 6-zero substitution (B6ZS) code is used to: **a.** eliminate dc wander in a line code. **b.** reduce the effects of line noise. **c.** increase "timing" information in a line code. **d.** avoid code violations. **e.** none of these.
20. When the performance of a digital transmission system is tested with an eye diagram, excessive deformation of pulses is indicated by a/an: **a.** wide-open eye. **b.** closing down of the eye. **c.** increased opening of the eye. **d.** complete absence of the eye. **e.** none of these.
21. A detection scheme that makes use of phase as well as amplitude information in a PCM signal is called: **a.** synchronous. **b.** incoherent. **c.** DPSK. **d.** coherent. **e.** none of these.
22. A space vehicle that appears to be stationary to an observer on the earth while in orbit around the earth is said to be in a/an _______________ orbit. **a.** coherent **b.** incoherent **c.** geosynchronous **d.** shuttle **e.** none of these.
23. In electron devices for the amplification and generation of extremely high frequencies, the so-called "transit-time" problem refers to the fact that the time required for electrons to travel across a finite space in a device is long compared to a single cycle of the signal being processed. **a.** True. **b.** False.
24. The traveling-wave tube is especially popular for applications in communications satellites because of its: **a.** low cost. **b.** small size. **c.** relatively long life expectancy. **d.** shape. **e.** none of these.

REVIEW QUESTIONS: ESSAY

1. In what way(s) is digital communication different from data communication?
2. Name and discuss two important advantages of the use of digital techniques for the communication of information signals as compared to analog techniques.
3. In general, what basic idea(s) does the term *multiplexing* imply when used to describe the characteristics of a communications system?
4. What is meant by *sampling rate*? Describe two ideas that are important in choosing a sampling rate for a particular application.
5. What is quantization noise?
6. Describe the difference between a priority encoder and an ordinary digital encoder.
7. Describe briefly the operation of an IC voltage comparator and its application in an analog-to-digital converter.
8. What is a codec?
9. Describe briefly the differences between FDM and TDM.

10. What is a typical method for cycling a multiplexer or demultiplexer through its various positions?
11. What is a serializer? Where might it be used in a digital communications system?
12. What is interleaving as applied to a TDM system? Discuss the difference between word interleaving and bit interleaving.
13. What are start and stop bits? What are they used for?
14. Discuss the significant ideas implied by the term *baseband transmission*.
15. What is meant by *timing information* in a line code? Why is adequate timing information a valuable characteristic of a satisfactory line code?
16. What is AMI, and why is it used?
17. Explain why the equivalent of an analog system's repeater can appropriately be called a "regenerator" in a digital communications system.
18. Name and briefly describe three schemes that can be used to modulate RF carriers with information signals in digital form.
19. What is a geosynchronous orbit? Why is it of importance to a satellite communications system?
20. Describe the principal functions of a satellite transponder.
21. Why is a low-noise amplifier (LNA) located immediately adjacent to the active element of an earth station's antenna?
22. What is a traveling-wave tube? Why is "traveling wave" used in the title?

EXERCISES

1. Sketch a digital waveform (i.e., a pulse stream with 1 represented by an "on" level and 0 by "off") to represent the binary value 1 0 0 1 1 1 0 1.
2. Over a span of eight sampling instants an analog waveform has the following amplitude values, in volts: 3, 7, 11, 15, 12, 8, 4, and 1. **(a)** Convert each sampled value to a 4-bit word. **(b)** Sketch a pulse stream to represent the eight words.
3. When decoded, a pulse stream is found to represent the following binary words: 1000, 1101, 1111, 1101, 1000, 0011, 0000, 0011, and 1000. **(a)** Assuming that the waveform is from a PCM system, what are the decimal values of the amplitudes represented? **(b)** Reconstruct (sketch) an analog waveform using the decoded values. (Assume that the values were obtained using a uniform time interval between sampling instants. Plot the values on a set of *X-Y* coordinates and sketch a smooth curve through the points plotted.)
4. Refer to the parallel coder of Fig. 12.8. Determine the binary output word for the priority encoder for the following values of input voltage, V_A: **(a)** 0.45 V; **(b)** 0.95 V; **(c)** 1.48 V; **(d)** 2.85 V; **(e)** 3.65 V.
5. Refer to the D/A converter of Fig. 12.16. Assume $V_s = 4.0$ V. Analyze the circuit for a switching condition of 0101_2; that is, determine V_{out} for this condition.
6. Given: the 8-bit word 1 0 1 1 0 1 1 0. **(a)** Sketch a simple, unipolar, pulse-stream waveform to represent the word. **(b)** Aligned with your waveform of part (a), sketch the AMI waveform for the same word.
7. The following binary code frames are to be transmitted over a digital communications system: (1) 1 1 1 0 0 0 0 0 0 1 0 0 0 0 0 0 1 0 1 0, and (2)

1 1 1 1 0 0 0 0 0 0 1 1 0 0 0 0 0 0 1 1. **(a)** Copy the code frames on your paper and directly beneath them indicate, using "+," "−," and "0," what the AMI code for these frames would be. **(b)** Beneath the AMI code indicate the B6ZS modified code for the frames.

8. Calculate the wavelengths for the following satcom operating frequencies: **(a)** C-band: 6-GHz uplink, 4-GHz downlink; **(b)** Ku-band: 14-GHz uplink, 12-GHz downlink; **(c)** K-band: 29-GHz uplink, 19-GHz downlink.

13

LIGHT-WAVE-CARRIER COMMUNICATIONS SYSTEMS

13.1 INTRODUCTION

In Chap. 12 we examined several aspects of digital communication: some of its advantages, how it is implemented, and so on. One of the major advantages of communications by digital means is the ease with which digital signals can be time-division multiplexed. This means that digital communications provide the potential for very intensive utilization of communications transmission facilities—terminal equipment and transmission-link media. Indeed, if the bandwidth capability of a transmission system is sufficiently great, there is a potential for the "simultaneous" transmission of enormous numbers of digital signals over a single system.

A communications transmission system that uses light waves as the carrier is a system that has a high bandwidth capability. To help in appreciating the significance of this assertion, consider the following reasoning. A communications system operating at 20 MHz with a bandwidth of 20 kHz has a bandwidth capability of 0.1% (the bandwidth is 0.1% of the operating frequency). A system operating at 20 GHz with the same relative capability has an absolute bandwidth of 20 MHz. Visible light is a form of electromagnetic energy like the emission from a radio antenna. The effective frequency of visible light is of the order of 20,000 to 100,000 GHz. A 0.1% bandwidth at 20,000 GHz is 20 GHz! Signal transmission by means of light waves, in combination with the techniques of digital communication, threatens to overcome and surpass the early popularity of satellite communications systems.

That light can be used to transmit information has been known for some time. Alexander Graham Bell, the inventer of the telephone, received a patent for a light-wave communication device called a "photophone" in 1880, four years after his invention of the telephone. The development of the laser, a high-intensity light source, and fiber-optic transmission lines have made the commercial exploitation of the advantages of light-wave transmission feasible only in the last 15 years, approximately. Up until recently, application of this technology was relatively slow. However, the construction of so-called optical communications systems is now accelerating rapidly. Such systems, in certain applications such as point-to-point transmissions, are likely to make satellite systems obsolete virtually in their infancy. It is easily predictable that optical systems will play a major role in the communications facilities of the future.

In this chapter we examine the most essential, fundamental concepts of optical communications systems. We want to find out how light can be guided around corners and over great distances by a tiny optical fiber; how light is generated, modulated, and input to such lines as a signal carrier; how such light carriers are "received" and demodulated; and finally, how a system can transmit several light-wave carriers simultaneously.

13.2 OPTICAL COMMUNICATIONS SYSTEMS

An optical communications system is primarily a "conventional" telecommunications system that utilizes light waves as a carrier in one or more transmission links (see Chap. 12 for an extensive discussion of transmission links). Although lightwaves can carry analog signals, optical systems are invariably also digital communications systems. There are several reasons why this is the case (none of the reasons is the result of a physical constraint): (1) optical communications techniques are used in new systems; (2) new systems are designed to utilize the most advanced technology, such as digital, to achieve optimum message-carrying capacity, efficiency, performance and so on; and (3) optical transmission is broadband; it works well with digital techniques in providing excellent communications system performance, especially in point-to-point transmission applications.

An optical communications system, then, is one in which the transmission link is an optical transmission line instead of a terrestrial metallic conductor transmission line, or microwave link or satellite link, etc. Such a system has terminal facilities incorporating typical digital communications functions: analog-to-digital and digital-to-analog converters, multiplexers and demultiplexers, carrier generators (light sources, in this case) and modulators, receivers and demodulators, and so on. What is really new and different to be learned about an optical communications system (as compared to most other electronic communications systems) is the operation of the optical transmission line. Let's proceed to an examination of the fundamentals of guided light transmission.

13.3 FUNDAMENTALS OF GUIDED LIGHT TRANSMISSION

A special tube made of glass or plastic for the purpose of guiding light is called an *optical fiber*. An optical fiber is a waveguide for light waves. The term "fiber" is appropriate because this tube or guide is a slender, thread-like structure. "Optical" means that it has to do with light. An optical fiber is able to guide or conduct light along a path that is not a straight line. It can accomplish this feat with only a minimum of attenuation of the light. Fibers with attenuation characteristics of the order of 0.2 dB/km (decibels per kilometer) have been demonstrated in the laboratory. By comparison, the attenuation of 19-gauge twisted-wire-pair transmission line (in a multipair cable) at voice frequencies is about 0.6 dB/km. Because attenuation on twisted-wire-pair line increases rapidly with frequency, operation is limited to approximately 1 MHz.

It is not difficult to conceive of a perfectly straight tube functioning as a light guide; light could simply shoot down the tube. Optical fibers, however, are seldom perfectly straight. By what principle are they able to conduct light around curves? The answer is *refraction*. (See Chap. 6 for a discussion of refraction in conjunction with a presentation on the propagation of radio waves.)

Refraction means bending of a wave-like entity such as light. The direction of a light wave—a light ray—is bent when the ray passes between two media in which the velocities of propagation (of light) are different. You have experienced this phenomenon if you have ever been puzzled when trying to locate something under water while looking at it from above the surface of the water. You will recall that the object (e.g., a bar of soap in a bathtub) was not where you "saw" it to be. The light rays from the object were refracted as they left the surface of the water. Light travels faster in air than in water.

Optical fibers guide light by refraction. In simplest form, a light "conductor" could consist of a solid glass or plastic rod surrounded by air (see Fig. 13.1). Since the solid and air have different propagation characteristics, light rays in the rod would be refracted from the sides of the rod and thus be guided through it.

For various reasons not of major concern to us at this point, optical fibers for communications applications have a form somewhat more complex than the simple "light pipe" of Fig. 13.1. As shown in Fig. 13.2, a typical optical fiber consists of three basic elements: a central *core* (a solid rod of glass or plastic) surrounded by a protective coating of a different material called a *cladding*,

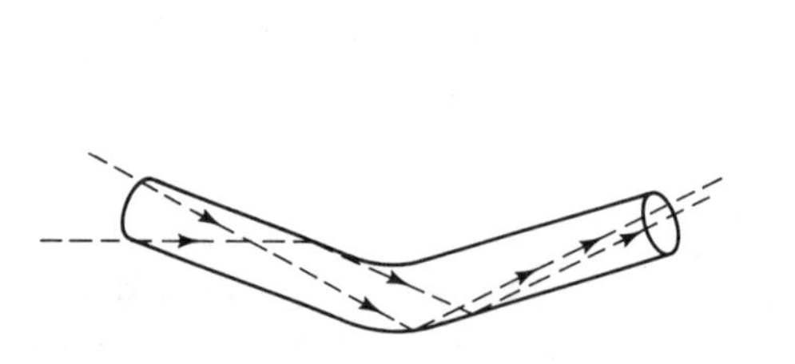

Figure 13.1 Concept of light travel through a light pipe.

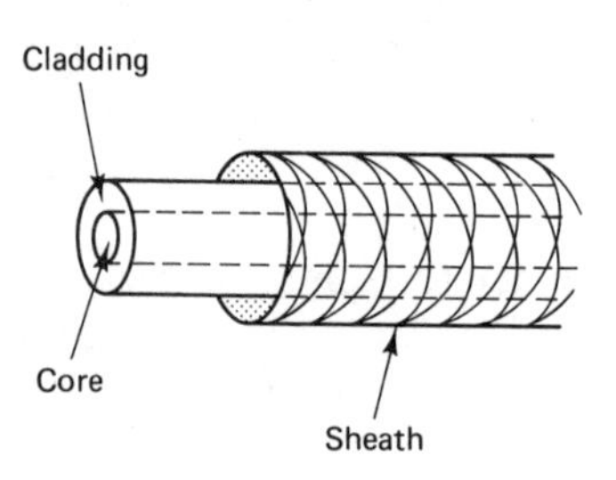

Figure 13.2 Three basic parts of an optical fiber.

which, in turn, is covered with a protective *sheath*. Before proceeding further with the specifics of optical fibers, let's examine briefly the basic "rules" of refraction and learn some terminology commonly used in discussions involving optical communication.

Snell's Law

The performance of light rays in refraction at the boundary of two light-conducting media is predictable from a principle known as Snell's law. Before looking at Snell's law, however, let's learn the meaning of basic terms associated with refraction. Refer to Fig. 13.3. Observe that Fig. 13.3 depicts the boundary between two media. Each medium is characterized by a property called its *refraction index, n*. The media in the diagram have indexes of n_1 and n_2; n_1 is greater than n_2. The index of refraction of a material is inversely proportional

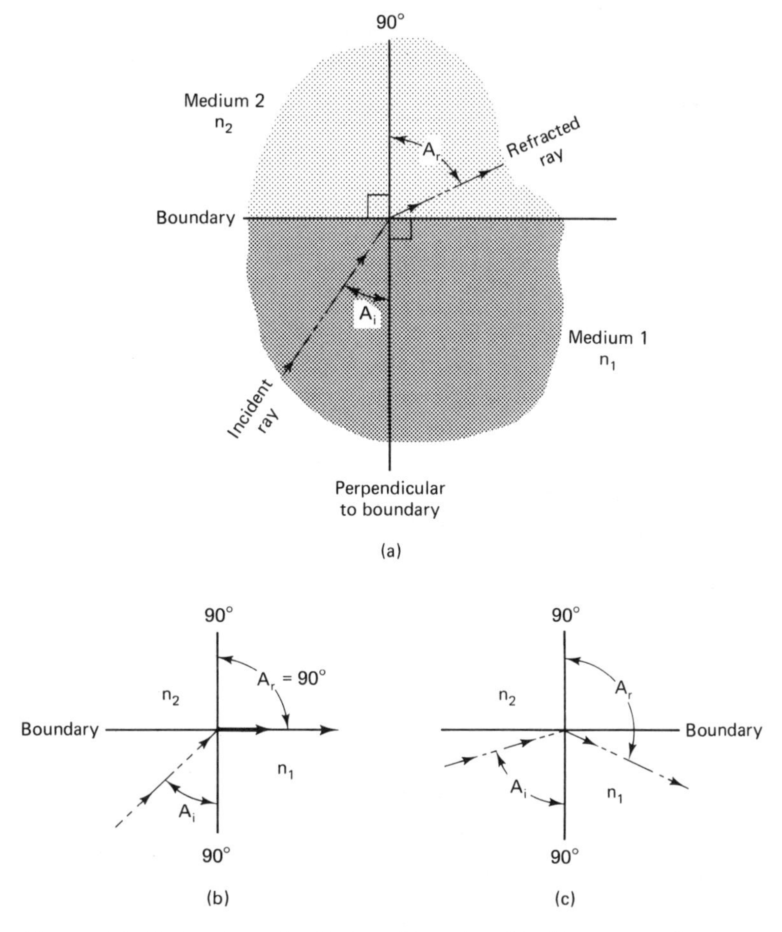

Figure 13.3 Refraction and reflection: (a) angles of incidence and refraction at boundary between media of different indexes of refraction; (b) A_i = critical angle; (c) $A_i >$ critical angle.

to the velocity of propagation of light in the material. For example, air has an index of refraction of approximately 1 (1.0002914); a typical n for glass is 1.5: light propagates more slowly through glass than through air. In Fig. 13.3, light travels faster in medium 2 than in medium 1.

Examine Fig. 13.3 further. Observe that three conditions of a light ray encountering a media boundary are shown. The conditions relate to different *angles of incidence*. The angle of incidence, A_i, of an arriving (incident) light ray is the angle the ray makes with a line perpendicular to the media boundary at the point where the ray meets the boundary. The *angle of refraction*, A_r, is the angle between the perpendicular to the boundary and the ray as it continues on its way. There is a consistent relationship between A_i and A_r, Snell's law:

> The ratio of the sine of the angle of incidence in medium 1 (of two media) to the sine of the angle of refraction in medium 2 is a constant, K, and equal to the ratio of the index of refraction n_2 of the second medium to that, n_1, of the first:
>
> $$\frac{\sin A_i}{\sin A_r} = \frac{n_2}{n_1} = K$$

Note from Fig. 13.3 that when A_i is relatively small, as in Fig. 13.3(a), the ray is bent but is able to exit medium 1; that is, it is able to cross the boundary and continue in the general direction of its original path. In Fig. 13.3(c), however, where A_i is quite large, the ray is reflected back into medium 1; it is not able to escape. Figure 13.3(b) illustrates what is called the *critical angle*. The critical angle is the incidence angle which produces a refraction angle of 90°. When the refraction angle is 90°, the ray neither exits the first medium nor is reflected back into it. Its direction is along the boundary; it is said to be absorbed.

When optical fibers are used as transmission lines for light in a communications system, it is important that they be operated so that most of the light that is introduced to the fiber remains in it until the destination is reached. That is, the condition of interest is when the angle of incidence is greater than the critical angle, the condition of Fig. 13.3(c).

In optical-fiber operation, the angle of incidence is set by the relationship of the light source to the source end of the fiber. Light from a typical source travels in all directions. The result is that all three of the conditions illustrated in Fig. 13.3 are likely to occur in the excitation of an optical fiber, as shown in Fig. 13.4. Only those rays that enter the fiber parallel to its axis or which have incidence angles greater than the critical angle will be propagated in the fiber.

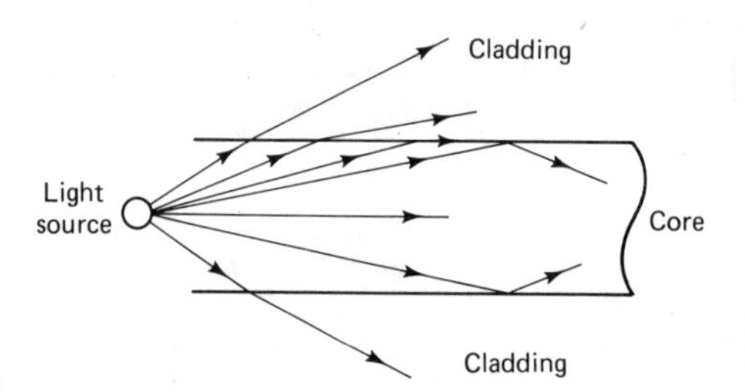

Figure 13.4 Light directions at input to optical fiber.

For a given source–fiber interface, there can be numerous ray directions which produce incident angles greater than the critical angle and which therefore result in propagation of light in the fiber. Each of these directions is referred to as a *propagation mode* or simply *mode*. Propagation (net distance traveled in a direction parallel to the axis of the fiber) is the result of the refraction of a ray from first one boundary of the fiber core and then the opposite boundary—a zigzag path.

For each propagation mode (initial direction of a light ray) there is a different zigzag path. The total distance traveled (from input to any other point on a fiber transmission line) on each path is different from that of every other path. The result is that communications signals carried by light waves when transmission is multimodal (many zigzag paths) will not have the same phase relationship along an optical-fiber transmission line that they had at its input. The energy of a signal in multimodal transmission may be dispersed into thousands of different paths (modes). The effect on the pulses of digital communications is to distort them by stretching them out, rounding the corners, etc. The appearance of transmitted pulses is similar to that of systems in which the bandwidth is inadequate. Multimodal transmission is therefore bandwidth limiting. Bandwidth limiting reduces to some degree the potential advantages of optical transmission.

There are at least two ways of minimizing the problem of bandwidth limiting associated with multimodal transmission. One technique is to use fibers whose core diameter is so small that only a single ray is propagated. That is, only a single path or mode is produced by light entering the fiber. The technique is called *single-mode propagation*. The second technique involves using a core medium that is not homogeneous in its refraction index across the diameter of the core. The fiber is said to have a *graded index*. A fiber with a homogeneous medium has a *step index*. Let's examine these concepts separately and in greater detail.

Single-Mode Propagation

Single-mode propagation is produced in an optical fiber when the diameter of its core is not more than about three times the wavelength of the light being transmitted. This means a core diameter of not more than 5 to 10 μm. The core diameters of multimode fibers are in the range of 50 to 100 μm. Again, the advantage of single-mode propagation is that it minimizes the distortion (in the form of spreading) of signal pulses caused by modal delay (different travel times caused by light traveling along different paths or modes).

Disadvantages of single-mode propagation include the fact that it is more difficult to inject light into the ultra-small-diameter fiber. In order to inject a satisfactory level of light, it is necessary to use an expensive laser source as compared to a more economical source such as an LED (light-emitting diode) usable with multimode fibers. Single-mode fibers are more difficult to splice and interconnect than the larger multimode fibers. Nevertheless, single-mode fibers appear to be the more popular choice for new optical transmission systems.

Graded-Index Fibers

It is now technologically possible to manufacture optical fibers with cores whose refraction indexes vary in some fashion across their diameters. The effect of this grading of the refractive index is to alter the velocity of light in relation to the region of the fiber in which the light is traveling. For example, grading typically causes light traveling near the boundary of the core and cladding to travel at a greater velocity than light traveling near the center of the core. The final result is that difference in travel delays caused by different modes is reduced almost to an insignificant minimum. Bandwidth limitation of multimodal transmission is significantly reduced in graded-index fibers.

Graded-index fibers are produced by adding suitable impurities (dopants) to the basic core material during the manufacturing process of the fiber. Precise control permits the production of fibers of various *index profiles*. An index profile diagram depicts the variation of the refractive index of a fiber with respect to its diameter. One common graded-index profile is shown in Fig. 13.5. A step-index profile is included for comparison.

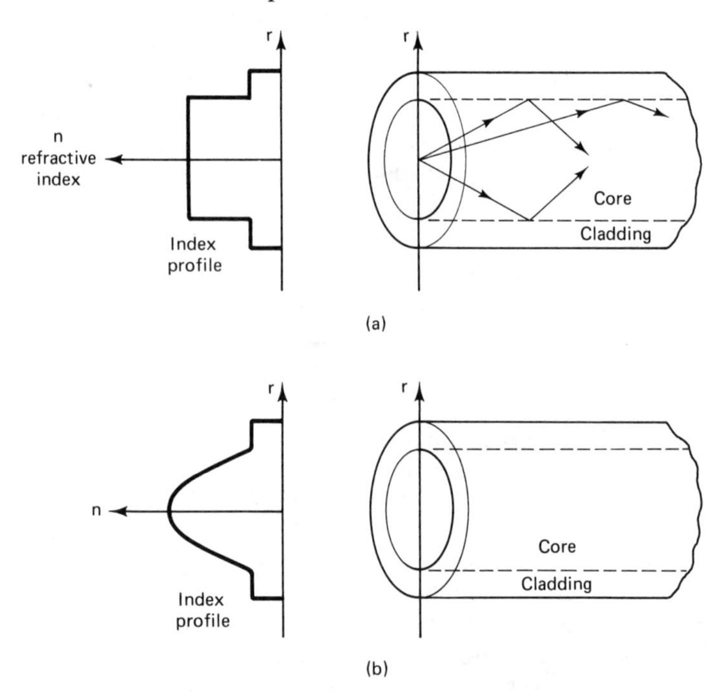

Figure 13.5 Optical-fiber index profiles: (a) optical fiber with step-index profile; (b) optical fiber with graded refractive index profile.

13.4 TRANSMISSION CHARACTERISTICS OF FIBER-OPTIC LINES

Optical fibers are now manufactured and packaged in forms suitable for use in operational environments similar to those of other forms of communications transmission lines. Other types of transmission lines include open-wire line (now virtually obsolete), twisted-wire pair lines, coaxial cables, etc. Practical considerations typically require that transmission-line packages include a sheath for environmental protection of the vital elements of the line. In addition, some method of providing physical strength to the package, so that it can be strung between poles, for example, is usually incorporated. Strength members include strands of glass yarn or even steel wire. Fiber-optic cables with two or more fibers in a single package are available. There is no standardization of optical-line packages; commercial products of numerous forms are available.

Attenuation

As with other forms of transmission lines, energy injected at the input end of a fiber-optic line diminishes with distance along the line. Light energy in an optical line is attenuated by four basic loss mechanisms: scattering, absorption, loss in connections, and loss due to fiber bending.

Scattering and absorption losses are related in that they are caused primarily by impurities or flaws in the medium of an optical-fiber core. Refer to Fig. 13.6. Observe that when a light ray encounters a flaw (an "air hole," for example) or an impurity in its path through a fiber, a refraction occurs. The ray may be deflected sufficiently to exit the core and be absorbed by the cladding or sheath of the fiber. Or, it may simply be absorbed by a particle of opaque impurity embedded in the core material. In either case, the energy involved in the particular event is lost to the system.

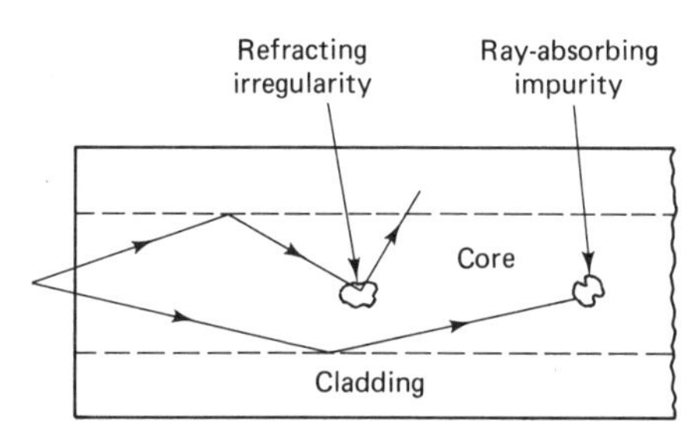

Figure 13.6 Refraction and absorption by irregularities and impurities in core material.

A continuing goal of companies that manufacture optical fibers is to find ways to reduce the number of flaws and impurities in a specified length of the fiber. Considerable progress has been made. As recently as 1968, attenuation in fibers amounted to 1000 dB/km. Even at that, fiber-optics pioneers were confident that optical transmission systems would someday be a popular reality. Today, for example, the Corning Glass Works with its Code 1505 and 1516 products can supply optical fibers with attenuation specifications of "2.9 dB/km at 850 nm"; and the price is competitive with copper-wire transmission lines. (The "850 nm" portion of the specification refers to the wavelength of the light to be propagated. The fibers are best suited to propagate this particular wavelength. A nanometer is 1/1000 of a micrometer.)

The manufacture of optical fibers in a gravity-free space station is considered an exciting possibility for the not-too-distant future. It has long been speculated that glass could be manufactured virtually impurity- and imperfection-free in a gravity-free environment. A recently conducted space manufacturing experiment by the Westinghouse Corporation appears to confirm the speculation. Expectations of further reductions in the attenuation of optical fibers have been heightened.

Practical optical transmission systems require that fibers be connected to energy sources, receivers, and to other sections of fiber transmission lines. There is a potential for energy loss at each connection point. Every system has some connections losses.

Connection losses are the result of light rays encountering imperfections in the boundaries of the transmission media where two media are joined to permit the passage of light from one to the other. For example, if two fibers to be connected together had perfectly smooth ends (a practical impossibility) and were perfectly aligned and of the same diameter, all the light from the sending fiber would be injected into the receiving fiber. However, if there is any surface roughness on the ends of the fibers where they are joined, some light will be refracted by the surface imperfections and lost to the system. If the two fibers are not of the same diameter or not perfectly aligned, some light may escape and represent a loss of energy to the system. Connection losses accumulate and can quickly become a major portion of the total losses of a system. Assuring that minimum-loss connections are obtained warrants significant effort in setting up and maintaining an optical transmission system.

Bending losses are the result of energy lost when light waves are required to make an excessively sharp bend. When analyzing the behavior of light as a wave phenomenon, we must remember that a wave has a "width" perpendicular to the direction of travel of the wave. When the wave bends around a corner, the outside edge of the wave must travel faster than the inside edge: If it doesn't, it isn't bending. (As an analogy, when a column of marchers goes around a corner, persons in the outside positions must step faster, or persons in the inside positions must mark time, in order for the line to remain straight during the turn.) If the bend is too sharp, part of the wave would have to travel faster than the speed of light, which it obviously cannot do. The result is that some of the light simply exits the fiber and is lost by absorption in the cladding or sheath.

The total losses of an optical communications system includes the sum of all the losses produced by the mechanisms described above. Design for maximum performance requires attention to assure minimization of each type of loss.

Numerical Aperture and Acceptance Angle

An important characteristic of an optical fiber is its ability to gather light. This property is called the *numerical aperture* (NA). Numerical aperture is defined in terms of another technical concept: the acceptance cone or acceptance angle.

Consider Fig. 13.7(a). It is apparent from the drawing that a fiber will accept light for propagation only within a certain angle with respect to its axis. If the angle of a light ray is too large, the angle of incidence at the boundary between the core and the cladding is less than the critical angle. The ray is not reflected from the boundary; it exits the core. Since the fiber is a cylinder, the

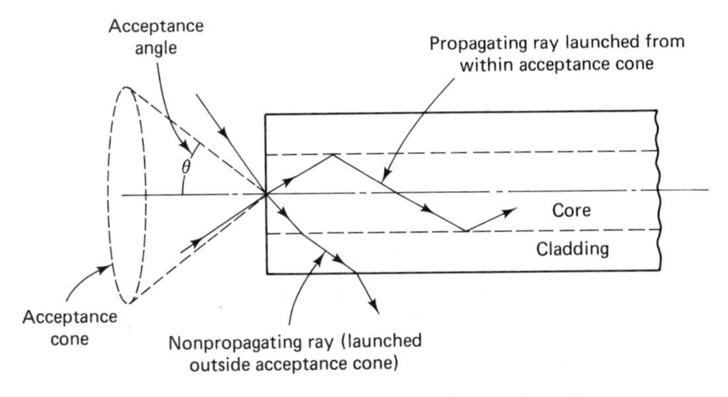

Figure 13.7 Acceptance cone of an optical fiber.

limiting angle of acceptance is a solid angle—a cone (see "acceptance cone" in Fig. 13.7). The numerical aperture is defined as the sine of half the angle of the acceptance cone, also called the acceptance angle, or

$$NA = \sin \theta$$

where θ is the acceptance angle or half the angle of the acceptance cone.

The NA is determined by the relative indexes of refraction of the core and cladding materials. It is not a function of the core diameter. For example, for a step-index fiber with a core index of refraction of n_1 and a cladding index of n_2, the formula for NA is

$$NA = \sqrt{(n_1)^2 - (n_2)^2}$$

Commercially available fibers for communications applications generally have an NA of less than 0.5. This corresponds to an acceptance angle of 30° and an acceptance cone angle of 60°. The numerical aperture specification of a fiber is of importance when the design details of the coupling between a light source and a fiber is under consideration.

Bandwidth

When light—visible or nonvisible—is being considered, discussion usually revolves around wavelength rather than frequency. Light of different colors has different wavelengths. Light of a single color—monochrome light—when used as a carrier in a communications system is analogous to a radio-frequency carrier of a single frequency. You will recall that when an RF carrier is modulated, side frequencies are produced. Similarly, when a light carrier is modulated to achieve the communication of information, the frequency of the light is changed. Or, more typically, it is said that the wavelength of the light energy is changed. A successful RF transmission system must have sufficient bandwidth to accommodate all significant side frequencies resulting from the modulation. In like manner, an optical transmission system must have sufficient bandwidth to carry all the wavelengths having significant energy produced by any modulation process.

The ability of an optical-fiber transmission line to transmit light energy of different wavelengths—its bandwidth, in other words—is limited primarily by two mechanisms. The two mechanisms are identified by the commonly used terms of *modal dispersion* and *chromatic dispersion*. In contexts related to the phenomenon of light, the term *dispersion* implies ideas such as "breaking up," "scattering," and/or "spreading about." These meanings are relevant in connection with a study of the light-propagating characteristics of optical fibers. Here dispersion relates to the spreading of the time of arrival at a destination of various components of transmitted light. The effect of this dispersion on transmitted information signal pulses is to attenuate them and spread or stretch them out over longer time intervals. The pulses are distorted; they become difficult, indeed, impossible to decode without error. Because they occupy more time, the maximum number that can be transmitted in a second is reduced—the effective bandwidth of the transmission system is reduced.

We have already examined the phenomenon of modal dispersion. We did not identify the mechanism with that term at the time. Modal dispersion leads to the distortion of a waveform described in earlier paragraphs: The distortion results from different propagation paths requiring different travel times from input to output of an optical transmission system. You will recall that using either single-mode or graded-index fibers helps to minimize this type of distortion (it is also referred to as modal delay spreading).

Chromatic dispersion, also sometimes called *material dispersion*, refers to the fact that different wavelengths propagate through the "material" of an optical fiber at different velocities. The refractive index is different for different wavelengths of light. If a light transmission is spread over a band of wavelengths, the relative phase of the signal components will be altered when the energy reaches the receiving end of the system. This "dispersion" of the phase of the components causes the information pulses to be attenuated and distorted by spreading. If the distortion is excessive, the information contained in the pulses cannot be "read" correctly by the receiving equipment. The spread of wavelengths—the bandwidth of the input signal—must be limited to a value that does not cause excessive distortion. The chromatic dispersion mechanism limits the usable bandwidth of the transmission medium.

The degradation, by the mechanisms just described, of a signal being transmitted over an optical transmission line increases with the distance traveled by the light. Furthermore, optical communications systems invariably transmit digitized signals. In this context, stating the bit-transmitting capacity (recall: a bit means a binary digit, a 1 or 0) of a subsystem—optical line, etc.—is the usual way of specifying its performance. Putting these two ideas together—bit "speed" and distance—leads to the usual specification format for optical lines: megabits or gigabits per second × kilometers. For example, a typical specification for a multimode fiber propagating an 0.8-μm emission is "1 Gb/s × km" ("one gigabit per second times kilometers"). This means that digital information could be transmitted on the fiber line at a rate of 1 Gb/s over a distance of 1 km without needing to be regenerated and reamplified in a repeater station.

The value of the "product" specification is that it provides for easy conversion for speed versus distance trade-offs. The 1 Gb/s × km is easily changed to 100 Mb/s × 10 km or 10 Mb/s × 100 km, etc. (To get a better handle on these figures, consider that the text in all the volumes of the *Encyclopaedia Brittanica* is equivalent to 100 Mb, approximately: The entire encyclopaedia could be transmitted in 1 s over a distance of up to 10 km!)

An interesting and fortunate characteristic of optical fibers is that chromatic dispersion "takes a holiday" at the wavelength of approximately 1.3 μm. On the other hand, fiber attenuation increases with wavelength but has a minimum or "null" at a wavelength of 1.5 to 1.6 μm. Major research and development efforts to improve the performance of fiber-optic systems are now concentrating on the development of so-called *single-mode* light sources. Specifically, these are lasers (more about lasers below) that will produce single-wavelength light at 1.55 μm. They will be used in conjunction with single-mode fibers. Consequently, they permit performance optimization of three factors: The single-mode fiber minimizes modal dispersion; the single-wavelength light virtually eliminates the problem of chromatic dispersion; and the 1.55-μm wavelength emission corresponds to the wavelength at which glass is most transparent. The result is that developers expect to achieve the amazing performance of at least 200 Gb/s × km from fiber-optic systems.

13.5 LIGHT SOURCES FOR OPTICAL COMMUNICATIONS SYSTEMS

When we consider an optical communications system as an analogy of an RF communications system, the optical fiber link is the equivalent of the transmission line of the RF system. Of course, we must also have the equivalent of the RF transmitter and the receiver. A light source capable of being modulated is the equivalent of the transmitter. Two sources currently used for this purpose are the surface light-emitting diode (LED) and the injection laser diode (ILD). Incandescent (filament-operated) light sources are not suitable sources for any but low-frequency information signals because of a characteristic called *thermal hysteresis:* A filament stores heat temporarily, and therefore changes in the intensity of its light output lags behind changes in its exciting current by a significant degree. Let us examine the operation and characteristics of LED and ILD sources in some depth.

Light-Emitting Diodes

The least expensive and simplest, in terms of theory of operation, of light sources used in fiber-optic systems is the light-emitting diode (LED). You are familiar with LEDs in their application as readouts for electronic calculators, digital watches, etc.

An LED is, first of all, simply a semiconductor diode. That is, it is made of a "sandwich" of two types of semiconductors. One layer is an *N*-type semiconductor: The current is carried by electrons contributed by an impurity added to a basic semiconductor material. The second layer is a *P*-type semiconductor:

The current is carried by a positive, electron-deficiency characteristic, the contribution of a different added impurity. This carrier bears a title familiar to all who have studied semiconductor devices: It is called a hole. The region of the sandwich where the two layers are joined together is called the junction of the diode. At the junction, some of the electrons of the N-type material fill the holes of the P-type material, thereby creating a "barrier" of negatively charged atoms on the P side of the barrier. Additional electrons from the N-type material will not move toward the P-type material unless this potential barrier is overcome by an external voltage applied to the sandwich: positive to the P-type material, negative to the N-type material. When such a voltage is applied, the junction is said to be forward biased: A current will flow in the external circuit.

When a diode is forward biased, electrons and holes move across the junction. There are now two types of current carriers on each side of the junction. Some of these carriers of opposite types recombine. Electrons and holes have different energy levels as a result of the nature of the atoms from which they come. The energy-difference phenomenon is called an energy band gap. The amount of energy of the band gap is determined by the materials used as the semiconductor. When the carriers recombine, the band-gap energy is released in the form of light *photons* (units of light energy). That is, light is radiated from the site of the recombination. The wavelength of the light, and whether it is visible or not, depends on the gap energy. Remember, this phenomenon occurs in all semiconductor diodes. A light-emitting diode is one in which the materials used provide light in the visible range, and are constructed so that that light can escape from the device to be seen.

Light-emitting diodes designed and fabricated especially for use in fiber-optic communications systems incorporate various modifications and refinements which distinguish them from the more common indicator-type LEDs. The thrust of such modifications, obviously, is to make the devices more effective as subsystems of the optical system. First, such devices are fabricated from semiconductor materials chosen to provide a desired emission wavelength. Although a number of materials have been used, the most popular new diodes use gallium arsenide indium phosphide (GaAsInP) as the material that determines their principal operating characteristics. This quaternary (four-element) compound emits radiation with a wavelength of 1.3 μm. This is a desirable wavelength because fibers have a loss null at 1.3 μm.

Second, the diode will be shaped geometrically to enhance its emission effectiveness and facilitate coupling the emitted energy to a fiber. Like a real sandwich, a junction diode package has a relatively large plane surface on two sides and a relatively narrow edge on its other sides (see Fig. 13.8). Although some light is emitted from a junction edge, most of the radiation is available from the upper surface of an LED. Hence LED sources are typically "surface" emitters. To improve the light-launching capability of the diode, much of one surface is etched away to provide a well. Etching the depression removes much of the light-absorbing material above the active region of the diode and provides a convenient contour for coupling a fiber to the diode, as shown in Fig. 13.8.

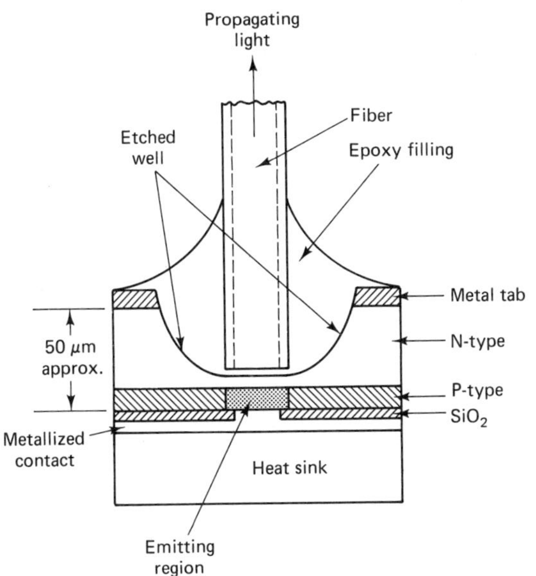

Figure 13.8 Light-emitting diode (LED) as source for fiber-optic system.

Currently, diode–fiber combinations are available in which several hundred microwatts at the most can be produced and injected into an optical fiber.

Injection Laser Diodes

A laser diode is a light-emitting diode with something extra. That something extra is described as an *optically resonant cavity*. A cavity is a space. An optically resonant space is a space, first of all, which is provided at two ends with some means of reflecting light. Second, the dimension of the space (cavity) between the reflecting ends has a precise relationship with the wavelength of the light to be produced in the device. Because of this relationship, some of the light in the space is reflected back and forth between the ends producing standing waves. The space is "resonant" to the light.

The real significance of an optically resonant cavity is that feedback (optical feedback in this case) can occur. With feedback, the optical equivalent of oscillation is possible. Optical oscillation produces a striking increase in the amount of light energy available from the device. Furthermore, this light is *coherent light*. Coherent light is not simply light of a single wavelength (monochromatic): The light from an LED is monochromatic. "Coherent" means that the waves are of the same phase, as well as of the same frequency. The result is that light emerges from a laser in a very narrow, intense beam. Let's examine lasing in greater detail.

The word "laser" is an acronym; it is formed as follows: *l*(ight) *a*(mplification by) *s*(timulated) *e*(mission of) *r*(adiation). The words whose first letters form "laser" indicate the factors required for lasing (emission of laser light). In the case of an LED that is caused to lase, we have already examined the mechanism of emission of radiation: the recombination of band-gap electrons and holes. If an LED is very carefully cleaved at appropriate planes of the semiconductor crystal, a cavity containing the sites of light emission is formed. (To "cleave" means to cut or separate. In this instance, however, "cleave" implies a more precise separation than just a "cut," as with a saw or knife. Rather, it means a separation at a natural separating point—a crystal "plane"—provided by the crystalline structure of the semiconductor itself.) The cleavage planes of the crystal form quite effective "mirrors" for the light emitted in the enclosed space. The separation of the mirrors is also right for resonant reflection of light of the wavelength produced by the diode.

What happens in a diode laser, then, is that when forced to conduct a current by forward biasing, an initial emission is produced. Some of this emitted light is reflected from a cleavage plane back to the opposite plane, thence back to the first plane, and so on. Each time the light passes through the cavity it *stimulates* more electron–hole recombinations, and, of course, more light radiation. If the primary excitation (diode current) is sufficient, the device will indeed produce super emission levels of coherent, monochromatic light—it will lase. The light beam from a laser is very narrow and has an extremely high energy density as compared to that from other sources.

Why is a diode laser of this type called an *injection laser diode?* The question cannot be answered with a single word, phrase, or even short sentence. Some background is required.

First, the light produced by either an LED or a diode laser is by a process called *electroluminescence*. Electroluminescence, in turn, refers to light produced by the effect of an electric field on the atomic structure of a material.

Electroluminescence is to be contrasted with incandescence. Incandescence is associated with heating to a high temperature. A burning ember and an incandescent light bulb produce light as a result of their high temperatures. Electroluminescence is a *cold* light.

When electroluminescence occurs, the atoms of a material are caused to release light energy packets—photons—simply as the result of the force of an electric field acting on them. In the case of LEDs and diode lasers, the electric field causing the luminescence is that produced by the forward-bias voltage across the diode junction. This voltage produces a current and the current is described as the result of *injecting* electrons into the *P*-type material and holes into the *N*-type material. Or, it is said that there is an *injection* of minority carriers across the junction barrier. It is this injection action that leads to the recombination of current carriers in diode lasers, and hence to the production of light. And, of course, it leads to the name given the device.

Injection laser diodes are significantly more expensive than the LEDs used for light sources in optical communications systems. This is because the structure of an ILD is more complex than that of an LED. A complex structure requires more, as well as more difficult, steps in the fabrication process. In many communications applications (e.g., long-distance trunk lines) the greater cost of the ILD is justified. The injection laser produces much higher levels of radiation than the LED. Optical output power levels of the order of milliwatts is possible from an ILD. You will recall that several hundred microwatts at the most can be produced and launched into an optical fiber by an LED. The inherent dimensions (microsmall) of an ILD match those of typical fibers closely. This minimizes the effect of material dispersion (see above). Injection laser diodes are well suited for application as the source for single-mode fiber-optic transmission lines which have several advantages over multimode fibers.

A single-mode fiber is such because of its extremely small diameter. Single-mode propagation is desirable because it minimizes modal dispersion. So-called single-mode lasers for use with single-mode fibers have been available for some time. Even though the output of these devices is basically of one wavelength, propagation of their output in a single-mode fiber is still subject to some chromatic dispersion.

As this is being written a new diode laser is emerging from several development laboratories around the world. This device is referred to as a *single-frequency laser*. It represents one step of improvement beyond that of the single-mode laser. Its operation provides an even purer and more oscillation-mode-limited output than that of previous lasers. (A full technical explanation of its design, fabrication, and operation is beyond the goals of this text.) The net result is that its output minimizes chromatic dispersion in single-mode propagation systems. Its potential emergence as a commercially available device portends great things for optical communications systems. As indicated in the preceding section, it is predicted that systems employing such devices will achieve bit rate × distance products of at least 200 Gb/s × km.

13.6 MODULATION OF LIGHT SOURCES

If you have studied other communications systems in this book or elsewhere, you are familiar with the concept of modulation. You understand that in order for a "carrier" to transport information, it must be changed (modulated) in some way. After modulation the information is contained either in the carrier or some product of the modulation process. A lightwave carrier can, of course, be modulated. Indeed, by its nature (high frequency or short wavelength) it possesses the potential for carrying greater quantities of information (or data) than any other type of communications carrier known. Let's examine the process of modulating a light-wave carrier.

The amount of light produced by either an LED or an ILD is proportional to the amount of injection current (i.e., the current produced when the diode junction is forward biased). Hence a light source could be modulated by an analog signal simply by driving the diode with a transistor whose operating current is controlled by the analog information signal. The concept is illustrated in the circuit of Fig. 13.9. The circuit shown would produce the equivalent of amplitude modulation. In the case of a light source, the effect is called *direct intensity modulation*. Although analog modulation is employed in some actual optical communications systems, its extent is very limited. One could expect to find the simple modulation scheme suggested by the circuit of Fig. 13.9 to be enhanced significantly in an actual installation.

The real potential of an optical system to provide high-density communications

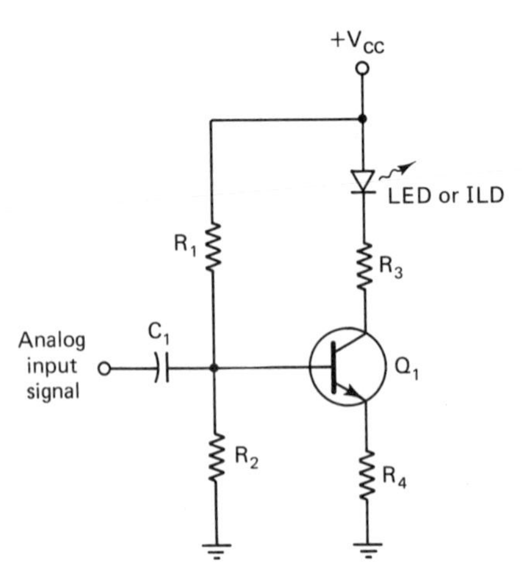

Figure 13.9 Simple circuit for direct intensity modulation of light source.

is best realized when information is transmitted in digital form. It is useful to consider an optical communications system as a digital communications system in which the transmission link is a fiber-optic subsystem (see Chap. 12). The significance of this statement is that the light source in the optical link is modulated in a manner very much like that of the RF carrier in RF systems: The light-source transmitter is fed numerous time-division-multiplexed digitized analog and digital data signals. These signals modulate the light source, typically using digital intensity modulation.

Digital intensity modulation is best compared with amplitude-shift keying in an RF system: A light source is turned on and off, or driven between two levels of intensity at a very high rate. Potentially, the rate can be in the gigahertz range. Of course, the modulation is in accordance with an appropriate digital code. The basic ideas involved are represented in a general way in Fig. 13.10(a). The more specific concept of a light-source drive circuit controlled by a digital signal is shown in Fig. 13.10(b).

The concepts illustrated in Fig. 13.10 are basically the same whether the source is an LED or an injection laser. In either case, enhancements of the simple circuit shown would include techniques to improve the switching speed of the drive circuit itself. However, the best performance of a laser source is obtained only with attention to the optimum control of several critical operating parameters. For example, a laser is sensitive to operating temperature. Laser drive circuits typically employ feedback circuits that monitor the output operating level and adjust drive conditions for changes in output caused by ambient temperature changes.

A laser light source is generally supplied with a bias current. This current, sometimes called a *prebias*, is maintained near but just below the "lasing" threshold of the device. That is, it is a bias current for the "off" condition of the laser. A prebias is used with a laser because it improves performance: It reduces switch-on delay. It makes compensation for aging or ambient temperature changes easier.

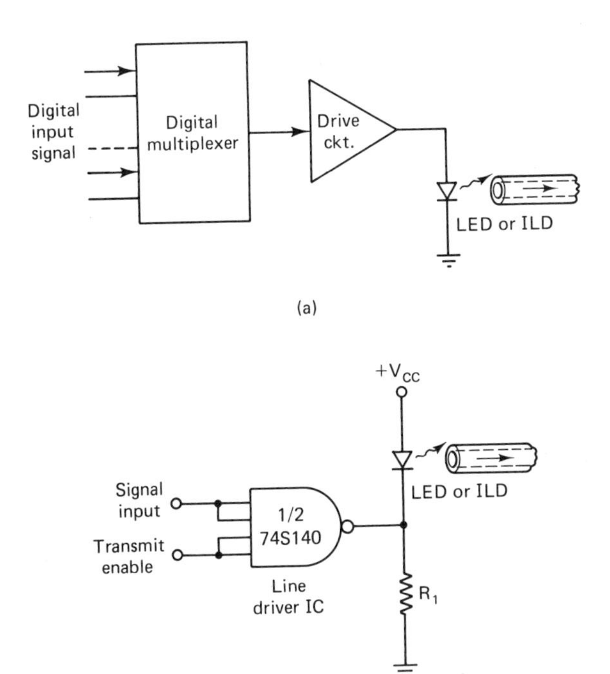

Figure 13.10 Light-source drive and modulation: (a) digital intensity modulation of light source; (b) possible TTL-compatible drive circuit.

13.7 LIGHT DETECTORS FOR OPTICAL COMMUNICATIONS SYSTEMS

After reaching its destination a modulated light beam must be demodulated in order for the system to recover the information being transported by the lightwave carrier. In current systems the light signal is first converted to an electrical signal and amplified before any action to recover the information content occurs. The device used for the purpose of converting the light carrier of an optical system into an electrical signal is called an *optical detector* or a *photodetector.* A photodetector is not a "detector" in the sense of a demodulator as the term is used, for example, in conjunction with radio receivers. It is more analogous to the antenna of a radio receiver.

In general, the "receiver" subsystem of an optical system includes a photodetector to convert the light signal into an electrical signal followed immediately by an electronic amplifier. Once amplified the signal is processed further by more or less conventional electronic communications circuits. The concept is illustrated in Fig. 13.11. Our purpose in this section, however, is primarily to gain knowledge about detector devices themselves. Let's have a look at the most important ones.

There is an assortment of devices used to detect light and/or convert it to an electrical signal in the numerous applications of optoelectronics. However, fiber-optic communications systems almost invariably utilize devices of the *photoemissive* type.

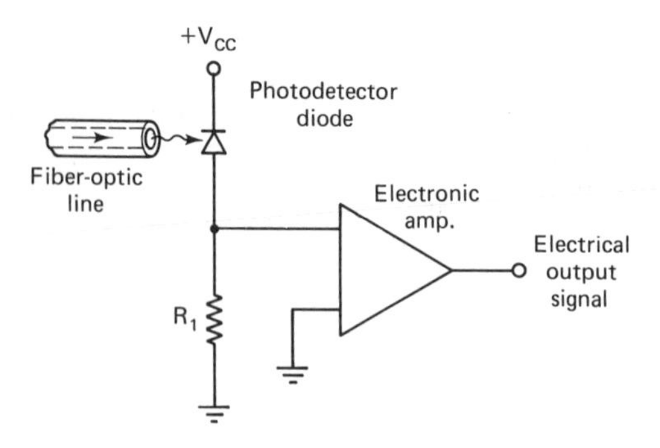

Figure 13.11 Optical system receiver.

Photoemission refers to the liberation of electrons from a solid material as the result of light striking the material. Historically, photoemission was the mechanism operating in the so-called "phototubes" of the vacuum-tube era of electronics. In those devices, the process was referred to as "external" photoemission: Electrons were liberated from a metallic photocathode and were attracted across an evacuated space within a glass envelope (a tube) by means of externally applied positive voltage.

In the present era of semiconductor electronics, there are available PN junction diodes, called photodiodes, which have the capability of photoemission. In these, electrons are liberated within the diode structure itself by the action of light striking the semiconductor material of the junction. Appropriately, they are referred to as internal photoemitters.

The basic process of the liberation of an electron by photoemission in a semiconductor diode is as follows. When struck by a photon (a unit energy bundle) of light, an electron in the valence band of an atom receives an increase in energy sufficient to boost it to the conduction band. That is, it becomes like the "free" electrons of a conductor. The freeing of the electron leaves a hole in the valence band; an electron–hole carrier pair is generated by the process and is available to augment a current in the circuit external to the diode.

Let's expand on this sketch of the basic process in several ways. First, the diode junction is reverse biased (a positive potential is connected to the *N*-type material, a negative potential is connected to the *P*-type material). Majority carriers are attracted back to their sides (holes to *P*-type material and electrons to *N*-type material). The so-called depletion region is created on either side of the junction. Majority carriers are prevented by the electric field (of the reverse bias) from crossing the region. Associated with the condition, however, is a current known as the reverse leakage current. The reverse leakage current of a diode is the result of minority carriers being attracted across the junction: holes in the *N*-type material attracted by the negative potential connected to the *P*-type material, electrons in the *P*-type material attracted to the positive potential connected to the *N*-type material.

When a light photon enters the depletion region described in the preceding paragraph it generates the electron–hole carrier pair described previously. The photon must possess a sufficient energy. It need not actually strike an electron but simply come near enough to excite it. The electron and hole so generated come immediately under the influence of the reverse bias field and as a result add to the leakage current of the diode. The reverse current in excess of the normal leakage current is proportional to the light energy striking the depletion region. It is called the photocurrent. The concept of electron–hole pair generation is illustrated in Fig. 13.12.

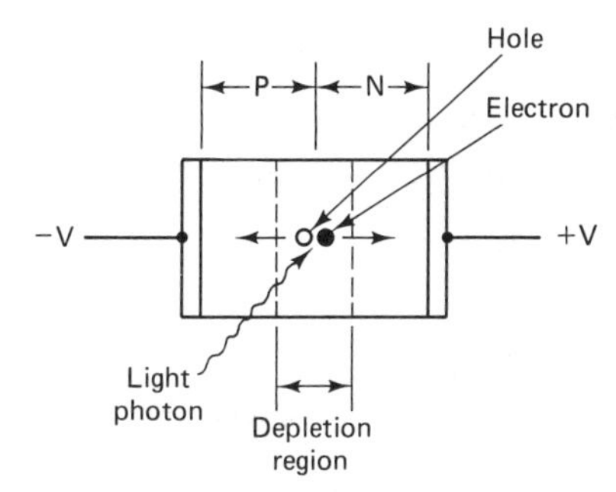

Figure 13.12 Creation of electron–hole pair by light photon entering depletion region of P-N junction diode.

The depletion region enumerated above is called an *intrinsic region*. The term "intrinsic" is used when carriers in a semiconductor are available only as the result of natural causes. Natural causes include elevated temperature, light, etc. Intrinsic carriers are distinguished from those which are "donated" by impurity atoms used to dope semiconductors. There is a type of specialized photodiode in which the intrinsic region surrounding the junction is increased in size by sandwiching a layer of pure (nondoped) semiconductor material between the *P*- and *N*-type materials. The device is called a PIN diode; the "I" calls attention to the fact that an intrinsic layer is present in its structure. The purpose of adding the intrinsic layer is to increase the size of the region that can be radiated by light. The consequence of an enlarged intrinsic region is an increase in photocurrent. A diagram illustrating the composition and structure of PIN photodiodes is given in Fig. 13.13. The schematic circuit symbol of a photodiode is represented in Fig. 13.14.

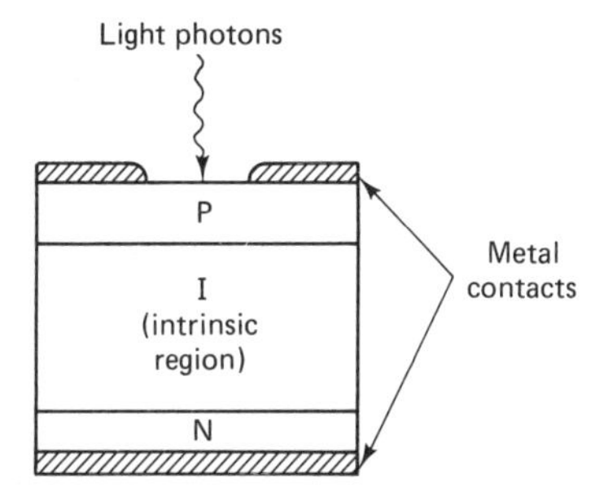

Figure 13.13 PIN photodiode.

Figure 13.14 Photodiode schematic circuit symbol.

The Avalanche Photodiode

Recall from your study of ordinary semiconductor diodes the phenomenon called avalanche breakdown. This is a process that produces a very high reverse current in a diode. It occurs when the reverse-bias voltage across a diode junction is excessive. A very high reverse bias imparts a high energy level to the relatively few electrons and holes of a reverse current. These high-energy carriers excite new carrier pairs. The process is called *ionization*. The additional carriers produce further ionization, and so on. The effect is a multiplication or amplification of the reverse current. The multiplication factor can be of the order of 1000 or more. The current is called an *avalanche* current. The process is utilized in the zener diode.

An avalanche current can also be produced in a photodiode. Photodiodes fabricated with special structures and materials to enhance this process are called *avalanche photodiodes* (APDs). An APD will have a large intrinsic region similar to that of a PIN diode. In addition, however, it will have a structure which, in conjunction with a higher-than-normal reverse voltage, will produce a region of an extremely high electric field. It is this high field that imparts sufficient energy to the electron–hole pairs produced by photoemission to also result in an ionization of the region.

The end result of ionization in the high-field region is an avalanche current. The current is initiated and controlled by the light striking the intrinsic region. An avalanche photodiode achieves internal amplification. It is useful in applications where the received light level in a fiber-optic transmission system is particularly low. A diagram of the structure of an avalanche photodiode is given in Fig. 13.15.

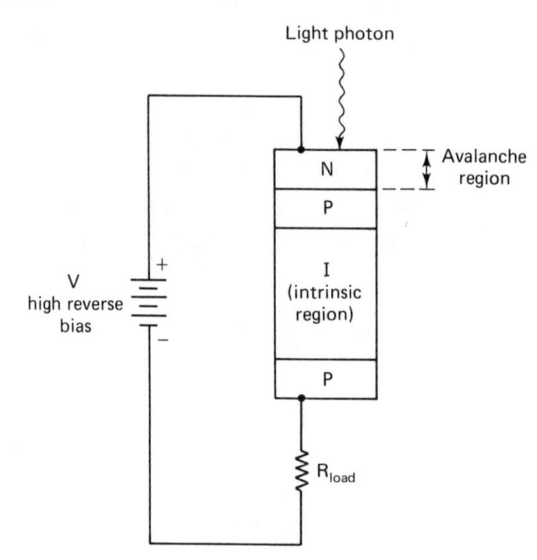

Figure 13.15 Structure of avalanche photodiode (APD).

Photodiode Circuits

Whatever its form, a photodiode is a device which when provided with a reverse bias voltage can produce a current proportional to the intensity of the light striking its active region. In a simple circuit containing an appropriate series resistor that current can be converted into a voltage which can be amplified and/or processed as any other electrical signal. A model circuit illustrating the generalized concept is given in Fig. 13.16(a). Please note that the rudimentary circuit incorporates the functions identified as a current-to-voltage converter and a linear amplifier. A simple implementation with an FET is shown in Fig. 13.16(b).

One of the most promising applications of fiber-optic transmission systems is as long-distance trunks (major transmission links) in transcontinental and transoceanic telecommunications systems. Such applications are designed to take maximum advantage of the broad-bandwidth capabilities of fiber-optic systems. Consequently, the electronics portions of the transmitting and receiving functions are required to operate at extremely high frequencies—in the gigahertz range. For example, the linear amplifier used in conjunction with the photodiode detector at the receiving end of a fiber-optic transmission line would be required to operate in the gigahertz range.

Considerable research and development effort has produced special field-effect transistor devices for broad-bandwidth applications. These are fabricated from gallium arsenide (GaAs) instead of silicon. Gallium arsenide devices, unlike

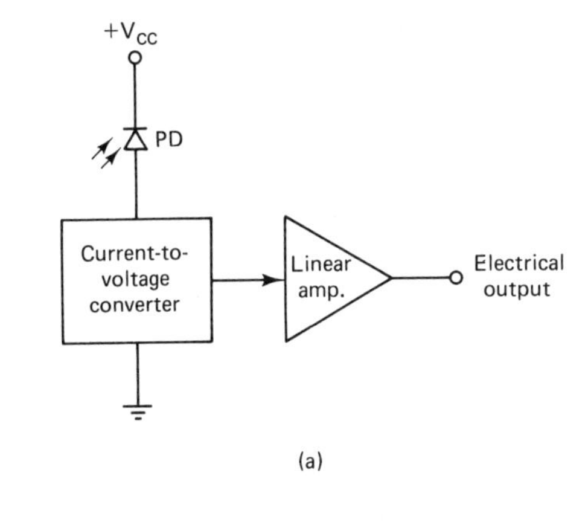

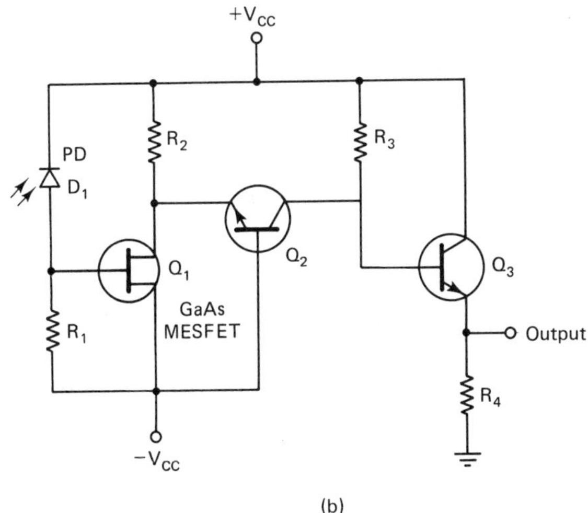

Figure 13.16 Receiver concepts for optical communications system: (a) optical receiver functions; (b) optical receiver implemented with GaAs MESFET.

silicon devices, have been found to operate with low noise and high gain at microwave frequencies (GHz range). One particular set of commercial FETs of this type are designated GaAs MESFETs. (The "M" is for microwave, the "E" is for enhancement mode, and the "S" is for Schottky barrier.)

13.8 OPTICAL COMMUNICATIONS SYSTEMS: A SECOND LOOK

At the beginning of this chapter we took a very brief, general look at optical communications systems. We noted that such a system is usefully considered as basically a conventional communications system, but one that uses an optical transmission line as the major transmission link. Being unfamiliar with optical transmission systems, we proceeded to examine the basic principles and operating characteristics of the subsystems and devices utilized in such systems.

It is time now to summarize and reinforce what we have learned about optical fibers, light sources, photodetectors, etc. Let's examine the block diagram of an optical communications system. We want to pay particular attention to the interrelationships of the devices/subsystems we have studied and how they work together to accomplish the desired result of information transmission.

A generalized model of a communications system utilizing a light-wave-carrier transmission link is shown in Fig. 13.17. The details of terminal installations are omitted. We will assume that the system is a digital system. All incoming information signals from users of the system are multiplexed and presented to the optical link in digitized form. Details of digitizing and time-division multiplexing are presented in Chap. 12.

Examine the diagram of Fig. 13.17 carefully. Observe the striking similarity of this system to the radio-frequency systems that have been presented in this book. The transmitter function consists, in part, of a source—a light source which is analogous to an oscillator and RF amplification in conventional transmitters. The light signal is modulated and fed to the fiber transmission line. The modulation is digital intensity modulation, as discussed in Sec. 13.6.

In an actual system regenerative repeaters would be located along the transmission line. The information signals being carried by the system are regenerated by the equipment at these locations: The light is "taken off" the line and detected. Then information signals are converted to electronic form and new digital pulses are produced under the control of the received pulses. The new signals are used to modulate a local light source. The regenerated light carrier is reinjected into the fiber line.

At the final destination of the system—the receiving terminal—the light carrier is detected in a photodetector (it is converted to an electrical signal). The electrical signal represents numerous digital information signals. These signals are demultiplexed and dispatched to their respective destinations—terminal facilities of users of the system. Some of these signals may be digital data signals; they will be used directly by computers. Other signals may be digitized analog signals. They will be reconverted to analog form and sent to telephone headsets, other audio facilities, or video facilities.

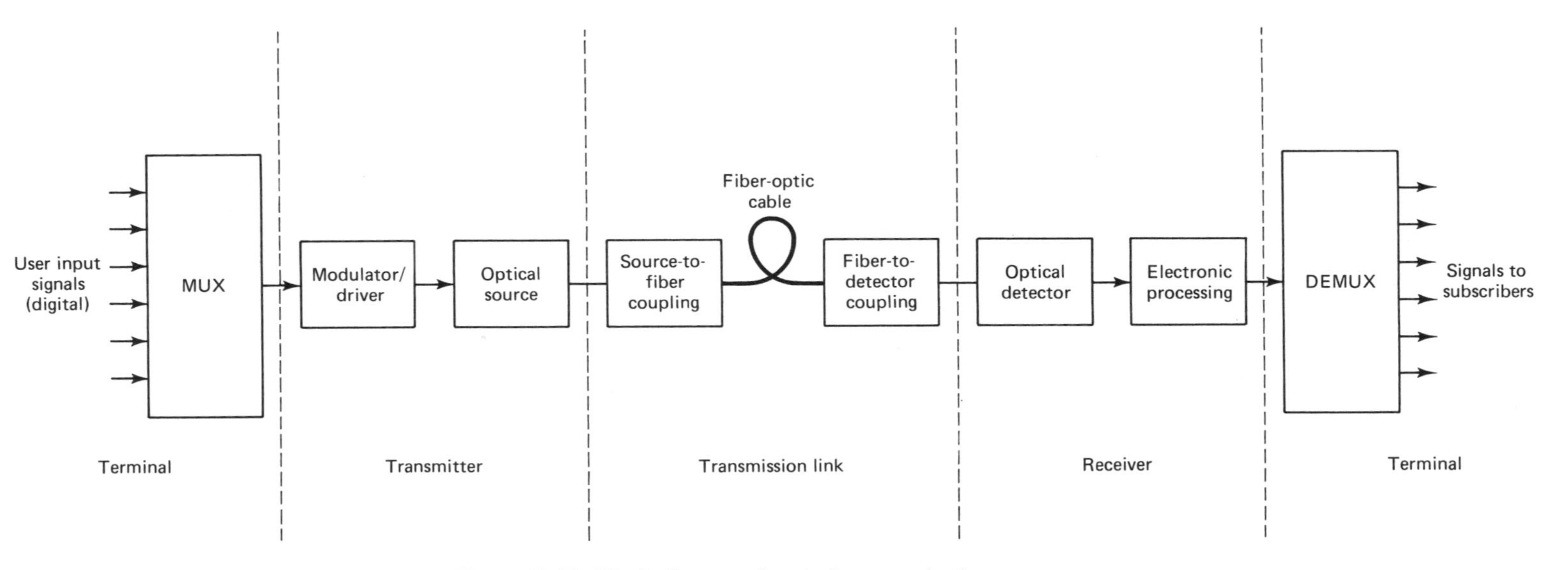

Figure 13.17 Block diagram of optical communications system.

ROMISE, AND FUTURE SYSTEMS

ear almost daily describing new advances in the technology of nication. Almost invariably, the articles recite already known technology and project a great future for it. As for the future, s "the capabilities of tomorrow's optical technology challenge or "the future capacity (of fiber systems) is going to be virtually equently seen. There is no evidence of any doubt that fiber- eventually predominate as the means of transmitting information cus on the advantages of fiber optics and then look at a few of the developments that make light-wave systems appear even more promising as the communications technology of the future.

Advantages of Optical-Fiber Communications Systems

Tremendous information-handling capability. Because optical-fiber transmission offers several significant advantages over "copper cable" or microwave technologies, one is reluctant to choose one advantage over another with which to start a discussion. Surely, however, the tremendous information-handling potential of fiber optics is one of its most important advantages. The prospect of almost unimaginably intensive utilization of communications facilities prompts the speculation that optical-fiber systems will someday be the least expensive communications technology by far.

Small size and weight. Optical-fiber cables are already relatively small in diameter and light in weight. They offer significantly greater information transmission capacity in significantly less space than do copper-based cables. However, improvements in optical-fiber cables are still being achieved. Considerable developmental effort is in process to achieve greater capacity in even less space.

Freedom from electrical noise. An optical fiber is made of glass, or, sometimes, plastic. These materials are insulators—dielectrics. As a consequence, a light-wave carrier is free from the potential influence of other forms of electromagnetic radiation. It is not possible to induce an interfering signal into a light-wave transmission from another communications source—an adjacent coaxial line, a microwave radiation field, or whatever. An optical-fiber transmission line may pass through or lie adjacent to regions of intense electromagnetic activity without having its signal polluted by electrical noise. It is also relatively simple to prevent a light-wave carrier from receiving noise from other light-wave sources.

Signal security. It is extremely difficult, indeed virtually impossible, to "steal" signal from a light-wave system without detection. It is literally necessary to "break into" a light-wave transmission system to obtain a signal. On the other hand, signals radiate easily from copper-based transmission lines. The

microwave signals of communications satellites, for example,
the taking by anyone with a "dish." Optical transmissions are rel
a distinct advantage to users such as the military, government, a

Low transmission loss. A typical, practical single-mode
be operated at a 1.31-nm wavelength with an attenuation of less th
This permits operation of lightwave systems with more than 30 km
19 mi) between regenerator-repeater stations. This is operation at rates of several hundred million bits per second. Developmental systems have been operated at a rate of over 445 Mb/s over a distance exceeding 130 km without a repeater.

Repeater stations (locations containing equipment for detecting, amplifying, regenerating, and retransmitting signals of a communications system) are expensive. They are also the sources of potential equipment failure and downtime for a system. Low transmission loss permits the use of links that have greater distances between repeaters, fewer repeaters for a given distance. Significant advances in lowering transmission losses in fiber systems still seem to be possible, indeed, probable, judging by reports in technical journals. Low transmission loss may prove to be one of the most significant advantages of fiber optics.

Mechanical properties of fiber cables. Steady effort on the improvement of fiber-optic cables has resulted in the availability today of cables that are strong, rugged, and surprisingly, very flexible. Cabled fibers can be bent in quite sharp curves. They are more flexible than metallic cables of the same bit-handling capacity. These characteristics, together with their smaller size and lower weight, give fiber cables a significant advantage over metallic cables of the same information-handling capacity. This advantage is of importance with regard to the storage, handling, transportation, and installation of the cables.

System reliability and maintainability. Because optical-fiber transmission systems require fewer repeaters, they are less subject to equipment breakdown. The fibers themselves are virtually immune to deterioration by natural environmental factors. In fact, a fiber cable is unlikely to be made inoperative except by accidental or malicious mechanical invasion. It is now predicted that the optical components of light-wave transmission systems have a practical lifetime of at least 20 to 30 years. All of these factors combine to make optical systems highly reliable and their maintenance simpler and less costly than communications systems of other types with comparable information-handling capability.

Lower cost per information bit transmitted. When all of the advantages cited just above have been refined to their maximum, the net result will be a communications system that costs less to transmit a bit of information than virtually any other system now known. That condition is rapidly being approached. In fact, in some applications, notably long-distance systems, the condition may have already been achieved. If the information in the numerous and frequent articles on fiber optics in newspapers and technical journals can be relied on, there is no room for doubt that optical communications systems will be the dominant communications technology sometime in the not-too-distant future.

Future Developments

Wavelength-division multiplexing. Ample bandwidth characteristics of communications systems provide for the simultaneous transmission of many channels of information by various means. We have already examined in other chapters the concepts of frequency-division multiplexing (FDM) and time-division multiplexing (TDM). You will recall that one of the advantages of digital communications systems is the ease with which channels can be time-division multiplexed when information is in digital form.

Both FDM and TDM can be utilized in optical communications systems. In FDM several baseband signals (signals before modulation of the primary carrier) are combined and the composite signal modulates the optical signal. In TDM, various signals are allocated time slots in which they can control the optical signal.

Wavelength-division multiplexing occurs when two or more light sources of different "colors" (different wavelengths) share the same fiber. The phenomenon is also called color multiplexing. The procedure is still experimental. It requires, of course, light sources of differing wavelength outputs. It also requires photodetectors capable of distinguishing between different wavelength inputs. Commercial application of wavelength-division multiplexing awaits the availability of suitable transmitting and receiving devices. Indications are that these are under development and that their availability is only a matter of time. The advent of wavelength-division multiplexing as a commercial reality will serve to increase further the attractiveness of optical communications systems by increasing their channel capabilities.

Optical amplification and switching. Although the mechanism of a laser light source involves the amplification of light, direct light amplification is not used in present optical communications systems. When amplification is required, the optical signal is converted to an electrical signal, amplified electronically, and reconverted to an optical signal. The switching of optical signals is at a somewhat similar stage of development. For example, if it is desired to switch from transmitting on one fiber to another, the result is typically accomplished by a mechanical change in the coupling between the fibers involved. However, both of these effects—amplification and switching—have been accomplished experimentally by purely optical means. Laser-like optical amplifiers have been developed and operated in experimental applications. True optical switches have also been constructed and operated experimentally.

The prospects of optical amplification and switching are exciting ones. If developed to the point of practical commercial availability, devices that could perform these functions would increase still more the advantages of optical communications systems. Direct optical amplification would simplify significantly the design and construction of repeaters. Optical switching might well lead to even greater reliability of optical systems by the elimination of mechanical-type switching devices.

The future of optical communications is indeed exciting to contemplate. It has been said that current optical systems, in comparison to their foreseeable sophistication and capabilities, are like radio communications systems when they were at the early radiotelegraph stage. In comparison to modern, two-way radio equipment, the first radio transmitters were simply broadband noise sources that were turned on and off.

GLOSSARY OF TERMS

Angle of incidence The angle an arriving light ray, at a boundary between two light-transmitting media, makes with a line perpendicular to the media boundary at the point where the ray meets the boundary.

Angle of refraction The angle a light ray makes with the perpendicular to a boundary between two light-transmitting media as it continues on its way after encountering the boundary. *See also* Angle of incidence.

Avalanche breakdown A process that produces a very high reverse current in a diode as a result of ionization in the junction region and which occurs in the presence of an excessive reverse bias voltage.

Avalanche photodiode A diode used to detect the intensity of a light source which makes use of the avalanche breakdown mechanism in conjunction with photoemission to achieve a marked increase in light sensitivity level. *See also* Avalanche breakdown and Photoemission.

Chromatic dispersion The dispersion that occurs in an optical fiber as a result of the different propagation times of light-beam components of differing wavelengths. *See also* Dispersion *and* Material dispersion.

Cladding The light-transmitting material that surrounds an optical fiber core and which has a refractive index different from that of the core.

Coherent light Light, such as is produced by a laser, in which all components are comparatively in phase producing an intense, narrowly focused beam.

Core (of an optical fiber) The central portion of an optical fiber through which light transmission occurs.

Critical angle The angle of incidence at the boundary of two light-transmitting media which results in a light ray neither exiting a source medium to a medium of different refractive index nor being reflected back into the source medium (i.e., angle of refraction = 90°).

Dispersion The separation or spreading of components of a light beam resulting from differing transit times of the components through a light-transmitting medium. Produces a stretching (distortion) of digital pulses transported by light beams used as carriers.

Electroluminescence The phenomenon of the production of a "cold" light which occurs when a strong electric field alters the energy relationships at the atomic level of a substance.

Graded-index fiber An optical fiber in which the refractive index across the radial cross section of its core varies (is graded) with distance from the center to minimize the effect of dispersion.

Injection laser diode A special light-emitting diode, fabricated so as to contain an optically resonant cavity, which will lase (produce a high-intensity, coherent light beam) when operated at a forward current level which injects current carriers in sufficient number to exceed a required threshold level.

Intrinsic region (of a semiconductor) The portion of a semiconductor in which current carriers are available only as a result of natural effects such as elevated temperature, incident light, etc. (as distinguished from current carriers donated by impurity atoms).

Lase To emit light in a narrow, very intense beam as a result of a process that includes the stimulation of the light-emitting mechanism by light amplified in an optically resonant cavity.

Laser [*l*(ight) *a*(mplification by) *s*(timulated) *e*(mission of) *r*(adiation)]. A device that produces a narrow, very intense beam of light by means of an electronic process which includes the stimulation of light emission by light amplified in an optically resonant cavity.

Material dispersion The dispersion (of a light beam) which occurs in an optical fiber as a result of the differing propagation times of light-beam components of different wavelengths. *See also* Chromatic dispersion.

Modal dispersion The dispersion (of a light beam) which occurs in an optical fiber as a result of the differing transit times associated with different propagation paths or modes. *See also* Dispersion.

Numerical aperture (NA) The sine of the acceptance angle of an optical fiber.

Optical detector An electronic device capable of converting a light-intensity level into an electric current proportional to the intensity level.

Optically resonant cavity A three-dimensional space in which light can be propagated, with at least partially reflecting surfaces on two opposing sides and with the dimensions of the space between the reflecting surfaces equal to a multiple of a half-wavelength of the light used in the cavity to make it resonant to that light.

Photon A unit quantity (a quantum) of electromagnetic energy exhibiting both wave and particle behavior and usually associated with light, X-rays, gamma rays, etc.

Refraction The bending of a light ray or wave as a result of its passing at an angle other than a right angle through the boundary between two different materials, or layers of the same material, at which the velocity of propagation changes.

Refraction index A quantitative measure, inversely proportional to its velocity of light propagation, of a medium's light-bending property; specifically, the ratio of the sine of the angle of incidence to the sine of the angle of refraction for a ray passing from one medium to another.

Sheath The protective covering of a device such as an optical fiber or coaxial transmission line.

Single-mode propagation The technique of transmitting light in an optical fiber so that there is but one path (mode) traveled by the light from input to output (as distinguished from numerous zigzag paths in multimode propagation).

Step index A description of the index of refraction of an optical fiber made of a homogeneous medium; a constant index of refraction across the cross-sectional diameter of an optical fiber.

Surface light-emitting diode A semiconductor diode designed and fabricated so that a comparatively large surface of its junction is exposed to facilitate the radiation of light produced by the recombination of electron–hole pairs in the junction region.

REVIEW QUESTIONS: BEST ANSWER

1. A unique characteristic of an optical communications system is that it makes use of: **a.** transmission lines. **b.** guided light. **c.** digital signals. **d.** time-division multiplexing. **e.** none of these.
2. The "waveguide for light" used in optical communications systems is called a fiber because it is: **a.** slender and threadlike. **b.** brittle. **c.** made from wheat. **d.** made from cloth. **e.** none of these.
3. An optical fiber can propagate a light beam or wave over great distances and around corners because of the physical principle called: **a.** reflection. **b.** photoemission. **c.** refraction. **d.** Snell's law. **e.** none of these.
4. According to Snell's law, at a boundary between two light-transmitting media of different indexes of refraction: **a.** the angle of incidence is equal to the angle of refraction. **b.** the ratio of the sine of the angle of incidence to the sine of the angle of refraction is a constant. **c.** refraction occurs only if the angle of incidence is 90°. **d.** the critical angle is always greater than 90°. **e.** none of these.
5. For light waves propagated in an optical fiber, the angle of incidence is: **a.** less than the critical angle (of refraction). **b.** equal to the critical angle. **c.** greater than the critical angle. **d.** greater than the angle of refraction. **e.** none of these.
6. In an optical fiber, each angle of incidence produces a different path through the fiber called a/an: **a.** index. **b.** dispersion. **c.** propagation mode. **d.** index profile. **e.** none of these.
7. Modal dispersion causes signal waveforms to be distorted by spreading. Consequently, dispersion is an effect that is: **a.** band limiting. **b.** amplitude limiting. **c.** of no significance. **d.** beneficial for analog signals. **e.** none of these.
8. A technique for eliminating model dispersion which requires fibers of very small core diameter is called: **a.** step index. **b.** graded index. **c.** monofiber. **d.** single-mode propagation. **e.** none of these.
9. A source of attenuation on an optical transmission line that has no analogy on a copper-wire line is the result of: **a.** heat loss. **b.** insulation loss. **c.** resistance loss. **d.** bending loss. **e.** none of these.
10. An optical transmission line that has a specification of 500 Mb/s × 10 km could transmit, without regeneration: **a.** 1000 Mb/s over a 100-km line. **b.** 5 Gb/s over a 1000-m line. **c.** 50 Mb/s over a 1000-km line. **d.** 5 Mb/s over a 10,000-m line. **e.** none of these.
11. An essential step in the fabrication of an injection laser diode from an ordinary light-emitting diode is that of making at least one of the dimensions of its active region be determined by the planes of the semiconductor crystal. The process step is called: **a.** cleaving. **b.** sawing. **c.** cutting. **d.** hacking. **e.** none of these.
12. When the optically resonant cavity of an LED amplifies the light initially produced by electron–hole recombinations and the amplified light stimulates more light emission, the device is said to be: **a.** lasing. **b.** emitting. **c.** propagating in single mode. **d.** injecting light. **e.** none of these.
13. The device used to convert a light-wave carrier into an electrical signal is called a/an: **a.** optoreceiver. **b.** optical detector. **c.** photodetector. **d.** either b or c. **e.** none of these.
14. A region of a semiconductor in which current carriers are liberated only by natural

means, such as elevated temperatures, light, etc., is called a/an ______________ region. **a.** *P*-type **b.** donor **c.** inherent **d.** intrinsic **e.** none of these.

15. In comparison with other communications systems, one of the most striking characteristics of an optical communications system is its: **a.** high cost. **b.** great complexity. **c.** tremendous information-handling capability. **d.** high energy requirement. **e.** none of these.

REVIEW QUESTIONS: ESSAY

1. What is refraction? What part does this phenomenon play in the behavior of an optical fiber?

2. Name and describe the three basic parts of an optical fiber for communications applications.

3. Give the meanings of *angle of incidence, angle of refraction*, and *critical angle* as these terms are used in conjunction with the phenomenon of refraction.

4. What is a propagation mode in a fiber?

5. What is the difference between a fiber with a step index and one with a graded index? Why is a graded-index fiber used?

6. What is bandwidth limiting, and why does dispersion in a fiber cause it?

7. Why are electroluminescent sources, as opposed to incandescent sources, most appropriate for high-performance optical communications systems?

8. What is an LED? an ILD? Discuss the use of these devices in optical communications systems.

9. What is an optically resonant cavity?

10. Describe briefly digital direct-intensity modulation.

11. Why is a PIN diode so-called?

12. Why is an avalanche diode so-called? What advantage(s) does an APD provide?

EXERCISES

1. What is the absolute bandwidth for each of the following carrier frequencies if the relative bandwidth for each is 0.25%? **(a)** 15 MHz; **(b)** 150 MHz; **(c)** 1.5 GHz; **(d)** 150 GHz; **(e)** 1500 GHz.

2. If the attenuation is 0.2 dB/km and 0.6 dB/km for fibers and twisted-wire-pair transmission lines, respectively, what length of each medium can be used before the signal is down by 10 dB?

3. Given: Light rays or waves traveling through medium 1 with an index of refraction of n_1 toward medium 2 with an index of n_2. Using Snell's law, determine the critical angle for each of these combinations of indexes of refraction (n_1 is given first in each case): **(a)** 1.6, 1.0; **(b)** 1.4, 1.38; **(c)** 1.5, 1.1; **(d)** 1.6, 1.4; **(e)** 1.9, 1.05.

4. Given a fiber with a propagation product specification of 1 Gb/s $\times$ km. Assuming that a communications system is operated at its maximum possible transmission rate, how long would be required to transmit 100 Mb of data over the following distances: **(a)** 5 km; **(b)** 25 km; **(c)** 100 km; **(d)** 1000 km.

ANSWERS TO REVIEW QUESTIONS

All answers represent the *best* answer, not the only answer possible.

CHAPTER 1

1. a
2. a
3. b
4. b
5. b
6. e

CHAPTER 2

1. b
2. d
3. b
4. e
5. d
6. a
7. b

CHAPTER 3

1. c
2. a
3. a
4. d
5. c
6. b
7. d
8. b
9. c
10. b
11. b
12. a
13. b
14. e
15. d
16. b
17. d
18. a
19. c
20. a
21. a
22. f
23. c
24. e
25. b
26. c

CHAPTER 4

1. b
2. b
3. a
4. c
5. d
6. b
7. c
8. a
9. a
10. c
11. d
12. a
13. d

CHAPTER 5

1. c
2. d
3. b
4. c
5. b
6. d
7. a
8. b
9. b
10. c
11. d
12. a
13. a
14. c
15. b
16. d
17. b
18. a
19. a
20. c
21. b
22. c
23. a
24. d
25. c
26. b
27. a

CHAPTER 6

1. d
2. a
3. b
4. a
5. e
6. b
7. c
8. e
9. a
10. d
11. e
12. d
13. a
14. d
15. a
16. c
17. d
18. a
19. b
20. c
21. c
22. a
23. b
24. c
25. d
26. c
27. d
28. a
29. d
30. d
31. b
32. d
33. d
34. c
35. e
36. d
37. b
38. b
39. d
40. b
41. a
42. c
43. c
44. b
45. a
46. b
47. c
48. b
49. a
50. c
51. d
52. c
53. a
54. c
55. d
56. b
57. d
58. b
59. e

CHAPTER 7

1. d
2. b
3. b
4. c
5. c
6. a
7. d
8. c
9. b

CHAPTER 8

1. d
2. a
3. b
4. b
5. b
6. d
7. a
8. d
9. c
10. d
11. a
12. d
13. c
14. b
15. d
16. c
17. b
18. a
19. d
20. c
21. d
22. c
23. a
24. c
25. b
26. c
27. a
28. b
29. a
30. d

CHAPTER 9

1. c
2. e
3. e
4. c
5. a
6. e
7. d
8. d
9. a
10. c
11. e
12. c
13. a
14. d
15. b
16. a
17. d
18. a
19. c
20. d
21. b
22. a
23. c
24. d
25. b
26. a
27. b
28. c
29. d
30. b

CHAPTER 10

1. e
2. c
3. b
4. c
5. b
6. c
7. e
8. d
9. b
10. a
11. d
12. b
13. d
14. b
15. b
16. c
17. a
18. d
19. a
20. d
21. a
22. c
23. b
24. b
25. d
26. d
27. b
28. d
29. b
30. a
31. a

CHAPTER 11

1. b
2. a
3. c
4. b
5. c
6. b
7. a
8. e
9. b
10. d
11. b
12. a
13. b
14. a
15. d
16. d
17. c
18. a
19. d
20. d
21. a
22. c
23. a
24. b
25. d
26. b

CHAPTER 12

1. d
2. b
3. d
4. a
5. d
6. b
7. c
8. d
9. b
10. a
11. d
12. a
13. a
14. d
15. b
16. b
17. d
18. a
19. c
20. b
21. d
22. c
23. a
24. c

CHAPTER 13

1. b
2. a
3. c
4. b
5. c
6. c
7. a
8. d
9. d
10. b
11. a
12. a
13. d
14. d
15. c

BIBLIOGRAPHY

BARKER, FORREST L., and GERSHON J. WHEELER, *Mathematics for Electronics* (2nd ed.). Menlo Park, Calif.: The Benjamin-Cummings Publishing Co., 1978.

BELL, TRUDY E., "Single-Frequency Semiconductor Lasers," *IEEE Spectrum,* Vol. 20, No. 12, 1983, pp. 38–45.

BELL, TRUDY E., "Technology 84, Communications," *IEEE Spectrum,* Vol. 21, No. 4, 1984, pp. 53–57.

BERLIN, HOWARD M., *Design of Phase-Locked Loop Circuits, with Experiments.* Indianapolis, Ind.: Howard W. Sams & Company, Inc., 1978.

BOWICK, CHRIS, *RF Circuit Design.* Indianapolis, Ind.: Howard W. Sams & Company, Inc., 1982.

BOYD, WALDO T., *Fiber Optics Communications, Experiments, and Projects.* Indianapolis, Ind.: Howard W. Sams & Company, Inc., 1982.

CANNON, DON L., and GERALD LUECKE, *Understanding Communications Systems.* Dallas: Texas Instruments, Inc., 1980.

CORWIN, WALTER L., "A Communications Network for the Summer Olympics," *IEEE Spectrum,* Vol. 21, No. 7, 1984, pp. 38–44.

DAVIS, DWIGHT B., "Making Sense of the Telecommunications Circus," *High Technology,* September 1985, pp. 20–29.

DEMAW, DOUG, *Practical RF Design Manual.* Englewood Cliffs, N.J.: Prentice-Hall, Inc., 1982.

DRENTEA, CORNELL, *Radio Communications Receivers.* Blue Ridge Summit, Pa.: TAB Books, Inc., 1982.

FEHER, KAMILO, *Digital Communications: Satellite/Earth Station Engineering.* Englewood Cliffs, N.J.: Prentice-Hall, Inc., 1983.

FELDT, TERRY, "Communications Satellites—The Push Is On," *ElectronicsWeek,* Vol. 58, No. 4, 1985, pp. 45–49.

FIKE, JOHN L., and GEORGE E. FRIEND, *Understanding Telephone Electronics.* Dallas: Texas Instruments, Inc., 1983.

FRIEND, GEORGE E., et al., *Understanding Data Communications*. Dallas: Texas Instruments, Inc., 1984.

HAYWARD, WES, and DOUG DEMAW, *Solid State Design for the Radio Amateur*. Newington, Conn.: American Radio Relay League, Inc., 1977.

KERSHAW, JOHN D., *Digital Electronics: Logic and Systems* (2nd ed.). North Scituate, Mass.: Breton Publishers (a division of Wadsworth, Inc.), 1983.

KRAUS, JOHN D., *Antennas*. New York: McGraw-Hill Book Company, 1950.

LANCASTER, DON, *Active-Filter Cookbook*. Indianapolis, Ind.: Howard W. Sams & Company, Inc., 1975.

LANCE, ALGIE L., *Introduction to Microwave Theory and Measurements*. New York: McGraw-Hill Book Company, 1964.

MALVINO, ALBERT PAUL, *Electronic Principles* (2nd ed.). New York: McGraw-Hill Book Company, 1979.

MARTIN, JAMES, *Future Developments in Telecommunications* (2nd ed.). Englewood Cliffs, N.J.: Prentice-Hall, Inc., 1977.

MASTEN, LARRY B., and BILLY R. MASTEN, *Understanding Optronics*. Dallas: Texas Instruments, Inc., 1981.

MIMS, FORREST M., III, *A Practical Introduction to Lightwave Communications*. Indianapolis, Ind.: Howard W. Sams & Company, Inc., 1982.

NABER (National Association of Business and Educational Radio), *Two-Way Radio Technician Certification Examination Handbook*. Washington, D.C.: National Association of Business and Educational Radio, 1985.

NOLL, EDWARD M., *Broadcast Radio and Television Handbook* (6th ed.). Indianapolis, Ind.: Howard W. Sams & Company, Inc., 1983.

NOLL, EDWARD M., *General Radiotelephone License Handbook* (7th ed.). Indianapolis, Ind.: Howard W. Sams & Company, Inc., 1982.

ORR, WILLIAM I., *Radio Handbook* (22nd ed.). Indianapolis, Ind.: Howard W. Sams & Company, Inc., 1981.

OWEN, FRANK F. E., *PCM and Digital Transmission Systems*. New York: McGraw-Hill Book Company, 1982.

PERSONICK, STEWART D., "Switches Take to Optics," *ElectronicsWeek*, Vol. 58, No. 11, 1985, pp. 55–58.

RAMO, SIMON, JOHN R. WHINNERY, and THEODORE VAN DUZER, *Fields and Waves in Communication Electronics*. New York: John Wiley & Sons, Inc., 1965.

RODEN, MARTIN S., *Digital and Data Communication Systems*. Englewood Cliffs, N.J.: Prentice-Hall, Inc., 1982.

RYDER, JOHN D., *Electronic Fundamentals and Applications: Integrated and Discrete Systems* (5th ed.). Englewood Cliffs, N.J.: Prentice-Hall, Inc., 1976.

SENIOR, JOHN M., *Optical Fiber Communications: Principles and Practice*. London: Prentice-Hall International, Inc., 1985.

SPILKER, JAMES J., JR., *Digital Communications by Satellite*. Englewood Cliffs, N.J.: Prentice-Hall, Inc., 1977.

STARK, HENRY, and FRANZ B. TUTEUR, *Modern Electrical Communications: Theory and Systems*. Englewood Cliffs, N.J.: Prentice-Hall, Inc., 1979.

TOCCI, RONALD J., *Digital Systems: Principles and Applications* (3rd ed.). Englewood Cliffs, N.J.: Prentice-Hall, Inc., 1985.

APPENDIX A
DECIBELS

The ability to use decibels is virtually a necessity for anyone working or desiring to work on the technical side of the electronics communications field. Let's look carefully at the what, why, and how of decibels.

Let's start by looking at the word "decibel" itself. It is a combination of "deci" and "bel." *Bel* comes from "bell." It was chosen to honor Alexander Graham Bell, the inventor of the telephone. *Deci* is a prefix meaning "one-tenth" (i.e., $\frac{1}{10}$ or 10^{-1}). Therefore, a *decibel* is literally one-tenth of a *bel*. What, then, is a bel?

Bel is the name of a unit used in conjunction with the comparison of power levels. Precisely,

$$N_B \text{ (number of bels)} = \log \frac{P_1}{P_2}$$

In words, "the number of bels is equal to the common logarithm (logarithm to the base 10) of a power ratio." Hence

$$N_{dB} \text{ (number of decibels)} = 10 \log \frac{P_1}{P_2}$$

The factors P_1 and P_2 represent the values of any two power levels that are to be compared. In general, power levels can be of any type: electrical, mechanical, thermal, etc. In this book we are concerned primarily with the power levels of communications signals at various points in a communications system. Power-level comparisons are most meaningful when the power levels are related in some way. For example, the power gain of an electronic amplifier—a comparison of the output power to the input power—is often expressed in decibels:

$$\text{gain (dB)} = 10 \log \frac{P_{out}}{P_{in}}$$

You may well be wanting to ask: "Why use the logarithm of the power ratio; why not just use the simple ratio?" A simple answer to this question is that human physiological response to sensory stimuli such as light and sound is approximately logarithmic. This means that, for example, if the sound power at your ear of an airplane flying overhead is 10,000 times the power, at your ear, of a person clapping hands close by, you do not perceive the airplane power as being 10,000 times as great. Your perception is more like 40 times since 10 log 10,000 = 40. Thus power ratios (and, as we shall see, voltage and current ratios) expressed in decibels provide information which is more closely related to real-life effects than do simple direct ratios. There are also advantages to working with gain and loss figures expressed in decibels. These will be made apparent as we learn more about the use of this special unit.

A.1 LOGARITHMS: A REVIEW

As we have seen, the mathematical function known as a *logarithm* is at the heart of the definition of a decibel. Therefore, in learning decibels it is useful to start with a review of common logarithms.

A logarithm, in simplest terms, is another way to express an exponent. A logarithm is the exponent that expresses the power to which a number, called the *base,* must be raised to equal a given number. When the base is 10, the logarithm is called a *common logarithm* or *base-10 logarithm*. For example, the common logarithm of 10,000 or 10^4 is 4. The common logarithm of an integral power of 10 is equal to the exponent used in powers-of-10 notation.

In general terms, if N is a given number, b is the base of the logarithm system being used, and L is the logarithm of N, then

$$L = \log_b N$$

or

$$N = b^L$$

The given number is called the antilog (antilogarithm) of L,

$$N = \text{antilog}_b\, L$$

Virtually all but the simplest hand-held calculators incorporate the capability for providing the common logarithms of numbers. Examine your calculator right now for this function. The key for the function is usually labeled "log." The logarithm is obtained by entering N and then pressing the *log* key. For example, the common logarithm of 1285 is 3.1089. For practice, verify for yourself the following:

$$\log 2.894 = 0.4615$$

$$\log 38.94 = 1.5904$$
$$\log 27{,}583 = 4.4406$$
$$\log 1{,}389{,}562 = 6.1429$$

The antilogs of the logarithms of these examples are the original given numbers. Antilogs can also be found on calculators. Remember,

$$\text{antilog}\, L = b^L = N$$

Thus

$$\text{antilog}\, 0.4615 = 10^{0.4615} = 2.894$$

The antilog_{10} can be obtained on calculators usually in one of two ways. On some calculators there is a key labeled "10^x." On calculators with this key, find the antilog by entering the logarithm and then pressing this key. On other calculators the same function is obtained by pressing first a key usually labeled "INV" (for inverse) and then the key labeled "log." Check your calculator now to see if it has the "10^x" function. If it does not, you must use "INV" with "log" to obtain antilogs. Verify the method of obtaining antilogs with your calculator by finding the antilogs of the examples above.

As you are undoubtedly aware, common logs are not the only logarithms used in practical mathematics and technology. Logarithms to the base e (e = 2.718281828) are also used extensively, especially in electricity and electronics. Logarithms to the base e are called *natural logarithms*. The abbreviation for natural logarithm is either $\log_e$ or *ln*. The latter is used most often. Calculators having the *log* function virtually always have the *ln* function also.

A.2 RULES FOR LOGARITHMS

Since logarithms are the equivalent of exponents, mathematical operations with logarithms follow the rules for exponents. We restate these here for your review and illustrate their application to logarithms.

1. *Products*. The exponent of the product of two or more factors of a common base is equal to the sum of the exponents of the factors.

   $$b^u \times b^v = b^{u+v} \qquad (b \neq 0)$$

   The logarithm of the product of two numbers is equal to the sum of the logarithms of the numbers.

   $$\log (N \times M) = \log N + \log M$$

2. *Quotients*. The exponent of the quotient of two numbers of a common base is equal to the exponent of the numerator minus the exponent of the denominator.

   $$\frac{b^u}{b^v} = b^{u-v} \qquad (b \neq 0)$$

   The logarithm of the quotient of two numbers is equal to the logarithm of the numerator minus the logarithm of the denominator.

$$\log \frac{N}{M} = \log N - \log M$$

3. *Powers.* The exponent of the power of a number raised to a power is equal to the product of the exponents.

$$(b^u)^v = b^{u \times v} \qquad (b \neq 0)$$

The logarithm of a power of a number is equal to the power times the logarithm of the number.

$$\log N^u = u \times \log N = u \log N$$

4. *Zero as an exponent.* Any quantity (other than zero) raised to the zero power is 1.

$$x^0 = 1$$

The logarithm of 1 to any base is zero.

$$\log_b 1 = 0 \qquad (b \neq 0)$$

Example 1

Using the rules for operations with logarithms, evaluate the following expressions: (a) log (273 × 546); (b) log (0.2789 × 35,482); (c) log (450/15); (d) log (2.485/39.62); (e) log $(376.8^{2.34})^{9.752}$; (f) log $(144^2)^{0.25}$; (g) Check your solutions by performing the operations first and then finding the logarithm of the result. (h) What is the significance of a negative logarithm?

Solution. (a) log (273 × 546) = log 273 + log 546 = 2.436 + 2.737 = 5.173
Check: 273 × 546 = 149,058 log 149,058 = 5.173
(b) log (0.2789 × 35,482) = −0.5546 + 4.550 = 3.995
Check: 0.2789 × 35,482 = 9896 log 9896 = 3.995
(c) log (450/15) = log 450 − log 15 = 2.653 − 1.176 = 1.477
Check: 450/15 = 30 log 30 = 1.477
(d) log (2.485/39.62) = 0.3953 − 1.598 = −1.203
Check: 2.485/39.62 = 0.06272 log 0.06272 = −1.203
(e) $\log (376.8^{2.34})^{9.752} = 2.34 \times 9.752 \times \log 376.8 = 58.79$
Check: $(376.8^{2.34})^{9.752} = 6.1098 \times 10^{58}$ $\log 6.1098 \times 10^{58} = 58.79$
(f) $\log (144^2)^{0.25} = 0.25 \times 2 \times \log 144 = 1.079$
Check: $(144^2)^{0.25} = 12$ log 12 = 1.079
(h) A negative logarithm indicates that the given number is less than 1.

Be sure to review logarithms and the rules for their application until you are confident of your skill with them. If necessary, obtain a text with a more extensive presentation on logs and do additional studying of this subject. Decibels are easy if you understand logarithms; they inevitably remain a great mystery if you do not understand logs.

A.3 DECIBELS

Converting a power ratio to decibel units is simply a matter of applying the formula for decibels:

$$N_{dB} = 10 \log \frac{P_1}{P_2}$$

That is, first evaluate the power ratio, P_1/P_2, and then find the logarithm of the result; multiply the logarithm by 10.

Example 2

Convert the following power ratios to decibels: (a) 392.6 W/2.761 W; (b) 0.8274 mW/0.07349 mW; (c) 345.8 μW/0.0389 mW; (d) 73.59 kW/0.0238 MW; (e) 23.78 μW/7.481 mW.

Solution. (a) $N_{dB} = 10 \log (392.6/2.761) = 21.53$ dB

(b) $10 \log (0.8274/0.07349) = 10.51$ dB

(c) $10 \log (345.8 \times 10^{-6}/0.0389 \times 10^{-3}) = 9.489$ dB

(d) $10 \log (73.59 \times 10^{3}/0.0238 \times 10^{6}) = 4.902$ dB

(e) $10 \log (23.78 \times 10^{-6}/7.481 \times 10^{-3}) = -24.98$ dB

A.4 DECIBELS FROM VOLTAGE OR CURRENT RATIOS

Although the decibel unit, by definition, pertains to a power ratio, with proper care it can be applied to voltage or current ratios. By way of review, recall the formulas for electrical power:

$$P = VI \qquad P = V^2/R \qquad P = I^2/R$$

Consider the effect on the calculation of decibels when we use the formula for P involving V and R:

$$P_1 = \frac{(V_1)^2}{R_1} \qquad P_2 = \frac{(V_2)^2}{R_2}$$

Hence

$$N_{dB} = 10 \log \frac{(V_1)^2/R_1}{(V_2)^2/R_2} = 10 \log \left[\left(\frac{V_1}{V_2}\right)^2 \times \frac{R_2}{R_1}\right]$$

$$= 2 \times 10 \log \frac{V_1}{V_2} + 10 \log \frac{R_2}{R_1}$$

or

$$N_{dB} = 20 \log \frac{V_1}{V_2} + 10 \log \frac{R_2}{R_1} \tag{A.1}$$

If the resistance is the same at the two points where power is being considered, that is, if $R_1 = R_2$, the second term drops out. (Recall: $\log 1 = 0$.). Equation (A.1) simplifies to

$$N_{dB} = 20 \log \frac{V_1}{V_2} \tag{A.2}$$

The practical significance of Eq. (A.2) is that only a voltmeter is needed to determine decibel values when the equivalent resistances at the points of measurements are the same. (As a matter of fact, many multimeters manufactured primarily for the use of field technicians have a scale calibrated in decibels. The scale can be used when the meter is measuring ac voltage. With most such meters, the scale is correct only for measurements across 600 Ω.) If the resistances are not the same, the decibel value obtained by voltage measurements must be corrected by the use of the second term, 10 log R_2/R_1, of Eq. (A.1). Of course, a decibel value can be obtained by measuring voltages at two points in a system and plugging the values into Eq. (A.2), even when R_1 and R_2 are not equal. This is often done in the practical world. The result is not a true decibel value, however.

It is possible to obtain equations similar to Eqs. (A.1) and (A.2) for current values. The results are

$$N_{dB} = 20 \log \frac{I_1}{I_2} + 10 \log \frac{R_1}{R_2} \qquad (A.3)$$

and when $R_1 = R_2$,

$$N_{dB} = 20 \log \frac{I_1}{I_2} \qquad (A.4)$$

Be sure you are aware that the positions of the factors R_1 and R_2 in the second terms of Eqs. (A.1) and (A.3) are reversed between (A.1) and (A.3). This, of course, occurs because of the different forms of the original equations for power.

Example 3

The signal voltage at the input of an audio amplifier measures 1.35 mV across an input of 760 Ω. The output is 0.54 V across an 8-Ω speaker load. (a) Calculate P_{in} and P_{out} and determine the gain of the amplifier in decibels. (b) What is the "gain" of the amplifier in decibels using voltage measurements only? (c) What is the true gain of the amplifier in decibels, using voltage measurements and known resistance values?

Solution. (a) $P_{in} = \frac{(V_{in})^2}{R_{in}} = \frac{(1.35 \times 10^{-3})^2}{760} = 2.398 \text{ nW}$

$$P_{out} = \frac{(V_{out})^2}{R_{out}} = \frac{(0.54)^2}{8} = 36.45 \text{ mW}$$

$$G_{dB} = 10 \log \frac{P_{out}}{P_{in}} = 10 \log \frac{36.45 \times 10^{-3}}{2.398 \times 10^{-9}}$$

$$= 71.82 \text{ dB}$$

(b) G_{dB} (using voltage measurements only) $= 20 \log \frac{V_{out}}{V_{in}}$

$$= 20 \log \frac{0.54}{0.00134}$$

$$= 52.04 \text{ dB}$$

(c) G_{dB} (true) $= G_{dB}$ (voltage only) $+ 10 \log \frac{R_{in}}{R_{out}}$

$$= 52.04 + 10 \log \frac{760}{8} = 52.04 + 19.78$$

$$= 71.82 \text{ dB}$$

A.5 ZERO AND NEGATIVE DECIBELS

What is the significance of the result when power, voltage, or current measurements and decibel calculations yield a result of 0 dB? To find the answer, let's recall the significance of log N:

$$L = \log_b N \qquad N = b^L$$

If L (or N_{dB}) $= 0$, then $N = 1$. And in the decibel formula, N is equivalent to the power (or voltage or current) ratio. In other words, if $N_{dB} = 0$, the ratio is 1: the values being compared are equal. If the gain of a black box is 0 dB, the output power is equal to the input power.

What is the significance of a negative decibel value? In order for a decibel value to be negative, the log of the ratio involved must be negative. And a logarithm is always negative for a number between 0 and 1 (i.e., for a fraction). Thus a negative decibel value indicates that the value in the numerator of the ratio being considered is smaller than the value in the denominator of the ratio. If the numerator is the P_{out} of a black box and the denominator is the P_{in} of that box, the box produces attenuation, rather than gain, in processing the signal. For example, many electrical filters have a gain characteristic of zero or negative decibels; they provide no gain or may actually attenuate a signal somewhat. When the gain is in negative decibels the filter is said to have an *insertion loss* (the filter puts in—inserts—a loss in the signal path).

A.6 OVERALL GAIN

In typical electronics systems, communications systems especially, a signal is processed in turn by several functions (subsystems and/or sections) in its passage between the input to the system and its output. The signal path is a "chain" of processing functions. Each function increases, decreases, or leaves the signal energy level unchanged. That is, each function has either a positive dB gain, a negative dB "gain" or a zero dB "gain." If we know each such gain, we can determine the overall system gain very easily: simply add the individual dB gains algebraically.

This simple technique of determining total dB gain—by summing the dB gains of individual sections—is a major attraction of the use of decibels in the analyses of electronic systems performance. The concept can be applied to a total system or to any two or more successive functions. The criterion for the application of the technique is that the output of one function serves as the input of the next function in the signal path, and so on. The process works because of the rule for the logarithm of products: The logarithm of a product is equal to the sum of the logarithms of the factors.

Consider the case of two cascaded amplifier stages: A_1 with a gain G_1 and A_2 with gain G_2 ("cascaded" means that the output signal of the first amplifier is the input signal of the second amplifier). Mathematically, for A_1,

$$G_1 = \frac{v_{1out}}{v_{1in}} \qquad \text{or} \qquad G_1(\text{dB}) = 20 \log \frac{v_{1out}}{v_{1in}}$$

(For simplicity we assume that $R_{1in} = R_{1out} = R_{2in} = R_{2out}$.) And for A_2,

$$G_2 = \frac{v_{2out}}{v_{2in}} \quad \text{or} \quad G_2(\text{dB}) = 20 \log \frac{v_{2out}}{v_{2in}}$$

For overall gain,

$$G_T = \frac{v_{2out}}{v_{1in}} \quad \text{or} \quad G_T(\text{dB}) = 20 \log \frac{v_{2out}}{v_{1in}}$$

But since

$$G_1 \times G_2 = \frac{v_{1out}}{v_{1in}} \times \frac{v_{2out}}{v_{2in}}$$

and

$$v_{2in} = v_{1out}$$

Then

$$G_1 \times G_2 = \frac{v_{1out}}{v_{1in}} \times \frac{v_{2out}}{v_{1out}} = \frac{v_{2out}}{v_{1in}}$$

That is,

$$G_T = G_1 \times G_2 \quad \text{or} \quad G_T(\text{db}) = 20 \log (G_1 \times G_2) = G_1(\text{dB}) + G_2(\text{dB})$$

A.7 DECIBEL REFERENCE LEVELS

Because the decibel formula contains a ratio—a comparison of two power levels—the decibel is said to be a *relative* unit. A relative unit is to be distinguished from an *absolute* unit. Examples of absolute units are amperes, volts, ohms, etc. However, absolute decibel units can be created by the use of set reference power levels against which to compare any other power level. For instance, 1.0 mW is the reference for the *dBm* unit, an absolute dB unit. (The "m" in *dBm* comes from *m*illiwatt.)

Example 4

The output of a microphone is 0.001 μW. What is the output in dBm?

Solution. $N_{dbm} = 10 \log \frac{P_0}{1 \text{ mW}} = 10 \log \frac{0.001 \times 10^{-3}}{1} = -60$

Decibel Meters

There are numerous forms of electrical meters which suggest that they measure decibels simply because they have at least one scale calibrated in decibels. In general, such meters are simply ac voltmeters. Further, readings obtained from them, when they are used correctly, are in one of the several types of *absolute* decibel units which are in common usage. The most common absolute decibel unit used with meters is the *dBm*, which we examined just above.

Recall that a dBm is based on 1 mW as the reference power level. However, because most dB meters measure ac volts and not power, there must be a second component to the condition under which the meter provides correct dBm values. That second component is the value of the resistance across which a reading is taken. If voltage readings are always taken across the same values of resistance, the meter can be calibrated in watts (the power unit) or in an absolute dB unit, such as dBm. The most common value of resistance specified is 600 Ω. That is, most volt-ohm-milliameter test meters (VOMs), for example, have a dB scale which is based on a reference power level of 1.0 mW and measurement across 600 Ω only. If a measurement is taken across a resistance significantly removed from 600 Ω, the dB reading is erroneous.

Consider the implications of measuring dBm across 600 Ω. If, when being used correctly, a dBm meter provides a reading of 0 dBm (across 600 Ω), what voltage is being measured?

$$V = \sqrt{PR} = \sqrt{1 \times 10^{-3} \times 600} = 0.7746 \text{ V (rms)}$$

That is, across a load of 600 Ω, a power level of 1 mW requires an rms voltage of 0.7746 V. A typical VOM dB scale (actually, dBm scale) is calibrated from approximately −20 dB to +10 dB. This corresponds to a full-scale voltage of approximately 2.5 V. Hence the dB scale generally corresponds to the 2.5-V ac scale of a meter.

Obviously, if the voltage at the point where a dB reading is to be taken is a very low voltage or is greater than 2.5 V, a 2.5-V full-scale function is not appropriate for the measurement. For example, can the dB scale be used for voltages in the range 10 to 50 V? The answer is "yes." Simply select a function of appropriate full-scale value; read the dB value from the 1-dB scale, and apply a correction to the dB value. The corrections (which are typically printed on a meter's face) for scales other than the basic scale used for the dB function can be obtained through the following reasoning process.

> A measurement taken on the X-volt scale of a VOM (with a dB scale) gives a reading of 5 dBm. The meter's dB scale corresponds to the 2.5-V scale. What correction should be applied to the dB reading to get the correct dBm reading? The value of any voltage measured on the X-volt scale is $X/2.5$ times the value of the voltage on the 2.5-V scale. Hence any value on the X-volt scale is greater in dB value by the amount $20 \log X/2.5$. The dB correction to be applied to the meter's dB reading is $+20 \log X/2.5$. Correction values for typical scales are as follows:

$$\text{0.5-V scale: } +20 \log \frac{0.5}{2.5} = -14 \text{ dB}$$

$$\text{2.5-V scale: } +20 \log \frac{2.5}{2.5} = 0 \text{ dB}$$

$$\text{10-V scale: } +20 \log \frac{10}{2.5} = +12 \text{ dB}$$

$$\text{50-V scale: } +20 \log \frac{50}{2.5} = +26 \text{ dB}$$

$$\text{250-V scale: } +20 \log \frac{250}{2.5} = +40 \text{ dB, etc.}$$

The correct dBm value for this sample measurement is

$$5 \text{ dBm} + 20 \log \frac{X}{2.5}$$

If $X = 50$, then

$$\text{correct dBm} = 5 + 26 = 31 \qquad \text{the correction is } +26 \text{ dB}$$

A.8 ATTENUATORS

Electronic instruments that provide audio- and/or radio-frequency signals for the testing and servicing of communications equipment are common. They are called signal generators and are almost indispensable to any activity involved in servicing communications equipment.

To be truly useful, signal generators must be designed and constructed so that the user can vary the output signal's amplitude as well as its frequency. Most commonly, the amplitude control of such equipment is by means of *calibrated attenuators*. An attenuator in such applications is usually a special switch-controlled network. The network is made up of a number of resistors. The resistors are so arranged that relatively precise amounts of attenuation of the output signal can be chosen by means of a switching circuit. The switching circuit may consist of a multiposition, multideck selector switch. Or, it may consist of several slide or toggle switches in combination with a variable resistance (a "pot") and a dB meter.

In almost all cases the attenuator function of signal generators is calibrated in decibels of attenuation rather than in voltage values. Switch positions, for example, are labeled with negative dB values: −6 dB, −10 dB, −20 dB, etc. On some instruments switch positions are labeled with both −dB and voltage values.

Example 5

A technician is testing a receiver with an RF generator that incorporates a calibrated attenuator controlled with slide switches, a variable "pot" FINE control, and a dB meter. The output of the generator when total attenuation equals 0 dB is 100 μV. To achieve a desired signal level, the technician has set the attenuator as follows: Slide switches labeled −6 dB, −10 dB, and −20 dB are set to "IN" and the FINE control is adjusted for a reading of +5 dB on the meter. (a) What is the total attenuation of the 100-μV signal? (b) What is the amplitude of the signal at the output of the attenuator in microvolts?

Solution. (a) One of the principal reasons for the popularity of decibels comes into play in determining total attenuation: We simply sum, algebraically, the dB values:

$$\text{total attenuation} = +5 + (-6) + (-10) + (-20) = -31 \text{ dB}$$

(b) To determine the amplitude of the signal, we start with the formula for decibels,

$$N_{dB} = 20 \log \frac{v_{out}}{v_{in}}$$

We are looking for v_{out}. We divide both sides of the equation by 20 and rearrange:

$$\log \frac{v_{out}}{v_{in}} = \frac{N_{dB}}{20}$$

Now, since the antilog log $N = N$, we take the antilog of both sides of the equation:

$$\text{antilog} \log \frac{v_{out}}{v_{in}} = \text{antilog} \frac{N_{dB}}{20}$$

$$\frac{v_{out}}{v_{in}} = \text{antilog} \frac{N_{dB}}{20}$$

$$v_{out} = v_{in} \text{ antilog} \frac{N_{dB}}{20}$$

Substituting values from the example gives us

$$v_{out} = 100 \text{ antilog} \frac{-31}{20} = 100 \times 0.02818\ \mu\text{V} = 2.818\ \mu\text{V}$$

A.9 DECIBEL CORRECTIONS FOR RESISTANCE

In solving Example 5 in the fashion demonstrated, we may have made a serious error: Although we did not state our assumption, we proceeded with the solution on the basis that the resistance across which the signal would be present—the "output impedance" of the attenuator—would be equal to that used in the design of the attenuator. The "calibration" of the attenuator (i.e., the dB markings on the switch positions, the reading of the dB meter, etc.) are correct only for a very specific equivalent resistance across the output of the attenuator. Recall: Decibels are based on power ratios, not voltage ratios. An attenuator is typically a form of voltage divider, not a power divider. Its markings (the power ratios involved) are correct only for the designed output resistance. The most common output resistance for signal generator attenuators, as well as many other applications, is 600 Ω.

Any time we use decibel units based on voltage ratios we run a risk of being in error if we do not consider the effect of resistance in the formula

$$N_{dB} = 20 \log \frac{v_1}{v_2} + 10 \log \frac{R_2}{R_1} = 20 \log \frac{v_1}{v_2} - 10 \log \frac{R_1}{R_2}$$

In practical work, we may become involved with voltage ratios in several ways. We may simply measure signal voltages at two points in a system and wish to express a comparison with decibels. We may use the decibel scale of a multimeter to express the signal level at a single point. (In this instance we are comparing our reading with the reference level designed into the meter scale.) Or we may be using a "calibrated" attenuator, as described above.

In the case of two voltage measurements, the equivalent resistances across which the voltage measurements are taken may be quite different. To convert the comparison of the voltages into decibel units correctly, the resistances should be determined and a correction factor calculated and applied to the dB value, using the equation above. In the latter instances—the dB meter and calibrated attenuator—it is necessary for one to know the design reference resistance. If our circuit of interest has a different resistance, we again correct by substituting resistance values into the complete formula.

Example 6

A technician takes a reading in a signal circuit with a multimeter and reports the signal level as "+8 dB." The equivalent resistance of the circuit at the measurement point is 2000 Ω. The meter is set on the 2.5-V range and the dB scale is a typical dBm scale referenced to 600 Ω. (a) Is the reported "+8 dB" a correct value? (b) If the value is not correct, what should it be?

Solution. (a) Since the meter is being used on the 2.5-V scale, the dB reading is correct for scale factor. However, since the dB scale is referenced to 600 Ω and the measurement is being taken across 2000 Ω, the reported reading is not a true dB value: A correction for the resistance offset is required for a true dB value.

(b) The correction factor for resistance is obtained as follows:

$$\text{resistance correction factor} = 10 \log \frac{R_R}{R_M}$$

where the subscript R resistance is the reference resistance and the M resistance is the resistance of the circuit across which the voltage is measured:

$$\text{resistance correction factor} = 10 \log \frac{600}{2000}$$

$$= -5.2 \text{ dB}$$

For the measurement taken,

$$\text{true value} = +8 + (-5.2) = +2.8 \text{ dB}$$

It is useful to have a reliable, intuitive (nonmathematical) method for determining whether a correction for resistance difference should increase a decibel value or decrease it. Consider the following reasoning. The basic formula for calculating power, given voltage and resistance, is $P = V^2/R$. Hence for a given amount of energy, a voltage reading will be higher across a larger resistance. Voltage readings across resistances larger than the reference resistance are always distorted to the high side: Related dB values must be corrected by applying correction factors that *reduce* the apparent dB value. Of course, the other side of the coin is that when output resistances are smaller than the input or reference resistance values, the dB values need to be *increased*.

APPENDIX B TELECOMMUNICATIONS AND GOVERNMENT

B.1 FEDERAL COMMUNICATIONS COMMISSION (FCC)

The field of human endeavor identified by the term "telecommunications" includes all forms of communications generally associated with the term "electronic communications." Specifically, telecommunications includes radio, television, telephone, telegraph, facsimile, data transmission, satellite communication, and so on. This list does not exhaust all present or potential techniques for using electronic and computer technology for the processing and transmission of information.

When information transmission takes place through the atmosphere (or space) instead of on wires, it uses a very limited "natural" resource—the electromagnetic frequency spectrum. Since this resource is limited, governments around the world have controlled its use in various ways virtually from the beginning of the development of wireless communication. We can assume that the exercise of such control has been as an effort to ensure the effective and efficient use of the resource so as to benefit the greatest number of its users and potential users. Whether or not the effort has succeeded is not the point here. For us, as workers and potential workers in this field, it is important to have a working knowledge of at least some of the elements of governmental regulations. In particular, it is important that we have a knowledge of those elements of regulation that affect the design and operation of communications equipment.

In the United States, laws and regulations affecting telecommunications are administered by the Federal Communications Commission (FCC), an agency of the executive branch of the federal government. Basic national policy regarding telecommunications is determined by international treaties and laws written by the U.S. Congress. These instruments are interpreted and executed by the FCC. The FCC also writes its own regulations to supplement the more basic policy instruments and provide details for the interpretation and administration of them. As communications technology changes, regulations must change to ensure that advancements can be utilized to greatest advantage for the country. Just as rapid changes have been occurring in technology in recent years, so have many changes been implemented in FCC regulations. Here we will be concerned with just three particular areas of regulations: (1) frequency-range categories in the radio-frequency spectrum, (2) emission-type designations, and (3) types of communications services.

Frequency-Range Categories

In its regulations for telecommunications services, the FCC divides the radio-frequency spectrum into a set of very explicit frequency ranges or categories. These categories, with their unique names and abbreviations, are universally used in the technical literature of the field of telecommunications. Needless to say, you will be handicapped if you do not learn the categories and their abbreviations. They are presented in Table B.1.

If you will study Table B.1 carefully you will see that there are two aspects to the way the ranges have been set up which will help you remember them more easily. First, and most obvious, is the fact that the numbers defining the ranges are all multiples of three: 3, 30, and 300. Second, each range is a frequency *decade;* that is, the upper limit in each range is 10 times the frequency of the lower limit of the range.

TABLE B.1 THE RADIO-FREQUENCY SPECTRUM

Frequency range	Designation	Abbreviation
30–300 Hz	Extremely low frequency	ELF
300–3000 Hz	Voice frequency	VF
3–30 kHz	Very low frequency	VLF
30–300 kHz	Low frequency	LF
300 kHz–3 MHz	Medium frequency	MF
3–30 MHz	High frequency	HF
30–300 MHz	Very high frequency	VHF
300 MHz–3 GHz	Ultra high frequency	UHF
3–30 GHz	Super high frequency	SHF
30–300 GHz	Extra high frequency	EHF

Emission-Type Designations

In the jargon of FCC rules and regulations, the energy radiated by an antenna is called an *emission.* This is an appropriate term since it comes from the verb "to emit," meaning "to send out." In its task of regulating telecommunications operations, the FCC is concerned not simply with the frequencies of the carriers of the various stations but also with the type of modulation used by each station. The type of modulation is important in regulation since the amount of frequency spectrum required for a given station is related to the type of modulation being used. The FCC's emission designations and their descriptions are listed in Table B.2.

TABLE B.2

Carrier	Description
	Amplitude-modulated
A0	Carrier carrying no modulation or information
A1	Telegraphy, on-off; no other modulation
A2	Telegraphy; on-off; amplitude-modulated
A3	Telephony; carrier with double sideband
A3A	Telephony; reduced carrier with single sideband
A3B	Telephony with two independent sidebands
A3H	Telephony, full carrier with single sideband
A3J	Telephony, suppressed carrier with single sideband
A3Y	Digital voice modulation
A4	Facsimile (slow-scan TV)
A5C	Television with vestigial sideband
A9B	Telephony or telegraphy with independent sidebands
A9Y	Nonvoice digital modulation
	Frequency-modulated
F1	Telegraphy; frequency-shift-keyed
F2	Telegraphy; on-off; frequency-modulated tone
F3	Telephony; frequency- or phase-modulated
F3Y	Digital voice modulation
F9Y	Nonvoice digital modulation
F4	Facsimile
F5	Television
F6	Telegraphy; four-frequency diplex
	Pulse-modulated
P0	Radar (pulsed carrier without information)
P1D	Telegraphy; on-off keying of pulsed carrier
P2D	Telegraphy; pulsed-carrier tone-modulated
P2E	Telegraphy; pulse-width tone-modulated
P2F	Telegraphy; phase or position tone-modulated
P3D	Telephony; amplitude-modulated pulses
P3E	Telephony; pulse-width-modulated
P3F	Telephony; pulses phase- or position-modulated

The designations in Table B.2 may be modified to provide even more encoded information about carriers. The allowable bandwidth in kilohertz of a carrier is indicated by writing a number in front of the emission designation. For example, 10A3 is used to specify an AM standard broadcast band carrier. It indicates a carrier with double sidebands and a bandwidth of 10 kHz, a typical bandwidth for the broadcasting of voice and music programs. Similarly, 150F3 might appear in the specification of a standard FM broadcast station carrier. The video carrier for a standard TV station would be designated 6000A5C.

Carrier specifications with the "P" designation have been appearing more frequently in the recent past, and will appear more frequently in the future as computer technology and data transmission become more important elements in telecommunications.

Designations with a "Y" suffix are relatively new, but are rapidly becoming common, again because of the rapidly increasing quantity of computer data being transmitted as well as the utilization of computer technology to achieve exciting new features and capabilities in communications hardware.

B.2 TYPES OF COMMUNICATIONS SERVICES

There are many regulations of electronic communication which are common to all types of services. On the other hand, because each service has its unique characteristics, there are numerous regulations which are specific to each type of service. By type of service is meant categories such as standard AM broadcast service, two-way radio communication, standard television broadcast, citizens' band radio, and so on. It is the goal of this section to provide only a brief listing and description of the most common services, services for which the technical topics covered in this book are most likely to be appropriate.

Standard AM Broadcast Service

This is very likely the service with which the greatest number of people in the United States are familiar. It is the service that most people know simply as "AM radio." The term "standard broadcast station" is defined by the FCC as a station licensed to transmit radiotelephone emissions intended to be received by the general public.

The AM broadcast band includes the frequencies between 535 and 1605 kHz. The band is divided into channels of 10 kHz each. The FCC has designated 107 AM channels. Transmitter power authorizations are in the range 250 W to 50 kW.

Because of the great number of AM stations in the United States, the chances of interference between stations are large. There are many restrictions on AM stations, all in an effort to minimize interference. Regulations require that a person with a commercial radio operator license of any class except a Marine Radio Operator Permit be on duty in charge of the transmitter during all periods of broadcast operation. The carrier frequency of an AM station must be stable to within ± 20 Hz of the assigned frequency.

International AM Broadcast Service

This is a category for stations that transmit programs intended for reception by the general public in foreign countries. Such stations are assigned carrier frequencies in the range between 5.95 and 26.1 MHz and within the following bands:

Band	Frequency (MHz)
A	5.95–6.2
B	9.5–9.775
C	11.7–11.975
D	15.1–15.450
E	17.7–17.9
F	21.450–21.750
G	25.6–26.1

The frequencies in the international AM service are higher than those of standard AM broadcast stations. The emissions, therefore, have much shorter wavelengths. Listeners to such broadcasts are called shortwave listeners (SWLs). Transmitters are not authorized with power ratings of less than 50 kW. Carrier frequencies must be held to within 0.0015% of assigned frequencies.

Shortwave stations generally use directional antenna systems to obtain coverage of specific countries or regions of the world. Reliance on ionospheric propagation is an integral part of their normal operation. Since inosopheric propagation is subject to many variables—time of day and year, sunspot cycle, etc.—international shortwave stations are assigned several frequencies in one or more bands to permit flexibility in matching optimum propagating conditions with desired coverage. Shortwave listeners generally refer to wavelength bands rather than frequency bands. For example, the 15.1- to 15.450-MHz band might be referred to as the "20-meter band" since an emission in that frequency range has a wavelength of approximately 20 m (300 megameters per second/15.1 MHz = 19.87 m).

FM Broadcast Service

The FM broadcast service includes stations whose transmissions are intended for the general public. FM stations are assigned frequencies in the range 88 to 108 MHz. An FM channel is 200 kHz in width. There are 100 available channels. Required stability of the carrier or so-called center frequency of FM stations with authorized output power of more than 10 W is ±2000 Hz. For stations of 10 W or less, the figure is ±3000 Hz.

Power assignments for FM stations are made not simply on the basis of transmitter power but on effective radiated power (ERP). Effective radiated power takes into consideration the antenna design. It is equal to the product of input power to the antenna and antenna gain:

$$\text{ERP} = \text{input power to antenna} \times \text{antenna power gain}$$

Noncommercial education FM stations have power ratings of 10 W. Highest-power stations are permitted to operate at a maximum ERP of 100 kW with an antenna height of 200 ft above average terrain. You will note that FM broadcast frequencies are in the VHF category.

Television Broadcast Service

Television broadcast stations are assigned frequencies in three ranges. These are called low-band VHF, 54 to 88 MHz; high-band VHF, 174 to 216 MHz; and UHF, 470 to 800 MHz. A television channel is 6 MHz in width. There are 82 in all: 12 VHF channels and 70 UHF channels. A TV station must broadcast two carriers: a *visual carrier* for picture information and an *aural carrier* for sound. The departure of the visual carrier from the assigned frequency must not exceed ± 1000 Hz. The aural carrier must remain stable within ± 1000 Hz and must be equal to the actual visual carrier plus exactly 4.5 MHz. Power assignments to TV stations are made with the following maximum power restrictions:

Channels	Effective radiated power
2–6	100 kW (20 dBk)
7–13	316 kW (25 dBk)
14–83	5000 kW (37 dBk)

The unit "dBk" appears in FCC regulations and means "decibels above 1 kW." The visual carrier is amplitude modulated but is radiated as a so-called *vestigial sideband emission,* A5C. In TV broadcasting, part of the lower sideband is attenuated before it is fed to the antenna. The aural carrier transmitted by TV stations utilizes frequency modulation.

Intimately associated with the television broadcast service is a category called Auxiliary Broadcast Services. This category includes broadcasting services with the special purpose of facilitating the pickup and relay of aural and visual signals. The emissions, except for those of television translator stations, are not intended for the general public. Television translator stations are special stations that rebroadcast TV programming into isolated or remote areas.

Amateur Radio Service

Amateur radio operating is a hobby. However, it is recognized by all governments as a valuable source of technically skilled personnel, a resource of great value, especially in times of emergency. Amateurs have made numerous contributions to the development of radio communications, and continue to do so. In recognition of the importance of this hobby, a share of the radio-frequency spectrum is allocated for use by amateur operators, or "hams" as they are called. Frequency allocations change from time to time, but as this is being written (1985), amateur frequency allocations include the following:

Frequency range	Popular name of band
kHz	
1800–2000	160-meter
3500–4000	80-meter
4383.8	
7000–7300	40-meter
14,000–14,350	20-meter
MHz	
21.000–21.450	15-meter
28.000–29.700	10-meter
50.000–54.000	6-meter
144–148	2-meter
220–225	
420–450	
1215–1300	
2300–2450	
3300–3500	
5650–5925	
GHz	
10.0–10.0	
24.0–24.25	
48–50, 71–76	
165–170, 240–250	

Operation on the amateur bands is governed by numerous regulations involving class of operator license, type of emission, type of carrier frequency control, and many other factors. The types of emission that can be transmitted by amateurs include the following: A0, A1, A2, A3, A4, A5, F0, F1, F2, F3, F4, F5, and P. Not all emissions may be used at all allocated frequencies, however.

Power restrictions vary with frequency of operation and type of emission. Unlike power restrictions for broadcast stations, which are based on actual RF output power from a transmitter, restrictions on power for amateur transmitters pertain to dc input power to the final stage of the transmitter. The maximum dc input power to the final stage for any type of amateur operation is 1 kW. Lower restrictions apply to certain operations; for example, a *novice operator* (a beginner) is restricted to use of 250 W.

The most popular types of amateur transmitters use carrier frequencies which are variable—the operator is free to vary the master oscillator of his/her transmitter in order to obtain an optimum operating frequency. Of course, the choice is restricted to the bands specified above and operation must not cause interference with other stations operating in the same band. Hams use many forms of emission, the two most popular being single-sideband radiotelephone and CW (continuous-wave) code. Proficiency in sending and receiving the International Morse code is still a requirement for licensing as an amateur radio operator.

Private Land Mobile Radio Services

This is a category whose rules and regulations pertain to the radio communications of a wide variety of agencies. Included under this title are services for users such as Public Safety (police, fire, highway maintenance, forestry-conservation, and local government), Special Emergency, Industrial, Land Transportation, and Radiolocation. The services are essentially two-way radio communications between fixed-base land stations and vehicular mobile stations, or between two or more vehicular mobile stations. The communications hardware generally consists of compact transceiver units—a combination of transmitter and receiver functions in one hardware package. Stations and their owners must be licensed. Operators of the stations are not required to have a license. The requirement that all persons undertaking to repair, service, or adjust the transmitters of such stations be licensed by the FCC was recently discontinued. Types of emissions used by stations include: A3 (AM), A3J (single-sideband), and F3 (FM). Many different frequency bands have been allocated for use in these services. A great variety of restrictions as to transmitter power, frequency stability, bandwidth, etc., are in effect. The users of mobile radio communications services provide employment for large numbers of electronics technicians in the maintenance of the communications hardware.

Marine Radio Services

All large commercial vessels are required to have one or more forms of radio communications facilities. Several hundred thousand small commercial and pleasure boats are also equipped with two-way radiotelephone hardware. Radiomarine equipment consists, most generally, of compact transceiver units. There are 96 operating channels in the 2- to 3-MHz and 156- to 158-MHz bands. Transmitter operating frequencies are crystal controlled. Emissions are either single-sideband, in the MF channels, or FM, in the VHF channels. Maximum permissible transmitted power is 25 W. However, many transceiver units provide for operation at reduced power to minimize interference. Regulations also provide for the installation and utilization of ship radar. Operators of marine radio equipment are not required to be licensed.

Aviation Radio Services

Aviation radio services include many radio navigation facilities for aircraft, as well as radiotelephone and radiotelegraph communications. Navigation aids are classified as communication, navigation, traffic control, and landing. Most aviation radio services use the portion of the spectrum between 108 and 136 MHz. Navigation services utilize the range 108 to 118 MHz, and traffic control utilizes frequencies in the range 118 to 136 MHz. Aircraft and aeronautical ground stations utilize frequencies spread throughout the 108- to 136-MHz band.

Although aeronautical radio equipment is small and compact, it is generally more specialized than other types of two-way radio equipment. Typical transmitter/receiver combinations may be operable on the same frequency, like transceivers in other services, or transmitters and receivers may be tunable to different frequencies to permit a mode of operation called *duplex*. In some cases, receivers can both receive radio navigational signals and permit voice communication with a ground station. The VHF omnidirectional range (VOR) navigational signals are utilized by all but the minimally equipped aircraft. The instruments of receivers capable of receiving VOR (or OMNI) signals can indicate bearing and aircraft positioning with respect to a particular VOR range station. Some aviation radio service stations require licenses for operators.

Citizens Radio Service or Citizens Band Radio (CB Radio)

Citizens Radio Service provides for use of the radio-frequency spectrum, in certain limited frequency bands, by those U.S. citizens who do not otherwise have access to this common resource. The rules and regulations of the service provide for four classes of service: A, B, C, and D. The class C service is intended for the remote control of objects or devices by radio (e.g., model airplanes). Classes A, B, and D provide for radiotelephone service utilizing amplitude modulation and/or single-sideband emission. However, the class D band, with 40 channels in the 27-MHz region, is by far the most popular band. Equipment invariably consists of compact, low-powered transceiver units, limited by law to 4 W output on AM and 12 W peak envelope power (PEP) on single sideband (SSB). All transmitters must be crystal controlled. Most transmitters are designed for operation on all of the 40 available channels. Operator licensing was never required; recently, even station licensing has been discontinued by the FCC.

B.3 OPERATOR AND TECHNICIAN CERTIFICATION AND LICENSING

Prior to December 31, 1981, licensing was required of many operators and virtually all transmitter maintenance personnel in the various radio services listed above. Before that date the FCC tested applicants for several different classes of operator licenses; license certificates were issued to those successfully completing examinations which were highly technical in nature. Among the licenses that were required for certain occupations were the First Class Radiotelephone Operator License, the Second Class Radiotelephone Operator License, and the Third Class Radiotelephone Operator License. (These licenses are generally referred to as "Commercial Operator Licenses.") The First Class License, for example, was required of all persons responsible for the maintenance of AM, FM, and television broadcast stations. A Second Class License was a minimum requirement for maintenance technicians in most other radio services.

In 1981, the FCC adopted several rule changes which have had a widespread effect on the licensing of operator and maintenance personnel of various radio services. First, the requirements for the FCC First Class Radiotelephone Operator License were eliminated and the examination and license were discontinued effective December 31, 1981. On the same date, the Second Class Radiotelephone Operator License was eliminated and effectively replaced by the General Radiotelephone Operator License. A Restricted Radiotelephone Operator Permit, which is issued without any examination, was established as part of the changes implemented.

In 1984, the FCC eliminated the rules that permitted only licensed commercial radio operators to perform certain duties in the Private Land Mobile and Fixed, Personal, and Domestic Public Fixed Radio Services. This change became effective November 11, 1984. The change effectively eliminated the requirement for a General Radiotelephone Operator License for technicians servicing most two-way radio equipment.

The net result of FCC rule changes in recent years is to effectively eliminate the requirements for licenses for operators in every radio service except amateur, aeronautical, international common carrier, and maritime. While eliminating licensing requirements for operator and maintenance personnel, the FCC has emphasized ever more strongly that the ultimate responsibility for the proper operation of a radio station is always that of the station licensee or owner. The FCC, in announcing the rule change, stressed that "the installation, service and maintenance of transmitter equipment should be performed by a qualified technician certified by organizations or committees representative of users in the private land mobile or fixed services."

The FCC has, recently, actively worked at reducing its role as a testing/certification agency. For example, it has asked educational and training institutions to review their practice of encouraging or requiring all telecommunications students to pass the FCC General Radiotelephone Operator License Examination. It seeks "cooperation in decreasing the demand for unnecessary FCC radiotelephone licenses." At the same time, the FCC encourages private organizations to establish technician examination/certification programs. And it advises educational institutions that "technical training for industry recognized certification programs will better qualify your students for jobs in the dynamic telecommunications marketplace."

Several private-sector organizations have announced intentions to set up programs to examine and certify telecommunications technicians. The following is an alphabetical listing of six organizations that offer nationwide certification programs:

Electronics Technicians Association, International (ETA)
Iowa State University Station
P.O. Box 1258
Ames, IA 50010
Attn: Director of Certification
Telephone: (515) 294-5060

Examination and certification is available from an associate level up to a master level.

International Society of Certified Electronic Technicians (ISCET)
2708 W. Berry Street

Fort Worth, TX 76109
Telephone: (817) 921-9101

Two levels of certification testing are available. An associate-level examination deals primarily with basic, generalized electronic theory. The journeyman-level examination consists of additional testing in one of several specialized options. Options include communications, industrial electronics, audio equipment servicing, and computer repair, among others.

National Association of Business and Educational Radio (NABER)
P.O. Box 19164
Washington, DC 20036
Attn: Certification Program Coordinator
Telephone: (202) 887-0920

NABER's program is entitled Two-Way Radio Technician Certification Examination. The aim of the program is to provide a means "by which technicians who are competent in the current technology of our (two-way radio) industry may be identified." Examinations for specialized endorsements in areas such as cellular and microwave technology are contemplated.

National Association of Radio and Telecommunications Engineers, Inc. (NARTE)
P.O. Box 15029
Salem, OR 97309
Attn: President, Certification and Endorsement
Telephone: (503) 581-3336

The NARTE program offers seven levels of certification, with endorsements covering specialized categories such as land mobile, broadcasting, aviation and marine, telegraphy, and radar servicing. NARTE enlists the services of nearly 400 technical/vocational colleges around the country for administering its tests.

National Institute for Certification in Engineering Technologies (NICET)
National Society of Professional Engineers
1420 King Street
Alexandria, VA 22314
Attn: General Manager of Certification Programs
Telephone: (703) 284-2835

NICET offers three levels of certification, which are tied to the number of years of applicable work experience—zero years for entry level, 5 years for the middle level, and 15 years for the senior level.

Society of Broadcast Engineers, Inc. (SBE)
P.O. Box 50844
Indianapolis, IN 46250
Attn: Certification Secretary
Telephone: (317) 842-0836

The SBE certification program is designed for broadcast engineers and technicians.

Certification programs typically have extensive study materials available.

Such materials may include sample tests, detailed outlines of each examination section, recommended study references, suggestions for preparing for the examination, and test-taking strategies. If you are interested in obtaining further information about a particular program, contact the appropriate organization directly. Each is interested in being of service to prospective employees of the telecommunications industry.

INDEX